6th International Conference on Fluid Mechanics and Industrial Applications (FMIA 2022)

Journal of Physics: Conference Series Volume 2280

Taiyuan, China
23 – 24 April 2022

ISBN: 978-1-7138-5530-9
ISSN: 1742-6588

Printed from e-media with permission by:

Curran Associates, Inc.
57 Morehouse Lane
Red Hook, NY 12571

Some format issues inherent in the e-media version may also appear in this print version.

Institute of Physics
Dirac House, Temple Back
Bristol BS1 6BE UK

Phone: 44 1 17 929 7481
Fax: 44 1 17 920 0979

techtracking@iop.org

Additional copies of this publication are available from:

Curran Associates, Inc.
57 Morehouse Lane
Red Hook, NY 12571 USA
Phone: 845-758-0400
Fax: 845-758-2633
Email: curran@proceedings.com
Web: www.proceedings.com

TABLE OF CONTENTS

Preface

2022 the 6th International Conference on Fluid Mechanics and Industrial Applications (FMIA 2022) included original and peer-reviewed research papers and held on April 23-24, 2022 in Taiyuan, China. This conference was sponsored by Asian Union of Information Technology, and was attended by academicians, researchers, entrepreneurs, government agencies and policy-makers, postgraduates.

FMIA is an annual international conference devoted to the discussion of recent developments and applications of parallel computing in the field of fluid mechanics, CFD, fluid power, fluid flow and related disciplines. The field of fluid mechanics is vast and covers numerous and diverse applications. The conference covered a wide range of topics, including basic formulations and their computer modelling as well as the relationship between experimental and analytical results. The meeting provided a forum for discussing new work on fluid mechanics and in particular for promoting the interchange of novel ideas and the presentations of the latest developments in this field. This conference would also provide an ideal environment to develop new collaborations and meet experts on the fundamentals, applications, and products of the mentioned fields.

This two-day conference was organized in technical association with the Asian Union of Information Technology and included invited keynote talks and oral paper presentations from both academia and industry. The topics covered in this FMIA 2022 are: 1. Aerodynamics; 2. Thermodynamics and Heat Transfer; 3. Fluid Theory and Related Algorithms. Each keynote speakers lasted for 30 minutes, including 5 minutes for Q&A. and for other sessions, there was 20 minutes for each presentation, including 5 minutes for Q&A. The book is useful to professionals and scientists involved or interested in fluid mechanics and industrial applications. Hopefully, all participants and other interested readers would benefit scientifically from the proceedings and also find it stimulating in the process.

We would like to thank all the author submitted papers to this FMIA 2022 and thank all the reviewers for their time and effort in reviewing articles. Especially we would like to thank the organizing committee for their valuable advices in the organization and helpful peer review of the papers.

Conference Organizing Committee,
Wuhan, China

Committees

Chairmen
Prof. Sanbao Hu, Wuhan University of Technology, China
Prof. Yangzhao Ma, Wuhan University, China

Members
Prof. Hengjia Zhu, Civil Aviation University of China, China
Prof. Jianjun Xu, Northeast Petroleum University, China
Prof. Dongtao Hu, China University of Geosciences, China
Dr. Xiaohe Deng, Wuhan University of Technology, China
Prof. Feng Liu, Huazhong University of Science and Technology, China
Dr. Qifei Du, Tiangong University, Tianjin, China
Dr. Jianghao Niu, Civil Aviation University of China, Tianjin, China
Dr. Jianpeng Shi, Xi'an Jiaotong University, China
Prof. Y.D. Shi, Zhejiang University, China
Dr. X.L. Li, Peking University, China
Dr. D.M. Li, Shandong University of Science and Technology, China
Prof. H. Chen, Shanghai University of Engineering Science, China
Associate Prof. S. Zhu, Wuhan University, China
Prof. Pin Wang, Guangxi college of Education, China
Dr. X.W. Tian, Xi'an University of Science and Technology, China
Dr. Y.H. Dou, Xi'an Shiyou University, China
Dr. Yuan Xi, Panjin Industrial Technology Institute of DUT, China
Dr. H. Ma, Qingdao University of Science and Technology, China
Dr. R.J. Ma, Tongji University, China
Dr. Zhigang Fang, Wuhan University of Technology, China
Dr. Yong Wang, Chongqing University, China

Peer Review Statement

All papers published in this volume have been reviewed through processes administered by the Editors. Reviews were conducted by expert referees to the professional and scientific standards expected of a proceedings journal published by IOP Publishing.

- **Type of peer review:** Double Anonymous
- **Conference submission management system:** Morressier
- **Number of submissions received:** 98
- **Number of submissions sent for review:** 62
- **Number of submissions accepted:** 62
- **Acceptance Rate (Submissions Accepted / Submissions Received × 100):** 63.3
- **Average number of reviews per paper:** 2.03
- **Total number of reviewers involved:** 13
- **Contact person for queries:**
  Name: Jie Wu
  Email: ICFMIA@163.com
  Affiliation: Wuhan University of Technology

Aero-acoustics Investigation of a Bionic Airfoil Horizontal Axis Wind Turbine Using LES-DCS Approach

Weimin Wu[1], Wenling Dong[2, *], Shiwei Li[3] and Suocheng Zhang[2]

[1]Chongqing University of Science and Technology, Chongqing, China
[2]Inner Mongolia Energy Investment Group New Energy Co., Ltd., Hohhot, China
[3]Chongqing Yongneng Industrial (Group) Co., Ltd., Chongqing, China

*Corresponding author e-mail: asataleso@163.com

Abstract. The wind turbine noise has become an important issue in the current green energy field, therefore, it is necessary to use new and advantageous methods to study the problem. This research employed the LBM-DCS method to investigate the wake sound radiation characteristics for a bionic airfoil horizontal axis wind turbine, combining the large eddy simulation (LES) model and the wall-adapting local eddy-viscity (WALE) model. The obtained results show that the aeroacoustic model used here is reasonable and effective, which are consistent with the relevant experimental measurement result. The appearance and changing characteristics of the acoustic radiation ring have been discussed and analyzed to a certain extent. Based on the accumulation of this research, we will combine more experimental data to carry out a further systematic investigation on the acoustic radiation characteristics to the wind turbine.

key word: acoustic radiation; wind turbine; LBM-DCS method; wake region.

1. Introduction

Wind energy is one of the fastest-growing clean energy sources among renewable energy sources. As an important equipment for utilizing wind energy, horizontal axis wind turbines have always been a research hot-spot. As the installed location of wind turbines is getting closer to the range of human activities, the noise problem has become increasingly prominent, for which the studies have shown that the noise generated by wind turbines can affect physical and mental health of human beings [1, 2] as well as the fatigue life of wind power equipment [3, 4].

For horizontal axis wind turbines, the noise can be divided into mechanical noise and aerodynamic noise according to the source. Among them, the mechanical noise mainly comes from the moving mechanical parts inside the nacelle, such as gearboxes, generators, cooling fans and other auxiliary equipment, and the mechanical noise can be improved by taking a better lubrication of rotating machinery and increasing the sound insulation performance of the engine room. However, the control of aerodynamic noise from blades is relatively difficult and thus the aerodynamic noise is the main noise source for wind turbines. Generally speaking, the aerodynamic noise of blades includes three types [5-7]: firstly, it is turbulent inflow noise, which is generated by the interaction between turbulent flow within the atmospheric boundary layer and the leading edge of the blade; secondly, it is blade self-excited noise that is generated by the air flow on the blade surface mainly in the trailing edge area of the blade; thirdly, it is low frequency noise which is the unsteady load noise generated by the velocity depletion zone with the blade rotation and the steady noise by the impeller cycle movement.

At present, there are mainly three research methods on the aerodynamic noise of wind turbines, including the experimental measurement, the theoretical prediction and CFD numerical simulation. Among them, for experimental measurement, the related scholars [8-12] have carried out a series of studies on the impact of aerodynamic noise of typical horizontal axis wind turbines on human health, the sound source characteristics of wind turbines of different scales, the impact of measurement environment and its coupling with other related factors (such as airfoil type, trailing edge shape, Lift-to-drag ratio, angle of attack change), etc. For theoretical prediction, the related scholars [13-18] proposed the BPM (Brooks, Pope, and Marcolini) model and the airfoil self-noise prediction model based on neural networks as well as a series of related improvement models. Due to the limitations of the experimental measurement and the theoretical prediction in many aspects and the shortcomings of flexibility of use, many scholars began to engage in the research on the CFD numerical simulation of wind turbine aerodynamic noise. The Ffowcs Williams and Hawkings (FW-H) acoustic analogy formulation became a major numerical simulation algorithm, for which the different turbulence models [19-21] (such as LES, DES and IDDES) are used to calculate the static pressure pulsation of the flow field.

However, the FW-H algorithm mainly uses array feature points to study the wake noise radiation of wind turbine, which lacks a more comprehensive and detailed analysis ability of acoustic radiation field. In view of this, the paper will use the direct computation of sound (DCS) [22, 23] method to study the acoustic radiation field of the horizontal axis wind turbine wake flow with the bionic airfoil, so as to find an efficient way to achieve a more in-depth investigation of the relevant characteristics. Additionally, the high-fidelity CFD algorithm uses the Lattice Boltzmann method (LBM), which unlike the finite volume method, the LBM algorithm [24, 25] has a strong adaptability to mesoscopic hydrodynamic behavior description. At the same time, it can avoid the huge consumption of computational resources such as molecular dynamics (MD) [26], so as to better meet the capture and description of important flow details such as blade tip wakes of wind turbines. Thus, it can provide a better basic data for aeroacoustic computation.

In this study, a CFD unsteady calculation based on the LBM algorithm of the bionic airfoil wind turbine is performed to provide static pressure pulsation data for aeroacoustic computation, in which the large eddy simulation (LES) model and the wall-adapting local eddy-viscity (WALE) model are employed to handle turbulent calculations and wall flow. The paper is organized as follows: Section 2 briefly describes the methodology of DCS; Section 3 describes the 3D solid model of the bionic airfoil impeller, the discrete layout of spatial lattice and conditions for CFD solving; Section 4 provides the results for aeroacoustic computation and gives the corresponding analysis and understanding; Section 5 summarizes the conclusions drawn in this work.

2. A Brief Description of the Direct Computation of Sound

The numerical solution process of fluid dynamics based on LBM algorithm can be adapted to the compressible situation, so that the algorithm can directly calculate the sound pressure fluctuation on the physical level. In addition, LES turbulence model is used in the calculation process so that the spatial resolution of important physical flow parameters in the computational domain can be determined by the physical planning of the lattice.

The acoustics approach used here is the direct computation of sound (DCS), which does not introduce any artificial shear viscosity or stabilization. Due to the inherent property of LBM algorithm is the calculation of transient and compressible flow naturally, and the function matching of the calculation code, the static pressure fluctuation directly appears to propagate at a numerical speed C_{num}

$$C_{num} = \frac{1}{\sqrt{3}}\frac{dx}{dt} \tag{1}$$

where dx is the resolution at a given lattice level in the specific computational domain, and dt refers to the associated time-step for the same lattice level.

For the ideal gas law for the single phase flows, there is a thermodynamic speed of sound which is calculated by

$$C_{thermo} = \sqrt{\frac{\gamma RT}{M}} \tag{2}$$

where γ is the adiabatic index of the flow medium, and R is the perfect gas constant, and M is the molecular weight.

To be consistent with the physical thermodynamic speed of sound, that is to say that it satisfies the following relationship

$$C_{num} = C_{thermo} \tag{3}$$

Therefore, the discrete time resolution dt by

$$dt = \frac{1}{\sqrt{3}} \frac{dx}{C_{thermo}} \tag{4}$$

One thermodynamic speed related attribute involved in the above calculation is the refractive index, in which a specific speed of sound in the fluid medium v_{medi} defined as following

$$C_{thermo} = n \cdot v_{medi} \tag{5}$$

where n is the refractive index that is a dimensionless parameter, and in this study it is set to 1. It is worth noting that the refractive index should always be equal to or greater than 1 to ensure the stability of the value, because the propagation of information can only be slowed down and cannot exceed the time-step limit.

There is a key index of acoustic analysis is sound pressure level (SPL), which is the deviation from the average pressure caused by sound waves. The SPL is a logarithmic measure of the effective sound pressure relative to the reference value, which is calculated here as

$$SPL = 10 \cdot \log_{10}\left(\frac{p_{st}^2}{p_{ref}^2}\right) \tag{6}$$

where p_{st} is the fluctuating gauge static pressure in the computational domain, and p_{ref} is the reference pressure equal to 2×10^{-5} Pa, which is the threshold of human hearing and measured in decibels (dB) above the reference level.

As for the relevant algorithms of LBM and LES, there is no specific description here. The relevant references can be found in the corresponding literature in the Introduction part, or just read the references [24-27].

3. CFD Model

The related visual descriptions are rendered in a unified way to provide readers with a better understanding for the CFD model of wind impeller, in which the wind turbine impeller part and other parts are used different colors rendering.

3.1. 3D solid model of the bionic airfoil wind turbine

The geometry model of the physical wind turbine is realized by a series of cross-sectional bionic airfoil matching the corresponding local chord length and twist angle, which is manifested in Figure 1.

The sky blue part is the wind turbine impeller, and other parts, such as alternator, tail rudder and tower, are rendered black, in which the gap between the impeller and the alternator is considered to be non-existent, and the relationship of relative motion exists. This relative motion relationship is matched with the re-planning of the dynamic orthogonal lattice at each time step to complete the CFD calculation of the impeller rotational motion, which is similar to the moving mesh algorithm (MMA) in the finite volume method (FVM), but has stronger adaptability and accuracy to complex geometric shapes than the MMA.

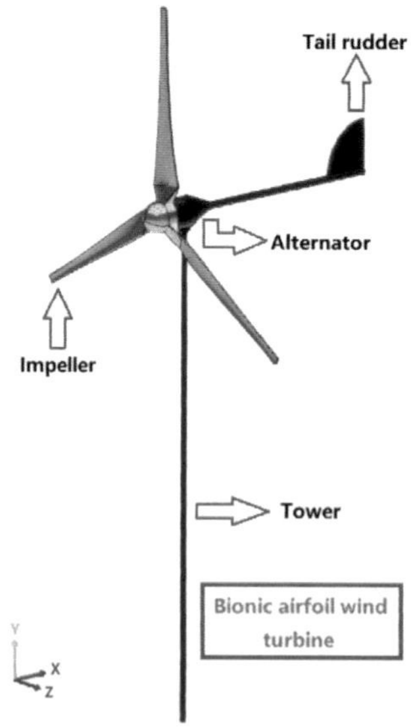

Figure 1. Bionic airfoil impeller for horizontal axis wind turbine.

The sweep surface diameter of the impeller is 1.4 m and the designed wind speed is 10 m/s, as well as the starting wind speed is 3 m/s and the rated power of the wind turbine is 300 W. Additionally, the data for standard airfoil was obtained by reverse engineering techniques scanning the sparrowhawk wing structure.

3.2. CFD domain setting and lattice layout in 3D space

The CFD domain of bionic airfoil wind turbine model is 9 (m) × 5 (m) × 6.95 (m) in the X, Y, and Z directions, respectively. There is an encrypted lattice area used in the model to better capture the fluctuations in static pressure for aeroacoustic computation, which is shown in Figure 2. The geometric size of the CFD model is established with reference to the corresponding open section wind tunnel experimental conditions to provide a reference for subsequent experimental research.

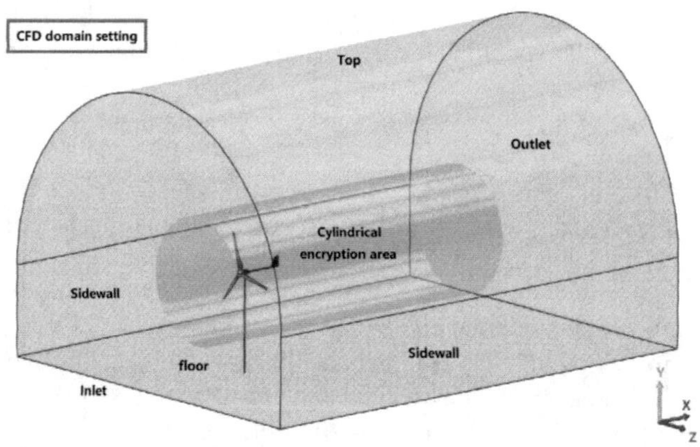

Figure 2. Calculation area setting for the CFD model.

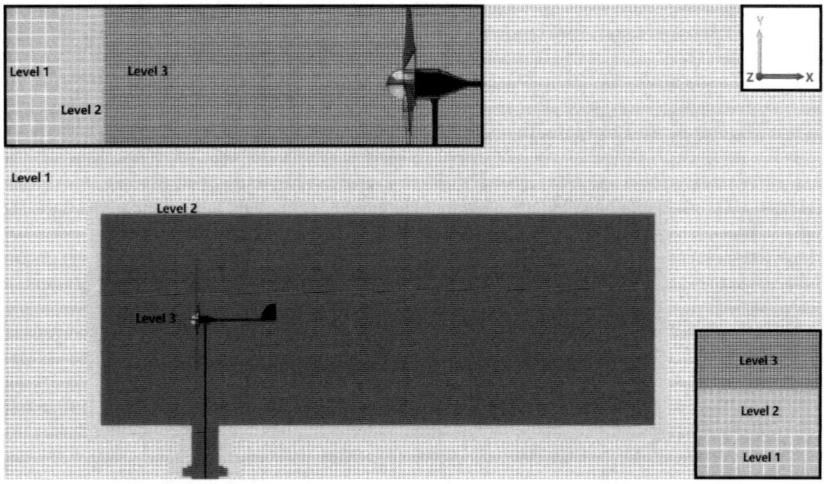

Figure 3. Lattice space discrete scheme in the CFD model.

The discrete scheme with the D3Q27of this study uses the octree lattice structure (OLS) algorithm, which offers the possibility to use non-uniform lattice structures and thus different spatial scales at different locations in the fluid domain of CFD model. Among them, different spatial scales are arranged hierarchically, and the spatial and temporal scales solved by each level are twice smaller than those of the previous level, thus forming the OLS of wind turbine CFD model in Figure 3. As the gradient of aerodynamic parameters near the wake area is at a high level, in order to better capture the detailed changes of important physical parameters, such as the static pressure value, such dense lattices are adopted here to discretize the area to provide a better basis for computational acoustics. The specific number of initial lattices in the discrete model is shown in Table 1.

Table 1. Details on the number of lattices.

Label / Time	Level 1	Level 2	Level 3	All levels
0.0 (s)	1,856,268	1,148,016	11,701,104	14,705,388
2.0 (s) roughly	1,897,712	636,971	13,137,048	15,671,731
3.9 (s) roughly	1,897,712	636,971	13,137,048	15,671,731
5.9 (s) roughly	1,897,712	636,971	13,137,048	15,671,731
7.9 (s) roughly	1,897,712	636,971	13,137,048	15,671,731

According to the analysis in Table 1, the total number of lattices is roughly at the order of 15.7 million, in which the Level 3 has the largest number of lattices at about 13.1 million.

3.3. *Solution Conditions*

The inflow boundary at the left of model is set to velocity-inflow type with the value of 10 m/s at the normal direction, matching a standard atmospheric pressure 101,325 Pa, and the related outflow boundary is specified as the pressure-outlet type. The density of the fluid medium is 1.225 kg/m^3 and the corresponding ambient temperature is 290 K with the turbulence intensity 0.5%.

The design tip speed ratio of impeller is designated as 5.5, corresponding to the rotation speed at about 750.3 r/min. Additionally, the turbulent flow characteristics are simulated by the LES model and the WALE model to promote computational efficiency and ensure computational accuracy and numerical stability.

4. Calculation Results

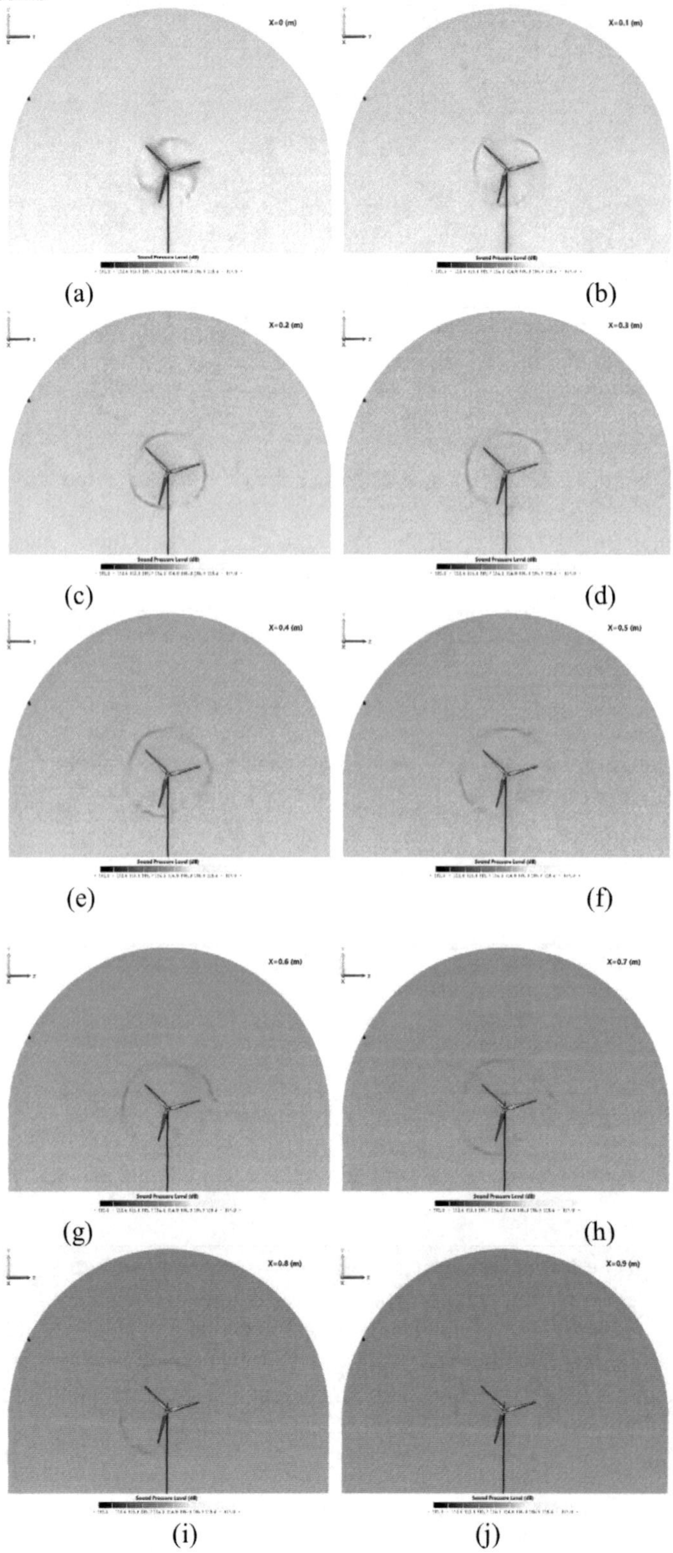

Figure 4. Variation of SPL on feature planes.

Due to the strong fluctuation of the initial wake flow field of the wind turbine, this study performed the aerodynamic acoustic calculation with the static pressure fluctuation data of CFD after the numerical flow has stabilized at 7.9 (s) approximately, in which the associated wind tunnel time-averaged background radiation value is removed. The study carried out the extraction of sound pressure level (SPL) through the feature planes, comprising the X=0 (m), X=0.1 (m), X=0.2 (m), X=0.3(m), X=0.4 (m), X=0.5 (m), X=0.6 (m), X=0.7 (m), X=0.8 (m) and X=0.9 (m). The specific calculation results are shown in Figure 4.

According to the analysis of the above calculation results, the aerodynamic acoustic characteristics of the ring appear obviously here due to the impeller rotation effect, especially for the tip vortex region. The SPL values on the series of characteristic surfaces are between 110 (dB) and 120 (dB), which are consistent with the relevant experimental measurement result [28], indicating that the calculation of the aeroacoustic model based on LBM-DCS is reasonable and effective. From Figure 4(a) to (j), the overall intensity of the acoustic radiation is decreasing gradually with the increase of the distance forward along the X-axis. The acoustic radiation intensity along the circumference of the tip vortex ring of this acoustic feature is not constant, but shows a certain random fluctuation, which should be mainly due to the asymmetric motion of the tip vortex. Furthermore, the radius of the acoustic radiation ring is gradually increasing, which is also related to the radial outward migration of the tip vortex and there is also a certain degree of weakening of the sound radiation intensity in the tower shadow effect area.

5. Conclusion

In this study, we employed the LBM-DCS method to investigate the wake sound radiation characteristics of a bionic airfoil horizontal axis wind turbine, in which the large eddy simulation model and wall-adapting local eddy-viscity are utilized to better obtain a basic CFD data.

The related calculation results show that the aeroacoustic model based on LBM-DCS is reasonable and effective. There is a aerodynamic acoustic characteristics of the ring that are clearly observed, and the acoustic radiation intensity along the circumference shows a certain random fluctuation due to the asymmetric motion of the tip vortex. As well as the tower shadow effect also reduces the intensity of the acoustic radiation in the near wake region of the wind turbine. In the follow-up research, we will combine more experimental data to carry out a further systematic investigation on the acoustic radiation characteristics of the bionic airfoil horizontal axis wind turbine.

Acknowledgments

This work has been financially supported by the Chongqing Natural Science Foundation of China (No. cstc2020jcyj-msxmX0314), the PhD Start-up Fund from Chongqing University of Science and Technology (No.181903017), and the Outstanding Post-doctoral Funding from Xinjiang Autonomous Region of China (No.2020816-2-3).

References

[1] Poulsen A H, Raaschou-Nielsen O, Peña A, et al. Impact of Long-term Exposure to Wind Turbine Noise on Redemption of Sleep Medication and Antidepressants: A Nationwide Cohort Study[J]. Environmental Health Perspectives, 2019, 127(3): 037005.

[2] Pohl J, Gabriel J, Hübner G. Understanding Stress Effects of Wind Turbine Noise-The Integrated Approach[J]. Energy Policy, 2018, 112: 119-128.

[3] Njiri J G, Beganovic N, Do M H, et al. Consideration of Lifetime and Fatigue Load in Wind Turbine Control[J]. Renewable Energy, 2019, 131: 818-828.

[4] Sun X, Dai Q, Bilgen O. Design and Simulation of Macro-Fiber Composite Based Serrated Microflap for Wind Turbine Blade Fatigue Load Reduction[J]. Materials Research Express, 2018, 5(5): 055505.

[5] Deshmukh S, Bhattacharya S, Jain A, et al. Wind Turbine Noise and its Mitigation Techniques: A Review[J]. Energy Procedia, 2019, 160: 633-640.

[6] Ocker C, Blumendeller E, Berlinger P, et al. Localization of Wind Turbine Noise Using a

Microphone Array in Wind Tunnel Measurements[J]. Wind Energy, 2021, 1-19.
[7] Li X, Xu Y, Jiang M. Research Status and Trend of WindTurbine Aerodynamic Noise[J]. Applied Mathematics and Mechanics, 2013, 34(10): 1083-1090.
[8] Møller H, Pedersen S, Staunstrup J K, et al. Assessment of Low-frequency Noise from Wind Turbines in Maastricht[J]. National Wind Watch Inc, 2012.
[9] Lee G S, Cheong C, Shin S H, et al. A Case Study of Localization and Identification of Noise Sources from a Pitch and a Stall Regulated Wind Turbine[J]. Applied Acoustics, 2012, 73(8): 817-827.
[10] Boorsma K, Schepers G. Wind Turbine Noise Measurements in Controlled Conditions[J]. International Journal of Aeroacoustics, 2017, 16(7-8): 649-665.
[11] Llorente E, Gorostidi A, Jacobs M, et al. Wind Tunnel Tests of Wind Turbine Airfoils at High Reynolds Numbers[J]. Journal of Physics: Conference Series, 2014, 524(1): 012012.
[12] Cheng J, Zhu W, Fischer A, et al. Design and Validation of the High Performance and Low Noise CQU-DTU-LN1 Airfoils[J]. Wind Energy, 2014, 17(12): 1817-1833.
[13] Brooks T F, Pope D S, Marcolini M A. Airfoil Self-noise and Prediction[M]. Washington, DC: National Aeronautics and Space Administration, Office of Management, Scientific and Technical Information Division, 1989.
[14] Moriarty P, Guidati G, Migliore P. Recent Improvement of a Semi-empirical Aeroacoustic Prediction Code for Wind Turbines[C]. 10th AIAA/CEAS Aeroacoustics Conference, 2004: 3041.
[15] Errasquin L. Airfoil Self-noise Prediction Using Neural Networks for Wind Turbines[D]. Virginia Tech, 2009.
[16] Pal P, Datta R, Rajbansi D, et al. A Neural Net Based Prediction of Sound Pressure Level for the Design of the Aerofoil[M]. Swarm, Evolutionary, and Memetic Computing and Fuzzy and Neural Computing, Springer, Cham, 2019, 2019: 105-112.
[17] Choi D K. Multilevel-Modeling Interpretation of Trailing-Edge Noise Models for Wind Turbines with NACA 0012 Airfoil[J]. International Journal of Precision Engineering and Manufacturing-Green Technology, 2021, 8(5): 1501-1514.
[18] Zhu W J, Shen W Z, Barlas E, et al. Wind Turbine Noise Generation and Propagation Modeling at DTU Wind Energy: A review[J]. Renewable and Sustainable Energy Reviews, 2018, 88: 133-150.
[19] Ghasemian M, Nejat A. Aero-acoustics Prediction of a Vertical Axis Wind Turbine Using Large Eddy Simulation and Acoustic Analogy[J]. Energy, 2015, 88: 711-717.
[20] Maizi M, Mohamed M H, Dizene R, et al. Noise Reduction of a Horizontal Wind Turbine Using Different Blade Shapes[J]. Renewable Energy, 2018, 117: 242-256.
[21] Benim A C, Diederich M, Gül F, et al. Computational and Experimental Investigation of the Aerodynamics and Aeroacoustics of a Small Wind Turbine with Quasi-3D Optimization[J]. Energy Conversion and Management, 2018, 177: 143-149.
[22] Colonius T. Direct Computation of Aerodynamic Sound[D]. Stanford University, 1994.
[23] Dorrego J R, Ríos A, Hernandez-Escobedo Q, et al. Theoretical and Experimental Analysis of Aerodynamic Noise in Small Wind Turbines[J]. Energies, 2021, 14(3): 727.
[24] Svyetlichnyy D. Neural Networks for Determining the Vector Normal to the Surface in CFD, LBM and CA Applications[J]. International Journal of Numerical Methods for Heat & Fluid Flow, 2018. DOI:10.1108/HFF-08-2017-0292
[25] Thibault S, Holman D, Garcia S, et al. CFD Simulation of a Quad-rotor UAV with Rotors in Motion Explicitly Modeled Using an LBM Approach with Adaptive Refinement[C]. 55th AIAA Aerospace Sciences Meeting, 2017, 2017: 0583.
[26] Chen C, Jiang X. Transport Property Prediction and Inhomogeneity Analysis of Supercritical n-dodecane by Molecular Dynamics Simulation[J]. Fuel, 2019, 244: 48-60.

[27] Zheng X, Montazeri H, Blocken B. CFD Simulations of Wind Flow and Mean Surface Pressure for Buildings with Balconies: Comparison of RANS and LES[J]. Building and Environment, 2020, 173: 106747.
[28] Wu W, Yao J, Dai Y, et al. Research on Effect of Fusion TIp Structure Change on Flow Field and Noise Field in Tip Region of Wind Turbine[J]. Acta Energiae Solaris Sinica, 41(12): 190-198.

Study on Rotating Stall of Multistage Compressor Using Wavelets

Ziwei Wang[1, 2], Shuai Ma[1], Jiantao Wang[1], Bin Li[1] and Pengcheng Cui[1, *]

[1] Computational Aerodynamic Institution, China Aerodynamics Research and Development Center, Mianyang, China
[2] National University of Defense Technology, Changsha, China

*Corresponding author e-mail: pccui@foxmail.com

Abstract. Rotating stall in the high speed compressor can lead to violent disruption of flow, damage to the blade structures and, eventually, engine shutdown. To investigate the mechanism of stall inception and stall propagation in multi-stage compressor, the onset and transient behavior of rotating stall in a transonic 2-stage compressor are simulated based on the in-house software ASPAC. Many sampling points are placed to monitor the flow variation. The analysis results of circumferential sampling points indicate that, the spectral characteristics of circumferential sampling points in steady state and unstable state are similar, the pressure pulsation is similar, but there exist phase difference. In the transition state, the main frequency is similar and the secondary frequency is slightly different, the 17th blade channel at the blade top of rotor 2 produces a disturbance with lower frequency and higher amplitude before other channels, which triggers the stall disturbance. The analysis results of axial sampling points show that the blade row interference frequency exists in the whole flow change process, and the disturbance of adjacent blade rows is dominant. The stall disturbance starts with rotor 2. This initial disturbance soon spreads to its adjacent stator 1 and stator 2. Rotor 1 is finally affected and enters into unstable state at about the 12th revolution.

1. Introduction

Aerodynamic stability of compressor is one of the key factors influencing performance, stability, and reliability of aero-engine. As one of the main forms of aerodynamic instability in compressors, rotating stall will cause high pressure pulsation, vibration, even structure resonance, so it should be avoided during the compressor operation. However, the flow mechanism of rotating stall is not fully understood [1]. Nowadays, experimental test and comprehensive measurement during rotating stall, especially for multi-stage high speed machines, remain a challenging task [2]. Simulating the whole process of rotating stall development based on CFD is an effective way to uncovering the flow mechanism.

3D unsteady Reynolds-averaged Navier-Stokes (URANS) methods have been successfully applied to simulation of rotating stall phenomenon in compressors [3]~[12], and most of the studies concentrated on rotors or single-stage compressors. It is pointed by Wu [12] that, flow characteristics of rotating stall in single-row environment are different from that in multi-row environment. However, very limited work can be found on multi-stage compressors [2].

Through suitably characterizing instability of compressors flow under rotating stall, the underlying physical mechanism can be partly revealed. Wavelet analysis was introduced as a very useful tool to extract features of unsteady disturbances, which can track the instability characteristics of compressors well [13] [14]. However, most of the analysis concentrates on regions near rotor tip, the propagation and influence of rotating stall in the whole compressor is rarely researched.

This paper presents an investigation into the behavior of rotating stall in a 2-stage high speed axial compressor with full-annulus unsteady method, the wavelet analysis is used to study the flow characters along the entire compressor flow path, both axially and circumferentially.

2. Methodology

2.1. Flow solver

The simulation is conducted based on the in-house code ASPAC. The 3-D compressible Navier-Stokes equations is solved by fully-implicit scheme with a cell-centered finite volume method. The inviscid flux is evaluated by Roe Scheme with the Van Albada limiter. The viscous flux is determined in a central differencing manner with Gauss's theorem. The second-order backward difference is applied to the temporal derivative and the inner iteration is conducted at each time step. The current computations use the one-equation Spalart-Allmaras turbulence model. The parallelization is conducted based on MPI.

Taking Rotor 37[15] as an example, comparison between simulation results of ASPAC and experiment results is shown in Fig. 1. ASPAC_RANS means the steady reynold averaged Navier-Stokes equations are solved, ASPAC_URANS means the unsteady reynold averaged Navier-Stokes equations are solved. The results of ASPAC match well with experimental results.

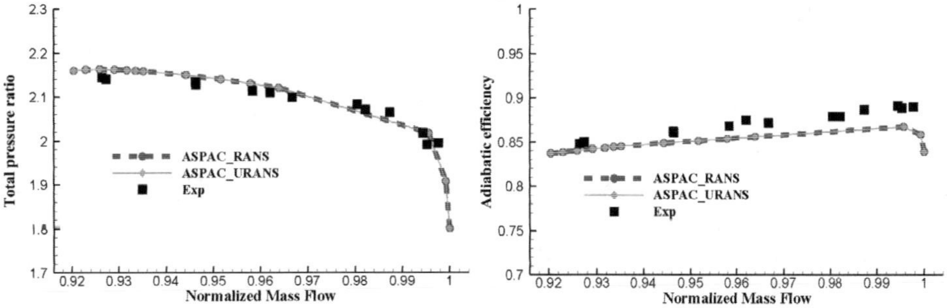

(a) total pressure ratio (b) adiabatic efficiency

Figure 1. Rotor 37 speedline

Taking Stage 35[16] as an example, comparison between simulation results of ASPAC and experiment results is shown in Fig. 2. In terms of ASPAC_RANS, the mixing plane method is applied. In terms of ASPAC_URANS, the dynamic overlapped grid technique is applied at rotor/stator interface. The results of ASPAC match well with experimental results.

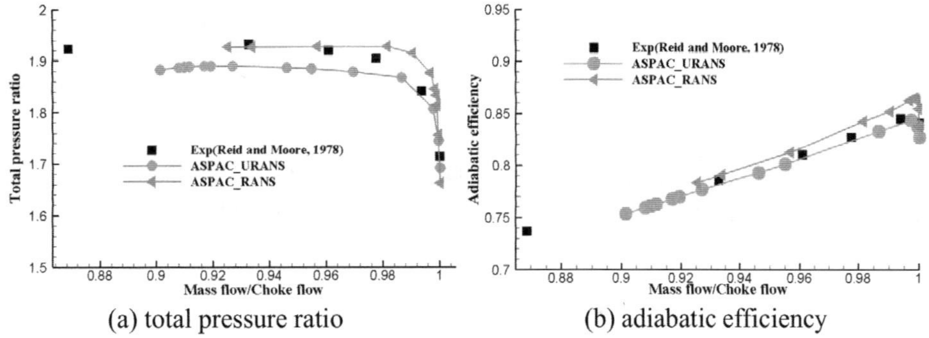

(a) total pressure ratio (b) adiabatic efficiency

Figure 2. Stage 35 speedline

2.2. Computation domain and boundary conditions

The 2-stage compressor contains four blade rows, which are rotor1, stator1, rotor2 and stator2, where rotori denoted rotor of ith stage and statori denoted stator of ith stage. The blade number of the rows are 23, 25, 29 and 41, respectively. The design speed is 33790rpm.

Numerical model of the 2-stage compressor is shown in Fig. 3(a). Full-annulus grid of the compressor is shown in Fig. 3(b). The space distance of the first layer of grid is 3e-6m and y+ is 1. Cell number of the full-annulus grid is 83 million.

(a) Numerical model

(b) Grid

Figure 3. Numerical model and grid of the 2-stage compressor

During the simulation, the inlet total pressure is 101325Pa, the inlet total temperature is 288.15K, the axial inflow is adopted. The rotational speed is set to 33790rpm.

The speedline of the compressor is simulated via mixing plane method (ASPAC_RANS), the static pressure at hub of outlet is set and the radial equilibrium equation is applied. When it comes to near stall condition, the steady flow is used as the initial flow field for unsteady computation (ASPAC_URANS), the boundary conditions remain the same.

Based on the convergent unsteady flow field, the development of rotating stall is simulated, where the boundary condition at outlet is changed to "choked" throttle model that specifies corrected mass flow at the exit is used. This boundary condition allows variation of exit static pressure to match the compressor exit mass corrected to the exit total condition. The formula is as follows:

$$P_c = P_a + \frac{\dot{m}_r^2}{k_t} \tag{1}$$

Where P_c is static pressure at hub of outlet, P_a is the environment pressure which is set as 101325Pa in current study, $\dot{m}_r$ is mass flow at outlet, K_t is a parameter to control the throttle area. After P_c is calculated, the radial equilibrium equation is applied at outlet.

When applying the choked throttle model, K_t is calculated at the first step, which is then multiplied by 0.95 and stay unchanged during the whole process of unsteady simulation, thus the rotating stall can be triggered more quickly this way.

The full-annulus simulation is conducted using 256 ARM V64 CPU cores. A temporal discretization level of 2000 physical time steps per rotor revolution and 50 inner iteration steps are used. The CFL number is set to 5.

3. Result analysis

3.1. Global parameters

The speedline of the compressor is simulated via mixing plane method (ASPAC_RANS). The result is shown as the red line in Fig. 4(a). When it comes to near stall condition, the steady flow is used as the initial folw field for unsteady computation (ASPAC_URANS), after the unsteady computation is converged, the choked throttle model is used. The onset and development of rotating stall in the compressor is shown as the green line in Fig. 4(a). After fully developed, the rotating stall is marked in balck circle as shown in Fig. 4(a).

Variation of mass flow during stall development is shown in Fig. 4(b). At about 8.5th revolution, the throttle is applied, the total pressure at outlet increases sharply, the mass flow decreases sharply. At about 16th revolution, the mass flow drops to the lowest point. The mass flow then increases gradually, after 6 revolutions, the mass flow reaches to the rotating stall condition as marked in Fig. 4(a). After entering into rotating stall state, the fluctuation range of inlet mass flow is larger than that of outlet mass flow, which indicates that, the stall cells have bigger influence on inlet of rotor than outlet of stator, due to the rectification of stator.

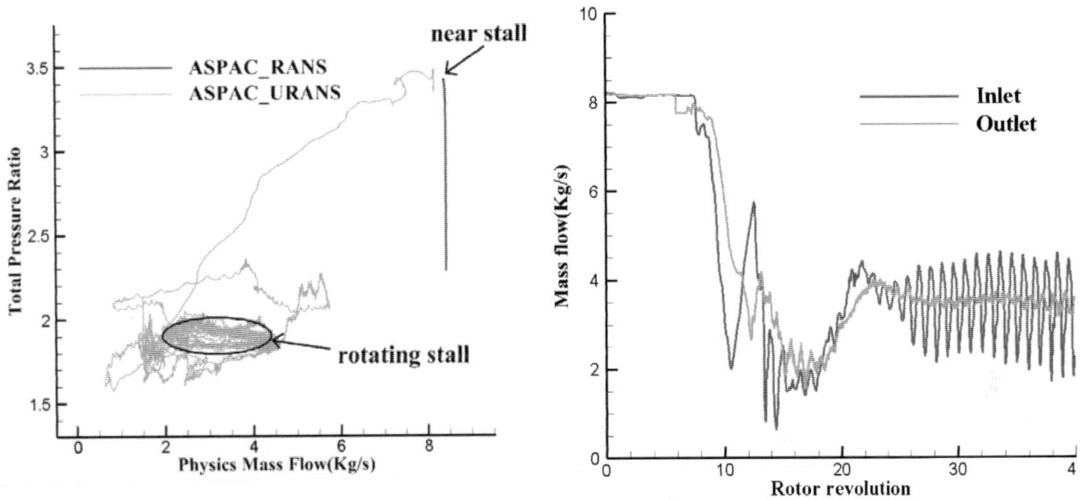

(a) Speedline (b) Mass flow

Figure 4. Global parameters variation during onset and development rotating stall

3.2. Wavelet analysis

To obtain high-resolution information regarding the unsteady flow field, a set of refined sample points were arranged in the computational domain, as shown in Fig. 5. In the rotor domain, the sample points are placed near the casing wall to sample the influence of rotor tip clearance on flow during rotating stall. In the stator domain, the sample points are placed near the middle span to sample the overall variation of flow in stator domain. In the blade domains (rotor or stator), the sample points are placed from inlet to outlet, which is used to analyse the propagation characters of rotating stall in the compressor. The sample points are placed in every blade passage to sample flow field in the whole annulus, the sample points are distributed along the positive direction of the rotating shaft. The sample points were arranged in the relative frames.

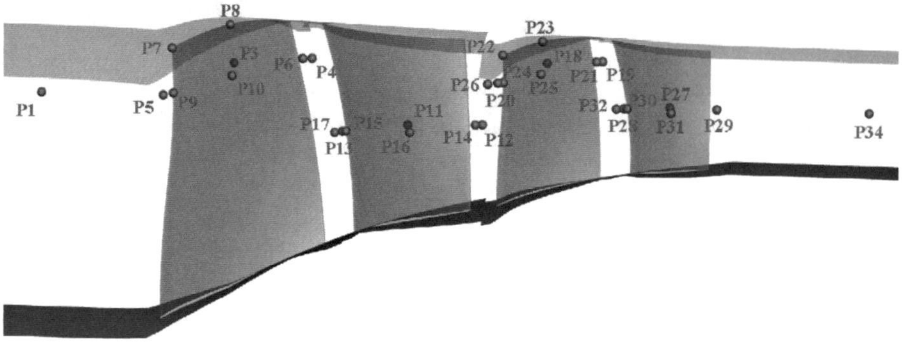

(a) Looking from the circumferential direction

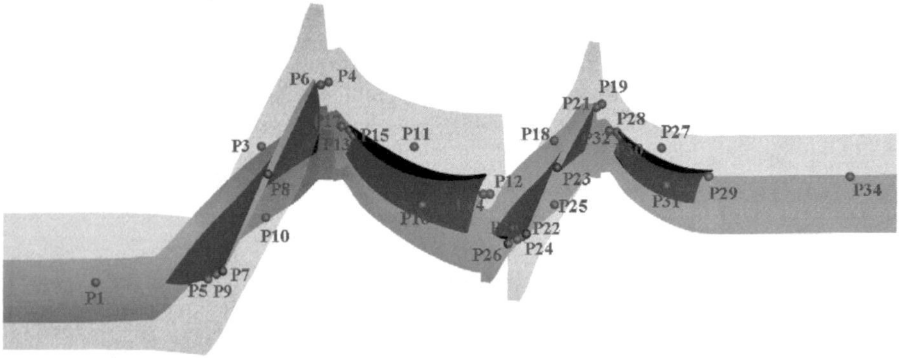

(b) Looking from above

Figure 5. Distribution of the sample points

It is generally believed that the flow in different blade channels is periodic circumferentially in the same blade row of compressor. However, in the process of compressor instability development, this periodicity will be destroyed. In order to explore how stall influence the flow periodicity, the pressure pulsations and corresponding wavelet analysis of P8 (rotor 1 tip) and P23 (rotor 2 tip) in all blade channels in the whole annulus are investigated respectively. This is because rotating stall often occurs near the blade tip, and the sampling points at the tip clearance can well reflect the characteristics of unstable flow.

Limited to the length of the article, one sampling point is selected for every four blade channels. It can be seen from Fig. 6 that the flow state can be divided into three parts, namely, The near stall steady flow state (steady state) before about the 7th revolution, the complete stall state (unstable state) after about the 20th revolution, and the transition state between the two states, i.e. stall development state (transition state). The instability disturbance is marked by a high amplitude region with a rotation frequency smaller than rotor rotation frequency, which is the same as the conclusion that the stall disturbance frequency is less than the rotation frequency in previous studies [1].

During steady state and unstable state, the spectral characteristics between different blade channels of the same blade row are very similar, and the pressure fluctuation curves differ by a certain phase, indicating the propagation of the disturbance along the circumferential direction.

In the transition state, the flow frequency characteristics between different blades can be summarized as follows: the main frequency is the same, and the secondary frequency distribution is different. The secondary frequency indicates the stall inceptions occur in one or more blade channels in the process of stall development. The 17th blade channel at the blade top of rotor 2 produces a disturbance with lower frequency and higher amplitude before others blade channels, which triggers stall disturbance.

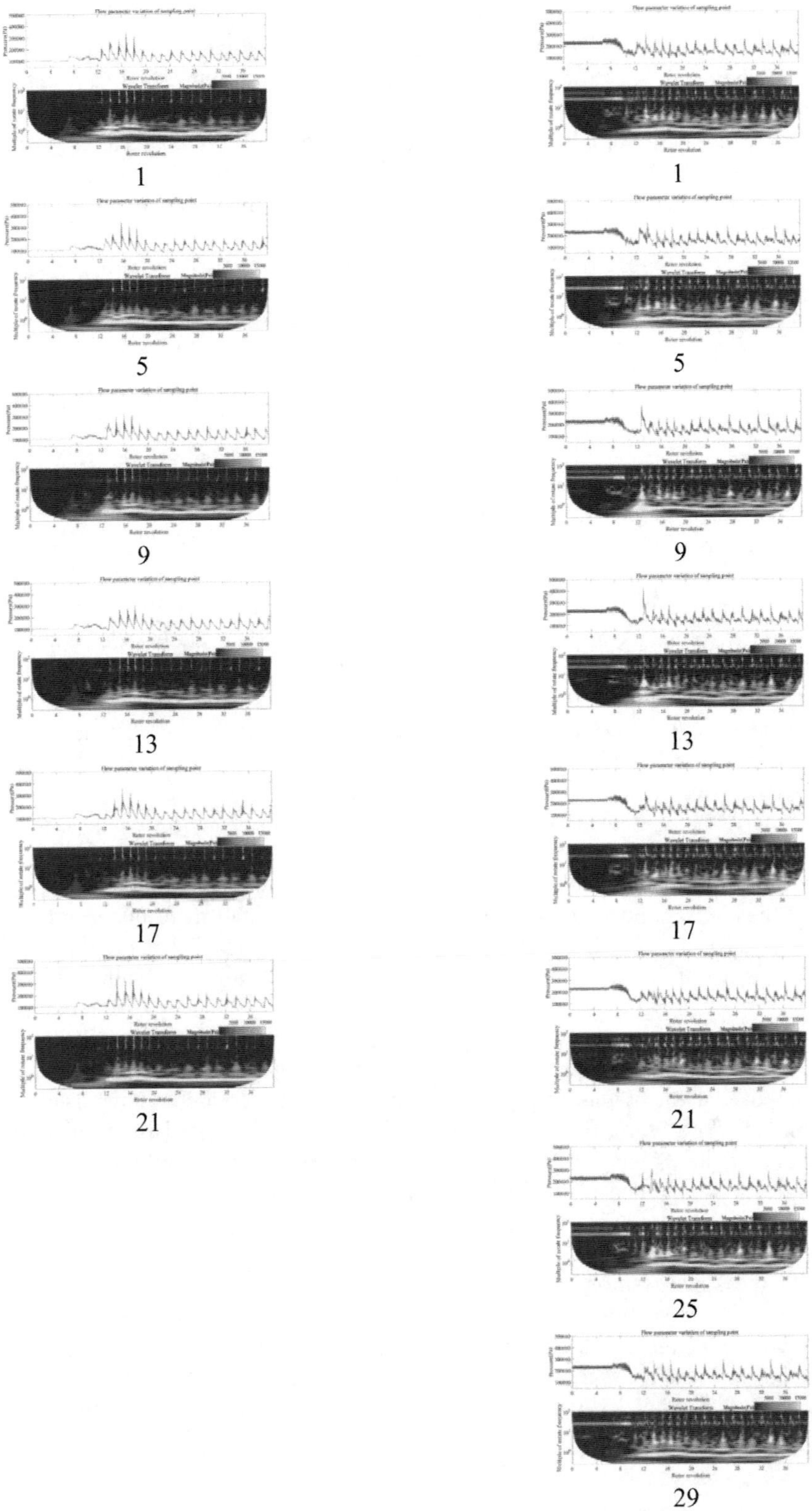

(a) P8 (rotor 1 tip) (b) P23 (rotor 2 tip)

Figure 6. Wavelet analysis results of P8 and P23

In conclusion, along the circumferential direction, the spectrum characteristics of sampling points between different blade channels are very similar in steady state and unstable state, and the pressure pulsation is similar, but there is phase difference; there is slight difference in the transition state, but the dominant frequency is the same. Based on the above analysis, the wavelet analysis results of a blade channel sampling point can roughly represent the main characteristics of flow at this axial position.

Therefore, when analysing the wavelet analysis results of sampling points at different axial positions, the sampling point of the first blade channel is selected, the axial sampling points are chosen as follows:

Table 1. Chosen axial sampling points

	Inlet of the blade row	Middle of the blade row	Outlet of the blade row
Rotor 1	P1	P3	P4
Stator 1	P17	P11	P12
Rotor 2	P26	P18	P19
Stator 2	P32	P27	P34

The wavelet analysis results of sampling points at different circumferential positions are shown in Fig. 7. The flow is still divided into stable state, transition state and unstable state. Frequencies can also be divided into three categories. They are the blade row interference frequency in the multiple of rotor frequency range [10, 100], which exists in all flow states. The stall cell frequency and its integer multiple or superimposed with other disturbances is the multiple of rotor frequency range [0, 10], this frequency remains stable in the instability stage and has high amplitude; finally, the frequency existing between the multiple of rotor frequency range [0, 1] in the transition state period is called the stall inception frequency. In the transition state period, the amplitude is high and basically disappears in the unstable state. It can be used to analyse the flow characteristics in the process of stall development.

The causes of blade row interference frequency are as follows:

Standing in a rotating coordinate system that is stationary relative to the current blade row (the rotation speed of the coordinate system is the same as that of the current blade row). Due to the relative rotation of other blade rows and the current blade row, the ratio of disturbance frequency to rotor frequency is the number of blades of other blade rows, the multiple of the number of blades, or the linear combination between the blade numbers of different rows. The wavelet analysis results clearly show the periodic disturbance induced by the number of blades of adjacent blade rows. The blade row interference disturbance frequency is about an integral multiple of the number of adjacent blades, which is the inherent characteristic of unsteady flow in the compressor. The blade row interference frequency exists in the whole flow process, and the amplitude at some time under unstable state is generally higher than that under steady state. It can also be seen from the pressure pulsation diagram that many high-frequency pulsations are superimposed in the pressure peak area under unstable state, which is the pulsation caused by blade row interference. In a word, the blade row interference frequency exists in the whole flow development process, and in the unstable state, the pressure fluctuation amplitude of blade row interference at some times is large due to the destruction of flow stability by stall cells.

As can be seen from Fig. 7(a) and Fig. 7(c), in the rotor domain, after about the 12th revolution, there is a stall disturbance with about 0.4 times of the rotor frequency, and the amplitude is relatively large. In the stator domain, the superposition of blade rotation disturbance and stall disturbance plays a dominant role. It should be noted that in the transition state between the 6th and 12th revolution, only the dominant frequency with large amplitude appears in the region of rotor 2, between [0, 1] of the rotor frequency, as shown in Fig. 7(c). This shows that the stall disturbance starts with rotor 2, and this initial disturbance soon spreads to its adjacent stator 1 (Fig. 7(b)) and stator 2 (Fig. 7(d)). As can be seen from Fig. 7(a), rotor 1 is finally affected and enters into unstable state at about the 12th revolution.

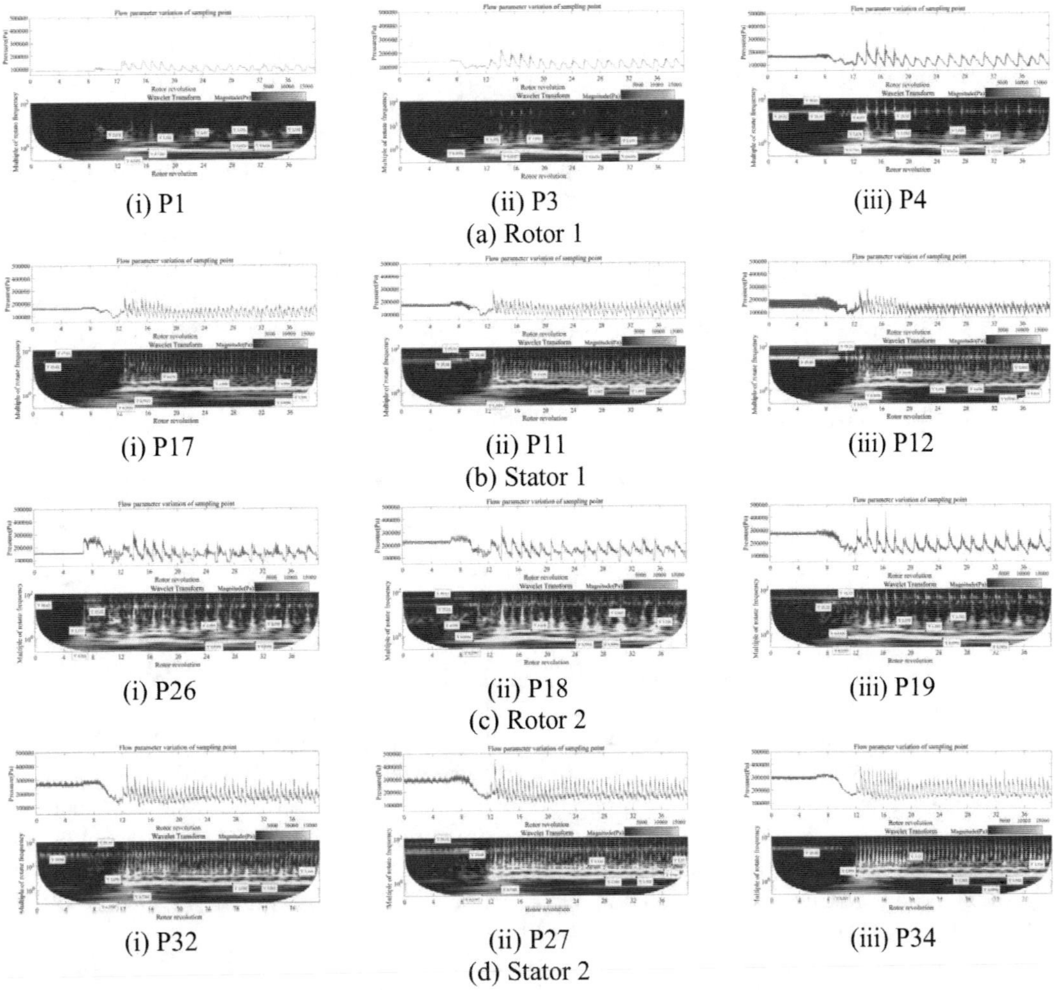

Figure 7. Wavelet analysis results of sampling points at different axial positions

4. Conclusion

After entering into rotating stall state, the fluctuation range of inlet mass flow is larger than that of outlet mass flow, which indicates that, the stall cells have bigger influence on inlet of rotor than outlet of stator, due to the rectification of stator.

The analysis results of circumferential sampling points show that the spectral characteristics of circumferential sampling points in steady state and unstable state are similar, the pressure pulsation is similar, but there exist phase difference. In the transition state, the main frequency is similar and the secondary frequency is slightly different. It can be seen that the 17th blade channel at the blade top of rotor 2 produces a disturbance with lower frequency and higher amplitude before other channels, which triggers the stall disturbance.

The analysis results of axial sampling points show that the blade row interference frequency exists in the whole flow change process, and the disturbance of adjacent blade rows is dominant. The stall disturbance starts with rotor 2. This initial disturbance soon spreads to its adjacent stator 1 and stator 2. Rotor 1 is finally affected and enters into unstable state at about the 12th revolution.

Acknowledgments
The authors would like to thank Dr. Junbing LIU and Dr. Jingfan CHEN of the National University of Defense Technology for their helpful discussing on the data processing technology.

References

[1] C.S. Tan, I. D., S. Morris, A.Wadia, Spike-Type Compressor Stall Inception, Detection, and Control. Annu. Rev. Fluid Mech. 42 (2010) 275-301.
[2] Fanzhou Zhao, J. D., Mehdi Vahdati, Post-stall Behaviour of a Multi-stage High Speed Compressor at Off-design Conditions. Proceedings of ASME Turbo Expo 2018. (2018).
[3] J.L. Lu, W. Chu, and K Peng, Numerical and experimental research of stall inception on subsonic axial-flow compressor rotor. J. Aerosp. 224 (2010) 1027-1039.
[4] J.A. Khan, K. P., S. Ahmad and A. Mushtaq, Effect of Circumferential Grooves and Tip Recess on Stall Characteristicistics of Transonic Axial Compressor Rotor, 49th AIAA Aerospace Sciences Meeting including the New Horizons Forum and Aerospace Exposition Orlando, Florida. (2011)
[5] X.Y. Yang, F.F. Ning, Numerical investigation of flow field characteristicistics at near stall condition in axial compressor, Gas Turbine Experiment and Research, 24 (2011) 29-34.
[6] P.F. Ju, F.F. Ning, Numerical study of near-stall flow feature on transonic compressor, Journal of Aerospace Power, 37 (2016) 1055-1064.
[7] S. Kim, G. P., C. A. Hall, R. P. Grewe, M. J. Wilson, E. Gunn, Stall Inception in Low-Pressure Ratio Fans. J Turbomach. 141 (2019) 071005.
[8] Q.P. Li, Y.H. Wu, W.L. Chu, Full-annulus simulations on axial compressor rotor at near stall condition, Journal of Aerospace Power, 26 (2011) 2330-2338.
[9] J.P. Chen, M. Hathaway, G.P. Herrick, Prestall Behavior of a Transonic Axial Compressor Stage via Time-Accurate Numerical Simulation. J Turbomach. 130 (2008) 041014.
[10] M. Zaki, L.N. Sankar, and S. Menon, Hybrid Reynolds Averaged NavierStokes KineticEddy Simulation of Stall Inception in Axial Compressors, Journal of Propulsion and Power, 26 (2010) 1276-1282.
[11] N. Gourdain, S. Burguburu, F. Leboeuf, G.J. Michon, Simulation of rotating stall in a whole stage of an axial compressor, Cumput Fluids, 39 (2010) 1644-1655.
[12] Y.H. Wu, X. Wang, F.Y. Lang, Numerical simulation of stall inception in single-stage axial-flow compressor, Journal of Aerospace Power, 29 (2014) 125-132.
[13] J. Schreiber, B. Paoletti, and X. Ottavy, Observations on Rotating Instabilities and Spike Type Stall Inception in a High-Speed Multistage Compressor, Int. J. Rotating Mach., 2017, pp. 7035870.
[14] F. Lin, J. Chen, and M. Li, Wavelet Analysis of Rotor-Tip Disturbances in An Axial-flow Compressor, J. Propul. Power, 20 (2004) 319–334.
[15] J. Dunham, CFD Validation for Propulsion System Components, AGARD-AR-335. (1998).
[16] L. Reid, R.D. Moore, Performance of single-stage axial-flow transonic compressor with rotor and stator aspect ratios of 1.19 and 1.26, respecitively, and with design pressure ratio of 1.82, Tech. Rep. TP-1338, National aeronautics and space administration. (1978).

The comparison of adjoint-based grid adaptation and feature-based grid adaptation method

Pengcheng Cui, Yongqian Zheng, Peimin Xu, Bin Li, Mingsheng Ma and Guiyu Zhou*

Computational Aerodynamics Institute, China Aerodynamics Research and Development Center, Mianyang 621000, China

* Corresponding author e-mail: pccui@foxmail.com

Abstract. Grid adaption is a popular method to enhance the resolution of flow field and the precision of numerical simulation, which automatically optimizes the grid distribution instead of manual complicated work. There exist usually two grid adaptation methods, the feature based grid adaption and adjoint based grid adaption, the former focuses on shocks, vortexes and other features of flow field, and the latter focuses on lift, drag and other aerodynamic characteristics. The comparison of adjoint based grid adaption and feature based grid adaption method is investigated in this paper. Numerical simulations show that both feature adaption and adjoint adaption could improve the resolution of flow field and the precision of numerical simulation such as lift and drag. As for the flow features, the feature adaptation could capture the obvious shock waves and vortexes in the flow field, the adjoint adaptation, however, only captures the flow features that are contributory to the accuracy of aerodynamic characteristics. As for the aerodynamic characteristics, some shock waves and vortexes have little influences to the forces, so the feature adaptation is not efficient as adjoint adaptation, which could greatly improve the accuracy of aerodynamic characteristics.

1. Introduction

Numerical simulations play an important role in the design process in aerospace industry recently, which provide more details of flow field than wind tunnel [1]. Grid is the basis of computational fluid dynamics, grid adaptation is a popular method to enhance the resolution of flow field and the precision of numerical simulation, which automatically optimize the grid distribution instead of manual complicated work [2-3]. Higher resolution flow field and more accurate aerodynamic characteristics can be obtained with less grids to maximize computational efficiency. At the same time, the mesh adaptation optimization reduces the dependence on the initial mesh quality, reduces the requirements for engineers' work experience, enables beginners to easily obtain high-quality CFD data, minimizes the intervention of "human" to the greatest extent, and frees engineers from the tedious work of repeatedly adjusting the mesh [4].

There exist usually two grid adaptation methods, the feature based grid adaption and adjoint based grid adaption, the former focuses on shocks, vortexes and other features of flow field, and the latter focuses on lift, drag and other aerodynamic characteristics. The feature-based grid adaptation tries to improve the identification of shock wave, vortex and other flow phenomenon, which automatically detect the flow characteristics and refine the mesh near flow features [5-8]. The adjoint equation establishes the mathematical relationship between aerodynamic force, flow variables and local residual. Through the adjoint theory, the calculation error of local residual on target output can be quantified, and the local grid regions with great influence on the calculation error of target output can be detected, providing theoretical guidance for grid adaptive optimization [9-12]. Adjoint method based on the error estimation and mesh adaptive is more suitable for the calculation of the aerodynamic characteristics of the aircraft.

The comparison of adjoint based grid adaption and feature based grid adaption method is presented in this paper. The structure of this article is as follows. Section 2 is introduction of numerical method, section 3 is the results and discussion, and finally section 4 is a brief conclusion.

2. Computational Method

The numerical computations in this paper are implemented by the China Aerodynamics Research and Development Center, where there is an inhouse flow solver NNW-FlowStar [13-14], which has developed feature-based grid adaptation and adjoint-based grid adaptation. FlowStar is an unstructured finite volume CFD solver, which attended several drag and hi-lift workshops, the results show that FlowStar behaved excellently in predicting aerodynamic characteristics [15-16].

2.1. Governing equation

The governing equation is Navier-Stokes equation, which are discretized by finite volume strategy.

$$\frac{\partial}{\partial t}\iiint_V Q dV + \oiint_{\partial\Omega}\left(H(Q)\cdot\hat{n} - Q\vec{v_g}\cdot\hat{n}\right) dS = \oiint_{\partial\Omega} H_v(Q)\cdot\hat{n}\, dS \tag{1}$$

2.2. Feature based grid adaptation

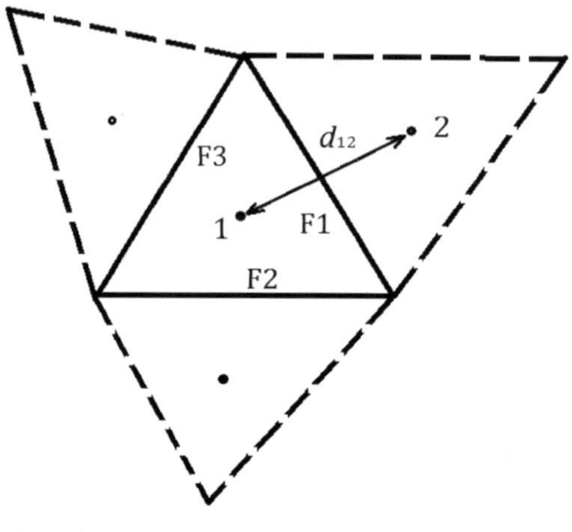

Figure 1. Templates of quasi gradient computing

Quasi-gradient method is used to capture the shocks in flow field. The theoretical basis of quasi-gradient adaptive criterion assumes that the calculation error will be larger in the region with larger flow gradient. In this paper, the density ρ, pressure p, velocity v, and temperature T are weighted to get a comprehensive gradient.

$$q = w_1\cdot\frac{\rho}{\rho_\infty} + w_2\cdot\frac{p}{p_\infty} + w_3\cdot\frac{|\boldsymbol{v}|}{|\boldsymbol{v}_\infty|} + w_4\cdot\frac{T}{T_\infty} \tag{2}$$

Where w_1-w_4 are the weight of density, pressure, velocity and temperature. The weight acts as a switch, when only one gradient of flow field variable is used, the weight of other variables is assigned to 0. Mesh adaptation also needs to consider the influence of mesh size. The pseudo-gradient value of weighted flow field variable in this paper is calculated as follows:

$$\nabla q = (q_1 - q_2)(d_{12})^P \tag{3}$$

In formula (3), subscripts 1 and 2 refer to the grid cells on both sides of the surface of a grid, as shown in Fig.1, and d is the body-center distance of the cell on both sides, and the index P is the distance influence factor.

2.3. Adjoint based grid adaptation
Adjoint equations can be expressed as following:

$$\left(\frac{\partial \boldsymbol{R}}{\partial \boldsymbol{U}}\right)^T \boldsymbol{\Psi} = \left(\frac{\partial f}{\partial \boldsymbol{U}}\right)^T \tag{4}$$

Where R is the residual of flow field equations, f is the objective function. The adjoint equation interpretates the relationship of residual, objective function and flow variables. Adjoint variables refer to the sensitivity of target function to local perturbation.

The solution strategy of adjoint equations is combined in our software. The LU-SGS method and multi grids are used to get the adjoint variable [17, 18].

The discretization error of target function can be calculated by adjoint equations. The estimation of objective function can be expressed as:

$$f_h(\boldsymbol{U}_h) \approx f_{est} = f_h(\boldsymbol{U}_h^*) - \boldsymbol{\Psi}_h^{*T} \boldsymbol{R}_h(\boldsymbol{U}_h^*) \tag{5}$$

Where f_{est} is the objective function, $f_h(\boldsymbol{U}_h^*)$ is objective function estimated by interpolation flow variable, $\boldsymbol{\Psi}_h^{*T}\boldsymbol{R}_h(\boldsymbol{U}_h^*)$ is the error correction of objective. $\boldsymbol{U}_h^*$ and $\boldsymbol{\Psi}_h^*$ are flow solution and adjoint variables.

The adaption parameter provides the guidance of arid adaptation [13]:

$$\varepsilon = \frac{1}{2}\left\{\left|\left[\boldsymbol{R}_h(\boldsymbol{\Psi}_h^Q)\right]\left[\boldsymbol{U}_h^Q - \boldsymbol{U}_h^L\right]^T\right| + \left|\left[\boldsymbol{\Psi}_h^Q - \boldsymbol{\Psi}_h^L\right]^T\left[\boldsymbol{R}_h(\boldsymbol{U}_h^Q)\right]\right|\right\} \tag{6}$$

2.4. Grid refinement method
A set of dense meshes is needed for error correction and mesh adaptive techniques based on adjoint theory. In this paper, the embedded fine meshes are obtained by global subdivision of coarse meshes, and the global objective function is modified and the adaptive parameters are calculated There are many methods to divide tetrahedron. In order to improve the quality of the fine mesh elements and simplify the methods [19]. The dense mesh tetrahedron obtained by the method of one-half tetrahedron and one-quarter tetrahedron is not similar to the original tetrahedron. If the original tetrahedron is poor, the quality of the dense mesh tetrahedron obtained after splitting will be worse In order to keep the shape of the original tetrahedron as much as possible, in this paper, the method of one-eight-point mesh division is adopted uniformly, that is, a new grid point is added on each side of the tetrahedron to divide the original tetrahedron into eight small tetrahedron, and the newly added grid point is located at the midpoint of the tetrahedron edge. There are three different subdivision methods of tetrahedron because of the different connection methods of points. This paper selects a subdivision method that minimizes the maximum dihedral Angle of the embedded fine mesh.

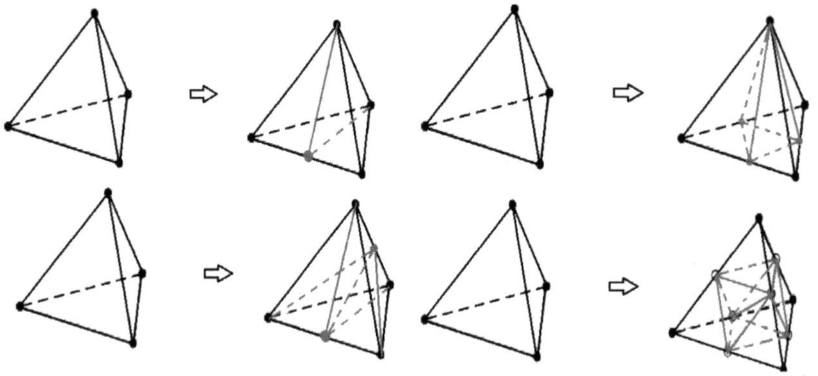

Figure 2. Tetrahedron embedding methods

3. Results and Discussion

Grid adaptation method is applied to the simulation of inviscid flows around two-dimensional and three-dimensional configurations, Euler equations and tetrahedron grids are employed. The difference of feature-based grid adaptation and adjoint-based grid adaptation is compared by same case.

The NASA0012 airfoil is used to study the difference of feature based grid adaptation and adjoint-based mesh adaption:

$$\begin{aligned} &M_\infty = 0.84,\ \ \alpha{=}3.06^\circ, \\ &T_\infty{=}255.56K,\ Re{=}1.172\times10^7 \end{aligned} \tag{7}$$

3.1. The differences of adapted grids

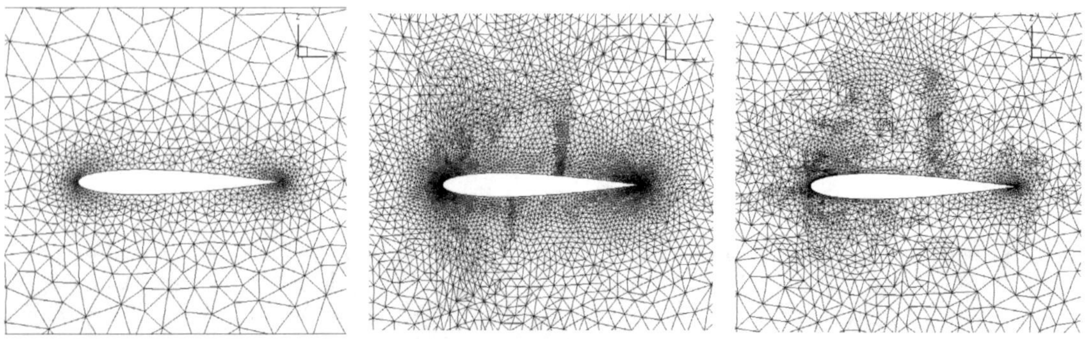

(a) Initial grid (b) Feature adaptation (c) Adjoint adaptation

Figure 3. Grids difference

The same initial mesh is optimized by feature based grid adaptation and adjoint-based grid adaptation method, Fig.3 shows the whole mesh of different methods, and the finally number of grids are almost same. Different mesh smoothness: the spatial distribution of feature adaptive mesh is smoother than that of adjoint adaptive mesh. Different degree of grid encryption: feature adaptive encryption is very fine at the shock position, while the encryption is coarse at the shock position accompanied by adaptive encryption; Different mesh encryption ranges: the adjoint adaptive mesh has a great degree of encryption near the front and rear edges of the wing, and a mesh similar to the boundary layer appears near the leading edge of the wing, and the encryption degree is relatively large at the trailing edge of the wing, while the feature adaptive mesh has a small degree of encryption near the front and rear edges of the wing.

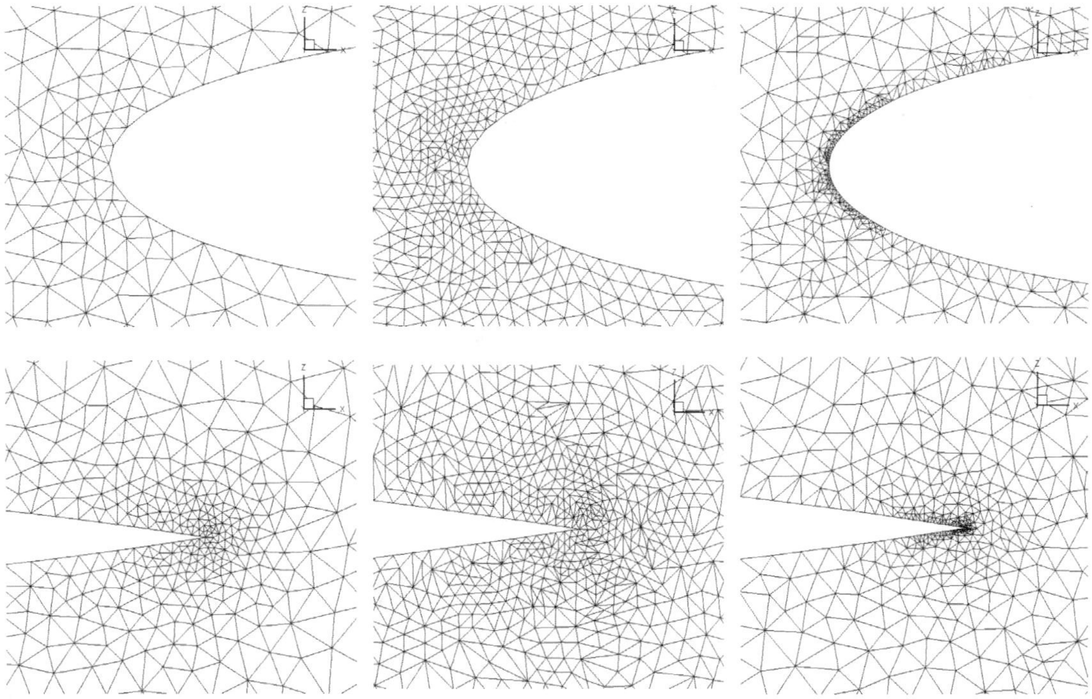

(a) Initial mesh (b) Feature adaption (c) Adjoint adaption

Figure 4. The detail of grids

There are other differences between the grids of different adaptation methods. Fig.4 shows the detail of grids difference in the leading edge and trailing edge. It can be seen that both two method refine the grids near the leading edge and trailing edge of airfoil, the grid distinguishability of initial mesh is not enough. However, the adjoint-based adaption refines the grids near the leading edge and trailing edge of airfoil more apparently. The grids near the leading edge and trailing edge of airfoil are considered more sensitive to aerodynamic characteristics such as drag and lift, these areas are more contributive to the accuracy of aerodynamic characteristics. And the grids near the weak shock wave in the lower surface is considered less contributive to the accuracy of aerodynamic characteristics.

3.2. The differences of flow field

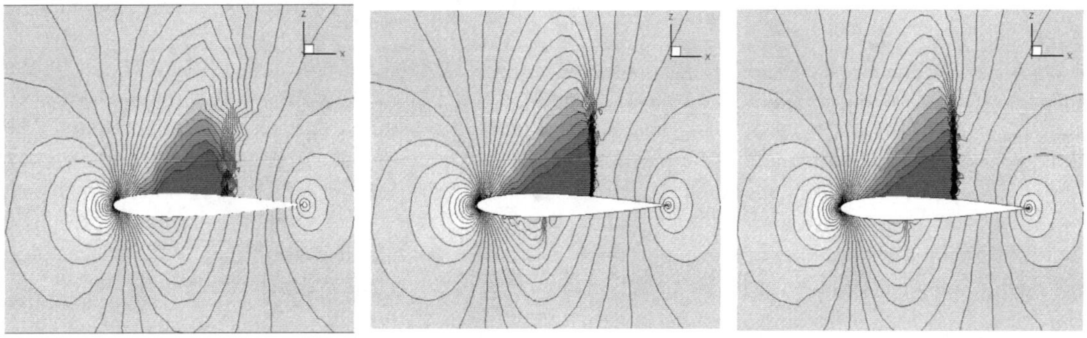

(a) Initial mesh (b) Feature adaptation (c) Adjoint adataption

Figure 5. Flow field

The grids are the basis of numerical simulation, and the different meshes could generate different flow fields. Fig.5 shows the flow field of initial grid and different grid adaptation method. The initial mesh is not enough to get an accurate flow field, the strong shock on the upper face is too wide, there is almost no weak wave on the lower surface. Two grid adaptation methods can both enhance the definition of flow field, and the feature adaptation could get more clear distinguishability of shock waves.

3.3. The differences of aerodynamic characteristics

We often care more about the whole aerodynamic characteristics in design procedure, and the differences on aerodynamic characteristics of feature-based adaption and adjoint-based adaption is compared in this section. In order to compare the effect of mesh adaptation, the curves of NACA0012 airfoil drag coefficient with the number of adaptive times and mesh quantity are given. The results obtained by adjoint adaptive meshwork are more accurate than those obtained by feature adaptive meshwork. The calculation accuracy of resistance coefficient obtained by adjoint adaptive meshwork is 10.1% higher than that of feature adaptive meshwork. The error correction effect based on adjoint equation is obvious, and the accuracy of high-order correction result is 7.5% higher than that of adjoint adaptive meshwork.

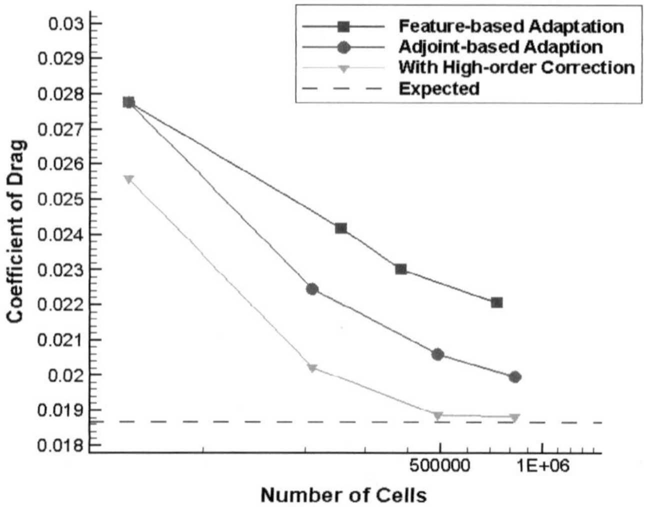

Figure 6. Adjoint drag adaption

4. Conclusion

The comparison of adjoint based grid adaptation and feature based grid adaptation method is investigated in this paper. Numerical simulations show that both feature adaption and adjoint adaption could improve the resolution of flow field and the precision of numerical simulation such as lift and drag. As for the flow features, the feature adaptation could capture the obvious shock waves and vortexes in the flow field, the adjoint adaptation, however, only captures the flow features that are contributory to the accuracy of aerodynamic characteristics. As for the aerodynamic characteristics, some shock waves and vortexes have little influences to the forces, so the feature adaptation is not efficient as adjoint adaptation, which could greatly improve the accuracy of aerodynamic characteristics.

Acknowledgments

This work was financially supported by National Numerical Wind Tunnel (NNW) and National Natural Sciences Foundation of China (12102453). Special thanks to Yueyue Yang, Jie Zhang, Xiaojun Wu for their suggestions on this article.

References

[1] D M Bushnell, M R Malik. Role of Computational Fluid Dynamics and Wind Tunnels in Aeronautics R&D: NASA/TP–2012-217602[R]. Washington, D. C.: NASA, 2012.
[2] J R Chawner, J F Dannenhoffer and N J Taylor. Geometry, Mesh Generation, and the CFD 2030 Vision: AIAA-2016-3485[R]. Reston: AIAA, 2016.
[3] A Quarteroni. Isogeometric analysis[M]. Numerical Models for Differential Problems. 2017.
[4] J Slotnick, A Khodadoust, J Alonso, et al. CFD Vision 2030 Study: A Path to Revolutionary Computational Aerosciences: NASA/CR-2014-218178[R]. Washington, D. C.: NASA, 2014.
[5] A Shih, Y Ito, R Koomullil. Solution Adaptive Mesh Generation Using Feature-Aligned Embedded Surface Meshes[R]. AIAA paper 2007-558, 2007.
[6] R E Bartels, V Vatsa, J R Carlson, et al. FUN3D Grid Refinement and Adaptation Studies for the Ares Launch Vehicle [R]. AIAA paper 2010-4372, 2010.
[7] Y L Moigne. Adaptive Mesh Refinement Sensors for Vortex Flow Simulations[C]// ECCOMAS2004-European Congress on Computational Methods in Applied Sciences and Engineering, 2004.
[8] M Jiang, R Machiraju, D Thompsony. Detection and Visualization of Vortices[R]. 2005.
[9] K K Chand, K D Lee. Adaptation of Structured Grids with Redistribution and Embedding [R]. AIAA paper 99-36515, 1999.
[10] A Choudhary. Verification of Compressible and Incompressible Computational Fluid Dynamics Codes and Residual-based Mesh Adaptation[D]. Virginia Polytechnic Institute and State University, 2014.
[11] D A Venditti, D L Darmofal. A Grid Adaptive Methodology for Functional Outputs of Compressible Flow Simulations [R]. AIAA paper 2001-2659, 2001.
[12] M A Park. Adjoint-Based, Three-Dimensional Error Prediction and Grid Adaptation[J]. AIAA Journal, 2002, 42(9):2002-3286.
[13] P. C. Cui, Y. Q. Deng, J. Tang, Adjoint-based grid adaptation and error correction, Acta Aeronautica et Astronautica Sinica, 37(10): 2992-3002, 2016.
[14] P. C. Cui, J. Tang, B. Li, A conservative interpolation method for overset mesh via supermesh, Acta Aeronautica et Astronautica Sinica, 39(3): 121569, 2018.
[15] D. Levy, K. Laflin, E. Tinoco, Summary of data from the fifth AIAA CFD drag prediction workshop, 51st AIAA Aerospace Sciences Meeting, DLR, 2013.
[16] P. L. Roe, Approximate Riemann Solvers, Parameter Vectors, and Difference Schemes, Journal of Computational Physics, 135(2):250-258, 1997.
[17] B Li, Y Q Deng, J Tang, et al. Discrete adjoint method for 3D on unstructured grid[J]. Acta Aeronautica et Astronautica Sinica, 2014, 35(3): 674-686.
[18] B Li, Y Q Deng, J Tang, et al. Application of par-allel multigrid algorithm to discrete adjoint optimiza-tion[J]. Acta Aeronautica et Astronautica Sinica, 2014, 35(8): 2091-2101.
[19] J Tang, Y Q Deng, M S Ma, et al. Parametrization and grid deformation techniques for fly-wing shape optimization[J]. Acta Aeronautica et Astronautica Sinica, 2015, 36(5): 1480-1490.

Analysis of Aerodynamic Characteristics of High-speed Train with Asymmetric Heads

Weibin Li[1], Yuejun Wang[1], Yunquan Hao[2], Wei Zhao[1,*], Yanlin Zhang[3]

[1] Computation Aerodynamics Institute, China Aerodynamics Research and Development Center, Mianyang 621000, China
[2] School of Sciences, Southwest Petroleum University, Chengdu 610500, China
[3] CRRC Zhuzhou Locomotive Co., Ltd., Zhuzhou 412001, China

*Corresponding author e-mail: ethnuaa@163.com

Abstract. High-speed transportation is closely related to people's livelihood and national economy, so it becomes an important development aspect to the whole country. The safety, comfort and economy of high-speed train are the key consideration during its design process. In the future, new trains and old ones will be used for a long time, and the performance of new trains which run on the old lines may be limited. These situations make the research of high-speed train with asymmetric heads much necessary. Aiming at the feasibility of the train with asymmetric heads, the heads of train CRH3 and another typical train with a similar speed are selected. Based on the two heads, the digital models of the traditional trains with symmetric heads and the new trains with asymmetric heads are built firstly. And then their aerodynamic characteristics are computed by the SIMPLE method. In the numerical part, the performances of the trains are compared on the condition without crosswind, which shows that one of the new trains with asymmetric heads is the best among them. Moreover, the aerodynamic characteristics of this new model under crosswind is analyzed, which demonstrates that the train with asymmetric heads is feasible. The research shows that the train with asymmetric heads can achieve better results through the proper aerodynamic shape optimization design, which presents an idea for the future high-speed train design.

1. Introduction

The development of high-speed railway has a far-reaching impact on improving people's livelihood, promoting national economic growth and enhancing international political status. By the end of 2018, 31 provinces and cities in China had opened high-speed railways and 3268 trains are employed. The total length of the railways is over 29000km and 2.005 billion passengers are passed throughout the year. These make our country with the longest high-speed railway mileage, the highest transportation ability and the most complex network of the railway in the world.

With the development of China's high-speed railway, the safety and stability of high-speed trains have become one of the key concerns [1]. For high-speed trains running at 300km/h, the aerodynamic drag force accounts for more than 80% of the total that of the whole train. And the aerodynamic drag force is approximately proportional to the square of speed. Therefore, the decrease of aerodynamic drag force is the primary consideration during the design process of high-speed trains [2, 3]. The train shape plays a decisive role in the aerodynamic performance of the train. Thus its optimization has always been an important means to reduce the aerodynamic resistance of the trains [4].

In aerodynamic tests of trains, references [5-7] show the aerodynamic performances of different trains. The aerodynamic characteristics under different conditions are also compared to get a summarization of the relationship between head shape and aerodynamic performance, which provides a basis for head shape optimization. In system optimization, reference [8-13] carried out a series of work on head, windscreen, air conditioning, underbody and other parts based on the aerodynamic characteristics simulations. In these works, experiments in the wind tunnel were used to demonstrate the optimization results. In the algorithm of optimization, reference [14] proposed a constrained multi-objective aerodynamic optimization design method based on adaptive nondominated sorting genetic algorithm, and reference [15] presented a multi-objective automatic optimization method with five variables for streamlined head design. All of the above research results provide strong support for the shape optimization design of high-speed trains in the future.

With the development of high-speed transportation, the complex running state and increasingly harsh environment of high-speed rail put forward a test to the existing lines. At the same time, due to the high cost and long period of new head design, the trains with new or old heads have been in a state of coexistence for a long time. In order to ensure the sustainable development of the new head and full utilization of the old head, it is necessary to carry out relevant research on trains with asymmetric heads.

Aiming at the feasibility of trains with asymmetric heads, the aerodynamic characteristics of the trains are present in this paper. In Section 2, four models of marshalling are designed based on the heads of CRH3 and another train with the same speed level. In Section 3, the grids and methods we used are briefly presented. In Section 4, the numerical simulations of the four models under situations without crosswind are carried out and compared. In Section 5, the feasibility of the train with asymmetric heads is analyzed. This paper is concluded in Section 6.

2. The model of trains

In this paper, we choose the head of train CRH3 (denoted as Type 1) as the reference and that with the same speed level (denoted as Type 2) as a combined head-type. These two head types are combined to get four groups of marshalling which are shown in Table 1. Trains of three vehicles with full scale is selected for calculation. The computational models include roadbed, track, cowcatcher, bogies and pantograph fairing, and windshield structure is adopted between two vehicles.

Table 1. Four groups of marshalling

Model	I	II	III	IV
Head	Type 1	Type 1	Type 2	Type 2
Tail	Type 1	Type 2	Type 1	Type 2

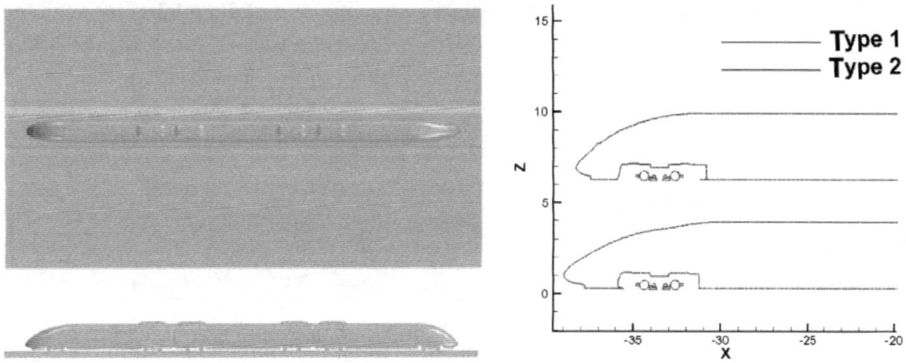

Figure 1. Model II with asymmetric heads.

For a clear understanding of high-speed train with asymmetric heads, Figure 1 shows the vertical view and side view of the three vehicles marshalling of Model II. The length of Type 1, Type 2 and the middle vehicle are 25.8m, 26.5m and 24.8m respectively. Furthermore, the comparison of the control line for the symmetry plane of the two head types are also given, we can see that the surface of Type 2 is smoother and the corresponding gradient of the head type section is better.

3. Grid and method

The characteristic scale of the train, L, is 3.7m and the reference area is 11.58m2. For the three vehicles model, the size of computational region is shown in Figure 2 (a). Especially, the distance between outlet and tail vehicle is expanded to about 40 times of L. In crosswind conditions, the leeward width is increased to about 35 times of L to ensure the full development of shedding vortices. Eight layers of prismatic grids are arranged near the surface of the train, where the minimum height of the first layer is 1mm and the corresponding Y+ is about 60. The grid partition strategy satisfies the requirements of non-equilibrium wall function and can get a better simulation of the pressure distribution.

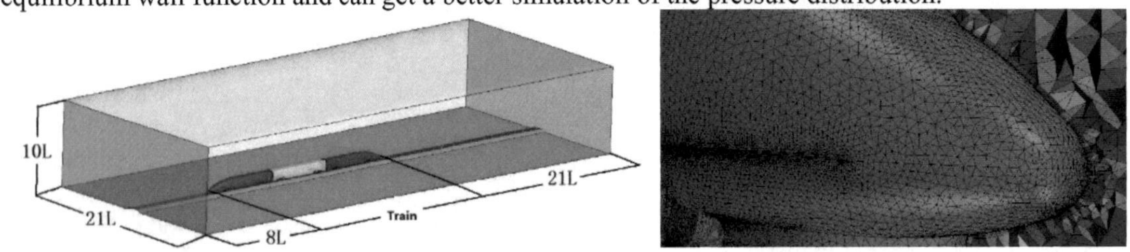

(a) (b)

Figure 2. Computational region and prismatic grids. (a) Computational region. (b) Prismatic grids.

The train runs at a speed of 300km/h, its aerodynamic is calculated by incompressible methods and the corresponding equation is incompressible N-S equations. And the k-ε two-equation turbulence model is applied. The inlet and outlet are set as the uniform inflow velocity entrance and pressure outlet, respectively. The ground and the train surface are set as the moving wall and non-slip wall, respectively. The far field is treated by symmetric boundary conditions. For crosswind conditions, the velocity inlet is set as components to specific the speed of the two orientations.

4. Aerodynamic performance without crosswind

In order to compare the aerodynamic performance of models I to IV, the flow field calculation was carried out under the condition without crosswind. The velocity of the train is 300km/h.

Figure 3 shows the surface pressure distributions of the head and tail vehicles for the two types, and it can be seen that the two type heads have the similar distribution law under the steady state. Concretely, the stagnation point is located at the nose cone of the head. When the air flow pass through the stagnation point, the speed is accelerated resulting in a rapid decrease of the surface pressure. Therefore, a negative pressure area is formed at the transition region of the head.

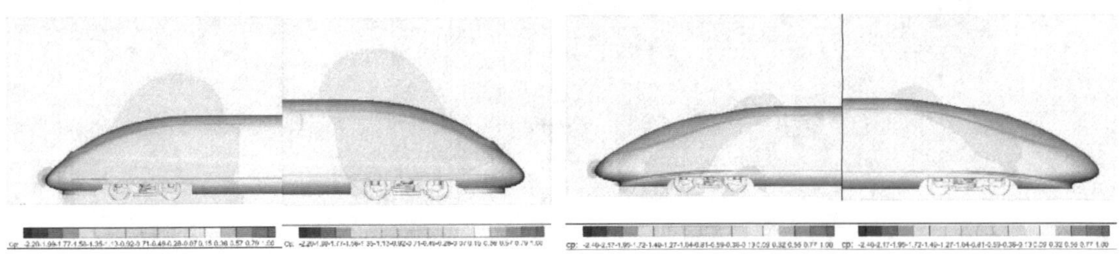

(a) (b)

Figure 3. Pressure distribution of the head and the tail vehicles. (a) Type 1. (b) Type 2.

In order to get quantitative comparisons of these two type heads, we give the pressure load on the symmetrical plane of the head and tail in Figure 4. It shows that the pressure coefficients of the two heads are quite different, which is caused by the aerodynamic shapes. Concretely, the minimum pressure of Type 1 is smaller than that of Type 2, and the peak position of negative pressure is in the front. This is caused by the large change of surface curvature in the transition zone between the head and carriage of type 1, the short head shape and the front position of wiper. In addition, the negative pressure area on the surface of Type 1 is greater than that of Type 2, which will cause large differential pressure resistance and affect the aerodynamic performance of the marshalling.

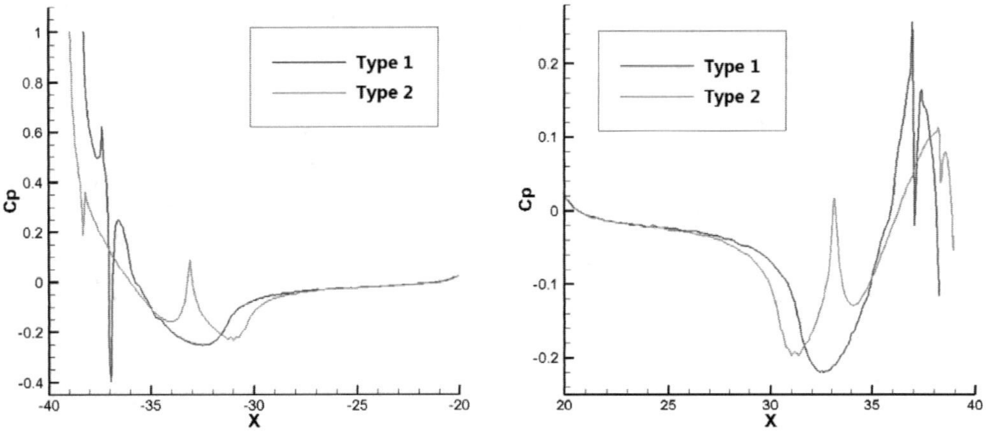

Figure 4. Pressure load distribution on the symmetric planes.

The drag force of the four models is calculated and shown in Figure 5. No matter which model, it can be seen that the drag summation of the head and the tail is about half of the whole train drag, which means shape optimization design of the head is important for the high-speed train. Among the four models, model II drag coefficient is the smallest and model III drag coefficient is the largest, which is mainly caused by the aerodynamic characteristic of the second vehicle. Thus the Type 1 and Type 2 are satisfied to assign as the head and the tail, respectively. From this comparison, we can see that the marshalling model with asymmetric heads performs better when the head and tail are well designed.

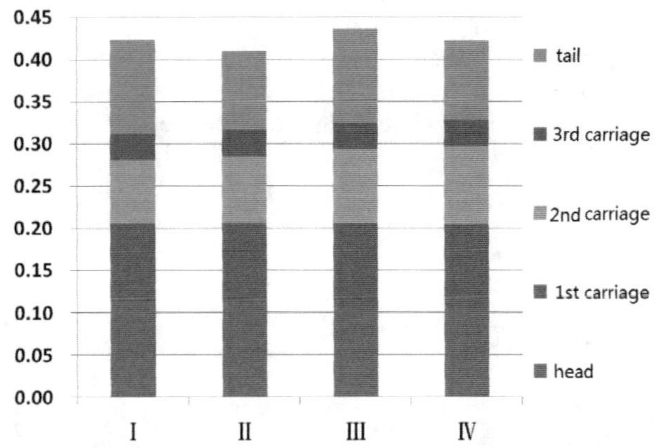

Figure 5. Drag coefficient of three vehicles marshalling.

5. Aerodynamic performance under crosswind

The surface pressure distribution under crosswind conditions is shown in Figure 6. Since the crosswind direction is perpendicular to the running direction of the train, the flow has actually become to flow around bluff body. This causes a wide range of separated flow on the leeside leading positive pressure on the whole windward side and negative pressure on the top of the vehicle body. Accordingly, the drag, lift and capsizing force of the whole vehicle will increase. For the tail, the originally symmetrically distributed tail vortex is rolling from the windward side to the leeside, resulting in drag increasing of the tail.

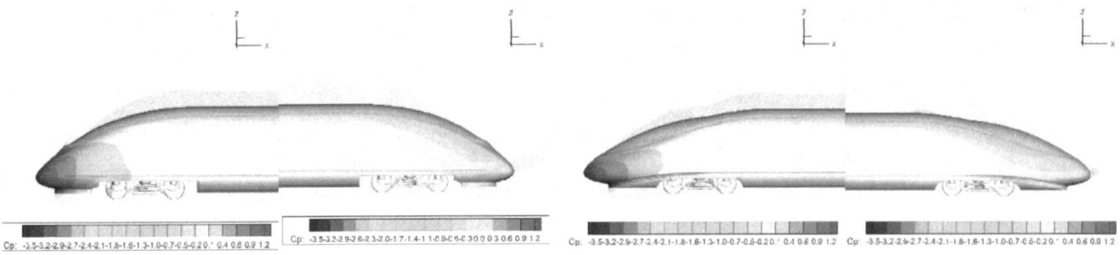

(a) Type 1 (b) Type 2

Figure 6. Pressure distribution on the upwind side of the train in crosswind condition

The results in Section 4 show that the marshalling model II with the asymmetric heads exhibits good aerodynamic performance in the conditions without crosswind. In order to verify its stability and safety, the flow field simulation under crosswind conditions is carried out and compared with model I. The calculation conditions are shown in Table 2.

Table 2. Conditions with crosswind

Case	Velocity (km/h)	Crosswind speed (m/s)
1	300	5
2	300	15
3	300	25
4	300	35

Figure 7 shows the aerodynamic calculation results of models I and II under cases in Table 2. It can be seen that the drag coefficient of the head decreases and that of other carriages increase with the increase of crosswind speed. This is because that difference of pressure coefficient on the head decrease when the crosswind speed increasing, but that on other part are increased. By contrast of Figure 7(a) and (b), we can see that the aerodynamic performance of model II is much better than that of model I. Thus the train with asymmetric heads is also feasible under crosswind conditions.

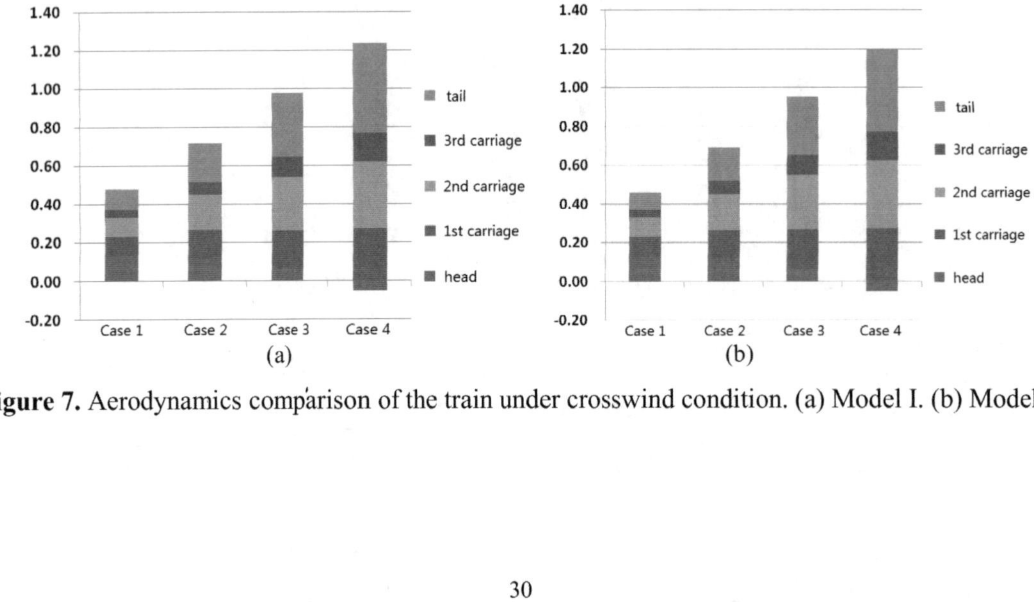

(a) (b)

Figure 7. Aerodynamics comparison of the train under crosswind condition. (a) Model I. (b) Model II.

6. Conclusion

Based on two typical types of high-speed train heads, four trains are established in the form of mutual combination. And their aerodynamics characteristic under the conditions without and with crosswind are analyzed and compared. Through the comparisons, it shows that the train with asymmetric heads shows better performance if the head and tail are well designed. And the train with asymmetric heads is feasible under the situations with crosswind. Thus the idea of asymmetric heads should be carefully considered in the design of the future high-speed train or pipeline train.

Acknowledgments

This work was supported by the National Numerical Windtunnel project. Special thanks to Shuai Ma for his suggestions on this article.

References

[1] Yang G W, Wei Y J, Zhao G L, et al. Research Progress on the Mechanics of High Speed Rails [J], Advances in Mechanics. 2015, 45: 201507. (in Chinese)

[2] Raghunathan R S, Kim H D, Setoguchi T. Aerodynamics of high-speed railway train [j]. Progress in Aerospace Sciences, 2002, 38: 469.

[3] Tian H Q. Development of research on aerodynamics of high-speed rails in China. Engineering Science, 2015, 17(4): 30. (in Chinese)

[4] Tian H Q. Railway train aerodynamic [M]. BeiJing: China Railway Press, 2007.

[5] Liu T H, T ian H Q. Comparison Analysis of the FullScale Train Tests for Trains with Different Shapes Passing Tunnel [J]. China Railway Science, 2008, 29(1): 51-55 (in Chinese)

[6] Cooper R K. Wind tunnel tests on a 1/25th scale APT-P model [J]. British Rail Research Division Technical Memorandum TMAERO, 1982.

[7] Kwak M, Yun S, Lee Y, et al. Optimum Nose shape of a Front –rear Symmetric Train for the Reduction of the Total Aerodynamic Drag [J]. Journal of Mechanical Science & Technology, 2013, 27(12): 3733-3743.

[8] Huang Z X, Chen L. The Wing Tunnel Test and Research on Aerodynamics Characeristics of High Speed Trains[J]. Rolling Stock, 2011, 49(12): 1-6. (in Chinese)

[9] Kong F B, Li M, Han L, et al. Optimization Design of Aerodynamic Shape for CRH380BL EMU [J]. Railway Locomotive & Car, 2012, 32(6): 9-12. (in Chinese)

[10] Yu Y Z, Jiang X D, Sun J. Study of Pressure Change of CRH3A High-Speed Train Passing a Tunnel[J]. Journal of DALIAN JIAOTONG University, 2015, 36(5): 46-49. (in Chinese)

[11] Yao S B, Guo D L, Yang G W, et al. Distribution of High-speed Train Aerodynamic Drag [J]. Journal of the China Railway Society. 2012, 34(7): 18-23. (in Chinese)

[12] Shan W, Deng H, Wang J T, et al. The Optimization Design of Aerodynamics Resistance Characteristics of CRH3 High Speed Multiple Units. Rolling Stock, 2012, 50(2): 1-32. (in Chinese)

[13] Deng H, Xia D W, Yin H, et al. Pneumatic Character Comparison and Research of Head Shapes Between New Generation High-speed EMU 380CL and Prototype Vehicle [J]. Journal of Dalian Jiaotong University. 2013, 34(2): 28-33. (in Chinese)

[14] Yao S B, Guo D L, Sun Z X, et al. Multi-objective optimization of the streamlined head of high-speed trains based on the Kriging model. Sci China Tech Sci, 2012, 55: 3495-3509. (in Chinese)

[15] Zhang L, Zhang J Y, Li T, et al. Multi-objective Optimization Design of the Streamlined Head Shape of Super High-speed Trains. Journal of Mechanical Engineering, 2017, 53(2): 106-114. (in Chinese)

Numerical simulation on static aerodynamic characteristics of supersonic cavities

Lei Luo[1], Hongyang Chen[1], Jie Zhang[1], Hongyin Jia[1,*] and Peihong Zhang[1]

[1]Computational Aerodynamics Institute, China Aerodynamics Research and Development Center, Mianyang, China

*Corresponding author e-mail: hongyinjia@foxmail.com

Abstract. A numerical study on supersonic cavity flow is carried out, where the influence law of the freestream parameters and configuration factors on static aerodynamic characteristics is investigated, focusing on the change of pressure distribution on the cavity walls at high and low Mach numbers. It is found that with the Mach number increment at high speeds (M>2), the pressure peak on the rear side of the cavity bottom floor rises, and the change in pressure distribution on the cavity walls shows a variating trend in flow type from a closed-cavity flow to an open-cavity flow, which is consistent with the law at low speeds (M≤2). At high speeds, the change in Reynolds number has a certain impact on the pressure distribution in the cavity, which is inconsistent with the law at low speeds. The increase of Reynolds number will increase the pressure on the rear side of the cavity bottom floor and on the rear wall, and the greater the Mach number, the greater the difference caused by Reynolds number variation. When the length-to-depth ratio increases at high speeds, the pressure peak on the rear side of the cavity bottom floor decreases, and the flow type is changing from an open-cavity flow to a transitional-cavity flow, which is consistent with the law at low Mach numbers, while the pressure variation is greater at high speeds. The influence of the length-to-width ratio on the pressure distribution is relatively small, and the increase of length-to-width ratio increases the pressure peak on the rear side of the cavity bottom floor slightly. Spanwise bending has a certain impact on the pressure distribution. The pressure peak on rear side of the bottom floor increases when the cavity is bending, which is consistent with the influence caused by length-to-width ratio increment.

1. Introduction

Cavity flow is one of the classical problems in fluid mechanics [1]. Because the research on cavity flow has great theoretical value and engineering practical significance, it has attracted extensive attention for a long time. In as early as 1960s, the United States began to conduct wind tunnel tests and engineering calculations research on cavity flow in different velocity regions [2, 3]. Subsequently, scholars carried out a lot of research on cavity flow characteristics and flow mechanism [4, 5, 6, 7, 8, 9, 10], including the influence law on static and dynamic characteristics of configuration factors and flow parameters, as well as different flow control strategies. Some basic conclusions on the pressure distribution, flow oscillation and spectrum characteristics are drew. An important application of the cavity static aerodynamic characteristics is the internal weapon bay design. The shear layer impacts the rear lip of an internal weapon bay strongly, resulting in the gradual pressure increase in the bay along the flow direction. This phenomenon can produce an additional pitching moment on the internal weapon and affect its release safety [6, 7]. Although the design of internal weapon bay at low Mach numbers (M≤2) has a few practical applications in nowadays, with the demand of high-speed internal weapon release, the research on the static aerodynamic characteristics of high Mach numbers (M>2) cavity becomes more and more important.

The static aerodynamic characteristics of the cavity is mainly affected by the freestream parameters and configuration factors. For the freestream parameters, it is pointed out that the static characteristics of subsonic and transonic cavities is barely affected by Reynolds numbers [11], but is closely related to Mach numbers [12]. The most significant effect of Mach number variation is the change of cavity flow type [2]. That is, with the increase of Mach number, cavity flow gradually changes from a closed-cavity flow to a transitional-cavity flow, and then to an open-cavity flow, resulting in significant changes in wall pressure. For the cavity configuration factors, the length-to-depth ratio is the most important factor affecting the static characteristics. At low Mach numbers, with the decrement in length-to-depth ratio, the separation areas near the front and rear wall in the cavity gradually approach and merge [8]. So that the high pressure on the rear wall is pulled down by the low pressure on the front wall, and the pressure distribution along the cavity bottom floor becomes gentle [2, 5]. Consequently, the flow type changes from a closed-cavity flow to an open-cavity flow. Moreover, the length-to-width ratio can also affect the pressure distribution in the cavity to a certain extent. Under different cavity flow types, the pressure peak at the cavity rear lip shows a nonlinear trend with length-to-width ratio variation [13]. From the conclusions of these mechanism studies, the static aerodynamic characteristics of low Mach number cavity is affected by the coupling of many factors, and its influence law is very complex. For the high Mach number cavity, due to higher energy in the shear layer, its impact on the cavity rear wall is stronger, conducting a block effect on the flows inside and outside the cavity [15]. As a result, the influence law of cavity static aerodynamic characteristics may be quite different from that at low Mach numbers, which needs to be further studied.

Therefore, a numerical study on supersonic cavity flow is presented, where the influence law of the freestream parameters and configuration factors on the static aerodynamic characteristics is investigated, focusing on the change of pressure distribution on the cavity walls at high and low Mach numbers.

2. Numerical simulation method and validation

2.1. Numerical simulation software

The NNW-Flowstar solver (the present version of the MFlow solver), which is developed by Computational Aerodynamics Institute of China Aerodynamics Research and Development Center, is used to carry out the numerical simulations. The solver is a large-scale parallel-computing flow field solver based on unstructured hybrid grids and Reynolds Averaged Navier-Stokes (RANS) equations. It can use arbitrary grid-cell shapes and has great flexibility. The calculation space is discretized by the finite volume method, and the unknown variable is located at the center of the grid cell. The discrete equations are solved by implicit LU-SGS method or explicit Rung-Kutta method. The NNW-Flowstar solver has various options to use, such as spatial discrete schemes for the convection and diffusion terms, time iterative methods, different turbulence models, etc. This software has now been widely used in engineering practices, and its calculation precision and accuracy are verified and validated by various classical examples [16, 17].

2.2. Calculation Model for Validation

The experimental cavity model in reference [5] is taking for the numerical validation. This wind-tunnel model is simplified into a calculation model composed of symmetrically arranged square cavities and octagonal plates. The shape is shown in Figure 1. The cavity is 12 inches long, 2.5 inches wide and 0.5 inches deep. The incoming Mach number is 2.86, the static pressure is 3114Pa and the static temperature is 123.3K. The computational grid is shown in Figure 2. The hybrid grid strategy is used to divide the computational space. Hexahedral and triangular prism elements are used in the boundary layer, and the far wall area is filled with pyramid and tetrahedral elements. The total number of grid elements is about 12 million.

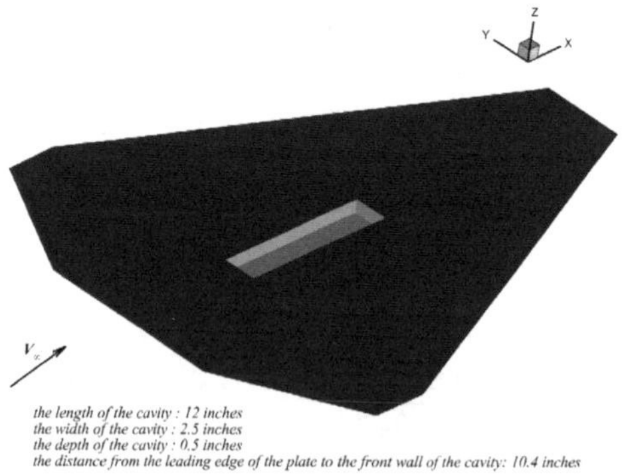

Figure 1. Calculation model for validation.

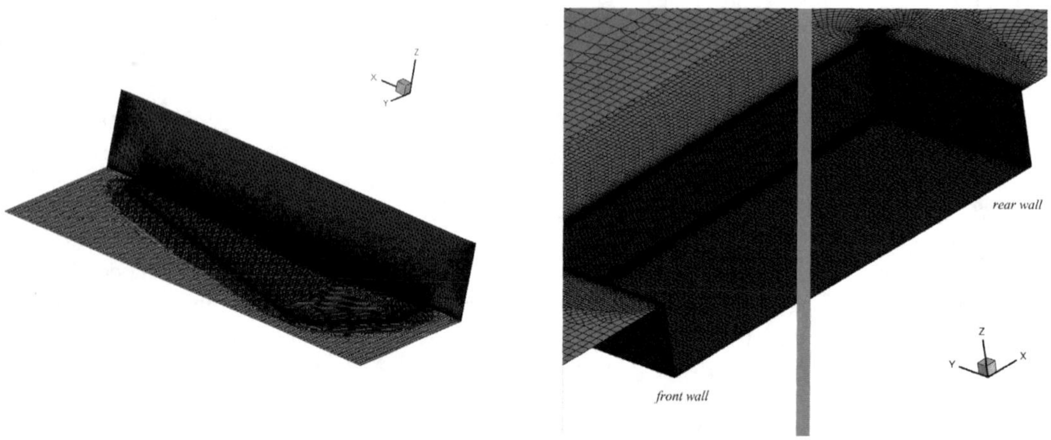

a. full view b. detail view

Figure 2. Mesh of calculation model.

As the length-to-depth ratio of the cavity is 24, the flow around the cavity is a typical closed-cavity flow. For a closed-cavity flow, similar to the classical back stage and front stage flow, there is a large-scale separation vortex at each side of the front and rear wall. On the opening side of the cavity, there are two shock waves between the separated vortices. From upstream to downstream, they are usually called the impingement shock and exit shock respectively [5]. The flow in the area between the two shock waves is almost parallel, and the pressure on the cavity bottom floor is close to a constant distribution.

2.3. Numerical Validation

A comparison study on the discrete schemes of inviscid convective flux [18, 19, 20] is proposed, implementing Roe, van Leer, HLLC+ and HLLE++ schemes. Among them, HLLC+ scheme is a mixture of HLLC and HLLE schemes, where HLLE scheme is used in strong discontinuous regions such as shock waves while HLLC scheme is implemented in other regions. The turbulence model adopts the k-ω SST two-equation model. Figure 3 shows the comparison of calculation results with different schemes. It can be seen that the calculation results of different schemes match well with each other, and the pressure coefficient on the rear plate is in good agreement with the experimental values.

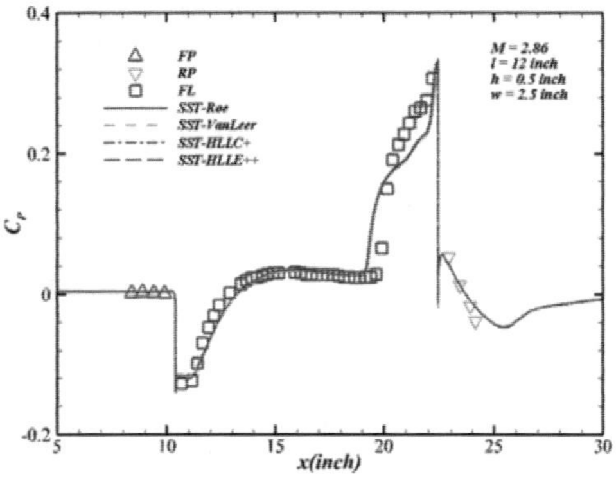

Figure 3. Calculation results of different schemes and experimental results (*Cp* distribution).

Figure 4 shows the calculation results using the Spalart-Allmaras (SA) one-equation model and *k-ω* Shear Stress Transfer (SST) two-equation model respectively. It can be seen that the performance of the two turbulence models on simulating the separation vortex scale of the front wall are close, and they are in good agreement with the experimental values. The two turbulence models have the same simulation results for the pressure platform between two separated vortices in the cavity. For the separation flow at the rear wall, there is little difference between the two models in predicting the position of the separation point. However, the calculation results of the two turbulence models after the separation point are quite different. The pressure peak calculated by the *k-ω* SST model are closer to the experimental one rather than that of the SA model. This indicates that there should be obvious differences between the two turbulence models in calculation results of the separation zone flow at the rear wall of the cavity. According to the results above, the HLLE++ scheme and *k-ω* SST turbulence model are selected for the following numerical research on the static aerodynamic characteristics of supersonic cavity.

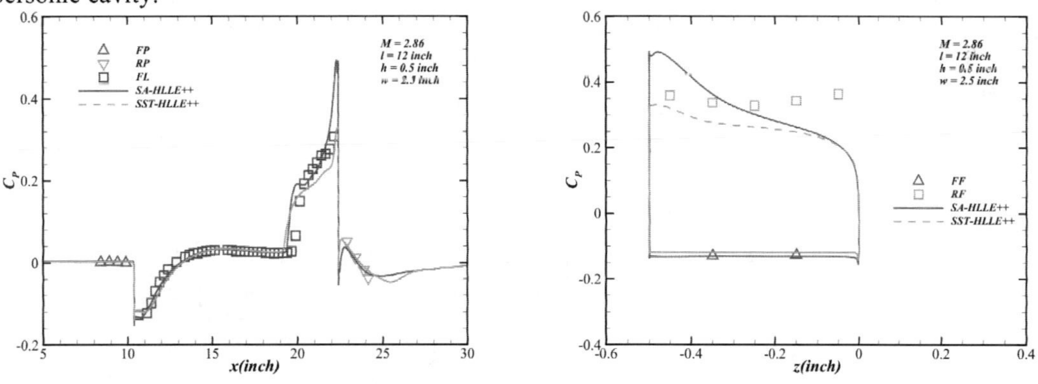

a. bottom floor b. rear wall

Figure 4. Calculation results of different turbulent models and experimental results (*Cp* distribution).

3. Influence of freestream parameters

3.1. Mach Number

There have been many studies on the static aerodynamic characteristics of cavities at low Mach numbers. Taking the experimental results in reference [7] as an example, Figure 5 shows the pressure distribution on the cavity bottom floor with length-to-depth ratio of 7.0, when the Mach numbers are not greater than 1.8. Due to the differences in static, dynamic and total pressure when Mach number varies, the pressure coefficient is not given here, but the variable P_u which is the local static pressure nondimensionalized by the freestream static pressure. It can be seen from the figure that the pressure peak on the cavity bottom floor increases gradually with the increase of Mach number. The literature points out that with the Mach number increment, the flow in the cavity changes from a transitional-cavity flow to an open-cavity flow, and the critical Mach number is between 0.9 and 1.2.

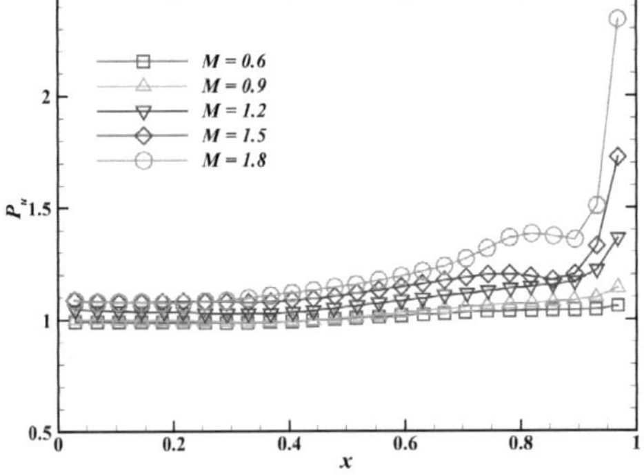

Figure 5. P_u distribution along the cavity bottom floor at low Mach numbers (recreated from ref. 7).

A rectangular cavity with length-to-depth ratio of 7.5 and length-to-width ratio of 5.0 is taken as the research object here. Figure 6 shows the comparison of P_u distribution on the bottom floor and rear wall at different Mach numbers, and Figure 7 shows the P_u distribution on the cavity surface at different Mach numbers. It can be seen that the pressure peak on the cavity bottom floor increases significantly as Mach number increases. When the Mach number is 1.5, $P_{u,max}$ is about 1.20, while when the Mach number is 4.0, $P_{u,max}$ is about 1.72, which is about 43.3% higher than the former. On the contrary, the pressure on the front side of the cavity bottom floor does not increase much as Mach number increases. It indicates that the pressure difference between the front and rear side of the cavity increases significantly with Mach number increment, which could result in additional pitching moment on the releasing object during delivery process. The influence of Mach number increment on the pressure change in the cavity is mainly on account of the rapid rise in dynamic pressure, which significantly enhances the impact effect of shear layer hitting on the rear wall.

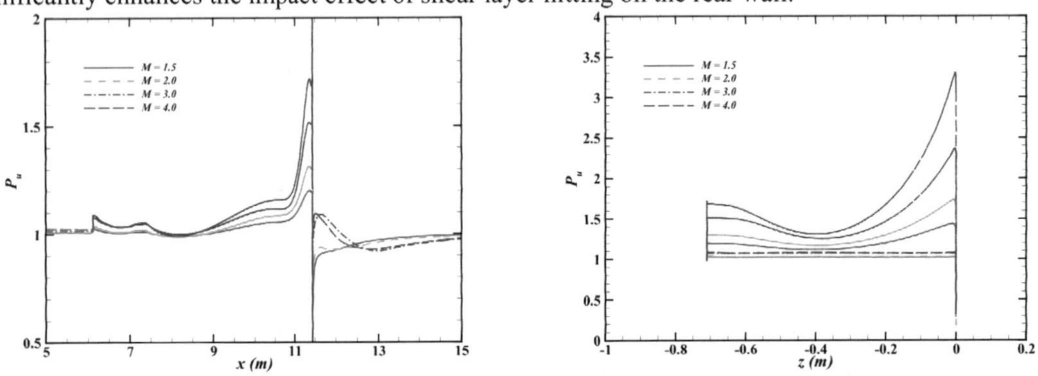

a. bottom floor b. rear wall

Figure 6. P_u distribution along the cavity bottom floor and rear wall at different Mach numbers

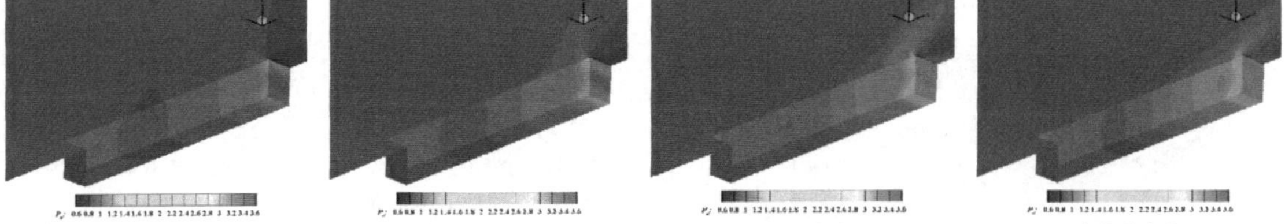

a. M 1.5 b. M 2.0 c. M 3.0 d. M 4.0

Figure 7. P_u distribution on the cavity walls at different Mach numbers

3.2. Reynolds Number

It is pointed out by other researches that the Reynolds number has little influence on the static aerodynamic characteristics of the cavity when the Mach number is low. However, for high Mach number cavity flow, the increase of Mach number will significantly enhance the shear effect. The change of Reynolds number will not only significantly change the boundary layer thickness at the front lip of the cavity, but also directly affect the viscosity transportation in the cavity. In order to compare the Reynolds number effects at low and high Mach numbers, the calculated Mach numbers are taken as 1.5 and 4.0.

Figure 8 and Figure 9 show the P_u distribution along the cavity bottom floor with different Reynolds numbers, at Mach number of 1.5 and 4.0 respectively. Figure 10 shows the P_u distribution on the cavity surfaces with different Reynolds numbers at Mach numbers 1.5 and 4.0. It can be seen that the pressure peaks on the bottom floor and rear wall increase with the increase of Reynolds number. As the viscosity transportation of the shear layer is limited to a smaller scale with Reynolds number increment, the shear-layer impact effect on the rear wall is stronger. It also can be seen from the plots that the pressure variation due to Reynolds number at a low Mach number is less than that at a high Mach number. It indicates the necessity of a flight-ground correlation study to take the influence of Reynolds number into account for high Mach number cavity flows. Moreover, it can be noticed that within the range of Reynolds numbers discussed in this section, the change of Reynolds number does not cause the change in cavity flow type. The character of an open-cavity flow is maintained within the Reynolds number and Mach number ranges.

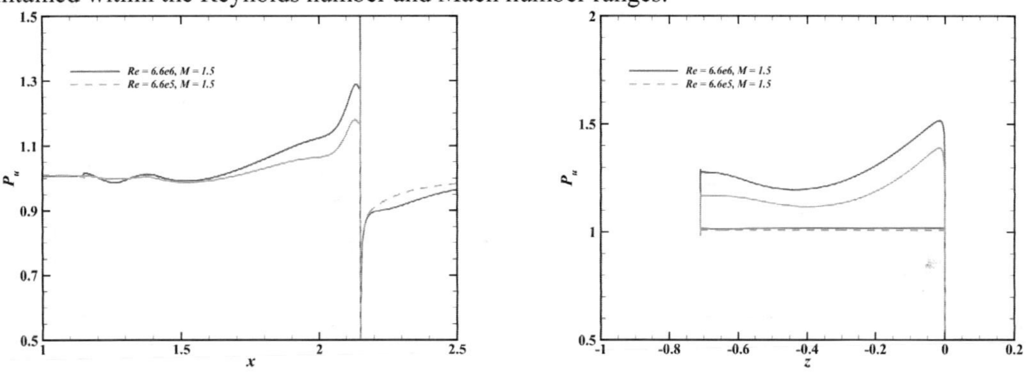

a. bottom floor b. rear wall

Figure 8. P_u distribution along the cavity bottom floor and rear wall at different Reynolds numbers (Mach number 1.5)

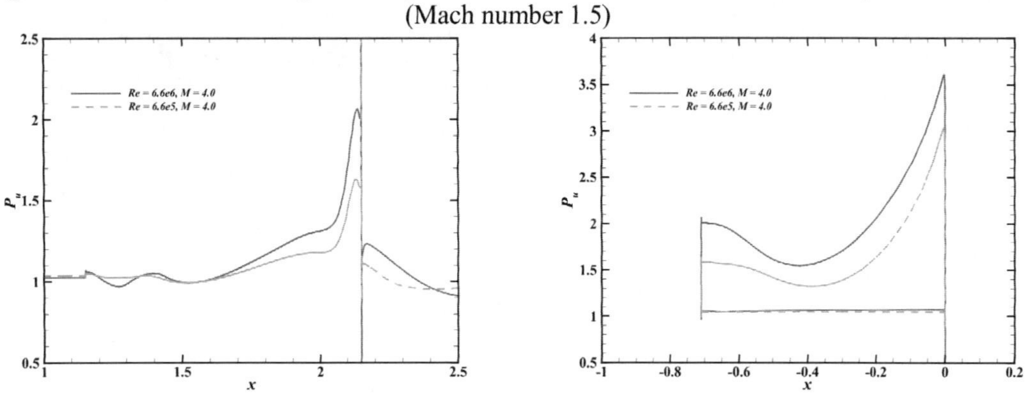

a. bottom floor b. rear wall

Figure 9. P_u distribution along the cavity bottom floor and rear wall at different Reynolds numbers (Mach number 4.0)

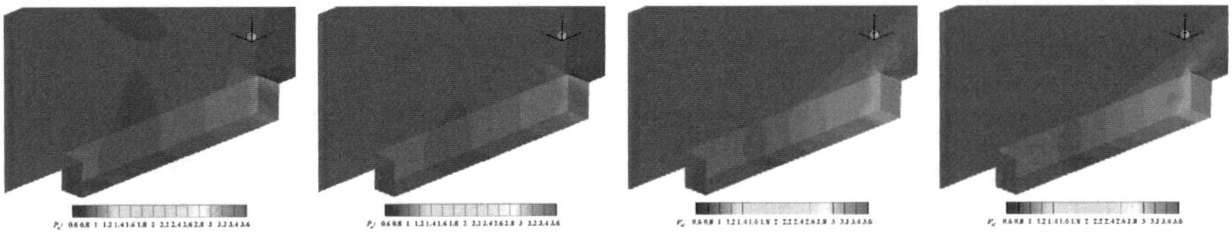

a. M 1.5, Re 6.6×10[5] b. M 1.5, Re 6.6×10[6] c. M 4.0, Re 6.6×10[5] d. M 4.0, Re 6.6×10[6]

Figure 10. P_u distribution on the cavity walls at different Reynolds numbers

4. Influence of configuration factors

4.1. Length-to-depth Ratio

The length-to-depth ratio is supposed to be the most important configuration factor in cavity flow and the main factor affecting the flow type. The influence law of length-to-depth ratio at low Mach numbers has been well concluded. However, the influence law of at high Mach numbers is still unclear. Therefore, an investigation is presented to compare the influence law of length-to-depth ratio at low and high Mach numbers in this section. The Mach numbers calculated here are still 1.5 and 4.0.

Figure 11 and Figure 12 show the P_u distribution along the cavity bottom floor and rear wall with length-to-depth ratio ranging from 6.0 to 9.0 and length-to-width ratio of 5.0, at Mach number of 1.5 and 4.0 respectively, while Figure 13 and Figure 14 show the P_u distribution on the cavity walls. It can be seen from the plots that the change of length-to-depth ratio will lead to significant changes in wall pressure distribution, and the pressure change at high Mach number is greater than that at low Mach number. At high Mach number, with the increase of length-to-depth ratio, the pressure peak on the cavity bottom floor increases while the pressure in the front side of the cavity decreases, and the pressure peaks downside and upside the rear wall is in a trend of decrease. It is noted that when the length-to-depth ratio is 9.0, the P_u value in the front side of the cavity has been less than 1, indicating that the flow type is turning into a transitional-open one. The above results indicate that the influence law of length-to-depth ratio on the static pressure distribution of the cavity at high Mach numbers is close to that at low Mach numbers.

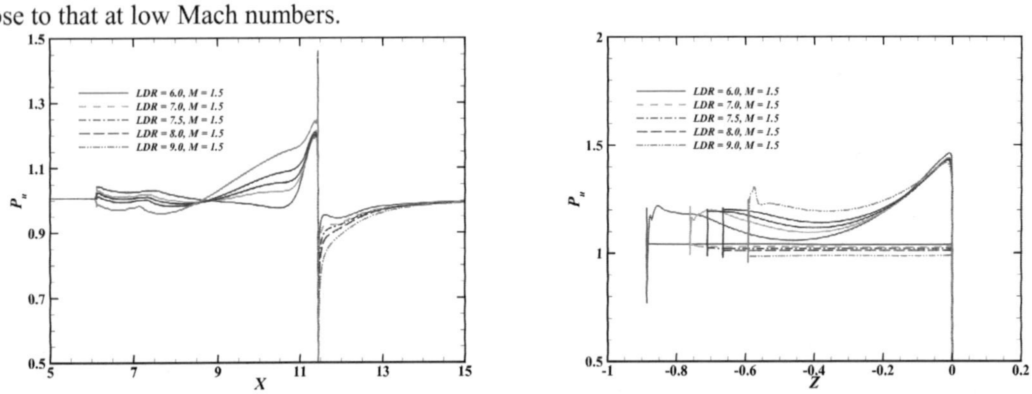

a. bottom floor b. rear wall

Figure 11. P_u distribution along the cavity bottom floor and rear wall with different length-to-depth ratios (Mach number 1.5)

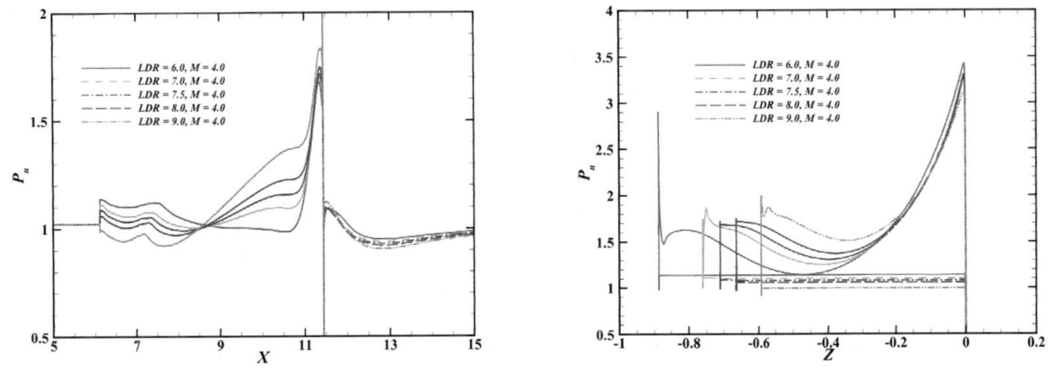

a. bottom floor b. rear wall

Figure 12. P_u distribution along the cavity bottom floor and rear wall with different length-to-depth ratios (Mach number 4.0)

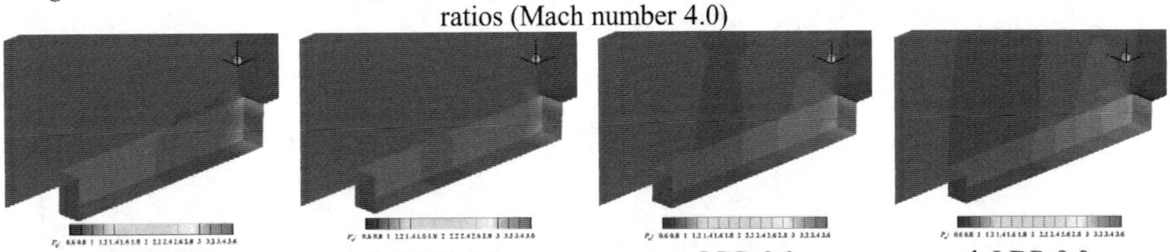

a. LDR 6.0 b. LDR 7.0 c. LDR 8.0 d. LDR 9.0

Figure 13. P_u distribution on the cavity surfaces at different length-to-depth ratios (Mach number 1.5)

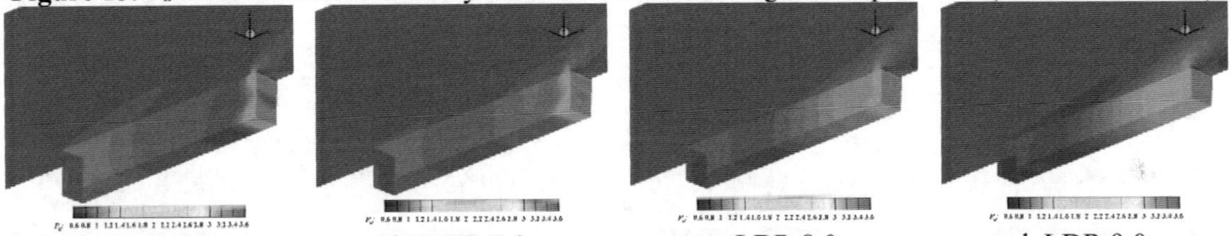

a. LDR 6.0 b. LDR 7.0 c. LDR 8.0 d. LDR 9.0

Figure 14. P_u distribution on the cavity walls at different length-to-depth ratios (Mach number 4.0)

4.2. Length-to-width Ratio

Due to the flow-separation zone around the front and rear wall, spanwise flow exists in the cavity, which is obviously affected by the parameter length-to-width ratio. Figure 15 and Figure 16 show the P_u distribution along the cavity bottom floor and rear wall with length-to-depth ratio of 7.5 and length-to-width ratio ranging from 4.0 to 6.0, at Mach numbers of 1.5 and 4.0 respectively, while Figure 17 shows the P_u distribution on the cavity walls. It can be seen that the length-to-width ratio has limited effect on the pressure. With the increase of length-to-width ratio, the pressure peaks on the cavity bottom floor and rear wall decreases slightly. Moreover, it can be noticed that the property of open-cavity flow does not change when the length-to-width ratio changes little. Consequently, the influence of length-to-width ratio on the static characteristics of cavity at high Mach numbers is still small, which is consistent with the conclusion at low Mach numbers.

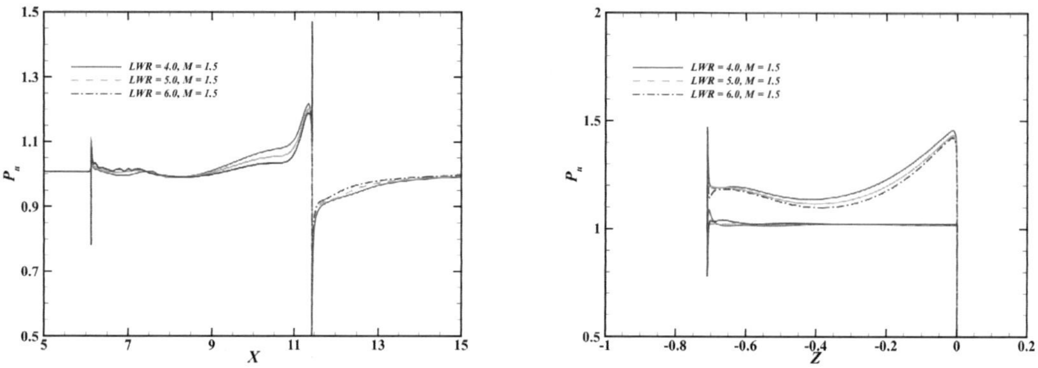

a. bottom floor b. rear wall

Figure 15. P_u distribution along the cavity bottom floor and rear wall with different length-to-width ratios (Mach number 1.5)

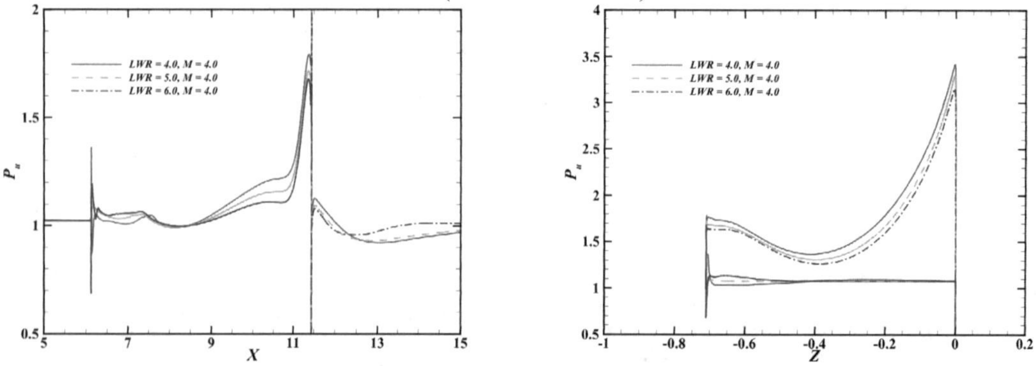

a. bottom floor b. rear wall

Figure 16. P_u distribution along the cavity bottom floor and rear wall with different length-to-width ratios (Mach number 4.0)

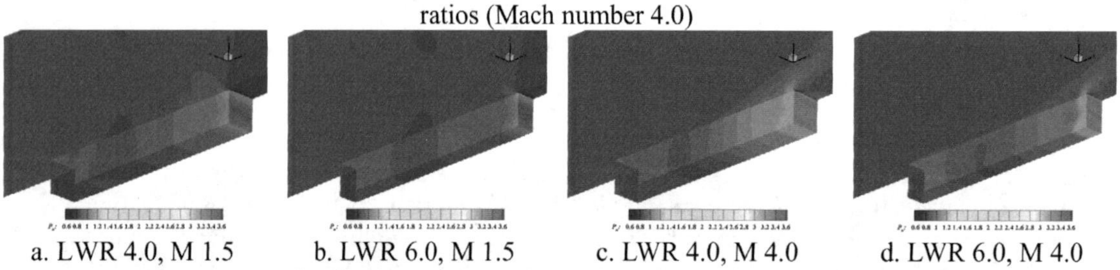

a. LWR 4.0, M 1.5 b. LWR 6.0, M 1.5 c. LWR 4.0, M 4.0 d. LWR 6.0, M 4.0

Figure 17. P_u distribution on the cavity walls at different length-to-width ratios

4.3. Spanwise Bending

In the internal weapon bay design, the fuselage form may be diverse. In order to investigate the influence of curved fuselage on the static characteristic, a spanwise-bending cavity shown in Figure 18b is considered for comparison with the straight cavity previously investigated.

Figure 19 shows the comparison of P_u distribution along the bottom floor and rear wall between the two kinds of cavities with length-to-depth ratio of 7.5 and length-to-width ratio of 5.0 at Mach number 4.0, while Figure 20 shows the P_u distribution on cavity walls. It can be seen that the pressure distribution along the bottom floor of the spanwise bending cavity is close to that of the straight cavity. The pressure peak value on the rear side of the former is slightly larger than that of the latter, while the pressure on the front side of the former is less than that of the latter. The non-uniformity of pressure distribution near and on the rear wall of the former turns out to be slightly stronger than that of the latter. Combined with the previous result, the influence of spanwise bending on pressure distribution is similar to that of length-to-width ratio, which indicates that the change in spanwise flow caused by spanwise bending is close to that caused by length-to-width ratio variation, within the parameter ranges discussed in this study.

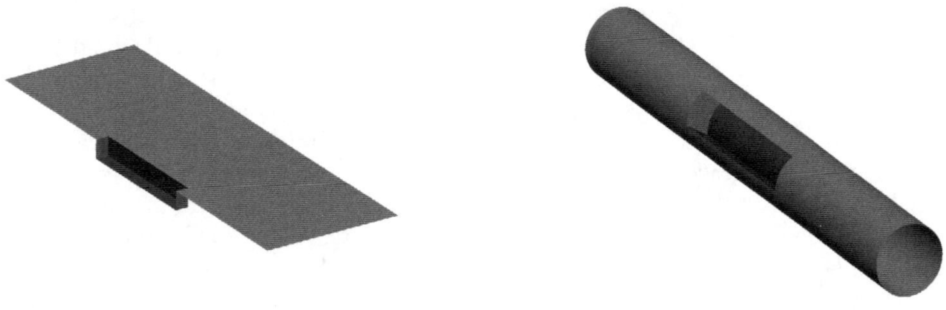

a. straight cavity b. spanwise-bending cavity

Figure 18. Configurations of the straight and spanwise-bending cavity

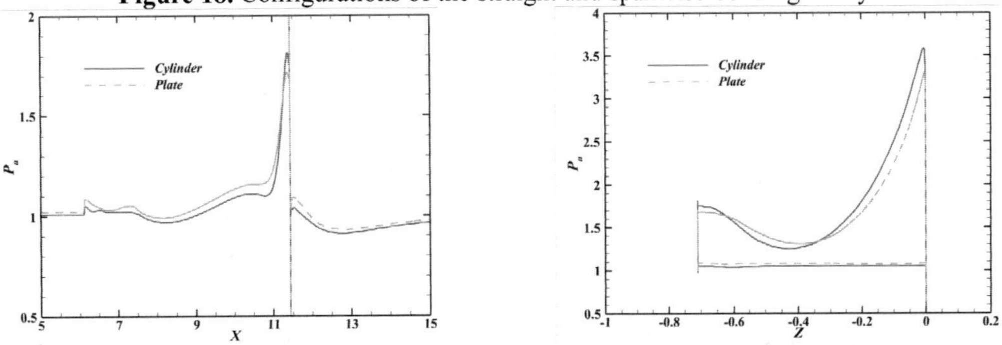

a. straight cavity b. spanwise-bending cavity

Figure 19. P_u distribution along the bottom floor and rear wall of the straight and spanwise-bending cavity

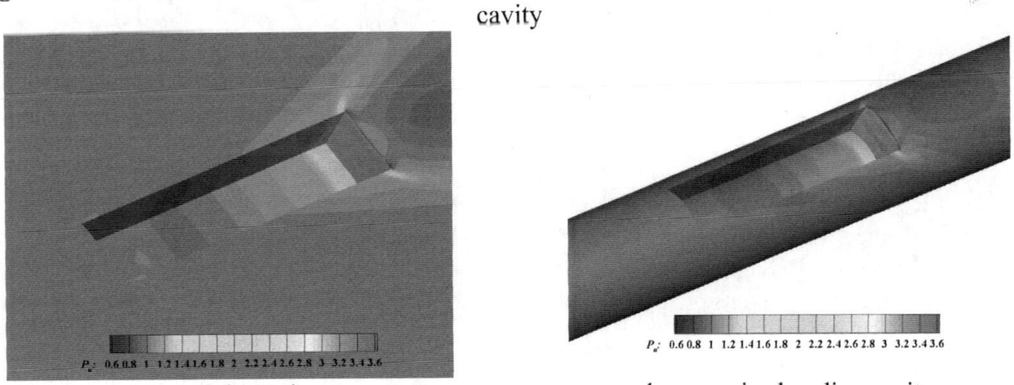

a. straight cavity b. spanwise-bending cavity

Figure 20. P_u distribution on the walls of the straight and spanwise-bending cavity

5. Conclusion

In this numerical investigation on supersonic cavity flow, the influence laws of two freestream parameters (Mach number and Reynolds number) and several configuration factors (length-to-depth ratio, length-to-width ratio and spanwise bending) on the cavity static aerodynamic characteristics are discussed. The conclusions of this study are drew as follows:

(1) At high speeds, the influence of Mach number on the pressure distribution in the cavity is consistent with that at low speeds. With the increase of Mach number, the pressure peak at the downstream of the bottom floor shows an upward trend, and the change in pressure distribution

characteristics indicates that the flow type is changing from a closed-cavity flow to an open-cavity flow.

(2) At high speeds, the change of Reynolds number will have a certain impact on the pressure distribution in the cavity, which is different from that at low speeds. The increase of Reynolds number will increase the pressure downstream the cavity bottom floor and on the rear wall, and the greater the Mach number is, the greater the difference will be caused by Reynolds number.

(3) The influence of length-to-depth ratio on the pressure distribution in the cavity at high speeds is close to that at low speeds. The increase of length-to-depth ratio reduces the pressure peak at the cavity rear lip, and the influence caused by the change of length-to-drag ratio at high speeds is greater than that at low speeds. In the parameter range investigated in this study, the increase of length-to-depth ratio leads to flow type change from open-cavity flow to transitional-cavity flow. In contrast, the influence of length-to-width ratio on the pressure distribution in the cavity is relatively small, and the increase of this factor will slightly increase the pressure peak downstream the cavity bottom floor.

(4) Spanwise bending will have a certain impact on the pressure distribution in the cavity. The enhancement of spanwise flow will slightly increase the pressure peak downstream the cavity bottom floor after spanwise bending, which is consistent with the length-to-width ratio increment, within the parameter range discussed in this study.

References

[1] S. J. Lawson, G. N. Barakos, Review of numerical simulation for high-speed, turbulent cavity flows, Prog. Aeosp. Sci. 47 (2011) 186–216.
[2] A. F. Charwat, J. N. Roos, F. C. Dewey, J. A. Hitz, An investigation of separated flows - part I: the pressure field, J. Aeosp. Sci. 28-6 (1961) 457–470.
[3] A. F. Charwat, J. N. Roos, F. C. Dewey, J. A. Hitz. An investigation of separated flows - part II: flow in the cavity and heat transfer, J. Aeosp. Sci. 28-7 (1961) 513–527.
[4] N. E. Suhs, Computations of three-dimensional cavity flow at subsonic and supersonic Mach numbers, AIAA 1987-1208 (1987).
[5] R. L. Stallings Jr., F. J. Wilcox Jr., Experimental cavity pressure distributions at supersonic speeds, NASA TP-2683 (1987).
[6] R. L. Stallings, Jr., F. J. Wilcox Jr., D. K. Forrest, Measurements of forces, moments, and pressures on a generic store separating from a box cavity at supersonic speeds, NASA TP-3110 (1991).
[7] J. F Wu, Investigation on aerodynamic characteristics of internal weapons bay system, Graduate Faculty of China Aerodynamics Research and Development Center (2012, in Chinese).
[8] J. E. Rossiter, Wind-tunnel experimental on the flow over rectangular cavities at subsonic and transonic speeds, DCA Report and Memoranda No. 3438 (1964).
[9] L. Larcheveque, P. Sagaut, Le Thien-Hiep, P. Comte, Large eddy simulation of a compressible flow in a three-dimensional open cavity at high Reynolds number, J. Fluid Mech. 516 (2004) 265–301.
[10] D. S. Dolling, S. W. Perng, Y. L. Leu, An experimental study of passive control of hypersonic cavity flow oscillations, AFRL-SR-BL-TR-98-0240 (1998).
[11] E. B. Plentovich, J. Chu, M. B. Tracy, Effects of yaw angle and Reynolds numbers on rectangular-box cavities at subsonic and transonic speeds, NASA TP-3099 (1991).
[12] L. G. Kaufman II, R. L. Clark, Mach 0.6 to 3.0 flows over rectangular cavities, AFWAL-TR-82-3112 (1982).
[13] J. Zhang, E. Morishita, Okunuki T, and et al. Experimental and computational investigation of supersonic cavity flows, AIAA 2001-1755 (2001).
[14] K. M. Chung, A study of transonic rectangular cavity of varying dimensions, AIAA 99-1909 (1999).
[15] O. H. Unalmis, N. T. Clemens, D. S. Dolling, Cavity oscillation mechanisms in high-speed flows, AIAA J., 42-10 (2004) 2035-2041.

[16] J. T. Chen, Y. Q. Deng, N. C. Zhou, Y. B. Zhang, CFD investigations of the HiLiftPW-2 configuration with the MFlow solver, AIAA 2014-2571 (2014).
[17] J. T. Chen, Y. B. Zhang, H. Zhao, G. Zhou. Numerical investigations of the NASA Common Research Model with aeroelastic twist, J. Aircr., 55-4 (2018) 1469-1481.
[18] P. L. Roe, Approximate Riemann solvers, parameter vectors and difference schemes, J. Comput. Phys., 43 (1981) 357-372.
[19] B. van Leer, Flux-vector splitting for the 1990s, NASA N91-21073 (1991).
[20] R. W. Tramel, R. H. Nichols, P. G. Buning, Addition of improved shock-capturing schemes to OVERFLOW 2.1, AIAA-2009-3988 (2009).

Control of Inner Flow Field Inside A Transonic Centrifugal Compressor with Splitter Through Chordwise Sweep

Dong Liang[1], Huixin Chen[1], Jun Ou[2], Donghai Jin[1, 3, *], Xingmin Gui[1, 3]

[1] School of Energy and Power Engineering, Beihang University, Beijing, China
[2] System Design Institute of Hubei Aerospace Technology Academy, Wuhan, China
[3] Jiangxi Research Institute of Beihang University, Jiangxi, China

*Corresponding author e-mail: jdh@buaa.edu.cn

Abstract. The strategy of sweep design should be based on the inner flow field of the compressor. In this paper, a transonic centrifugal compressor within the splitter is discussed in detail, and both forward sweep and backward sweep are studied with numerical simulation for optimization. Splitter introduces extra leakage and it's worth noting that the high-entropy region is shifted from the suction side of the main blade to the suction side of the splitter. Different from the transonic axial rotor, the leakage of the main blade mainly occurs after the passage shock and will be mixed with the leakage of the splitter in the suction side of the splitter. Therefore, the flow control on the middle and rare part of the compressor is more essential for the performance of the stage. In this regard, the backward sweep of the main blade exhibits a good effect by reducing the mixing loss of secondary leakage. Due to the redistribution of the airflow, the mainstream velocity in the tip area is accelerated in the forepart and reduced in the rare part, the aerodynamic loss in the high-entropy zone is then improved, and the flow capacity of the compressor is also increased. The best performance is achieved by an 8° backward sweep for the centrifugal compressor in this work, the pressure ratio is increased by 2.84% and isentropic efficiency is increased by 1.40%, compared to the prototype.

1. Introduction

The structure of the inner flow field inside the centrifugal compressor is three-dimensional, the direction of the pressure gradient in the flow field is highly related to the meridian location of airflow. This makes the secondary flow characteristics in the centrifugal compressor more complicated, resulting in the actual performance of the centrifugal compressor often being lower than the design value. Therefore, researchers concentrated on optimization of design factors to control the development of secondary flow and wake inside the impeller, thereby obtaining a better flow field distribution and also providing better incoming flow conditions for the downstream radial diffuser [1].

The study of the internal flow of centrifugal compressors has been held for a long time. In the 1950s, some scholars have studied the flow nonuniformity caused by rotating and radial pressure differences with the help of potential flow theory [2]. In 1960, Dean [3] summarized this nonuniformity at the impeller outlet as a jet-wake model and discussed the mixing process of wake in the downstream diffuser. This classical model is still used by many engineers. According to the location of secondary flow,

Eckardt [4] modified the Dean jet-wake model and classified flow into three categories: 1. End wall secondary flow near casing and pressure surface; 2. End wall secondary flow near hub and suction surface; 3. Clearance leakage flow. These secondary flows finally converge to form the exit wake. Hathaway [5] pointed that the leakage jet in the clearance of centrifugal impeller is composed of two streams of fluid from the pressure surface and suction surface of the main blade. The incidence angle of incoming flow will affect the intensity of different secondary flows and influence the position and intensity of the wake. Zageneh [6] carried out the reverse three-dimensional optimization design of the SRV2 impeller and considered that the impeller loss was mainly determined by secondary flow, inlet shock, shock-leakage interference, flow separation at blade surface, and other factors.

Splitter is first introduced by Wennerstrom [7] and then wildly used in centrifugal compressors for higher solidity in the rare part of the impeller, the capacity of pressure rise is strengthened, and flow separation is also suppressed. But in the meanwhile, the mechanism of loss generation is different in the two channels. The higher entropy region is often located on the suction side of the splitter and is close to the shroud, which is significantly related to the leakage of both blades. Kaneko and Tsujita [8] have studied the inner flow field of the impeller, the leakage of the main blade will migrate to the suction side of the splitter, and mix with the leakage generated by the splitter clearance in the same channel. Ibaraki [9] states that the leakage of the main blade is relatively steady within the passage shock, but the leakage of the splitter is sensitive to the leading edge shock, the break down of leakage vortex devotes to the high entropy in this channel. Erdmmenger [10] also presents the leakage trajectory in the two channels, the interaction of the leakage generated by the main blade and the leading edge of the splitter almost decides the stall point of the impeller. Therefore, the development of flow loss in the two channels is not balanced, which means that the flow control in the high-entropy area should be focused on.

The control of secondary flow in centrifugal impeller has gone through three stages of development, mainly involving three aspects of geometric parameters: meridional channel geometry, blade loading along the flow direction, and swept shape along blade spanwise. With the deepening of research, the technology of using sweep to control radial flow distribution is gradually developed. At first, the sweep was applied in axial flow compressor based on the design idea of "shock free", back swept was used to control shock. Rodgers [11] points out that the change of sweep angle of the centrifugal impeller will change the meridional velocity distribution at the inlet, and then change the distribution of the internal flow field inside the impeller passage. The impeller designed by DFVLR [12] adapts the aft-loaded design to reduce loss of inlet shock and uses the backward sweep and backward lean. Compared with the previous high loading impeller developed in 1988, the internal flow field of the new impeller is improved, the outlet uniformity is increased, and the stage pressure ratio, efficiency and choke flow range are improved. Hazby [13] takes high loading centrifugal impeller as a research object, the forward sweep can further reduce the load on the leading edge to reduce the shock wave loss and leakage loss and is beneficial to restrain the spanwise migration of secondary flow. Ganesh's [14] study shows that forward sweep can suppress the flow separation on the casing surface and weaken the shock-boundary layer interference in the tip region, thus increasing the working stability and flow range. Erdmmenger studies the sweep patterns of the main blade and splitter blade of a centrifugal compressor. It demonstrates that blade sweep can change the interaction between the leakage and the splitter leading edge, and it's possible to maintain the operating range of the impeller while achieving a higher pressure ratio and efficiency with aft sweep of the leading edge.

In the research of axial flow compressor, sweep angle has a strict definition, which can be divided into a chordwise sweep and axial sweep. However, for centrifugal compressors, the design of sweep is limited by the three-dimensional characteristics of the impeller. Thus, the sweep angle of the centrifugal impeller is generally defined based on the leading edge of the impeller, and the definition methods are divided into the following two types as summarized by Hazby. In the first definition, the forward sweep is to extend the tip of the impeller forward while keeping the metal angle of the extended section unchanged, and the backward sweep is to truncate the blade based on the original impeller. In the second definition, the position of the tip leading edge is changed, based on keeping the distribution of metal angle and thickness along the relative chord length unchanged. In the latter definition, the change of blade tip leads to the change of whole channel geometry, so the load distribution at different meridional positions will also be changed; while the first definition will only change the geometry of blade leading edge, and the geometry at the middle and rear section of the impeller is almost unchanged, which is more suitable for the evaluation of swept as a single variable. Therefore, the first definition is adopted in this paper.

2. Design of Centrifugal Compressor

2.1. Design Method

A high loading centrifugal compressor within the splitter was designed. The transonic centrifugal compressor is composed of a single-stage centrifugal impeller, a radial diffuser, and an axial diffuser. The design parameters of this high loading compressor are shown in Table 1. The parameters in the table are designed for the maximum climbing condition.

Table 1. Design parameters of centrifugal compressor

Parameters	Value	Units
Shaft speed	29000	rpm
Design mass flow rate	3.608	kg/s
Stage pressure ratio	4.91	
Stage efficiency	0.807	
Flow coefficient	0.064	
Load coefficient	0.705	
Impeller count full/splitter	14/14	
Radial diffuser count	20	
Axial diffuser count	56	
Leading edge hub radius	41	mm
Leading edge tip radius	109	mm
Impeller tip radius	190	mm
Impeller tip speed	577	m/s
Impeller tip clearance	0.3	mm

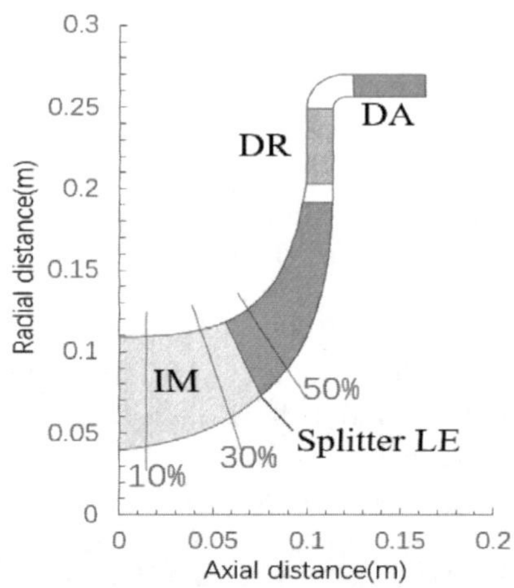

Figure 1. The meridian view of centrifugal compressor

The design of the centrifugal compressor adopts the three-dimensional inverse design method, which is wildly used by modern designers [15]. At first, the meridional positions of the flow channel and blades should be given as shown in Fig. 1. Within the geometric and thermodynamic parameters, the circulation distribution of the tip and the root along the flow direction is controlled. Combined with the streamline curvature method, the airflow angle distribution of the blade can be calculated. And then the arc of the blade profile is fitted, based on the results, a three-dimensional blade can be obtained. Both the centrifugal impeller and the diffuser adopt this design method. The centrifugal impeller adopts the traditional ruled surface design, and the elementary blade shape of the radial diffuser along spanwise direction is consistent.

2.2. *Design for Aft-loaded Impeller*

The load distribution of the impeller is generally divided into three types: fore-loaded, uniform loaded, and aft-loaded. Some articles believe that the use of fore-loaded at tip and aft-loaded at root is more conducive to the performance of the centrifugal compressor [16]. However, due to the high load coefficient of the centrifugal compressor, the impeller abandons the fore-loaded design at the tip, and uses the aft-loaded design instead. The inlet flow field of the impeller is transonic, and a backswept design is adopted to control the exit wake. These two reasons also make the aft-loaded design better. The impeller adopts an additional power reduction design at the inlet to control the intensity of the shock.

The exit wake of the centrifugal impeller is strong, and the backswept design of the impeller can reduce the expansion angle. Therefore, the load on the rear of the impeller is reduced, and flow separation is inhibited; in addition, the axial vortex in the backswept impeller and the speed difference in the passage can cancel out a part of each other. This mechanism helps to suppress the separation of the boundary layer and the formation of secondary vortices, which in turn helps to improve the uniformity of the flow field at the impeller exit. In the process of three-dimensional design, the impeller is given a negative deviation angle and has the characteristic of a large backswept angle.

Due to the high flow coefficient and load coefficient of the impeller, the splitters are used to reduce the blockage between the main blades and increase flow range. The leading edge of the splitter is located at the 40% meridian chord length of the main blade, and the circumferential position is at the 50% pitch between the two main blades. The shape of the splitter is consistent with the main blade, which can be regarded as a partial truncation of the main blade. Both the main blade and the splitter are not designed with sweep angle or lean angle and adopt trailing edge stacking.

2.3. *Radial Diffuser and Axial Diffuser*

High loading centrifugal compressors often use vaned diffusers to improve stage efficiency and pressure ratio. Vaned diffusers are commonly divided into wedge-shaped vanes and cambered vanes. Due to the high kinetic energy at the exit of the impeller, both the radial diffuser and the axial diffuser are designed with cambered vanes, to ensure the static-pressure rise capacity. And the radial diffuser is designed with a blunt trailing edge. The diffusing effect is mainly achieved in the radial diffuser section, which adopts the design of weak rear loading and a thin leading edge to adapt to the supersonic condition at the inlet. While the axial diffuser mainly plays the role of rectification, the bending angle is relatively large.

3. Numerical Methods

3.1. *Simulation Settings for Centrifugal Compressor*

To evaluate the performance of the designed centrifugal compressor and analyze the flow field, a three-dimensional constant numerical simulation is conducted using NUMECA. Autogrid5 is used to generate a structured grid of fluid channels, and the compressor inlet and outlet are appropriately extended to ensure calculation convergence. Because the impeller has multiple blades in a single channel, H&I grid topology is selected to obtain better grid quality. The radial diffuser uses O4H grid topology; the geometry of the axial diffuser channel is relatively simple. Therefore, a HO type grid is selected and the topology of the one-single passage of the compressor is shown in Fig. 2. The flow boundary is extended outwards, the lengths of the inlet section and exit section are both about the axial length of the impeller.

The working substance used for calculation is actual gas. Spalart-Allmaras model is adopted here, which is wildly used because of easy convergence and fast calculation speed. The multi-grid method is used to accelerate the convergence. Set the total temperature, total pressure, and flow angle as an inlet boundary condition, and set the static pressure as the exit boundary condition. The interfaces between rotor and stators are combined with a mixing plane approach. Use the last calculation result as the initial field, and gradually increase exit static pressure during calculation. There are still some controversies about the stall characteristics of centrifugal compressors, and it is not easy to accurately capture them in numerical simulations. In this paper, until the exit static pressure increase causes the calculation to diverge, at which point the centrifugal compressor is considered to be in a stall state.

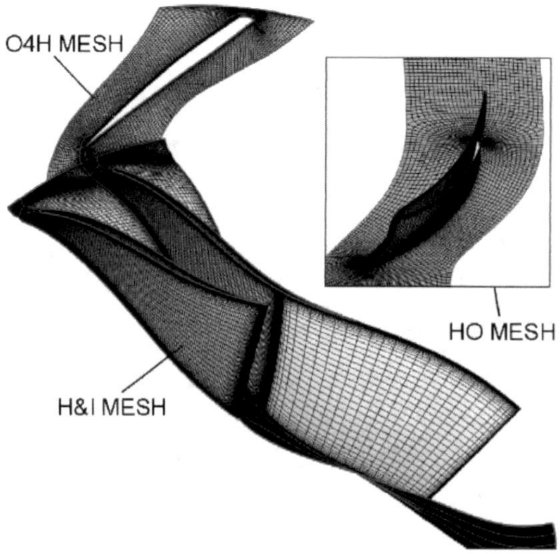

Figure 2. Topology of grid mesh

3.2. Mesh Independence Validation

To eliminate the influence of grid differences on the calculation results, four mesh with different grid numbers are set to verify the grid independence. The topology of the four grids remains unchanged. The total number of grids in coarse mesh C is 0.8 million. Based on mesh C, the number of grids on the streamwise, spanwise, and circumferential of the entire stage compressor is evenly increased. The total number of grids gradually increases from 2.1 million to 3.1 million and then to 5 million, respectively represented by mesh M1, mesh M2 and mesh F. Fig. 3 shows the performance line of four different mesh calculated under the same boundary condition. It can be seen that the difference in performance parameters between M2 and F is within 0.5%. From the perspective of saving time of calculation, M2 is chosen for subsequent numerical calculations.

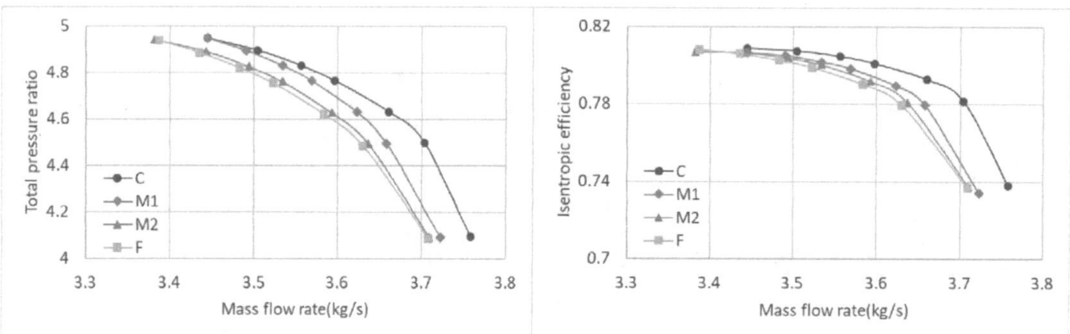

(a) Total pressure ratio-mass flow rate (b) Isentropic efficiency-mass flow rate

Figure 3. Performance line of four mesh

4. Numerical Results

4.1. Overall Performance

In this paper, the angle between the new leading edge and the original radial leading edge is defined as the sweep angle. The leading edge of the original impeller is along the radial direction. The root of the blade stays the same, the leading edge of the tip moves forward or backward while keeping the metal angle unchanged. The ratio of blade thickness to chord length remains constant along the chord, so the throat area will change linearly with sweep angle. Obviously, the sweep angle defined here is the relative sweep angle, the reference plane is formed by the chord line of the tip and the leading edge of the original impeller. The stagger angle of the main blade is 60°, so the leading edge is no longer perpendicular to the hub when the tip moves, that is, the circumferential lean is introduced through the sweep of ruled lines. Fig. 4 shows the leading edges of five impellers used in research, while the leading edges of splitters in all designs are kept straight. Chordwise sweep angle is defined as positive when swept forward and is negative when swept backward. The sweep angle is varied from -8° to +8°and represented by FSW-8, FSW-4, STR, BSW-4, and BSW-8 respectively.

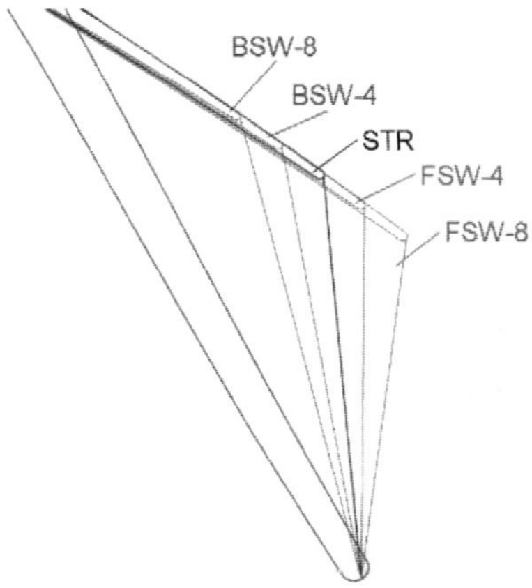

Figure 4. Definition of sweep angle

Fig. 5 gives the performance map of five centrifugal compressors. The increase of sweep angle will reduce the choke flow and the flow margin is slightly improved. The decrease in choke flow is mainly due to the increase of sweep angle reduces the throat area of the impeller channel, it can be seen that the choke flow of the compressor is almost linearly related to the sweep angle.

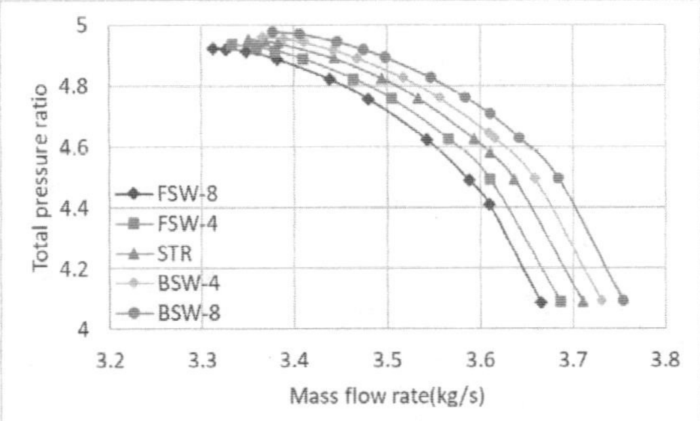

(a) Total pressure ratio-mass flow rate

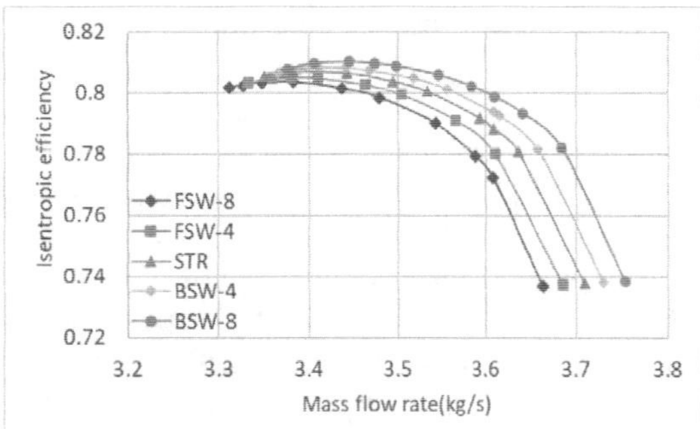

(b) Isentropic efficiency-mass flow rate

Figure 5. Comparison of calculated performance

The comprehensive surge margin (SM) is defined as:

$$SM = \left(\frac{(\pi_{tt}^*/\dot{m})_{stall} - (\pi_{tt}^*/\dot{m})_{design}}{(\pi_{tt}^*/\dot{m})_{design}} \right) \quad (1)$$

where the π^* and the $\dot{m}$ represent the total pressure ratio and the mass flow rate, respectively.

It can be seen SM continues to increase with the sweep angle. This is due to the lower flow rate at the stall point in the forward-swept design, and the second reason is that the pressure ratio at the design flow condition continues to decrease and the choke point moves to the left.

Table 2. Performance at design condition

	FSW-8	FSW-4	STR	BSW-4	BSW-8
$\dot{m}$(kg/s)	3.609	3.610	3.610	3.609	3.610
π_{tt}^*	4.41	4.49	4.58	4.64	4.71
η_{tt}^*	0.772	0.780	0.788	0.794	0.799
π_{IM}^*	5.18	5.25	5.33	5.39	5.46
η_{IM}^*	0.877	0.882	0.887	0.891	0.895
$\sigma_{DR,DA}$	0.851	0.855	0.859	0.861	0.863
SM	21.6%	19.1%	16.5%	14.6%	13.1%
OR	9.6%	9.6%	9.7%	9.8%	10.0%

Operating range (OR) is defined by Erdmenger as:

$$OR = \frac{\dot{m}_{choke} - \dot{m}_{stall}}{\dot{m}_{choke}} \quad (2)$$

OR will not be affected by the design mass flow rate, and can reflect the width of flow range compared to the choke mass flow rate. In this definition, backward sweep even slightly increased the flow range as shown in Table 2.

However, it can be found that the isentropic efficiency of the compressor decreases with the sweep angle in the whole working range, which means that backward sweep can improve performance for the aft-loaded centrifugal compressor.

4.2. Structure of Shocks and Flow Field

The distribution of Mach number at blade tip is shown in Fig. 6. There are shocks on the suction side of the main blade and the splitter, which are referred to by shock 1 and shock 2 respectively. The inlet of the main blade has a pre-compression design and the blade bending angle is small. Therefore, the strength of shock 1 is limited, which is mainly expressed as an expansion wave. Between the two shocks, the main flow accelerates due to the contraction of the flow channel, and the supersonic range is located between 20%-30% meridian position.

Along with the increase of sweep angle from -8° to +8°, the relative Mach number of shock 1 decreases from 1.28 to 1.23. At the same time, the relative Mach number on the pressure surface has increased. This is mainly because the chord length of the blade tip increases with sweep angle, while the metal angle of the extension section remains unchanged, so the pre-compression of the blade tip is strengthened, which will cause the airflow on the pressure surface to accelerate. Affected by the acceleration of the pressure surface of the adjacent blades, the relative Mach number near the suction surface also increases significantly, which will increase pressure difference across the gap. Since the leading edge of the splitter is located at 40% meridian position, it also leads to an increase in Mach number at the leading edge of the splitter. The Mach number of shock 2 increases from 1.48 to 1.53 when the sweep angle increases, while the incidence angle decreases. Earlier studies have shown that it can be approximately considered that the shock loss is proportional to $(M^2 - 1)^3$, where M is the vertical component of the Mach number. Therefore, the increase of shock 2 causes more loss. The low momentum fluids gradually mix up the downstream, the flow field is smoother at the exit than 80%

meridian position. Flow on the pressure side of the main blade is forced by the blade, which has a higher pressure and only receives a little impact from the secondary flow, acts as the jet at the exit.

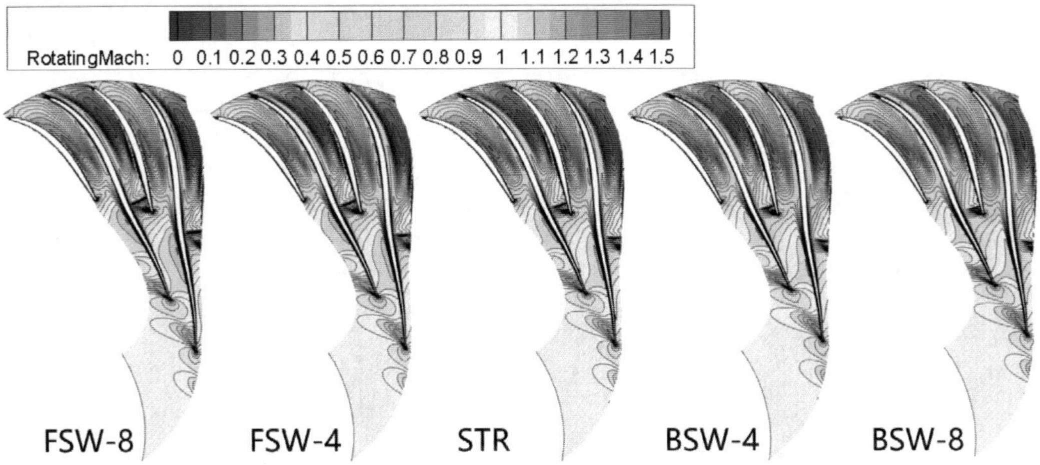

Figure 6. Relative Mach distribution at design operation condition at 90% span

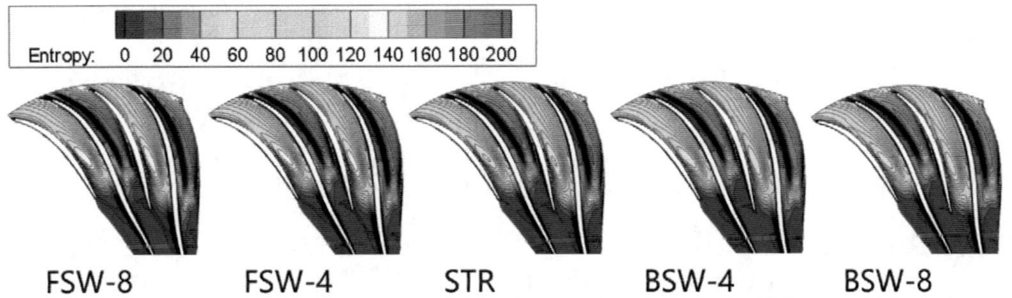

Figure 7. Entropy distribution at design operation condition at 90% span

The factor that has more impact on the performance of the impeller is the accumulation of the low momentum fluid in the downstream passage, corresponding to the high entropy region. Fig. 7 shows the entropy distribution at the same span position. It can be seen that the entropy in both channels increases with the sweep angle. There is less loss in the front part of the impeller, no obvious separation is generated and the flow field is satisfactory. Aerodynamic loss is formed from 50% meridian position and is primarily located in the suction side of the splitter, where the tip leakage flow and main flow is mixing. A part of the leakage of the main blade mixed in the suction side of the main blade, and the other part cross the tip of the splitter as secondary leakage, creating stronger mixing in the suction side of the splitter.

The tip leakage loss is related to the intensity of leakage, the velocity of the main stream, and the leakage angle. The intensity of leakage flow is mainly determined by the static pressure difference between the suction surface and the pressure surface near the blade tip. Fig. 8(a) shows the distribution of static pressure distribution at 90% span of the main blade surface. Only three designs are shown in the figure to make it easier to distinguish, the conclusion is the same in the other two designs. The abscissa is the meridian position, and the ordinate Cp is defined by the following formula:

$$C_p = \frac{p - p_{inlet}}{p^*_{inlet} - p_{inlet}} \tag{3}$$

where p^* and p represent the total pressure and the static pressure, respectively.

The position of shock 1 and shock 2 is marked by the black lines. The relative position of shock 1 is basically constant in different designs, absolute movement of shock 1 in the passage is mainly due to the movement of the leading edge. Forward sweep reduces the static pressure on the pressure surface, consequently, the pressure difference across the blade is reduced. At this position, the load on the tip decreases, and the tip leakage intensity will also be weakened. Static pressure on the surface of the splitter is affected by shock 2, which has a lower value due to stronger shock at the splitter leading edge as shown in Fig. 8(b). Obviously, shock 2 is stronger than shock 1 and will increase the pressure difference across the splitter gap, leading to leakage loss and leakage-shock interaction in the suction side of the splitter. Forward sweep reduces the load on the front of the impeller, but accelerates the meridional velocity in the impeller, especially at the leading edge of the splitter, increasing shock 2 and higher leakage loss, ultimately reducing the performance of the impeller.

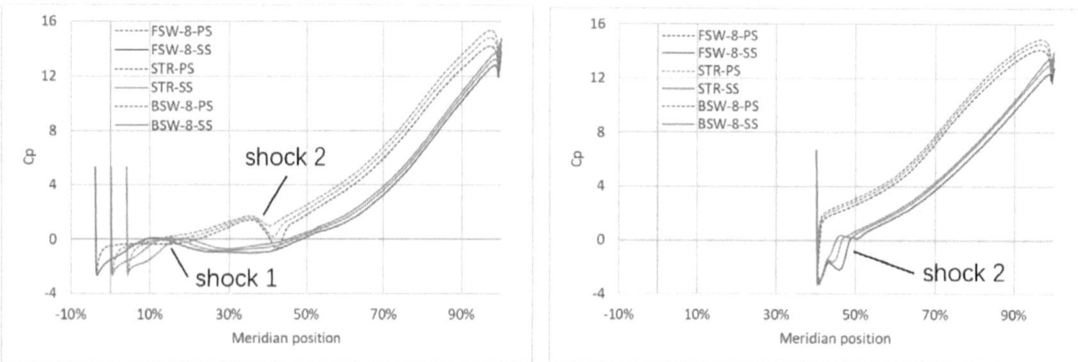

(a)Main blade (b)Splitter

Figure 8. Static pressure distribution of blade at 90% span

4.3. Mechanism of Loss Generation

To discuss the influence of shock and secondary flow on the development of entropy more clearly, Fig. 9 shows the distribution of circumferentially averaged meridional velocity and entropy at 90% span along the streamwise direction. Meridional velocity is accelerated before 40% position and gradually decreases as leakage flow accumulated downstream of the splitter. In FSW-8, meridional velocity is higher than other designs in all positions, especially at the leading edge of the splitter, and leads to higher losses after 40% meridian consequently. In Fig. 9(b), before 10% meridian, the entropy increase in FSW-8 is lower than other designs, which indicates that forward sweep is beneficial to reduce the loss at the leading edge of the impeller. At 10%-40% meridian, the mixing loss of the leakage flow and the main flow increases slightly as the intensity of shock 1 is weak. The process of reducing the performance mainly occurs in the channel after 40% meridian position for forward sweep designs. The meridional velocity increases with sweep angle and leads to an increase in the low momentum area downstream, and a higher mixing loss is produced.

The high entropy zone is mainly located in the corner formed by the suction surface of the splitter and the casing. Fig. 10 shows the distribution of entropy of S3 surface at 30% and 50% meridian position respectively. Apparently, the leakage flow injects into the suction side of the blade, and the mixing loss is introduced near the shroud. The interaction between the main stream and the leakage is strengthened with the increase of sweep angle as mentioned above. In the case of FSW-8, there is the largest area of low momentum fluid near the tip and the high-entropy area near the shroud increases. Additionally, a higher entropy area of the wake is formed in the suction side of the splitter due to secondary leakage as shown in Fig. 10(b).

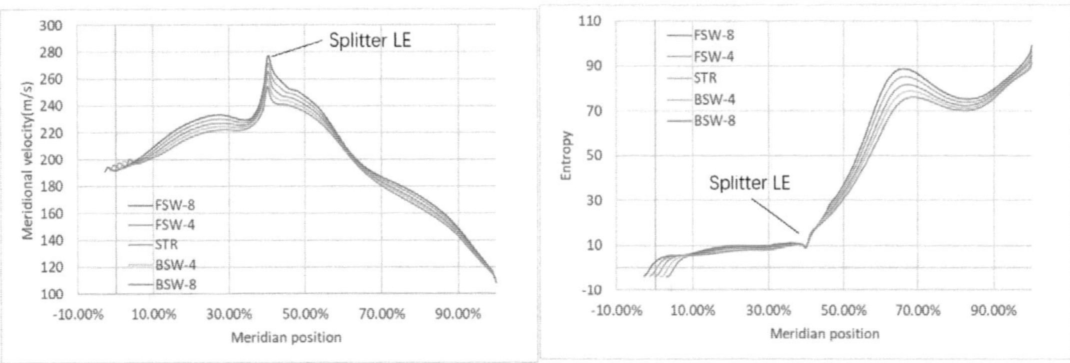

(a) Meridional velocity (b) Entropy

Figure 9. Distribution of meridional velocity and entropy at 90% span

According to the location of the high entropy region, leakage generated by the splitter contributes the most in this process, the leakage generated by the main blade is already mixed out at 50% meridian. Consequently, in the forward sweep design, the entropy in both passages increases, along with the increase of unevenness at the exit, especially in the suction side of the splitter. On the contrary, the backward sweep design suppresses the meridional velocity and improves the flow field near the shroud. The mixing loss of leakage is controlled by backward sweep, and the impeller performances better at the design condition.

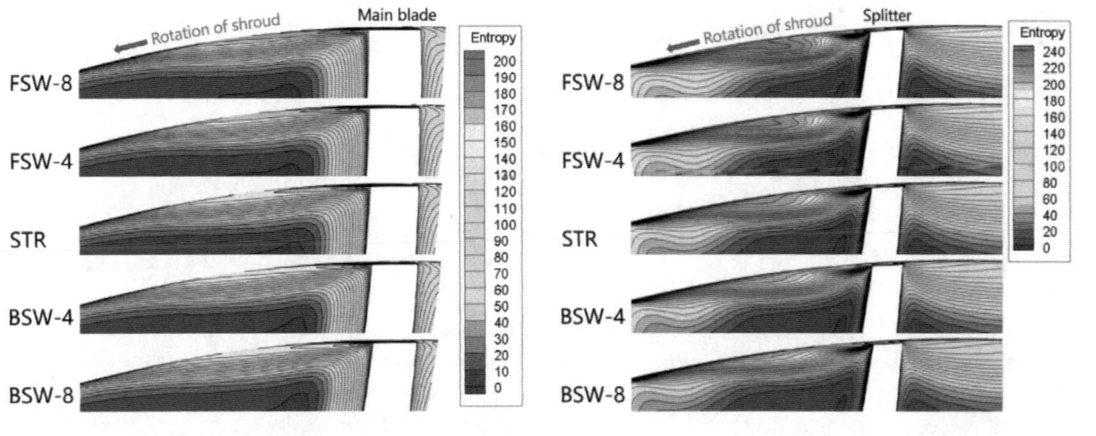

(a) 30% meridian in the suction side of the main blade (b) 50% meridian in the suction side of the splitter

Figure 10. Entropy distribution

4.4. Spanwise Distribution

Due to the design of sweep, the blade force along spanwise is un-uniform, forward sweep will initiate the radial migration of airflow from hub to shroud. In the inlet of the impeller, the static pressure difference to maintain radial balance will then be reduced. Fig. 11 gives the distribution of static pressure between the hub and shroud at 5% and 30% meridian position. It can be seen that the spanwise static pressure gradient of the forward sweep design is the smallest, which is conducive to suppressing the spanwise migration of secondary flow. Below 30% span position, the static pressure gradient is almost the same in all designs, this means the flow field at root is not wildly affected by sweep. At 30% meridian position, the spanwise static pressure gradient increases in all designs compared to upstream. However, the pressure gradient of FSW-8 is the highest at this position, spanwise migration of secondary flow is strengthened by forward sweep. This is one of the reasons why the forward-swept design has not significantly improved the performance of the impeller.

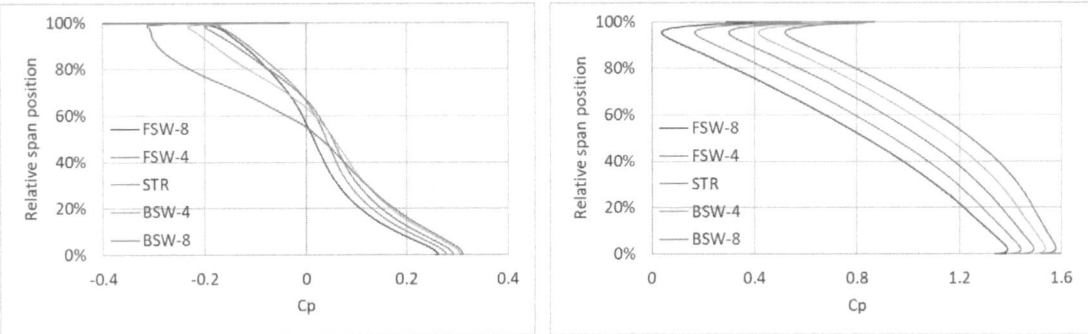

(a) 5% meridian position (b) 30% meridian position

Figure 11. Static pressure distribution

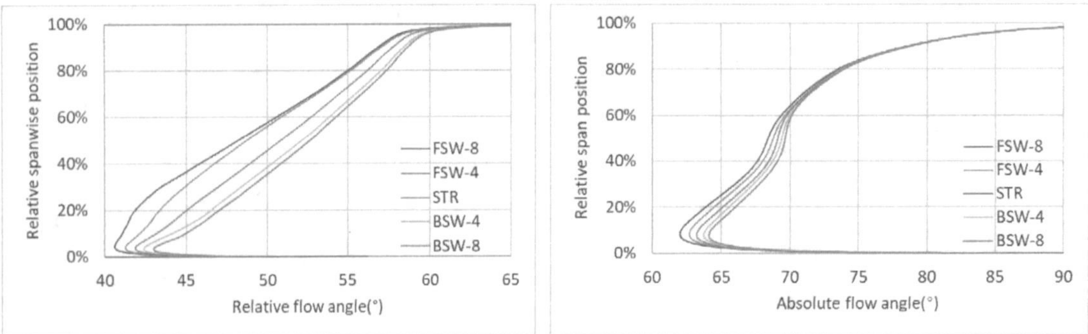

(a) IM inlet (b) IM exit

Figure 12. Absolute flow angle distribution

Redistribution of mainstream occurs at the inlet of the impeller, averaged relative flow angle changes with the sweep angle as shown in Fig. 12(a). For the STR, the relative flow angle is linear to the spanwise position apart from the near-wall region, indicating that the inlet flow is evenly distributed. For the forward sweep designs, extra blade force causes the mainstream to turn to the axial direction in the whole spanwise, and then the meridional velocity is accelerated. For the backward designs, the incidence of the blade increases, therefore shock 1 is strengthened by flow acceleration in the suction surface, while the meridional velocity is suppressed. Fig. 12(a) also indicates that the rate of increase will decrease as the backward sweep angle increases, thus there might be an optimal sweep angle for each compressor.

As the flow develops, the effect of sweep on the mainstream velocity gradually weakens, but the leakage flow generated at the tip of the splitter gradually accumulates and expands toward the hub, and the secondary flow component in the channel increases. Comparing the distribution of absolute flow angle at impeller exit as shown in Fig. 12(b), it can be found that the spanwise distribution is uneven due to the existence of low momentum fluids at the rare part of the channel. Low momentum fluids generated in the tip region gradually affect the whole passage and mix up in this process. Secondary flow in the channel is suppressed in the backward sweep design, so the spanwise flow distribution is improved, and the outlet flow field is relatively more uniform, the efficiency of the downstream diffuser is also improved.

5. Conclusion

This paper discusses the design and characteristics of the aft-loaded centrifugal compressor, and chordwise sweep is adapted to control the internal flow field, especially the suction side of the splitter. Many studies hold the idea that the forward sweep of the blade will reduce the loss at the tip, increase the efficiency and flow range, but the actual effects of the sweep are related to the mechanism of loss generation. Inside the impeller with splitter, the high entropy region locates in the suction side of the splitter, secondary leakage, instead of leakage at the main blade tip, dominates the performance of the stage.

(1) The decrease in the chord length of the tip of the blade will lead to the weakening of the tip pre-compression feature, which will lead to the decrease of the relative Mach number near the pressure surface. The velocity of main stream is slowed down, the expansion wave characteristics will be suppressed, and the intensity of shock at splitter leading edge is then reduced. The aerodynamic loss of leakage and secondary leakage is decreased due to lower meridional velocity and weaker leakage flow.

(2) Sweep forward reduces the tip load of the blade, and the capacity of pressure rise is weakened, while backward sweep has the opposite effect. The total pressure ratio of backward sweep is higher than the prototype in the whole flow range.

(3) Backward sweep increases the throat area of the impeller, the radius of the main blade tip is larger, and choke flow capacity is increased. The comprehensive surge margin will decrease with a higher choke mass flow rate and higher stall mass flow rate, while the operating range is hardly changed.

Acknowledgements

This study was financially supported by the National Science and Technology Major Project (2017-|-0005-0006 and 2019-II-0020-0041).

References

[1] Krain H. (1998). “Swirling impeller flow.” Journal of Turbomachinery. 110, 122-128.

[2] Michel D. J., Ginsburg A., Mizisin J. (1951). “Experimental investigation of flow in the rotating passages of a 48-inch impeller at low speed.” NACA, Report No.: RM E51D20.

[3] Dean, Senoo Y. (1960). “Rotating wakes in vaneless diffusers.” Journal of Basic Engineering, 563-574.

[4] Eckardt D. (1976). “Detailed flow investigations within a high-speed centrifugal compressor impeller.” Journal of Fluids Engineering, 390-399.

[5] Hathaway M. D., Chriss R. M., Wood J. R., Strazisar A. J. (1993). “Experimental and computation of the NASA Lewis Low-Speed Centrifugal Compressor flow field.” Journal of Turbomachinery, 115, 527-541.

[6] Zangeneh M., Amarel N., Daneshkhah K., Krian H. (2011). “Optimization of 6.2:1 pressure ratio centrifugal compressor impeller by 3D inverse design.” Proceedings of ASME Turbo Expo 2011. Vancouver, Canada.

[7] Wennerstrom A. J., Frost G. R. (1974). “Design of a Rotor Incorporating Splitter Vanes for a High Pressure Ratio Supersonic Axial Compressor Stage.” Aerospace Research Laboratories Wright-Patterson Air Force Base Ohio, Report No.: ARL-74-0110.

[8] Kaneko M., Tsujita H. (2015). “Numerical investigation of influence of tip leakage flow on secondary flow in transonic centrifugal compressor at design condition.” Journal of Thermal Science, 24(2), 117-122.

[9] Ibaraki S., Furukawa M., Iwakiri K., et al. (2007). “Vortical Flow Structure and Loss Generation Process in a Transonic Centrifugal Compressor Impeller.” ASME Turbo Expo: Power for Land, Sea, and Air; 2007, 1089-1098.

[10] Erdmenger R. R., Michelassi V. (2014). “Impact of main and splitter blade leading edge contour on the performance of high-pressure ratio centrifugal compressors.” Proceedings of ASME Turbo Expo 2014: Turbine Technical Conference and Exposition, Dusseldorf, Germany.

[11] Rodgers C. (1998). “The centrifugal compressor inducer.” ASME International Gas Turbine &

Aeroengine Congress & Exhibition, Stockholm, Sweden.
[12] Krain H., Hoffmann B., Rohne K. H., Eisenlohr G., Richter F. (2007). "Improved high pressure ratio centrifugal compressor." Proceedings of ASME Turbo Expo 2007: Power for Land, Sea and Air; Montreal, Canada.
[13] Hazby R., Xu L. (2007). "Effect of leading edge sweep on performance of a transonic centrifugal impeller." ISABE-2007-1225.
[14] Ganesh C. S., Nagpurwala Q. H., Dixit. (2010). "Effect of leading edge sweep on the performance of a centrifugal compressor impeller." SASTECH Journal, 9(2), 55-62.
[15] Ji Guofeng, Yin Mingxia, Gui Xingmin. (2010). "Aerodynamic design of a small centrifugal compressor." Journal of Aerospace Power, 25(3), 557-564.
[16] Zangeneh M., Schleer M., Ploger F., etc. (2004). "Investigation of an inversely designed centrifugal compressor stage—part I: design and numerical verification." Journal of turbomachinery, 126(1), 82-90.

Analysis of the aerodynamic characteristics of the outflow field of an SUV under the effect of side wind

Yang Zhou, Qingyun Guan, Ling Xiao, and Xiayi Yuan

Guangzhou Automobile Group Co., Ltd Automotive Engineering Institute, Guangzhou, China

*Corresponding author e-mail: yun.913@163.com

Abstract. A domestic SUV model was selected, and two different side wind simulation methods of "partial car" and "partial wind" were analyzed to numerically simulate the flow field of SUV under different intensity of side wind, and the velocity-pressure distribution and aerodynamic force of the flow field under different side wind intensity were obtained. coefficients under different side wind strengths. The wind tunnel test was conducted at the same time, and the simulation results were compared and analyzed with the test results. The analysis results show that: the results of the two simulation methods are close to each other, and the trend is the same as that of the test, which is that the aerodynamic drag, lateral force and lift force gradually increase with the increase of the side yaw angle; considering the operation and difficulty of model building, the side wind simulation method is recommended to "deflect the wind"; the pressure maps of different parts of the body are analyzed to qualitatively The pressure maps of different parts of the body are analyzed to qualitatively explain the effect of side wind on the aerodynamic characteristics of SUV, which provides a reference for further research on the aerodynamic characteristics of SUV under the influence of side wind.

1. Introduction

Aerodynamic drag is critical to vehicle fuel economy and is an important indicator for evaluating the aerodynamic performance of a vehicle. Most of the aerodynamic drag at this stage is developed under the angle of zero yaw angle, while the results of a literature study show that zero degree only accounts for less than 50% of the weight in the real driving environment [1]. Therefore, it is necessary to consider the aerodynamic resistance under side-wind conditions, and then make a comprehensive assessment of the optimization scheme of aerodynamic resistance by the weighting coefficient method, so that the aerodynamic resistance development evaluation method can be more relevant to the actual needs of users.

Secondly, as people pay more attention to the safety of automobile, it is prompted to continuously reduce the sensitivity of automobile to side wind and improve the driving stability of automobile in the process of automobile development and design, so it is important to study the aerodynamic characteristics of automobile under the action of side wind and find out the change law of aerodynamic characteristics of automobile with the action of side wind for the design and development of automobile [2-3].

Many scholars have conducted a series of studies on the effect of side wind on the aerodynamic performance of automobiles. The literature [4] simplified a domestic car to explore the effect of different intensity of side winds on the external flow field of the car and its related aerodynamic forces. In the papers [5-6], CFD software was used to simplify the car, three types of side wind simulation analysis methods were given, and a 1:4 scaled-down model was made for wind tunnel tests, and the simulation results were compared with the tests. The aerodynamic and flow field structures of a simple car-like form under the influence of transient winds were studied in [7] and [8], respectively. Most of the current domestic and foreign research objects are simplified car models, and the bottom of the car is flat, ignoring the bottom structure of the car, which is very different from the actual situation, and the bottom structure has a large impact on the aerodynamic characteristics of the car when it operates in the state of no side wind [9]. In this paper, an SUV, for example, contains detailed chassis data, and analyzes the influence of different intensity side winds on the aerodynamic characteristics of SUV through two different simulation methods, while conducting wind tunnel tests on the whole vehicle, and comparing the simulation results with the tests.

2. Side wind simulation method

2.1. Lateral yaw angle

Vehicle in the process of driving encounter side wind situation as shown in Figure 1, the figure V_x for the vehicle travel speed, V_y for the lateral wind speed, V_R for the vehicle travel speed and lateral wind speed of the synthetic speed, α for the side deviation angle, the vehicle right deviation for positive.

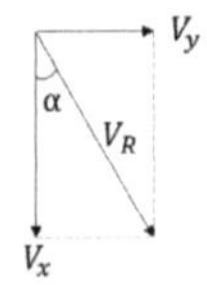

Figure 1. Speed diagram.

According to Jaguar Land Rover, the side yaw angles for side wind conditions are 5, 10 and 15 degrees according to the UK Met Office data [10]. As there is no relevant research in China, this paper combines Jaguar Land Rover and the development needs of the project itself, the side yaw angles under side wind conditions are selected as -2, -5 and -10 degrees, and in order to compare with the test data, this paper selects the vehicle left deflection, i.e., the side yaw angle is negative.

2.2. Simulation methods

The side wind simulation method mainly has two methods: "partial car" and "partial wind", and the method of "partial car" is similar to the test method of Tongji wind tunnel test, that is, the vehicle is deflected by the corresponding angle α, as shown in Figure 2.

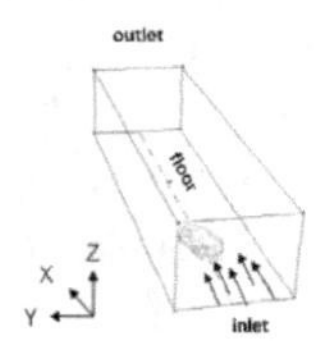

Figure 2. " partial car" diagram.

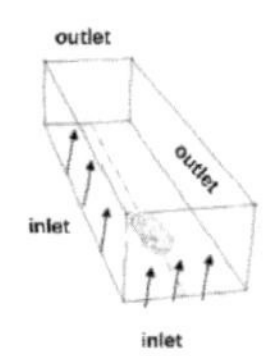

Figure 3. "partial wind" diagram.

The "partial wind " approach is to set the vehicle speed and side wind speed at the entrance instead of deflecting the vehicle during modeling, as shown in Figure 3.

In this paper, both methods are simulated and then compared with the test data to compare the accuracy of the two methods.

3. Calculation working condition setting

3.1. Geometric model establishment

The simulation model is selected for a mass-produced SUV, because the chassis has a large impact on the aerodynamic drag of the side wind condition, the model contains detailed chassis data, and the dimensions of the vehicle (L×W×H) are 4695mm×1885mm×1726mm, and the model is shown in Figure 4.

Figure 4. Schematic diagram of the simulation model

3.2. Simulation conditions and computational domain settings

The calculation domain of both methods is a rectangular body, as shown in Fig. 2 and 3, where the length direction is parallel to the X-axis, the width direction is parallel to the Y-axis, and the height direction is parallel to the Z-axis, and the dimensions of the calculation domain are set according to Table 1.

Table 1. Dimensions of the calculation domain

X (Length) Direction /m		Y(width) Direction /m		Z(height) Direction /m	
Front	Rear	Left	Right	Top	Bottom
-15	40	-7	7	10	Ground line

Because the aerodynamic drag mainly has a significant effect on the reduction of fuel consumption at high speed, the synthetic speed of this simulation is chosen to be 120km/h. According to the synthesis and decomposition of the speed, combined with the previous selection of the side yaw angle, the final simulation conditions are shown in Table 2.

Table 2. Simulation working condition table

Swing angle /°	Travel speed /km/h	Lateral wind speed /km/h	Synthetic speed /km/h
-2	119.93	-4.19	120
-5	119.54	-10.46	120
-10	118.18	-20.84	120

3.3. Mesh division and boundary condition setting

The mesh division of the model is shown in Fig. 5. In order to simulate more accurately, the bottom and the rear part of the vehicle are encrypted in the mesh division.

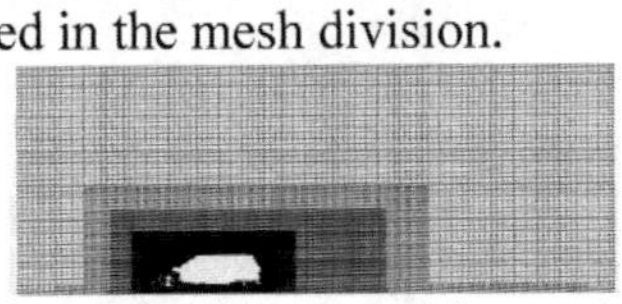

Figure 5. Schematic diagram of the model body mesh

For the two different simulation methods, the boundary conditions are also set differently, as shown in Table 3.

Table 3. Boundary condition settings for the two simulation methods

Boundary type	Method 1 " partial car"	Method 2 " partial wind"
Inlet boundary	Velocity inlet， Vx=VR，Vy=Vz=0	Velocity inlet， Vx=Vx，Vy=Vy，Vz=0
Right wall	Wall boundary	Velocity inlet， Vx=Vx，Vy=Vy，Vz=0
Outlet boundary	Pressure outlet	Pressure outlet
Left wall	Wall boundary	Pressure outlet
Upper wall	Wall boundary	Wall boundary
Ground	Moving ground	Moving ground

3.4. Turbulence model
The flow field outside the vehicle is generally a constant, isothermal, incompressible three-dimensional flow field, which should be set up according to turbulence because of the complex shape of the vehicle that can easily cause separation. The turbulence model is selected as SST K-Omega model. The SST K-Omega model integrates the advantages of K-Omega model in the near-wall area and K-Epsilon model in the far-field calculation, while adding the transverse dissipation derivative term and considering the transport process of turbulent shear stress in the definition of turbulent viscosity, which is more widely applicable and can be better for the area of easy separation of the body. The flow field around the body can be better simulated.

4. Comparison of simulation analysis and test
In order to compare the accuracy of the two simulation methods, the aerodynamic test of the SUV was conducted in the wind tunnel laboratory of Tongji University, and the vehicle side deflection was achieved by turning the balance turntable, and the test results were compared with the simulation results.

4.1. Aerodynamic drag
The simulation and test results of aerodynamic drag are shown in Fig. 6. The results of both simulation methods are close to each other, and the trend is the same as that of the test, which is that the aerodynamic drag gradually increases with the increase of the side yaw angle. Among them, the aerodynamic resistance increases less when the lateral yaw angle is smaller (-2 degrees), and increases significantly when the lateral yaw angle is larger (-5 degrees and -10 degrees).

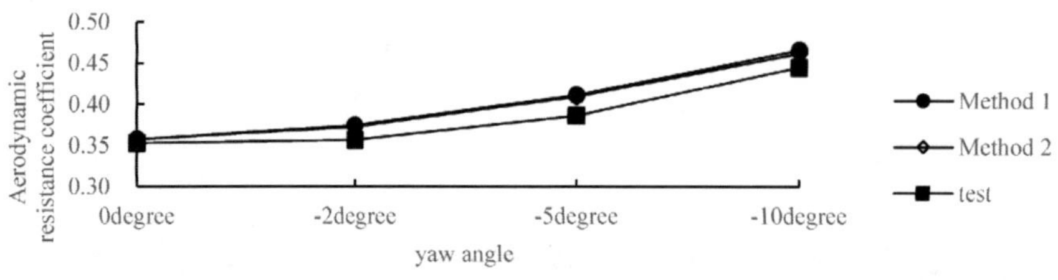

Figure 6. Variation of aerodynamic drag coefficient with side yaw angle

From the tail pressure map, it can be seen that the simulated tail pressure is symmetrical on the left and right sides when the side yaw angle is zero, as shown in Figure 7. The wind tunnel test is shown in Figure 8 after the pressure measurement points are arranged in the tail section and made into a pressure map. Comparing the two tail pressure maps, it can be found that the tail pressure measured in the test is higher than the simulation, which indicates that the tail pressure recovery is better in the test, which is also consistent with the relatively low aerodynamic drag coefficient measured in the test.

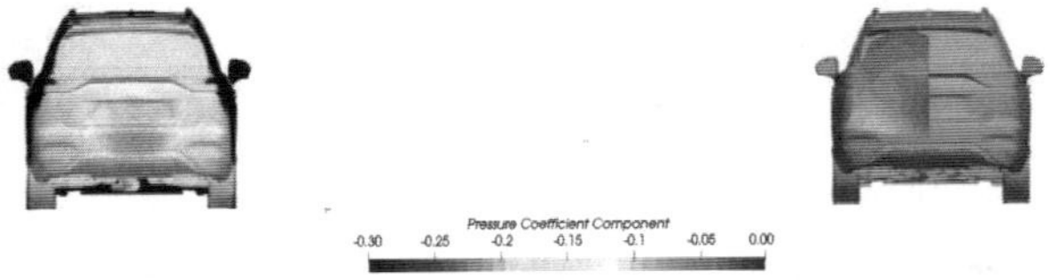

Figure 7. 0 degree simulation tail pressure map **Figure 8**. 0 degree test tail pressure map

Figure 9 shows the corresponding simulation and test tail pressure maps for a lateral yaw angle of -2 degrees, from left to right, the simulation results of method 1, simulation results of method 2, and test results. Comparing the simulation and test tail pressure maps at -2 degree and 0 degree, we can see that the change of tail pressure is very small when the side yaw angle is small, so the change of aerodynamic drag is also small, which is consistent with the conclusion in Figure 6.

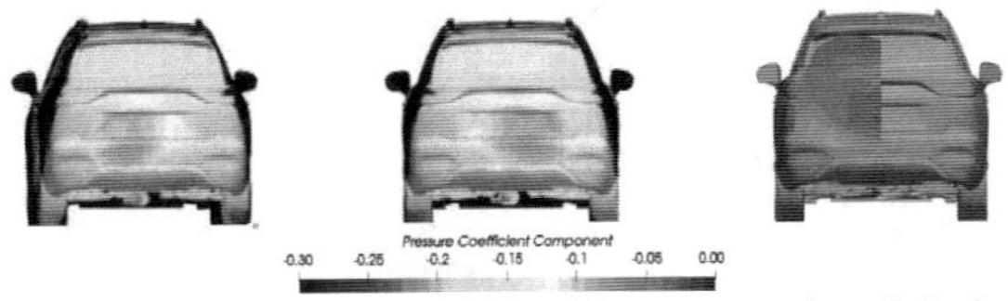

Figure 9. Tail pressure map with side yaw angle of -2 degrees

Fig. 10 and Fig. 11 show the corresponding simulation and test tail pressure maps when the side yaw angle is -5 degrees and -10 degrees, respectively, and the simulation results of method 1, simulation results of method 2 and test results are shown from left to right. When the side yaw angle increases significantly, it can be seen that the separation on the left side of the vehicle is serious and the tail pressure becomes significantly worse, which leads to a significant increase in aerodynamic drag.

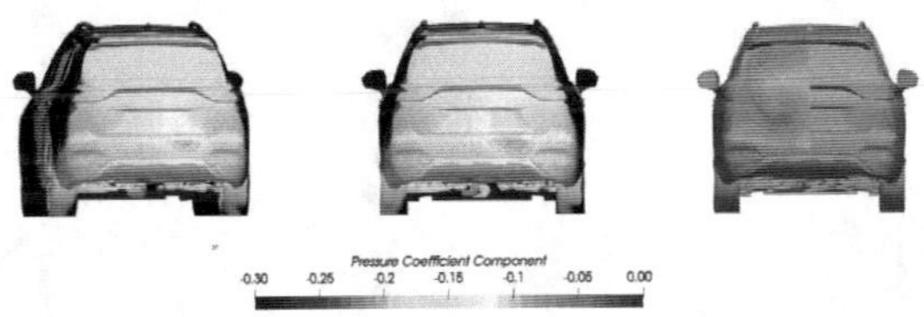

Figure 10. Side yaw angle -5 degrees tail pressure map

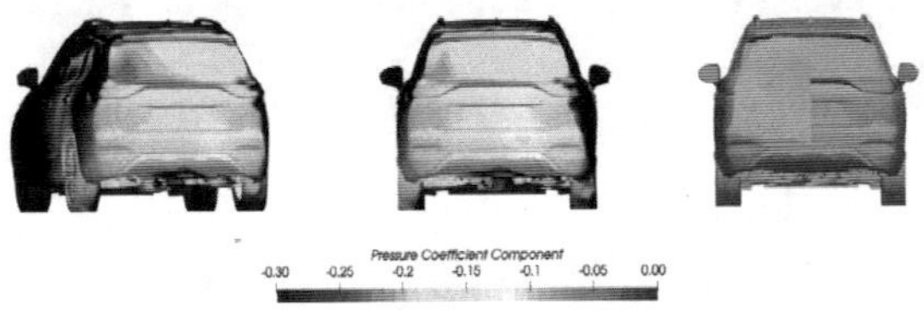

Figure 11. Lateral yaw angle -10 degree tail pressure map

4.2. Lateral force

Under the action of side wind, the airflow flows from the right side of the body to the left side of the body through the roof, the pressure in front of the right side of the vehicle rises, and the pressure difference exists between the two sides due to the low airflow separation pressure on the left side, and the lateral force is generated, as shown in Figure 12, and with the enhancement of the side wind, the pressure difference between the two sides becomes larger, and the lateral force also increases, and the lateral force coefficient rises (absolute value), as shown in Figure 13.

Because the pressure on the right side in front is obviously higher than the pressure on the left side, the pressure difference between the two sides is large, while the pressure difference between the two sides in the rear is small, the body will be subject to the action of the transverse moment around the Z axis in the counterclockwise direction, which is not conducive to driving safety.

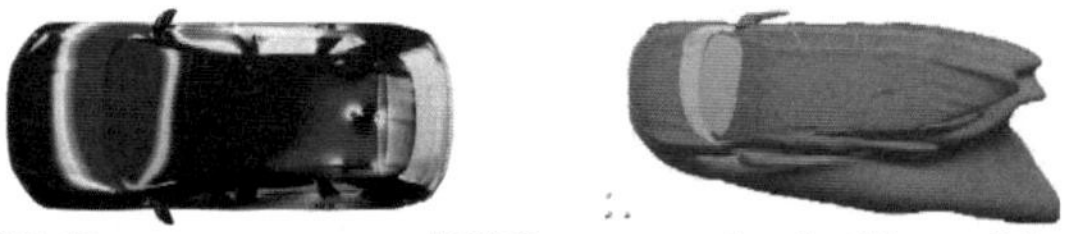

Figure 12. Pressure map and ISO map on both sides of the body

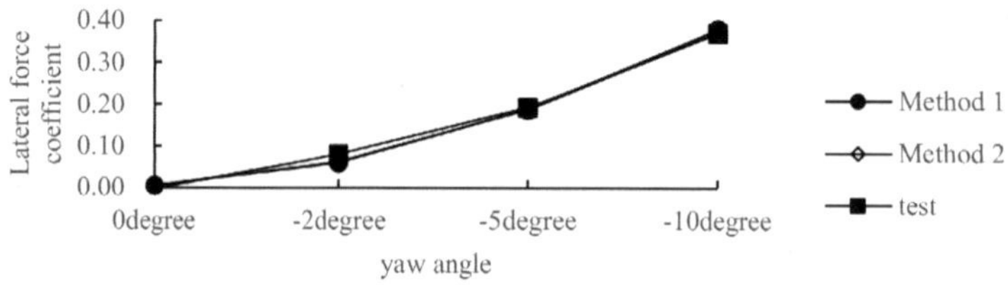

Figure 13. Variation of lateral force coefficient with lateral yaw angle

4.3. Lifting force

At zero yaw angle, the body is well streamlined, with more airflow flowing away from the roof and sides of the car and relatively less airflow entering the chassis. By the effect of side wind, the windward area of the car increases, the streamline degree of the body becomes worse, and the airflow into the chassis increases, resulting in increased resistance to airflow at the bottom of the body, lower speed and higher pressure. With the increase of side deviation angle, the pressure of the chassis rises, see Figure 14.

Side yaw angle-2 degrees Side yaw angle-5 degrees Side yaw angle -10 degrees

Figure 14. Sump pressure map with different side yaw angle

At the same time, as the airflow to the top flows over a long distance before it merges with the airflow at the bottom, the speed has to become larger and the pressure decreases. Therefore, the pressure difference between the top and bottom surfaces of the body becomes larger, and the lift force increases with the increase of the side yaw angle, see Figure 15.

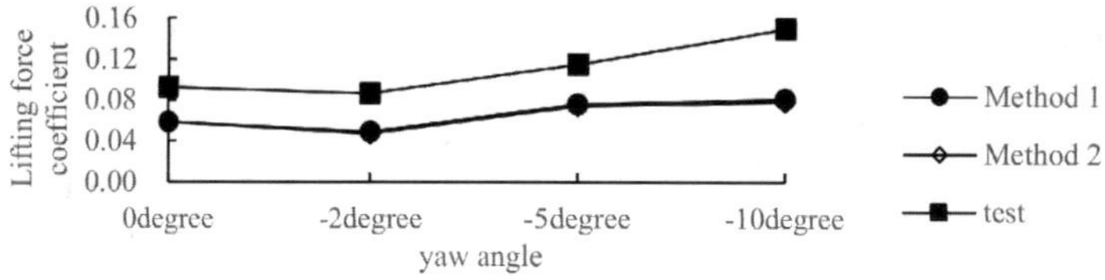

Figure 15. Variation of lift coefficient with side yaw angle

5. Summary

1) The trend of both the 2 simulation methods and the test are consistent, but in terms of operation and difficulty of model building, simulation method 2 is simpler to operate because it does not need to deflect the car and only changes the inlet wind speed, so the side wind simulation method is recommended to "partial wind".

(2) Under the condition of side wind, the flow field outside the car becomes asymmetric, the pressure at the rear decreases, the pressure difference between the two sides of the body increases, the chassis pressure increases and the roof pressure decreases, causing the aerodynamic drag, lateral force and lift of the car to increase with the increase of side wind intensity.

3) Relative to the aerodynamic drag and lift, the lateral force of this SUV changes more significantly under the side wind, which seriously affects the driving stability of the car.

(4) The pressure changes on each surface of the SUV body under the influence of different side wind strengths and the mechanism of aerodynamic force changes were analyzed to provide relevant reference for further in-depth research on the aerodynamic characteristics of the SUV under the influence of side wind.

References

[1] Gu zhegnqi, Jiang lehua, Wu jun, Numerical analysis and computer simulation of automobile bypass flow [J]. Journal of Aerodynamics. 2000,18(2):188-193.

[2] Howe,M.s.,On the infrasound generated when a train enters a tunnel,Journal of Fluids and Structures,Volume 17,Issue 5,April 2003,Pages 629-642.

[3] Chen hongye, Wang ailing, Zheng zhizhen, Simulation study on the aerodynamic characteristics of a car under different side winds[J]. Mechanical management and development. 2010,25(2):81-82

[4] WANG Fuliang. Numerical simulation of the crosswind effects on the aerodynamic performance of a sedan[D]. Changchun:Jilin University， 2005,38 (8): 1255-1258

[5] GONG Xu,GU Zhengqi,LI Zhenlei,et al. A study on the numerical simulation of car aerodynamic characteristics under cross wind conditions[J],Automotive Engineering,2010,32(1):13-16.

[6] GONG Xu. Analysis and optimization of high speed aerodynamic stability of tractor trailer vehicle in cross winds[D]. Changsha:Hunan University,2011.

[7] VOLPE R, FERRAND V,SILVA A D,et al. Forces and flow structures evolution on a car body in a sudden crosswind[J].Journal of Wind Engineering and Industrial Aerodynamics, 2014, 128(5):114-125.

[8] FERRAND V. Forces and flow structures on a simplified car model exposed to an unsteady harmonic crosswind[J] . Journal of Fluid Engineering, 2014,136(1) : 249- 256.

[9] Yang yi,Xu yongkang,Shen xiawei,et al. Aerodynamic lift-based bottom flow field improvement for automobiles[J].Journal of Central South University (Natural Sciences Edition) 2013, 44 (10):4063 -4068

[10] Howell,J., Passmore,M., and Windsor,S.,"A Drag Coefficient for Test Cycle Application," SAE Technical Paper 2018-01-0742, 2018, doi:10.4271/2018-01-0742.

The influence of wind turbine blade rotation on anemometer

Yaqiang Zhou*, Lizhu Tian, Zhiwen Jiang, Yapeng Li, Zhaohe Wu, Chenglong Qi, Yanxu Gou, Yonghe Xu, Dayu Du, Bao Wang, Yuan Wu, Wei Feng, Peng Li

Datang Bayannur Wind Power Generation Co., Ltd, Bayannur Inner Mongolia, China

*Corresponding author e-mail: 34080575@163.com

Abstract. In this thesis, numerical simulation modeling is used to calculate the influence of wake on the wind speed anemometer. The input conditions include wind speed, wind direction, wind density, etc., and a matrix database is formed, which is used to call data with lidar. Evaluation of the actual deviation of wind speed and direction at different wind speeds. Evaluation of the actual deviation of the anemometer under different installation positions. Through the results of numerical analysis, improve the wind speed and direction collection accuracy of the unit.

1. Introduction

It is impossible to accurately assess the status of the anemometer simply by comparing the Doppler lidar equipment with the anemometer. It is necessary to further clarify other interference factors, including the impeller rotation in the running state and the impeller structure in the Y-shaped stop state. The influence of the wind direction indicator is used to analyze the initial yaw value of the wind direction indicator and optimize the operation of the unit [1-3].

Create a geometric model of the wind turbine, input initial data including wind turbine wind, nacelle tower, geometric dimensions of the installation position of the wind turbine, geometric dimensions of the installation position of the wind turbine, and installation position of the wind turbine; use the model for numerical simulation calculation, and the input condition is wind speed ; The output content is the modification factor of the generator wind direction instrument being disturbed by the impeller; including: in the non-stop state, the wind turbine wind direction meter corresponding to different wind speeds is affected by the modification factor of the impeller movement; the blade Y-shaped stop state, corresponding to different wind speeds The wind power direction indicator is modified by the impeller movement [4-5].

2. Model description

This project uses domestic QFLUX software, physical simulation and analysis research work. The software uses finite volume method and arbitrary polyhedral unstructured network system. Various types of turbulence models have been developed (such as SA, Standard/SST k)-RANS models such as ω, Standard/RNG/Realizable ke, or DES, DDES, IDDES, etc. Hybrid model).

This project focuses on the study of the influence of rotation on the location results of wind speed and direction collection, and the numerical simulation will directly adopt the unsteady calculation mode.

The external flow field is established with the surface of the blade, the hub, and the nacelle as the flow wall boundary. Considering that the focus area is the root of the blade and the nacelle, the blade is retained from the root to the maximum chord length. According to previous project experience and the analysis of the flow of the whole machine blade, the flow in the upper part of the nacelle is mainly affected by the root of the blade and the upper part of the nacelle. If all the blades of the whole machine are modeled, the overall grid size will increase accordingly, which may reduce the overall calculation accuracy. Therefore, the calculation in this project is based on the method of using partial blades.

Blades and hubs are rotating mechanisms throughout the analysis, and a rotating grid is established in this area. The three-dimensional geometric model of the whole machine is shown in Figure 1.

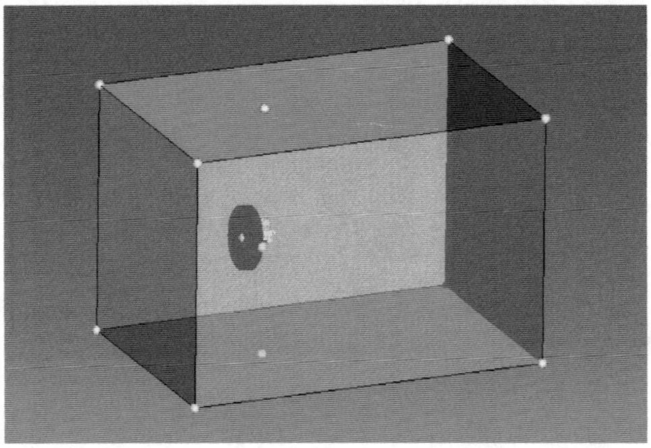

Figure 1. Three-dimensional geometric model of the flow field of the whole machine

The computing domain of the whole machine is divided into two parts, the first part is the rotation area of the wind wheel, and the second part is the flow area of the external flow field.

3. Calculate parameters and boundary conditions

The fluid medium adopts incompressible air, the density ρ=1.004kg/m^3, and the dynamic viscosity coefficient $\mu=1.7894\times10^{-5}$ kg/(m·s). According to different inflow boundaries, specify the inflow velocity (wind profile) at the same time, and the wind shear coefficient is 0.131.

Figure 2. Type A unit calculation model monitoring points

4. Results

It can be seen from the side-view velocity cloud diagram of the flow that there is an obvious vortex area on the upper part of the nacelle. This flow state exists under the action of blade rotation. Under the influence of this state, the wind speed in the upper part of the nacelle is unstable. That is, the wind speed exhibits this fluctuation with time, and this fluctuation affects the measurement result of the anemometer.

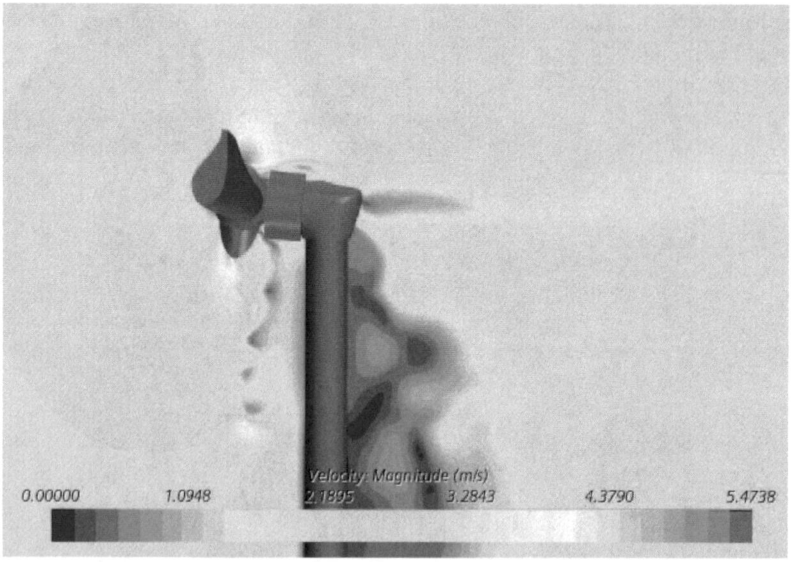

Figure 3. Side view of velocity cloud map (3m/s)

The most important thing is that this rotation has a greater impact on the wind direction. Since the rotation of the blades disturbs the flow in the upper part of the nacelle, the direct effect of this disturbance is the horizontal tangential direction, which strengthens the back-and-forth swing of the wind direction. The specific flow characteristics are shown in the Figure 3.

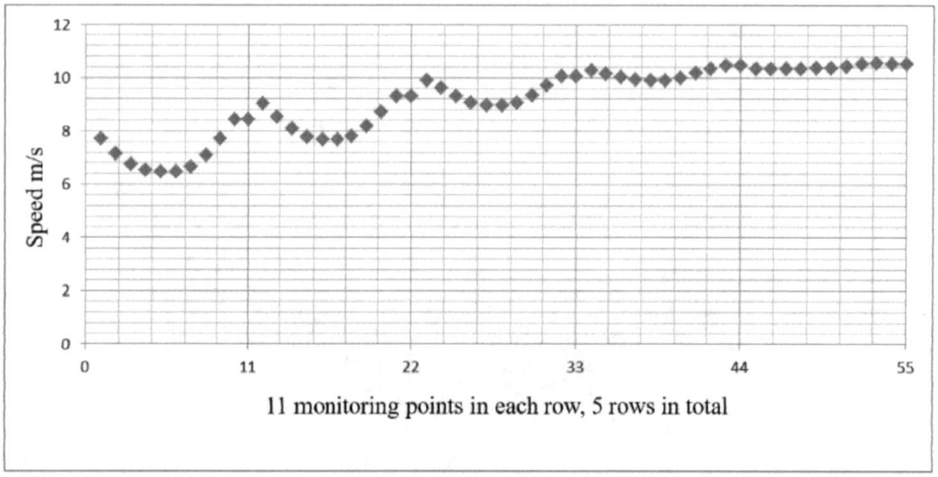

Figure 4. Average wind speed at different monitoring points (11m/s)

It can be obtained regularly from the above figure that as the number of layers increases, the wind speed of the anemometer gradually approaches the true value. In the highest row, wind speed fluctuations are reduced and are basically unaffected by rotation. In the low-level position (rows 1-3), the wind speed changes regularly on both sides and in the middle.

Therefore, it is preliminarily inferred that after the height of the anemometer, the current height is 1.2m, and the accuracy of wind speed collection will be greatly improved after the height of 0.4m.

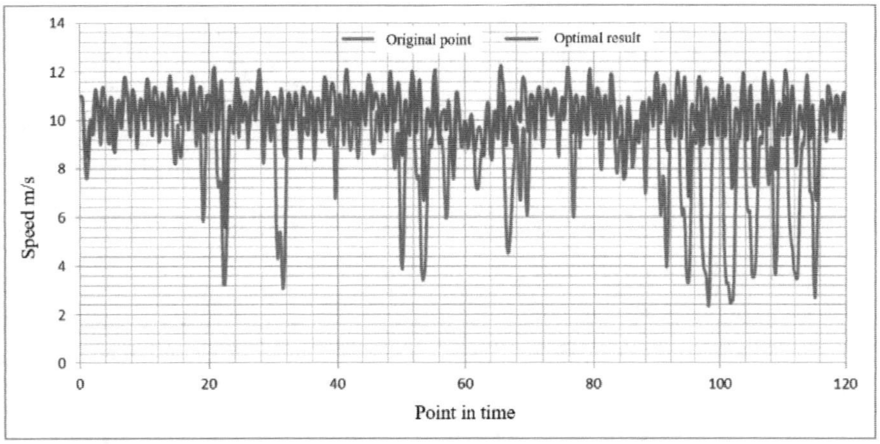

Figure 5. Comparison of optimization results of anemometer (11m/s)

Table 1. Wind speed optimization results

Inlet wind speed (m/s)	Average anemometer (m/s)	Average wind speed after optimization (m/s)	Deviation after optimization
3	2.23	2.69	11.4%
5	4.07	4.62	7.6%
7	5.82	6.46	7.7%
9	7.69	8.56	4.9%
11	9.65	10.37	5.8%

5. Conclusion

In summary, after numerical analysis and evaluation of the anemometer, the wind speed results at different locations can be displayed intuitively, and the optimization results can be found within a feasible range through data analysis. After heightening, the accuracy of the anemometer's collection results is greatly improved.

References

[1] U. Datta, J. Shi, A. Kalam. Primary frequency control of a microgrid with integrated dynamic sectional droop and fuzzy based pitch angle control. Int. J. Electr. Power Energy Syst, 111, 248-259 (2019).

[2] H. Iwasawa, N. Sakitani, M. Nomura et al. The effects of heat, icing, and stretching on knee joint capsule in rat contracture model of spinal cord injury[J] Physiotherapy, 2015, 101

[3] J. Jin, M. Virk. Effect of Wind Turbine Blade Profile Surface Roughness on Ice Accretion——A Numerical Case Study. Proceedings of 2020 3rd International Conference on Energy and Power Engineering（EPE2020）. 2020

[4] G. Peeters, W. Paepegem, J. Degroote. Effect of rotor–tower interaction, tilt angle, and yaw misalignment on the aeroelasticity of a large horizontal axis wind turbine with composite blades[J] . Wind Energy . 2020 (7)

[5] C. Qin, A. Mulroney, M. Gupta. Anti-icing epoxy resin surface modified by spray coating of PTFE Teflon particles for wind turbine blades. Materials Today Communications . 2020

Flow Characteristic Study of High-speed Cavity Based on Detached-eddy Simulations

Yongqian Zheng, Jie Zhang, Huan Li, Xiaojun Wu and Hongyin Jia *

China Aerodynamics Research and Development Center, Mianyang, Sichuan, China

*Corresponding author e-mail: jiahongyin@163.com

Abstract. The cavity is widely used in the advanced flight vehicle. However, there are also some new aerodynamic characteristic when cavity is exposed in high-speed flow. The flow characteristic of high-speed cavity is studied in this paper based on detached-eddy simulation method. The M219 cavity benchmark under transonic flow is used as the validation case for numerical methods and grid strategy. Comparison of predicted sound pressure levels and spectra against the wind tunnel experiment has proved that the numerical methods capture the major oscillation dynamics inside the cavity. The aerodynamic characteristic of cavity with a store is investigated in this paper, there are three vortexes in the cavity and the main vortex in the middle rotates around the store. The shear layer breaks earlier compared with clean cavity due to the turbulent flow around the store. However, the small store has little influence to the sound pressure level of cavity.

1. Introduction

The cavity is widely used in the advanced flight vehicle. However, there are also some new aerodynamic characteristic when cavity is exposed in high-speed flow, the flow inside the cavity is complicated and include some serious phenomenon such as the flow separation, wave/vortex interference, shock/shear interference, which may generate aerodynamic noise and break the structure of the cavity and flight vehicle [1-3]. So the research of high-speed cavity is extremely important to the cavity design.

In recent years, the detached-eddy simulations (DES) method has been applied to the study of cavity flow [4, 5]. Shieh and Morris investigated the two-dimensional and three-dimensional flow in a subsonic cavity with A variety of different slenderness ratios and specific turbulent boundary layer conditions [6, 7]. Hamed studied the cavity flow at supersonic velocity and discussed the influence of Re number on the flow-field [8], the results show that DES method has obvious advantages in the prediction of cavity noise, especially in the broadband noise spectrum. Shoeb and Klaus studied the characteristic of rectangular open cavity [9], the results are compared with the sound pressure level test results and the root mean square pressure test results. Bidur analyzed the unsteady flow characteristics of three-dimensional cavities, including cavities and cavities with missiles [10]. Davis presented the numerical calculation results of unsteady flow characteristics in the case of three-dimensional cavities and cavities containing plugins [11].

The aerodynamics characteristics of high-speed cavity is investigated by numerical simulation. The structure of the article is as follows. Section 2 is introduction of numerical method, section 3 is the results and discussion, and finally section 4 is a brief conclusion.

2. Numerical Method

2.1. Flow Solver

The numerical simulations in this paper were conducted by the China Aerodynamics Research & Development Center (CARDC) NNW-Flowstar software [12-14], which is an unstructured finite volume cell-center CFD solver and has participated in the 5th and 6th CFD drag prediction workshop. Roe scheme is used for inviscid flux computations [15]. The unsteady turbulence is predicted by improved delayed detached-eddy simulation (IDDES) method [16], based on Spalart-Allmaras (SA) turbulence model.

2.2. Validation of numerical method

European DESider project designed the M219 cavity, which is a benchmark and has full and accurate wind-tunnel data, the M219 cavity often used to validate the cavity flow predictions [17-19]. Fig.1 shows the size of M219 cavity, and the length-to-depth ratio is 5.

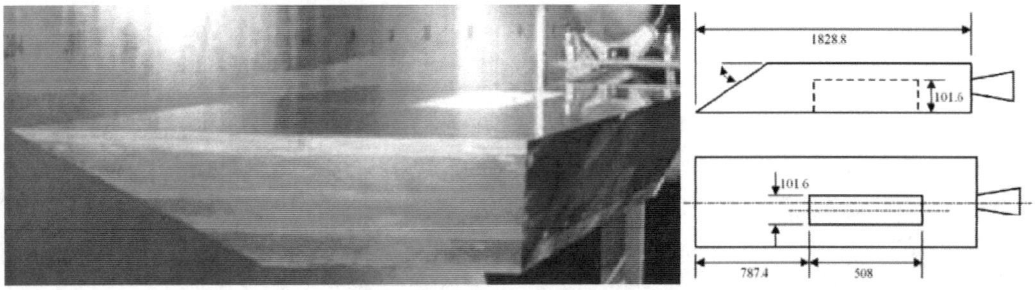

Figure 1. M219 cavity

Fig.2 shows the computational domain of three-dimensional cavity flow and computational mesh, the hybrid mesh is used to simulate the cavity flow, combining structured mesh and unstructured mesh, and the total number of grids is 10 million. Fig.3 shows the comparison of sound pressure level of wind tunnel and numerical simulation. The results show that the numerical results in this paper are qualitatively and quantitatively comparable to the wind tunnel experiment.

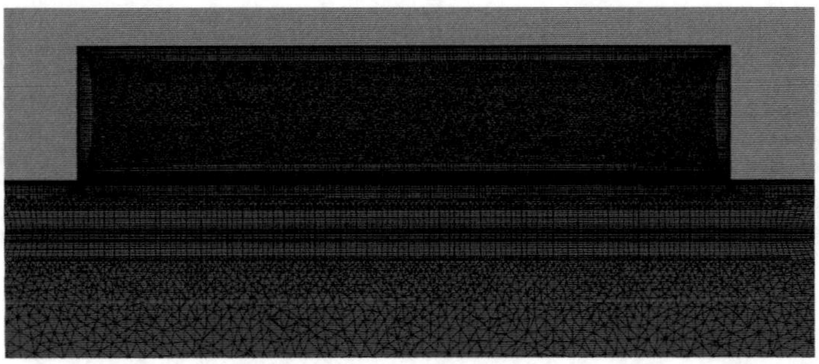

Figure 2. Local mesh of three-dimensional cavity flow

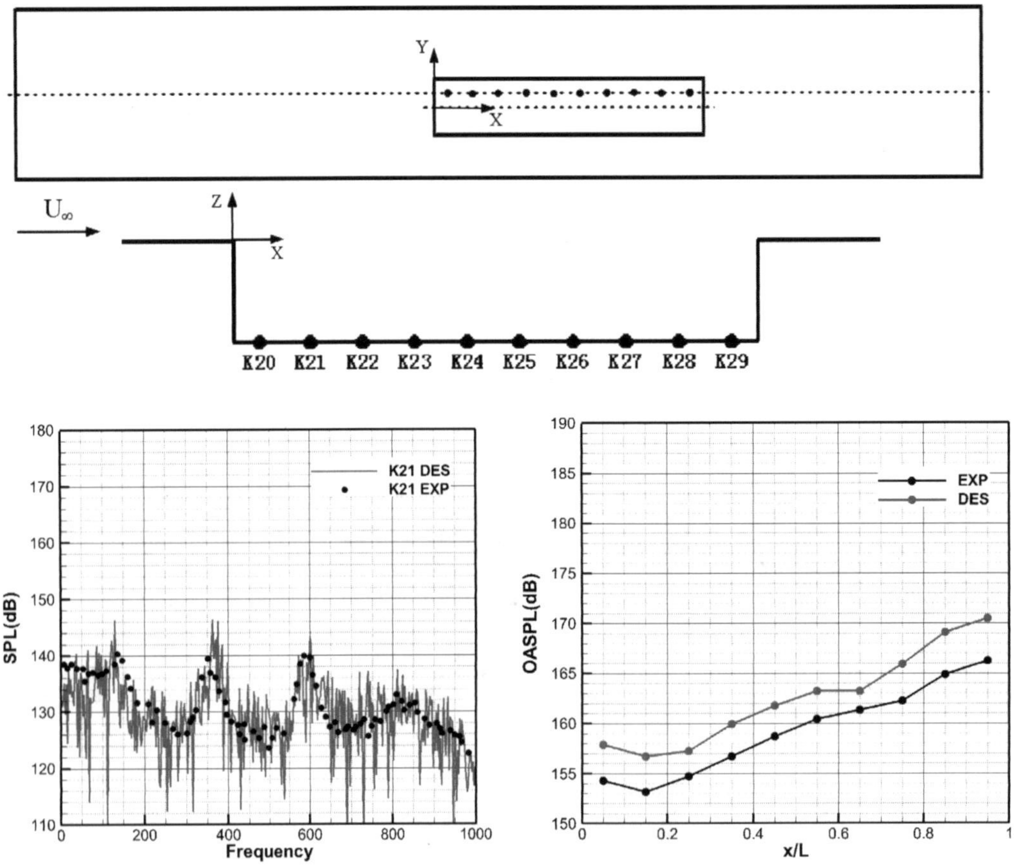

Figure 3. Comparison of sound pressure level

2.3. Geometry and Mesh

(a) Flying-wing standard model (b) Simplified cylindrical store

Figure 4. M219 standard model and simplified cylindrical store

Fig.4 shows the grids of M219 cavity and a simplified cylindrical store, the computational grids are unstructured hybrid grids, including tetrahedral, pyramid, and triangular prism grid. The triangular prism grid and tetrahedron grid are used to simulate the boundary layer and the isotropic region of the spatial flowfield, respectively, and the pyramid grid is used to transition the different cells. The grids inside the cavity are specially refined to simulate the vortexes and shear layer, and grids where the shear layer may go through are locally refined, which can make an accurate simulation of aerodynamic noise, and the number of grid cells is 10 million.

The validation of M219 cavity benchmark under transonic flow and the hybrid unstructured grid strategy indicate that the numeric method of this paper is capable to capture the major oscillation dynamics inside the cavity is appropriate for the investigation of the flow characteristic of cavity.

3. Results and Discussion

3.1. Flow Characteristic of clean cavity

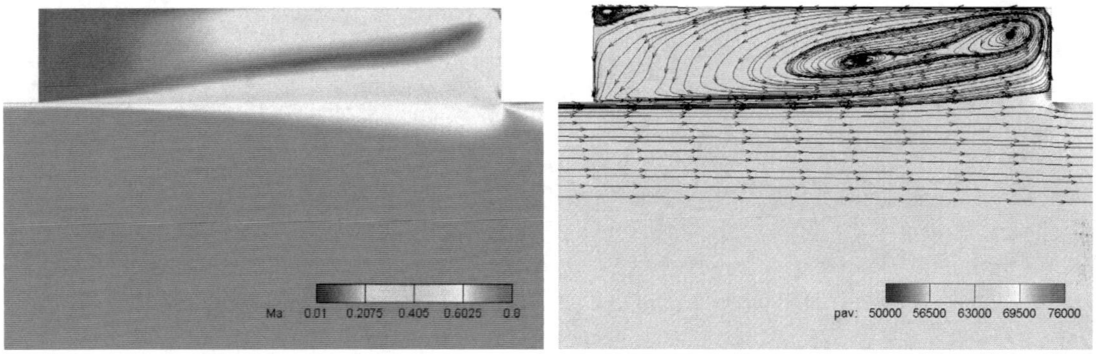

(a) Mach number distribution (b) Pressure and streamline distribution

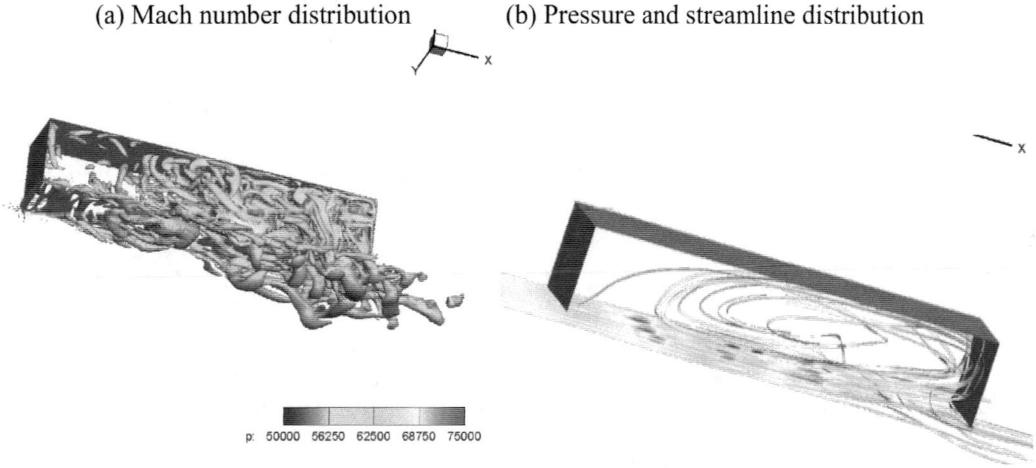

(c) Q-criterion of clean cavity (d) Spatial streamline of clean cavity

Figure 5. Flow characteristic of clean cavity

Fig.5 shows the flow characteristic of clean cavity, including Mach number distribution, pressure distribution, streamline distribution and Q-criterion distribution. There are three vortexes in the cavity, distributed in the front, middle and tail of cavity. The high-speed airflow passes though the cavity and impacts on the trailing edge of cavity, then the airflow flows back to the leading edge of cavity and forms the main vortex in the middle of cavity. There are also two small vortexes at the corner of cavity.

The high-speed flow out of cavity and the low-speed flow inside the cavity cause an obvious shear layer, which has great influence to the dynamic characteristics of cavity. It can be seen that the shear layer is stable at the front of cavity, and starts to break up and becomes unstable at middle of cavity. Q-criterion distribution shows that the shear layer becomes oscillating at the tail of cavity, and cause high pressure and high pressure fluctuation. The high pressure and high pressure fluctuation may damage the structure of cavity.

3.2. Flow Characteristic cavity with a store

As is usually the case there is an object inside the cavity, such as the embedded store. So a simplified cylindrical store is set inside the cavity to investigate the flow characteristic, the computational conditions are as the same as the case of clean cavity.

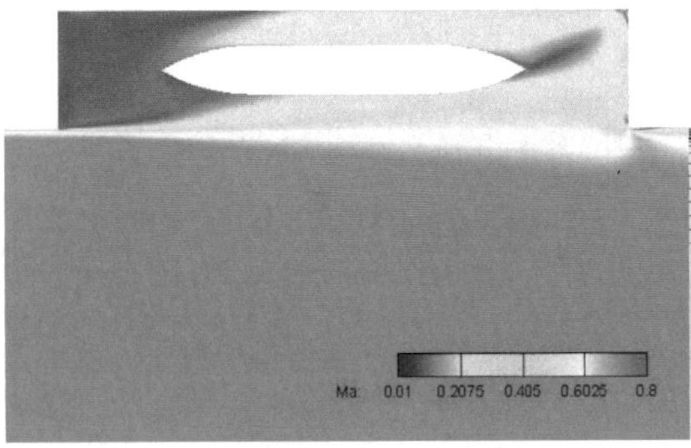

(a) Mach number distribution

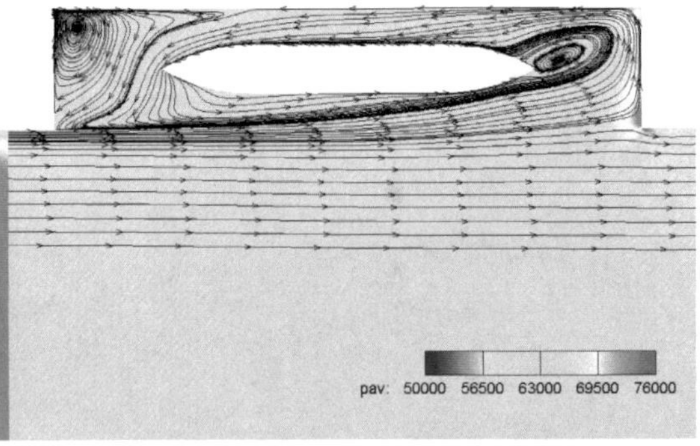

(b) Pressure and streamline distribution

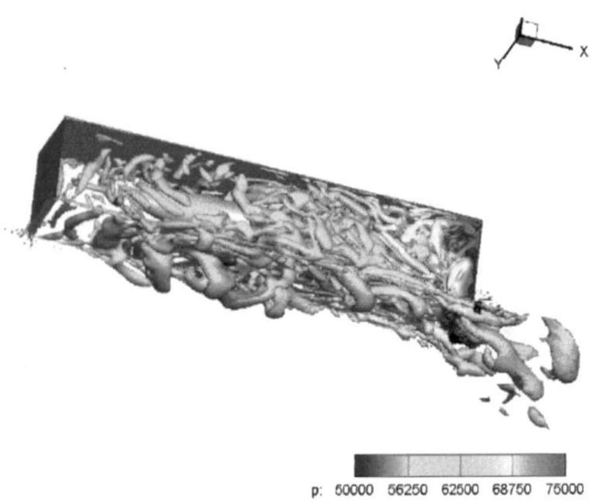

(c) Q-criterion of clean cavity

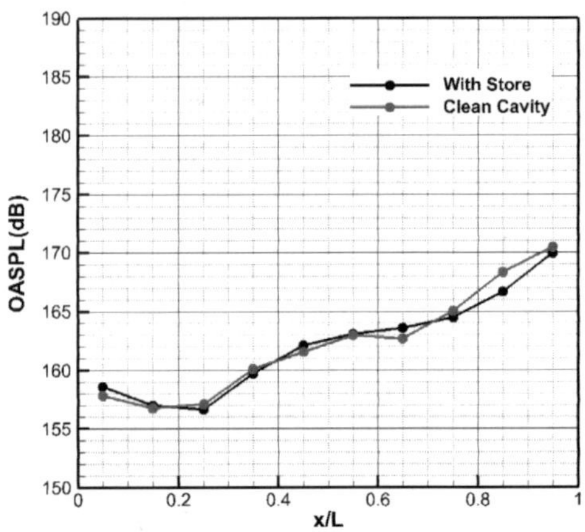

(d) Comparison of aerodynamic noise

Figure 6. Flow characteristic of cavity with a store

Fig.6 shows the flow characteristic of cavity with a store, including Mach number distribution, pressure distribution and Q-criterion distribution. There are also three vortexes in the cavity, distributed in the front, middle and tail of cavity. The main vortex in the middle rotates around the store, and the vortex in the front is much larger than the case of clean cavity.

There is also an obvious shear layer which is caused by the high-speed flow out of cavity and the low-speed flow inside the cavity, Q-criterion distribution shows that the shear layer starts to break up and becomes unstable at middle of cavity. Compared with the case of clean cavity, when there is a store inside the cavity, the shear layer breaks earlier. It can be seen from the streamline distribution that airflow is forced flow back to the front of cavity, and the airflow of the main vertex meet the shear layer earlier and breaks the shear layer.

10 monitoring points in the base of cavity are set to investigate the aerodynamic noise of cavity, the sound pressure level (SPL) in the tail of cavity is much larger than the SPL in the front. The high-speed

airflow strikes in the tail, which makes high pressure and high pressure fluctuation, and results in laeger SPL. Although the flow characteristic of clean cavity and with a store is totally different, the vortexes and shear layer vary larger in two cases, but the SPL of two cases is almost same, which means the simplified cylindrical store has little influence to the sound pressure level of cavity.

4. Conclusion

The M219 cavity benchmark under transonic flow is used as the validation case for numerical methods and grid strategy. Comparison of predicted sound pressure levels and spectra against the wind tunnel experiment has proved that the numerical methods capture the major oscillation dynamics inside the cavity. The aerodynamic characteristics of cavity with a store is investigated in this paper, there are three vortexes in the cavity and the main vortex in the middle rotates around the store. The shear layer breaks earlier due to the turbulent flow around the store. However, the small store has little influence to the sound pressure level of cavity.

The store release characteristics of embedded cavity and more detail of aerodynamic noise of high-speed cavity will be investigated in the future.

Acknowledgments

This work was financially supported by National Numerical Wind Tunnel (NNW) and National Natural Sciences Foundation of China (12102453). Special thanks to Pengcheng Cui for his suggestions on this article.

References

[1] P. C. Cui, B. Li, J. Tang. Investigation of store separation characteristics by leading edge spoilers before supersonic cavity. Journal of Physics. 2020, 1600 012017.
[2] C. Coley, A. Lofthouse, Correlation of Weapon Bay Resonance and Store Unsteady Force and Moment Loading, 50th AIAA Aerospace Sciences Meeting including the New Horizons Forum and Aerospace Exposition, 2012.
[3] S. J. Lawson, G. N. Barakos, Assessment of Passive Flow Control for Transonic Cavity Flow Using Detached-Eddy Simulation, Journal of Aircraft, 46(3):1009-1029, 2009.
[4] C. Lada, K. Kontis, Experimental Studies on Transitional and Closed Cavity Configurations Including Flow Control, Journal of Aircraft, 47(2):723-729, 2010.
[5] S. J. Lawson, G. N. Barakos, Assessment of Passive Flow Control for Transonic Cavity Flow Using Detached-Eddy Simulation, Journal of Aircraft, 46(3):1009-1029, 2009.
[6] C. M. Shieh and P. J. Morris. Comparison of two- and three-dimensional turbulent cavity flows [C]. 39th AIAA Aerospace Sciences, Reno, Nevada, January, 2001. AIAA Paper 2001-0511.
[7] C. M. Shieh and P. J. Morris. Parallel computational aeroacoustic simulation of turbulent subsonic cavity flow [C]. 6th AIAA/CEAS Aeroacoustics Conference and Exhibit, June 12 14 2000. AIAA Paper 2000-1914.
[8] A. Hamed, D. Basu and K. Das. Effect of Reynolds number on the unsteady flow and acoustic fields of supersonic cavity [C]. 4th ASME/JSME Joint Fluids Engineering Conference, Proceedings of FEDSM03, Honolulu, Hawaii, USA, July 6 11 2003. ASME. Paper FEDSM 2003-45473.
[9] K. A. H. Syed, Numerical investigation of 3-D open cavity with and without cover plates [C]. 47th AIAA Aerospace Sciences Meeting Including the New Horizons Forum and Aerospace Exposition, Orlando, Florida. 2009.
[10] K. K Bidur, Alistair Saddington. Computational study of cavity flow field at transonic speeds [C]. 47th AIAA Aerospace Sciences Meeting Including the New Horizons Forum and Aerospace Exposition. Orlando, Florida. 2009: 18. 2009.
[11] M. B. Davis, P. Y., B. R. Smith, K. M. Chankaya, R. A. Johnson. Store trajectory response to unsteady weapons bay flow fields [C]. 47th AIAA Aerospace Sciences Meeting Including the New Horizons Forum and Aerospace Exposition. Orlando, Florida. 547: 35. 2009.

[12] P. C. Cui, Y. Q. Deng, J. Tang, Adjoint-based grid adaptation and error correction, Acta Aeronautica et Astronautica Sinica, 37(10): 2992-3002, 2016.
[13] P. C. Cui, J. Tang, B. Li, A conservative interpolation method for overset mesh via supermesh, Acta Aeronautica et Astronautica Sinica, 39(3): 121569, 2018.
[14] D. Levy, K. Laflin, E. Tinoco, Summary of data from the fifth AIAA CFD drag prediction workshop, 51st AIAA Aerospace Sciences Meeting, DLR, 2013.
[15] P. L. Roe, Approximate Riemann Solvers, Parameter Vectors, and Difference Schemes, Journal of Computational Physics, 135(2):250-258, 1997.
[16] L. Xiao, Z. Xiao, Z. Duan, et al. Improved-Delayed-Detached-Eddy Simulation of cavity-induced transition in hypersonic boundary layer [J]. International Journal of Heat & Fluid Flow. 51(2):138-150, 2015.
[17] L. Temmerman, B. Tartinville, C. Hirsch. URANS Investigation of the Transonic M219 Cavity [J]. Springer Berlin Heidelberg, 2012.
[18] S Lawson, G N Barakos, P Nayyar, et al. Appendix for M219 cavity flow: Computations with and without bay doors. 2009.
[19] Y P X, B J Q, G B Z , et al. Suppression Effect of Jet Flow on Aerodynamic Noise of 3D Cavity[J]. Applied Mechanics and Materials, 2014, 444-445(5):588-595.

Simulation Research on the Effect of Reynolds Number on the Aerodynamic Characteristics of High Lift Device

Hui Zhao *, Yueyue Yang, Yaobing Zhang and Weibin Li

Computational Aerodynamics Institute, China Aerodynamics Research and Development Center, Mianyang 621000, China

*Corresponding author e-mail: zbhui@cardc.cn

Abstract. Based on the consideration of grid factors and turbulence model factors, the confluence boundary layer and aerodynamic characteristics of the NHLP-2D airfoil are studied. Compared with the experimental results, the grid distribution near the confluence boundary layer is important for calculating the width of the wake area. The calculation results of the structured and unstructured grids are basically the same when the surface grid distribution is the same. On the basis of the mesh refinement in the wake area, the aerodynamic characteristics calculated by the SA turbulence model and the distribution of the total pressure coefficient in the confluence boundary layer are in good agreement with the experimental results. The width of the confluence boundary layer and the wake strength decrease with the increase of Reynolds number, the lift coefficient increases with the increase of Reynolds number, but the drag and pitching moment have opposite pattern. The maximum lift of the airfoil increases as the Reynolds number increases. As the Reynolds number increases, the stall angle of attack of the airfoil gradually increases and then remains stable with the increase of the Reynolds number.

1. Introduction

At present, wind tunnel testing is still the main tool for obtaining actual aircraft aerodynamic characteristics and verifying CFD methods. However, due to the limitation of domestic wind tunnel size and test conditions, the current domestic wind tunnel test Reynolds number is 1-2 orders of magnitude lower than flying. Therefore, when the wind tunnel test results are extrapolated to flight conditions, the Reynolds number influence correction is required. In order to make better use of the CFD method to modify and extrapolate the wind tunnel test results, and to provide a reference for the analysis of the influence of the typical aircraft Reynolds number in the future. Due to the complexity of high-lift multi-segment airfoil flow, high-efficiency and accurate numerical simulation research based on the CFD method has been ongoing. In order to ensure the validity of the calculation results, the typical characteristics of the flow must be captured in the numerical calculation. The influence of Reynolds number on aerodynamic characteristics, confluence boundary layer, stall angle of attack, etc. under the condition of fixed Mach number are researched in this paper.

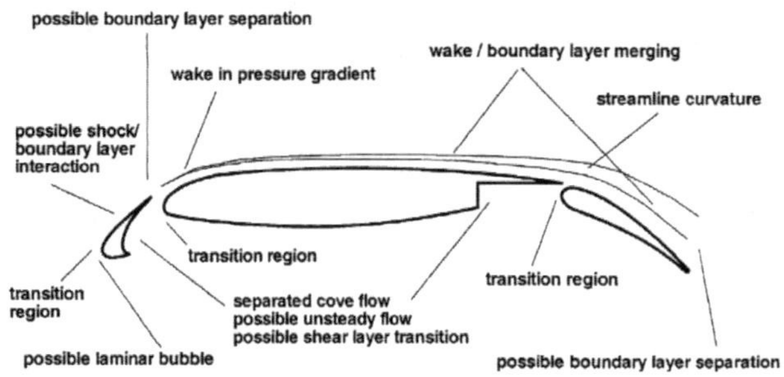

Figure 1. Flow characteristics around multi-segment airfoil

2. Simulation method and grids

The calculations in this paper mainly involve steady calculations, all of which are calculated using the sub-transsonic flow field solver NNW-FlowStar independently developed by CARDC. The NNW-FlowStar solver is a massively parallel flow field solver based on the unstructured hybrid grid of the grid and the Reynolds average NS equation. It can use grid cells of any shape and has greater flexibility. The calculations in this paper mainly adopt the implicit LU-SGS method, the space convection term adopts the Roe format, and the turbulence model includes SA and SST turbulence models.

In this paper, two sets of structured and unstructured grids are used for calculation. The grid distribution of structured and unstructured grids on the object surface is consistent. In order to simulate the wake area well and calculate the drag correctly, the far field boundary is close to 90 times of aerodynamic reference length away from the airfoil. The aerodynamic reference length and the grid information are shown in Table 1 below, and the grid comparison is shown in Figure 2 below. Because the topological structure of the structural grid is easy to control, the grid distribution strategy of grid refine around confluence boundary layer area is adopted. The impact of this grid refine on the calculation results will be analyzed later.

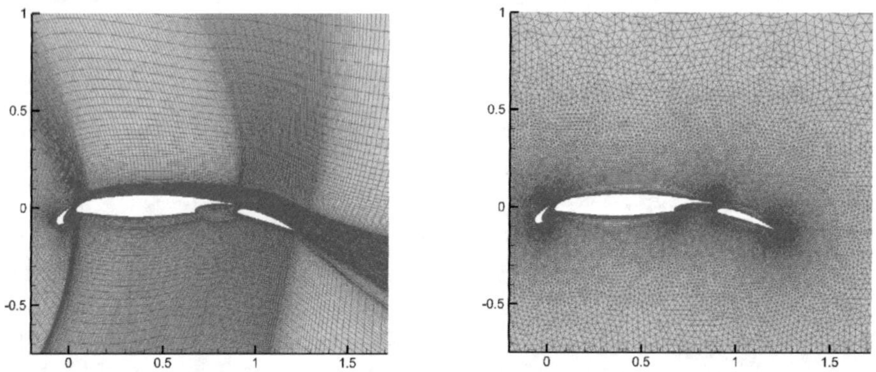

Figure 2. Grid distribution of NHLP-2D [1] (Left: structured, Right: unstructured)

Table 1. Grid information.

Grid Type	Number of grid cells	Thickness of first layer grid
Structured	501, 000	2E-5
Unstructured	309, 810	2E-5

Three kinds of boundary conditions are mainly used. The surface of the model is solid wall boundary, the longitudinal plane is a symmetrical boundary, and the rest are far-field boundaries. The reference information for aerodynamic calculation is shown in Table 2 below.

Table 2. Reference Information.

Reference area (m^2)	Reference length (m)	Reference point (m)
1	1	(0.25,0,0)

3. Simulation and analysis of results

3.1. Comparison of calculation and experiment results

The aerodynamic characteristics of the airfoil are calculated using the structural and unstructured grids introduced earlier, with Mach number of 0.197 and Reynolds number of 3.52×10^6. The influence of factors such as structured and unstructured grids are compared with SST and SA turbulence models. Figure 3~5 respectively shows the comparison of the aerodynamic characteristics of the NHLP-2D airfoil calculated in this paper and the experiment results [2]. It can be seen that the calculation results of the SA model are closer to the experimental values than the SST model. The calculation results of the structured and unstructured grids are very close, especially under the SA turbulence model condition.

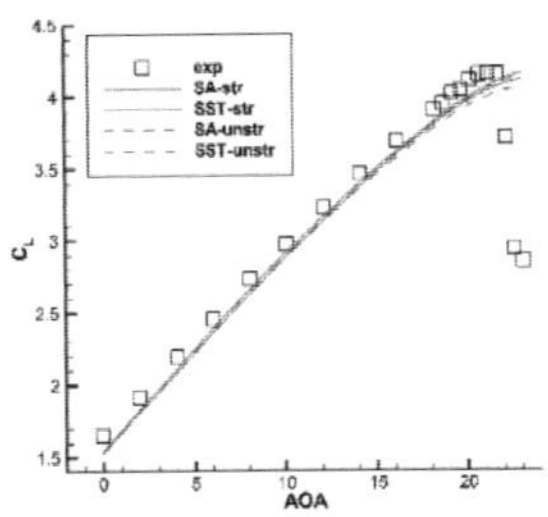

Figure 3. Comparison of C_L

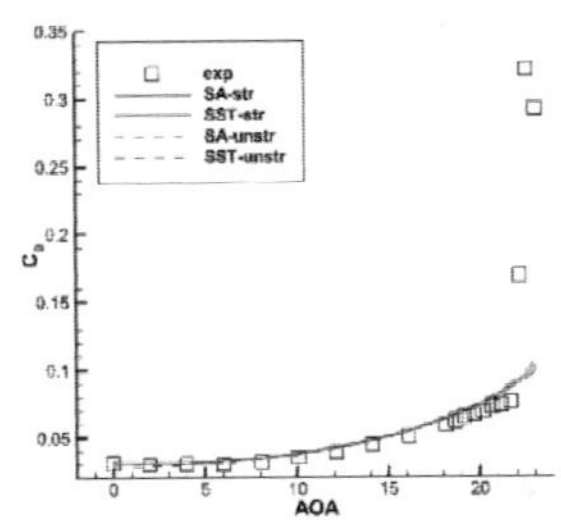

Figure 4. Comparison of C_D

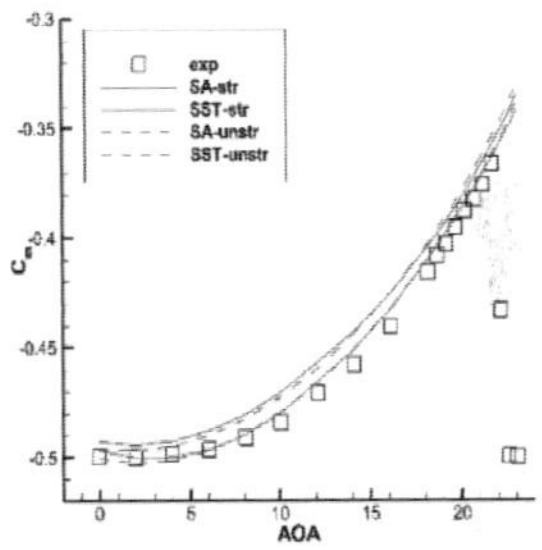

Figure 5. Comparison of C^m

The NHLP-2D airfoil flow field contains the phenomenon of mixing the wake and the local boundary layer. In order to facilitate the observation and analysis the change of the boundary layer, the total pressure coefficient (Cp_{tot}) distribution perpendicular to the airfoil surface was measured in the wind tunnel experiment. The measurement position is shown in the figure 6. Figures 7~8 show the comparison between the airfoil surface Cp_{tot} and the experiment results under different angles of attack. The different calculation results are basically consistent with the experiment results. The calculation results of structured and unstructured grids are different to a certain extent. The wake width calculated by structured grids is closer to the experiment. The calculation results of SA and SST turbulence model are consistent. The calculated wake width of the structural grid is closer to the experimental value, the calculated results of SA and SST are nearly the same, and the predicted wake width of the structural grid is closer to the experimental.

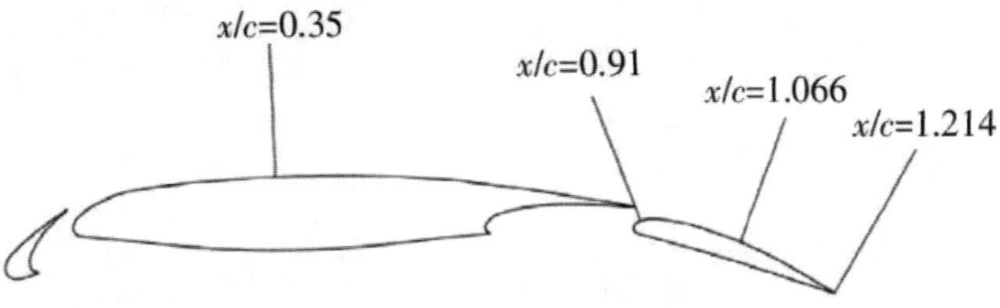

Figure 6. Detection point location of Cp_{tot}.

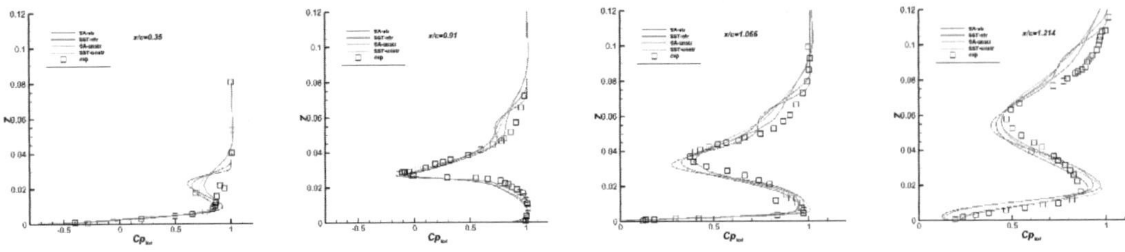

x/c=0.35 x/c = 0.91 x/c = 1.066 x/c=1.214

Figure 7. Comparison of Cp_{tot} (a=4.01°)

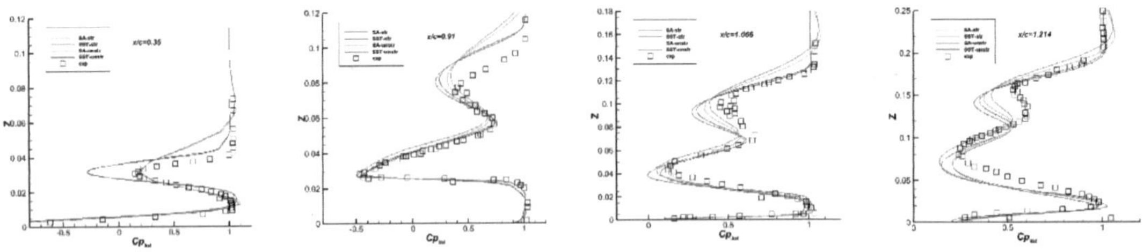

x/c=0.35 x/c = 0.91 x/c = 1.066 x/c=1.214

Figure 8. Comparison of Cp_{tot} (a=20.18°)

For the prediction of the slat and main wing wake width, the calculation results of the SA model and SST model based on the structural grid can achieve the predictive effect. At the periphery of the entire confluence boundary layer, the incoming flow is basically free from the interference of the boundary layer and wakes. Cp_{tot} should be close to 1. The results of this study and the test results both reflect this trend, but the calculation results of the structured grid are much better than unstructured grids, especially the wake width prediction, considering that the structured grids are refined in the wake area, in order to analyze the impact of the grid refine in the wake area, special treatment is performed in the wake area. As shown in Figure 9. The refinement unstructured grid is calculated with the SA turbulence model. The comparison result is shown in Figure 10. The width is consistent with the structural grid, which is very close to the experimental value. It shows that the mesh refinement near the confluence boundary layer has a great influence on the calculation results, especially the prediction of the width of the wake area.

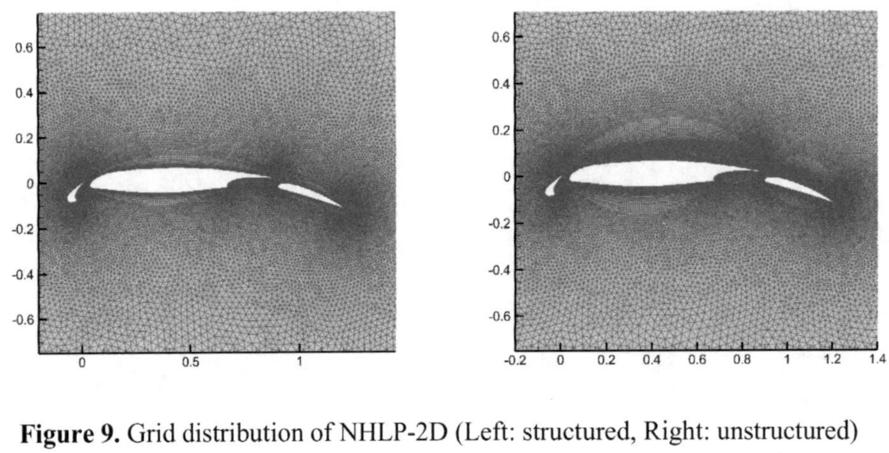

Figure 9. Grid distribution of NHLP-2D (Left: structured, Right: unstructured)

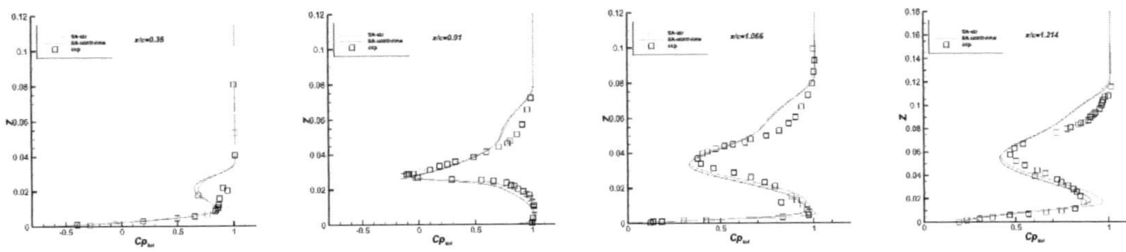

x/c=0.35 x/c = 0.91 x/c = 1.066 x/c=1.214

Figure 10. Comparison of Cp_{tot} between new grid results and experiment results(a=20.18°)

SA turbulence model is more suitable for aerodynamic calculation of high-lift configurations. The grid distribution near the confluent boundary layer has a direct influence on the calculation results of the wake area. SA turbulence model is applied to study the influence of Reynolds number on the aerodynamic characteristics of multi-section airfoil with the refine unstructured grid.

3.2. Influence of Reynolds number on boundary layer

The influence of the Reynolds number on the confluence boundary layer can be reflected by the distribution of the total pressure coefficient. When a = 20.18°, the change curve of the total pressure coefficient under different Reynolds numbers is shown in Figures 11. It can be seen that as the Reynolds number increases, the thickness of the entire confluence boundary layer gradually decreases, and the strength of the wake decreases as the Reynolds number increases [3].

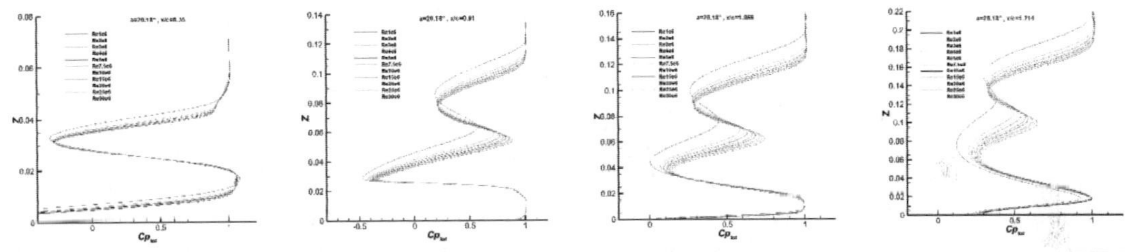

x/c=0.35 x/c = 0.91 x/c = 1.066 x/c=1.214

Figure 11. Cp_{tot} change with Reynolds number (a=20.18°)

3.3. Influence of Reynolds number on aerodynamic coefficient

When a = 4.01° and a = 20.18°, the influence curve of Reynolds number on the aerodynamic coefficient is shown in Figure 12. It can be seen from the content of the previous section that the Reynolds number increases, the thickness of the confluence boundary layer decreases, and the lift coefficient increases with the increase of the Reynolds number. Drag coefficient and pitching moment coefficient have the opposite pattern.

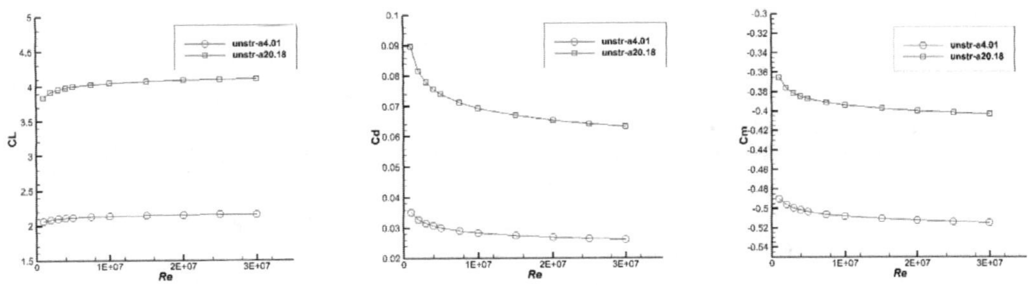

Figure 12. Influence of Reynolds number on aerodynamic coefficient

The change rule of airfoil lift curve with Reynolds number is shown in Figure 13. In the linear section, the lift is not much different, and the maximum lift increases with the increase of Reynolds number. As the angle of attack continues to increase, the airfoil "stalls". It can be seen from the figure that as the Reynolds number increases, the airfoil stall angle of attack shows a gradually increasing trend, and the Reynolds number continues to increase. The angle no longer increases and remains relatively stable.

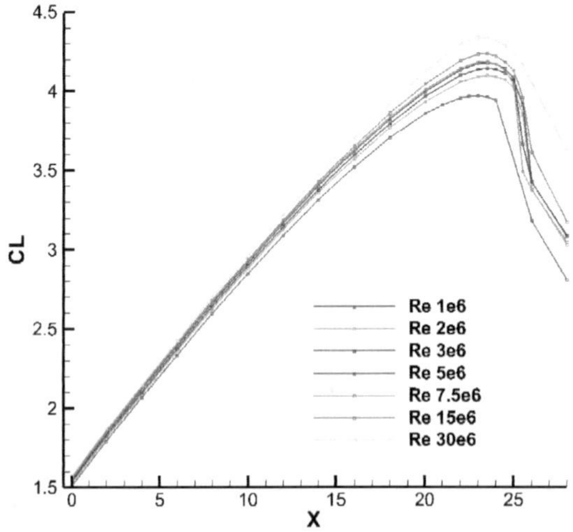

Figure 13. Change rule of airfoil lift curve with Reynolds number

4. Conclusion

Based on the consideration of grid factors and turbulence model factors, the confluence boundary layer and aerodynamic characteristics of the NHLP-2D airfoil are studied in this paper. Compared with the experimental results, it is shown that the grid distribution near the confluence boundary layer is important for calculating the width of the wake area. When the surface grid distribution is the same, the calculation results of the structured and unstructured grids are basically the same. On the basis of the mesh refinement in the wake area, the aerodynamic characteristics calculated by the SA turbulence model and the distribution of the total pressure coefficient in the confluence boundary layer are in good agreement with the experimental results. The width of the confluence boundary layer and the wake strength decrease with the increase of Reynolds number, the lift coefficient increases with the increase of Reynolds number, but the drag and pitching moment have opposite pattern. The maximum lift of the airfoil increases as the Reynolds number increases. As the Reynolds number increases, the stall angle of attack of the airfoil gradually increases and then remains stable with the increase of the Reynolds number.

Acknowledgments
Special thanks to Pengcheng Cui for his suggestions on this article.

References

[1] G. Ian,Fejtek, CFD 96 Computer Code Validation Challenge—multiple Element Airfoil, the Forth Annual Conference of the CFD Society of Canada, 1996
[2] M.Burt, A selection of experimental test cases for the vali-dation of CFD codes: Chapter 5-summaries of the test cases, AGARD, 1994.
[3] Liu Y, Gao Y, Wang J, et al. Calculation Research on the Effect of Reynolds Number on the High Lift Device Flow Characteristcs I—Aerodynamic Characteristics and Confluent Boundary Layer [J]. Civil Aircraft Design & Research, 2018.

The influence of wind turbine blade rotation on Wind direction meter

Lizhu Tian*, Yaqiang Zhou, Zhiwen Jiang, Yapeng Li, Zhaohe Wu, Chenglong Qi, Yanxu Gou, Yonghe Xu, Dayu Du, Bao Wang, Yuan Wu, Wei Feng, Peng Li

Datang Bayannur Wind Power Generation Co., Ltd, Bayannur Inner Mongolia, China

*Corresponding author e-mail: 34080575@163.com

Abstract. In this paper, a numerical analysis of the influence of the wind turbine impeller on the wind direction indicator is carried out. The layout position of the wind direction indicator and the wind speed change are analysed respectively, and the current equipment position is analysed whether it is suitable for the installation of wind measurement equipment, and the best layout is given.

1. Introduction

Wind turbines rely on dozens of micro-sensors to monitor their operating conditions and monitor the structure, internal sensitive locations and system components of the wind turbine at any time to ensure efficient and safe operation. The wind direction indicator is a very important component in the entire power generation system. Inaccurate or incorrect sensor signals will not only make the wind turbine inefficient, but also affect the safe operation of the wind turbine and the life of other components [1-3].

The wind vane is installed on the nacelle frame on the top of the nacelle and is easily disturbed by the rotation of the blades. Therefore, the actual power curve of the equipment is seriously affected, the output power of the equipment cannot meet the design requirements, and the volume of wind energy and the efficiency of the machine are reduced. The purpose of this paper is to select the best installation position on the nacelle through numerical simulation to reduce the interference of blade rotation on the wind vane. Ensure the safe operation of wind turbines [5-9].

2. Model description

The surface mesh is divided by a combination of tetrahedral and triangular meshes. The division requirements are: no free edges, no repeating units, and the minimum angle of the mesh is greater than 20°; the specific setting of the surface mesh size is shown in the table below.

Table 1. Mesh size of each part

Part name	Surface mesh size (m)
Engine room, wheel hub	0.1
Blade	0.1
Bracket	0.08
Rotation domain	0.2
Wind tunnel	2

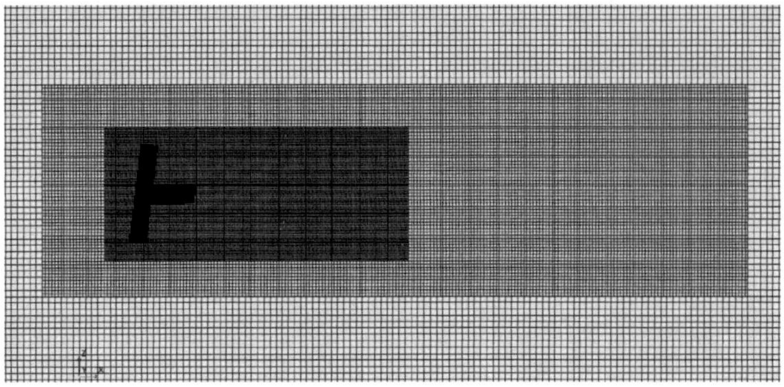

Figure 1. Overall grid

3. Calculate parameters and boundary conditions

Hyper mesh is used for geometric cleaning and mesh generation, and the STAR-CCM+ solver is used for calculation. Establish a geometric model of the wind turbine, input initial data including wind turbine blades, nacelles, towers, the geometric dimensions of the wind vane installation location, and the installation location of the wind vane; use the model for numerical simulation calculation, the input condition is wind speed; the output content is wind power generation The correction coefficient of the wind vane perturbed by the impeller; including: when the wind vane is not stopped, the wind vane corresponding to different wind speeds is affected by the impeller. Correction factor for impeller disturbance.

Figure 2. Type A unit calculation model monitoring points

4. Results

The unit is equipped with two anemometers symmetrically on both sides of the center line. First, the left and right anemometers are compared. The definition of left and right is shown in the sampling position.

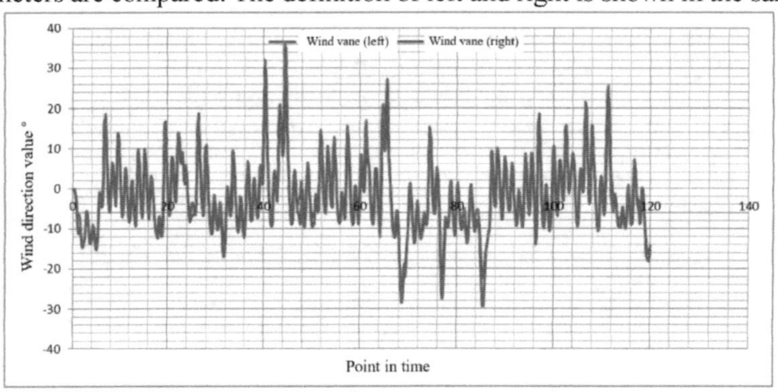

Figure 3. Comparison of wind vane results (9m/s)

Figure 3 is a comparison of the fluctuations of the wind vane on both sides of the 9m/s wind direction. The collection results of the two anemometers are consistent and there is not much deviation. Therefore, the two can be regarded as the same result at the current position. On the whole, the wind direction fluctuation range of the wind vane is (between -10 degrees and 10 degrees).

After numerical averaging, the average wind direction approached 0. Obviously, this numerical averaging method cannot show the characteristics of wind direction collection results.

Therefore, we transform the wind direction angle to the tan value. First compare the wind direction points on the same floor. Select two sides and the current wind direction value points, a total of three positions for comparison.

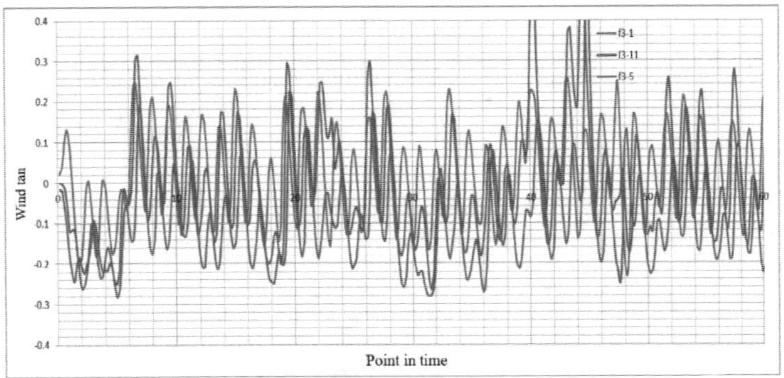

Figure 4. Comparison of results of collection points in the same layer (9m/s) (f-1 is the leftmost collection point, f-11 is the rightmost collection point, f-5 is the position of the wind vane)

From the comparison of the results, the collection position of the same layer is not much different, and the far right side is slightly better than the current collection position. Continue to compare the collection results of collection points at different heights.

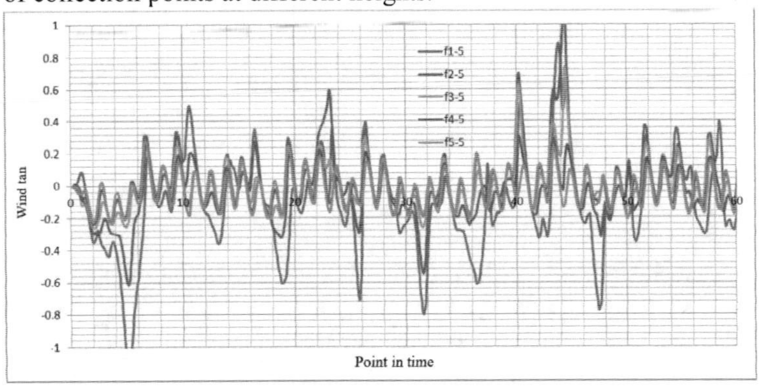

Figure 5. Comparison of results of collection points in different layers (9m/s) (f1-5 is the lowest position collection point (-0.4m), f3-5 is the wind vane position, f5-5 is the highest position collection point (+0.4m))

Comparing the results of the collection points on the 5th floor, the height difference of each floor is 0.2m, and f3-5 is the current position of the wind vane. It can be seen intuitively from the result that as the height increases, the collected fluctuation value decreases, and the collected result of the wind vane fluctuates around a value of 0.

At the same time, considering the heightening of the right wind vane, the comparison of the three is as follows:

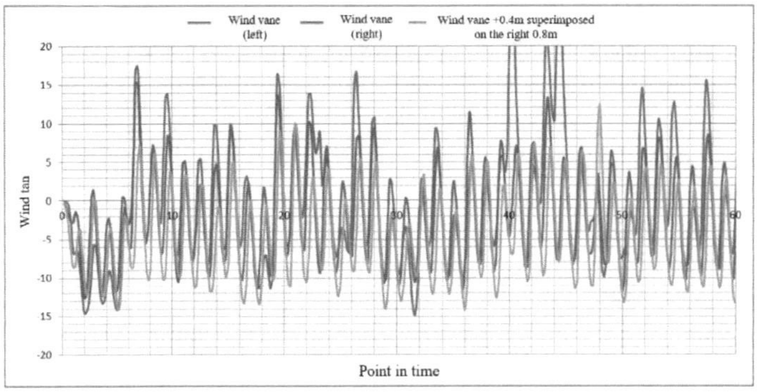

Figure 6. Comparison of optimization evaluation results (9m/s)

After comparing the three, the fluctuation of the wind direction is significantly reduced after 0.4m is increased, the overall fluctuation range is reduced from (-10 to 10 degrees) to (-10 to 5 degrees), and the fluctuation is reduced again after the right position.

The result of the same comparison under the wind speed of 11m/s is as follows:

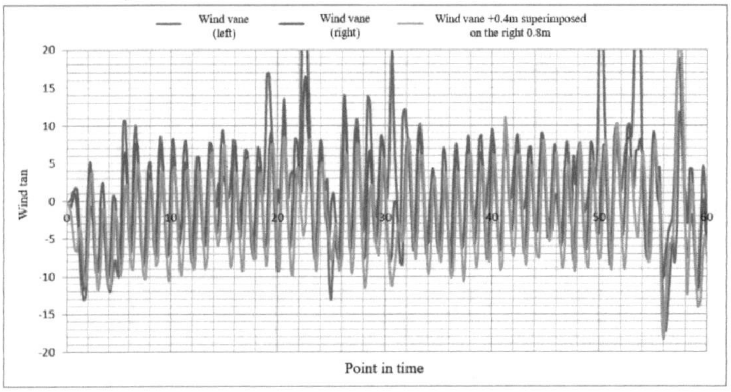

Figure 7. Comparison of optimization evaluation results (11m/s)

The optimal evaluation results of the wind vane under the same wind speed of 11m/s show the same characteristics.

5. Conclusion

After establishing a calculation model for the wind turbine 1:1, taking into account wind shear, blade rotation, different wind speeds and other factors, the wind speed and direction meter collection results of the unit were compared and optimized and evaluated.

After calculation, due to the rotation of the impeller, the flow wake of the basis has a certain impact on the downstream collection instrument. The rotating wake strengthens the swing of the left and right winds, making the collection results of the wind vane present a fluctuating state. After optimization and evaluation, the layout position of the contrast wind vane needs to be heightened, and at the same time, it needs to be placed on the right, but the effect of the right placement is limited. Taking into account the actual on-site installation situation. The wind vane can also be used to increase the height of 0.4m.

References

[1] H. Sun, Y Luo, Z. Zhao, R. Chang. The Impacts of Chinese Wind Farms on Climate. Journal of Geophysical Research. Atmospheres . 2018 (10).

[2] I. Balog, P. Ruti, I. Tobin, V. Armenio, R. Vautard. A numerical approach for planning offshore wind farms from regional to local scales over the Mediterranean. Renewable Energy. 2016.

[3] S. Kalantar, M. Ahmad. Thermal Analysis of 10-MW-Class Wind Turbine HTS Synchronous Generator. IEEE Transactions on Applied Superconductivity: A Publication of the IEEE Superconductivity Committee . 2014 (2).
[4] A. Tahi, C. Massfon. Thermal Analysis of Wind Turbine Nacelle Operating in Algerian Saharan Climate. Energy Procedia . 2012.
[5] J. Zhang, L. Song. Stress State Analysis on a Rock Anchor RC Foundation of Wind Driven Generator. Applied Mechanics and Materials . 2014 (448).
[6] U. Bhardwaj, A. Teixeira, C Soares. Reliability prediction of an offshore wind turbine gearbox. Renew. Energy, 141, 693-706 (2019).
[7] S. Bhattacharya, N. Nikitas, J. Garnsey, N.A. Alexander, J. Cox,D. Lombardi, D. Muir Wood, D.F.T. Nash. Observed dynamic soil–structure interaction in scale testing of offshore wind turbine foundations. Soil Dynamics and Earthquake Engineering . 2013.
[8] I. Lotsberg. Structural mechanics for design of grouted connections in monopile wind turbine structures. Marine Structures . 2013.
[9] Zhipeng Xue, Jie Fu, Minmin Zhao. Numerical Simulation and Analysis of the Arrangement of Wind Direction Meter for Wind Turbine[J]. E3S Web of Conferences, 2019, 118:02040.

Numerical analysis of horizontal bearing capacity of large-diameter implanted rock-socketed piles for offshore wind turbines

Zhaoqi Wu[1,]*, Xingcheng Zhang[1], Qiming Song[2], Wei Liu[2] and Zhibing Chen[2]

[1] College of Civil Engineering, Fuzhou University, Fuzhou 350108, China
[2] Fujian Yongfu Power Engineering Co., Ltd., Fuzhou 350108, China

*Corresponding author e-mail: zhaoqi_wu@163.com

Abstract. There is no standard for rock-socketed pile foundation of offshore wind farm at China and abroad. According to the practical geological and environmental conditions of offshore wind power plant, response of rock-socketed pile under horizontal load was analyzed by using a finite element software, ABAQUS. Then, the parameters of rock-socketed pile, i.e. depth, pile diameter, ratio of pile diameter to thickness, and grouting thickness were investigated by orthogonal analysis and parametric analysis. It is found that the thickness of grouting has little influence on the displacement of pile head. The diameter and the ration of diameter to thickness of pile, rock-socketed depth have significant influence. The displacement of pile tip tends to stable and the bending moment at bottom of pile become zero when the rock-socketed depth reaches $3d_p$. The maximum Tresca stress of grouting body and 1st principal stress of rock mass decrease with an increase of rock-socketed depth.

Keywords: offshore wind farm, rock-socketed piles, horizontal bearing capacity, numerical simulation, parametric analysis.

1. Introduction

Wind energy is highly valued as a clean and renewable energy source with huge reserves, and offshore wind farms are being constructed on a large scale in last 2 decades. Mono-pile foundation is the most widely used for its convenience and efficiency in fabrication and construction. In most southeast sea areas of china, there is a geological condition of thin overburden layer. Consequently, it is necessary to implant the mono-pile into the rock layer to meet the requirements of its bearing capacity.

However, there is no standard for rock-socketed pile of offshore wind farm at China and abroad. It is often designed by referring to the relevant specifications for rock-socketed pile foundation of port engineering or highway bridges [1, 2]. There are great differences between them in pile foundation size, construction mode and load action mode. The above mentioned design code has been unable to meet the requirements of the rock-socketed pile with large diameter, large depth and high horizontal bearing capacity.

In this paper, according to the practical geological and environmental conditions of offshore wind power plant, response of rock-socketed pile under horizontal load was analyzed by using a finite element software, ABAQUS. Then, the parameters of rock-socketed pile, i.e. depth, pile diameter, ratio of pile diameter to thickness, and grouting thickness were investigated by orthogonal analysis and parametric analysis.

2. Establishment and validation of numerical model

2.1. Numerical Model

The large commercial finite element software ABAQUS was used to analyze the behavior of rock-socketed pile. The analytical model includes three parts: steel pipe pile, grouting body and rock mass. In order to reduce the time-consuming of calculation, a finite element simulation was carried out by using one-half model. The rock diameter is defined as 20 times the pile diameter (d_p), and the soil height is 2 times the pile depth (h_p), in order to reduce the influence of boundary constraints on pile stress.

Pile, rock mass and grouting body were modeled by a 3D solid element C3D8R. The wall of pile and grouting body were divided into two elements in order to capture their bending behaviors [3]. The rock mass mesh near the pile hole was refined and encrypted. The mesh of the model is shown in Fig. 1.

Contact pairs were arranged between the interfaces of steel pile, grouting body and rock mass. A hard contact is set in the normal direction, and Coulomb friction with penalty function was used in the tangent direction. And the corresponding friction coefficient was defined as 0.4 according the literature [4].

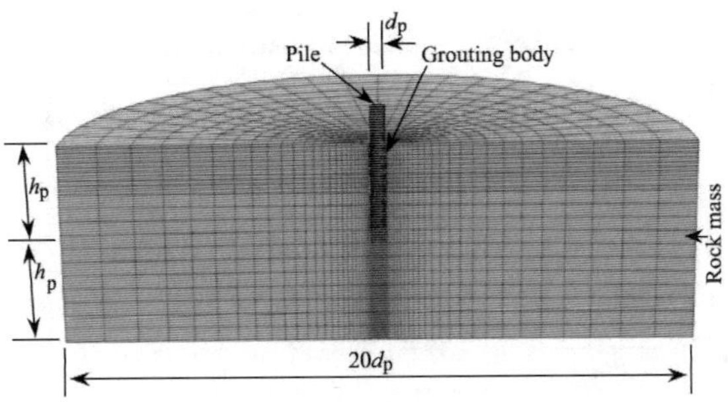

Figure 1. Finite element model of rock-socketed pile.

Steel pipe pile adopts double bilinear isotropic follow-up strengthening constitutive model. Strain-hardening modulus was defined as 1% of elastic modulus (E_s). The E_s and Poisson ratio were specified as 2.1×10^5 MPa and 0.30. Mohr-Coulomb strength criterion was use to simulate the behavior of rock and the performance parameters (Table 1) came from the geological investigation report. The material properties of grouting body was simulated by CDP model of concrete, which can fully consider the stiffness degradation and damage of materials and reflect the failure phenomenon of grout materials [5]. The stress-strain relationship of grouting body under uniaxial tension and compression is shown in Fig. 2.

Bottom and sides of rock mass were fixed. Symmetric constraints were used on symmetric surface of the model. Before horizontal force was applied at top of the pile, rock stress were balanced under gravity.

2.2. Validation

To validate the numerical model, field horizontal test of large diameter steel pipe pile in Donghai Bridge Wind Power Plant [6] was simulated. The comparison of tested and finite element results is shown in Fig. 3. It is found the tested and finite element results compare well, which indicates that the above mentioned numerical model is effective.

Table 1. Values of the parameter in Mohr-Coulomb failure criterion

Name	t / m	γ / kN/m^3	φ / °	c / kPa
Artificial fill	0.3	18.0	25	5
Fragmentary strongly differentiated granite	4.9	21.0	35	10000
Meso-differentiated granite	5.4	24.0	45	40000

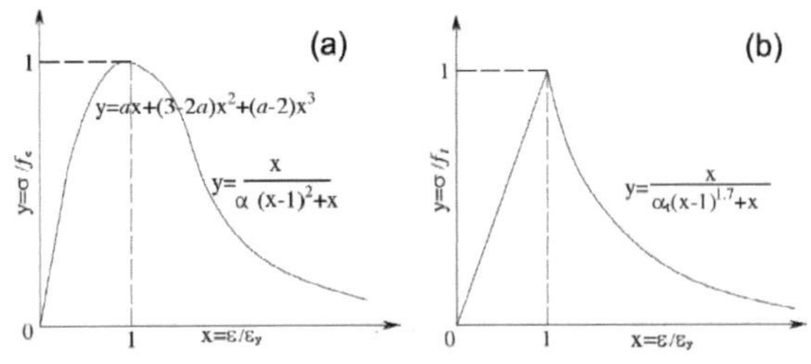

Figure 2. Stress-strain relationship of grouting body: (a) compression; (b) tension.

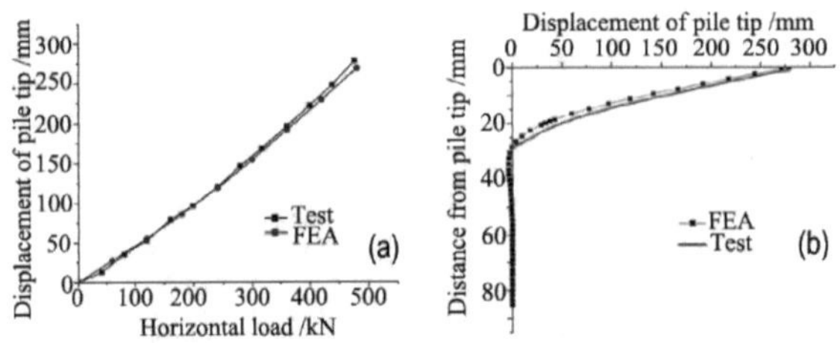

Figure 3. Comparison of tested and finite element results: (a) load-displacement relationship; (b) displacement of pile under force equal to 480kN

3. Mechanism of rock-socketed piles

Horizontal load-tip displacement of a typical rock-socketed pile is shown in Fig. 4a. According to Technical Code for Building Pile Foundation (JGJ 94-2008) [7], the load at the first stage with sudden change on the load displacement curve is taken as the horizontal critical load. According to Fig. 4a, when the load is before 6400kN, the overall growth is linear. When the load was increased to 7200kN, the first mutation occurred. When the load was applied to 8000kN, the pile tip displacement reached about 249mm, and the pile was greatly bent and could not continue to bear the load. To sum up, 6400kN is taken as the horizontal critical load value of the rock-socketed pile.

For the steel pipe pile bearing horizontal load, the bending moment of the pile is also one of the key indexes, especially the maximum bending moment of the pile must be checked and judged. Firstly, the rock-socketed piles below the mud surface were partitioned, and a section was defined for every 0.5m of filled soil and broken rock, and a section was defined for every 1m of moderately differentiated rock. A total of 16 bending moments were defined along the pile body to extract sections.

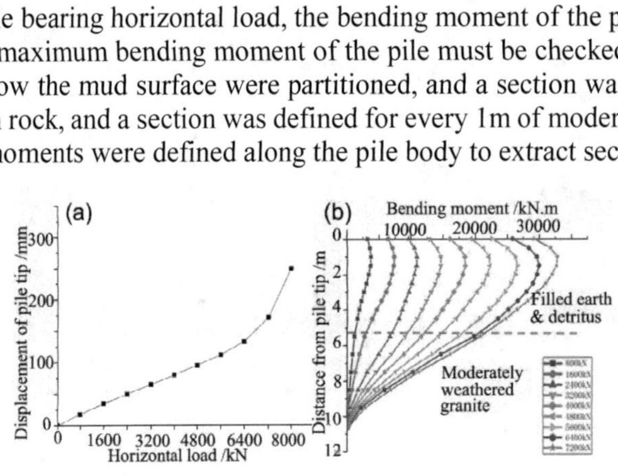

Figure 4. Displacement of a typical pile: (a) load-displacement relationship; (b) displacement of pile.

Bending moment extraction results of test pile are shown in Fig. 4b. Under the action of graded load, the maximum bending moment appears in the depth range of 1.0-1.5 m below mud surface, and with the increase of load, the maximum bending moment of pile body moves down slightly. After entering the moderately differentiated rock strata, the bending moment value drops sharply, and the bending moment value at the pile bottom position can be ignored, indicating that the embedment effect of moderately differentiated rock is obvious. However, for the steel pipe pile, the position of the maximum bending moment is the dangerous point of the structure, that is, the most unfavorable section, which needs to be observed in the stress process.

As for the peripheral grout, the standard 28-day compressive strength of high-strength grout material greater than 70MPa was selected, the corresponding stress requirements for the grouting material of offshore fan foundation are proposed in DNV-OS-J101 specification. For rock-socketed piles, grouting materials with slightly lower strength are generally selected. The compressive strength of grouting material selected is 70MPa. The material coefficient is recommended to be 3.0 in the specification when calculating the ultimate stress. After calculation, the rock-socketed pile. The Tresca stress cloud diagram of peripheral grout under critical load is selected for analysis, as shown in Fig. 5a.

According to Fig. 5a, the maximum Tresca stress of grout does not exceed the set strength requirement of 23.3MPa, which is in line with the stress index proposed for grout in DNV specification. At the same time, the maximum value of the Tresca stress of the grout is 4.9~5.7m below the mud surface under each grade of load, which includes the boundary between the highly weathered rock and the moderately differentiated rock (5.2m below the mud surface), indicating that the surrounding grout is prone to the phenomenon of high Tresca stress near the position where the rock properties change. Due to the different properties of the two rock strata, the stress is not uniform, and the phenomenon of stress concentration is easy to occur.

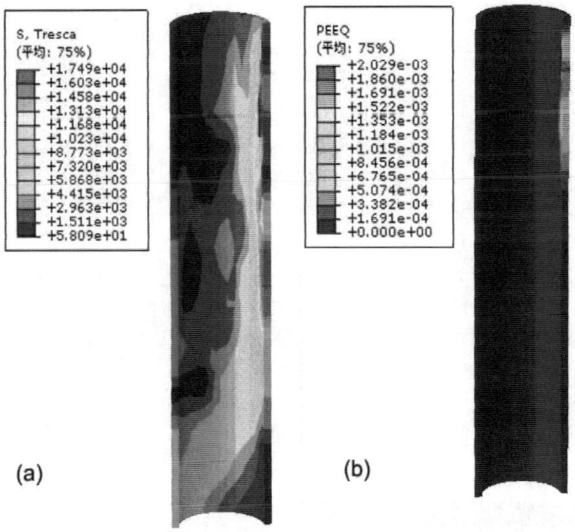

Figure 5. Contours of grouting body: (a) Tresca stress; (b) plastic strain.

Based on the peripheral grouting, CDP damage criterion [5] was adopted to list the equivalent plastic strain cloud map under the corresponding load to analyze the failure modes of grouting at different stages. As shown in Fig. 5, when it reaches 6400kN, the plastic strain reaches 2.029E-3, which is located about 1.2m below the mud surface, corresponding to the maximum bending moment of pile body under relevant load.

With the crack propagation of grouting, the plastic zone develops to a range of 4.0m below the mud surface, but no obvious plastic strain appears in the rest of grouting. When the horizontal load

Journal of Physics: Conference Series **2280** (2022) 012012 doi:10.1088/1742-6596/2280/1/012012

increases, there will be more serious crushing phenomenon in a certain part of the grout, and finally through cracking. His position corresponds to the maximum bending moment of the steel pipe pile in the process of loading, and they are correlated.

Fig. 6a shows the displacement distribution of rock and soil mass under the load of 6400kN. It can be seen that the deformation of rock and soil mass is concentrated in the filling layer above -4.7m and the fragmentary granite layer. Due to the large elastic modulus and buried depth, the deformation of the moderately differentiated granite is small. Fig. 6b shows the contour of the maximum first principal stress of moderately weathered granite. When the horizontal force reaches the critical load, the maximum principal stress was transferred from the pile bottom to the interface between broken rock and moderately weathered granite. Under the action of the previous load, the rock was tensile failure. The internal forces of the rock were redistributed and the tensile stress of the interface of the rock becomes larger. The Tresca stress of the adjacent grouting body was also the largest, so it can be considered that the failure of the grouting has a certain relationship with the tensile failure of the rock mass.

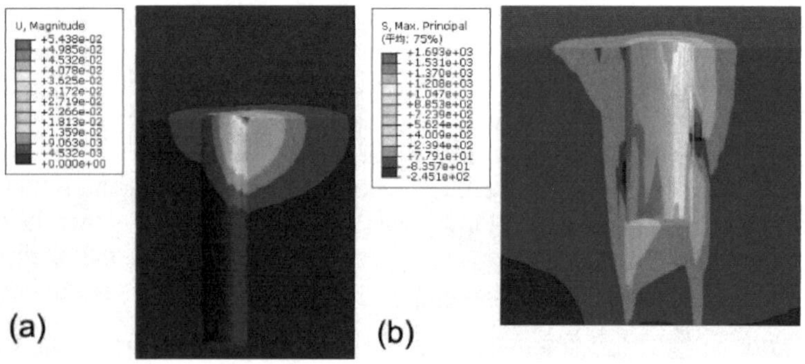

Figure 6. Displacement and stress contours of rock mass: (a) dis placement; (b) first principal stress of moderately weathered granite.

4. Orthogonal test analysis

According to the design concerns, four factors of rock-socketed depth, pile diameter, pile diameter to thickness ratio and peripheral grouting thickness were selected for study. Rock-socketed depth is determined by integer multiple of pile diameter. Diameter thickness ratio is related to pile diameter. So the three factors are involved in mutual influence. The maximum rock-socketed depth was recommended to be $5d_p$, so the rock-socketed depth of this time is d_p and $5d_p$. For the pile diameter, 1.8m and 4.5m were selected as horizontal factors, and the peripheral grouting thickness was 5cm and 25cm. As for the diameter thickness ratio of pile, 20 and 60 are taken as parameters to study based on the recommended value of current offshore wind power specifications.

The interaction of factors was considered in this test, so orthogonal test table $L8(2^7)$ was selected for test arrangement and result analysis. The horizontal displacement data of pile head under the condition of horizontal critical load of 6400kN were selected and summarized, as shown in Table 2.

According to the intuitive analysis method, the range of each factor is obtained, as shown in Table 3. It can be found that the influence degree of each factor on the result is pile diameter > diameter thickness ratio > interaction between pile diameter and diameter thickness ratio > rock-socketed depth > interaction between pile diameter and rock-socketed depth > grouting thickness.

It can be concluded that pile diameter and diameter to thickness ratio have great influence on displacement results, while peripheral grouting thickness has little influence on displacement results. Therefore, in the subsequent analysis, the influence of peripheral grouting thickness changes on displacement is ignored, and the concrete influence of pile diameter, diameter-thickness ratio and rock-socketed depth on static response is only discussed.

Table 2. Orthogonal test displacement of pile tip

	A	B	A*B	C	A*C		D	δ_p
	1	2	3	4	5	6	7	/mm
1	1(1.8)	1(20)	1	1(1D)	1	1	1(50)	88.21
2	1	1	1	2(5D)	2	2	2(250)	66.64
3	1	2(60)	2	1	1	2	2	139.01
4	1	2	2	2	2	1	1	121.13
5	2(4.5)	1	2	1	2	1	2	5.16
6	2	1	2	2	1	2	1	3.65
7	2	2	1	1	2	2	1	11.47
8	2	2	1	2	1	1	2	9.36

Table 3. Visual analysis table

Test	Factors						
	A	B	A*B	C	A*C	Blank	D
K_1	414.99	163.66	175.68	243.85	240.23	223.86	224.46
K_2	20.28	280.97	268.95	200.78	204.40	220.77	220.17
R	394.71	117.31	93.27	43.07	35.83	3.09	4.29

5. Parametric analysis

The influence of pile diameters, diameter-thickness ratios and rock-socketed depths on the static response of rock-socketed piles was further studied by taking the top displacement, bending moment, maximum Tresca stress of grouting body and the first maximum principal stress of rock as indexes. Four pile diameters, d_p = 1.8m, 2.7m, 3.6m and 4.5m. Five diameter-thickness ratios, δ = 20, 30, 40, 50 and 60; Five rock-socketing depth, $H_p = d_p$, $2d_p$, $3d_p$, $4d_p$ and $5d_p$. A total of 100 finite element models were established for parametric analysis.

5.1. Displacement of pile tip

Fig. 7 shows the tip displacements of the piles with varied diameters and rock-socketed depths. It is found that the displacement decreases with an increase of rock-socketed depth, and tends to be stable when the rock-socketed depth is greater than $3d_p$. With the increase of the ratio of diameter to thickness, the displacement of the pile increases.

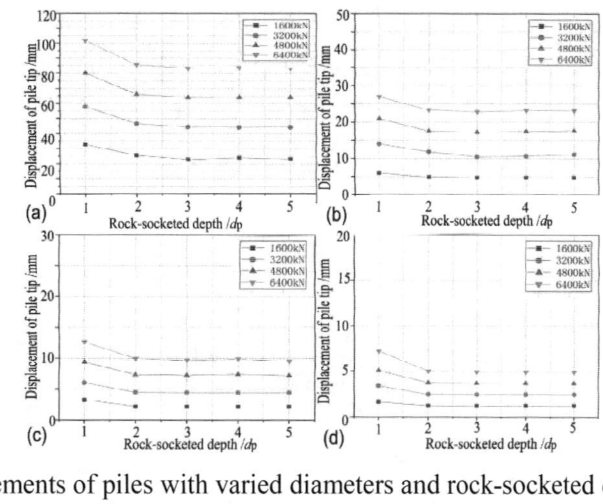

Figure 7. Tip displacements of piles with varied diameters and rock-socketed depth: (a) d_p = 1.8m; (a) d_p = 2.7m; (a) d_p = 3.6m; (a) d_p = 4.5m.

5.2. Bending moment of pile

Fig. 8 hows the comparison of pile bending moment with different pile diameter and rock-socketed depth. It can be found that with the increase of the ratio of diameter to thickness, the bending moment value presents a decreasing trend, however, the position of the maximum bending moment is relative stable. The properties of piles tend to be from rigid piles to flexible piles.

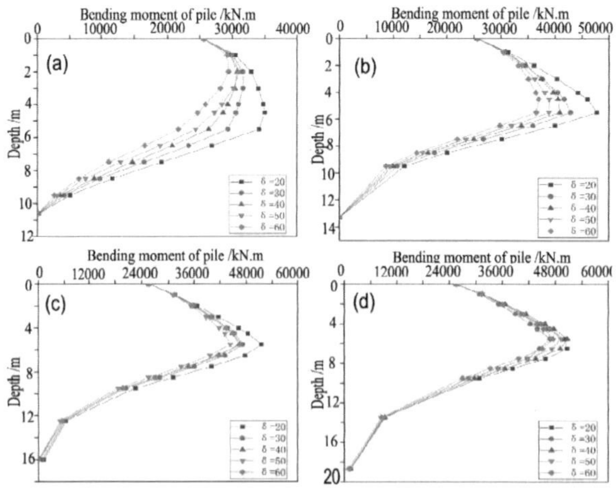

Figure 8. Bending moment of piles with varied diameters and rock-socketed depth: (a) d_p = 1.8m; (a) d_p = 2.7m; (a) d_p = 3.6m; (a) d_p = 4.5m.

5.3. Tresca Stress of grouting body

Fig. 9 shows bending moment of pile with varied diameters and rock-socketed depth. It can be found that the maximum Tresca stress of grout shows a law of fluctuation and decrease with an increase of rock-socketed depth. The maximum Tresca stress is smaller than the compression strength of grouting body.

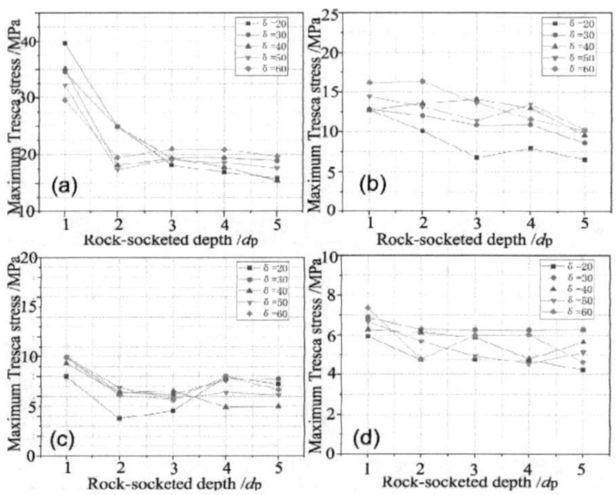

Figure 9. Maximum Tresca stress of grouting with varied pile diameters and rock-socketed depth: (a) d_p = 1.8m; (a) d_p = 2.7m; (a) d_p = 3.6m; (a) d_p = 4.5m.

5.4. 1st principal stress of rock mass

The maximum 1st principal stress of rock with different pile diameters and rock-socketed depths is shown in Fig. 10. It can be found that with an increase of rock-socketed depth, the maximum first principal stress of rock shows a law of fluctuation and decline. After the rock-socketed depth reaches

$2d_p$ or $3d_p$, the first principal stress of rock mass becomes stable.

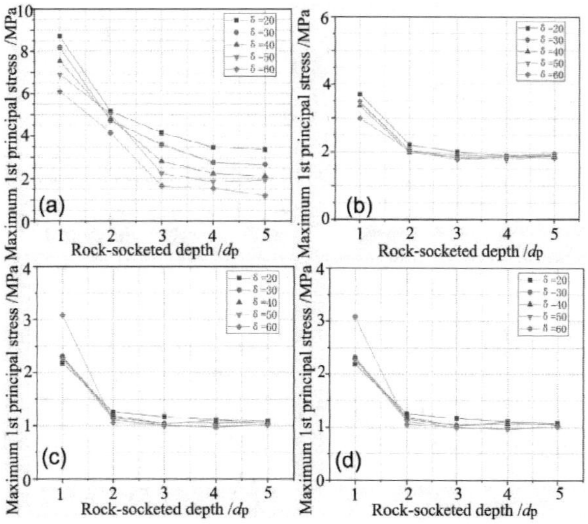

Figure 10. Maximum 1st principal stress of rock with varied pile diameters and rock-socketed depth: (a) $d_p = 1.8$m; (a) $d_p = 2.7$m; (a) $d_p = 3.6$m; (a) $d_p = 4.5$m.

6. Conclusions

(1) The tested and finite element results compare well. The numerical model built in this paper is effective to analyze behavior of rock-socketed pile under horizontal loads.

(2) The orthogonal test shows that the thickness of grouting has little influence on the displacement of pile head. The diameter and the ration of diameter to thickness of pile, rock-socketed depth have significant influence.

(3) The displacement of pile tip tends to stable and the bending moment at bottom of pile become zero when the rock-socketed depth reaches $3d_p$.

(4) The maximum Tresca stress of grouting body and 1st principal stress of rock mass decrease with an increase of rock-socketed depth.

Acknowledgments

This work was supported by Scientific Research Project of Fujian Yongfu Power Engineering Co., Ltd., China.

References

[1] Industrial standard of the People's Republic of China (2000). Design and Construction Code of Rock-socketed Piles for Harbour Engineering (JTJ285), Beijing, China. (in Chinese)

[2] Industrial standard of the People's Republic of China (2007). Code for Design of Ground Base and Foundation of Highway Bridge and Culverts, (JTG D63), Beijing, China. (in Chinese)

[3] DNV GL (2019). Analysis of Grouted Connections Using the Finite Element Method (DNVGL-RP-0419), Denmark.

[4] Lotsberg, Inge (2013). Structural mechanics for design of grouted connections in monopile wind turbine structured. Marine Structures, 32(7): 113-135.

[5] Lee J H, Fenves G L (1998). Plastic-damage model for cyclic loading of concrete structures. Journal of Engineering Mechanics, 124(8): 892-900.

[6] Gong W, Huo S, Yang C. et al (2015). Experimental study on horizontal bearing capacity of large diameter steel pipe pile for offshore wind farm. Shuili Xuebao, 46(S1): 34-39. (in Chinese)

[7] Industrial standard of the People's Republic of China (2008). Technical code for building pile foundations, Beijing, China. (in Chinese)

Study on the effect of inflow conditions on wind flow acceleration in mountainous micro-topography

Yuan Yin[1], Yangsen Li[1], Ruizong Lin[1], Zhiwei Liu[2], Benzhao Fu[2], Guo Huang[3, *], Hongjie Zhang[3]

[1] State Gird Fujian Electric Power Co. Ltd, Fuzhou 350003, China
[2] State Grid Fujian Economic Research Institute, Fuzhou 350003, China
[3] China Electric Power Research Institute, Beijing, 100055, China

*Corresponding author e-mail:huangguocn@126.com

Abstract. The acceleration of wind flow in mountainous areas can cause the destruction or collapse of transmission towers. The wind speed and direction of inflow have a significant impact on the flow acceleration in mountainous areas. The mountain terrain in the coastal area of Fujian was selected to study the influence of the inflow wind speed and direction on the acceleration effect. By analyzing the radar chart of the wind speed acceleration ratio of the monitoring points under 24 wind direction angles, the variation law of the wind speed acceleration ratio is consistent with the trend of the influence law obtained by the analysis of the surrounding terrain. The obtained simulation results are correct and credible, which can be used to correct the design wind speed of transmission lines in mountainous areas.

1. Introduction

China is a country dominated by mountains, and the mountainous area accounts for more than 65% of the country's land. Especially in the western region, the terrain is undulating and continuous. In the mountain areas, wind flow will accelerate or change direction because of rough terrain. Due to heat exchange and terrain interference, the characteristics of wind flow in the mountain areas are very different from those of the flat ground. The annual wind disaster will cause a large number of buildings and transmission towers to be damaged or collapsed. Therefore, the study on near-ground wind flow in mountainous areas has significant value for the design and safety evaluation of large-scale structures sensitive to wind loads such as transmission lines. Numerous scholars have carried out numerical simulations of wind flow in mountainous areas and achieved a series of research results.

Dhunny studied the key parameters in CFD numerical simulation, such as boundary conditions, grid independence test, discretization scheme, turbulence model and iterative convergence criteria [1]. Liu simulated the drag effect of vegetation by adding a negative scour term to the momentum equation [2]. Ishihara established a transition area around complex terrain to play a smooth transition role [3]. Bitsuamlak simulated a simple two-dimensional and three-dimensional wind flow. By comparing the results of wind tunnel tests and field measurements, it was found that the numerical simulation results on the windward side were closer to the observation data than on the leeward side [4]. Kikuchi used the Monte Carlo simulation method to propose a terrain factor for considering the influence of complex terrain on the wind direction. The accuracy of the CFD simulation and the Monte Carlo simulation were verified by the wind tunnel test data and the observation results of weather stations [5]. Tse conducted field measurements, scaled wind tunnel tests and mesoscale meteorological simulations (WRF) on the complex terrain wind flow, respectively. It was found that the error between the wind tunnel test results and the field measurement results was mainly due to the difference between the inlet wind speed profile of the wind tunnel test and the real terrain. The difference between the WRF simulation results and the field measurement results was due to the lower grid resolution of the WRF [6]. Abd-Elaal carried out numerical simulations of downburst winds on two real terrains and studied the variation of horizontal and vertical wind speeds [7].

The wind speed and direction of inflow have a significant impact on the acceleration effect of the mountain micro-topography. In this paper, numerical simulation methods are used to study the influence of the inflow condition on the acceleration effect.

2. Computational Model

The mountainous terrain of about 10km×10km was selected as the simulation object in the coast of Fujian, as shown in Figure 1. Ten points are selected in the terrain to analyze the wind flow. These points are located near the ridges, in order to show the shielding effect of hillsides on wind flow. The coordinates of the extracted points are shown in Table 1.

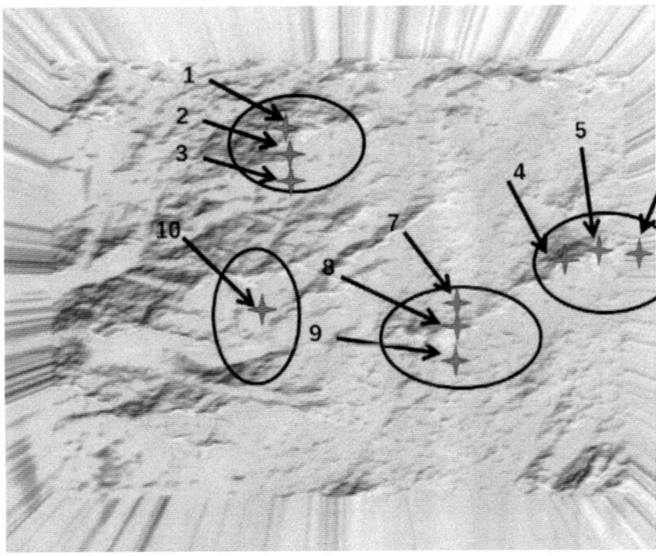

Figure 1. Ten measuring points in mountainous areas

Table 1. The latitude and longitude coordinates of the measuring points

No.	North Latitude	East Longitude
1	24°18.787′	188°1.979′
2	24°18.508′	188°1.979′
3	24°18.277′	188°1.982′
4	24°17.126′	188°5.052′
5	24°17.277′	188°5.384′
6	24°17.335′	188°5.605′
7	24°16.401′	188°3.802′
8	24°16.239′	188°3.805′
9	24°15.930′	188°3.802′
10	24°16.151′	188°1.860′

In order to study the influence of the inflow wind speed and wind direction on the wind speed acceleration ratio, the combined working conditions of wind speed and wind direction are set. Under the condition of type-A inflow profile, the inflow wind speed was set as 10m/s, 20m/s and 30m/s, respectively, and the wind direction angles ranged from 0° to 345°.

3. Simulation calculations

A square computational domain is established, and the wake region has a distance of 10 times the characteristic length to avoid the effect of the outlet. The velocity condition is set for the inlet, where the velocity equals to 10m/s. The pressure outlet condition is set for the outlet. The wall condition is set for the surface, and the symmetrical boundary condition is set for other boundaries. Figures 2(a)~(j) show the radar charts of wind speed acceleration ratio of 10 monitoring points at 24 wind direction angles.

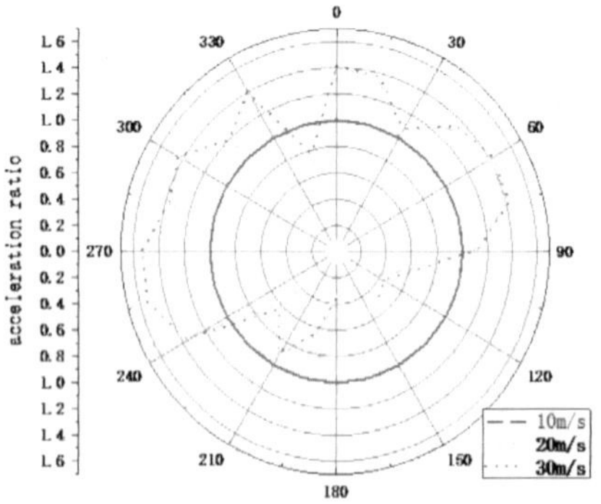

(*a*) Point 1

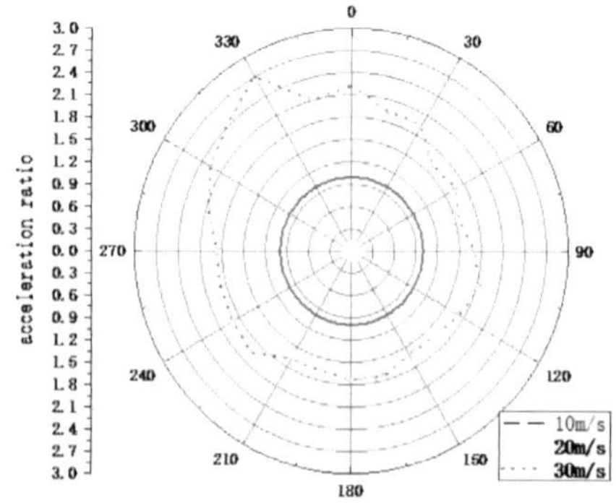

(*b*) Point 2

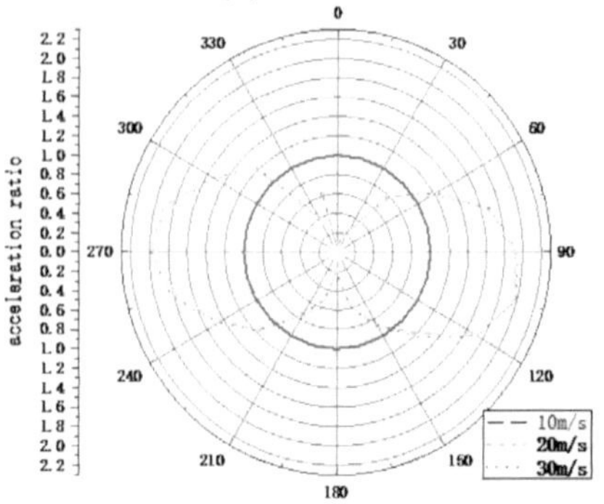

(*c*) Point 3

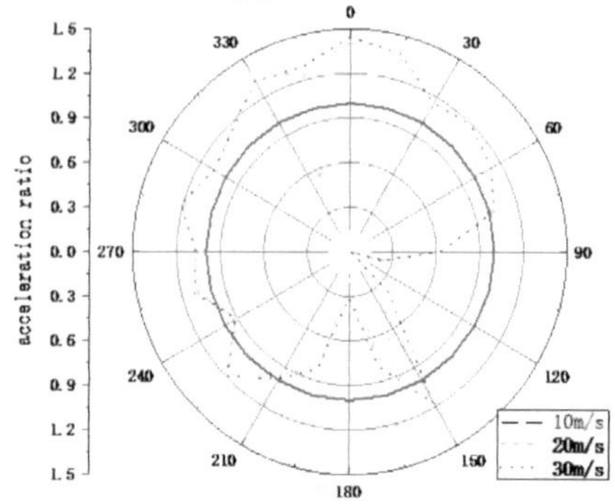

(*d*) Point 4

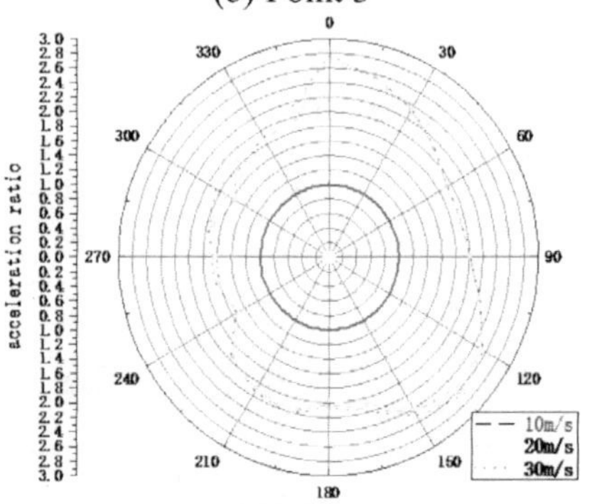

(*e*) Point 5

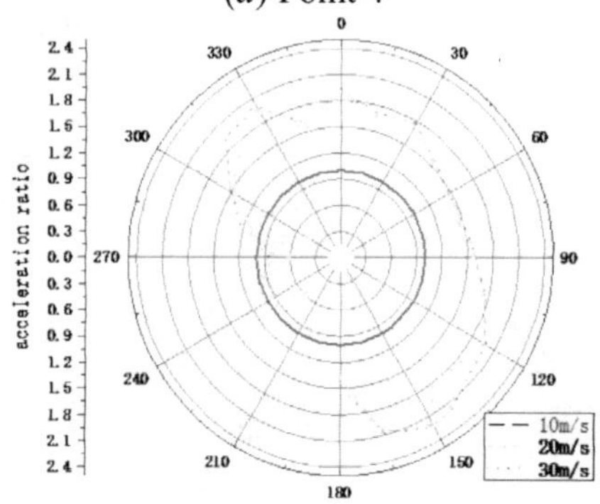

(*f*) Point 6

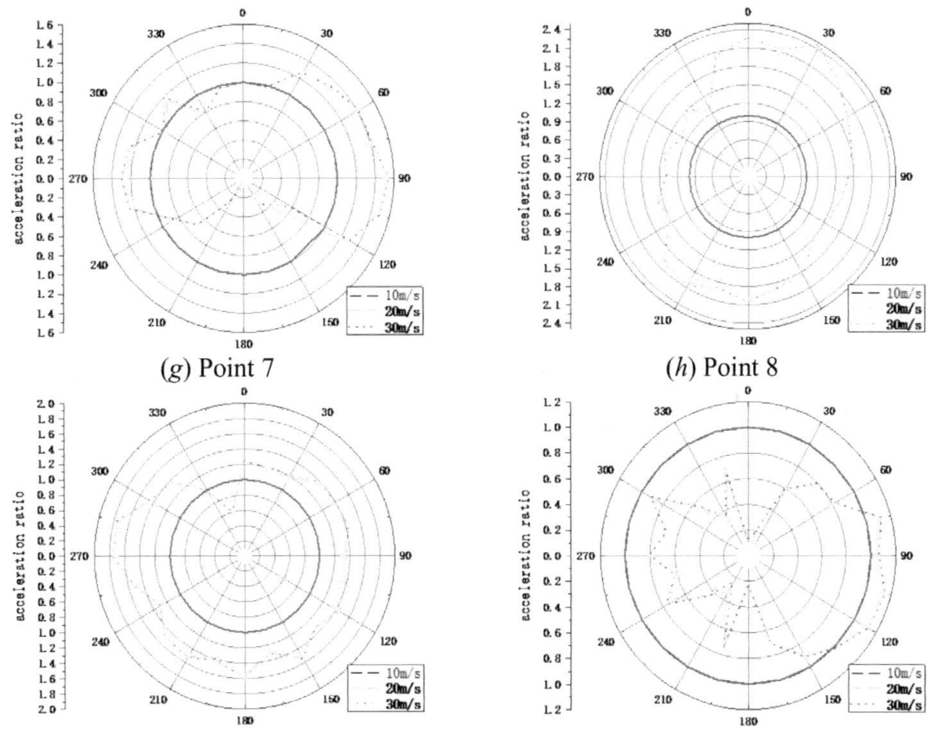

(*g*) Point 7 (*h*) Point 8

(*i*) Point 9 (*j*) Point 10

Figure 2. Radar chart of wind speed acceleration ratio

The monitoring point 1 showed in Figure 2(a) is on the windward side with the wind from the northwest, northeast and north, and the acceleration effect is significant. The acceleration ratio of wind speed is greater than 1, and the maximum value is close to 1.4. The point 1 is on the leeward side with the wind from the southwest, southeast and south, and the shading effect is significant. The wind speed acceleration ratio is much less than 1, and the minimum value is close to 0.2. The data analysis shows that the point 1 is located on the hillside of the hilly terrain, which is consistent with the actual terrain where the measurement point is located.

The monitoring point 2 showed in Figure 2(b) is on the windward side with the wind from all directions, and the acceleration effect is significant. The wind speed acceleration ratio is greater than 1, and the maximum value is close to 2.7. The data analysis shows that the point 2 is located at the top of the hill, which is consistent with the actual terrain where the measuring point is located.

The monitoring point 3 showed in Figure 2(c) is on the windward side with the incoming flow from a small range between the southwest and northeast, and the acceleration effect is significant. The wind speed acceleration ratio is greater than 1, and the maximum value is close to 1.8. The data analysis shows that the point 3 is in a canyon extending from east to west. According to Figure 1, it can be seen that it is consistent with the actual terrain where the measuring point is located.

The monitoring point 4 showed in Figure 2(d) has mountains near the east, west and south sides. The wind shielding effect from these three directions is significant, and the wind speed acceleration ratio is less than 1. The point 4 is still slightly shielded because of the hill from a distance, and the wind speed acceleration ratio is only slightly greater than 1. There is a crack in the mountain at the 4 o'clock direction, and the wind speed acceleration ratio in this direction is slightly greater than 1. The data analysis shows that the point 4 is located in the hillside area surrounded by mountains on three sides. According to Figure 1, it can be seen that it is consistent with the actual terrain where the measuring point is located.

The monitoring point 10 showed in Figure 2(j) is blocked by the nearby mountains in most of the incoming flow directions. Most of the wind speed acceleration ratios are less than 1, and the minimum wind speed acceleration ratio is 0.12. Only in the southeast the shielding effect of the wind speed is weakened, and the maximum wind speed acceleration ratio is 1.16. The data analysis shows that the measuring point 10 is located in a valley basin surrounded by mountains. It can be seen from Figure 1 that it is consistent with the actual terrain where the measuring point is located.

From the analysis of Figure 2(a) to 2(j), it can be seen that:

(1) Under the action of incoming flow from different wind directions, the laws of wind speed acceleration and shading effect of the 10 monitoring points are consistent. The variation law of the wind speed acceleration ratio with the wind direction angle at each monitoring point is consistent with the influence law obtained from the analysis of the surrounding terrain, which proves that the simulation analysis results are correct and credible.

(2) The acceleration ratio of the wind speed obtained at each monitoring point is little affected by the inflow wind speed.

(3) The radar map of wind speed acceleration ratio can clearly and quantitatively shows the influence of surrounding terrain and incoming wind direction on each monitoring point, which is very convenient for designers to correct the design wind speed in mountainous areas.

4. Conclusions

The radar chart of the wind speed acceleration ratio at the monitoring points under 24 wind direction angles are analyzed. The variation law of the wind speed acceleration ratio with the wind direction angle is consistent with the influence law obtained by the analysis of the surrounding terrain. The simulation analysis results are correct and credible, and simulation method which can be used to correct the design wind speed of transmission lines in mountainous areas.

Acknowledgments

This work was financially supported by the State Grid Fujian Economic Research Institute. The project name is Research on assessment technology of mechanical state of towers based on numerical wind tunnel of atmospheric boundary layer (52130N210007).

References

[1] Dhunny A. Z., Loll chund M. R., Rughooputh S. D. D. V. Wind energy evaluation for a highly complex terrain using Computational Fluid Dynamics (CFD) [J]. Renewable Energy, 2017, 101: 1-9.

[2] Liu Z., Ishihara T., He X., et al. LES study on the turbulent flow fields over complex terrain covered by vegetation canopy [J]. Journal of Wind Engineering and Industrial Aero dynamics, 2016, 155: 60-73.

[3] Ishihara T. A Nonlinear Wind Prediction Model MASCOT: Development and Application (Numerical Prediction Technique of Airflow Over Complex Terrain) [J]. Journal of Japan Society of Fluid Mechanics, 2003, 22: 387-396.

[4] Bitsuamlak G. T., Stathopoulos T., Bédard C. Numerical Evaluation of Wind Flow over Complex Terrain: Review [J]. Journal of Aerospace Engineering, 2004, 17(4): 135-145.

[5] Kikuchi Y., Ishihara T. A study of topographic multiplier considering the effect of complex terrains and tropical cyclones [J]. Journal of Wind Engineering and Industrial Aerodynamics, 2012, 104-106: 558-564.

[6] Tse K. T., Li S. W., Fung J. C. H. A comparative study of typhoon wind profiles derived from field measurements, meso-scale numerical simulations, and wind tunnel physical modeling [J]. Journal of Wind Engineering and Industrial Aerodynamics, 2014, 131: 46-58.

[7] Abd-Elaal E.-S., Mills J. E., Ma X. Numerical simulation of downburst wind flow over real topography [J]. Journal of Wind Engineering an d Industrial Aerodynamics, 2018, 172: 85-95.

Overload Response Analysis of UAV in Severe Weather

Ting Zhu [a], Shenglin Xia, Shiwei He, Kunshan Huang

Chengdu Aircraft Industrial (Group) CO., Ltd, Chengdu, Sichuan, China

[a] icfmia@163.com

Abstract. Turbulence will occur when the aircraft passes through the gust area, which will affect the handling quality and ride quality of the aircraft, increase the bending moment of the wing root and shorten the fatigue life of the structure. At present, in engineering calculation, when the gust load of UAV is considering, the UAV is usually regarded as a rigid body according to the general specification, and the overload response is obtained by solving the dynamic equations of rigid body. The flexibility of UAV is not considered, and the more realistic harsh wind field environment is not considered, so there is a certain deviation between the results and the actual values. Therefore, the gust load was studied considering the aircraft elasticity and the more realistic harsh wind field environment, and compares the calculation results with that considering the standard wind field, and makes a full assessment of the gust overload of UAV. It provides the basis for the design of aircraft structure strength and flight safety, and has important engineering significance.

Key words: severe weather, UAV, elastic aircraft, gust overload.

1. Introduction

When unmanned aerial vehicles (UAV) fly in the atmosphere, they often encounter gusts, that is, airflow disturbances of varying severity, which are called turbulence and gust in engineering applications. The gust will change the effective angle of attack of the aircraft's aerodynamic lifting surfaces, causing a sudden change in lift that will produce an aircraft dynamic response involving elastic deformation. The light impact of the gust on the aircraft will make it throwing up and down, shaking from side to side, causing it difficult to control, inaccurate instrument data and so on and reduce flight quality. In severe cases, it will lead to structural damage and even disintegration of the aircraft, seriously endangering flight safety. In summary, when aircraft encounters with severe weather, it has a significant impact on mission performance and flight safety, and therefore the response of aircraft to gust excitation must be considered in the process of aircraft design.

At present, in the calculation of UAV gust overload, two models of discrete gust and continuous gust used in engineering are generally considered for the calculation of aircraft gust load. The strength of these two wind fields is generally more than the actual wind field. In other words, the design overload calculated by these two wind fields is often too conservative. When the wind field is unknown, the two wind field models are used to calculate the gust overload, and the larger one being used as the basis for the structural strength design is appropriate. When the wind field is known, still using the engineering wind field model will often cause redundancy in aircraft structure design. Therefore, it is necessary to calculate the gust overload of the aircraft according to the real wind field in the real aircraft flight mission, and the results being as the basis of aircraft structure design is of great significance to improve the utilization rate of aircraft structure.

With the continuous development of aviation science and technology, the performance requirements of aircraft continue to increase, which makes the aircraft develop towards the trend of low structural weight and high flexibility, thereby intensifying the coupling effect of structure and aerodynamics, and the phenomenon of aeroelasticity is becoming more and more obvious. The gust load is often greater than the maneuvering load for the aircraft with the large aspect ratio layout, and the structure should be designed according to the most severe gust load, and most of the fatigue damage of the structure is caused by the gust load. At present, when calculating the gust of the UAV in engineering, the aircraft is regarded as a rigid body, and the overload response is obtained by solving the dynamic equations, without considering the elasticity of the aircraft, and the result has a certain deviation from the actual value. Therefore, the gust overload calculation of elastic aircraft is considered in this paper, which provides a basis for the design of aircraft structural strength, which has important engineering significance.

2. Gust response modeling

The aeroelastic equation of motion based on generalized coordinates can be written as formula (1) [1]:

$$M_q\ddot{q} + C_q\dot{q} + K_q q = \frac{1}{2}\rho V^2 Q_q q + \frac{1}{2}\rho V^2 Q_g w_g \tag{1}$$

Where, q is the generalized coordinates corresponding to the rigid body mode and the elastic mode of the aircraft, w_g is vertical gust velocity, ρ is atmospheric density, V is flight speed, M_q is the generalized mass matrix, C_q is the damping matrix, K_q is the generalized stiffness matrix, Q_q is the generalized unsteady aerodynamic coefficient matrix, Q_g is the generalized unsteady aerodynamic coefficient matrix corresponding to the gust.

The unsteady aerodynamic rational function is fitted by the minimum state (MS) method. The rational function fitting formula [2] based on the MS method is written as formula (2):

$$Q(s) = Q_0 + Q_1 s + Q_2 s^2 + \sum_{j=1}^{N} \frac{D_j E_j^T}{s+r_j} s \tag{2}$$

Where, Q_0, Q_1 and Q_2 are real coefficient matrix, D_j And E_j are real column vector, r_j is a real constant.

Transform the aeroelastic motion equation (1) excited by the gust into the Laplace domain, and replace Q_q and Q_g in equation (1) with Q(s) in equation (2), the state space equation of the elastic aircraft can be obtained as equation (3).

$$\begin{cases} \dot{x} = Ax + Bw \\ y_k = Cx + Dw \end{cases} \tag{3}$$

Where, $x = [q^T \quad \dot{q}^T \quad x_a^T]^T; w = [w_g \quad \dot{w}_g]^T; q = [T_z \quad R_y \quad q_1 \quad \cdots \quad q_n \quad x_a^T]^T; C = [\boldsymbol{\Phi}_k K_\varphi \quad \boldsymbol{\Phi}_k C_\varphi \quad \mathbf{0}]$. T_z denotes the generalized coordinates corresponding to the ups and downs mode of rigid body; R_y denotes the generalized coordinate corresponding to the pitching rigid body mode; q_i denotes the generalized coordinate corresponding to the elastic mode; N denotes the number of elastic modes; w_g denotes a gust entry; Φ_k denotes modal matrix of the node of interest; K_φ denotes the corresponding stiffness matrix; C_φ denotes corresponding damping matrix; A, B And D denotes the correlation matrix of the state-space equation; y_k denotes the acceleration response of the node of interest.

3. Modeling of severe wind field

3.1. Shape of wind field
The shape of the actual wind is ever-changing, but the configuration of discrete gust commonly used in engineering is "1-cos", such as the UAV specification [3], CCAR-23 [4], CCAR-25 [5], so the "1-cos" type is adopted as the configuration of the wind field in severe weather in this paper. It is converted to a form used by Nastran as formula. (4):

$$U = \frac{U_{de}}{2}\left[1 - \cos\left(\frac{\pi V t}{H}\right)\right] \tag{4}$$

Where V is the flight speed of the aircraft; t is the flight time; H is the gust gradient, which is the distance the aircraft travels when the gust velocity reaches its maximum value from zero.

3.2. Intensity of Wind field
Based on the detected typhoon weather data, the polyval function of MATLAB is used to model the original data reasonably, and the wind field intensity models at different altitude are obtained, as shown in the figure 1.

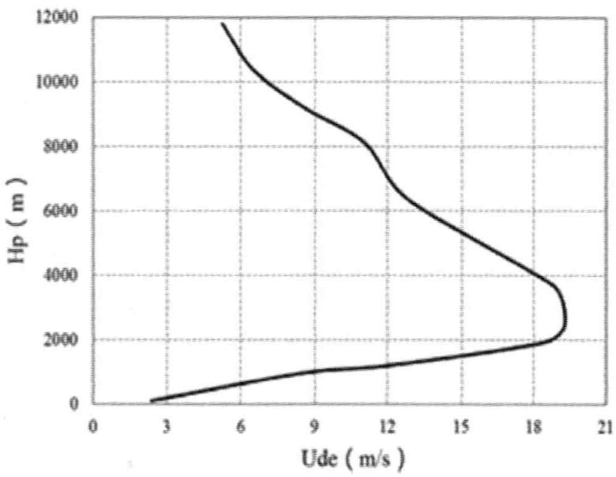

Figure 1. Variations of typhoon model intensity with altitude

3.3. Scale of wind field
At present, the scale of the wind field data obtained from typhoon detection is in the order of kilometers, which is not suitable for the calculation of the gust load of the UAV. analyze. In order to determine the appropriate wind field scale, the sensitivity analysis of wind field scale of UAV is carried out in this paper. In the CCAR-25 specification, the description of the wind field scale is: the gust scale must be adequately analyzed and calculated within the range of 9.1 m to 106.7 m, and the most severe load response is used as the gust design load. When determining the scale of the wind field model, this paper refers to this regulation and selects a series of scales: 9.1, 20, 30, 40, 50, 60, 70, 80, 90, and 106m. The comparison of the result curves is shown in Figure 2. The abscissa in the figure represents the scale of the wind field, and the ordinate represents the overload increment at the center of gravity of the UAV. The legend consists of 7 digits, representing the flight point, of which the first 3 digits represent the Mach number, the last 4 digits represent the flight height, such as 3670000, which means the flight state is "Ma=0.367, flight height=0m". It can be seen from the calculation result diagram that when the gust scale is 20m, the overload increment is the largest.

Therefore, 20m is used as the scale of the severe weather wind field to evaluate the gust overload response.

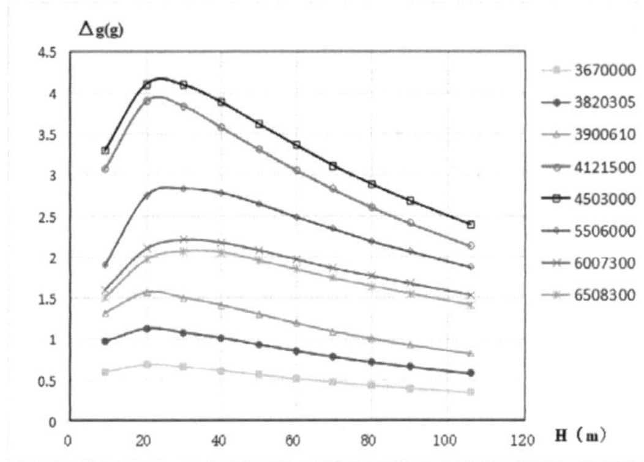

Figure 2. Variation of overload increment with the scale of wind field

4. Results and Analysis

4.1. Model

According to the above principles, the structural model and aerodynamic model of a large aspect ratio UAV are established. The structural model is a finite element model, as shown in Figure 3. In the finite element model, an aerodynamic mesh is established which is shown in Figure 4 and the dipole mesh method is used for unsteady aerodynamic calculations. In addition, it is necessary to filter out the wing surface nodes in the structural model, and establish the coupling relationship between the aerodynamic and structural models for data exchange. The filtered surface node is shown in Figure 5. Finally, the aerodynamic grid and the structural surface nodes are mapped to each other to establish an interpolation relationship. The interpolation method adopts the IPS method in Patran, as shown in Figure 6, and the exchange of coupling data between aerodynamic model and structural model can be realized. Thereafter, the aerodynamic mesh is output to the "aero.dat" file, the finite element model is output to the" stru.dat "file, and then the generated calculation control file "Discrete-Gust.bdf" is filled in according to the calculation conditions. Then, according to the gust file" CF. dat" corresponding to different gust configurations, the calculation is started.

Figure 3. Structural model

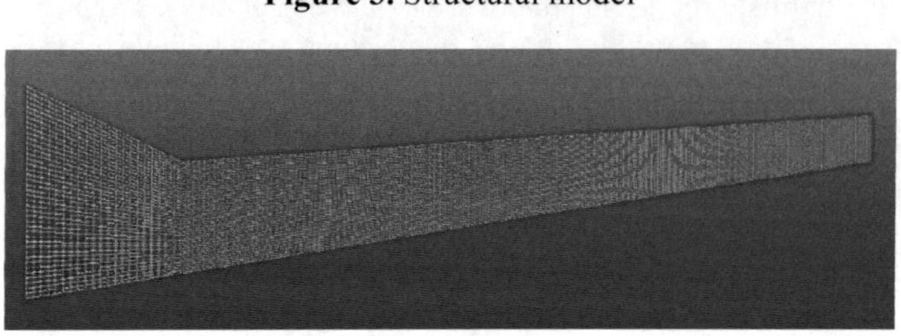

Figure 4. Aerodynamic grid

Figure 5. Surface node

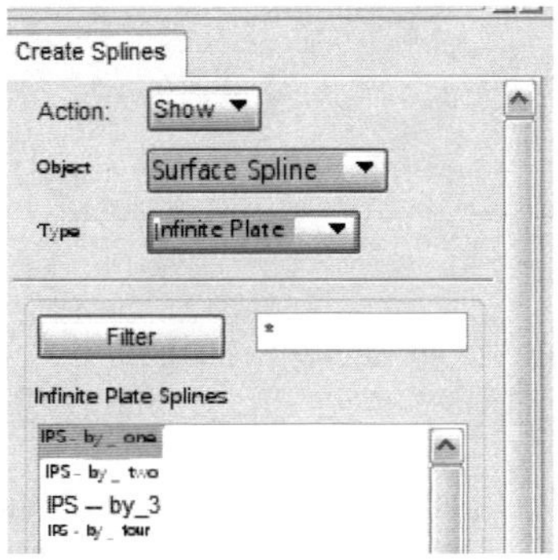

Figure 6. The interpolation relation is established

4.2. Calculation state

The gust load of the UAV is calculated and analyzed within the entire flight envelope. The calculation status points are shown in Figure 7 below. The abscissa in the figure represents the flight Mach number, the ordinate represents the flight altitude, and the legend represents the flight indicated airspeed (IAS) in km/h.

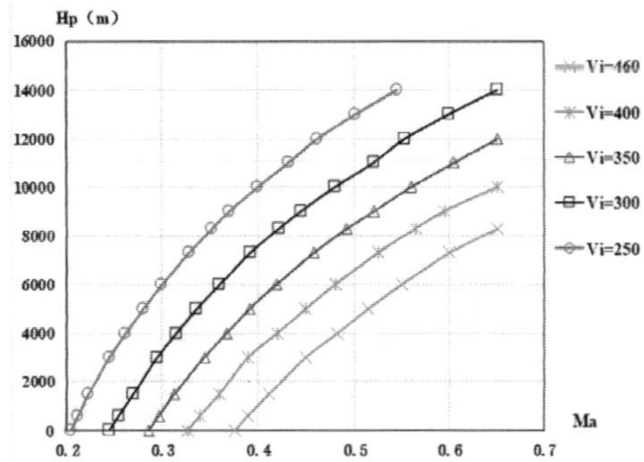

Figure 7. Calculated state point

4.3. Results

Based on the established severe wind field model, the overload increment at the center of gravity of the UAV in the full-envelope is shown in Figure 8. It can be seen that:

Under the same IAS, within the flight altitude range of 0~3000m, the overload increment increases with the increase of altitude. Near the height of 3000m, the overload increment reaches the maximum, and the maximum overload increment exceeds 4g, that is, the overload at the center of gravity exceeds 5g. Above 3000m, the overload increment decreases with the increase of height. After more than 8000m, the overload increment is less than 2g, that is the overload at the center of gravity is less than 3g.

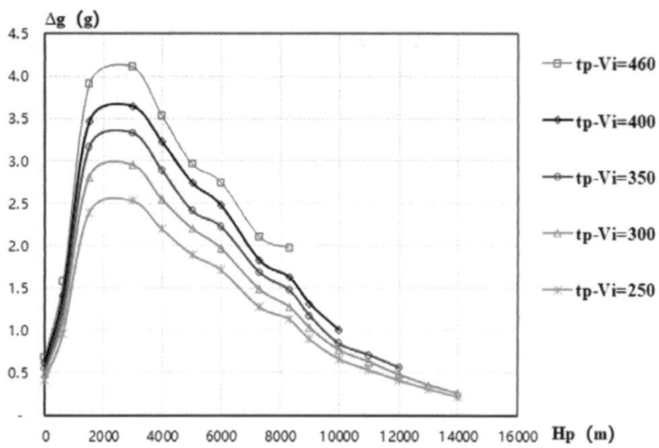

Figure 8. Maximum overload increment under different flight conditions

4.4. Comparative analysis
The calculation results based on the severe weather wind field model are compared with the results based on the general specification CCAR-23, as shown in Figure 9. It can be seen that the overload increment of the severe weather wind field model in the condition of the 14-level typhoon at the height of 1km~4km is greater than that of CCAR-23.

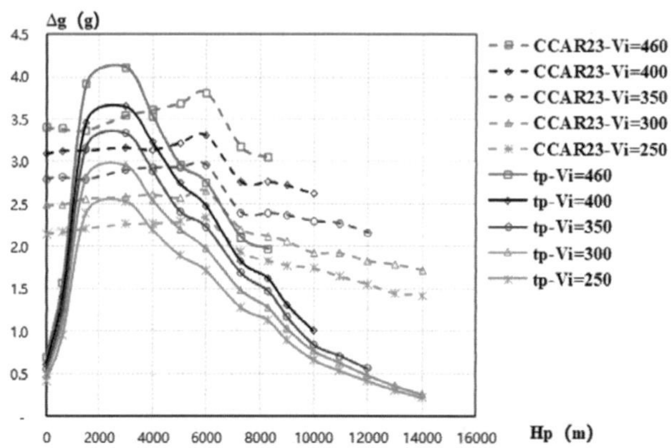

Figure 9. Comparison of overload increment

5. Conclusion
In this paper, the gust response calculation method considering aerodynamic elasticity and a wind field model for severe weather is completely established, and more realistic and accurate gust overload is obtained. Compared with the calculation results of the general specification CCAR-23, it shows that if the UAV needs to perform special tasks, such as typhoon detection, a more realistic wind field model needs to be used for further load evaluation to ensure flight safety. This method has great engineering application value.

References

[1] M Karpel, B Moulin, E Presente et. Al. Dynamic gust loads analysis for transport aircraft with nonlinear control effects [C]. 49th AIAA/ASME/ASCE/AHS/ASC Structures, Structural Dynamics, & Materials Conference 16th AIAA/ASME/AHS Adaptive Structures Conference 10t, 2008.

[2] D. Azoulay. M. Karpel. Characterization of methods for computation of aeroservoelastic systems response to gust excitation [C]. 47th AIAA/ASME/ASCE/AHS/ASC Structures, Structural Dynamics, and Materials Conference, 2006.

[3] General Armament Department of the Chinese People's Liberation Army. GJB5435.10-2005 Unmanned Aerial Vehicle Strength and Rigidity Specification Flight Load. 2005.10.

[4] Civil Aviation Administration of China. China Civil Aviation Regulation Part 23 Airworthiness Standard for Normal, Utility, Aerobatic and Commuter Aircraft CCAR-23.2005.01.

[5] Civil Aviation Administration of China. China Civil Aviation Regulation Part 25 Airworthiness Standard for Transport Aircraft CCAR-25.2001.10.

Design of Ducted Fan Based on Optimal Circulation Distribution

Zilun Zhang[1,*], Mingmin Bao[2], Nanxuan Qiao[3] and Tielin Ma[4]

[1] Shanghai Electro-Mechanical Engineering Institute, Shanghai, China
[2] Key Laboratory of Rotorcraft Aeromechanics, China Helicopter Research and Development Institute, Jingdezhen, China
[3]School of Aeronautic Science and Engineering, Beihang University, Beijing, China
[4]Institute of Unmanned System, Beihang University, Beijing, China

*Corresponding author e-mail: zzl666@buaa.edu.cn

Abstract. The ducted fan lift system has been widely used in vertical takeoff and landing (VTOL) unmanned aerial vehicle (UAV). According to the overall design requirements of a VTOL UAV, based on the optimal circulation distribution condition without blade element drag and combined with strip theory, the blade torsion angle distribution and chord length distribution are given. The parameters of ducted fan are designed, and the preliminary design of ducted fan lift system of this UAV is completed. The aerodynamic characteristics of the isolated fan and the ducted fan are compared by numerical simulation method, and the thrust characteristics of the ducted fan at different flight speeds are calculated. The results show that the duct increases the induced velocity at the propeller. Compared with the isolated fan, the power load and hover efficiency of the ducted fan are increased by 36.8% and 36.3% respectively. The hovering efficiency of the ducted fan designed by this method reaches 83%, which meets the design requirements and proves the effectiveness of the design method.

1. Introduction

VTOL UAV combines the characteristics of multi rotor aircraft and fixed wing aircraft and shows unique advantages. For example, it doesn't need a special runway, which reduces the requirements of takeoff and landing site. It doesn't need a launch recovery system, whicth reduces the use cost. It overcomes the disadvantages of slow flying speed, poor wind resistance and short range of multi rotor aircraft, and expands its application range. According to the current market demand, VTOL UAV can be applied in both military and civil fields.

Ducted fan refers to a propulsion device composed of an open propeller surrounded in the duct[1]. Compared with an isolated propeller of the same diameter, due to the enclosing effect of the duct, the formation of blade tip vortex is restrained and the wake energy loss is reduced. At the same time, the ducted lip also makes a certain contribution to the thrust, so that it can produce greater thrust and have higher aerodynamic efficiency when consuming the same power. It also has the advantages of low aerodynamic noise, compact structure and good safety. Therefore, as a propulsion device, it has been applied in VTOL aircraft since 1950s. The ducted VTOL UAV is to use the duct to ring the lifting components of the UAV, so that the UAV can realize VTOL and hovering through the thrust provided by the ducted fan device. Because the ducted fan is used as the lift system, this kind of UAV can reduce the size of the whole machine and has a compact structure. At the same time, the rotating parts have ducted protection and high safety. In the process of VTOL and hover, the thrust is generated by ducted fan. There are two main ways to generate lift in level flight. One is that the ducted fan deflects the thrust axis to the fuselage axis close to the horizontal attitude to provide forward power, and the lift is generated by the wing. The other is that the ducted fan doesn't deflect at a large angle, but still generates all the lift, and the forward power is provided by another separate propulsion device, or the thrust axis of the ducted fan tilts forward, and the forward power is generated by the horizontal component of the thrust.

As early as the 1960s, the United States and France developed two ducted VTOL aircraft x-22a and Nord 500 respectively[2, 3]. The ducted fan can tilt, which can not only generate the thrust required for VTOL, but also provide the pull force required for horizontal flight. Martin UAV has developed a V-bat VTOL fixed wing UAV[4]. A ducted fan propulsion device is installed at the tail of its fuselage, which can provide vertical upward thrust during takeoff. At the same time, a control surface is installed at the outlet of the ducted fan, which can be used to change the direction of thrust to realize the transition from VTOL to horizontal flight. In the VTOL project of the Defense Advanced Research Projects Agency(DARPA)[5, 6], the Phantom Swift of Boeing and the Lightning Strike of Aurora adopt the ducted fan as the lift system. The Phantom Swift is equipped with two ducted fans in the middle of the fuselage and at the wing tip, The fan on the fuselage is used to provide thrust during VTOL and the fan on the wing tip is used to provide the control force during transition. The Lightning Strike embeds 24 ducted fans into the wing and canard which can tilt to change the direction of thrust. Finally, Lightning Strike of Aurora won the project bidding. At present, there have been a lot of research on the design and aerodynamic characteristics of ducted fan. Fan Zhongyun et al.[7] proposed a propeller design method which can arbitrarily give the circulation distribution. Angelo et al.[8] determined the chord length distribution and torsional angle distribution of the blade by giving the blade working state and profile airfoil distribution. Serdar et al.[9] conducted experimental research on the performance and flow field of ducted fan in hovering and horizontal flight. Graf[10] studied the effects of different lip shapes and control rudders on the aerodynamic characteristics of ducted fan through wind tunnel test. Chen Tianyu et al.[11] studied the aerodynamic characteristics of tilt rotor aircraft in transition state based on the momentum source method.

As the lift system of ducted VTOL UAV, the performance of the ducted fan greatly affects the flight performance of the whole aircraft. How to design the ducted fan lift system accurately and efficiently according to the overall design requirements is an important content of the overall design of ducted VTOL UAV. Combined with the overall design requirements of a VTOL UAV, this paper gives the effective angle of attack distribution that makes all blade elements reach the maximum lift-to-drag ratio, and combined with the optimal circulation distribution based on no blade element drag, and gives the blade torsion angle distribution and chord length distribution through the strip theory. Finally, the preliminary design of the ducted fan lift system of the VTOL UAV is completed, and its aerodynamic characteristics are analyzed by numerical simulation.

2. Blade design method based on optimal circulation distribution

The blade design needs to design the geometric parameters according to the given thrust demand and power limit, mainly including the radial distribution of blade airfoil, chord length and torsion angle. In this paper, the optimal circulation distribution without blade element drag under the given conditions is designed, and then the blade geometric parameters are obtained by combining the strip theory.

According to the blade element theory, ignoring the friction generated by the blade element, the efficiency of the fan can be expressed as:

$$\eta = \frac{TV_0}{2\pi n_s M} \quad (1)$$

In this equation, these symbols are respectively expressed as: the thrust generated by the fan T, the incoming flow speed V_0, the fan speed n_s and the fan torque M. When there is a small increment $\Delta\Gamma$ in the circulation of the blade element at the blade diameter, the force generated by the blade element will change, and the thrust and torque of the whole blade will also increase. Therefore, the useful work will increase and the power absorbed by the blade will increase. The ratio of the two is expressed by:

$$k = \frac{\Delta TV_0}{2\pi n_s \Delta M} \quad (2)$$

When the position of blade element increasing circulation is different, k will also be different, In order to keep the added energy constant and obtain better efficiency, it is necessary to increase the circulation where the k is large and reduce the circulation where the k is small. Therefore, the optimal distribution of blade circulation along the radial direction is to make the k constant, which is the condition for the maximum efficiency of Betz propeller[12]. The stress of blade element is shown in the following figure:

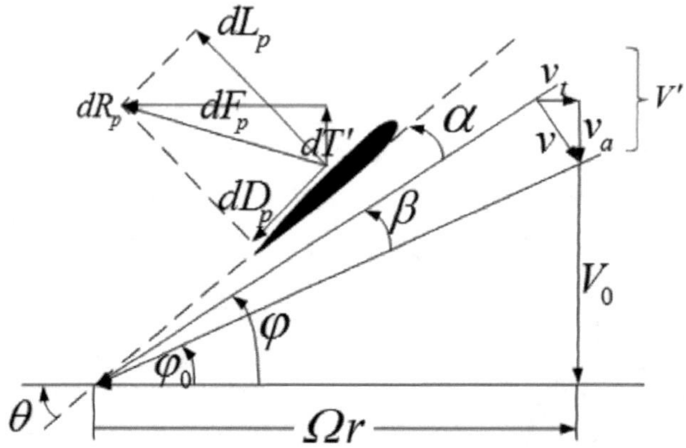

Figure 1. Stress analysis of leaf element

In the figure, these symbols are respectively expressed as: the blade element installation angle θ, the interference angle β generated by the blade tail vortex, the blade element effective angle of attack α, the induced pitch increment V', the axial induced velocity va at the propeller, the circumferential induced velocity vt at the propeller, the angular velocity Ω. The thrust and torque generated due to the increase of circulation can be expressed as:

$$\begin{cases} \Delta T = \rho \Delta \Gamma (\Omega r - vt) dr \\ \Delta M = \rho \Delta \Gamma (V_0 + va) r dr \end{cases} \tag{3}$$

Substituting equation 3 into equation 2 and combining with the geometric relationship in Figure 1, we can get:

$$V' = \frac{va + V_0}{\Omega r - vt} vt + va \tag{4}$$

Assuming that the number of blades is N_B, the relationship between induced velocity and circulation at the propeller is as follows[13]:

$$\begin{cases} va = \dfrac{N_B \Omega \Gamma(r)}{4\pi (V_0 + V')} \\ vt = \dfrac{N_B \Gamma(r)}{4\pi r} \end{cases} \tag{5}$$

Substituting equation 5 into equation 4, after simplification, the expression of circulation can be obtained as follows:

$$\Gamma(r) = \frac{4\pi (V_0 + V') V' \Omega r^2}{N_b \left[(V_0 + V')^2 + \Omega^2 r^2 \right]} \tag{6}$$

When V' is given, the radial distribution of circulation can be obtained, and the circulation distribution meets the maximum efficiency condition of Betz propeller.

According to the strip theory, the blade element circulation can also be expressed as[12]:

$$\Gamma(r) = \frac{1}{2} V_{real} C_L(r) b(r) \tag{7}$$

V_{real} is the actual air velocity of blade element. The implicit equation for solving the interference angle is:

$$C_L \sigma = \frac{4 \sin(\phi_0 + \beta) \tan \beta}{1 - \tan \gamma \tan \beta} \tag{8}$$

In this equation, these symbols are respectively expressed as: the blade compactness σ, the geometric angle of attack ϕ_0, the blade drag lift angle γ.

At the beginning of the design, it is necessary to select the appropriate blade airfoil, effective angle of attack distribution and circulation distribution according to the incoming flow environment, fan disk area and other parameters. Therefore, the blade chord length distribution can be obtained through equation 7, which can be substituted into equation 8, the interference angle distribution can be obtained through Newton iteration method, and then the blade torsion angle distribution can be obtained to determine the blade shape, The blade obtained by this method meets the optimal circulation distribution condition without blade element drag.

3. Parameter design of ducted fan

ARA-D is selected as the blade airfoil. According to the overall design requirements of a VTOL UAV in hovering state[14], the design constraints are as follows: the fan disk radius is 0.315m, the incoming flow velocity is 0, the hovering efficiency is not less than 75%, and the target thrust is not less than 2500N. Given the effective angle of attack distribution that makes all blade elements reach the maximum lift-to-drag ratio, the optimal circulation distribution without blade element drag shown in Figure 2 can be obtained according to the constraints, and then the chord length distribution and torsion angle distribution of the blade can be obtained, as shown in Figure 3.

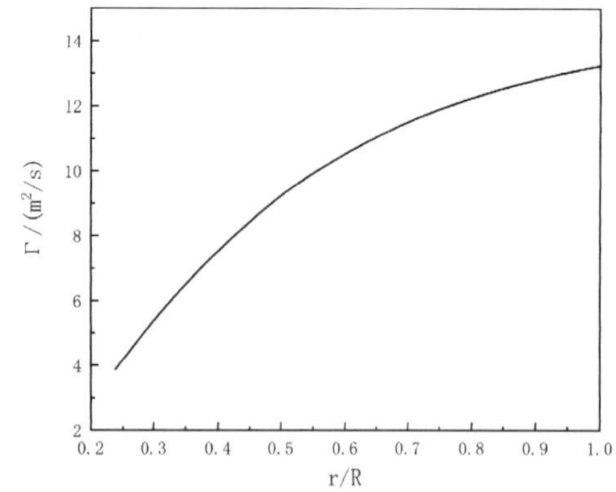

Figure 2. Optimal circulation distribution

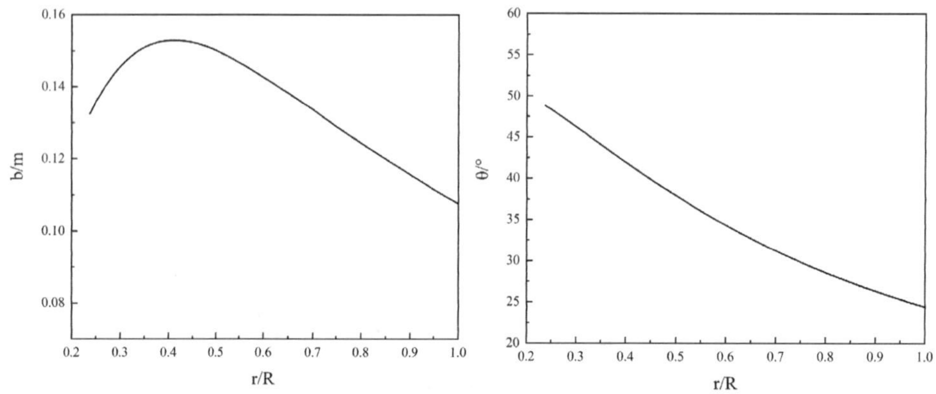

(a) chord length distribution (b) torsion angle distribution

Figure 3. Blade geometric parameters

Figure 4 shows the geometric parameters of ducted fan. These symbols are respectively expressed as: the ducted lip radius R, the tip clearance d_{tip}, the duct inner diameter D, the duct height C, the distance h between the propeller and the duct lip, and the diffusion angle β. The parameters related to duct are shown in the table below:

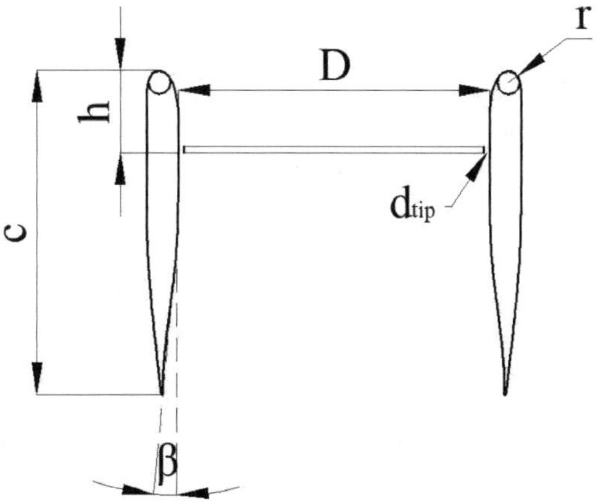

Figure 4. Geometric parameters of ducted fan

Table 1. Geometric parameters of duct

D(m)	c(m)	d_{tip} (mm)	r/c	h/c	β(°)
0.63	0.46	3	0.047	0.08	3

4. Analysis of aerodynamic characteristics of ducted fan

Different from the open environment of the isolated fan, the air flow in the ducted fan first passes through the rectification of the duct before reaching the fan disk. Therefore, the flow field characteristics at the fan are different from those of the isolated fan. The aerodynamic characteristics of the isolated fan and the ducted fan are compared by numerical simulation. The computational domain of the ducted fan is divided into three basins: blade, duct and outflow field, and the structural mesh is divided respectively. The flow field information of each field is coupled and exchanged through the interface boundary conditions. The multiple reference frame(MRF) model is used for steady simulation. The fan speed in hovering state is 7000rpm, and the calculation model is shown in Figure 5. The overall mesh volume is 12 million.

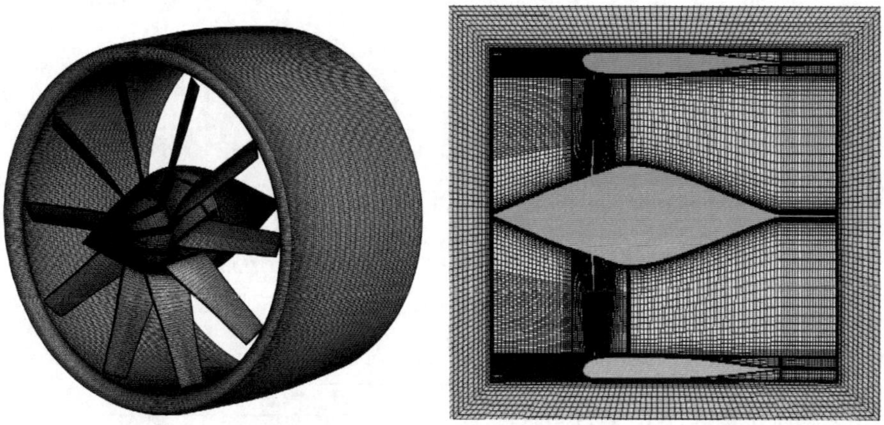

（a）surface mesh of global model （b）lap mesh of each field

Figure 5. Mesh model

Figure 6 shows the comparison of the velocity changes before and after the isolated fan and the ducted fan. It can be seen that the existence of the duct increases the induced velocity at the fan disk, especially near the fan tip. The increase of the induced velocity reduces the effective angle of attack of the blade resulting in the decrease of thrust. Figure 7 shows the relationship among thrust, power and speed of ducted fan under the incoming flow speed of 0 ~ 200km/h. At the same time, the performance of ducted fan in this state is compared with that of isolated fan. As shown in Table 2, it is found that the power load of ducted fan is increased by 36.8% and the hover efficiency is increased by 36.3%. Although the thrust decreases, the additional thrust generated by the duct increases the total thrust, so it has higher power load and hover efficiency. It can be seen from the data in Figure 7 and Table 2 that the hover efficiency of the ducted fan is 83% and the maximum thrust is 2593N, which meets the design requirements and proves the effectiveness of the design method.

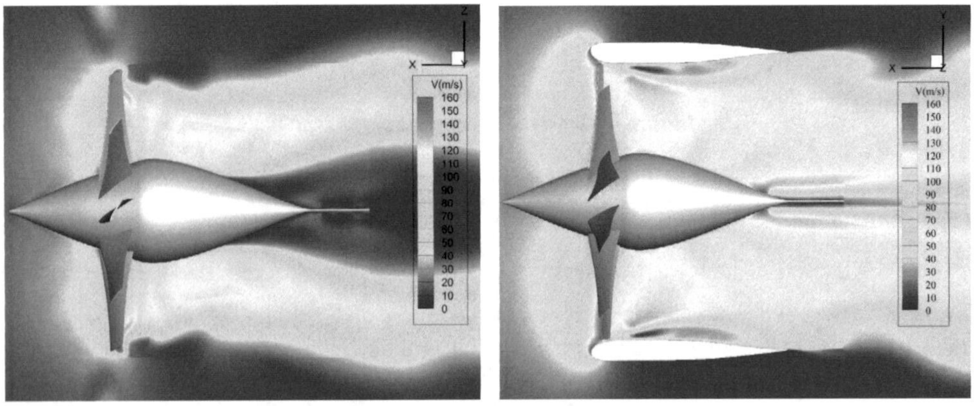

（a）isolated fan （b）ducted fan

Figure 6. Comparison of velocity changes before and after isolated fan and ducted fan

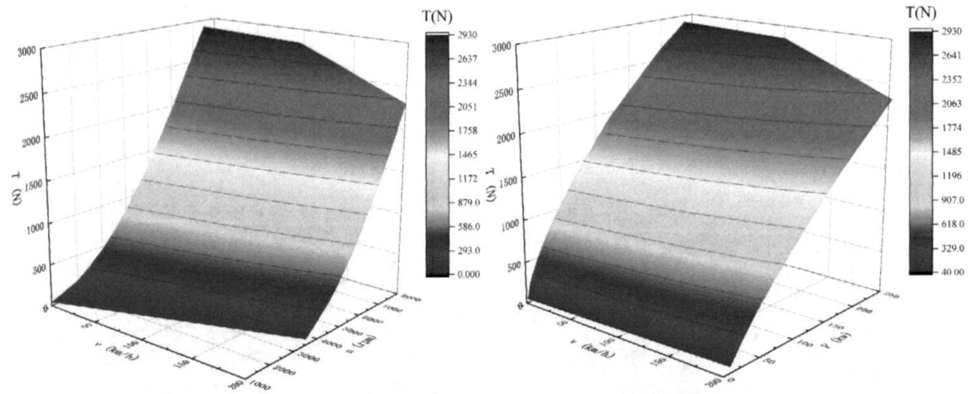

（a）thrust, incoming flow speed, rotating speed （b）thrust, incoming flow speed, power

Figure 7. Thrust characteristics of ducted fan

Table 2. Comparison of calculation results between isolated fan and ducted fan

	T(N)	M(N·m)	P(kw)	Power load(kg/kw)	η
Isolated fan	2534	327	240	1.06	0.609
Ducted fan	2593	248	182	1.45	0.83

5. Conclusion

Based on the optimal circulation distribution condition without blade element drag and combined with strip theory, this paper completes the preliminary design of the ducted fan lift system of a VTOL UAV. The aerodynamic characteristics of ducted fan are analyzed by numerical simulation, and its thrust performance is calculated. The results show that the duct increases the induced velocity at the fan disk and the fan thrust decreases, but the additional thrust generated by the duct increases the overall thrust. Compared with the isolated fan, the power load of the ducted fan is increased by 36.8% and the hover efficiency is increased by 36.3%. The maximum thrust and hover efficiency of the ducted fan can meet the overall design requirements, which proves the effectiveness of the design method. The research results can provide relevant reference for the design of the lift system of VTOL UAV.

References

[1] Xiao Senlin, Qiao Hongxin, Xu Guanxiong, et al. Optimized Matching Lift Unit Transmission Ratio of Engine Driven Ducted Fan[J]. MATEC Web of Conferences, 2018, 166: 04003.

[2] V.B. Paxhia, E.Y. Sing. X-22a Design Development[J]. Journal of Aircraft, 1965, 2(1): 2-8.

[3] H.A. Smith. French Nord Aviation VSTOL Aircraft[R]. 1969.

[4] P.R. Newswire. Martin UAV to Demonstrate Their Upgraded V-BAT at the Army Expeditionary Warrior Experiment [J]. PR Newswire US., 2021.

[5] T. Yasmin. X-Plane may impact future vertical lift program[J], National Defense, 2016, C(750): 18.

[6] B. Mark, J. Mike, W. David. DARPA begins VTOL X-Plane competition[J], Air international, 2014, 5, 26

[7] Fan Zhongyun, Zhou Zhou, Zhu Xiaoping. A Design Method for Propeller with Arbitrary Circulation Distribution[J]. Journal of Aerospace Power, 2019, 34(2): 434-441.

[8] S.D. Angelo, F. Berardi, E. Minisci. Aerodynamic Performances of Propellers with Parametric Considerations on the Optimal Design[J]. The Aeronautical Journal, 2002, 106(1060): 313-320

[9] S. Yilmaz, D. Erdem, M. Kavsaoglu. Effects of Duct Shape on a Ducted Propeller Performance[C]//Aiaa Aerospace Sciences Meeting Including the New Horizons Forum & Aerospace Exposition. 2013.

[10] W.E. Graf. Effects of Duct Lip Shaping and Various Control Devices on the Hover and Forward Flight Performance of Ducted Fan UAVs[D]. USA: Virginia Polytechnic Institute and State University，2005.

[11] Chen Tianyu, Xu Yi, Liang Zhuang, et al. Momentum-source Method and Analysis on Aerodynamic Characteristics for Quad Tilt Rotor Aircraft in Conversion State[J]. Air & Space Defense, 2020, 3(1): 17-22.

[12] Liu Peiqing. Theory and Application of Air Propeller[M]. Beijing: Beihang University Press, 2006.

[13] Cao Yihua. Modern Helicopter Rotor Aerodynamics[M]. Beijing: Beihang University Press, 2015.

[14] Zhang Zilun, Ma Tielin, Hao Shuai, et al. Design of A Distributed Propulsion VTOL UAV[C]. IEEE International Conference on Unmanned Systems. IEEE, 2019

Application research on matching design of suspended AUV propeller based on CFD

Pengxing Wu*, Xianlong Liu, Zhengcheng Ling and Cheng Chen

Kunming Ship Equipment Research and Test Center, Kunming, China

*Corresponding author e-mail: wupengxing@nuaa.edu.cn

Abstract. Combined with practical engineering application cases, this paper introduces a matching design method of suspended AUV propeller based on CFD. From theoretical design to practical engineering application, it is described from five aspects: demand analysis, design process, CFD simulation and experimental verification. The final results show that the designed propeller fully meets the application requirements, and this method is fully suitable for engineering practical application.

1. Introduction

At present, with the in-depth development and utilization of marine resources, AUVs play an increasingly important role as a key tool, and how to ensure long-endurance and low-energy operation of AUVs has become one of the main development directions [1]. Among them, the fixed-depth suspension of AUV is the depth control at low speed or zero speed, which enables the AUV to realize virtual anchoring at a certain depth for long-term fixed-point observation. At present, there are two main ways to achieve fixed-depth suspension: the propeller adjustment system and the buoyancy adjustment system [2]. The buoyancy adjustment system maintains the depth for a long time, but the structure is relatively complex, and it takes a certain time to change the buoyancy, resulting in a slow dynamic response. Relying on thrusters to achieve depth determination has the advantages of fast dynamic response and simple structure.

Aiming at the specific AUV fixed depth suspension problem, this paper selects the propeller adjustment method, and studies the application of propeller matching design of suspended AUV Based on CFD. The final test results show that the propeller designed in this paper can well meet the requirements of AUV fixed depth suspension.

2. Demand analysis

The diameter of the AUV used in this paper is about 324mm and the length is 2400mm. The head and tail lines are similar to streamline. The tail is equipped with a full-motion rudder, and the rudder spread exceeds the maximum outer diameter of the AUV. The AUV cabin is equipped with electronic equipment such as acoustics and communications. The overall structure is shown in Figure 1. It only needs to be suspended for a short time during use.

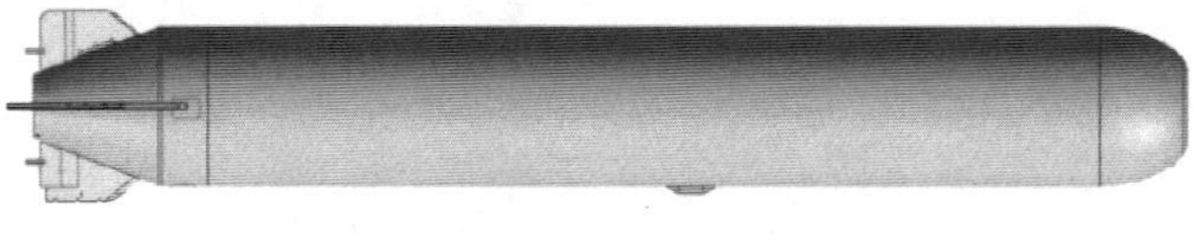

Figure 1. Overall outline of AUV.

During the design process, the restoring force generated by the center of gravity and the center of buoyancy of the AUV is particularly important for the attitude adjustment of the AUV center of gravity vector $f_G = J_1^{-1}(\eta_2)\begin{bmatrix}0 & 0 & W\end{bmatrix}^T$, buoyancy vector $f_F = -J_1^{-1}(\eta_2)\begin{bmatrix}0 & 0 & F\end{bmatrix}^T$, then its restoring force and moment vector can be expressed as:

$$g(\eta) = -\begin{bmatrix} f_G(\eta) + f_F(\eta) \\ r_G \times f_G(\eta) + r_F \times f_F(\eta) \end{bmatrix} \tag{1}$$

$$\begin{cases} r_F = \begin{bmatrix} x_F & y_F & Z_F \end{bmatrix}^T \\ r_G = \begin{bmatrix} x_G & y_G & Z_G \end{bmatrix}^T \\ W = mg \\ F = \rho g v \end{cases} \tag{2}$$

Among them, r_F and r_G respectively represent the coordinates of the center of buoyancy and the center of gravity of the AUV in the motion coordinate system. Combining the above results:

$$g(\eta) = -\begin{bmatrix} (W-F)\sin\theta \\ -(W-F)\cos\theta\sin\phi \\ -(W-F)\cos\theta\cos\phi \\ (y_g W - y_f F)\cos\theta\cos\phi + (z_g W - z_f F)\cos\theta\sin\phi \\ (z_g W - z_f F)\sin\theta + (x_g W - x_f F)\cos\theta\cos\phi \\ -(x_g W - x_f F)\cos\theta\sin\phi - (y_g W - y_f F)\sin\theta \end{bmatrix} \tag{3}$$

In the early overall design process of the AUV in this case, through the design analysis of the overall shape, material, center of gravity and buoyancy in the 3D software, it is ensured that the overall AUV is negatively buoyant, with a size of 40 ± 5N, and the restoring torque $g(\eta)$ is sufficient to meet the needs of use. The propulsion motor used has been determined, rated power 550W, rated speed 3000rpm. In the above case, in order to meet the suspension requirements, the propeller should be designed according to the actual situation.

3. Demand analysis

Through the analysis of specific requirements, the thrust generated by the propeller in this case is mainly used to overcome gravity. The propeller design in this paper starts from the initial propeller scheme. According to the initial propeller diameter D and pitch P/D, the propeller hydrodynamic simulation is carried out, and the thrust and resistance balance is achieved through multiple iterative design, and then the motor matching is carried out. When everything meets the conditions, The matching design process is over, and the overall process is shown in Figure 2.

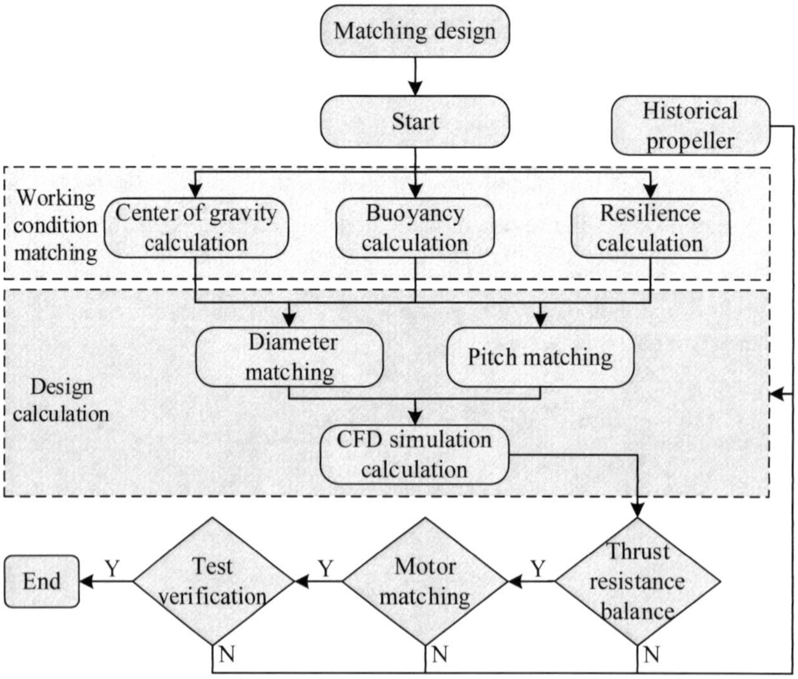

Figure 2. Matching design process

4. CFD simulation

4.1. Physical model

In this paper, the Ka type propeller is used. Due to the lack of relevant model data verification, in order to reduce the risk of design, the propulsion method of a single propeller is selected. The main parameters of propeller are shown in Table 1 [3].

Table 1. Main parameters of propeller

D(m)	$P_{0.7}/D$	A_B/A_0	D_h/D	Z	Turn around	Side angle
0.15	1.2	0.5	0.32	3	Right	0°

4.2. Watershed and meshing

This paper divides the watershed into dynamic domain and static domain, and adopts the numerical calculation method of multiple reference frame MFR [4]. Among them, the area where the propeller is located is the dynamic domain, and the area outside the propeller is the static domain. Both the watershed where the propeller is located and the external watershed are divided into cylindrical regions, the length and scale of which are shown in Figure 3. The whole is divided into 3 parts for block meshing. Due to the complex propeller model, it is difficult to generate high-quality structured meshes. Unstructured meshes are easier to achieve local refinement. Therefore, tetrahedral elements are selected for grid division in areas 1 and 2, and hexahedral elements are selected for grid division in areas 3. At the same time, the mesh refinement process is carried out on the leading edge, the trailing edge, the blade tip and the blade root of the blade.

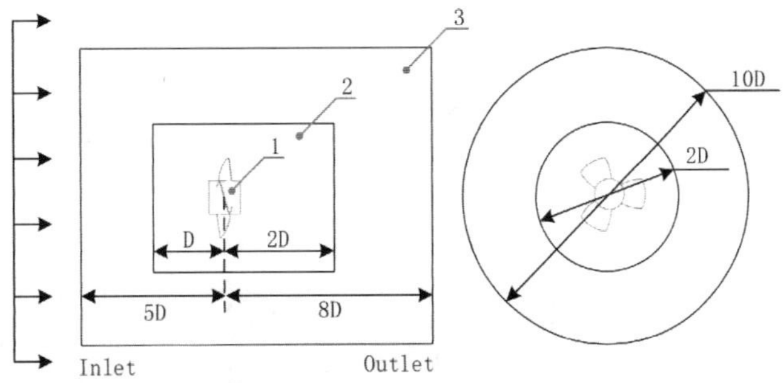

Figure 3. Watershed division

The final mesh model division is shown in Figure 4. There are about 300,000 nodes and 1.26 million elements on the wall, and the average quality of the elements is greater than 0.90.

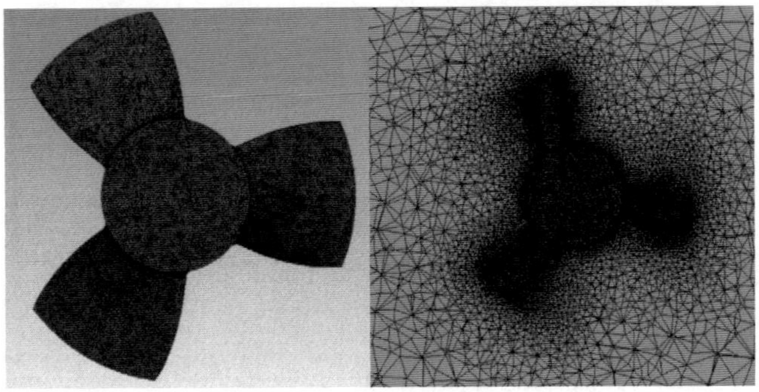

Figure 4. Propeller meshing

4.3. Boundary conditions
When setting the boundary conditions, only considering the open water test of the propeller in steady state, the numerical simulation calculation of the hydrodynamic performance of the propeller is carried out. Among them, cavitation, impurities, heat transfer and other factors are not considered; the fluid is assumed to be a pure, constant temperature, incompressible water medium; the influence of gravity, water surface wave making, pool wall and other factors is not considered, and it is treated as the problem of infinite water. The turbulence model adopts the SST k-ω model, which is a hybrid model, which combines the advantages of the k-ω model and the k-ε model, and makes up for the shortcomings of the two models [5]. Its equation can be expressed as:

$$\frac{\partial}{\partial t}(\rho k)+\frac{\partial}{\partial x_i}(\rho kui)=\frac{\partial}{\partial x_j}\left(\Gamma_k\frac{\partial k}{\partial x_j}\right)+G_k-Y_k \tag{4}$$

$$\frac{\partial}{\partial t}(\rho\omega)+\frac{\partial}{\partial x_j}(\rho\omega uj)=\frac{\partial}{\partial x_j}\left(\Gamma_\omega\frac{\partial \omega}{\partial x_j}\right)+G_\omega-Y_\omega+D_\omega \tag{5}$$

Where G_K is the turbulent kinetic energy generation term calculated according to the average velocity gradient; ρ is density; k is turbulent kinetic energy; u_i is the average turbulent velocity; x_j is coordinate component; ω is turbulence and special dissipation; D_ω is cross diffusion term.

The numerical calculation adopts the same boundary conditions: the inlet is uniform flow, the outlet adopts the static pressure condition, and the blade and the hub adopt the non-slip smooth adiabatic wall condition.

4.4. Analysis of simulation results

On the basis of the above preparations, the performance of the propeller is simulated and calculated, in which the propeller speed vector and streamline diagram are shown in Figure 5. In the case of different rotational speeds, the thrust conditions are shown in Table 2. It can be seen that when the rotational speed is greater than 1100, the propeller theoretically meets the requirements.

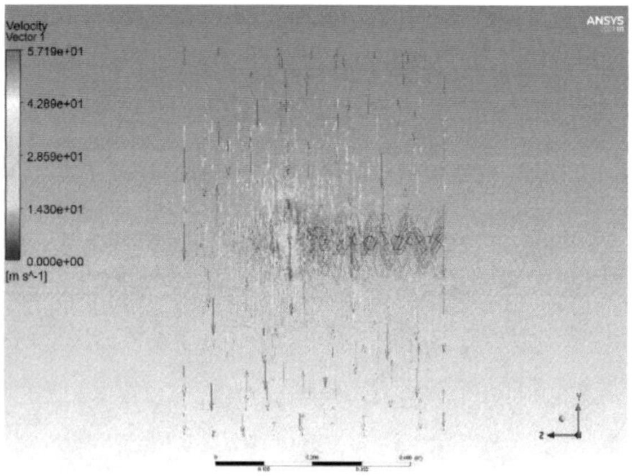

Figure 5. Velocity vectors and streamlines

Table 2. Propeller thrust simulation table

Rotating speed (rpm)	500	600	700	800	900	1000	1100	1200	1300	1400
Thrust(N)	14.2	18.1	25.0	29.3	37.8	42.2	55.1	63.8	76.5	91.3

5. Test verification

The designed propeller is processed into a real propeller through 3D printing technology, and the bollard thrust test is carried out in the pool, as shown in Figure 6. The motor speed is set from 500rpm. It can be seen that when the speed is 1100rpm, the propeller thrust is greater than 45N, and the error with the CFD simulation result is less than 5%, as shown in Table 3.

Figure 6. Pool thrust test.

Table 3. Actual thrust of propeller

Rotating speed (rpm)	500	600	700	800	900	1000	1100	1200	1300	1400
Thrust(N)	12.9	16.8	24.3	28	35.5	40.2	52.4	60.8	75.5	88.8

6. Conclusion

In this paper, from the perspective of engineering application, based on CFD technology, the floating AUV propeller is matched and designed, and the designed propeller is tested in the pool by 3D printing technology. The results show that the proposed method and design model fully meet the actual needs, and can be used for engineering practice. for reference.

References

[1] Zheng Z, Lian L, Karl S, Fangpo H, Youhong T, Andrew L. A survey on path planning for persistent autonomy of autonomous underwater vehicles, Ocean Engineering. 110 (2015) 303-313.

[2] Wilde P, Berry W. The role of oceanographic factors in the generation of global bio-events, Springer Berlin Heidelberg. 8(1986)75-91.

[3] Felicjancik J, Kowalczyk S, Felicjancik K, et al, Numerical Simulations Of Hydrodynamic Open-Water Characteristics Of A Ship Propelle, Polish Maritime Research. 23(2016) 16-22.

[4] Franzke R, Sebben S, Bark T, et al, Evaluation of the Multiple Reference Frame Approach for the Modelling of an Axial Cooling Fan, Energies. 12(2019) 29-34.

[5] Pang A, Skote M, Lim S Y, Modelling high Re flow around a 2D cylindrical bluff body using the k-ω (SST) turbulence model, Progress in Computational Fluid Dynamics An International Journal. 16(2016) 48-57.

Numerical investigation of the aerodynamic interference in 2-aircraft formation flight

Xingda Cui, Yonggang Yu, Shuai Ma, Zhongyun Xiao*, Lu Zhang and Fang Liu

Computational Aerodynamics Institute, China Aerodynamics Research and Development Center, Mianyang, China

*Corresponding author e-mail: xiaozy@cardc.cn

Abstract. Due to obvious reduction in drag, formation flight attracts wide attention of major powers in the world. Based on wild geese flying in a 'V' formation, a formation composed of two flying-wing aircrafts is designed. The aerodynamic interference in 2-aircraft formation flight is studied by CFD. Results show that 2-aircraft formation flight has little impact on leading aircraft, but has an evident impact on following aircraft. At the same AoA, the effect of increment in lift and reduction in drag of following aircraft is large. As considering the balance of gravity and lift, the reduction in drag of following aircraft is much larger, that is up to 25.1%. Under the aerodynamic interference effect, the following aircraft has a nosedown pitching moment increment, and becomes lateral static instability, which is opposite to the characteristic of solo flying-wing aircraft. The aerodynamic interference mainly comes from upwash flow induced by wingtip vortex of leading aircraft, which reduces the actual AoA and induced drag of following aircraft, and changes the lift distribution of both sides of following aircraft.

1. Introduction

The performance and efficiency of aircraft in formation flight for complex tasks are much higher than that of solo aircraft, which can make up for the deficiency of solo aircraft in the implementation of reconnaissance, search, battle, defense and other tasks. Therefore, formation flight has attracted wide attention from military powers around the world [1]. In addition, formation flight, especially in close proximity, can bring considerable aerodynamic advantage and save energy with the relatively minimal change to the current infrastructure [2].

However, due to the short distance between the aircrafts, there exists serious aerodynamic interference in the close formation flight, which not only brings aerodynamic advantage, but also has great impact on the aircraft attitude control [3]. There are many studies on the aerodynamic interference of multi-aircrafts including theoretical analysis [4-6], numerical calculation [7, 8], wind tunnel experiment [9-12], and formation flight experiment [13-15].

Hansen et al. [16] conducted formation flight experiments using two F/A-18 aircrafts. The results show that the drag of trailing aircraft decreases about 22% at the optimum position. Pahle et al. [17] carried out a two-aircraft formation flight experiment for the C-17 large transport-class vehicles, the flight data suggests that under the influence of the lead's vortex, formation flight saves more energy than single flight, and engine thrust and fuel flow are reduced by more than 10%. These formation flight experiments are costly and dangerous in reality. Liu et al. [18] carried out wind tunnel experiments on a two-aircraft formation consisting of a wing-body aircraft and a flying-wing aircraft. The results show that the following aircraft can obtain the maximum lift-drag ratio when the wing-tip vortex of leading aircraft is near the wingtip of following aircraft. However, there exists support interference in the wind tunnel experiment. In addition, when the number of formation aircraft is large, the implementation of wind tunnel experiment is difficult, and even impossible to complete. Yuan et al. [19], Liu et al. [20] adopted CFD method based on RANS equation to study the aerodynamic interference and obtained the aerodynamic benefits of two-aircraft formation flight.

Most of the above studies on formation flight mainly focus on the aerodynamic benefits, such as drag reduction, instead of the attitude of aircraft, such as pitch and roll. In addition, the balance of gravity and lift is not considered in these studies. In order to solve the above problems, in this paper the aerodynamic interference of a two-aircraft formation flight composed of flying-wing aircrafts is studied by CFD numerical simulation. Considering the balance of gravity and lift of the following aircraft, the drag reduction and lateral aerodynamic characteristics of the following aircraft are analyzed. The research results have certain guiding significance and reference value for optimizing the formation of aircrafts and elevating the flight range of aircrafts.

2. Numerical method

In this paper, the numerical simulation is conducted by a self-developed CFD solver named PMB3D2.0, which is a large-scale parallel CFD program based on multi-block structured grids including overlapping grid. PMB3D2.0 is suitable to evaluate aerodynamic characteristics of aircraft under the subsonic, transonic and supersonic flows, and has been widely used in some aerospace engineering projects. Compressible Navier-Stokes equations in a general curvilinear coordinate system are governing equations expressed as

$$\frac{\partial \tilde{Q}}{\partial \tau}+\frac{\partial \tilde{E}}{\partial \xi}+\frac{\partial \tilde{F}}{\partial \eta}+\frac{\partial \tilde{G}}{\partial \zeta}=NVIS\times\left(\frac{\partial \tilde{F}_v}{\partial \xi}+\frac{\partial \tilde{G}_v}{\partial \eta}+\frac{\partial \tilde{H}_v}{\partial \zeta}\right) \tag{1}$$

The above governing equations are discretized by finite volume method. The LU-SGS implicit time marching method is used for the time discretion of the equations. ROE which is a flux difference splitting scheme with second order upwind is used for spatial discretization. In this paper, the k-ω SST turbulence model is used. k-ω SST two-equation model combines k-ε and k-ω two-equation model by mixing function, which makes full use of the advantages of k-ε model in dealing with free flow and k-ω model in dealing with confined flow on the wall, and can simulate large separation flow. In order to improve the convergence speed of the flowfield for 2-aircraft formation flight, the multi-grid technology [21] is applied.

3. Formation model and grid

The object of this paper is a flying-wing aircraft [22] obtained by multi-iteration optimization design. The aircraft with 20 meters wingspan (b) has the advantages of low drag, high lift-drag ratio and better stealthy specialization. The aerodynamic advantage of formation flight is obvious and the aircraft in formation flight has the aerodynamic performance comparable to that of aircraft with high aspect ratio [23]. Studies have shown that wild geese flying in a 'V' formation can fly 70% more than those flying alone with the same energy [24-26]. The main reason is that the following wild geese can utilize aerodynamic advantages in the formation flight, and obtain additional lift by the upwash flow generated by other birds in the formation. According to the study [25, 26], when the geese flock is flying in formation with 11% span overlap of wingtips, the following geese can obtain the maximum lift from the upwash flow. Based on the above studies, in this paper we have designed a two-aircraft formation, as shown in Figure 1. The leading aircraft in the formation flight is represented by 'Leader', and the following aircraft is represented by 'Follower', which is in the left and rear direction of the leading aircraft. In the longitudinal direction $\triangle X = 1.75b$, in the lateral direction $\triangle Z = 0.89b$ and in the vertical axis $\triangle Y = 0$, where $b = 20$ meters.

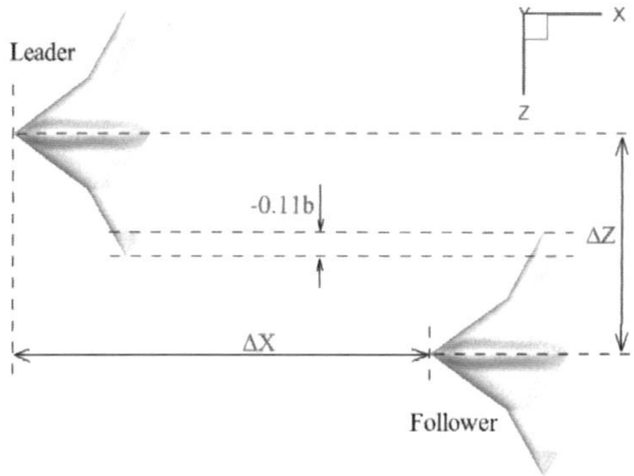

Figure 1. 2-aircraft formation model.

In order to adjust the aircraft's angle of attack for convenience, the overlapping grid technology is used to generate the structured computational grid for 2-aircraft formation flight. The grid of 2-aircraft formation consists of one background grid and two sub-grids, as shown in Figure 2. The background grid is a cuboid domain which size is more than 50 times the average aerodynamic chord length of the aircraft. In order to simulate the wingtip vortex of aircraft accurately, local mesh refinement for the wake field of wingtip vortex of aircraft in the background grid is conducted in this paper. The number of background grid cells is nearly 34 million. The sub-grid is the computational grid generated around the aircraft as shown in the Figure 3. The grid topology of the sub-grid is O-type. The normal first-level grid spacing is about 0.002mm, and y^+ could satisfy O(1) under the conditions of Mach 0.6 and flight Reynolds number. The number of grid cells in one sub-grid is nearly 20 million.

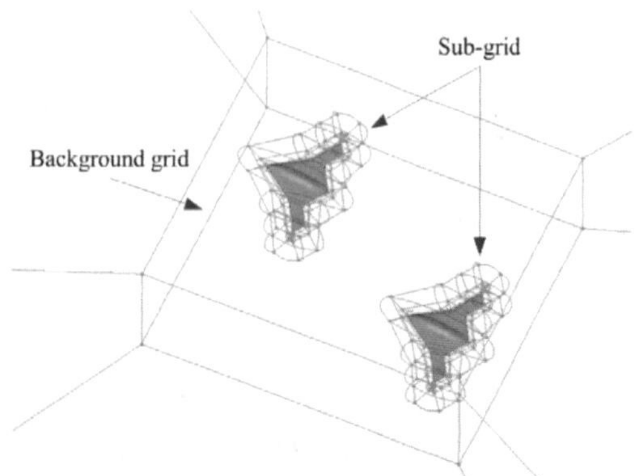

Figure 2. Overlapping grid of 2-aircraft formation flight.

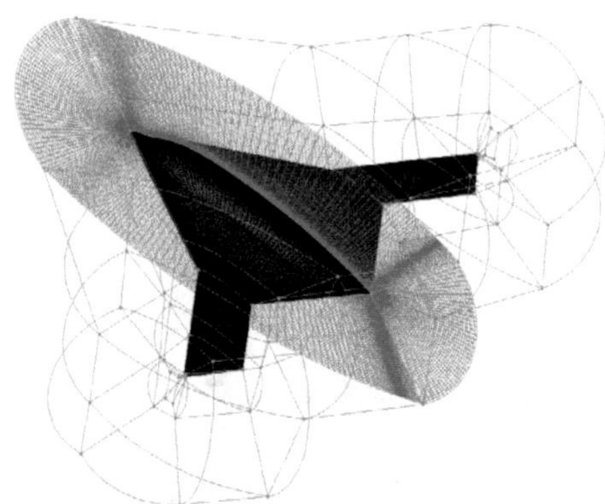

Figure 3. Sub-grid of solo aircraft in computational grid.

4. Validation of numerical method

In order to verify the reliability of the numerical method used in this paper, the wind tunnel test data of the flying-wing aircraft are used to assess and verify the numerical method. Test wind speed is about 40m/s, and Re is about 360 million while the scale of the test model to the real model is 1:5. Full turbulence simulation is used for CFD. Computational grid of the test model is overlapping grid as well with about 19.5 million grid cells. Figure 4 compares the CFD calculation of the aircraft with the wind tunnel test results. It can be seen that the CFD computation values of aerodynamic coefficient are in good agreement with the wind tunnel test results. The trend of variation of each curve is consistent in general, and the numerical calculation is reliable. The comparison results show that the numerical method and grid generation strategy adopted in this paper are reasonable and credible.

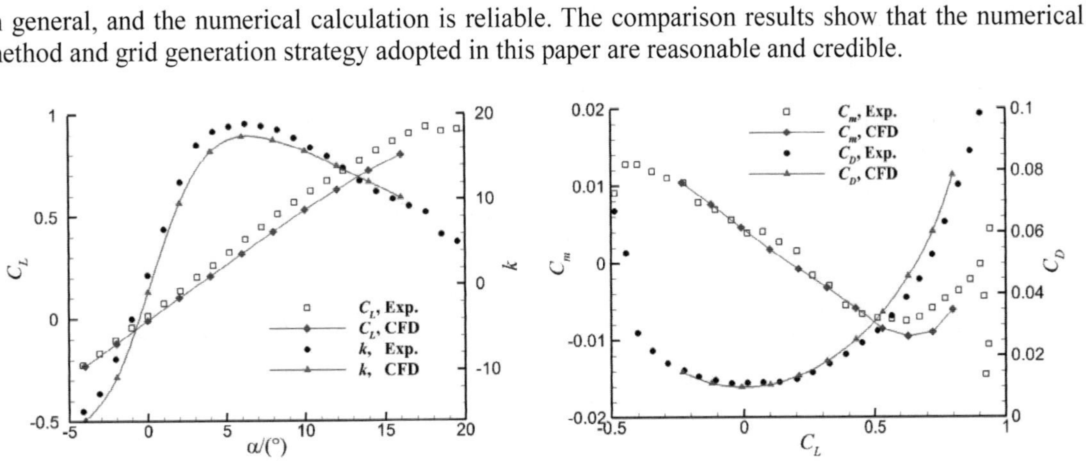

Figure 4. The aerodynamic coefficient of computation vs. experimental results.

5. Simulation results and analysis

In this paper, the aerodynamic interference of the two-aircraft formation flight is studied under the condition of Mach 0.6 and an altitude of 10 km.

5.1. *Impact on longitudinal aerodynamic characteristic*

Under the flight conditions of Mach 0.6 and an altitude of 10 km, the cruise AoA of solo flying-wing aircraft is about 5.15°, and the corresponding lift coefficient is almost 0.315. In this paper AoA is the nomenclature of angle of attack for the aircrafts. Table 1 shows the CFD results of longitudinal aerodynamic data for different aircrafts including solo aircraft and formation aircrafts at β=0° and α=5.15°. It can be seen that the two-aircraft formation flight designed in this paper has little impact on the aerodynamic characteristic of the leading aircraft, but increases the lift and reduces the drag of the following aircraft at the same AoA, namely α=5.15°. Compared with solo aircraft, C_L of the following aircraft increases about 14%, and C_D of the following aircraft decreases about 14%. Therefore, the lift-drag ratio k of the following aircraft increases about 32.6%.

Table 1. CFD results of longitudinal aerodynamic data for different aircrafts at β=0° and α=5.15°.

Name of aircraft	AoA (α/°)	Lift coefficient (C_L)	Drag coefficient (C_D)	Pitching moment coefficient (C_D)	Lift-drag ratio (k)
Solo	5.15	0.314827	0.016500	-0.007397	19.08
Leader	5.15	0.315641	0.016469	-0.007452	19.17
Follower	5.15	0.358919	0.014190	-0.011183	25.29

It should be noted that the above CFD results in Table 1 do not consider the balance of gravity and lift for the following aircraft. In order to accurately measure the aerodynamic disturbance caused by the two-aircraft formation flight to the following aircraft, the effect of gravity should be further considered. Then we keep the AoA of the leading aircraft equal to 5.15°, and gradually change the AoA of the following aircraft relative to free flow. The lift coefficient curve of the following aircraft is shown in Figure 5. It is confirmed by calculation that the AoA of the following aircraft in formation is about 4.477°, and the lift coefficient of the following aircraft at α_F=4.477° can be maintained at about 0.315, which is balanced with gravity of the following aircraft. In Figure 5 α_L is the AoA of the leading aircraft in formation.

Table 2 shows the CFD results of longitudinal aerodynamic data for different aircrafts under the balance of gravity and lift, namely $C_L \approx 0.315$. It can be seen that under the balance of gravity and lift, the drag reduction of the following aircraft in formation is further increased. According to Table 2, as the C_L of the following aircraft is about 0.315, the C_D of the following aircraft compared with that of solo aircraft decreases about 41cts (cts: C_D unit, 1cts = 0.0001) with a reduction of about 25.1%.

Table 2. CFD results of longitudinal aerodynamic data for different aircrafts at β=0° and $C_L \approx 0.315$.

Name of aircraft	AoA (α/°)	Lift coefficient (C_L)	Drag coefficient (C_D)	Pitching moment coefficient (C_D)	Lift-drag ratio (k)
Solo	5.15	0.314827	0.016500	-0.007397	19.08
Leader	5.15	0.315484	0.016473	-0.007441	19.15
Follower	4.477	0.315369	0.012360	-0.009408	25.52

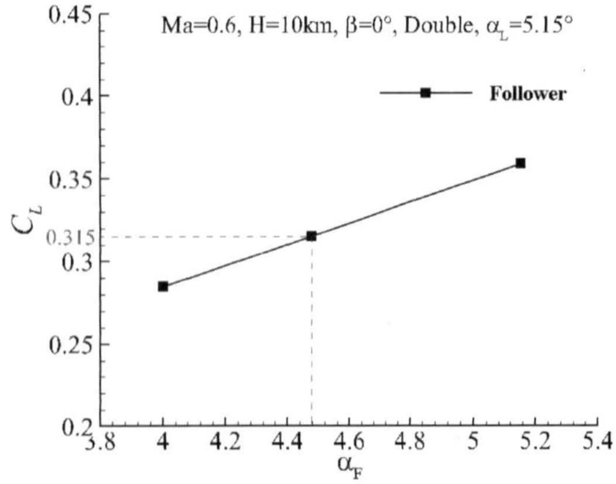

Figure 5. Lift coefficient curve of the following aircraft at α_L=5.15°.

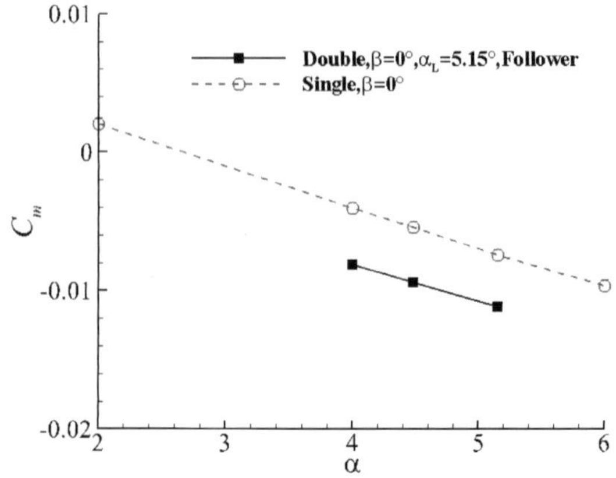

Figure 6. Pitching moment coefficient curves of the solo and following aircraft.

Figure 6 shows the pitching moment coefficient curves of the solo and following aircraft. It can be seen from Figure 6 that the two-aircraft formation flight has little impact on the longitudinal static stability of the following aircraft, but has great impact on the value of pitching moment. Under the aerodynamic interference of the leading aircraft, the following aircraft produces a significant nosedown pitching moment increment. The analysis shows that under the aerodynamic interference, the lift of the following aircraft increases significantly, while the main region of lift increased on the following aircraft is behind the center of mass (i.e. the moment reference point in this paper) of the aircraft, which results in the increase of the nosedown pitching moment.

5.2. Impact on lateral aerodynamic characteristics

As CFD method is used to calculate the flow field around an aircraft, the AoA of aircraft is usually achieved by fixing the aircraft and changing the angle of inflow. In this paper, the AoA of aircraft in two-aircraft formation flight can be obtained by turning the aircraft's sub-grid around the pitch axis without changing the angle of inflow. The aerodynamic moment directly obtained by CFD is equivalent to that under the body axis coordinate system with the AoA of 0°, which needs to be converted to that under the body axis coordinate system with the AoA of α as shown in Figure 7.

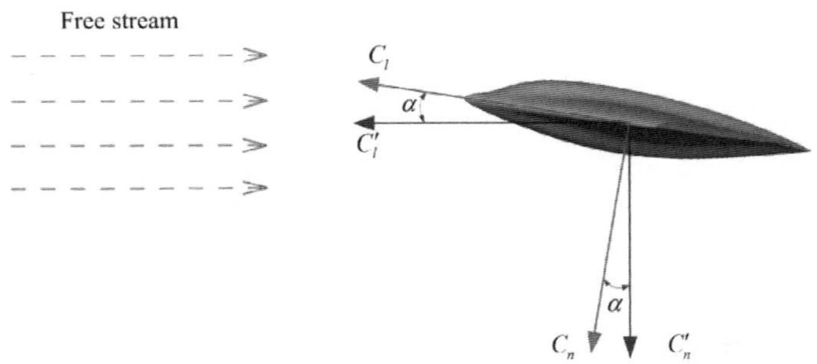

Figure 7. Schematic diagram of coordinate axis transformation of aircraft.

The conversion formulas for rolling and yawing moment are as follows.

$$\begin{bmatrix} C_l \\ C_n \end{bmatrix} = \begin{bmatrix} \cos\alpha & -\sin\alpha \\ \sin\alpha & \cos\alpha \end{bmatrix} \begin{bmatrix} C_l' \\ C_n' \end{bmatrix} \tag{2}$$

Figure 8 and Figure 9 show the rolling and yawing moment coefficient curves of the solo and following aircraft, respectively. The abscissa is the sideslip angle β of the aircraft. In formation flight, the AoA of the leading aircraft is 5.15°, and the AoA of the following aircraft is 4.477°. It can be seen that the two-aircraft formation flight has an evident impact on the lateral aerodynamic characteristics of the following aircraft. As β = 0°, the rolling moment and yawing moment are non-zero values under the aerodynamic interference, and the following aircraft will roll to the left and yaw to the left. For the solo aircraft, since the sign of $C_{l\beta}$ is negative, the aircraft has lateral static stability. Meanwhile, the solo aircraft has directional static instability due to the negative $C_{n\beta}$, which is consistent with the stability characteristics of flying-wing aircraft [27]. It should be noted that $C_{l\beta}$ is the lateral static stability derivative, and $C_{n\beta}$ is the directional (yaw) static stability derivative. Different from the conventional aircraft with vertical tail, flying-wing aircraft without vertical tail is generally directional static instability, which needs to be focused by strengthen control measures and additional stabilization system based on control rudder [28]. It is a major challenge to aircraft designers. However, the sign of $C_{l\beta}$ for the following aircraft becomes positive due to the aerodynamic interference. Then the following aircraft has lateral static instability, which needs to be considered in the flight control. Interestingly, the sign of $C_{n\beta}$ for the following aircraft is positive in the range of β = 0° ~12°. A positive $C_{n\beta}$ implies that the following aircraft has the directional static stability, which reduces the difficulty of the yawing control of the flying-wing aircraft and is favour to the yawing control.

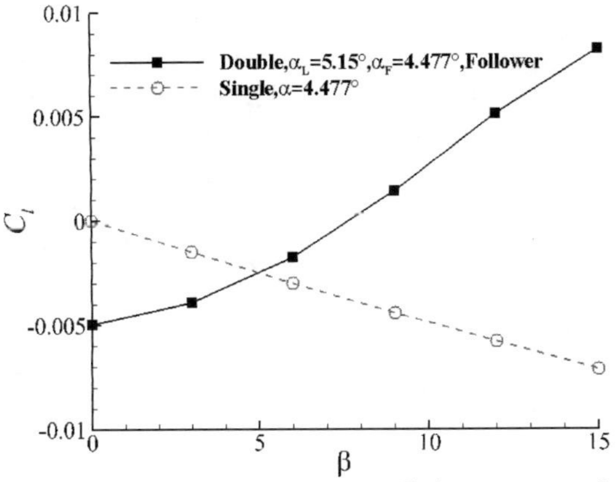

Figure 8. Rolling moment coefficient curves of the solo and following aircraft.

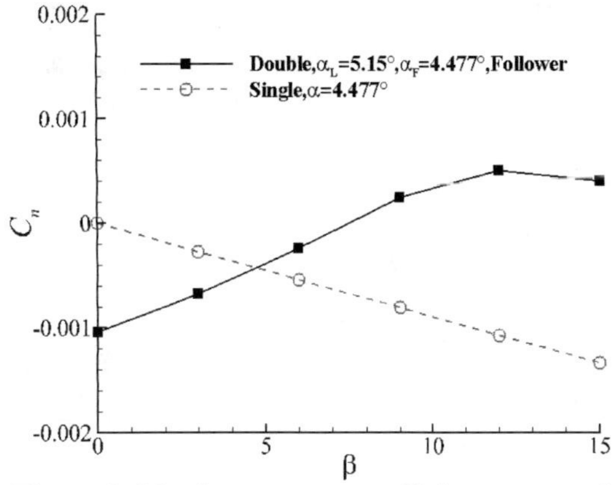

Figure 9. Yawing moment coefficient curves of the solo and following aircraft.

5.3. Impact of wake vortex and analysis

When an aircraft is flying at an angle of attack and flight speed, air flows through the wing of aircraft and then creates a pressure difference between the upper and lower surfaces of the wing, which causes the airflow on the lower surface of the wing to roll over the upper surface of the wing and produces an inward curling vortex from the wingtip. This is known as wingtip vortex. The whole aircraft generates two wake vortex zones with a certain width in the downstream of the aircraft, forming downwash flow between the two wake vortex cores and upwash flow on the outside. The schematic diagram is shown in Figure 10.

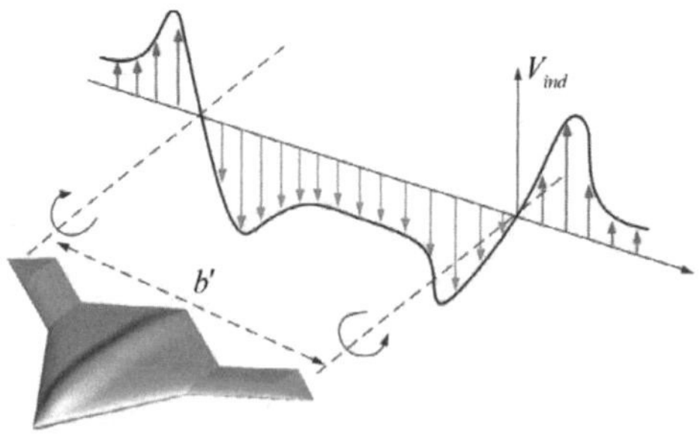

Figure 10. Schematic diagram of wingtip vortex induced velocity distribution.

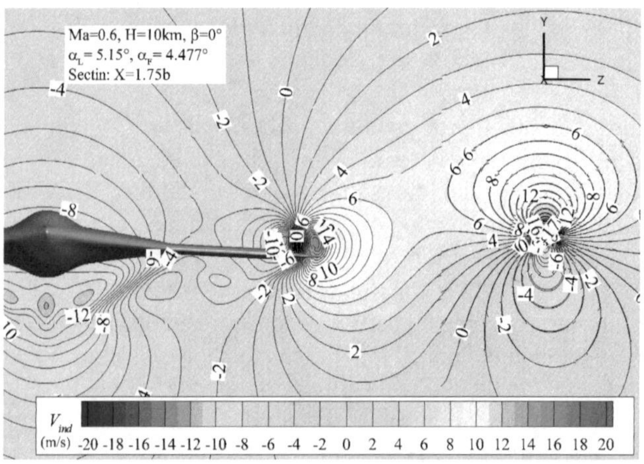

Figure 11. The distribution of partial velocity in the Y-direction on typical section (X=1.75*b*).

Figure 11 shows the distribution of partial velocity in the Y-direction on a typical section (X=1.75*b*). It can be seen that there does exist obvious upwash flow in the downstream wake field behind the left side of the leading aircraft. The behavior of placing the following aircraft in the wingtip vortex-induced upwash flow field generated by the leading aircraft is also known as Surfing Aircraft Vortices [17, 29]. It is considered that an upwash flow added to the free flow can increase the actual angle of attack of the aircraft, generates induced lift and reduces induced drag, as shown in Figure 12.

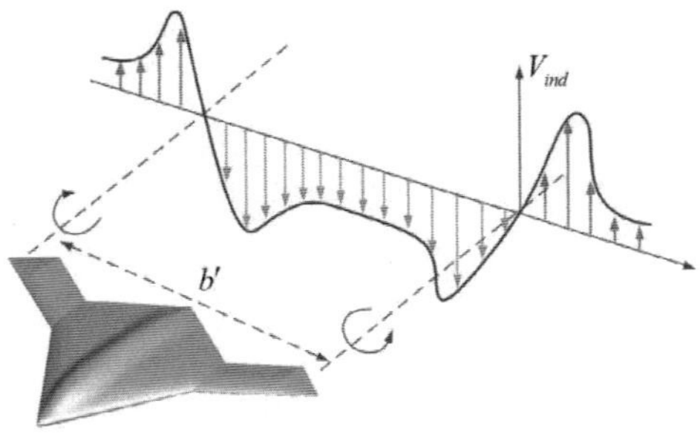

Figure 12. The diagram of lift and drag increment under the upwash flow.

Figure 13 shows the distribution of upwash angle $\Delta\alpha$ on different spatial sections of the two-aircraft formation flight. It can be seen that the positive area of the upwash angle $\Delta\alpha$ shown in red gradually increases with the increase of the flow direction position, which may be related to the increase of the influence range of the wake vortex field due to vortex dissipation. The positive upwash angle will have an important effect on the aerodynamic characteristics of the right wing of the following aircraft. In addition, along the wingspan direction of the wing, gradually approach the wingtip and gradually increase the wash angle. Obviously, this corresponds to the variation rule of upstream speed along the wing spread position shown in the Figure 11.

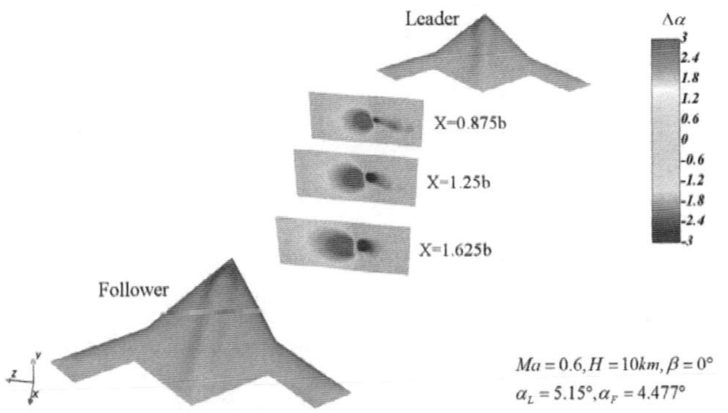

Figure 13. The distribution of upwash angle on different spatial sections.

Different spanwise sections on the solo aircraft and the following aircraft are further intercepted. $\eta > 0$ represents the right spanwise section of the aircraft and $\eta < 0$ represents the left spanwise section. Figure 14 shows the distribution of C_p at different sections of the solo and following aircrafts at $\beta = 0°$. The abscissa X is the ratio of the flow direction position and the wingspan length of aircraft. Compared with solo aircraft, the lift provided by the wing of the following aircraft is increased due to the influence of the upwash flow. As η increases, the lift increment provided by the wing of the following aircraft gradually increases, which is mainly due to the gradually increasing influence of the upwash flow near the wake vortex core.

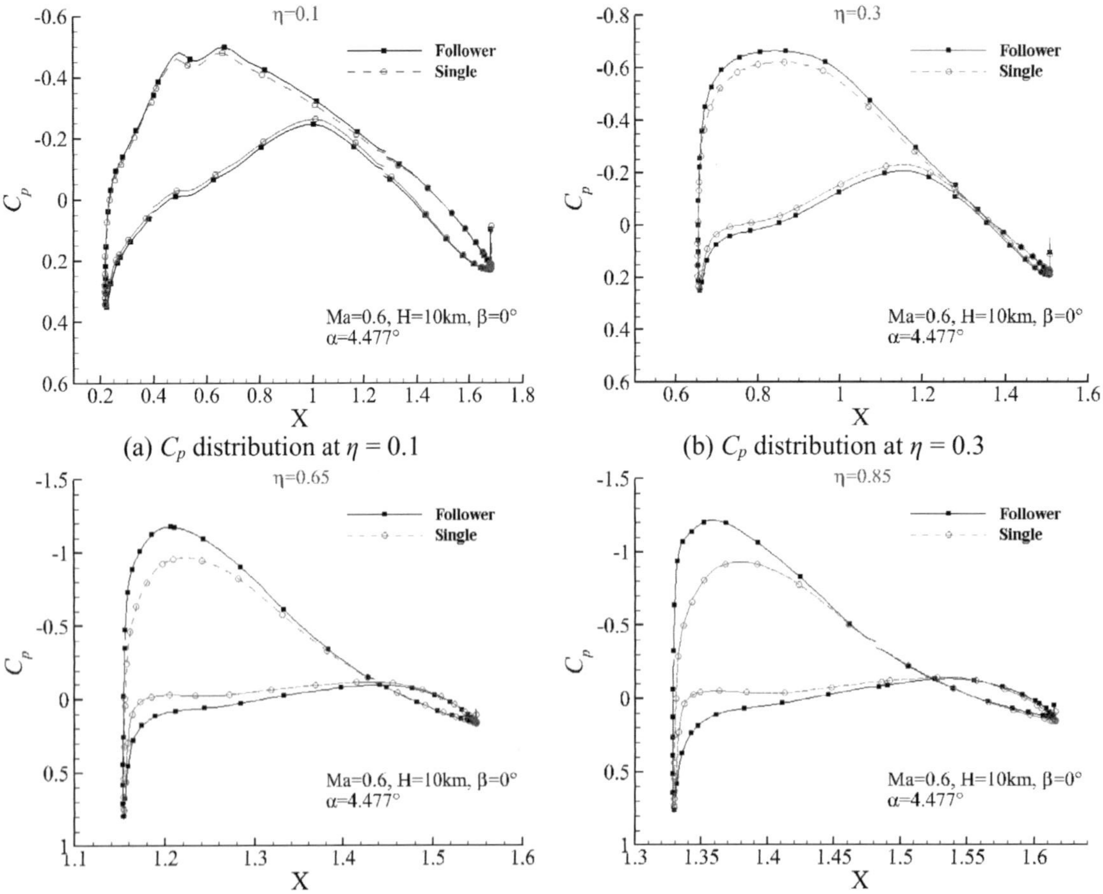

(a) C_p distribution at $\eta = 0.1$ (b) C_p distribution at $\eta = 0.3$

(c) C_p distribution at $\eta = 0.65$ (d) C_p distribution at $\eta = 0.85$

Figure 14. The distribution of C_p at different sections of the solo and following aircrafts (β = 0°).

Figure 15 shows the distribution of C_p at symmetrical sections of the following aircraft at β = 0°. It can be seen that the distribution of C_p at η= 0.1 and η= -0.1 sections is basically same. Due to close distance between the two sections, there is little difference of influence of upwash flow on the distribution of C_p at the two sections. While there is obvious difference in the distribution of C_p between η= 0.65 and η= -0.65 sections. That is because the C_p at η= 0.65 section is greatly affected by the upwash flow due to close distance from the wake vortex field, and the C_p at η= -0.65 section is little affected by the upwash flow due to long distance from the wake vortex field. It can be inferred that there is obvious asymmetry of lift distribution on the left and right wings of the following aircraft under the aerodynamic interference from the leading aircraft. The relatively large lift on the right wing of the following aircraft will result in a negative rolling moment that tends to roll the aircraft toward the left side.

Figure 16 shows the distribution of C_p at symmetrical sections of the following aircraft at β = 6°. It can be seen that if the sideslip angle β is positive such as β = 6°, there begins to exist difference in the distribution of C_p between η= 0.1 and η= -0.1 sections. That is because wake vortex field generated by the leading aircraft moves toward the left side from the right side of the following aircraft at positive sideslip angle, since the leading aircraft is located in front of the right side of the following aircraft. So the influence by the upwash flow on the C_p at η= 0.1 and η= -0.1 sections is different. Meanwhile, the influence area of the wake vortex field induced by the leading aircraft on the wing of the following aircraft increases gradually with the increase of β. The lift provided by the left wing mainly affected by the upwash flow on the outside of wake vortex core gradually increases, while the lift provided by the right wing affected by some upwash flow and most downwash flow on the inside of wake vortex core gradually decreases. The tendency of the following aircraft to roll toward the left side decreases, and the following aircraft begins to roll toward the right side gradually, which explains the lateral static instability of the following aircraft shown in Figure 8.

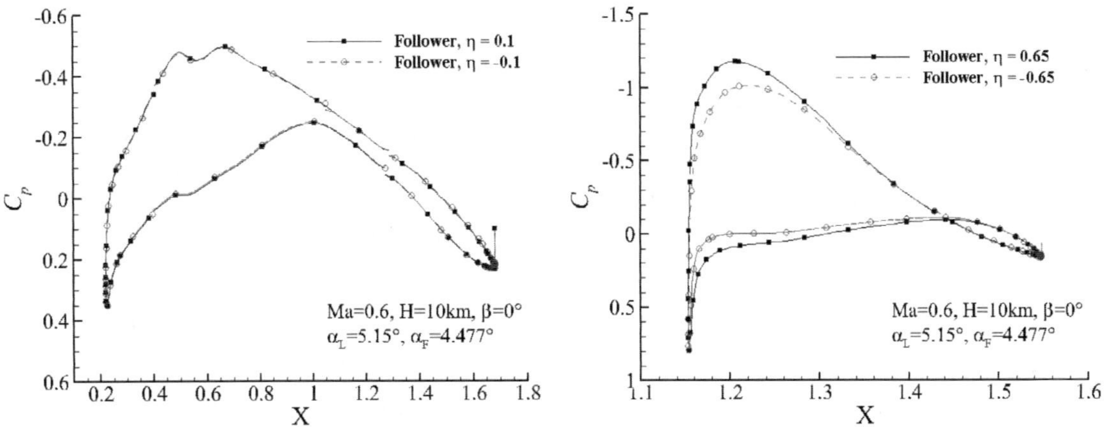

(a) C_p distribution at η = 0.1 and η = -0.1 (b) C_p distribution at η = 0.65 and η = -0.65

Figure 15. The distribution of C_p at symmetrical sections of the following aircraft (β = 0°)

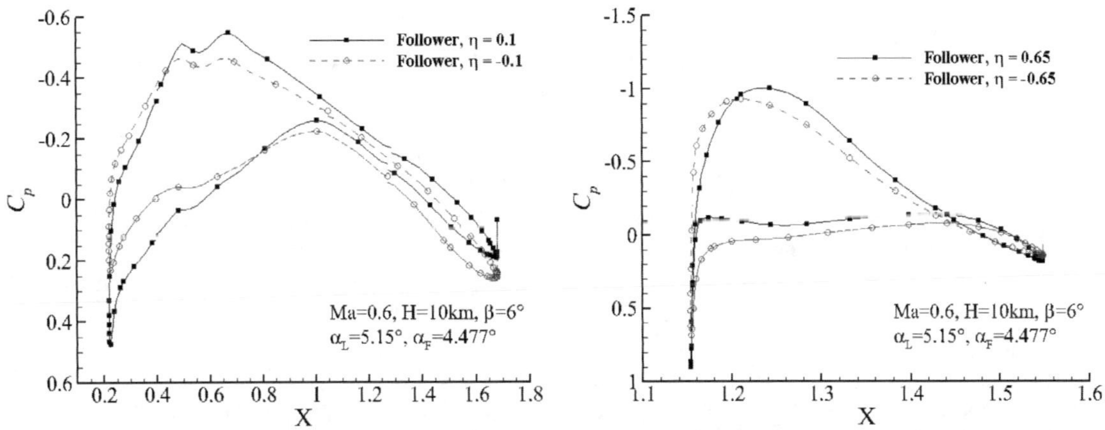

(a) C_p distribution at η = 0.1 and η = -0.1 (b) C_p distribution at η = 0.65 and η = -0.65

Figure 16. The distribution of C_p at symmetrical sections of the following aircraft (β = 6°).

6. Conclusion

Based on wild geese flying in a 'V' formation, a two-aircraft formation composed of flying-wing aircrafts is designed. The numerical simulation of the aerodynamic interference in two-aircraft formation flight is carried out by CFD.

The two-aircraft formation flight has little impact on the aerodynamic characteristics of the leading aircraft, but has an evident great impact on the aerodynamic characteristics of the following aircraft. The drag reduction of the following aircraft is particularly large, as high as 25.1%, especially considering the balance of gravity and lift. Under the aerodynamic interference effect, the following aircraft has a nosedown pitching moment increment, and becomes lateral static instability, which needs to be considered in flight control. The aerodynamic interference mainly comes from upwash flow induced by wingtip vortex of the leading aircraft. The research results have certain guiding significance and reference value for optimizing the formation of aircraft and elevating the flight range of aircraft.

References

[1] Q.J. Fan, Z. Yang, T. Fang, et al, Research status of coordinated formation flight control for multi-UAVs, Acta Aeronautica et Astronautica Sinica, 2009, 30(4): 683-691.
[2] Y. Tao, N. Xiong, X.B. Wang, et al, Experimental and computational investigation of hybrid formation flight for aerodynamic gain at transonic speed, Chinese Journal of Aeronautics, 2021, 34(1): 32-43.
[3] M. Pachter, J. J. D'Azzo, A.W. Proud, Tight formation flight control, Journal of Guidance, Control, and Dynamic, 2001, 24(2): 246-254.
[4] W. Blake, D. Multhopp, Design, performance and modeling considerations for close formation flight, AIAA-1998-4343.
[5] A. Dogan, S. Venkataramanan, W. Blake, Modeling of aerodynamic coupling between aircraft in close formation, Journal of Aircraft, 2005, 42(4): 941-955.
[6] J. Shan, H. Liu, Close formation flight control with motion synchronization, Journal of guidance control, and dynamics, 2005, 28(6): 1316-1320.
[7] Z.C. Zheng, Y. Xu, D.K. Wilson, Behaviors of Vortex Wake in Random Atmospheric Turbulence, Journal of Aircraft, 2009, 46(6): 2139-2144.
[8] Z.Y. Liu, Investigation on the mechanism of drag reduction and positional parameters optimization in close-formation flight, Hefei: University of Science and Technology of China, 2016.
[9] Z.A. Bangash, R.P. Sanchez, A. Ahmed, et al, Aerodynamics of formation flight, Journal of Aircraft, 2006, 43(4): 907-912.
[10] W.B. Blake, D.R. Gingras, Comparison of predicted and measured formation flight interference effects, Journal of Aircraft, 2004, 41(2): 201-207.
[11] E. Romander, M. Betzina, M. Silva, et al, Investigating tiltrotor formation flight via 1/48-sacle wind tunnel experiment, The American Helicopter Society 62nd Annual Forum, Phoenix AZ, 2006.
[12] I. Ayumu, M. Fumihide, A. Masahito, Detailed observations of interactions of wingtip vortices in close formation flight, Journal of Aircraft, 2012, 49(1): 206-213.
[13] M.J. Vachon, R.J. Ray, K.R. Walsh, et al, F/A-18 Aircraft Performance Benefits Measured During the Autonomous Formation Flight Project, AIAA-2002-4491, 2002.
[14] M.G. Wagner, L.D. Jacques, W. Blake, et al, Flight test results of close formation flight for fuel savings, AIAA-2002-4490, 2002.
[15] R.J. Ray, B.R. Cobleigh, M.J. Vachon, et al, Flight test techniques used to evaluate performance benefits during formation flight, AIAA-2002-4492, 2002.
[16] J. Hansen, B. Cobleigh, R. Ray, et al, Vortex-induced aerodynamic effects on a trailing F-18 aircraft flying in close formation, AIAA-2002-3432, 2002.
[17] J. Pahle, D. Berger, M. Venti, et al, An initial flight investigation of formation flight for drag reduction on the C-17 aircraft, AIAA-2012-4802, 2012.
[18] Z.Y. Liu, Y. Tao, Z.W. Shi, et al, Investigation on formation flight in wind tunnel, Journal of Experi-ments in Fluid Mechanics, 2016, 30(4): 20-25.
[19] Y. Yuan, J.G. Yan, Y.H. Qu, The aerodynamic coupling simulation in the formation flight of multi-UAV, Flight Dynamics, 2013, 31(1): 29-32.
[20] Y. Liu, X.X. Zhao, Z.M. Yuan, et al, The calculation on the aerodynamics of two aircrafts in proximity based on RANS method, The 4th China Aviation Science and Technology Conference, 2019.
[21] B. Mou, Z.Y. Xiao, Z. Zhou, et al, Application and investigation of multiblock multigrid method in complicated viscous flow fields calculation, Acta Aerodynamica Sinica, 2006, 24(1): 51-54.
[22] Y.G. Yu, Y. Huang, Z. Zhou, et al, Aerodynamic design of a flying-wing aircraft, Acta Aerodynamica Sinica, 2017, 35(6): 832-836, 878.
[23] G. Zhan, J. Wan, UAV formation flight control based on aerodynamic coupling effects, Journal

of Fudan University (Natural Science), 2019, 58(1): 14-18, 24.
[24] J.R. Speakman, D. Banks, The function of flight formations in Greylag Geese Anser anser; energy saving or orientation?, IBIS, 2008, 140: 280-287.
[25] P.B.S. Lissaman, C.A. Shollenberger, Formation flight of birds, Science, 1970, 168: 1003-1005.
[26] Z.W. Zhou, H.B. Duan, Y.M. Fan, et al, Unmanned aerial vehicle close formation control based on the behavior mechanism in wild geese, SCIENTIA SINICA Technologica, 2017, 47(3): 230-238.
[27] L. Li, C. Ma, L.X. Wang, Stability Features of Low Aspect-Ratio Flying Wings, Acta Aeronautica et Astronautica Sinica, 2007, 28(6): 1312-1317.
[28] G. Liu, Y.X. Qiu, H. Chen, Investigation of the direction control for the tailless aircraft configuration, Experiments and Measurements in Fluid Mechanics, 2003, 17(4): 1-9.
[29] S. Andrew Ning, Tristan C. Flanzer, Ilan M. Kroo, Aerodynamic Performance of Extended Formation Flight, Journal of Aircraft, 2011, 48(3): 855-865.

Study on effect of boundary conditions on simulation of wind flow in mountainous area

Yuan Yin[1], Yangsen Li [1], Ruizong Lin[1], Zhiwei Liu[2], Benzhao Fu[2], Guo Huang[3, *], Hongjie Zhang[3]

[1] State Gird Fujian Electric Power Co. Ltd, Fuzhou 350003, China
[2] State Grid Fujian Economic Research Institute, Fuzhou 350003, China
[3] China Electric Power Research Institute, Beijing, 100055, China

*Corresponding author e-mail: huangguo2010@163.com

Abstract. The simulation analysis of wind flow in mountainous area can assist the design of the transmission line. The conventional square computational domain is not conducive for the condition of different wind directions. In this paper, a circular computational domain is proposed for the solution of wind flow in mountainous area in the condition of different wind directions. A square computational domain and a circular one are used to solve the wind flow around the standard hill model. The simulation accuracy of the circular computational domain is verified by the wind profiles at the windward slope, the top of the slope and the leeward slope.

1. Introduction

Transmission lines often pass through mountainous areas, and the hills will accelerate and change the direction of the wind flow. The characteristics of wind flow in the mountains is very different from that in the flat ground. Research on wind flow in the mountains is helpful for the design of transmission lines. Many scholars have carried out a series of numerical simulations, and have achieved many research results. The research work mainly focuses on the key parameters in the simulation [1], especially on the turbulence model [2-4]. Liu simulated the drag effect of vegetation by adding a negative scour term to the momentum equation [5]. Ishihara established a transition area around complex terrain to play a role of smooth transition [6].

In the simulation, the terrain area of interest is firstly determined by the latitude and longitude coordinates. Then, a computational domain is established according to the wind direction and divided into grids. The computational domain is usually square. Finally, the boundary conditions is set to complete the calculation. The current simulation technology can provide the complex wind flow in the mountains, but it is very inefficient. This is because the computational domain must be re-established at every wind direction angle. In this paper, a circular computational domain is proposed, by which the computational domain is no need to be re-established for different wind directions. This method can greatly improve the simulation efficiency and facilitate the engineering designers.

2. Computational Domain Settings

2.1. Hill Model

The hill model is often used to verify the accuracy of the algorithm, which is adopted in the paper, and the maximum slope of this model is about 32°. The contour expression is shown in formula (1). The hill model is shown in Figure 1.

$$z(x,y)=h\cos^2\left(\frac{\pi\left(x^2+y^2\right)^{1/2}}{2L}\right) \tag{1}$$

Where h = 0.4m, L = 1.0m.

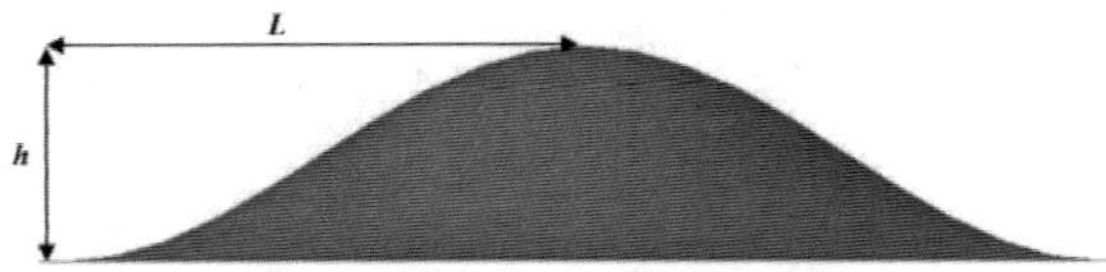

Figure 1. Standard hill model

2.2. Meshing

The structured mesh is used for mesh division, and the surface mesh of the hill model is shown in Figure 2.

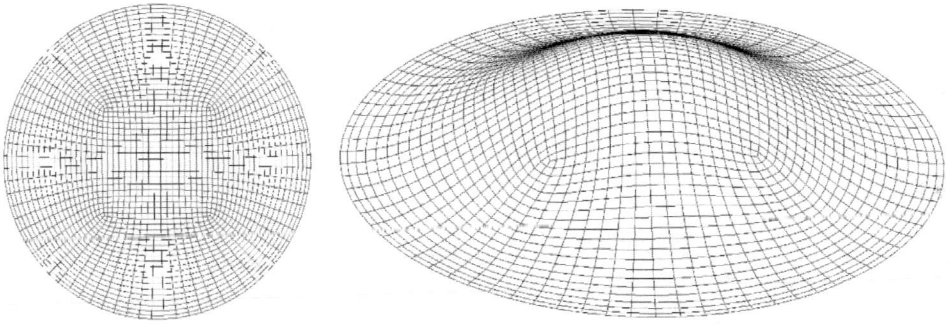

(a) Top view (b) Oblique view

Figure 2. Surface mesh of mountain model

For the simulation of wind flow, a square computational domain is established in usual, which is shown in Figure 3. The wake region usually has a distance of 10 times the characteristic length to avoid the effect of the outlet. The velocity condition is set for the inlet, where the velocity equals to 10m/s. The pressure outlet condition is set for the outlet. The wall condition is set for the surface, and the symmetrical boundary condition is set for other boundaries. The mesh is shown in Figure 3.

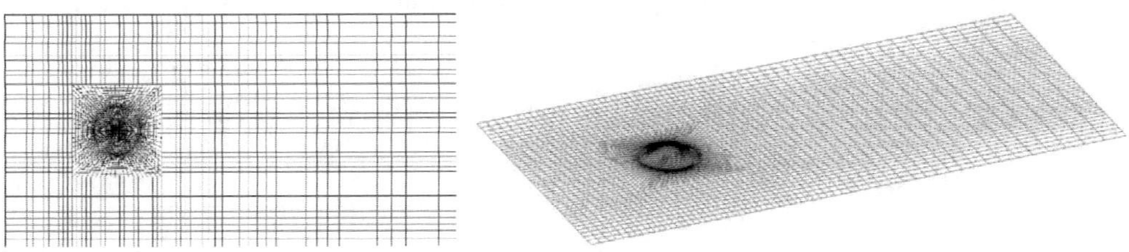

(a) Top view (b) Oblique view

Figure 3. Square computational domain

For the square computational domain, the computational domain needs to be re-established under the corresponding wind direction angle if the wind direction angle changes. Some computational software supports the General Grid Interface (GGI) Connections algorithm, by which a circular sub-computation domain can be set around the mountain model. By rotating the circular sub-computation domain, the cases with different wind direction angles can be simulated Although the GGI algorithm can improve the simulation efficiency, it is not conducive for the analysis of results after the sub-computation domain is rotated. In this paper, a circular computational domain is proposed. The radius of the computational domain is set to 10 times the characteristic length. The velocity condition is set for the outer boundaries which are both the entrance and the exit, where the velocity equals to 10m/s. The mesh is shown in Figure 4.

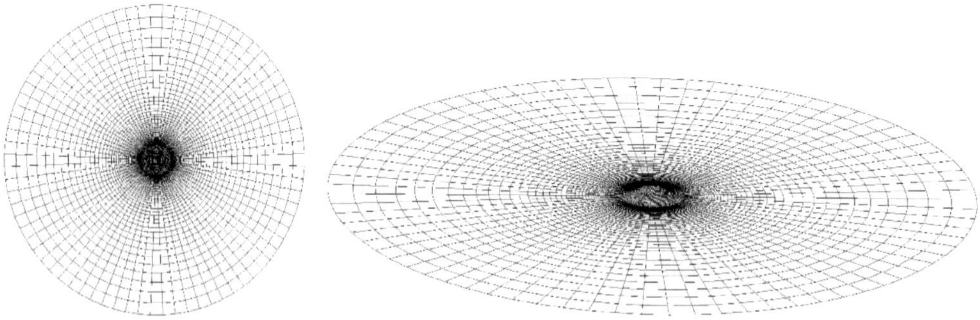

(a) Top view (b) Oblique view

Figure 4. Circular computational domain

2.3. Turbulence model

The N-S equation can be closed by the standard k-ε turbulence model, and the standard wall function is adopted for the flow near the wall with low Reynolds number. The governing equations of the standard k-ε turbulence model are as follows:

$$\frac{\partial(\rho k)}{\partial t}+\frac{\partial}{\partial x_i}(\rho k u_i)=\frac{\partial}{\partial x_j}\left[\left(\mu+\frac{\mu_t}{\sigma_k}\right)\frac{\partial k}{\partial x_j}\right]+G_k-\rho\varepsilon \tag{2}$$

$$\frac{\partial(\rho \varepsilon)}{\partial t}+\frac{\partial}{\partial x_i}(\rho \varepsilon u_i)=\frac{\partial}{\partial x_j}\left[\left(\mu+\frac{\mu_t}{\sigma_\varepsilon}\right)\frac{\partial \varepsilon}{\partial x_j}\right]+C_{1\varepsilon}\frac{\varepsilon}{k}G_k-C_{2\varepsilon}\rho\frac{\varepsilon^2}{k} \tag{3}$$

Where $\mu_t=\rho C_\mu\frac{k^2}{\varepsilon}$, $G_k=\mu_t S^2$, $S\equiv\sqrt{2S_{ij}S_{ij}}$, $S_{ij}=\frac{1}{2}\left(\frac{\partial u_j}{\partial x_i}+\frac{\partial u_i}{\partial x_j}\right)$. The model constants are as follows: C_μ=0.09, $C_{1\varepsilon}$=1.44, $C_{2\varepsilon}$=1.92, σ_k=1.0, σ_ε=1.3.

3. Results

In order to compare the effects of different computational domains on the simulation accuracy, slices are made along the symmetry plane of the computational domain, and the downwind wind speed is displayed in the slices, as shown in Figure 5. After comparing Figure 5(a) and 5(b), the velocity contours are basically same.

On the slices shown in Figure 5, five points are taken in turn along the downwind direction to extract the wind profile in the downwind direction, as shown in Figure 6(a)~(e). The extraction points are located on the windward slope, the top of the slope and the leeward slope, respectively. The coordinates of the extraction points are x = -1m, -0.5m, 0m, 0.5m, and 1m. The wind profiles at the extraction positions in the square computational domain and the circular one are basically coincide, which indicates that the different settings of the computational domain will not affect the simulation results. The circular computational domain proposed in this paper is suitable for the simulation of wind flow in the mountains.

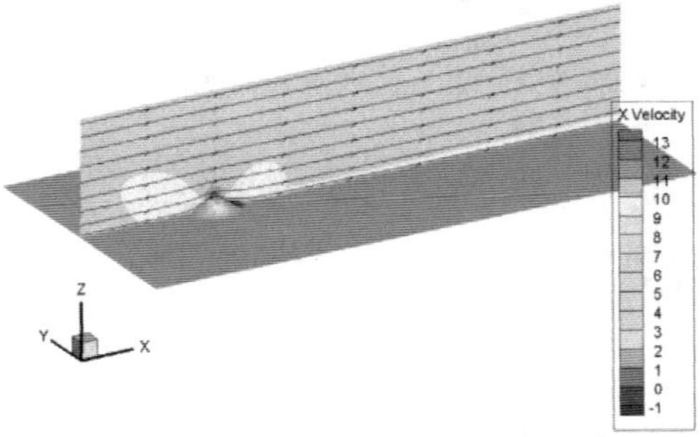

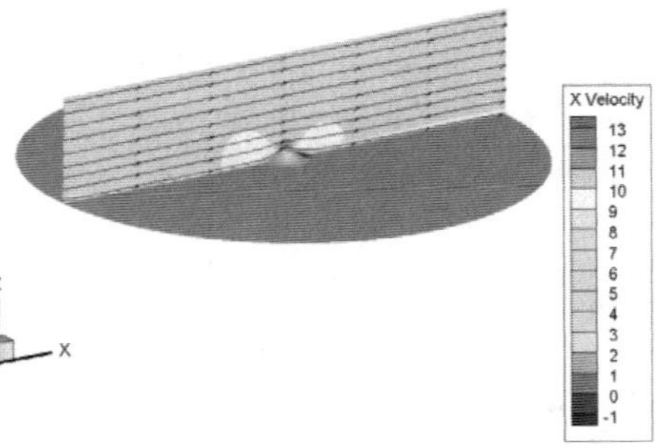

(a) Square computational domain (b) Circular computational domain

Figure 5. Wind speed downwind (m/s)

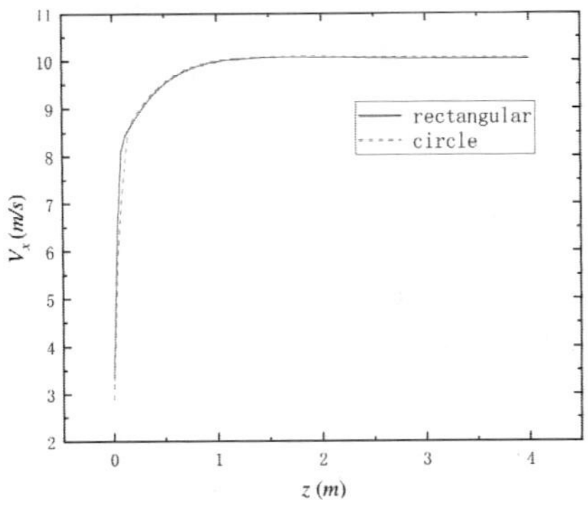

(*a*) *x* = -1m

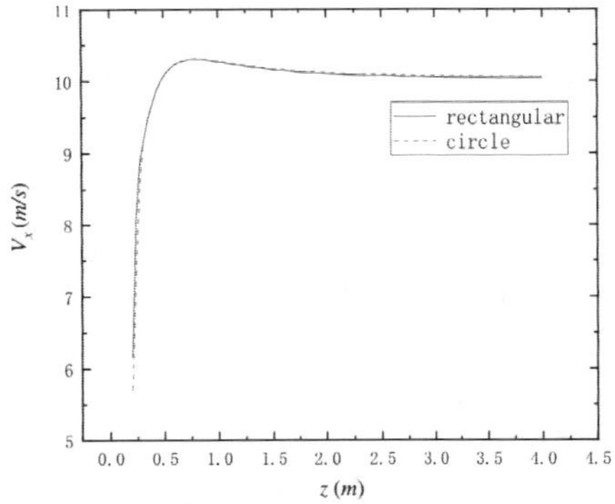

(*b*) *x* = -0.5m

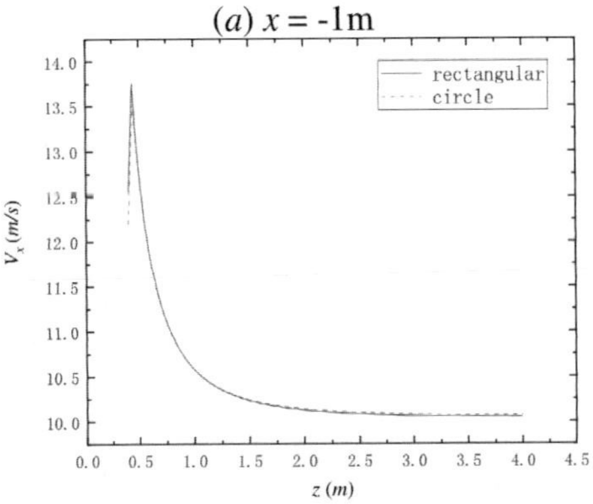

(*c*) *x* = 0m

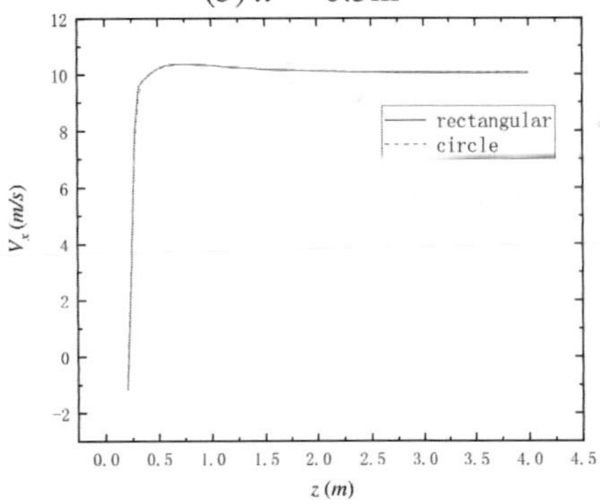

(*d*) *x* = 0.5m

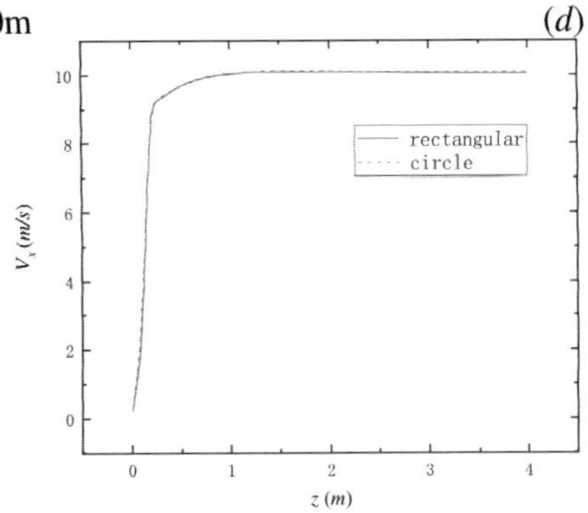

(*e*) *x* = 1m

Figure 6. Comparison of downwind speed profiles

4. Conclusion

In this paper, a circular computational domain is proposed. The square computational domain and the circular one are used to solve the wind flow around the standard hill model. The wind profiles at the windward slope, the top of the slope and the leeward slope are extracted. By comparison, it is found that the wind profiles obtained from the two computational domains are basically coincident, indicating that different settings of the computational domain will not affect the simulation results. The circular computational domain proposed in this paper is suitable for the calculation of wind flow in the mountains.

Acknowledgments

This work was financially supported by the State Grid Fujian Economic Research Institute. The project name is Research on assessment technology of mechanical state of towers based on numerical wind tunnel of atmospheric boundary layer (52130N210007).

References

[1] Dhunny A. Z., Loll chund M. R., Rughooputh S. D. D. V. Wind energy evaluation for a highly complex terrain using Computational Fluid Dynamics (CFD) [J]. Renewable Energy, 2017,101: 1-9.

[2] Blocken B., van der Hout A., Dekker J., et al. CFD simulation of wind flow over natural complex terrain: case study with validation by field measurements for Ria de Ferrol, Galicia Spain [J]. Journal of Wind Engineering and Industrial Aerodynamics, 2015, 147: 43-57.

[3] Risan A., Lund J. A., Chang C.Y., et al. Wind in Complex Terrain-Lidar Measurements for Evaluation of CFD Simulations [J]. Remote Sensing, 2018, 10(1): 59.

[4] Abdi D. S., Bitsuamlak G. T. Wind flow simulations on idealized and real complex terrain using various turbulence models [J]. Advances in Engineering Software, 2014, 75: 30-41.

[5] Liu Z., Ishihara T., He X., et al. LES study on the turbulent flow fields over complex terrain covered by vegetation canopy [J]. Journal of Wind Engineering and Industrial Aero dynamics, 2016, 155: 60-73.

[6] Ishihara T. A Nonlinear Wind Prediction Model MASCOT: Development and Application (Numerical Prediction Technique of Airflow Over Complex Terrain) [J]. Journal of Japan Society of Fluid Mechanics, 2003, 22: 387-396.

Influence of converging slot-holes (consoles) with different transition surfaces on film cooling performance

Manshuang Yu*, Dong Xuan

State Key Laboratory of Civil Aircraft Flight Simulation, Shanghai Aircraft Design and Research Institute, Shanghai, China

*Corresponding author email: yumanshuang201970@163.com

Abstract. To investigate the influence of the transition surfaces of consoles on film cooling performance, a series of numerical simulations were performed under two typical blowing ratio (M). Four different consoles were designed by changing the transition surface of the console, and their film cooling performance was studied under the same boundary conditions. The purpose of the present work is to determine the best configuration of the transition surface. The results show that whether at M=0.5 or M=1.0, compared with the reference case, the console with anti-s-shaped transition surface has the best spatially-averaged adiabatic film cooling effectiveness ($\eta_{ad,av}$), and the $\eta_{ad,av}$ is improved by 6.4% and 7.0% respectively, meanwhile, the discharge coefficient (C_d) is decreased by 2.9% and 3.2% respectively. Besides, the detailed flow and heat transfer mechanism of consoles with different transition surfaces under different blowing ratios are also investigated.

Keywords: film cooling; converging slot-holes; transition surfaces; film cooling effectiveness; discharge coefficient.

1. Introduction

With the increasing temperature of the turbine engine inlet, higher and higher requirements are put forward for the high-temperature resistance of turbine blade and other hot end components [1, 2]. Therefore, it is necessary to take efficient cooling measures to decrease the heat load of the turbine blade to ensure that the turbine blade could work regularly and reliably. As an efficient cooling method, film cooling is widely used in turbine blades and other hot end components [3-5]. The film cooling not only cools high temperature in the gas side but also forms a film between the high-temperature gas and the outer wall of the blade to play a thermal protection role.

In recent decades, scholars have carried much fundamental research on the interaction between film jet and mainstream, and the film cooling technology has been developed continuously. The innovation and application of shaped holes are the most noticeable research [6-10]. The mechanism is to control the flow between the film jet and the near-wall primary flow by optimizing the structure of film holes and taping the potential of film cooling further. Compared with conventional cylindrical holes, shaped holes could improve the film cooling effectiveness in both streamwise direction and span-wise direction, and make the turbine blades have better cooling uniformity. It is essential to reduce the amount of coolant air and the thermal stress of the turbine blades.

As a kind of shaped hole, the console was first developed and experimentally explored by Sargison [11, 12]. The transition surface of the film hole expands from the conventional round entrance along the lateral direction and shrinks along the streamwise direction, and then gradually transiting to rectangular slot outlet. Sargison et al. [13] studied the flow and cooling characteristics of the console in detail. They believed that the overall cooling effect of console was close to that of two-dimensional slot film hole, which could form a better span-wise film coverage on the downstream surface, and effectively suppress the normal lift of the film, thus significantly weakening the influence of film outflow on the downstream boundary layer and reducing the mainstream aerodynamic loss. Azzi et al. [14] and Yao et al. [15-17] found that the anti-kidney vortices formed downstream of the console are the key factors to improve film cooling effectiveness. Liu et al. [18-19] investigated the influence of the different divergence angles and area ratios of exit-entry on the film cooling performance of the console. The results show that adjusting the geometric parameters of console properly could make the film cooling effect improved.

Due to the interaction between the cooling jet and mainstream, film cooling performance is affected by many film hole geometry parameters. Changing and optimizing the geometry parameters of shaped holes is an important method to make the film cooling performance of shaped holes improved. Lee et al. [20-22] have employed the single objective surrogate model approximation method with the film cooling effectiveness as the objective function to optimize the geometry of fan-shaped holes and laidback fan-shaped holes on the flat surface. Furthermore, they have employed the RSM surrogate model coupling with a multi-objective genetic algorithm to optimize the geometry of laidback fan-shaped holes with the film cooling effectiveness and aerodynamic loss as the objective function. Wang et al. [23] and Huang et al. [24] have employed radial basis function neural network (RBFNN) coupling with the genetic algorithm to optimize the geometry of laidback fan-shaped hole and round-to-slot hole on the flat surface respectively. Furthermore, Huang et al. [25] have employed RBFNN coupling with a variant of non-dominated sorting genetic algorithm (NSGA-II) with the film cooling effectiveness and discharge coefficient as the objective function to optimize the geometry of laidback fan-shaped hole on the suction surface of turbine vane. Jiang et al. [26] have employed response surface approximation coupled with NSGA-II to perform a multiple objective optimization on the high pressure turbine vane to optimize the geometry of multi-rows film cooling holes with the injection angle and compound angle as the design variables.

To our knowledge, the console has excellent film cooling performance. However, few researchers devote to reveal the effect of console transition surfaces on film cooling performance. In the current research, to study the influence of different transition surfaces of film holes on film cooling performance, four consoles with different transition surfaces are selected and designed for numerical investigation and comparative analysis under two types of blowing ratio.

2. Computational Procedures

2.1. Physical model

For the current research, the reference case is selected from the experimental model of console [16]. The computational domain is established according to the geometrical configuration of the reference case, as is shown in Figure 1(a). Due to the periodical distribution feature of consoles, the computational domain only includes one pitch of console. The computational domain and geometry of the console mainly include the mainstream flow channel, the coolant flow channel, and the console. The slot length of console (l) is 10 mm, the circular entry diameter of the console (d) is 5 mm, and the pitch of the console (P) is 12mm. The origin of coordinates is located at the center of the console outlet. As is shown in Figure1 (b), the length and the height of the coolant flow channel are 80mm and 20mm, respectively. In the x-direction, the center of the console is 100mm away from the mainstream channel inlet and 80 mm away from the mainstream channel outlet, and the height of the mainstream channel is 60 mm.

The geometry of the console is shown in Figure 1(c). After the gradual transition, the transition surface of the console changes from the round at the film hole entry to the rectangular slot at the console outlet. The geometric parameters of the console mainly include the inclination angle (θ), the slot width

(s), the slot length (l) and the hole height (t). The hole height (t) and the inclination angle (θ) are fixed at 6 mm and 35°.

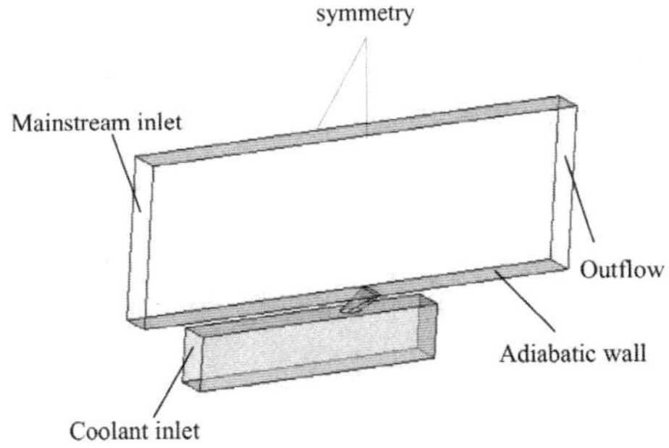

(a) Three-dimensional view

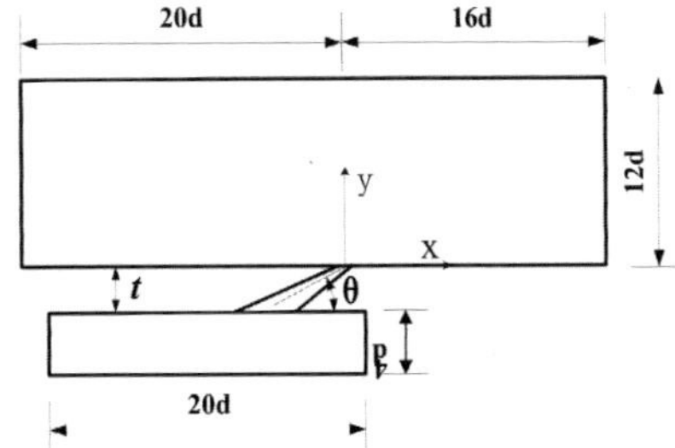

(b) Front view of computational domain

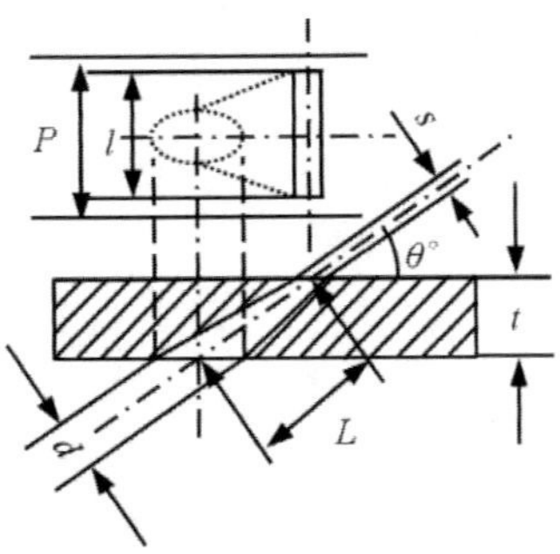

(c) Geometry of console

Figure 1. Schematic of computational model

To study the effect of different transition surfaces of the consoles on film cooling performance, four consoles with different transition surfaces are designed by changing the lateral transition surfaces, which are the outer-arc console, the inner-arc console, the s-shaped console, and the anti-s-shaped console respectively. The central sections of the reference and four designed consoles are shown in Figure 2.

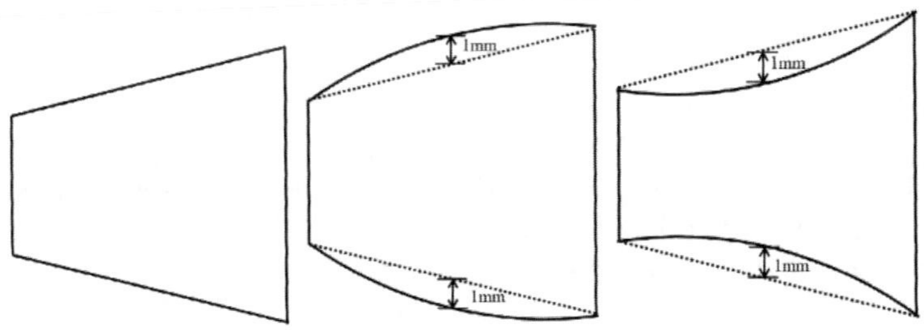

(a) Reference console (b) Outer-arc console (c) Inner-arc console

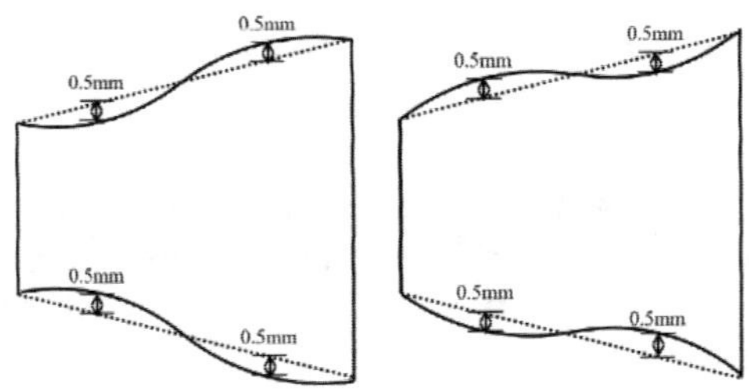

(d) S-shaped console (e) Anti-s-shaped console

Figure 2. The central sections of the reference and four designed consoles

2.2. Boundary condition and computational scheme

The boundary conditions of the calculation domain are set as follows: primary flow inlet and coolant flow inlet are velocity inlet. The temperature of mainstream and coolant flow are 335 K and 300 K respectively, and the primary flow velocity is 25 m/s. The primary flow outlet is pressure outlet, and the outlet static pressure is 0 Pa. The symmetry boundary condition is used for the two sides of the primary flow channel, and the adiabatic and no slip velocity condition is employed in the other wall surfaces. The turbulence intensity of free-stream in the primary flow channel is 2%.

It is noted that changing film hole exit shape could result in a change of average flow velocity, while the aerodynamic parameters in the film hole are not uniform. Therefore, the definition of the blowing ratio is based on the average parameters of the console inlet and primary flow entrance.

$$M = \frac{(\rho_c u_c)_{\text{inlet}}}{\rho_\infty u_\infty} \quad (1)$$

Where u_∞ and ρ_∞ are the velocity and density of primary flow entrance, respectively. u_c and ρ_c are the velocity and density of film hole inlet, respectively.

The adiabatic film cooling effectiveness is a key characteristic parameter for evaluating film cooling performance and it is defined as follows.

$$\eta_{ad} = \frac{T_\infty - T_{aw}}{T_\infty - T_c} \quad (2)$$

Where T_c presents the coolant flow temperature, T_{aw} presents the adiabatic wall temperature, T_∞ presents the primary temperature.

To better evaluate the performance of consoles with different transition surfaces, the spatially-averaged film cooling effectiveness in the range of x/d=1 and x/d=12 is another important evaluation parameter, which is defined as follows:

$$\eta_{ad,av} = \frac{1}{11d}\int_{d}^{12d} \eta_{ad,avx}\,dx \quad (3)$$

$$\eta_{ad,avx} = \frac{1}{2.4d}\int_{-1.2d}^{1.2d} \eta_{ad}\,dz \quad (4)$$

Where $\eta_{ad,avx}$ presents the laterally-averaged film cooling effectiveness in the range of z/d=-1.2 and z/d=1.2.

In order to evaluate the aerodynamic performance of consoles with different transition surfaces, the discharge coefficient is introduced. The discharge coefficient (C_d) is the ratio of the actual mass flow rate to the ideal mass flow rate through the film hole. A greater C_d implies a smaller aerodynamic loss in the console. C_d is defined as:

$$C_d = \frac{m_c}{A_{c,inlet}\sqrt{2\rho_c(p_c^* - p_c)}} \quad (5)$$

Where ρ_c and m_c are the density and the actual mass flow rate of the coolant flow. $A_{c,inlet}$ is the inlet area of film hole. p_c is static pressure of the film hole exit; p_c^* is the total pressure of the film hole inlet.

The film cooling performance and the flow field of the consoles are calculated using commercial software ANSYS FLUENT 14.5 based on Reynolds time-average method. Details of numerical simulation, computational meshes, independence verification of meshes, computational methods, and numerical verification have been described in detail in our previous work [24], and they are not presented here.

3. Results and Analysis

Table 1 presents the comparison of different transition surface film holes at M=0.5 and M=1.0. The results show that the transition surfaces on different film holes have a significant influence on the $\eta_{ad,av}$ downstream of the console exit between the ranges of s/d=1 and s/d=12. At M=0.5 and M=1, compared with the reference case, the $\eta_{ad,av}$ of inner-arc and anti-s-shaped console is improved, while that of outer-arc and s-shaped consoles is sharply deteriorated. At M=0.5, the $\eta_{ad,av}$ of inner-arc and anti-s-shaped consoles is improved by 5.8% and 6.4% respectively, while that of outer arc and s-shaped consoles is deteriorated by 13.8% and 7.2% respectively. At M=1, the $\eta_{ad,av}$ of inner-arc and anti-s-shaped consoles is improved by 1.0% and 7.0% respectively, while that of outer arc and s-shaped consoles is deteriorated by 24.5% and 12.1% respectively. The results also show that the transition surfaces on different film holes have a significant influence on the C_d. Compared with the reference console, the outer-arc and S-shaped consoles have larger discharge coefficients, i.e. smaller aerodynamic losses, and their discharge coefficients respectively increase by 5.0% and 3.4% at M=0.5 and respectively increase by 4.6% and 3.5% at M=1. The inner-arc and anti-s-shaped consoles have smaller discharge coefficients, i.e., more greater aerodynamic losses, and their discharge coefficients respectively decrease by 5.4% and 2.9% at M=0.5 and respectively decrease by 5.6% and 3.2% at M=1.

Table 1. Comparison of different transition surface consoles at M=0.5 and M=1

Different transition surface consoles	$\eta_{ad,av}$ (M=0.5)	C_d (M=0.5)	$\eta_{ad,av}$ (M=1.0)	C_d (M=1.0)
Reference	0.484	0.439	0.587	0.460
Outer-arc	0.417	0.461	0.443	0.481
Inner-arc	0.512	0.414	0.593	0.433
S-shaped	0.449	0.453	0.516	0.475
Anti-s-shaped	0.515	0.426	0.628	0.445

3.1. *Flow fields and dimensionless temperature distribution*

Figure 3 and Figure 4 present the dimensionless temperature (defined as $\Theta=(T_{aw}-Tc)/(T_\infty-Tc)$) contours on the film-hole centreline plane at M=0.5 and M=1. As is shown in figure 5 and figure 6, at M=0.5, the cooling jet flows out from the film hole, and a good film being oppressed by the primary flow can be formed and attached to the cooling wall. With the greater blowing ratio, the film hole exit momentum of cooling jet increases, the expansion ability along the downstream and the penetration ability to the mainstream increase. The thicker cooling film layer formed by the coolant air could have a higher influence range on the primary flow. Under the low blowing ratio, there is no significant difference in the dimensionless temperature contours on film-hole centreline planes with different transition surfaces. However, with the higher blowing ratio, the dimensionless temperature contours on film-hole centreline planes with different transition surfaces have an apparent difference. In Figure 4, compared with the reference case, the film layer in Figure 4(b) is the thickest and the outer-arc console has the best film coverage, while the film layer in Figure 4(c) is the thinnest and the inner-arc console has the worst film coverage in the y-direction, this is due to the difference of film hole transition surface. In Figure 4(b), the outer-arc transition surface guides the secondary flow to converge at the centre of the primary channel, while in Figure 4(c), the inner-arc transition surface guides the secondary flow to flow at both ends of the primary channel. In Figure 4(d) and Figure 4(e), for the s-shaped and anti-s-shaped transition surfaces, they all have a part of 'outer-arc transition surface' and a part of 'inner-arc transition surface' to guide the secondary flow, so the consoles with s-shaped and anti-s-shaped transition surfaces have similar film coverage in the y-direction.

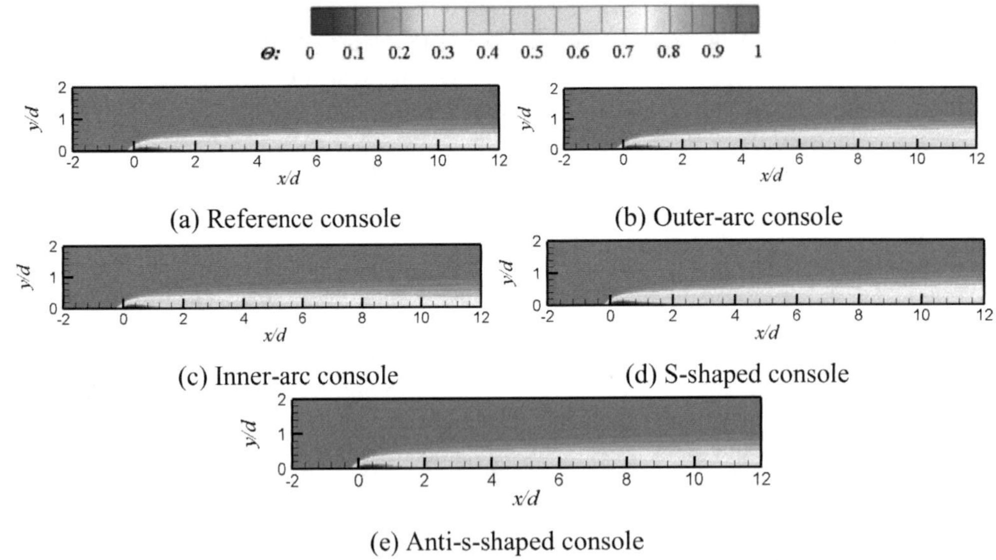

Figure 3. Dimensionless temperature contours on film-hole centreline plane at M=0.5

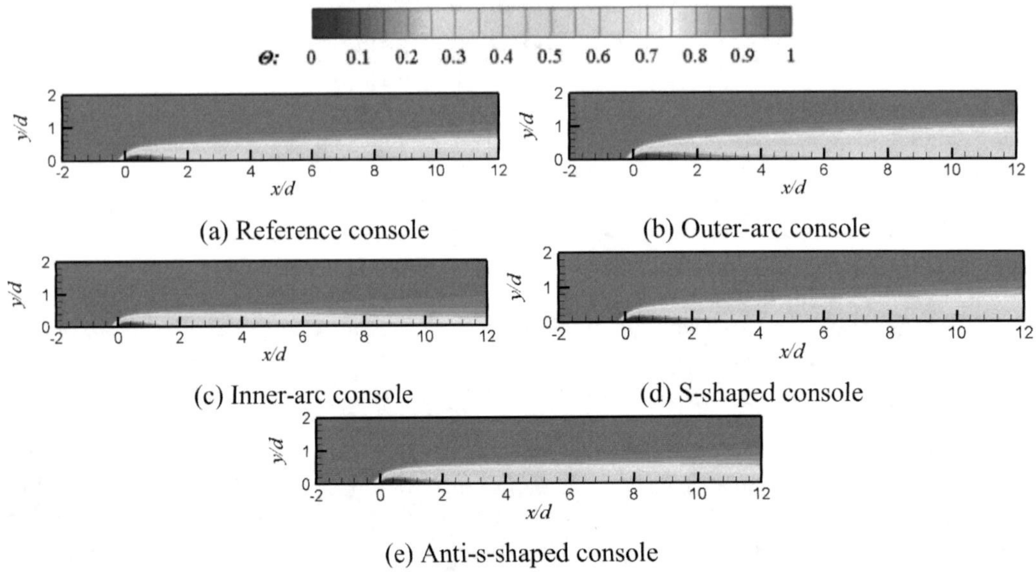

Figure 4. Dimensionless temperature contours on film-hole centreline plane at M=1

Figure 5 and Figure 6 present the dimensionless temperature contours at the symmetric plane at M=0.5 and M=1. Compared with the reference case, it can be seen that the dimensionless temperature of the inner-arc console and anti-s-shaped console have better laterally coverage, while the dimensionless temperature of out-arc console and s-shaped console in Figure 5 and Figure 6 have worse laterally coverage. This is because the inner-arc console and the anti-s-shaped console are with inner-arc transition surfaces near the film hole outlets, which are more conducive to guiding the secondary flow to both ends of the mainstream channel. The outer-arc console and s-shaped console are with outer-arc transition surfaces near the film hole outlets, which could lead the secondary flow to converge to the centre of the mainstream channel, and is not conducive to the lateral coverage of the cooling jet.

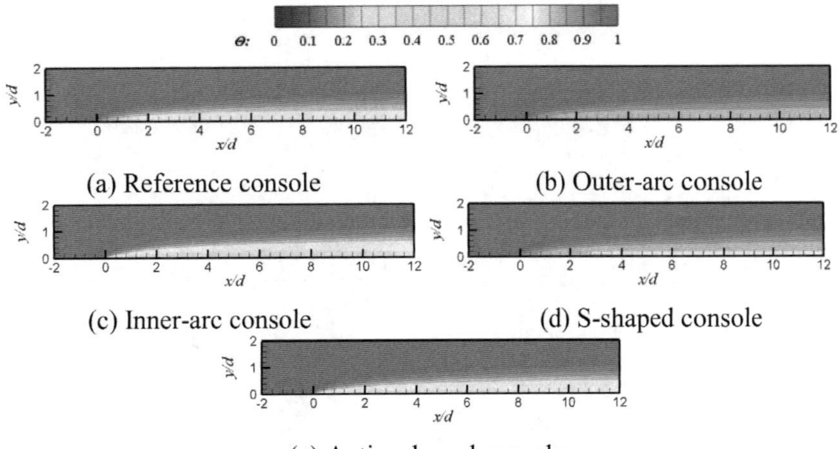

(a) Reference console (b) Outer-arc console

(c) Inner-arc console (d) S-shaped console

(e) Anti-s-shaped console

Figure 5. Dimensionless temperature contours at the symmetric plane at M=0.5

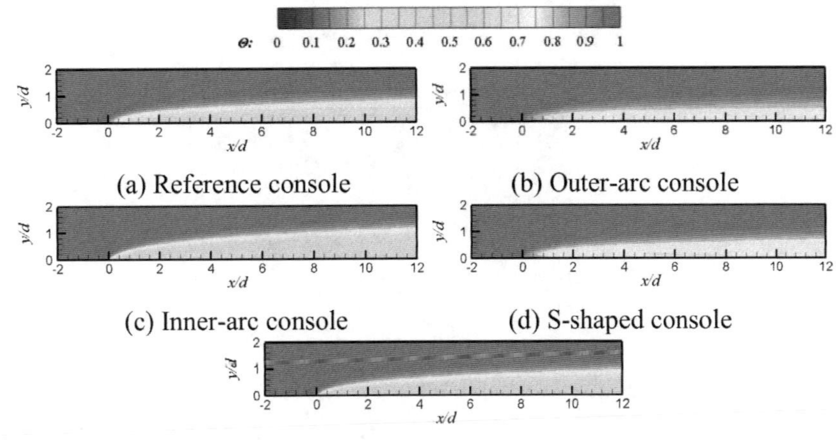

(a) Reference console (b) Outer-arc console

(c) Inner-arc console (d) S-shaped console

(e) Anti-s-shaped console

Figure 6. Dimensionless temperature contours at the symmetric plane at M=1

Figure 7 and Figure 8 present dimensionless temperature distribution and streamlines on the y-direction cross section at x/d=2 at M=0.5 and M=1. The consoles form reverse rotating vortex pairs which are opposite to the rotating direction of the traditional cylindrical holes in the downstream of the console outlet. This reverse rotating vortex pairs can effectively prevent the high-temperature primary flow from being sucked into the cooling wall below the cooling jet. The formation of these unique vortex pairs is closely related to the unique structure of consoles. From the five consoles with different transition surfaces, small vortex pairs are only formed at both ends of the reference console. Compared with the reference console, the inner-arc console and anti-s-shaped console have larger reverse rotating vortex pairs, which is more conducive to the lateral expansion of cooling jet. The outer-arc console and s-shaped console have relatively smaller reverse rotating vortex pairs. As a result of the guide effect of the outer-arc transition surface near the exit of the film, the vortex pairs core become closer to the cooling wall, inhibiting the lateral expansion of the cooling jet to a certain extent. The inner-arc transition surface near the outlet of the console is beneficial to the film cooling performance on the downstream near the outlet of the console.

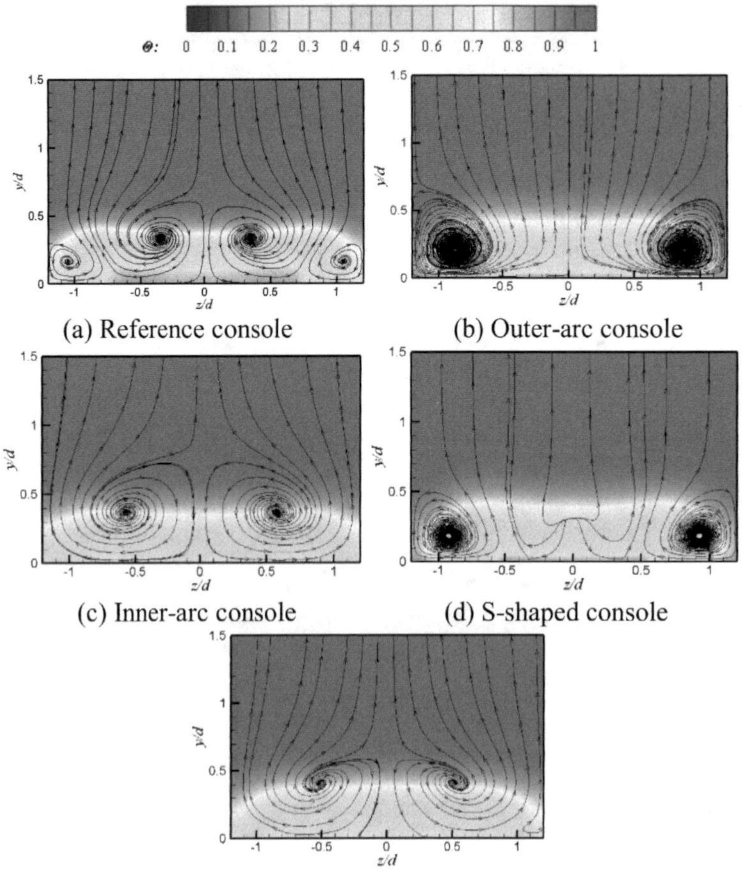

Figure 7. Dimensionless temperature distribution and streamlines on the y-direction cross section at x/d=2 at M=0.5

3.2. Film cooling effectiveness

Figure 9 and Figure 10, respectively, present the distribution of local η_{ad} at M=0.5 and M=1. At M=0.5, compared with the reference case, the local η_{ad} of inner-arc console and anti-s-shaped console is obviously improved, while the local η_{ad} of outer-arc console and s-shaped console is becoming worse. At M=1, compared with the reference case, similarly, the local η_{ad} of anti-s-shaped is obviously improved, the local η_{ad} of outer-arc console and s-shaped console is becoming worse. While the local η_{ad} of outer-arc console downstream near the console outlet is improved, and the local η_{ad} becomes worse downstream far away from the film hole exit. The reason is that the inner-arc transition surface near the film hole exit is more conducive to guide the secondary flow to both ends of the mainstream channel, while the outer-arc transition surfaces near console outlet tend to guide the secondary flow to converge at the center of the mainstream channel near the cooling wall surface. At M=1, the outer-arc transition surface guides the excessive secondary flow to the two ends of the cooling wall surface, which leads to reduced local η_{ad} in the center of the mainstream channel near the cooling wall surface.

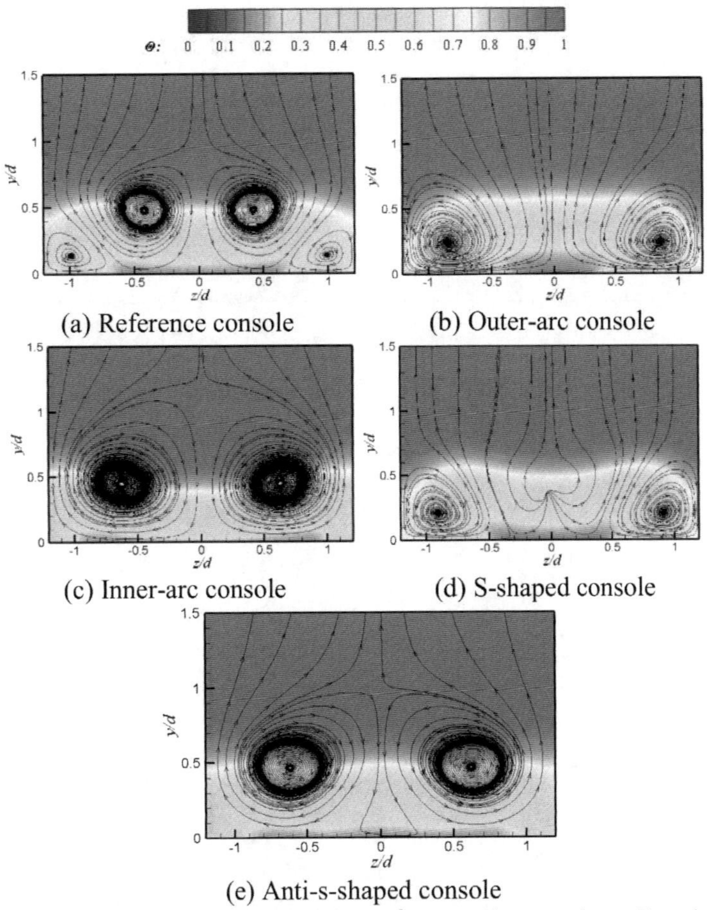

Figure 8. Dimensionless temperature distribution and streamlines on the y-direction cross section at x/d=2 at M=1

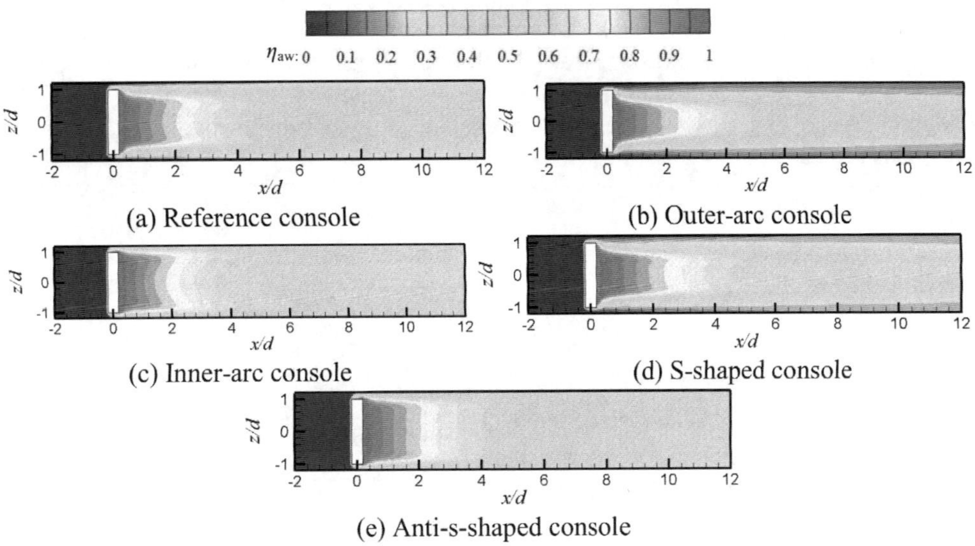

Figure 9. Distribution of local η_{ad} at M=0.5

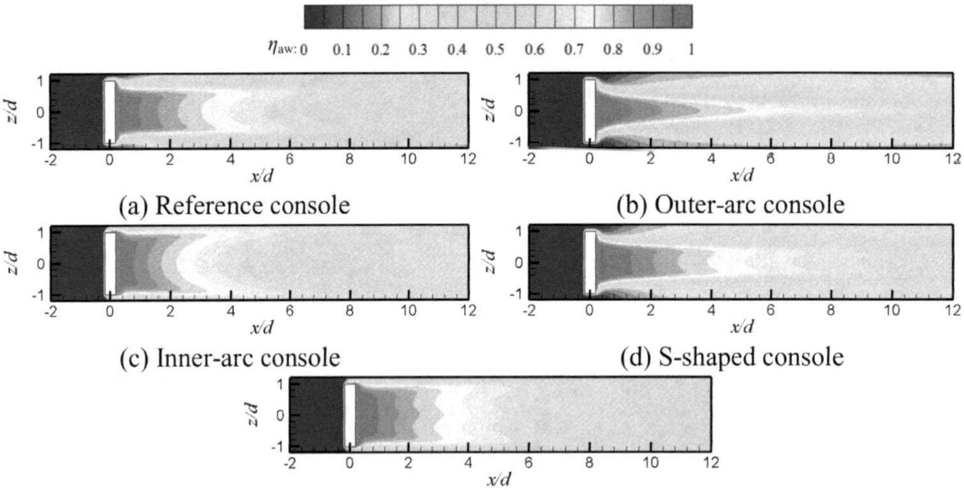

(a) Reference console (b) Outer-arc console

(c) Inner-arc console (d) S-shaped console

(e) Anti-s-shaped console

Figure 10. Distribution of local η_{ad} at M=1.0

Figure 11 and Figure 12, respectively, present the $\eta_{ad,avx}$ at M=0.5 and M=1. At M=0.5, compared with the reference case, the consoles with inner-arc and anti-s-shaped transition surfaces have better $\eta_{ad,avx}$, and these two types of film holes have similar $\eta_{ad,avx}$. In the range of x/d > 6, the $\eta_{ad,avx}$ of the console with anti-s-shaped transition surface is slightly higher than that of the console with inner-arc transition surface. The film hole with s-shaped and outer-arc transition surfaces have reduced $\eta_{ad,avx}$. Near the console exit at the position of x/d=1, the $\eta_{ad,avx}$ of consoles with inner-arc and anti-s-shaped transition surfaces has got the most significant improvement and is improved by 12.1% and 10.9% respectively. However, at the positon of x/d=1, the $\eta_{ad,avx}$ of consoles with outer-arc and s-shaped transition surfaces decreases the most, and is decreased by 15.1% and 8.0% respectively. At M=1, only the overall $\eta_{ad,avx}$ of the console with anti-s-shaped transition surface is better than that of the reference console. The console with inner-arc transition surface has better $\eta_{ad,avx}$ in the range of x/d < 6, while its $\eta_{ad,avx}$ becomes worse in the range of x/d > 6. Near the film hole outlet at the position of x/d=1, the $\eta_{ad,avx}$ of consoles with inner-arc and anti-s-shaped transition surfaces has got the most significantly improved and is improved by 18.0% and 12.1% respectively. The consoles with s-shaped and outer-arc transition surfaces have reduced $\eta_{ad,avx}$. Near the film hole exit at the position of x/d=4, the $\eta_{ad,avx}$ of console with outer-arc transition surface has decreased the most and is decreased by 25.2%. Near the film hole exit at the position of x/d=2, the $\eta_{ad,avx}$ of console with s-shaped transition surface has decreased the most and is decreased by 14.3%.

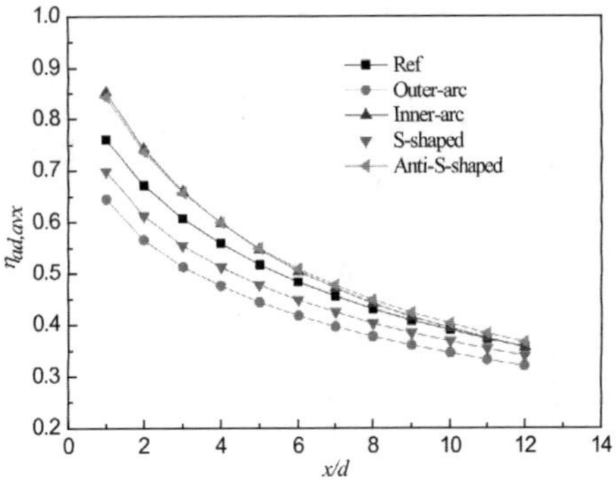

Figure 11. The $\eta_{ad,avx}$ at M=0.5

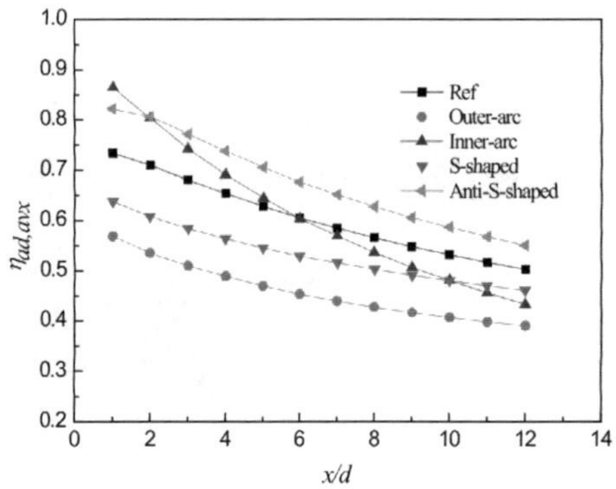

Figure 12. The $\eta_{ad,avx}$ at M=1

3.3. Discharge coefficient

In order to compare the pressure drop in the film hole more intuitively, a pressure drop ratio is introduced, which is defined as:

$$\psi = \frac{\left(p_c^* - p_c\right)}{\left(p_c^* - p_c\right)_{\text{ref}}} \tag{6}$$

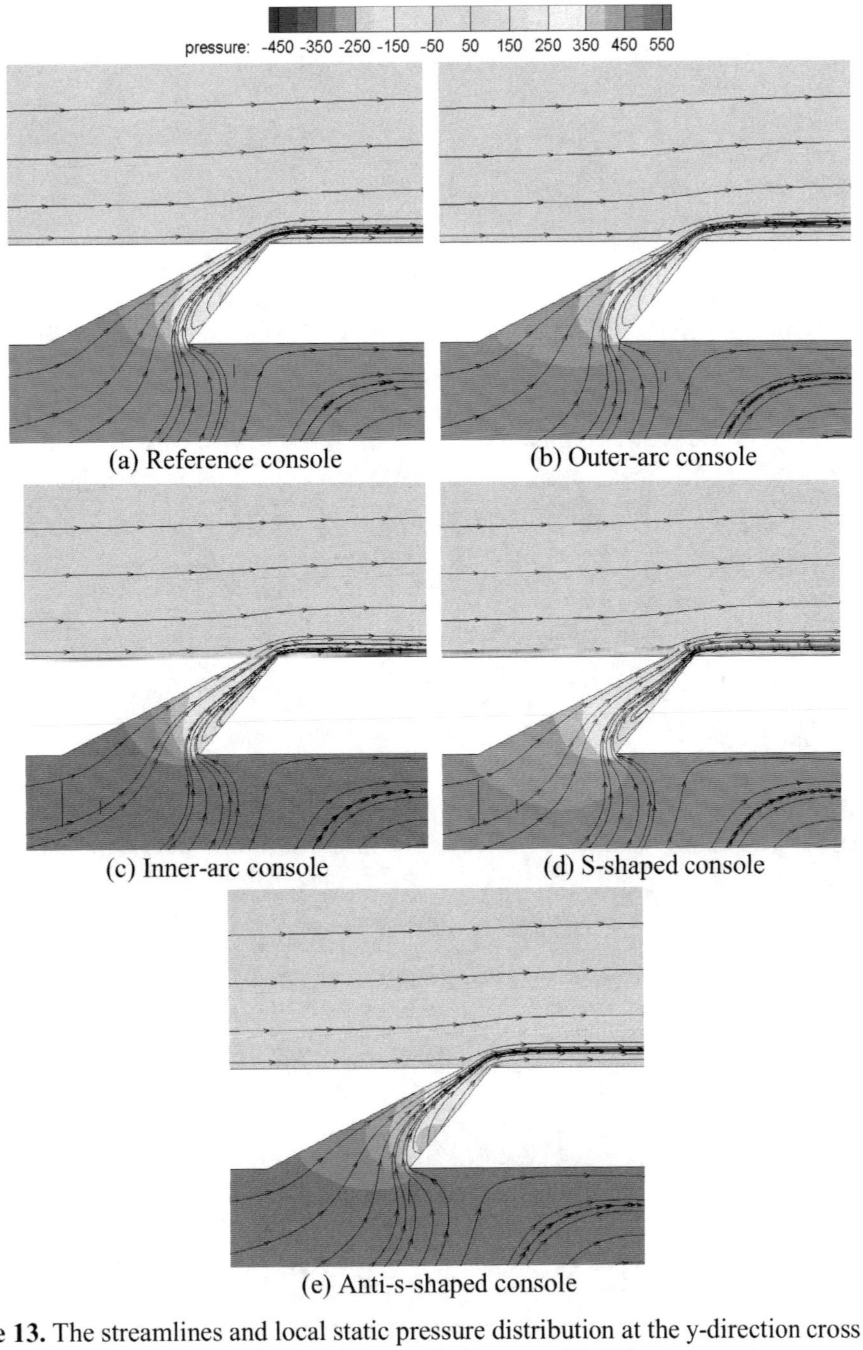

Figure 13. The streamlines and local static pressure distribution at the y-direction cross section through the console center at M=0.5

Table 2. The pressure drop ratios for consoles with different transition surfaces related to the reference case

	Outer-arc	Inner-arc	S-shaped	Anti-s-shaped
$\Psi(M=0.5)$	0.909	1.126	0.938	1.064
$\Psi(M=1.0)$	0.912	1.125	0.938	1.067

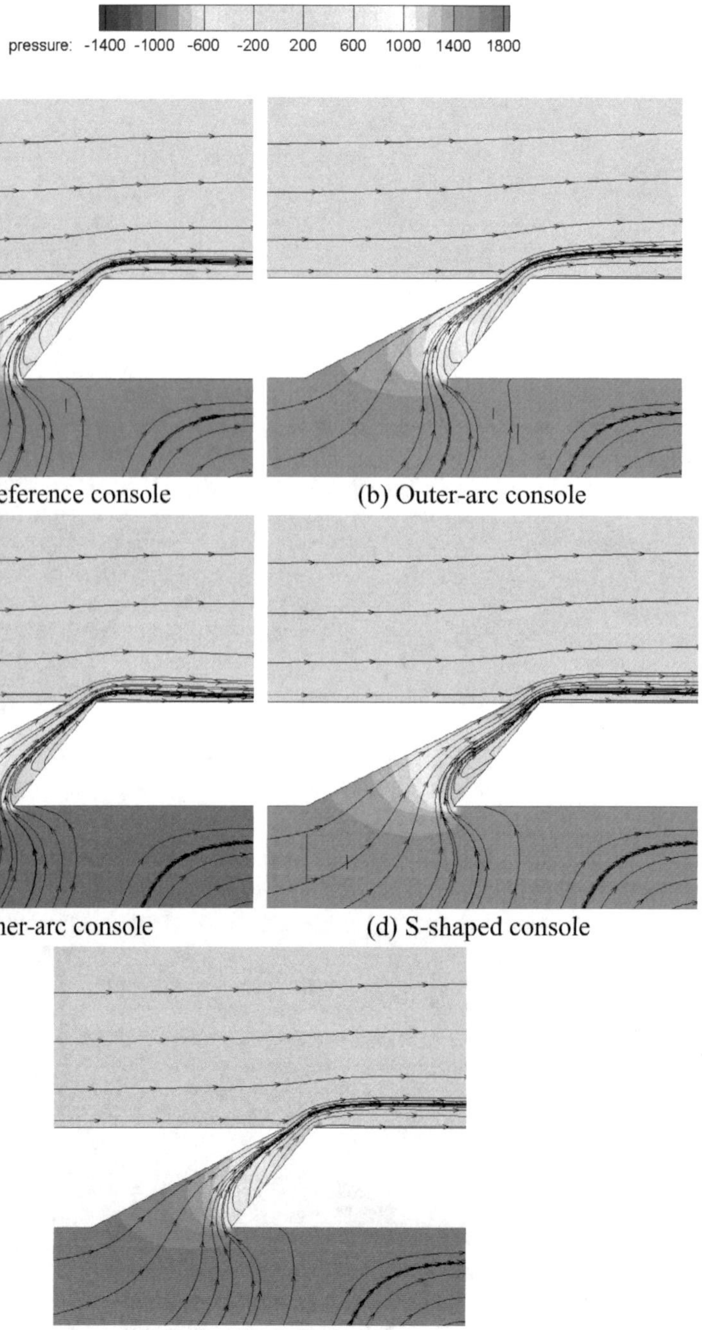

(a) Reference console (b) Outer-arc console

(c) Inner-arc console (d) S-shaped console

(e) Anti-s-shaped console

Figure 14. The streamlines and local static pressure distribution at the y-direction cross section through the console center at M=1

The pressure drop ratio for the consoles with different transition surfaces related to the reference case is shown in table 2. Figure 13 and Figure 14 show the streamlines and local static pressure distribution at the y-direction cross section through the console centre at M=0.5 and M=1. Compared with the reference console, the consoles with inner-arc and anti-s-shaped transition surfaces need larger bleeding pressure to overcome the pressure drop in film holes. The main reason is that the secondary flow guided by inner-arc and anti-s-shaped transition surfaces accelerates near the outlet of the console, resulting in greater flow loss.

4. Conclusions

To study the influence of transition surfaces of the console on film cooling performance, four transition surfaces are designed, including outer-arc transition surface, inner-arc transition surface, s-shaped transition surface, and anti-s-shaped transition surface. A series of numerical simulations were carried out under two types of blowing ratios of M=0.5 and M=1, and their heat transfer performance and flow field are investigated in detail. The present study is to determine the best configuration of transition surface of the console under different blowing ratios and to study the influence mechanism of the consoles with different transition surfaces on film cooling performance. The main conclusions are as follows:

1). At M=0.5 and M=1, the console with anti-s-shaped transition surface has the best $\eta_{ad,av}$. Compared with the reference console, the $\eta_{ad,av}$ is improved by 6.4% and 7.0% respectively, while the discharge coefficient is decreased by 2.9% and 3.2% respectively under the two blowing ratios.

2). At M=0.5 and M=1, the console with outer-arc transition surface has the worst $\eta_{ad,av}$. Compared with the reference console, the $\eta_{ad,av}$ is decreased by 13.8% and 24.5% respectively, while it has best dynamic performance, and the discharge coefficient is increased by 5.0% and 4.6% respectively under the two blowing ratios.

3). Compared with the reference case, at M=0.5 and M=1, the console with inner-arc transition surface has a better $\eta_{ad,avx}$, and its performance is improved by 5.8% and 1.0% respectively, while it has the worst dynamic performance and the discharge coefficient is decreased by 5.4% and 5.6% respectively.

Acknowledgments

Special thanks to Ying Huang for his suggestions on this article.

References

[1] D. Zhao, E. Gutmark, P. de Goey, A review of cavity-based trapped vortex, ultra-compact, high-g, inter-turbine combustors, Progress in Energy and Combustion Science 66 (2018) 42-82.
[2] L. Li, D. Zhao, X. Sun, Nonorthogonality analysis of acoustics and vorticity modes: Should thermoacoustic energy norm be time-invariant?, Aerospace Science and Technology 77 (2018) 149-155.
[3] R. S. Bunker, A review of shaped hole turbine film-cooling technology, Journal of heat transfer 127 (2005) 441-453.
[4] Y. J. Zheng, Hassan I, Experimental flow field investigations of a film cooling hole featuring an orifice, Applied Thermal Engineering 62 (2014) 766-776.
[5] D. G. Bogard, K. A. Thole, Gas turbine film cooling, Journal of propulsion and power 22 (2006) 249-270.
[6] M. Gritsch, W. Colban, H. Schair, et al, Effect of hole geometry on the thermal performance of fan-shaped film cooling holes, Journal of Turbomachinery 127 (2005) 718-25.
[7] H. H. Cho, D. H. Rhee, B. G. Kim, Enhancement of film cooling performance using a shaped film cooling hole with compound angle injection, JSME International Journal Series B Fluids and Thermal Engineering 44 (2001) 99-110.
[8] G. Barigozzi, G. Benzoni, G. Franchini, A. Perdichizzi, Fan-shaped hole effects on the aero-thermal performance of a film-cooled endwall, Journal of turbomachinery 128 (2006) 43-52.

[9] Y. Okita, M. Nishiura, Film effectiveness performance of an arrowhead-shaped film-cooling hole geometry, Journal of Turbomachinery 129 (2007) 331-339.
[10] C. L. Liu, H. R. Zhu, J. T. Bai, Study on the physics of film-cooling effectiveness enhancement by the converging-expanding hole, Journal of Aerospace Power 23 (2008) 598-604.
[11] J. E. Sargison, S. M. Guo, M. L. G. Oldfield, et al, A Converging Slot-Hole Film-Cooling Geometry—Part 1: Low-Speed Flat-Plate Heat Transfer and Loss ASME J Turbomach 124 (2002) 453-460.
[12] J. E. Sargison, S. M. Guo, M. L. G. Oldfield, et al, A converging slot-hole film-cooling geometry - part 2: transonic nozzle guide vane heat transfer and loss, ASME J Turbomach 124 (2002) 461-471.
[13] J. E. Sargison, M. L. G. Oldfield, S. M. Guo, et al, Flow visualisation of the external flow from a converging slot-hole film-cooling geometry, Experiments in Fluids 38 (2005) 304-318.
[14] A. Azzi, B. A. Jubran, Numerical modelling of film cooling from converging slot-hole, Heat & Mass Transfer 43 (2007) 381-388.
[15] Y. Yao, J. Z. Zhang, Investigation on film cooling characteristics from a row of converging slot-holes on fiat plate, Science China Technological Sciences 54 (2011) 1793-1800.
[16] Y. Yao, J. Z. Zhang, L. P. Wang, Film cooling on a gas turbine blade suction side with converging slot-hole, International Journal of Thermal Sciences 65 (2013) 267-279.
[17] Y. Yao, J. Z. Zhang, X. M. Tan, Numerical study of film cooling from converging slot-hole on a gas turbine blade suction side, International Communications in Heat & Mass Transfer 52 (2014) 61-72.
[18] C. L. Liu, H. R. Zhu, J. T. Bai, et al, Experimental research on the thermal performance of converging slot holes with different divergence angles, Experimental thermal and fluid science 33 (2009) 808-817.
[19] C. L. Liu, H. R. Zhu, J. T. Bai, et al, Film cooling performance of converging-slot holes with different exit-entry area ratios, Journal of Turbomachinery 133 (2011) 011020.
[20] K. D. Lee, K. Y. Kim, Shape optimization of a fan-shaped hole to enhance film-cooling effectiveness, International Journal of Heat & Mass Transfer 53 (2010) 2996-3005.
[21] K. D. Lee, K. Y. Kim, Surrogate based optimization of a laidback fan-shaped hole for film-cooling, International Journal of Heat & Fluid Flow 32 (2011) 226-238.
[22] K. D. Lee, A. Husain, K. Y. Kim, Multi-objective Optimization of a Laidback Fan Shaped Film-Cooling Hole Using Evolutionary Algorithm, International Journal of Fluid Machinery & Systems 3 (2010) 150-159.
[23] C. H. Wang, J. Z. Zhang, J. H. Zhou. Optimization of a fan-shaped hole to improve film cooling performance by RBF neural network and genetic algorithm, Aerospace Science & Technology 58 (2016) 18-25.
[24] Y. Huang, J. Z. Zhang, C. H. Wang, Shape-optimization of round-to-slot holes for improving film cooling effectiveness on a flat surface, Heat and Mass Transfer 54 (2018) 1741-1754.
[25] Y. Huang, J. Z. Zhang, C. H. Wang, et al, Multi-objective optimization of laidback fan-shaped film cooling hole on Turbine Vane Suction Surface, Heat and Mass Transfer 55 (2018) 1181-1194.
[26] Y. T. Jiang, H. F. Lin, G. Q. Yue, et al, Aero-thermal optimization on multi-rows film cooling of a realistic marine high pressure turbine vane, Applied Thermal Engineering 111 (2017) 537-549.

Aerodynamic Characteristic Analysis of V-22 Tilt-Rotor Aircraft in Hover

Peimin Xu, Yueyue Yang, Jie Zhou, Guiyu Zhou*

China Aerodynamics Research and Development Center, Mianyang, Sichuan, China

*Corresponding author e-mail: zhouguiyu29@163.com

Abstract. Tilt-rotor aircraft combines the vertical take-off and landing function of helicopters with the high-speed cruise function of fixed-wing aircraft, which has given rise to extensive attention in the aviation field, but the complex aerodynamic characteristics of the rotor flow field and the interference of rotor/fuselage have always been tough problems. In recent years, the continuous development of high-performance computer and the combination of high-precision CFD method and dynamic overset grid technology, make it possible to perform fine numerical simulation of tilt-rotor aircraft's flow field. In this paper, the low-dissipation fifth-order Roe-WENO scheme and overset grid method are used to solve the unsteady N-S equation, and the hovering state of tilt-rotor aircraft is numerically simulated. The result shows: under the interference of ground effect and fountain effect, the lift coefficient of blade when the azimuth angle is 180° to 360° is obviously smaller than that when the azimuth angle is 0° to 180°. The vortex structure of the hovering flow field of V-22 tilt-rotor aircraft is mainly composed of the fountain effect recirculation zone, the tip vortex wake and the root vortex wake. The tip vortex wake develops downwards, shrinks inward and approaches the fuselage. After the root wake falls off, it converges and fuses on the surface of nacelles. The root vortex wake is strong and moves towards the front of the fuselage and the plane of symmetry.

1. Introduction

Tilt rotor aircraft is the product of the combination of helicopter and fixed wing aircraft. It is mainly composed of rotor, wing, fuselage and tail wing. Rotor and engine nacelle are installed at the two wing tips of its wing. When the axis of the engine nacelle is in the vertical position, the tilt rotor aircraft is equivalent to a twin rotor transverse helicopter, which can take off and land vertically, hover and fly forward and backward; When the engine nacelle axis is in the horizontal position, the tilt rotor aircraft is equivalent to a fixed wing aircraft, which can fly at high speed and long range.

Due to the combination of the advantages of helicopter and fixed wing aircraft, tilt rotor aircraft involves a wide range of disciplines and many key technologies, and the rotor/wing/fuselage interference in aerodynamics is particularly complex.

2. Numerical Method

The turbulence model used in the numerical simulation is one-equation SA (Spalart-Allmaras) model. Roe scheme is used for spatial discretization, and the fifth order WENO scheme [1] is used to improve the accuracy of numerical simulation. LU-SGS implicit solution method is used for time discretization, and the time accuracy is improved by double time-step algorithm.

2.1. Governing equation

Based on the element center finite volume method, the integral form of the N-S equation in the conservative form is as follows:

$$\oiiint_{\Omega}\frac{\partial\bar{Q}}{\partial t}dV+\oiint_{S}(\bar{E}-\bar{E}_v)\bullet dS+\oiint_{S}(\bar{F}-\bar{F}_v)\bullet dS+\oiint_{S}(\bar{G}-\bar{G}_v)\bullet dS=0 \qquad (1)$$

Where Ω represents control volume, S represents control surface.

$$Q=\begin{pmatrix}\rho\\ \rho u\\ \rho v\\ \rho w\\ \rho e\end{pmatrix}\quad E=\begin{pmatrix}\rho u\\ \rho u^2+p\\ \rho uv\\ \rho uw\\ (\rho e+p)u\end{pmatrix}\quad F=\begin{pmatrix}\rho v\\ \rho vu\\ \rho v^2+p\\ \rho vw\\ (\rho e+p)v\end{pmatrix}\quad G=\begin{pmatrix}\rho w\\ \rho uw\\ \rho vw\\ \rho w^2+p\\ (\rho e+p)w\end{pmatrix}$$

$$E_v=\begin{pmatrix}0\\ \tau_{xx}\\ \tau_{xy}\\ \tau_{xz}\\ u\tau_{xx}+v\tau_{xy}+w\tau_{xz}-q_x\end{pmatrix}\quad F_v=\begin{pmatrix}0\\ \tau_{xy}\\ \tau_{yy}\\ \tau_{yz}\\ u\tau_{xy}+v\tau_{yy}+w\tau_{yz}-q_y\end{pmatrix}\quad G_v=\begin{pmatrix}0\\ \tau_{xz}\\ \tau_{zy}\\ \tau_{zz}\\ u\tau_{xz}+v\tau_{zy}+w\tau_{zz}-q_z\end{pmatrix}$$

2.2. Overset grid

Mesh generation is one of the keys of CFD calculation. Mesh quality directly affects the accuracy of numerical simulation [2]. Different from the fixed wing, the rotor has different forms of movements in the flight process. Therefore, in the grid generation, it is required not only to have good orthogonality and transition, but also to adapt to the large rotation movement.

The moving overset grid [3] does not need to be regenerated with the movement of the model. After each physical time step, the hole is dug again to generate the calculation grid at this time. During the calculation, the grid at zero time is used as the initial grid, and the Euler angle at current N time is used to calculate the rotation matrix. When calculating the grid at N+1 time, the grid at N time needs to be restored to the grid at zero time first. After the next displacement operation, the following relationship exists between the current grid and the initial grid:

$$X_{next}=M_{next}\bullet M_{cur}^{-1}\bullet\left(X_{cur}-X_{cgcur}\right)+X_{cgnext} \qquad (2)$$

Where M_{next} represents the coordinate transformation matrix at next time, M_{cur} represents the coordinate transformation matrix at current time, M_{cgcur} represents the centroid displacement at current time in inertial coordinate system, M_{cgnext} represents the centroid displacement at next time in inertial coordinate system.

2.3. Method verification

A symmetrical blade model with an airfoil of NACA0012 is selected, which is derived from the Caradonna-Tung rotor model experiment [4]. In the 1980s, a series of wind tunnel tests on this airfoil conducted by Caradonna and Tung at NASA Ames Research Center obtained a large number of valuable experimental data. Now these results have been widely used in the reliability test of rotor numerical program.

Fig. 1 shows the schematic diagram of the rotor experimental device of Caradonna-Tung. The rotor system consists of two straight blades with NACA0012 airfoil and no torsion. The blade tip rotation radius r=1.143m and the blade chord length C=0.1905m. In the numerical simulation, the blade root is undercut by 0.1R, and the blade hub and rotating shaft are ignored. Fig. 2 shows the Schematic diagram of Caradoona blade grid.

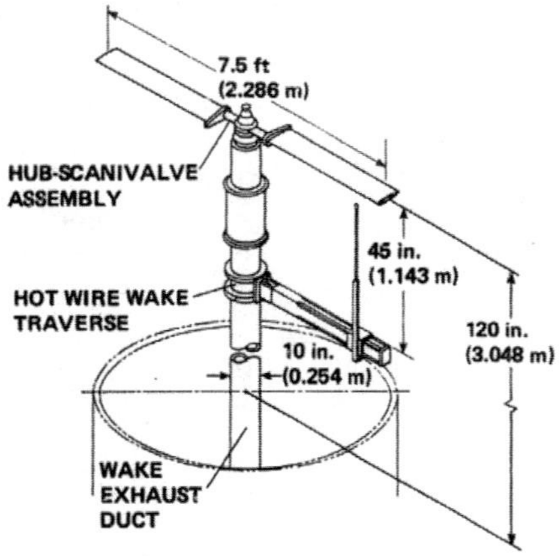

Figure 1. Schematic diagram of the rotor experimental device of Caradonna-Tung

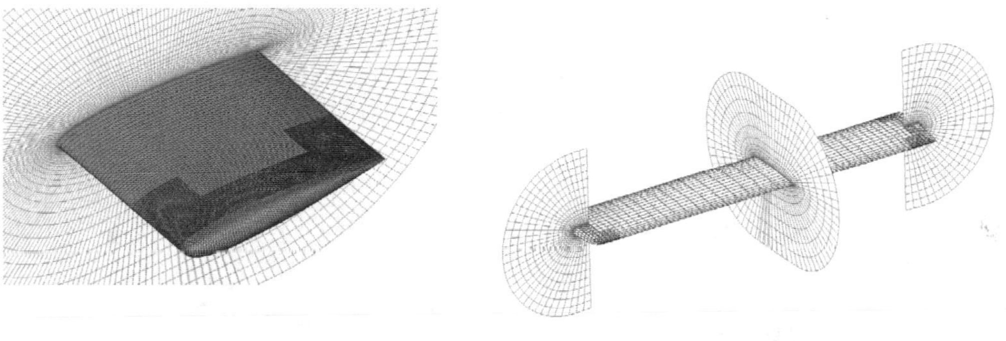

Figure 2. Schematic diagram of Caradoona blade grid

The incoming flow parameters in the numerical calculation are as follows:

Table 1. Calculation conditions of zero lift hovering state of Caradonna rotor

Height(km)	U_∞(m/s)	Ω(rpm)	θ_0(°)	Ma_{tip}
0	0	1500	0	0.52

Fig. 3 compares the pressure coefficient distribution of each section of the blade with the experimental value, in which the pressure coefficient is defined as follows:

$$C_p = 2(p - p_\infty)/\rho_\infty(\Omega r)^2 \tag{3}$$

Where p is the static pressure on the blade surface, is the rotational angular velocity, and r/R is the spanwise position of the blade profile. It can be seen that the pressure coefficient in this paper is in good agreement with the experiment, and the error is small.

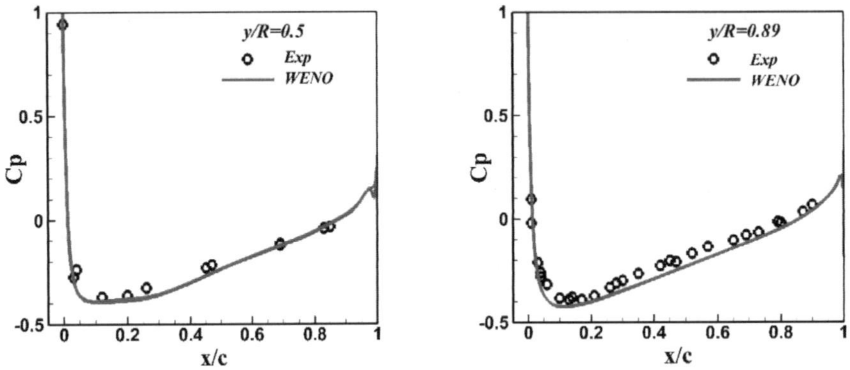

Figure 3. Comparison of pressure coefficient distribution of blade profile

3. Simulation and Discussion

3.1. Geometry and Computational grids

Since the shape parameters of V-22 tilt rotor aircraft are not fully disclosed, the modeling of its shape is mainly based on known pictures, data and some wind tunnel test references [5]. The V-22 tilt rotor aircraft is mainly composed of a horizontal twin rotor system composed of left and right blades, engine nacelles at both ends of the wing and the whole fuselage (also divided into wing, fuselage components and tail wing). In order to realize the given blade rotation and engine nacelle tilting motion in the numerical simulation, the root of the blade and the connection between the wing and the engine nacelle are cut off respectively.

In order to facilitate the given calculation parameters, the calculation model is a scaled model of V-22 tilt rotor aircraft with a scale of 0.25 [6, 7].

Fig. 4 shows the V-22 tilt rotor aircraft and its calculation model and Fig. 5 shows three views of calculation model.

Figure 4. V-22 tilt rotor aircraft and its calculation model

(a) Front view (b) Side view (c) Top view

Figure 5. Three views of calculation model

According to the literature investigation [8], taking the compressible viscous fluid as the wind tunnel test medium, when the 0.25 scale model is used for hover and forward flight experiments, the wind tunnel speed in helicopter mode is 62 knots, the given rotor pitch angle is 6°, and the forward ratio is 0.15. Therefore, the calculated rotation speed of the blade is 1377 rpm. The specific calculation parameters are shown in Table 2 below. Where the pitch angle refers to the angle of attack at the blade position of 0.75r.

Table 2. Numerical simulation parameters of V-22 tilt rotor aircraft in hover

Height(km)	U_∞(m/s)	Ω(rpm)	θ_0(°)	Ma_{tip}
0	0	1377	6	0.625

Considering the calculation efficiency and accuracy, in the first six blade rotation cycles, the weekly physical time step is set to 240 steps, that is, one-step rotation of 1.5° and sub iteration of 30 steps, so that the wake vortex can fully develop downward and the flow field can reach preliminary stability. After that, the physical time step is modified to 360 steps, that is, one step is rotated by 1° and the sub iteration is 40 steps. After 8 cycles, the flow field converges and the force coefficient changes periodically.

3.2. Blade Aerodynamic

Fig 6 and Fig 7 shows the variation curves of lift coefficient and resistance coefficient of single blade and left rotor in a rotation cycle in hover.

It can be seen that the lift coefficient of single blade is strongly influenced by coupled effect of ground effect and fountain effect. When the blade is at azimuth angle of 0° to 180° (outside the rotation axis of the rotors, away from the coupling interference zone), the lift coefficient is relatively high. While the blade is at azimuth angle of 180° to 360° (inside the rotation axis of the rotors, near the coupling interference zone), the lift coefficient is significantly reduced.

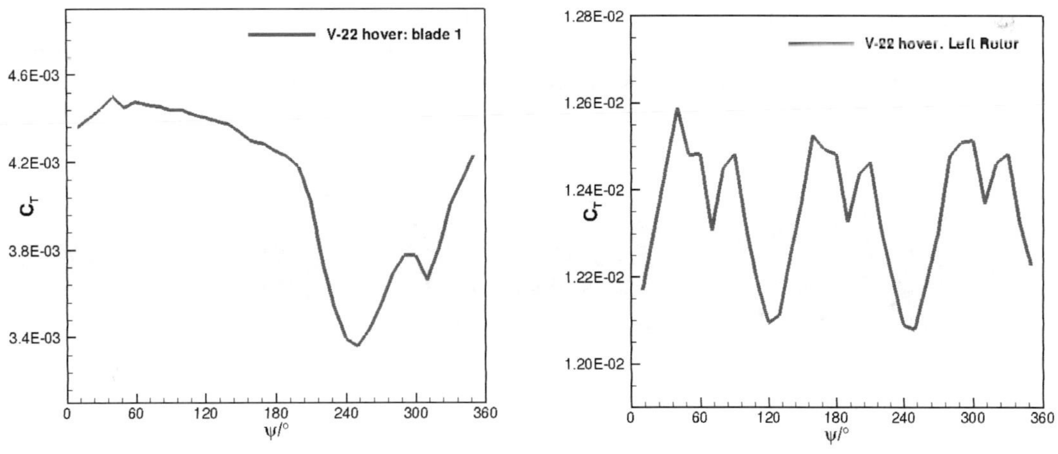

Figure 6. Lift coefficient of single blade and left rotor

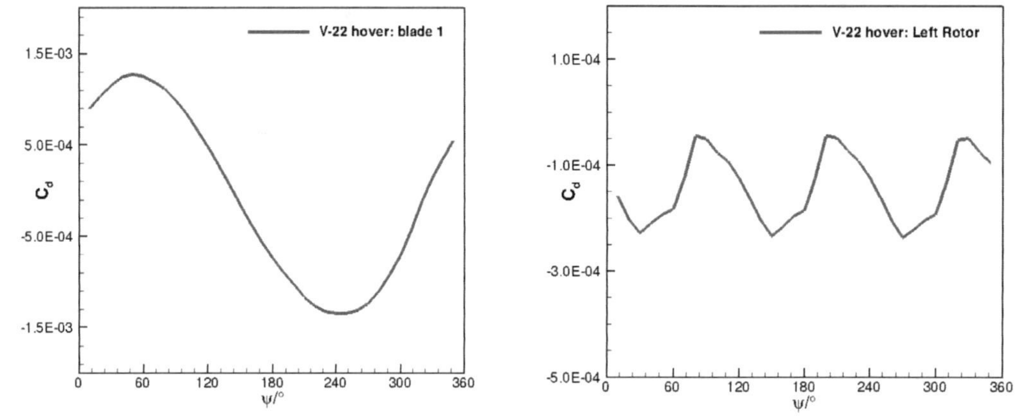

Figure 7. Resistance coefficient of single blade and left rotor

3.3. Flow Field and Vortex Structure

Fig. 8 shows different views of vorticity Q isosurface at the instantaneous moment of hovering flow field of V-22 tilt rotor aircraft, in which vorticity is calculated by Q criterion and colored with Mach number. Under the high-order scheme and huge amount of grid, the vortex structure of the whole aircraft is captured clearly, and the wake vortex develops downward to a long distance away from the fuselage.

From the front view of the vorticity isosurface, it can be seen that the propeller tip vortex gradually shrinks inward in the process of downward movement. After encountering the wing, the continuity of the wake vortex is interrupted. The vortex structures above the wing continue to accumulate to form the wake vortex surface and move towards and below the symmetry plane. Therefore, some vortex structures meet near the symmetry plane and roll up to form a fountain effect, some form a wake and move downward over the wing. After the continuity is interrupted by the wing, the outer tip vortex structure maintains the downward movement trend, but the inward contraction movement trend gradually disappears. At the same time, the tip vortex wake on both sides shows a trend of approaching the symmetry plane in the downward movement.

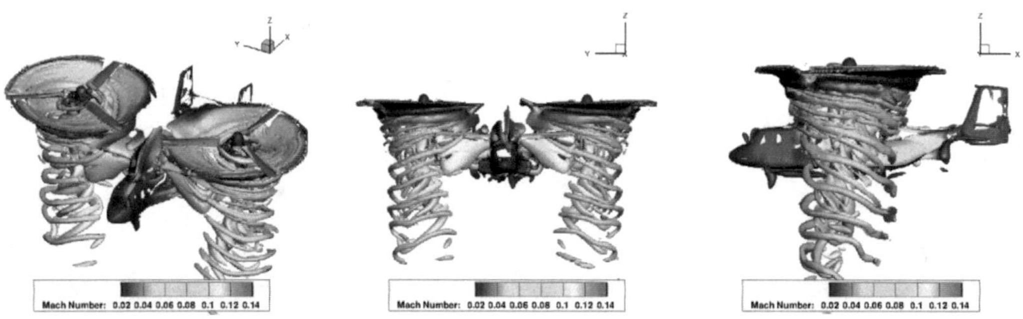

(a) Main view (b) Front view (c) Side view

Figure 8. Different views of vorticity Q isosurface of V-22 in hover

In order to analyze the asymmetry of tip vortex development in the X direction under the V-22 whole aircraft model, the X section(x=1.87m, where x=0m represents the head of V-22) passing through the wing is selected for vortex structure analysis. Fig. 9 shows the section vorticity contour map of this section at different times, where ψ refers to azimuth angle.

From the vorticity contour map across the wing section, it can be seen that the outer vortex structure of the rotor maintains a stable trend of inward contraction and downward movement. After the root vortex falls off, the root vortex on the outside of the rotor moves downward along the surface of the engine nacelle. The root vortex on the inside of the rotor is obstructed by the wing, and the gap is narrowed. When moving downward through the gap between the engine and the wing, the vortex structure is elongated and the motion trend is close to the symmetrical surface.

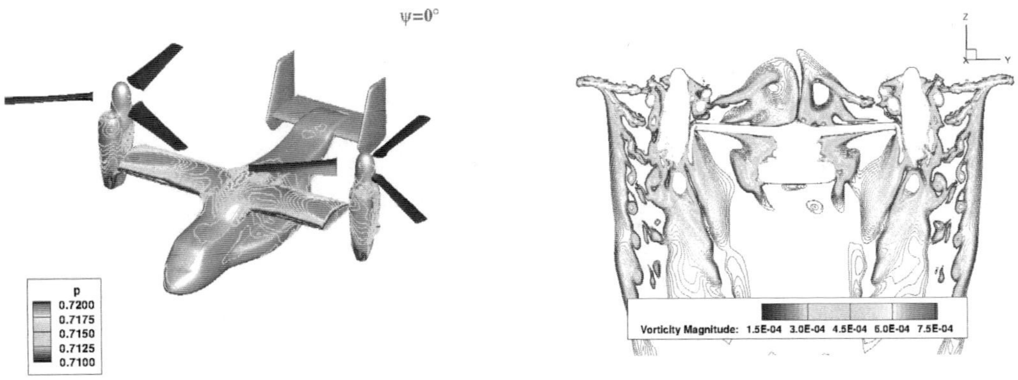

(a) ψ=0°

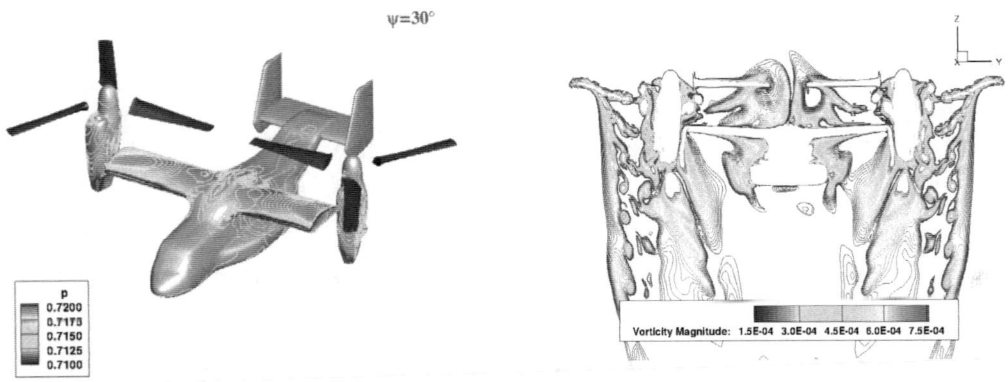

(b) ψ=30°

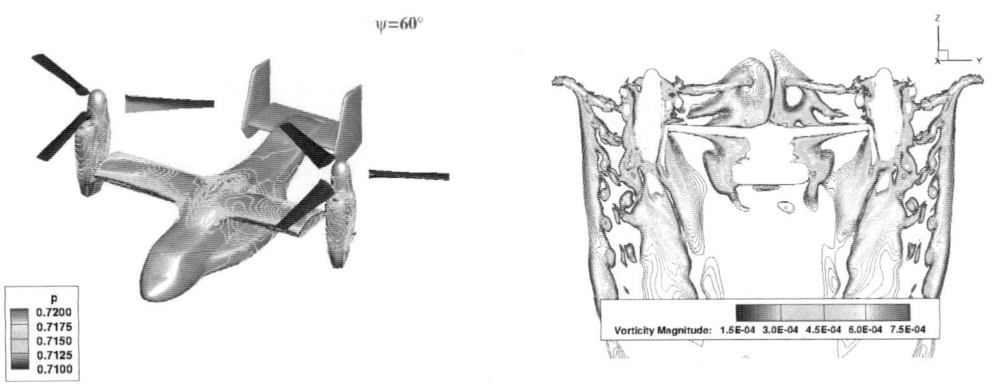

(c) ψ=60°

Figure 9. Vorticity contour map of X section of V-22 at different times in hover

4. Conclusion

The numerical simulation of hovering state is carried out for the V-22 Osprey tilt rotor aircraft model with a scale of 0.25, and the following preliminary conclusions are obtained: The vortex structure of hovering flow field of V-22 tilt rotor aircraft is mainly composed of fountain effect recirculation zone, tip vortex wake and root vortex wake. The tip vortex wake shrinks inward in the downward development and moves towards the symmetrical plane of the fuselage. After the propeller root vortex falls off, it converges and merges on the surface of the engine nacelle, and the strong vortex track inclined to the front of the fuselage and the symmetrical plane.

Acknowledgments

His work is supported by National Numerical Windtunnel (NNW) and national computational fluid dynamics laboratory of Beihang University. Special thanks to Pengcheng Cui for his suggestions on this article.

References

[1] Shu C.W. Essentially non-osillatory and weighted essentially non-oscillatory schemes for hyperbolic conservation laws[R]. NASA/CR-97-206253, 1997.
[2] Steger J L, Dougherty F C, Benek J A. A chimera grid scheme[C]. ASME Mini-Symposium on Advances in Grid Generation, Houston, TX, June 1982.
[3] Sturat E R, Norman E S, William E D. PEGASUS 5: An Automated Preprocessor for Overset-Grid Computational Fluid Dynamics[J]. AIAA Journal, 2003, 41(6).
[4] Caradonna F. X., Tung C. Experimental and Analytical Studies of a Model Helicopter Rotor in Hover[R]. NASA-TM-81-232, 1981.
[5] Young L A, Lillie D, McCluer M, et al. Insights into airframe aerodynamics and rotor-on-wing interactions from a 0.25-scale tiltrotor wind tunnel model[R]. NATIONAL AERONAUTICS AND SPACE ADMINISTRATION MOFFETT FIELD CA AMES RESEARCH CENTER, 2002.
[6] Mark A. Potsdam, Roger C. Strawn, "CFD Simulations of Tiltrotor Configurations in Hover"[J]. American Helicopter Society 58th Annual Forum, June 11-13, 2002.
[7] Freeman C. E., Mineck R. E. Fuselage surface pressure measurements of a helicopter Wind-Tunnel Model with a 3.15-Meter Diameter Single Rotor[R]. NASA TM 80051, 1979.
[8] Mineck R. E., Gorton S. A. Steady and periodic pressure measurement on a generic helicopter fuselage model in the presence of a rotor[R]. NASA TM-2000-210286, 2000.

Research on Building Ventilation Condition and Numerical Simulation Based on Fluid Mechanics Model

Huijuan Hao*

Henan Technical College of Construction, zhengzhou 450064, China

*Corresponding author e-mail: hhj11092022@163.com

Abstract. The article is based on the theory of computational fluid dynamics. We qualitatively analyse the difference between the total pressure on the windward side and the leeward side by numerical method, and compare and analyse the wind environment around the ground floor non-overhead courtyard building with the same width and height. The distribution of the velocity field, pressure field and trace map around the courtyard building with the ground floor fully overhead and the courtyard building with the ground floor not completely overhead are obtained. By introducing the CFD numerical simulation method, introducing finite element, Navier-Stokes equation, k-ε equation turbulence model, etc., combined with traditional ventilation design, the advantages of artificial ventilation and natural ventilation are integrated to obtain the optimal ventilation design scheme.

Key words: Fluid mechanics model; building ventilation; numerical simulation; indoor air flow field; thermal comfort.

1. Introduction

Green buildings are in the stage of rapid development in China, saving energy and improving the quality of indoor and outdoor environment of buildings are the basic requirements of green building design. Natural ventilation plays a vital role in energy saving and emission reduction, improving the comfort of the building environment and improving indoor air quality. Good natural ventilation can reduce the use time of air-conditioning equipment in transitional seasons, and it is the most effective for the Lingnan region with a humid and hot climate. passive design strategy [1]. Therefore, when air-conditioning technology is popularized and mechanical ventilation is widely used, architects should pay more attention to natural ventilation technology. Using Fluent software based on CFD simulation technology, the author analyses the distribution law of indoor flow field and temperature field, introduces indoor thermal comfort index, and proposes a simulation method of indoor natural ventilation flow field and thermal comfort. It can be seen from this that we should accurately evaluate and predict the outdoor ventilation of the building, and analyse the impact of the outdoor ventilation of the building on the indoor ventilation, especially the impact on the building's windproof measures and natural ventilation performance and the energy consumption of the building itself. There is no report on the ventilation performance prediction of hotel indoor and outdoor buildings in areas with hot summer and cold winter [2]. Therefore, this paper takes a hotel construction project as an example and uses the Swell Vent software to study the indoor and outdoor ventilation performance, the wind speed and flow of the surrounding site and the room. Field distribution in order to obtain the impact of natural ventilation on building energy consumption, and maximize the use of the natural environment around the building to achieve the purpose of building energy conservation.

2. Basic methods of CFD numerical simulation

There are three common methods to measure the wind environment of buildings: field measurement, wind tunnel experiment and CFD numerical simulation [3]. With the rapid development of computer performance, numerical simulation methods are widely used in the determination of wind environment. As shown in Figure 1, first, the finite volume method is used to obtain the calculation grid to generate calculation points to determine the boundary conditions of the object (the picture is quoted from An Overview of Experiments and Numerical Simulations on Airflow and Aerosols Deposition in Human Airways and the Role of Bioaerosol Motion in COVID-19 Transmission). The discrete equation is established and the test is repeated until the solution value converges, and the calculation result is finally displayed and output. This is the solution process of a brief CFD numerical simulation.

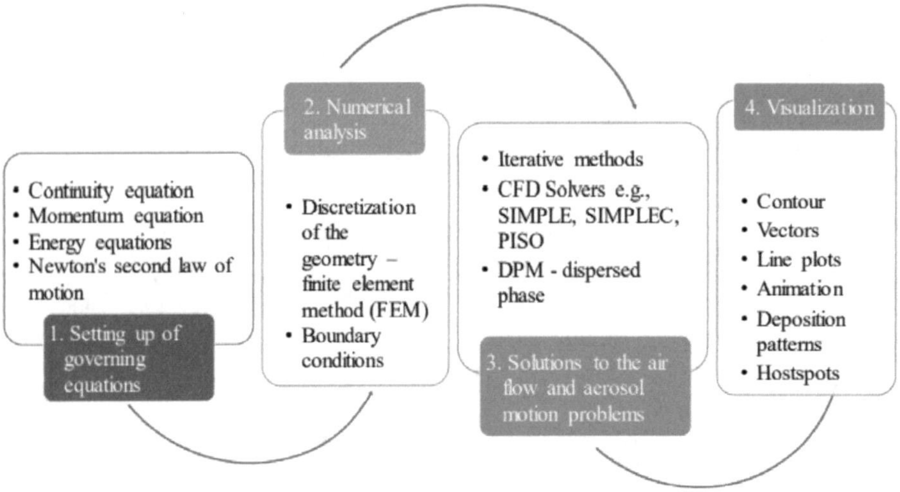

Figure 1. The solution process of CFD numerical simulation

Then, the flow chart of natural ventilation optimization design of green buildings based on CFD simulation is shown in Figure 2 (the picture is quoted from A CFD-Based Optimization of Building Configuration for Urban Ventilation Potential). First, the wind environment of the basic site is obtained by inputting the surrounding environment of the terrain and surrounding environment to obtain the wind environment of the basic site, which is the design basis of the overall layout, and then repeated CFD simulation optimization to obtain the indoor and outdoor wind environment, and then design the building shape, the enclosure interface, and improve the relationship between the building itself and the foundation, surrounding buildings, wind direction, and building layout, make full use of natural resources, obtain better natural ventilation conditions, and reduce the impact of unfavourable wind environment.

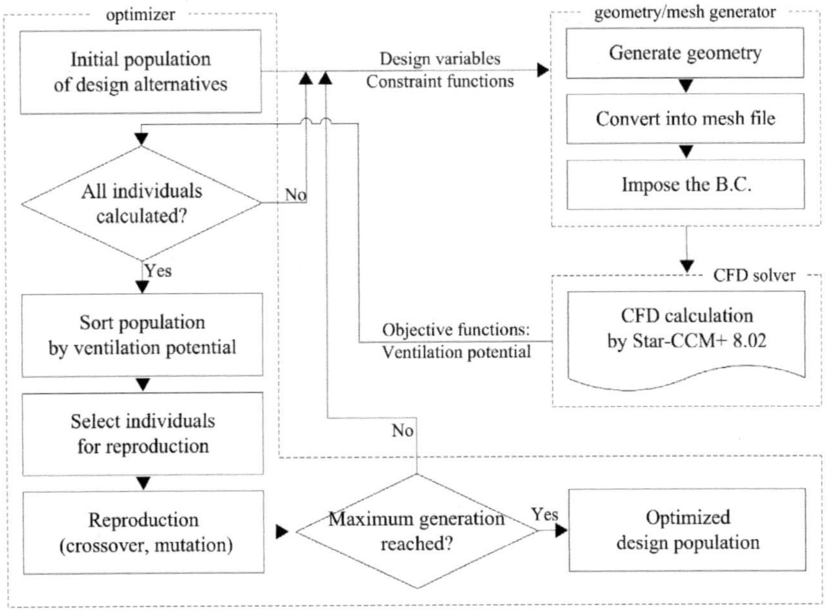

Figure 2. Optimization design process of natural ventilation in green buildings based on CFD simulation

3. Mathematical model

For the convenience of calculation, the following assumptions are made when establishing the mathematical model: it is assumed that the air in the office is incompressible and conforms to the Bossiness assumption; the air tightness of the indoor wall is considered to be good and the effect of air leakage is not considered [4]. The heat conduction and convection of the air in the office include two aspects: on the one hand, the heat conduction between the windows, walls and indoor air; on the other hand, the convection diffusion between the door and the indoor air. In this paper, the k-ε turbulence model combined with the energy conservation equation with temperature T as the variable is used to simulate the convection and heat transfer in the office under different air outlet positions. The governing equations for turbulent flow are the Reynolds-averaged Navier-Stokes equations in tensor form. Conservation of mass:

$$\frac{\partial U_i}{\partial X_i} = 0 \tag{1}$$

Where: U_i, X_i are dimensionless velocity and dimensionless length. Momentum equation:

$$\frac{\partial (U_i U_j)}{\partial X_i} = -\frac{\partial P}{\partial X_i} + \frac{\partial}{\partial X_i}[(\frac{1}{\mathrm{Re}} + \nu_t)\frac{\partial U_i}{\partial X_i} - \frac{2}{3}k] \tag{2}$$

Where P is the dimensionless pressure; Re is the Reynolds number; ν_t is the dimensionless air viscosity; k is the dimensionless turbulent kinetic energy. The turbulent kinetic energy transport equation:

$$\frac{\partial (U_i k)}{\partial X_i} = \frac{\partial}{\partial X_i}[(\frac{1}{\mathrm{Re}} + \frac{\nu_t}{\sigma_k})\frac{\partial k}{\partial X_i}] + P - \varepsilon \tag{3}$$

In the formula: σ_k is the turbulent model parameter, taking 1.0; ε is the turbulent kinetic energy dissipation rate [5]. The turbulent kinetic energy dissipation rate transport equation:

$$\frac{\partial(U_i\varepsilon)}{\partial X_i}=\frac{\partial}{\partial X_i}[(\frac{1}{\text{Re}}+\frac{\nu_t}{\sigma_\varepsilon})\frac{\partial\varepsilon}{\partial X_i}]+c_1 f_1\frac{\varepsilon}{k}P-c_2 f_2\frac{\varepsilon^2}{k} \tag{4}$$

In the formula, σ_ε is the turbulence model parameter, 1.3; the coefficient c_1, f_1, c_2, f_2 is 1.44, 1.0, 1.92, 1.0 respectively. The average shear stress generation term $\tau=\nu_t\frac{\partial U_i}{\partial X_j}[\frac{\partial U_i}{\partial X_j}+\frac{\partial U_j}{\partial X_i}]$, ν_t is calculated by the following formula

$$\nu_t=c_\mu f_\mu\frac{k^2}{\varepsilon} \tag{5}$$

In the formula: the coefficient c_μ, f_μ is taken as 0.09 and 1.0 respectively. The definition of the above dimensionless quantity is

$$X_i=\frac{x_i}{L_0}, U_i=\frac{u_i}{U_0}, P=\frac{p}{\rho U_0^{\ 2}}, k=\frac{k^*}{U_0^{\ 2}}, \varepsilon=\frac{\varepsilon^*}{U_0^{\ 2}}, \text{Re}=\frac{U_0 L_0}{\mu} \tag{6}$$

$x_i, u_i, p, k^*, \varepsilon^*$ represents the dimensional coordinate tensor, velocity tensor, pressure, turbulent kinetic energy and turbulent kinetic energy dissipation rate, respectively, and μ is the dynamic viscosity [6]. The finite volume method is used to discretize the above set of partial differential equations. Among them, the diffusion term uses the central difference method, while the QUICK difference method is used for the convection term. The SIMPLE algorithm and the staggered grid technology are used to realize the decoupling of pressure and velocity, so as to avoid the appearance of unreasonable pressure and velocity fields. The average wind speed changes along the height according to the exponential rate form commonly used in wind engineering is given by:

$$U=U_b(\frac{Z}{Z_b})^\alpha \tag{7}$$

U_b is the horizontal direction fractional velocity at the reference height Z_b, $Z_b=10m$; α is the roughness index, and α is taken as 0.30. After analyzing the meteorological data in Shenyang in recent years, Z_b three representative values U_b (Table 1).

Table 1. Selection of wind speed sample points

Serial number	Wind speed ($m\cdot s^{-1}$)	Remark
1	3.22	Average wind speed in transition season
2	1.95	Monthly mean wind speed minimum
3	3.93	Monthly mean wind speed maximum

The boundary conditions of the upper surface, the side and the outflow surface are much larger than the actual size of the calculation area, so it is considered that the flow at the outflow surface has been fully developed, and the normal gradient of each variable is zero. There is no actual physical boundary, and the assumption of boundary conditions is required [7]. Now the calculated control area is larger, the upper surface and the side are far away from the building wall, and the incoming flow is in the horizontal wind direction, it can be considered that the speed along the tangential direction The gradient is zero, the value along the normal direction is zero, and for other scalars such as k, ε, the gradient along the normal direction is taken to be zero.

4. Experimental simulation design

Using Fanger's thermal comfort equation to calculate the PMVη values of room types A and B under various working conditions, seven parameters need to be determined [8]. According to the actual living conditions of residents in Building 4. The values of the parameters are shown in Table 2.

Table 2. Parameter values

Metabolic rate M/ (W m^{-2})	70
acting	0
humidity/%	80
Clothing thermal resistance/(m^2·K·W^{-1})	0.08
Air velocity	Simulation calculated value
Radiation temperature/°C	30.4
air temperature	Simulation calculated value

According to the indoor three-dimensional flow velocity field and temperature field obtained by numerical simulation, combined with the thermal comfort parameters in Table 2, the Fanger equation was used to calculate the indoor thermal comfort index PMVη distribution under four meteorological conditions [9]. It can be regarded as the conclusion of thermal comfort environment within the range, and the thermal comfort conditions of the two types of units are compared and analysed as shown in Figure 3 (the picture is quoted from Efficiency evaluation of a floor standing air conditioner with different installation positions and air supply parameters applied to a large laboratory).

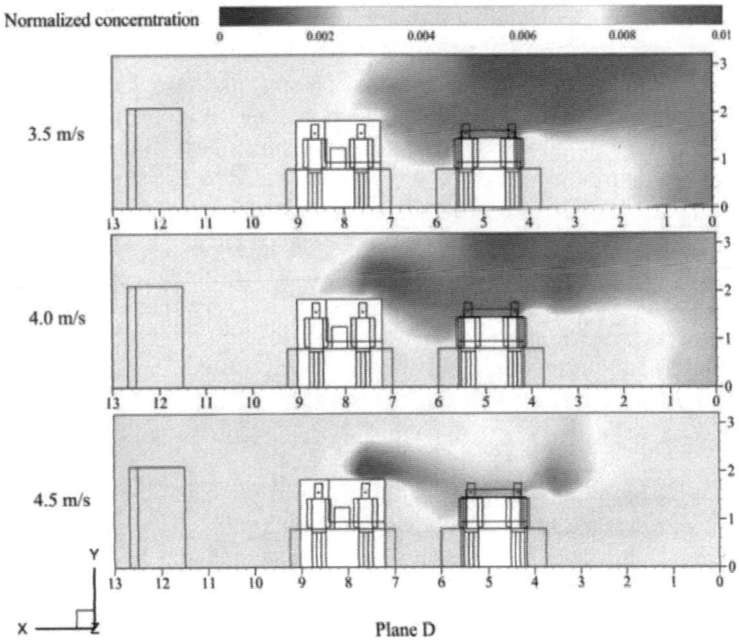

Figure 3. Distribution of thermal sensation index under different meteorological conditions (y=1.1m)

In order to judge the indoor thermal comfort, it is proposed that the ratio R2 of the area where the PMVη value is between -1 and +1 in the plane at the main activity height of 1.1, m accounts for the total room plane area as a measure of thermal comfort under natural ventilation. According to. The ratio of the area occupied by the indoor PMVη value between -1 and +1 under the four conditions is shown in Table 3.

Table 3. Area Ratio R_2

Weather condition	R_2/%	
	Type A	Type B
Weather conditions 1	32	24
Weather conditions 2	3	2
Weather conditions 3	23	4
Weather conditions 4	18	0

The comparison results show that the area ratio of the area with PMVη value between -1 and +1 in the type A room under the four conditions is also larger than that of the type B room, that is, the area that meets the comfort requirements of the type A room is more than that of the type B. Therefore, under the typical climatic conditions in summer, the natural ventilation effect and thermal comfort of type A room is better than that of type B room when the air conditioner is not turned on. Good natural ventilation conditions can effectively improve the thermal comfort environment. For Building 4, when the wind direction is south, the south wall is on the windward side, and part of the kinetic energy of the wind becomes static pressure, so that the pressure on the windward side is greater than the atmospheric pressure, forming a positive pressure area, and the pressure on the leeward side is less than the atmospheric pressure, forming a negative pressure area. Pressure zone. From the room type plan 2, room type A has direct air convection due to windows 1, 2, 4, 5, and 6 on the north and south sides; while room type B has windows on one side, so it is completely on one side. The role of wind pressure. For any room type, air temperature has a great influence on thermal comfort.

5. Conclusion

We use Swell Vent software to simulate the working conditions in different seasons and the indoor and outdoor ventilation of buildings in transition seasons. The results show that: in the outdoor pedestrian height area of the building, under the influence of the dominant wind in summer, a narrow tube effect occurs between the building complex and the south annex, and a large floc flow occurs at the passage, which has a great impact on travel. Therefore, it is recommended to Tall broad-leaved trees are planted on the outside of the site for proper shielding to reduce the impact of flocculation in the wind environment; the ventilation effect is good in winter, because the narrow pipe wind generated by the surrounding buildings has a small impact area, and there is no large flocculation in the site. The leeward side of the building forms a certain area of wind shadow area, which basically has no effect on the surrounding wind environment; the area with indoor ventilation times greater than 2/h accounts for 47.45% of the total area, and the natural ventilation effect is good, and the wind speed in most areas of the room and the ventilation times can better meet the indoor comfort.

References

[1] Yan Kai. Implicit Monte Carlo method in numerical simulation of radiation hydrodynamics. Atomic Energy Science and Technology, 55(3) (2021) 8.

[2] Jin Haian, Wang Dongxiang, Yu Jianfeng. Numerical simulation of computational particle fluid dynamics for the minimum spout velocity of spouted bed. Light Industry Machinery, 38(2) (2020) 7.

[3] Liu Yan, Qin Pan, Hong Wenpeng. Numerical Simulation of Rough Wall Grain-trending Sedimentation Process Based on Coupling of Computational Fluid Dynamics and Discrete Element Method. Science Technology and Engineering, 21(13) (2021) 17-22.

[4] Guo Ang, Zhang Jingkun, Hou Xiaojun, et al. Numerical Simulation of Ventilation System in Ship Silent Generator Room and Improvement of Real Ship. Ship Science and Technology, 42(9) (2020) 54-59.

[5] Kang Ahzhen, Yin Ruitao, Zhu Bing, et al. Numerical simulation of cofferdam wave and flow force during construction of cross-sea bridge based on LES. Journal of Southwest Jiaotong University, 55(3) (2020) 99-104.

[6] Sun Zhaoqiang, Zhang Xing, Liu Baoyu, et al. Optimal Design of Jet Pig Nozzle Based on Flow Field Numerical Simulation. Oil and Gas Storage and Transportation, 39(2) (2020) 87-92.
[7] Zeng Jia, Liu Binbin, Wang Jiamiao, et al. Numerical simulation and experimental verification of traditional Chinese medicine fluidized bed granulation process based on computational fluid dynamics. China Journal of New Medicine, 30(1) (2021) 11-18.
[8] Ding Mingsong, Fu Yang Aoxiao, Gao Tieshuo, et al. Hypersonic magnetohydrodynamics controls Hall effect. Acta Physica Sinica, 69(21) (2020) 18-22.
[9] Xu Baiping, Liu Biao, Liu Yao, et al. Mixed numerical simulation of distribution under the action of triangular rotors with different staggered angles. Chinese Journal of Chemical Engineering, 71(8) (2020) 101-114.

Study on analytical model of gas diffusion in underground pipe network

Tian Zhou*, Fan Yang, Wei Liu, Yue Tan

Beijing Institute of Radio Metrology and Measurement, China

*Corresponding author e-mail: zhoutians88@163.com

Abstract. Through the analysis of gas pipeline leakage causes and gas leakage diffusion model, the preventive measures for gas leakage accidents are put forward, which is of great significance for the harm of gas leakage and explosion accidents. In this paper, the mathematical model of gas leakage is established through hydrodynamics and pipeline gas motion equation, and the leakage process is analyzed by means of numerical simulation, the change law of gas dynamic leakage is obtained from the aspects of gas leakage conditions, gas movement and diffusion law.

1. Introduction

Urban gas generally includes natural gas, liquefied petroleum gas and artificial gas. The density of natural gas and artificial gas is higher than that of air, and it is passive diffusion of floating gas after leakage; The gaseous density of LPG is heavier than that of air, and it belongs to heavy gas diffusion after leakage. Different diffusion models need to be applied to gases with different properties. In the process of natural gas transmission and use, sometimes there will be gas leakage, which will produce serious potential safety hazards [1]. If measures are not taken in time, it will cause major safety accidents and economic losses. In recent years, gas poisoning and natural gas explosion accidents have occurred all over China. In daily life, natural gas leakage, poisoning and casualties caused by improper use of gas are common [2]. Therefore, how to realize the rapid and accurate detection of natural gas leakage has become a common concern [3].

2. Establishment of gas diffusion model

The neutral buoyancy diffusion model is used to estimate the concentration of the released gas mixed with air after the release, resulting in the downwind of the mixed gas cloud with neutral buoyancy. The gas pipeline leakage is generally described by the smoke plume model. Assuming that the wind direction is along the X axis, the wind speed is constant u, and the concentration of plumes in the continuous steady state source described by the Pasquill-Gifford model at H above ground level.

The gas whose density is greater than that of the ambient air is called heavy gas. Most of the existing heavy gas diffusion models are established on the basis of box model. Gaussian model and heavy gas diffusion box model have great limitations. In order to simulate the diffusion of heavy gas accurately and close to the reality, computational fluid dynamics (CFD) method is used to simulate the three-dimensional unsteady turbulent flow process. The leaked gas is regarded as incompressible fluid, and its temperature is regarded as the same as the ambient temperature, so K is selected- ε Based on the two equation model, the movement and diffusion law of leaked gas can be described by the control equation.

The control equation is discretized by a certain discretization method, and the boundary conditions are constructed according to the specific situation to solve the control equation, so that the gas concentration distribution and velocity distribution at the leakage site can be obtained. The geometric model of the leakage site is established in fluent and other software, and the momentum equation model, boundary conditions and other known parameters can be calculated automatically.

The determination of leakage rate is the basis of analyzing leakage diffusion and evaluating accident consequences. The main models are small hole leakage model, pipeline model and large hole leakage model according to the size of leakage aperture. In general, corrosion perforation is prone to small aperture leakage accidents, while for external damage (such as construction, natural disasters, etc.), large aperture leakage accidents are more likely to occur.

The small hole leakage model is applicable to the case of perforation leakage. Perforation leakage refers to the small hole formed by pipeline or equipment due to corrosion and gas leakage from the small hole. For the mass flow rate at any point of isentropic expansion of the ideal gas [4]:

$$Q = C_0 AP_2 \sqrt{\frac{2M}{RT^2}\frac{k}{k-1}\left[\left(\frac{P_a}{P_2}\right)^{\frac{2}{k}}\left(\frac{P_a}{P_2}\right)^{\frac{k+1}{k}}\right]}$$

Where, Q is the mass of gas passing through the hole; C0 is the gas leakage coefficient, generally ranging from 0.6 to 1.0. When the crack shape is round, take 1.00; when the crack shape is triangular, take 0.95; when the crack shape is rectangular, take 0.90; when the hole orifice is a tapered hole formed by inner corrosion (obtuse angle inlet), take 0.9 < C0 < 1.0; when the hole orifice is a tapered hole formed by external mechanical damage (obtuse angle inlet), take 0.6 < C0 < 0.9; A is the hole area, m2; P2 is the pressure at the central point of the corresponding pipeline at the leakage port, PA; M is the molar mass of gas, kg / mol; K is the hot melt ratio of gas, k = CP / CV; R G is the ideal gas constant, 8.314 kJ / (KMOL · K); T2 is the gas temperature at the central point of the corresponding pipeline at the leakage port, K; PA is the ambient pressure, PA [5].

From the perspective of safe work, the maximum flow of steam out of the hole is required. The pressure ratio causing the maximum flow rate is:

$$\frac{P_c}{P_{2c}} \leq \left(\frac{2}{k+1}\right)^{\frac{k}{k-1}}$$

The plug pressure PC is the maximum downstream pressure (critical pressure) leading to the maximum flow of hole or pipe flow, and its leakage is:

$$(Q_m)_c = C_0 AP_2 \sqrt{\frac{Mk}{RT^2}\left[\left(\frac{2}{k+1}\right)^{\frac{k+1}{k-1}}\right]}$$

The pipeline leakage model is suitable for the case of crack leakage. The cause of cracking and leakage is usually due to external force interference or overpressure rupture, which belongs to large-area leakage. The area of leakage port is usually 80% - 100% of the cross-sectional area of the pipeline. The instantaneous leakage of cracking leakage is large, resulting in a significant reduction in the pressure in the pipeline [6]. For ideal gas adiabatic pipe flow, the mass flux at sound velocity and non-sound velocity can be expressed by Darcy formula:

$$= \frac{m}{A} = Y_g \sqrt{\frac{2g_c\rho_1(P_1 - P_2)}{\sum K_f}}$$

Where G is the mass flux; M is the mass flow rate of gas, kg / S; A is the hole area, m2; YG is the gas expansion coefficient; GC is the gravity constant; ρ Is the upstream gas density, kg / m3; P1 is the upstream gas pressure, PA; P2 is the downstream gas pressure, PA; Σ KF is the difference loss term, which can be replaced by 4 fL/ D, f is the friction coefficient, D is the inner diameter, l is the pipe length, M [7].

For high-pressure gas transmission pipeline, there are two main flow states of gas [8].

When the hole diameter is small, the central point pressure P2 of the leakage port is slightly less than the starting point pressure P1, and far greater than the critical pressure PC. At this time, the leakage process is an isentropic expansion process with subcritical flow in the pipeline and critical flow in the leakage hole. The solution equation is:

$$\begin{cases} T_2 = \langle\frac{Y_1}{Y_2}\rangle T_1, P_2 = \langle\frac{Ma_1}{Ma_2}\rangle P_1\sqrt{\frac{Y_1}{Y_2}} \\ = \langle\frac{A}{A_0}\rangle P_2\sqrt{\frac{Mk}{RT^2}\left[\langle\frac{2}{k+1}\rangle^{\frac{k+1}{k-1}}\right]} = Ma_1P_1\sqrt{\frac{Mk}{RT_1}} = Ma_2P_2\sqrt{\frac{Mk}{RT_2}} \\ \frac{k+1}{2k}\ln\langle\frac{Ma_2^2}{Ma_1^2}\frac{Y_1}{Y_2}\rangle + \frac{(A_0/A)^2}{k\left[2/_{k+1}\right]^{(k+1)/(k-1)}}\langle 1-\frac{Ma_2^2}{Ma_1^2}\rangle + \langle\frac{4fL_e}{D}\rangle = 0 \\ Q = AP_2\sqrt{\frac{Mk}{RT^2}\left[\langle\frac{2}{k+1}\rangle^{\frac{k+1}{k-1}}\right]} \end{cases}$$

Where, f is the pipeline friction coefficient (dimensionless); L is the equivalent length of pipeline, m; D is the pipe diameter, m; A0 is the cross-sectional area of the pipeline, m2; G is the mass flow rate of gas, kg / S; 1 is the starting point of leakage, and 2 is the subscript; Other symbols are the same as above [9].

With the further increase of pore diameter, P2 is far less than P1, but it is still greater than the critical pressure PC. at this time, the leakage process becomes an equal line expansion process in which both the pipe and the leakage hole are critical flow. The solution equation is:

$$\begin{cases} \frac{k+1}{2k}\ln\langle\frac{2Y_1}{(k+1)Ma_1^2}\rangle + \frac{(A_0/A)^2}{k\left[2/_{k+1}\right]^{(k+1)/(k-1)}}\langle 1-\frac{1}{Ma_1^2}\rangle + \langle\frac{4fL_e}{D}\rangle = 0 \\ Q = AMa_1P_1\sqrt{\frac{Mk}{RT_1}\left[\langle\frac{2}{k+1}\rangle^{\frac{k+1}{k-1}}\right]} \end{cases}$$

In practice, when the leakage accident occurs for a certain time, the air source will be manually closed by the automatic control valve or the management personnel, so that the leakage presents a steady-state leakage process of gradually reducing and finally stopping. After the gas source is cut off, there is still a certain volume of gas in the pipe. The expression of dynamic leakage rate can be obtained according to the mass conservation equation [10].

When the air source is closed, the pressure at the starting point and along the way will gradually decrease with time. Even if the initial state is critical flow, the orifice leakage will become subcritical flow after a certain time. The criterion for changing from critical flow to subcritical flow is.

3. Summary

This paper mainly studies the gas leakage diffusion analysis model. The small hole leakage model, pipeline model and large hole leakage model can be used as a reference for small hole leakage caused by pipeline corrosion and pipeline cracking caused by external interference or overpressure. In particular, the monitoring of the output leakage rate of the model can effectively reflect the serious situation of leakage diffusion.

References

[1] Xue Kaixi, Hu Yanxiang, pan Guisheng, et al. Development trend of China's natural gas industry [J]. Gas and heat, 2011, 31 (12): b32-b37.
[2] Li Peiming, Jiao Wenling, song Hancheng, et al. Application and promotion of liquefied natural gas in China [J]. Gas and heat, 2008, 28 (1): b25-b28.
[3] Chen Li. Current situation and development trend of urban gas technology [J]. Gas and heat, 2007, 27 (12): 56-58.
[4] G. K. Batchelor, introduction to fluid dynamics, Cambridge University Press, London, 1970.
[5] L. Prandtl et al., translated by Guo yonghuai and Lu Shijia: introduction to fluid mechanics, Science Press, Beijing, 1981. (L. Prandtl, et al., Führer Dvrch die Strömungslehre, Friedr.

Vieweg und Sohn, Braunschweig, 1969.)
[6] Fluid mechanics, edited by Wu Wangyi, Peking University Press, Beijing, 1982.
[7] Xiao Xianglin Introduction to tribology [M] Tongji University Press, 1990, 5.
[8] Department of theoretical mechanics, Harbin Institute of technology Theoretical mechanics [M] Beijing: Higher Education Press, 2009, 7 [3] Tang Yanan Synchronous guidance of mechanical principle and complete solution of exercises [M] Xuzhou: China University of mining and Technology Press, 2008,3 [4] Sun Huan, Chen zuomo, Ge Wenjie Mechanical principle [M]. Version 8 Beijing: Higher Education Press, 2013, 5.
[9] Irrigation Association, Predicting and Estimating Landscape Water Use [Z]. Falls Church, VA USA: October 2001.
[10] James Barret ... , Golf Course Irrigation Environmental Design and Management Practices [M]. John Wiley & Son, Inc, USA: 2002.

Numerical simulation of sound scattering by a convected vortex

Ruixuan Ma[1,2,*], Conghai Wu[1], Shuaibin Han[1], and Shuhai Zhang[1]

[1]State Key Laboratory of Aerodynamics, China Aerodynamics Research and Development Center, MianYang, China
[2]Key Laboratory of Aerodynamic Noise Control, China Aerodynamics Research and Development Center, MianYang, China

*Corresponding author e-mail: maruixuan@cardc.cn

Abstract. The sound scattering by a convected isentropic vortex is numerically studied by solving the linearized Euler equations (LEE) with high-fidelity finite-difference methods and non-reflecting boundary conditions. Our computations are in good agreement with previous studies. Simulations for the scattering from a Taylor vortex at a range of convected velocity reveal that the convection of the vortex leads to the spectral broadening of the incident waves. The power spectral density analyses of the scattered pressure in the frequency domain show that the range of the spectral broadening is strongly influenced by convected velocity. This fact can be explained by the relative motion between the vortex and observer associated with the Doppler factor. The sidebands originating from the scattered fields at a fixed point in the frequency domain can be associated with the scattered beams observed in the case of scattering by a steady vortex.

1. Introduction

The non-uniform flow field results in not only the generation of the scattered waves, but also an obvious alteration of the amplitude and frequency of the incident waves. As one of the most significant problems in the area of aeroacoustics, the sound scattering from a vortical flow plays a principal role in a variety of fields of industrial [1-3] and scientific interest [3-7], including recognition and detection of acoustic targets, active noise reduction based on a vortex array, providing a benchmark used for testing high order schemes for aeroacoustics and cognition of interaction of complicated vortical flows with acoustic waves.

Research on sound scattering by a vortex-dominated flow attracts attention in numerous fields, such as superfluids, and plasma [8]. Originally from Lighthill [9,10], many scholars have carried out in-depth research in this field [11-14]. In recent years, several researchers have developed parabolic approximation [15] for long-range scattering and numerical approaches [16] to capture the details of the scattered field.

Most of the analytical and numerical studies on the scattering caused by turbulence were lack of accurate physical models to describe turbulent fields. In addition, turbulent flows, containing a large number of successive evolving vortices with varying scales and multiple physical processes of vortex motion, are so complicated that it is difficult to figure out precisely what role is played by a given character of turbulence in acoustic scattering. It is necessary to start from the research on a simplified two-dimensional vortical flow with an isolated vortex to bring further mechanical mechanisms responsible for acoustic scattering from turbulent flows. Utilizing spatial-scale separation between the sound wavelength λ and the vortex core size R, the vast majority of studies on this problem have been focused on theoretical efforts. Based on two limits of λ / R, two main approaches distinguished were presented. The first one, making use of the ray-tracing techniques [17, 18] from geometrical acoustics, holds in cases where the wavelength of the sound is very much smaller than the core size of the vortex, i.e. $\lambda / R \ll 1$. Originating from Born approximation, the second approach, which mostly combined with Lighthill's acoustic analogy [1, 19, 20], matched asymptotic expansions [21-24], theory of vortex sound [25], and partial harmonic expansion [26] in previous studies, is used in this regime where the condition $\lambda / R \gg 1$ is satisfied. Due to huge practical difficulties in solving the governing equations for acoustic scattering theoretically, including Navier-Stokes equations and Euler equations, theoretical studies had to apply numerous simplifications and approximations. As a result, these studies sometimes gave an irrational answer [1, 19, 27] and lost some significant properties of scattered fields, including phase, which were often not discussed in any detail. Limited by kinds of conditions for theoretical considerations, it is impossible to investigate wide-ranging effects of critical quantities on the scattered waves, such as the characteristic velocity of a vortex, the acoustic wavelength, and the vortex core size. While experimental studies [28-30] provided a wide-ranging frequency of incident waves, there are some new difficulties, such as the acoustic measurements of the near scattered field, precise control of flow properties, and desired incident waves, which make it hard to figure out concisely what properties are responsible for a scattering phenomenon observed in an experiment. Over the past few decades, rapid developments of computer technologies and numerical methods have enabled simultaneously resolved the flow and acoustic fields, which brought a huge opportunity to improve our understanding of wide-ranging scale effects and scattering mechanisms, by directly solving Navier-Stokes equations [18, 24] and linearized Euler equations [31].

In this paper, following the previous work [31], we focus on the acoustic scattering by a convected vortex. Unlike [31] which was focused on the development of a semianalytical model, our own numerical efforts are to investigate the effects of convected velocity on the spectral characteristics of scattered fields in detail by solving the linearized Euler equations with high-fidelity methods. The corresponding results bring further physical insights into the acoustic scattering by shear flows.

For brevity, the quantities in this paper are all normalized by the density ρ_∞ and the acoustic velocity c_∞, where the subscript ∞ represents the quantity in the far-field. The rest of paper is organized in the following way. Firstly, we present physical models and numerical methods in section 2. Section 3 is devoted to verifying the accuracy of the current numerical method. Section 4 investigates the sound scattering by a convected Taylor vortex, and then related discussions are presented. Finally, we summarize and conclude the current work in section 5.

2. Methodology

2.1. Physical model

Following Colonius et al. [18], Taylor vortex is considered in this paper. In the following, we give its expressions,

$$\begin{cases} u_r = 0 \\ u_\theta(r) = M_v r \exp\left[\left(1-(r/R)^2\right)/2\right] \\ \rho(r) = \left[1 - \dfrac{\gamma-1}{2} M_v^2 \exp\left(1-(r/R)^2\right)\right]^{1/\gamma-1} \\ p(r) = \dfrac{1}{\gamma}\left[1 - \dfrac{\gamma-1}{2} M_v^2 \exp\left(1-(r/R)^2\right)\right]^{\gamma/\gamma-1} \end{cases}, \tag{1}$$

where M_v denotes the vortex strength, and R denotes the vortex core size.

Next, following Clair & Gabard [31], we consider the Taylor vortex with a given convected velocity M_c in the x-direction, which are given by

$$\begin{cases} u_0 = M_c + M_v \dfrac{y-y_c}{R}\exp\left[\left(1-r^2/R^2\right)/2\right] \\ v_0 = -M_v \dfrac{x-x_c}{R}\exp\left[\left(1-r^2/R^2\right)/2\right] \\ \rho_0 = \left[1-\dfrac{\gamma-1}{2}\exp\left(1-r^2/R^2\right)\right]^{1/(\gamma-1)} \\ p_0 = \dfrac{1}{\gamma}\left[1-\dfrac{\gamma-1}{2}\exp\left(1-r^2/R^2\right)\right]^{\gamma/(\gamma-1)} \end{cases} . \quad (2)$$

In the above expressions, $r=\sqrt{(x-x_c)^2+(y-y_c)^2}$, and (x_c, y_c) is the center of the Taylor vortex at the time t, specifically $x_c = x_{c0} + M_c t$, $y_c = y_{c0}$, where (x_{c0}, y_{c0}) is the position of the vortex center at the initial moment.

2.2. Governing equations

We adopted a two-dimensional linearized Euler equations (LEEs) with the conservative form in the current work:

$$\frac{\partial \mathbf{U}}{\partial t}+\frac{\partial \mathbf{E}}{\partial x}+\frac{\partial \mathbf{F}}{\partial y}+\mathbf{H}=0 \quad (3)$$

with

$$\mathbf{U}=\begin{pmatrix}\rho' \\ \rho_0 u' \\ \rho_0 v' \\ p'\end{pmatrix}, \mathbf{E}=\begin{pmatrix}\rho_0 u' + \rho' u_0 \\ \rho_0 u_0 u' + p' \\ \rho_0 u_0 v' \\ u_0 p' + \gamma p_0 u'\end{pmatrix}, \mathbf{F}=\begin{pmatrix}\rho_0 v' + \rho' v_0 \\ \rho_0 v_0 u' \\ \rho_0 v_0 v' + p' \\ v_0 p' + \gamma p_0 v'\end{pmatrix}, \mathbf{H}=\begin{pmatrix} 0 \\ -\dfrac{\rho'}{\rho_0}\dfrac{\partial p_0}{\partial x} + \rho_0 u'\dfrac{\partial u_0}{\partial x} + \rho_0 v'\dfrac{\partial u_0}{\partial y} \\ -\dfrac{\rho'}{\rho_0}\dfrac{\partial p_0}{\partial y} + \rho_0 u'\dfrac{\partial v_0}{\partial x} + \rho_0 v'\dfrac{\partial v_0}{\partial y} \\ (\gamma-1)\left[p'\left(\dfrac{\partial u_0}{\partial x}+\dfrac{\partial v_0}{\partial y}\right)-\left(u'\dfrac{\partial p_0}{\partial x}+v'\dfrac{\partial p_0}{\partial y}\right)\right]\end{pmatrix}.$$

Here, ρ', u', v' and p' represent the acoustic fluctuations, namely density, velocity in the x-direction, velocity in the y-direction, and pressure; γ denote the ratio of the specific heats (taken to be 1.4 for air in what follows); ρ_0, u_0, v_0, and p_0 denote the base flow properties, which are given by (2).

2.3. Computation setup

An enhanced optimized schemes proposed by Wu [32] is adopted in this paper. A monochromatic plane incident wave is assumed to be generated at the bottom computational boundary, propagating along the $+y$ direction, as shown below

$$\begin{pmatrix}\rho' \\ u' \\ v' \\ p'\end{pmatrix} = \varepsilon \begin{pmatrix}1 \\ 0 \\ 1 \\ 1\end{pmatrix} \sin\left[\frac{2\pi}{\lambda}(y-t)\right], \quad (4)$$

where λ denotes the wavelength of the incident waves; ε is taken to be $10^{-5} \ll 1$ to ensure that the sound waves are well with the small perturbation zone and the interaction of the incident acoustic wave with vortex flow can be regarded as an isentropic process. To minimize the reflection of outgoing waves, a buffer zone proposed by [33, 34] is used at the boundaries to absorb sound waves. Moreover, extrapolation boundary conditions are applied on the last grid points. Specific computation settings are see Ref. [35] for details.

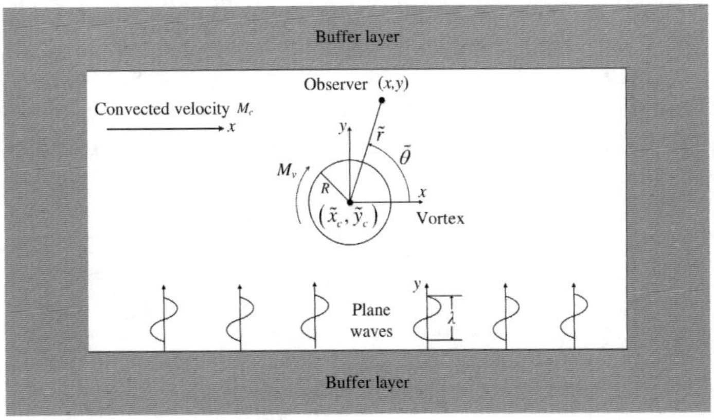

Figure 1. Schematic diagram of computation configuration.

3. Verification

Fig. 2 shows the scattered pressure at $r = 8R$. Our computations are in good agreement withprevious studies [18, 31].

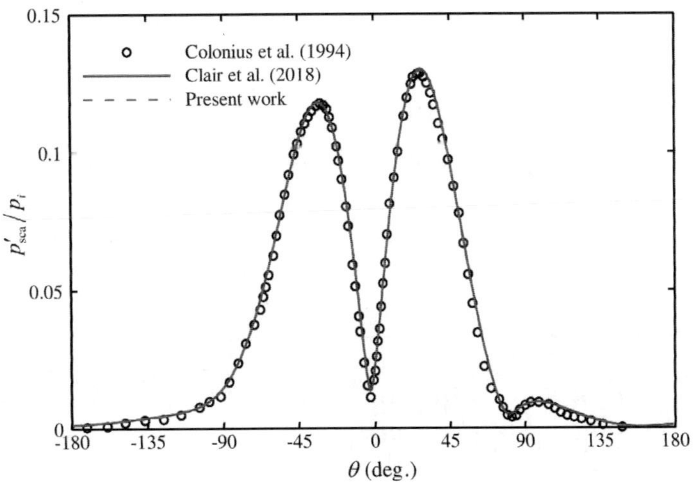

Figure 2. Comparison of root-mean-square pressure of the scattered field between the present work and the previous studies.

4. Results and Discussions

For the problem of the scattering of a plane wave from a Taylor vortex with a uniform convected velocity, the scattered fields depend on the three crucial dimensionless parameters: the length-scale ratio λ/R, the vortex Mach number M_v and the convected Mach number M_c. We maintain our focus on the roles played by the convected Mach number M_c in this paper. The computational results and related analyses, which involve the scattering of an incoming plane acoustic wave with a dimensionless wavelength λ/R=4 by a Taylor vortex with a Mach number M_v=0.125 at a convected Mach number from 0.1 to 0.5 are shown below in this section.

Fig. 3 presents the examples of the base pressure, the acoustic pressure, and the scattered pressure with $M_c = 0.2$ at the moment when the vortex moves to the origin of coordinates. In this case, the scattering is similar to that observed for a steady vortex in Fig. 2 and our previous study [35]. An alteration of the wavefronts due to the existing vortex. The scattered field is characterized by two beams on either side of the incident wave direction around $\theta = 60^{\circ}, 120^{\circ}$ (hereafter referred to as the first beam and the second beam respectively). A sight asymmetry between the first beam and the second beam is visible. The second beam is slightly stronger than the first beam. The two beams are in antiphase, and there exists a phase dislocation around $\theta = 0^{\circ}$. Due to an existing convected velocity M_c, the scattered field is moving in the x direction.

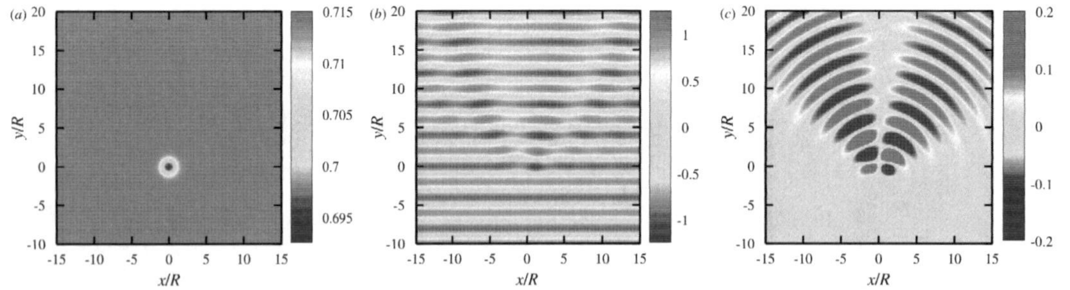

Figure 3. Snapshot of the pressure of (*a*) base flows, (*b*) total acoustic fields, and (*c*) scattered fields.

Fig. 4 shows the total acoustic pressure for a plane wave scattered by a vortex convected at a velocity $M_c = 0.2$ recorded at a fixed point (0,10*R*). Fig. 4(*a*) presents the total acoustic pressure signal in the time domain. Unlike the case of a steady vortex, the amplitude of the total acoustic pressure produced by acoustic scattering from a convected vortex no longer stays the same over time. Fig. 4(*b*) presents the Power Spectral Density (PSD) of the total acoustic pressure signal in the frequency domain, where ω denotes the angular frequency perceived by the observer at a fixed point (0,10*R*), and ω_0 the angular frequency of the incident waves, defined by $\omega_0 = 2\pi / \lambda$. When propagating through a convented vortex, a monochromatic plane wave is scattered over a range of frequencies, resulting in typical spectra with broadband sidebands on both sides of the tone ω_0 due to the motion of the vortex relative to the observer.

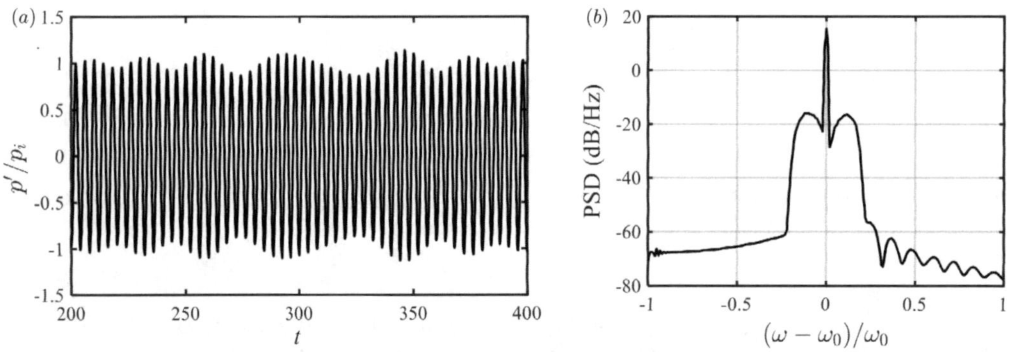

Figure 4. The total acoustic pressure recorded in the (*a*) time domain and (*b*) frequency domain.

Fig. 5 shows the scattered pressure for a plane wave scattered by a vortex convected at a velocity $M_c = 0.2$ recorded at a fixed point (0,10*R*). Fig. 5(*a*) presents the scattered pressure signal in the time domain, which is characterized by two wave packets corresponding to the two primary scattered beams observed in Fig. 3(c). Fig. 5(*b*) presents the Power Spectral Density (PSD) of the scattered pressure signal in the frequency domain, which is characterized by two beams located on both sides of the tone ω_0 associated with the two wave packets as well as the two primary scattered beams. A number of small sidelobes are visible for $\omega > \omega_0$.

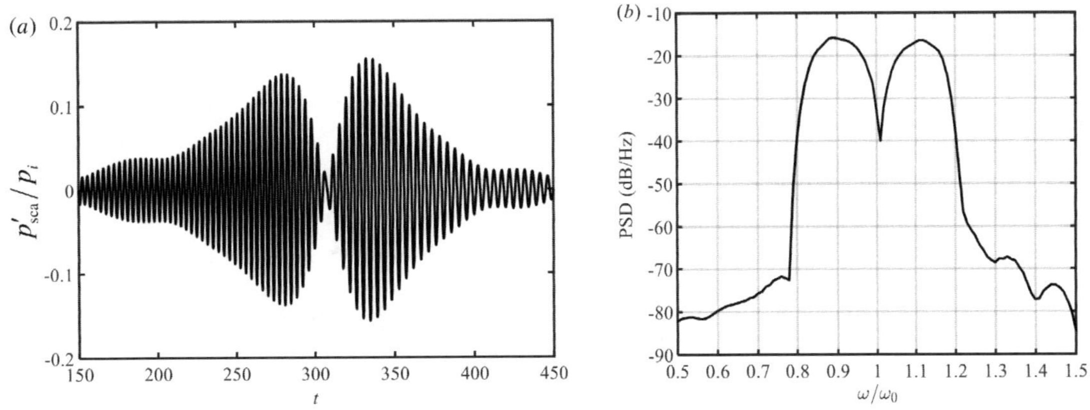

Figure 5. The scattered pressure recorded in the (*a*) time domain and (*b*) frequency domain.

The scattered waves perceived by the fixed observer are directly derived from the scattering by a convected vortex. Even though the incident field is monochromatic, the scattered pressure signal exhibits a number of frequencies. It can be described by the formula of the Doppler shifts

$$\omega(t) = \omega_0 \left(1 + M_c \cos\theta(t)\right), \tag{5}$$

where $\theta(t)$ denotes the relative azimuth angle of the fixed observer in the moving frame fixed to the vortex. It is easy to see that this Doppler shift is due to the relative motion between the vortex and the observer. Therefore, $\frac{\omega}{\omega_0} = \left(1 + M_c \cos\theta(t)\right)$ is called as Doppler factor. When the vortex is moving towards the fixed observer (at the time that $\cos\theta(t) > 0$), the scattered signal perceived by the observer is higher than the incident frequency ω. When the vortex is moving away from the fixed observer (at the time that $\cos\theta(t) < 0$), the scattered signal perceived by the observer is smaller than the incident frequency ω. The transition occurs at the time that $\cos\theta(t)=0$, and then the vortex passes just below the observer. The presence of the side beams on both sides of the incident frequency ω_0 in the frequency domain can be corresponding to the scattered beams observed in the case of scattering by a steady vortex with the same parameters, i.e M_v =0.125and λ / R=4 as employed in Fig. 2. Each scattered beam can be related to one side beam in the frequency domain for the case of scattering by a convected vortex. In Fig. 3(c), the center of the two primary scattered beams is located around $\theta = 60^\circ, 120^\circ$, corresponding to the Doppler factor of 0.9 and 1.1, as shown in Fig. 5(b). The second beam observed in Fig. 3(c) is responsible for the left side beam $\left(\omega < \omega_0\right)$. The second first beam observed in Fig. 3(c) is responsible for the right side beam $\left(\omega > \omega_0\right)$.

Fig. 6 shows the PSD of the scattered pressure at a range of velocity recorded at a fixed point (0,10R). The phenomenon that the scattered fields produced by a convected vortex in the frequency domain show the spectral broadening is more obvious as the convected velocity is increased, which is similar to the fact predicted by the formula (5). The strength of the side beams at all convected velocities does note exhibit significant differences. The width of side beams is scaled as the convected velocity M_c.

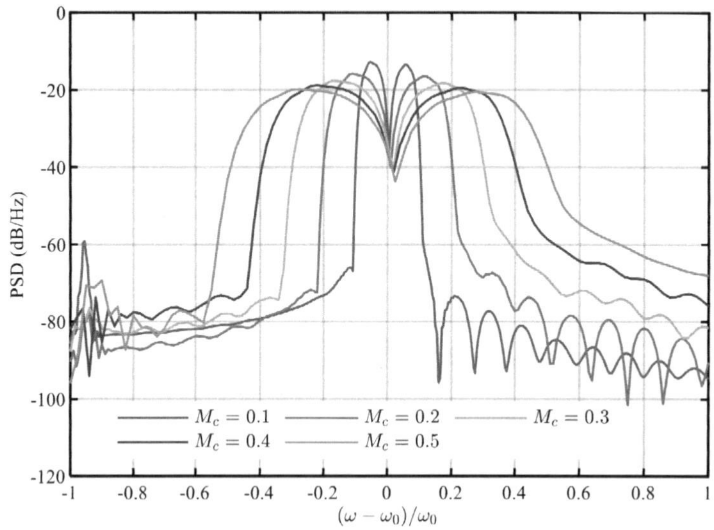

Figure 6. PSD of the scattered pressure p'_{sca} for different convected velocities M_c.

5. Conclusion

The computations and related PSD analyses, which involve the scattering of an incoming plane acoustic wave with a dimensionless wavelength λ/R=4 by a Taylor vortex with a Mach number M_v=0.125 at a convected Mach number from 0.1 to 0.5 are conducted. When propagating through a convented vortex, a monochromatic plane wave is scattered over a range of frequencies. This fact can be described by the formula of the Doppler shifts $1+M_c\cos\tilde{\theta}(t)$, which accounts for the relative motion between the vortex and observer. The spectra with broadband sidebands on both sides of the incident frequency are observed. The spectrum of the scattered signal is closely related to the directivity of scattered fields produced by a steady vortex. In the present cases, the scattered signal recorded at a fixed point in the frequency domain exhibits the two primary side beams on either side of the incident frequency, which can be attributed to the two primary beams observed in the case that the incident wave is scattered by a steady vortex with the same parameters, i.e M_v =0.125 and λ/R=4. Each scattered beam generates one frequency side beam. The range of the spectral broadening is strongly influenced by convected velocity. The width of primary beams is scaled as the convected velocity M_c.

It should be noted that the present study is focused on the parameters M_v =0.125 and λ/R=4. This choice is based on our previous parameter study [36], which has shown that significant and regular acoustic scattering occurs under these two parameters.

Acknowledgments

The authors gratefully acknowledge the support for this research provided by National Natural Science Foundation of China (Grant Nos. 11732016 and 12172374), Sichuan Science and Technology Program (Grant No. 2018JZ0076), CARDC Fundamental and Frontier Technology Research Fund (Grant No. PJD20200185), and National Numerical Windtunnel project.

References

[1] J.H. Ferziger, Low-frequency acoustic scattering from a trailing vortex, J. Acoust. Soc. Am., 56 (1974) 1705-1707.
[2] F. Lund, C. Rojas, Ultrasound as a probe of turbulencE, Physica D, 37 (1989) 508-514.
[3] D. Tschumperle, R. Rajaona, Non-intrusive vortex measurement by acoustic scattering, 4th AIAA/CEAS Aeroacoustics Conference, 1998.

[4] R. Labbe, J.F. Pinton, Propagation of sound through a turbulent vortex, Physical Review Letters, 81 (1998) 1413-1416.
[5] M. Oljaca, X. Gu, A. Glezer, M. Baffico, F. Lund, Ultrasound scattering by a swirling jet, Physics of Fluids, 10 (1998) 886-898.
[6] R.H. Kraichnan, The scattering of sound in a turbulent medium, J. Acoust. Soc. Am., 25 (1953) 1096-1104.
[7] R.K. Amiet, Refraction of sound by a shear-layer, J Sound Vib, 58 (1978) 467-482.
[8] Y. Ohno, Z. Yoshida, Nonlinear ion acoustic waves scattered by vortexes, Communications in Nonlinear Science and Numerical Simulation, 38 (2016) 277-287.
[9] M.J. Lighthill, On the energy scattered from the interaction of turbulence with sound or shock waves, Mathematical Proceedings of the Cambridge Philosophical Society, 49 (1953) 531-551.
[10] M.J. Lighthill, On sound generated aerodynamically I. General theory, Proceedings of the Royal Society of London. Series A. Mathematical and Physical Sciences, 211 (1952) 564-587.
[11] G.W. Ford, W.C. Meecham, Scattering of sound by isotropic turbulence of large reynolds number, J. Acoust. Soc. Am., 32 (1960) 1668-1672.
[12] W. Baerg, W.H. Schwarz, Measurements of scattering of sound from turbulence, J. Acoust. Soc. Am., 39 (1966) 1125-1132.
[13] M.S. Howe, Multiple-scattering of sound by turbulence and other inhomogeneities, J. Sound Vib., 27 (1973) 455-476.
[14] S.F. Clifford, E.H. Brown, Acoustic scattering from a moving turbulent medium, J. Acoust. Soc. Am., 55 (1974) 929-933.
[15] V.E. Ostashev, E.M. Salomons, S.F. Clifford, R.J. Lataitis, D.K. Wilson, P. Blanc-Benon, D. Juva, Sound propagation in a turbulent atmosphere near the ground: A parabolic equation approach, J. Acoust. Soc. Am., 109 (2001) 1894-1908.
[16] A. Attili, F. Bisetti, Statistics and scaling of turbulence in a spatially developing mixing layer at Re_{λ}=250, Physics of Fluids, 24 (2012).
[17] V. Pagneux, A. Maurel, Irregular scattering of acoustic rays by vortices, Phys Rev Lett, 86 (2001) 1199-1202.
[18] T. Colonius, S.K. Lele, P. Moin, The scattering of sound-waves by a vortex - numerical simulations and analytical solutions, Journal of Fluid Mechanics, 260 (1994) 271-298.
[19] S. Oshea, Sound Scattering by a Potential Vortex, J Sound Vib, 43 (1975) 109-116.
[20] T. Kambe, U.M. Oo, Scattering of Sound by a Vortex Ring, J. Phys. Soc. Jpn., 50 (1981) 3507-3516.
[21] R. Ford, S.G.L. Smith, Scattering of acoustic waves by a vortex, Journal of Fluid Mechanics, 386 (1999) 305-328.
[22] S.G. Llewellyn Smith, R. Ford, Three-dimensional acoustic scattering by vortical flows. I. General theory, Physics of Fluids, 13 (2001) 2876-2889.
[23] S.G. Llewellyn Smith, R. Ford, Three-dimensional acoustic scattering by vortical flows. II. Axisymmetric scattering by Hill's spherical vortex, Physics of Fluids, 13 (2001) 2890-2900.
[24] Y. Hattori, S.G. Llewellyn Smith, Axisymmetric acoustic scattering by vortices, Journal of Fluid Mechanics, 473 (2002) 275-294.
[25] M.S. Howe, On the scattering of sound by a rectilinear vortex, J Sound Vib, 227 (1999) 1003-1017.
[26] V.F. Kopiev, I.V. Belyaev, On long-wave sound scattering by a Rankine vortex: Non-resonant and resonant cases, J Sound Vib, 329 (2010) 1409-1421.
[27] A.L. Fetter, Scattering Sound by Classical Vortex, Physical Review a-General Physics, 136 (1964) 1488-1493.
[28] R. Labbe, J.F. Pinton, Propagation of sound through a turbulent vortex, Physical Review Letters, 81 (1998) 1413-1416.
[29] S. Manneville, P. Roux, M. Tanter, A. Maurel, M. Fink, F. Bottausci, P. Petitjeans, Scattering of sound by a vorticity filament: an experimental and numerical investigation, Phys Rev E Stat

Nonlin Soft Matter Phys, 63 (2001) 036607.
[30] J.F. Pinton, G. Brillant, Sound and vorticity interactions: Transmission and scattering, Theor. Comput. Fluid Dyn., 18 (2005) 413-433.
[31] V. Clair and G. Gabard, Spectral broadening of acoustic waves by convected vortices, Journal of Fluid Mechanics, 841 (2018) 50-80.
[32] C. Wu, R. Ma, Y. Wang, S. Zhang, Application of enhanced optimized difference schemes in acoustic scattering, Kongqi Donglixue Xuebao, 38 (2020) 1120-1128.
[33] D.J. Bodony, Analysis of sponge zones for computational fluid mechanics, Journal of Computational Physics, 212 (2006) 681-702.
[34] A. Mani, Analysis and optimization of numerical sponge layers as a nonreflective boundary treatment, Journal of Computational Physics, 231 (2012) 704-716.
[35] R. Ma, S. Zhang, Y. Luo, S. Han, Y. Wang, X. Wang, Numerical study on multiple acoustic scattering by a vortex array, J Sound Vib, 527 (2022) 116815.
[36] R.X. Ma, Y.M. Wang, S.H. Zhang, C.H. Wu, X.N. Wang, Numerical investigation of scale effect on acoustic scattering by vortex, Acta Physica Sinica, 70 (2021) 190-199.

Numerical Simulation of Orbit Controlled Hot-gas Lateral Jet of the triangular plate with rudder

Zhao Wan, Yangaoxiao Fu, Xinguang Wang *, and Qi Chen

China Aerodynamic Research and Development Center, Mianyang, China

*Corresponding author e-mail: wangxinguang@cardc.cn

Abstract. Hypersonic interception weapons have high requirements for the efficiency and accuracy of the control system. As an effective control approach, the thermochemical non-equilibrium effect of jet flow must be considered in order to accurately predict the aerodynamic characteristic. At present, relevant researches are mostly focused on the low-flow rate hot jet, while numerical simulation and wind tunnel test studies on high-flow rate hot jet are rare. In this paper, the interaction flow field of the rail-controlled jet at the altitude of 60Km and 70Km has been simulated by solving the three-dimensional viscous fluid governing equations. The influence of the thermalchemical non-equilibrium effect of the jet interaction on the flow field characteristics and aerodynamic characteristics of the aircraft was analyzed. The primary results show that: (1) The flow fields of the cold-gas jet, frozen-gas jet and hot-gas jet have significant differences under the high-flow rate jet condition, which is mainly reflected the size and the pressure value of the separation region before the nozzle. The separation region length of the cold-gas jet is approximately twice that of the hot-gas jet and the length of the frozen-gas jet is between them; (2)Jet interaction will increase the surface pressure of the aircraft and reduce the skin-friction. When Ma=25, the aerodynamic interaction of hot jet is only about 30% of that of cold jet; (3) Jet interaction on aerodynamic characteristics increases with altitudes. With the increase of Mach number, the difference between the predicted results of hot jet and cold jet is enlarged because of the enhancement of the thermochemical non-equilibrium effect.

1. Introduction

Hypersonic interceptor weapons are near space vehicles with strong defense capabilities. The interception intersection time is extremely short, which requires the interceptor itself to have rapid response ability and extremely high control accuracy. When the interceptor is flying at high speed in the near space environment, the thin atmosphere and the low incoming flow pressure lead to insufficient rudder control efficiency, making it difficult to meet the needs of maneuvering and ballistic control, so the jet control or the jet/rudder surface composite control must be used [1-5].

Jet interaction researches first began in the late 1950s as a qualitative study of interaction mechanisms and flow field structures based on the US space shuttle. In the late 1980s, stimulated by the demand for advanced anti-missile interceptors and space-to-earth transporters, a boom in jet research was set off. The research methods, and objects were gradually diversified, complicated, and practical. However, until 1997, AGARD still listed jet research and applications as "one of the key technologies for several major problems that missile aerodynamics is facing in the new century".

Up to now, the research on jet interaction has focused on the jet flow research of attitude control engines with low flow rate [6-13]. The influence of incoming flow parameters, jet parameters, jet similarity criteria and nozzle shape on the flow field characteristics and aerodynamic characteristics of the aircraft has been studied. For the hypersonic interceptor weapon, the attitude jet thrust is about 500N with flow rate about 0.2kg/s, while the orbital jet thrust is about 5000N with flow rate about 2kg/s. The rail control jet has significant influence on the flow field due to the large thrust and flow rate. The high-temperature thermochemical non-equilibrium effect of the gas and incoming flow becomes nonnegligible, and the jet interaction mechanism are more complex. The research on orbit-controlled hot jets is relatively limited, such as AIT interceptor, THAAD interceptor [14] and the experiments on orbit-controlled transient jet interaction by Arnold Engineering Development Center. The lack of systematic understanding of the mechanism and the aerodynamics influence of hot jets has led to the inability to assess the deviation caused by the cold jet prediction, which is commonly used in engineering. Therefore, the control accuracy of hypersonic interceptor weapons is limited to a certain extent.

Three methods are mostly used in the study of jet interactions: (i) an equivalent specific heat ratio model that simplifies the gas into one component, named as two component jet; (ii) a frozen flow model in which no chemical reaction occurs between the gas and air, named as frozen jet; (iii) a finite rate chemical reaction model including the secondary combustion of jet gas and air, named as hot jet. Comparative studies of the three methods have been carried out. Gnemmi et al. [15~17] have conducted the jet experiment of the TZN warhead in the ISL shock wave tunnel. The numerical results show that the cold jet separation zone is greater than that of the hot jet, and the flow field structure of the hot jet fits the experimental results better. Yang et al. [18] have studied the two-phase flow, component diffusion, secondary combustion and other factors of the jet interaction. The conclusion is that the effect of the two-phase flow on the aerodynamic characteristics is the smallest, followed by the component diffusion effect, and the largest impact is the secondary combustion. Li et al. [19,20] have simulated the jet interaction flow field for the separation between rocket stages using the cold jet, two component jet and hot jet approach. The results show that the length of the flow separation zone obtained from the three jet approaches increased sequentially. Chamberlain et al. [21] have studied the effect of wind tunnel scale, experiment model scale, wind tunnel test conditions and flight conditions on the normal force coefficient and pitch moment coefficient with the conclusion that the effect of cold and hot jets on force characteristics is not significant at Mach number of 3 and attitude of 25km. From the literature, it is clear that the difference of hot and cold jets on the flow field and aerodynamic characteristics exist. Most research focus on low flow rate attitude jet interaction, and these conclusions may decrease the accuracy of the design of orbit-controlled jet systems.

In this paper, the numerical simulation of jet flow field is carried out by using three methods named as the cold/frozen/hot jet. The influence of hot jet interaction on flow field characteristics and aerodynamic characteristics is studied, which can provide theoretical guidance for the design of orbit controlled jet system of hypersonic interception weapons in near space.

2. Numerical methods

2.1. Governing equations and discrete methods

Three-dimensional thermochemical non-equilibrium Navier-Stokes equations are used in the dimensionless form [22] as:

$$\frac{\partial \boldsymbol{Q}}{\partial t}+\frac{\partial \boldsymbol{F}}{\partial x}+\frac{\partial \boldsymbol{G}}{\partial y}+\frac{\partial \boldsymbol{H}}{\partial z}=\frac{1}{Re}\left(\frac{\partial \boldsymbol{F}_V}{\partial x}+\frac{\partial \boldsymbol{G}_V}{\partial y}+\frac{\partial \boldsymbol{H}_V}{\partial z}\right)+\boldsymbol{W} \tag{1}$$

Where $\boldsymbol{Q}=(\rho_i,\rho,\rho u,\rho v,\rho w,\rho E_t,\rho E_V)^T$. ρ represents the mixing gas density, ρ_i represents the density of species i, u, v, w represent the velocity in three different directions, E_t is the gas internal energy, E_V is total vibrational energy, Re is the Reynolds number, $\boldsymbol{F,G,H}$ and $\boldsymbol{F_V,G_V,H_V}$ are the convection terms and viscous terms, and W is thermochemical non-equilibrium source term. The governing equations are discretized by the finite difference method with the convective terms by AUSMPW+, the viscous terms by the central difference method, and the time integration by an implicitly LU-SGS method. In order to overcome the rigidity problem, the non-equilibrium source, convective and viscous terms are treated in a fully implicit manner as detailed in the literature [22].

2.2. Physicochemical model of jet gas and air mixings

The main components of the jet gas are CO_2, H_2O, CO, H_2, N_2, etc. After being ejected from the nozzle, a complex chemical reaction process with air needs to be considered. On the basis of the air reaction system, the chemical reactions involving C and H elements are further considered, which together constitute the C-H-O-N reaction system. To simplify the calculation, the ionization reaction of air is neglected here. A simpler 12-component reaction model (N_2, O_2, NO, O, N, CO_2, H_2O, CO, H_2, OH, C, H) is used, and most of the reaction rates are also from the study of Park et al. Table 1 gives the nozzle gas-air mixing chemical reaction equations and their forward and reverse reaction rate constants, and other parameters could be found in the literature [23-27]. The thermodynamic model is a one-temperature or two-temperature vibrational nonequilibrium model. The energy relations, vibrational nonequilibrium source term, and transport coefficients are detailed in the literature [22].

Three approaches of jet calculation are used in the simulation. The cold jet calculation is based on the calorically perfect gas assumption. The jet gas is air and the thrust outlet coefficients of momentum ratio and mass flow rate follow the similarity criterion to the experiment jet gas. The frozen jet gas is the experiment gas considering the component diffusion effect without chemical reaction. The hot jet calculation includes a finite rate chemical reaction model of the experiment jet gas and air.

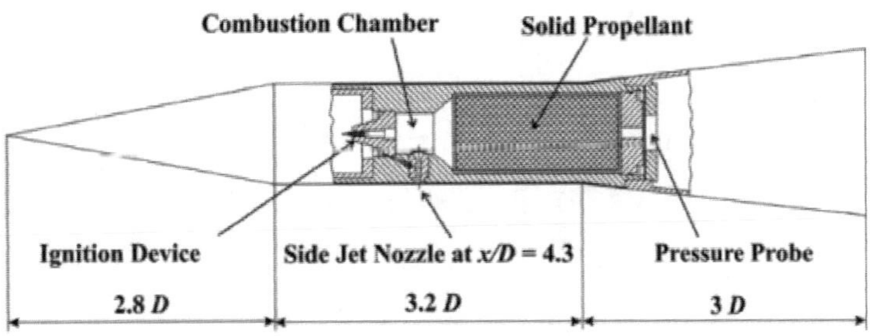

Figure 1. Sketch of the experiment model.

3. Numerical methods validation

For the conical hypersonic lateral jet test given in the literature [28], numerical simulations of the lateral cold/hot jet interaction flow field of the test model were carried out and compared with the experiment results to verify the numerical method. The shape of the test model is shown in Figure 1, with the diameter of the missile body D= 90mm and the diameter of the nozzle outlet $d_J = 4.6mm$. Air is used for the inflow of the wind tunnel experiments with $P_\infty = 0.923Bar$, $T_\infty = 105K$, $M_\infty = 3.0$, $\alpha = 0°$, total pressure of the jet $P_{0J} = 120Bar$, ratio of jet pressure $PR = P_{0J} / P_\infty = 130$. When the jet gas is air, the total temperature of the jet $T_{0JC} = 293K$, the temperature of the jet outlet $T_{JC} = 244K$, while the jet gas is gas in the hot jet simulation with the mass fraction of the gas component at the jet outlet shown in Table 2, and the model of the chemical reaction of the gas and the air mixture shown in Table 1, the total temperature of the jet $T_{0JC} = 2300K$, the temperature of the jet outlet $T_{JH} = 2058K$, and the specific heat ratio of the jet outlet $\gamma_H = 1.233$. The wall surface adopts adiabatic wall with zero pressure gradient, no slip condition, non-catalytic wall.

Table 1. Chemical reaction model of jet gas and air mixture

No.	chemical reaction	Positive reaction rate coefficient			Inverse reaction rate coefficient		
		Af	Bf	Cf	Ab	Bb	Cb
1	CO2+M1↔CO+O+M1	6.9015	-1.5	63,275	1.145	-0.75	0
2	H2O+M2↔H+OH+M2	6.0615	-1.31	59,400	2.309	-1.5	0
3	CO+M3 ↔C+O+M3	2.3014	-1.0	129,000	5.136	-1.0	0
4	N2+M4 ↔N+N+M4	7.0014	-1.6	113,200	4.278	-1.6	0
5	O2+M5 ↔O+O+M5	2.0014	-1.5	59,500	2.006	-1.0	0
6	NO+M6 ↔N+O+M6	5.0014	0.0	75,500	4.273	0.0	0
7	H2+M7 ↔H+H+M7	2.208	0.0	48,300	3.082	-0.18	0
8	OH+M8 ↔O+H+M8	6.967	0.21	5,100	3.002	0.0	0
9	O+NO ↔N+O2	8.406	0.0	19,450	2.405	0.5	0
10	O+N2 ↔N+NO	6.4011	-1.0	38,370	1.3911	-1.0	0
11	NO+CO ↔ CO2+N	4.592	0.5	12,070	9.906	-0.25	0
12	CO2+O ↔ O2+CO	2.107	0.0	27,800	4.114	0.5	23,800
13	CO+CO ↔ CO2+C	2.333	0.5	65,710	4.606	-0.25	0
14	CO+O ↔ O2+C	3.907	-0.18	69,200	1.348	-0.43	0
15	CO+N ↔NO+C	2.865	0.5	53,630	2.604	0.5	0
16	OH+CO ↔CO2+H	1.505	0.0	107	1.4110	-0.80	1,240
17	OH+H2 ↔H2O+H	5.006	0.0	2,320	1.857	0.01	9,800
18	H+O2 ↔OH+O	5.008	0.0	8,430	4.176	0.27	49
19	O+H2 ↔OH+H	4.007	0.0	5,140	1.297	0.03	4,160
20	OH+OH ↔H2O+O	7.606	0.0	504	8.657	-0.02	8,930

Figure 2 shows the comparison of pressure coefficient distribution along the central line of the jet. It can be seen that the numerical results by the method established in this paper fit the experimental data well, which means that these methods are suitable for the calculation of the interaction flow field of cold/hot jets.

Table 2. The mass fraction of hot gas species

	CO_2	CO	H_2O	H_2	O_2	N_2
c_i	0.3821	0.3547	0.1047	0.0173	0.0	0.1412

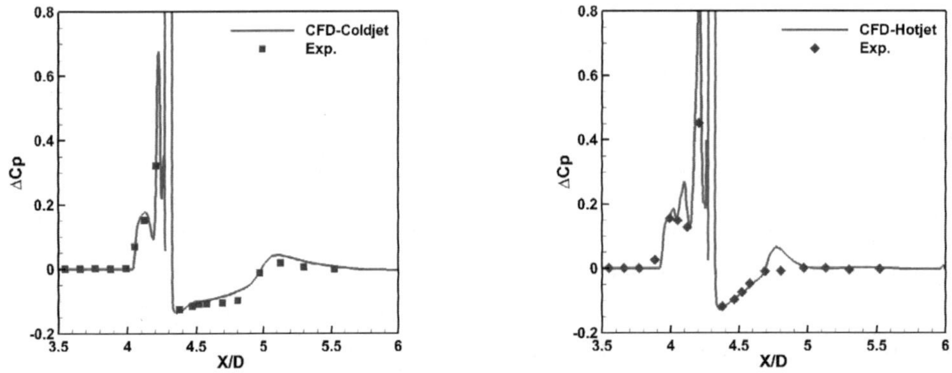

(a) Cold jet (b) hot jet

Figure 2. Comparison of surface pressure with tunnel test results.

Table 3. Nozzle parameters

Name		Jet parameters
Outlet diameter d/mm		234
Outlet Mach number/M_j		4.19
Outlet pressure P/Pa		2432.7
Outlet temperature T/K		1088.5
Outlet gas specific heat ratio		1.2524
Gas constant /(J/(kg*K))		397.3
Thrust F/N		2405
Mass rate Q/(kg/s)		0.746
Gas components /(mol/kg)	CO	0.07068
	N2	0.31344
	H2	0.22026
	H2O	0.29465
	CO2	0.10095

4. Numerical results

The shape of the triangular plate with rudder is shown in Figure 3. A C-type and H-type structural grid is used in the calculation with fine mesh points around the nozzle as shown in Figure 4. The flight altitude is 60Km, Mach number is 25, and angle of attack is 0°. The jet simulation start from the nozzle with the nozzle parameters detailed in Table 3.

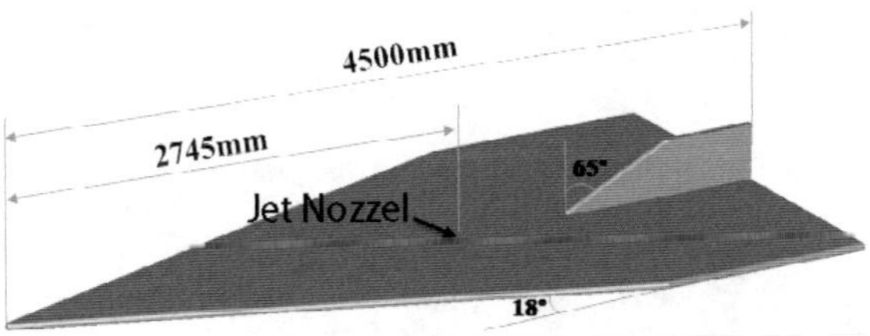

Figure 3. Shape parameters of flat rudder surface.

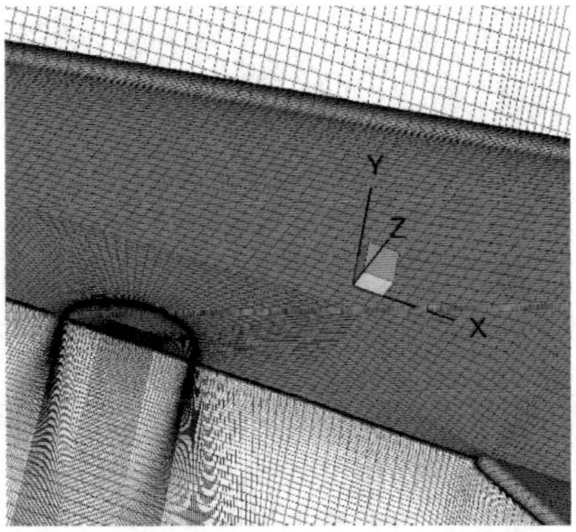

Figure 4. Grid diagram near the nozzle.

4.1. Effect of hot jets on field filed characteristics

Figure 5 shows the pressure distribution and streamlines in the symmetrical plane. The topology of the jet field obtained by the three approaches is basically the same. After the inflow and the jet gas meet in zone 1, part of it flows downstream with the jet gas, and part of it flows down to the wall surface because of the reverse pressure gradient, and the speed is increased with compression, and finally a three-dimensional compression wave is formed in zone 2. After the compression wave in zone 2, the flow becomes subsonic flow, and hits the wall area 3, forming a stagnation point where a surface high pressure zone exist (as shown in Figure 6 and the pressure peak is also marked). Simultaneously, part of the gas flows backward from zone 3, and the other part of the gas flows upstream resulting in an extremely complex separation structure (5, 6, 7 structures marked in the Figure 5).

Figure 6 shows the separation and reattachment upstream and downstream of the nozzle, where 'S' represents the separation line and 'A' represents the reattachment line. The separation area of the cold jet nozzle can clearly observe 4 separation lines and 5 singularities behind the nozzle. From the comparison of central line pressure before the nozzle as shown in Figure 7, the peak pressure point position for the three jets varies, as also seen from Figure 6. The frozen jet and hot jet have a weaker blocking effect on the inflow compared to the cold jet and the length of the separation area is significantly reduced, which is 1.6m for the cold jet, 1.3m for the frozen jet and 0.8m for the hot jet. Also, the structure of the flow field after the nozzle become simpler with less wall streamline singularities.

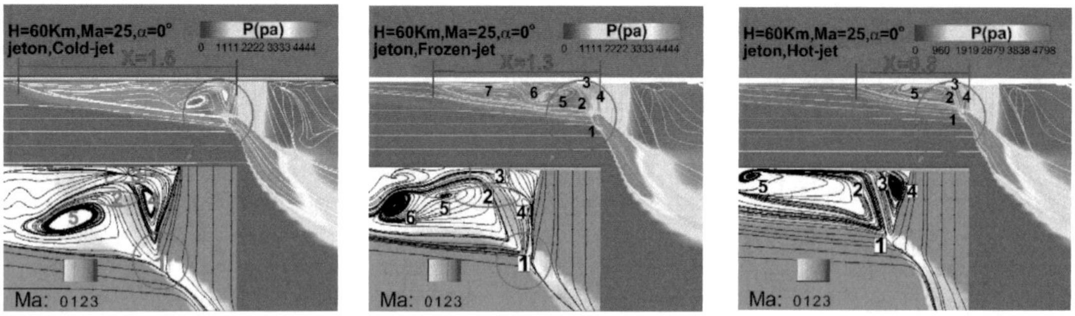

(a) cold jet (b) frozen jet (c) hot jet

Figure 5. Pressure distribution and streamlines in the symmetry plane.

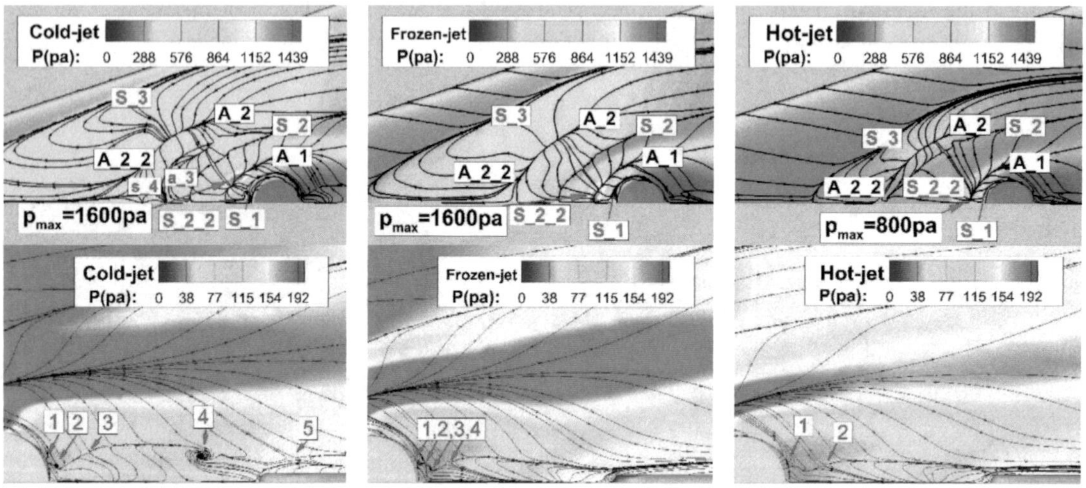

(a) cold jet (b) frozen jet (c) hot jet

Figure 6. Pressure distribution and wall streamline near nozzle. (up: upstream streamlines of nozzle; bottom: downstream streamlines of nozzle;)

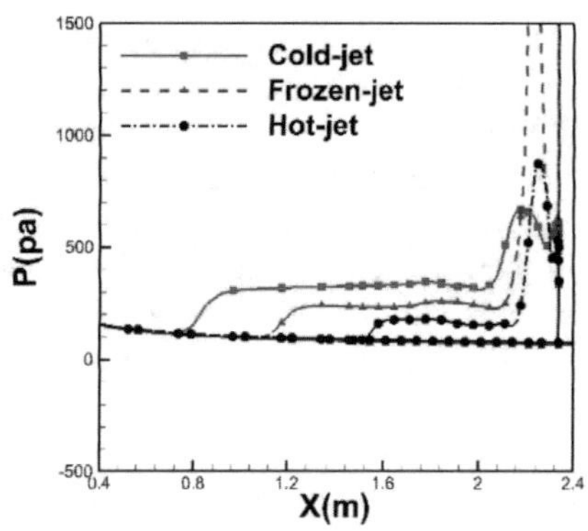

Figure 7. Central line pressure distribution of the three jets.

The main reasons for the difference between the separation zone of the three jets interaction are: (1) the boundary layer thickness of the hot jet is much thinner when considering the thermochemical non-equilibrium effect, which generally return smaller separation zone; (2) in terms of gas energy, according to the jet similarity criterion of static pressure consistency and momentum ratio consistency, the static pressure, density, and velocity of gas and cold jet air are consistent, because the air gas constant is smaller than the jet-gas constant, resulting in the temperature of cold jet air being higher than the temperature of gas (as shown in Figure 8a, the hot jet flow is 870K, Cold jet 1100K). In this temperature range, the gas reaction is weak and most of the energy is manifested as translational kinetic energy and rotational energy, which means that the energy of cold spray air collides with the incoming flow is greater; (3) within the gas temperature range about 4000K-8000K in the separation zone, the secondary combustion of the components happens and the gas components and the inflow air components have dissociation reaction with components variation (as shown in Figure 9) and temperature reduction. Overall the viscosity coefficient of the mixed gas is increased (as shown in Figure 8b) which means that the gas viscosity is enhanced. The combined effect of these three aspects makes the separation zone of the hot jet significantly smaller than that of the cold jet.

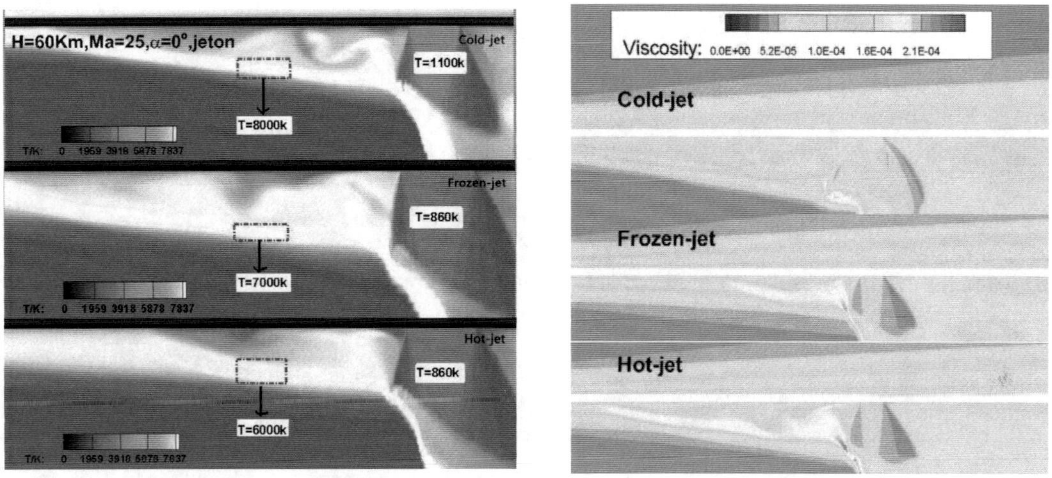

(a) temperature distribution (b) viscosity coefficient distribution

Figure 8. Temperature and viscosity coefficient contour in the symmetry plane.

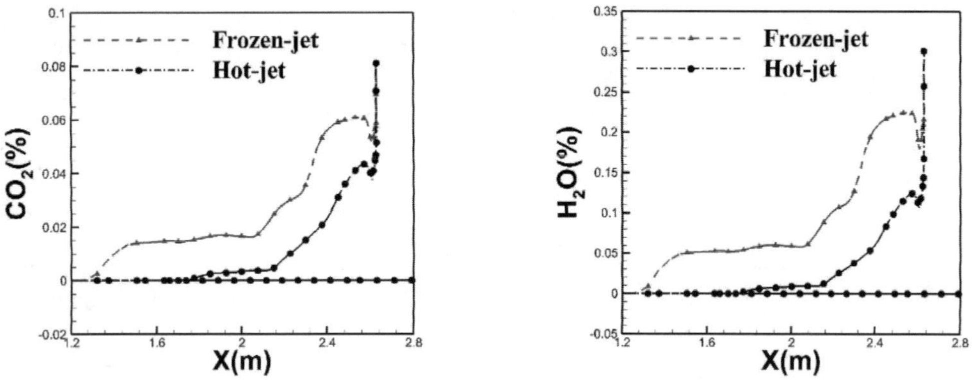

(a) CO_2 (b) H_2O

Figure 9. The distribution of components on the symmetry wall.

4.2. Effect of hot jets on aerodynamic characteristics

As can be seen from Figure 7, the pressure distribution around the nozzle and rudder surface of the three jets are significantly different, which means that the jet has an impact on the pressure, skin-friction and pitch moment. Taking the center point of the nozzle X=2745mm as the cross-section, the plate is divided into two parts, and the influence of cold/hot jets on the aerodynamic characteristics of the area before and after the nozzle is investigated with reference length Lref = 1.0m, reference area Sref = 1.0m^2, and moment reference point (nozzle center point) (2.745m, 0, 0).

4.2.1. Effect on pressure. In order to quantitatively investigate the effect of cold/hot jets on pressure, there is a surface pressure integral on the jet side of the plate, where '1' denotes the front wall value, '2' denotes the rear wall value in Table 4, and 'whole' denotes the aerodynamic integral of the entire wall surface. Jet interaction amount = jet on - jet off. When the jet is on, the pressure before and after the nozzle is increased about 1-3 times. For the whole and part pressure increment, the cold jet prediction is the highest, the frozen one is the second, and the hot one is the smallest with the amount of 30% of the cold jet.

Table 4. Jet interaction on the pressure.

	Jet-on		Jet-off		Jet-interaction		
	P1	P2	P1	P2	P1	P2	Whole
Cold	0.043224	0.036215	0.013286	0.011670	0.029938	0.024545	0.043406
Frozen	0.032771	0.030185	0.019210	0.016687	0.013561	0.013498	0.025791
Hot	0.028753	0.025297	0.019011	0.016557	0.009742	0.008740	0.011323

4.2.2. Effect on skin-friction. When the jet is on, the skin-friction between the front and rear of the nozzle is reduced by about 20%-50%. The result of the cold jet is significantly higher than that of the frozen and hot jet and the amount of skin-friction interaction predicted by the hot jet is only about 10% of the cold jet as shown in Table 5.

Table 5. Jet interaction on the skin-friction.

	Jet-on		Jet-off		Jet-interaction		
	f1	f2	f1	f2	f1	f2	Whole
Cold	0.004755	0.003574	0.009489	0.007954	-0.004734	-0.004380	-0.015901
Frozen	0.006004	0.004917	0.007111	0.004969	-0.001107	-0.000052	-0.001766
Hot	0.006632	0.004947	0.007201	0.005090	-0.000569	-0.000143	-0.001379

4.2.3. Effect on pitch moment. The head-up moment generated before the nozzle is slightly greater than the head-down moment generated after the nozzle (the positive value is the head-up moment and the negative value is the pitch-down moment), resulting in a smaller pitch-up moment generated by the jet interaction. The hot jet result is greater than the cold jet result. It should be particularly known that the pitch moment coefficient reflects the distribution of the force and is closely related to the flow field structure, and the relevant conclusion about the pitch moment coefficient is closely related to the incoming flow conditions (such as altitude, Mach number, angle of attack, etc.).

Table 6. Jet interaction on the pitch moment

	Jet-on		Jet-off		Jet-interaction		
	Mz1	Mz2	Mz1	Mz2	Mz1	Mz2	whole
Cold	0.036269	-0.030722	0.017018	-0.012369	0.019251	-0.018353	0.000099
Frozen	0.034644	-0.024657	0.020847	-0.013885	0.013797	-0.010772	0.003552
Hot	0.032935	-0.021457	0.020745	-0.013826	0.012190	-0.007631	0.006161

4.3. *Effect of inflow conditions*

In order to examine the effects of the flight altitude and Mach number on the flow field and aerodynamic characteristics, additional calculation at altitude of 60, Mach number of 15 and altitude of 70, Mach number of 25 are processed to compare with the previous results at altitude of 60, Mach number of 25.

4.3.1. Effect of flight altitude. The pressure and temperature distribution of the cold/hot jet at the altitude of 70 km in Figure 10 is compared with those of 60 km in Figure 5. It is clear that the length of the cold jet separation zone is increased, while the length of the hot jet separation zone does not change much. The reason is that there is little difference in temperature within the flow field under the same Mach number.

With the increase of altitude, the amount of jet interaction increases as listed in Table 7, the 70 km jet interaction is as twice as that of 60 km and the difference between the cold/hot jet interactions is still significant.

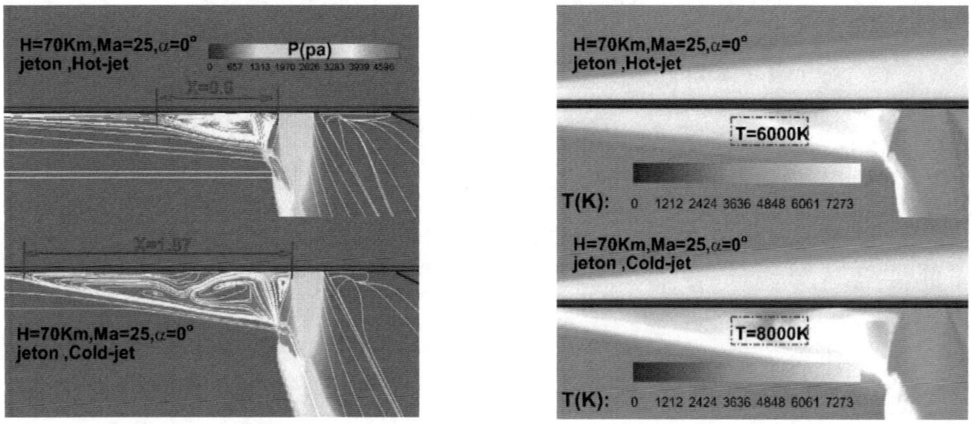

(a) pressure distribution (b) temperature distribution

Figure 10. Flow field of 70km on the symmetric plane (70km).

Table 7. Effect of altitude on jet interaction

	H=60km		H=70km	
Aerodynamics	Δ(Cold-jet)	Δ(Hot-jet)	Δ(Cold-jet)	Δ(Hot-jet)
P1	0.029938	0.009742	0.054672	0.017870
P2	0.024545	0.008740	0.056617	0.017772
f1	-0.004734	-0.000569	-0.005300	-0.002738
f2	-0.004380	-0.000143	-0.006130	-0.001898
Mz1	0.019251	0.012190	0.046920	0.020579
Mz2	-0.018353	-0.007631	-0.035592	-0.016246
Pressure	0.043406	0.011323	0.098498	0.029035
Skin-friction	-0.015901	-0.001379	-0.009031	-0.003080
Pitch moment	0.000099	0.006161	0.010008	0.024709

4.3.2. Effect of the inflow Mach number. Figure 11 shows pressure and temperature distribution of the inflow Mach number 15 in the symmetrical surface. Compared with the previous results of Ma=25 in Figure 5, the length of the cold jet separation zone increases by 20%, while the length of the hot jet separation zone increases by 100%. The flow field structure of the hot jet changes greatly, mainly because the flow field temperature is reduced (Ma=25 is 6000K, Ma=15 is 3000K), resulting in a significant weakening of the thermochemical non-equilibrium effect, and the difference between the hot and cold jet flow fields is reduced.

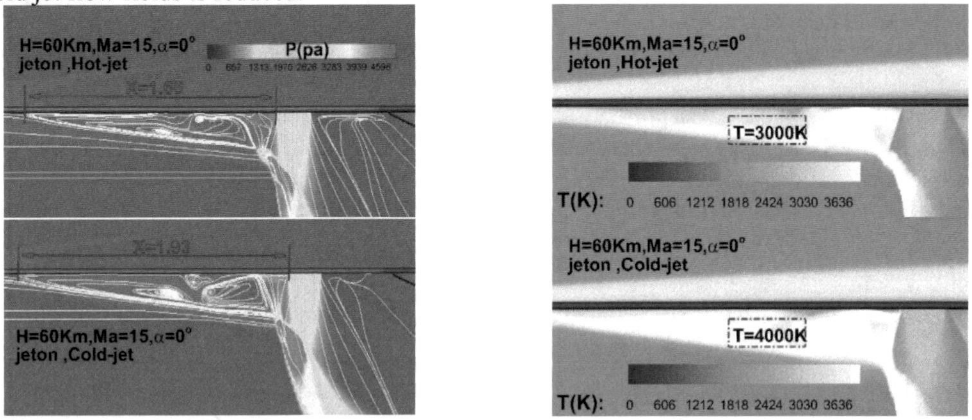

(a) pressure distribution (b) temperature distribution

Figure 11. Flow field of Ma=15km on the symmetric plane.

Table 8. Effect of Mach number on jet interaction

	Ma=25		Ma=15	
Aerodynamics	Δ(Cold-jet)	Δ(Hot-jet)	Δ(Cold-jet)	Δ(Hot-jet)
P1	0.029938	0.009742	0.030150	0.017591
P2	0.024545	0.008740	0.032816	0.024738
f1	-0.004734	-0.000569	-0.001912	-0.001644
f2	-0.004380	-0.000143	0.000972	0.000794
Mz1	0.019251	0.012190	0.030813	0.022355
Mz2	-0.018353	-0.007631	-0.028659	-0.020831
Pressure	0.043406	0.011323	0.062081	0.034327
Skin-friction	-0.015901	-0.001379	-0.001165	-0.002663
Pitch moment	0.000099	0.006161	0.009097	0.011376

Aerodynamics of different Mach number jets is given in Table 8. The interaction value of Ma=15 increases compared to the results of Ma=25, mainly because of the increase of the jet/inflow pressure ratio. The difference between the hot/cold jet interaction of Ma=15 is significantly reduced, and the interaction of the hot jet gas is greater than that of Ma=25. It is also further illustrated that the thermochemical non-equilibrium effect has a greater impact on the jet interaction.

5. Conclusion

Through the numerical simulation of the jet flow field using the cold/frozen/hot jet methods, the influence of jet interaction effect on flow field characteristics and aerodynamic characteristics is studied. The following conclusions are drawn:

(1) Under the condition of high-flow rate jet at high-altitude high-Mach number, the difference between the influence of cold/hot jet on the flow field characteristics and aerodynamic characteristics of the aircraft is significantly increased, which is different from the conclusion of low-flow rate attitude jet at the altitude of 25km.

(2) The flow topology of the separation area obtained by the cold/frozen/hot jet is basically the same. However, the size of the separation zone and the pressure value in this zone are quite different. The order of influence from large to small is the cold jet, frozen jet and hot jet.

(3) The jet interaction will increase the pressure near the nozzle and reduce the skin-friction. The amount of interaction predicted by the cold jet is significantly higher than that of the hot jet. At the altitude of 60-70km, the jet interaction amount of the hot jet is only about 30% of that of the cold jet.

(4) The inflow condition has a great impact on the aerodynamics. The jet interaction amount increases with the altitude, and rises with the Mach number, because of the enhancement of the thermochemical non-equilibrium effect. Also, the hot jet prediction is more sensitive to the Mach number than the Reynolds number.

From the simulation, it should be noticed that the jet flow have a significant impact on the rudder, which will affect the jet/rudder control efficiency. The following research will focus on the combination control of the jet and rudder. An experiment study of the hot jet under Mach number of 8 is also on the schedule, which will be helpful to fully understand the hot jet interaction.

Acknowledgments
This work was financially supported by 2019-JCJQ-ZD-047 fund and National Numerical Windtunnel.

References

[1] S. Li, Progress in aerodynamics of combination control for vehicles at high speed, Advances in Mechanics, 2009, 39(6) 740 -755.(in Chinese)
[2] Z. Tang, Y. Yang, J. Liu etc. The investigation and expectation on lateral jet interaction/control, Journal of Experiments in Fluid Mechanics, 2010, 24(4) 1-6.(in Chinese)
[3] C. Xu, C. Jiang, Z. Gao etc. The jet interaction effects of reaction control systems in hypersonic vehicles, Mechanics in Engineering, 2014, 36(2) 147-160.(in Chinese)
[4] M. Wei, Summary of blended control technology for missiles with lateral jets and aerodynamic surfaces, Modern Defence Technology, 2012, 40(1) 52-76.(in Chinese)
[5] Q. Jia, M. Wei, D. Guo, Orbital Lateral Thrust/Aerodynamic Force Blended Control Methodin High Altitude, Modern Defence Technology, 2015, 43(6) 61-67.(in Chinese)
[6] R.J. Margason, Fifty years of jet in cross flow research: computational and experimental assessment of jets in cross flow, Moffett Field, NASA Ames Research Center, Nov 1993, Report No: AGARD-CP-534.
[7] G. Partick, Computational comparisons of the interaction of a lateral jet on a supersonic generic missile, AIAA-2008-6883.Reston, 2008, VA:AIAA.
[8] D. James, Lateral jet interaction on a finned projectile in supersonic flow, AIAA-2012-0413. Reston, 2012, VA:AIAA.
[9] J. Chen, Q. Chen, Y. Xie etc. Numerical Study on the Interaction of Lateral Jet and Rudder Movement, Journal of Astronautics, 2014, 35(5) 515-520. (in Chinese)
[10] Y. Liu, W. Xu, J. Wu, Numerical simulation of flow field of lateral jet interaction in supersonic flow for a missile-like wing-body configuration, Acta Armamentarii, 2007, 28(8) 965-969.(in Chinese)
[11] J. Lai, Z. Zhao, Y. Li etc. Transverse jet interaction characteristics on rear section of missile model, Journal of Aerospace Power, 2019, 34(2) 469-478.(in Chinese)
[12] S. Wang, J. Hu, J. Wu etc. Numerical simulation of interaction between supersonic incoming flow and lateral jets from rotating missile, Acta Armamentarii, 2008, 29(10) 1220-1226.(in Chinese)
[13] H. Zhao, A. Gong, Z. Liu etc. Numerical study of lateral jet interaction at high altitude, Acta Aerodynamica Sinica, 2020, 38(5) 996-1003. (in Chinese)
[14] R. Chamberlain, D. McClure, A. Dang, CFD Analysis of Lateral Jet Interaction phenomena for the THAAD interceptor, AIAA-00-0963.
[15] P. Gnemmi, F. Seiler, Interaction of a lateral jet with the projectile external flow, AIAA-2000-4196. Reston, 2000, VA:AIAA.
[16] P. Gnemmi, R. Adeli, J. Longo, Computational comparisons of the interaction of a lateral jet on a supersonic generic missile, Presented at the AIAA Atmospheric Flight Mechanics Conference, Aug 2008, Honolulu, HI.
[17] P. Gnemmi, H.J. Schafer, Experimental and numerical investigations of a transverse jet interaction on a missile body, Presented at the 43rd AIAA Aerospace Sciences Meeting, Jan 2005. Reno, NV.

[18] Y. Yang, Z. Tang, J. Liu, Study on real gas effect in transverse jet interference, Acta Aerodynamica Sinica, 2006, 24(1), 28-33.(in Chinese)
[19] L. Yi, R. Bodo, E. Thino, Coupled simulation of CFD-flight-mechanics with a two-species-gas-model for the hot rocket staging, AIAA-2014-2371. Reston, 2014, VA: AIAA.
[20] L. Yi, R. Bodo, E. Thino, Numerical investigations on the aerodynamics of SHEFEX-III launcher, Acta Astronautica, 2014, vol.97, pp. 99-108.
[21] R. Chamberlain, A. Dang, D. Mcclure, Effect of Exhaust Chemistry on Reaction Jet Control, AIAA-1999-0806. Reston, 1999. VA:AIAA.
[22] W. Dong, Numeical Simulation and Analysis of Thermochemical Nonequilibrium Effects at Hypersonic Flow, Beijing University of Aeronautics and Astronautics, 1996.(in Chinese)
[23] C. Park, Review of Chemical-Kinetic Problems of Future NASA Missions I: Earth Entries, Journal of Thermophysics and Heat Transfer, 1993, 7(1) 385-398.
[24] C. Park, J.T. Howe, R.L. Jaffe et al. Review of Chemical-Kinetic Problems of Future NASA Missions II Mars Entries, Journal of Thermophysics and Heat Transfer, 1994, 8(1) 9-23.
[25] C. Park, R.L. Jaffe, H. Partridge, Chemical-kinetic parameters of hyperbolic Earth entries, AIAA-2000-021. Reston, 2000, VA:AIAA.
[26] C. Park, Assessment of two-temperature kinetic model of ionizing air, Journal of Thermophysics and Heat Transfer, 1989, 3(3) 233-244.
[27] C. Park, Nonequilibrium hypersonic aerothermodynamics, New York, John Wiley &Sons, 1990, pp. 27-41.
[28] B. Stahl, H. Emunds, A. Gülhan, Experimental investigation of hot and cold side jet interaction with a supersonic cross-flow, Aerospace Science and Technology, 2009, 13 488-496.

Research on dust suppression technology of fully mechanized mining workers by section control of airflow

Fang Chen*

China Coal Technology and Engineering Group Chongqing Research Institute, Chongqing, China

*Corresponding author e-mail: my-cafe@163.com

Abstract. Taking the 20106 fully mechanized mining face of Wangjialing coal mine as the engineering background, this paper analyzes the action relationship between the dusty air flow and fresh air flow formed by the coal caving and dust production of the fully mechanized mining machine, and through the establishment of mathematical model and field test, the best matching parameters of the relative position between the axial and radial air outlet of the segmented air control and dust control system and the head-on of the working face are obtained. The field application shows that after the optimization of the parameters of the wind and dust control system, the dust reduction efficiency of the total dust concentration at the driver's position and each transfer point can be increased by about 80%, and the dust reduction efficiency at the driver's position can reach 91.8%. The high concentration dust is controlled within 2m in front of the work, which can prevent the escape and diffusion of dust to the rear, and the working environment is significantly improved.

1. Introduction

The proportion of dust production of underground excavation face in the total underground dust production is second only to that of fully mechanized mining face, and the development and excavation of shaft and roadway has become the focus of dust prevention because of its characteristics of many processes, scattered dust sources and high dust dispersion [1-3]. At present, the main measures for dust control in fully mechanized working face are mainly the internal and external spraying of roadheader, air curtain dust control, wet dust removal, retransmission point spray and roadway air flow purification water curtain[4-6]. However, there are some unfavorable factors in the implementation: first, the active ventilation and passive ventilation mode are adopted in the fully mechanized face, which causes the gas flow in the working face to be atomized and affects the operational line of sight of the workers. Second, the above measures do not control the air flow of the fully mechanized excavation face, and the direct dust removal method is adopted, so the dust removal efficiency is relatively low [6-8].

2. General situation of Engineering

The 20106 return air roadway of Wangjialing coal mine is excavated along the 2# coal seam floor. The inclination angle of the coal seam is 2 ~ 4 °. The 2# coal belongs to dry and fragile (original moisture content 0.85%, firmness coefficient f = 0.46) and difficult to wet (U20 is about 0). The working face adopts ebz-220 cantilever roadheader to cut the coal and select Φ 1000 double opposite side soft air ducts supply air, and kcs-500 wet dust collector is equipped at the same time. The treated air volume is 450-550 m^3/min, and the measured suction air volume is 480 m^3/min.

The roadway excavated in 20106 working face has large section (s= 5.2×3.6m=18.72m^2), The characteristics of large air supply (500~550 m^3/min), high outlet wind speed (8~10 m/s, large air flow pressure head and large disturbance to the head-on flow field), high coal cutting intensity (daily footage of 15~20 m), serious coal crushing and large dust production lead to high dust production intensity, difficult treatment and high process requirements of dust prevention system. Therefore, it is urgent to carry out systematic research on the treatment of flour dust in comprehensive excavation, In particular, the ventilation and dust control technology of high-yield dust fully mechanized mining face is studied.

3. Determination of parameters of segmented wind and dust control system

In order to control the diffusion of flour dust from 20106 return air roadway excavation to the return air side, the fully mechanized excavation face is equipped with a segmented wind and dust control system, which is mainly composed of wall attached air duct, air regulating valve, rectifier air duct, etc. the system composition is shown in Figure 1. When the working face works normally, the fresh air flow moves forward and the dusty gas flows backward. There is an interaction surface between the fresh air flow and the dusty air flow in the front area of the working face, which is the critical interface (dust control surface) to control the outward diffusion of dust. In order to ensure that the dust control surface is always located in front of the driver of the fully mechanized excavator, The position of the dust control surface can be controlled by adjusting the distance between the axial air outlet and the radial air outlet and the coal wall of the working face, so as to ensure that the driver's position is in the fresh diversion, so that the dust control area is kept in front of the driver.

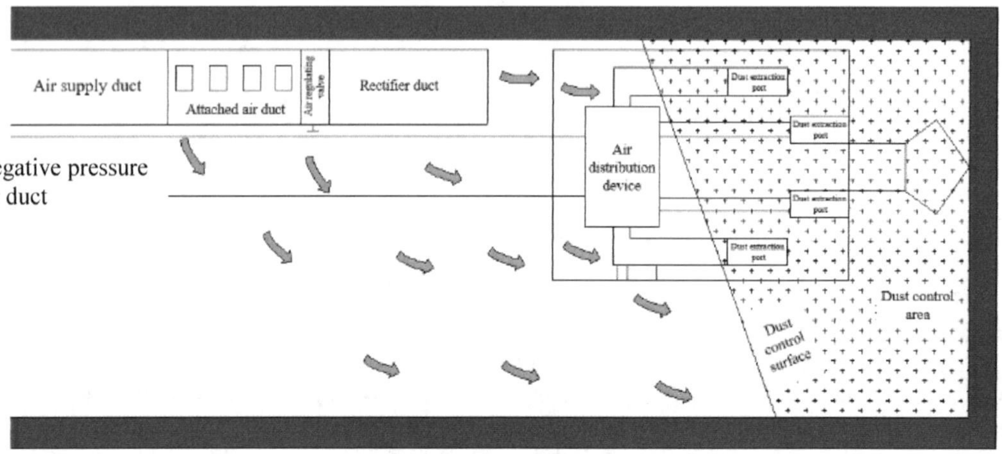

Figure 1. System composition

3.1. Determination of distance between axial air outlet and head-on

According to the relationship between the air distribution volume of the working face and the air extraction volume of the dust collector, Q_y=(1.2~1.3)Q_c [9-10], in the dust control area in front of the work, the air volume not extracted by the dust collector per unit time will diffuse disorderly with the dust generated by coal cutting in the dust control area, forming a high concentration dust storage area. During the normal production of the working face, the coal cutting is intermittent. When cutting from the left side to the right side, the excavation opportunity stops briefly. After the dust collector removes most of the dust in the dust control area, continue to cut the coal; Therefore, during coal cutting, the optimal distance between the axial air outlet and the coal wall of the working face is calculated by establishing a mathematical model (as shown in Figure 2).

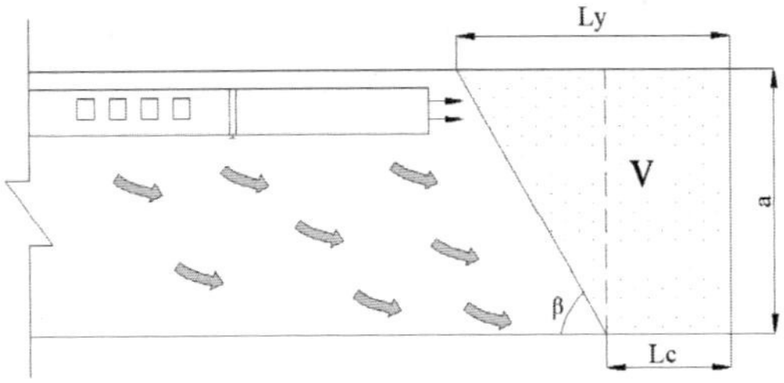

Figure 2. Calculation Model

The position of the dust control surface and the size of the angle β are related to the wind speed of the air flow in front of the driver. When the axial air output is large and the radial air output is small, the wind speed on the right side of the driver's position is relatively small. At this time, the position of the dust control surface will face the driver. The rear moves, the angle β increases; when the radial air output increases and the axial air output is small, the wind speed on the right side of the driver's position is larger, and the dust control surface moves relatively forward, so that β decreases.

In the high concentration dust storage area, the residual air volume of the pressure pumping system and the volume of the dust control area always exist $\Delta Q = V$;

$$V = \frac{ab}{2} \bullet (L_y + L_c), \quad L_c = L_y - a \bullet ctg\beta \qquad (1)$$

Ly is the distance between the axial air outlet of the air control system and the head-on; L_C is the distance between the air outlet of the dust suction hood and the head-on; a is the roadway width; b is the roadway height; β Is the included angle between the dust control surface and the roadway.

After simplification, the following formula is obtained.

$$L_y = \frac{\Delta Q}{ab} + \frac{a \bullet ctg\beta}{2}, \quad L_c = \frac{\Delta Q}{ab} - \frac{a \bullet ctg\beta}{2} \qquad (2)$$

On the face of the roadheader, since the position of the dust control surface needs to be controlled at the front of the driver and the rear of the dust hood, the angle β should vary between β' and β'', as shown in Figure 3.

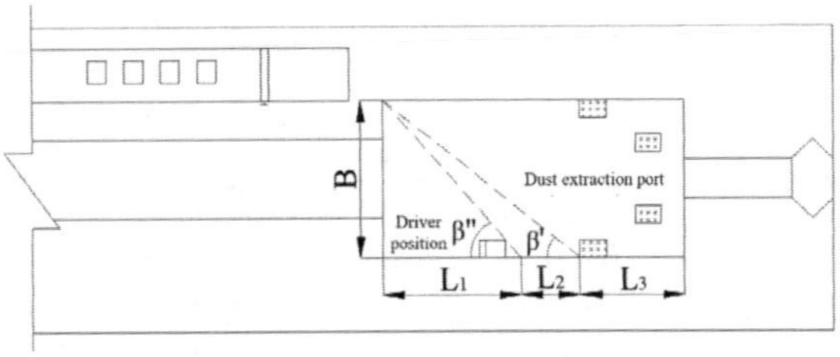

Figure 3. Schematic diagram of the position change of the dust control surface

Therefore, the following formula can be obtained.

$$\frac{\Delta Q}{ab} + \frac{a \bullet ctg\beta''}{2} \leq L_y \leq \frac{\Delta Q}{ab} + \frac{a \bullet ctg\beta'}{2} \qquad (3)$$

$$ctg\beta'' = \frac{L_1}{B}, \quad ctg\beta' = \frac{L_1 + L_2}{B}$$

After the formula (3) is simplified, the calculation formula (4) of the optimal matching parameter for the distance between the axial air outlet and the head is obtained.

$$\frac{\Delta Q}{ab} + \frac{aL_1}{2B} \leq L_y \leq \frac{\Delta Q}{ab} + \frac{a(L_1 + L_2)}{2B} \qquad (4)$$

In the formula, L_1, L_2, and L_3 are the distances between the driver and the tail, the driver and the nearest vacuum hood, and the vacuum hood and the head, and B is the width of the roadheader.

Through the dimensional analysis of the EBZ series fully mechanized excavator, the relationship curve between the axial air outlet distance L_y and the dust control surface angle β is obtained, as shown in Figure 4.

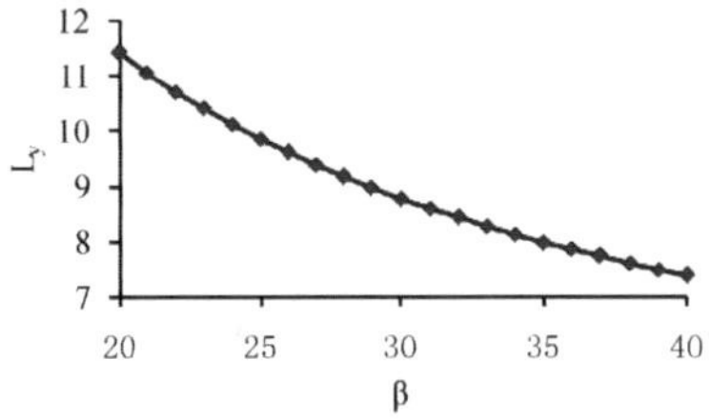

Figure 4. Relationship between axial air outlet distance and horizontal angle of dust control surface

According to the overall size of the fully mechanized excavator, it is calculated that 25°≤β≤35°. It can be seen from Fig. 4 that the optimal distance between the axial air outlet of the air control system and the head-end should be 8~10m, within this range, the shaft can be guaranteed The best matching relationship between the air outlet and the headwind speed.

According to the empirical formula (5) of the long-pressure short-extraction ventilation and dust removal system, it is calculated that the distance Ly≤21.6m from the outlet of the press-in air duct to the head, and the distance Lc ≤ 6.5m from the inlet of the dust extraction air duct to the head, the deduced air control The axial air outlet of the system is 8~10m away from the head, which meets the design requirements of the air outlet.

$$\begin{Bmatrix} L_y \le 5\sqrt{S} \\ L_c \le 1.5\sqrt{S} \end{Bmatrix} \tag{5}$$

According to the site conditions, when the distance from the air outlet to the head is ≤10m, the air outlet is located in the front of a running load point, and the air volume of the soft air duct will not bring the dust generated by a running load point into the front of the work front. Therefore, it is best when the axial air outlet is 8~10m away from the head.

3.2. Determination of distance between radial air outlet and head-on

In order to realize the dust control of the fully mechanized excavation face in 20106, a field test was carried out on the radial air outlet of the ventilation and dust removal system, that is, the distance between the wall-mounted air duct and the head-on to determine the best matching parameters.

In the entire ventilation and dust removal system, the position of the wall-mounted air duct is directly related to the dust control effect of the working face. In order to obtain the best installation position of the wall-mounted fan Field tests were carried out on the effect of dust control under three conditions: 15m, 20m and 25m, and passed the continuous test the dust concentration at the driver's position. The variation of average dust concentration from wall attached duct to head-on is shown in Figure 5.

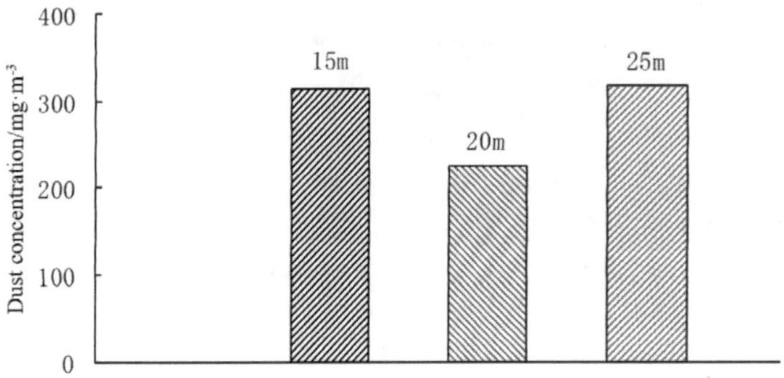

Figure 5. Variation of average dust concentration from wall attached duct to head-on

Through the field test, it can be seen that when the wall-mounted air duct is located at the rear of the second operation load point, the second-run load point is closer to the wall-to-wall air duct, where the wind speed is higher, and a large amount of dust on both sides of the transfer point is brought into the work by the wind flow. Therefore, the best position of the wall-mounted air duct should be 20m away from the head, and close to the load point of the second operation, and the best position is 5m before the load point of the second operation.

4. Effect analysis of wind and dust control

After field test, the original dust concentration at the driver's position on the fully mechanized excavation face was 2744 mg/m^3 without any dust reduction measures. Before the ventilation and dust removal system was improved and optimized, the average total dust concentration at the driver's position on the driving face of the return air road in 20106 was 1224.8 mg/m^3, and the average total dust concentration at the first operating load point of the fully mechanized excavator was 489.7 mg/m^3. The average total dust concentration at the operating load point is 253.2 mg/m^3.

After the improvement and optimization of the ventilation and dust removal system, compared with before optimization, the dust concentration at the driver's position decreased from 1224.8 mg/m^3 to 225.3 mg/m^3, the dust concentration decreased by 81.6%, and dust reduction efficiency was 91.8%; The dust concentration decreased from 489.7mg/m^3 to 146.8 mg/m^3, and the total dust reduction efficiency was 84.0%; the dust concentration at the second operation load point decreased from 253.2mg/m^3 to 99.0 mg/m^3, the total dust reduction efficiency was 86.4%; the ventilation and dust removal effect of the fully mechanized excavation working face of the return air lane was significantly improved, and the dust concentration of the working face was greatly reduced. The Comparison of dust concentration before and after optimization of dust control system is shown in Table 1, and the comparison effect of dust control is shown in Figure 6.

Table 1. Comparison of dust concentration before and after optimization of dust control system

Location	Average dust concentration /mg·m^{-3}			Dust reduction efficiency		Remark
	Original value	Before optimization	After optimization	Before optimization	After optimization	
Driver	2744	1224.8	225.3	55.4%	91.8%	Top coal cutting Collector on
Transport transfer 1#	916	489.7	146.8	46.5%	84.0%	
Transport transfer 2#	732	253.2	99.0	64.9%	86.4%	

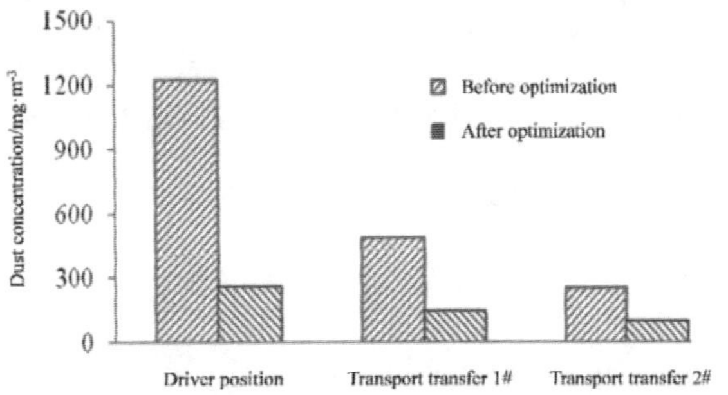

Figure 6. Comparison of dust concentration before and after optimization

In a word, after optimizing the sectional air outlet parameters of the original ventilation and dust removal system, the dust control and reduction effect of the ventilation and dust removal system is improved by about 80%, which can effectively reduce the dust concentration of tunneling flour. Under the best matching parameters, the average dust concentration at the driver's position of the system can be reduced to 225.3mg/m^3, and the dust concentration is stable at about 200mg/m^3 most of the time of coal cutting, The dust reduction efficiency can reach 91.8%. The high concentration dust is controlled within 2m in front of the work, and the working environment has been significantly improved. The field sampling test is shown in Figure 7.

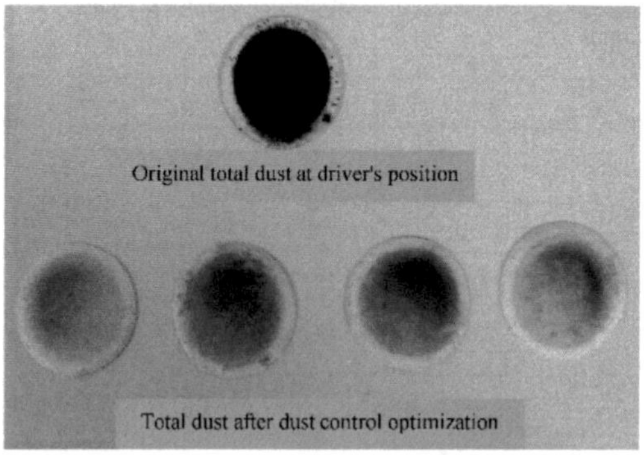

Figure 7. Field sampling comparison test

According to the field sampling test and analysis, when the ventilation and dust removal system of the working face is located at the best matching parameters (the axial outlet of the air supply duct is 8 ~ 10m away from the head-on, and the wall attached duct is 20m away from the head-on of the working face), the effect of wind and dust control is the best.

5. Conclusions

By analyzing the characteristics of coal cutting and dust production in the fully mechanized excavation working face, in order to obtain the best matching parameters of the segmented air control and dust control system, a corresponding mathematical model was established, and the axial air outlet distance of the segmented air control and dust control system was calculated. The best head-on distance is 8~10m.

Using the field test method, it is obtained that the optimal distance between the radial air outlet and the head of the segmented air and dust control system is 20m.

Through the on-site test, when the coal is cut normally at the working face, after the parameters of the segmented air control and dust control system are optimized, the dust control efficiency can be increased by about 80%, and the total dust reduction efficiency can reach 91.8%, and the high concentration dust is controlled in front of the work. The working environment has been significantly improved within the range of 2 m.

Acknowledgements

This work was supported by the Chongqing Natural Science Foundation of China (General Program) (cstc2021jcyj-msxmX1184).

References

[1] Shi Fuchuan, Kang Juding. Comprehensive mine dust prevention technology and management [M]. Beijing Coal Industry Press, 1994.

[2] Yang Shengqiang. Dust control theory and technology [M]. Xuzhou: China University of Mining and Technology Press, 2007.

[3] Wang Hong. The 40 years developmental review of the fully mechanized mine roadway heading technology in China[J]. Journal of China Coal Society, 2010, 35(11): 1815-1820.

[4] Gang Liu, Yan Xi, Zhang Rengui. Research and application of dust removal technology in the mechanized excavation face [J]. Industrial Safety and Environmental Protection, 2012, 33(2): 16-18.

[5] Liu Jian, Yao Haifei,Wei Chuanguang,et al. The structural optimization and numerical simulation

of wet centrifugal precipitator at working face of diving airway [J]. Journal of China Coal Society, 2010, 35(3): 424-428.
[6] He Qinghai. High-efficiency and comprehensive dust removal technology in fully mechanized excavation face of all-rock roadway [J]. Coal Science and Technology, 2013, 41(8): 81-83.
[7] Cheng Weimin, Liu Xiangsheng, Ruan Guoqiang, etc. The closed dust control theory and technical process for fast construction of coal roadway anchor excavation [J]. Chinese Journal of Coal, 2009, 34(2): 203-207.
[8] Zhang Sheji, Liu Yong, Zhou Runjin, et al. Test on dust distribution degularity and dust suppression and control technology in heading face [J]. Mining safety and environmental protection, 2010, 37(2): 30-33.
[9] Nie Wen,Cheng Weimin,Yu Yanbin,et al. The research and application on whole-rock mechanized excavation face of pressure ventilation air curtain closed dust removal system [J]. Journal of China Coal Society, 2012, 37(12): 1165-1170.
[10] Chen Fang,Zhang Sheji,Ma Wei,et al. Research and application of the technology of forced ventilation diversion to control and reduce dust in fully mechanized excavation face[J]. Journal of China Coal Society, 2018, 43(S2): 483-489.

The axial fan design and commissioning test with nonuniform inlet flow

Xiaoli Qu, Zebin Ren*, Wenguo Yang and Yuanqiang Luo

Facility Design and Instrument Institute, China aerodynamics Research and Development Center, Mianyang, Sichuan 621000, China

*Corresponding author e-mail: Renzebin401@163.com

Abstract. According to the two-dimensional flow theory of the axial fan rotor blades and the aerodynamic characteristics of the low speed wind tunnel, combined with the fan aerodynamic efficiency and wind tunnel pressure loss coefficient, a new equation which points out the inherent relationship of the fan blade setting angle, fan rotating speed and flow rate in wind tunnel circuit is derived. So a new method for fan rotor blade setting angle adjustment to satisfy the fan performance at off-design point by getting the test results of fan operating parameters but without the fan total pressure rise in the low speed wind tunnel is developed. Following the new method, the fan rotor blade setting angle adjusting value was provided directly only with the fan rotating speed and flow velocity in the wind tunnel test section, the adjusting target was achieved successfully by the new blade setting angle, the cost of the wind tunnel commissioning test were saved. The test results show that, after increasing the fan rotor blade setting angle by 4.5 degrees, when the flow velocity in the wind tunnel test section reaches 60m/s, the fan rotating speed is 570rpm, the deviation from the predicted fan rotating speed value of 575rpm is 0.9%. For the same test section flow velocity, the predicted value and the real value of the fan rotating speed are in good agreement, it proves that this method is reliable and accurate in practical application.

1. Introduction

Axial fans are widely employed for civil, military industrial, and transportation applications. Much work has been carried out in axial fan design in recent years, largely driven by high aerodynamics performance and low noise requirements. Jin-Hyuk Kim [1] develops an optimization procedure for axial fan design and increase the fan efficiency of 1.8% and a 31.4% improvement in the pressure rise. Sarraf [2] studies the effects of blade thickness on the performances of a controlled vortex designed axial fan and concludes that the aerodynamic efficiency of thick blades is lower than thin blades. Spasic [3] designs and tests the performance of reversible fans with two blade configurations, finds that reversible axial fan with the slightly doubly curved profile of blades can achieve a greater increase in pressure and flow than the fan with the straight profiles. Lee [4, 5] use a computerized preliminary design system to construct the basic 3-D blade geometry for axial fan and predict its aerodynamic performance, and the results of its predictions are compared with corresponding experimental results to verify the prediction accuracy of the design system. Lotfi [6] optimized the total pressure loss and tangential load of a low-speed axial fan cascade with variables related to blade configuration by using a genetic algorithm and

3-D RANS analysis. In recent years [7-10], several design methods have publicized for the purpose of axial fan noise analysis and reduction. Moreau [11-12] optimized the fan rotation speed, rotor-stator separation distance, rotor-stator blade count, the lean and sweep angle of the rotor blades and stator vanes to minimize the tones and broadband noise of a low noise fan in the DNW-NWB wind tunnel [13], which contributed much to make the wind tunnel to be a high quality aeroacoustic facility in Europe.

Many experiment researches and numerical simulations are carried out to verify and optimize the performance of the axial fan, many works are focus on the tip-gap and blade angle. Pogorelov [14] simulated the axial fan flow field of different, it was found that the tip-gap vortex wandering and as such the region of influence of the turbulent wake decreased and the dominant modes frequencies increased by reducing the tip-gap width. Bo Luo [15] made a systematic analysis on axial fans with different tip-clearance sizes to gain a clear understanding of the characteristics of tip leakage flow and investigate the generation mechanism of the fan noise of certain frequencies. Wilkinson [16, 17] test a scaled fan to validate the numerical models as well as investigating the aerodynamic performance of the M-fan under different operating, and the optimized design result for the M-fan was obtained. Weifeng He [18], Lei Zhang [19], Jiadong Tang [20] and Yan Xiao-kang [21] analyzed the static and dynamic characteristics of axial fans for different flow conditions, blade angle was adjusted to optimize the fan performance. Ye Xuemin [22-25] experimentally and numerically investigated the effect of the abnormal blade angle on axial fans aerodynamic and aeroacoustic performance,

This paper presents the design methods and commissioning test for the axial fan in a low speed wind tunnel in SHANGHAI JIAO TONG UNIVERSITY (SJTU). To obtain steady flow in the test section of the wind tunnel (Figure 1), an axial fan (Figure 2) is set between the 1st and the 2nd corner with a nonuniform inflow condition. Arbitrary vortex method is employed to optimize the geometry of fan blades and stators, for the purpures to achieve high aerodynamic efficiency and uniform flow at the fan outlet. In the early stage of fan commissioning, a new method for fan blade angle adjustment to satisfy the fan performance at off-design point by getting the test results of fan operating parameters but without the fan total pressure rise in the low speed wind tunnel is developed. Flowing the new method, the fan blade angle adjusting value was provided directly only with the fan rotating speed and flow velocity in the test section of the low speed wind tunnel in SJTU before the fan blade angle altered, the adjusting target was achieved by the new blade setting angle successfully and the time and cost of the wind tunnel commissioning test were saved.

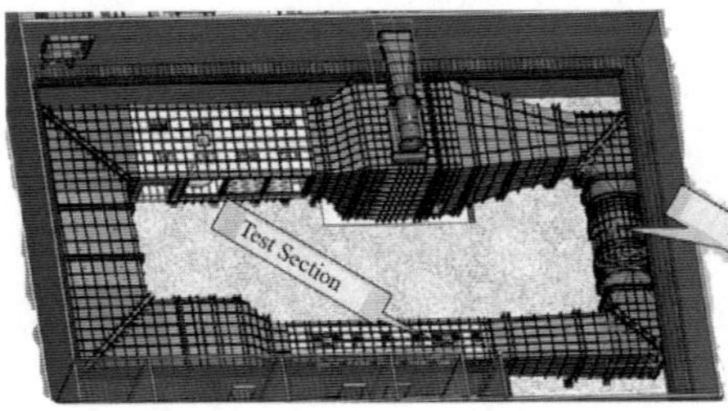

Figure 1. The wind tunnel in SJTU

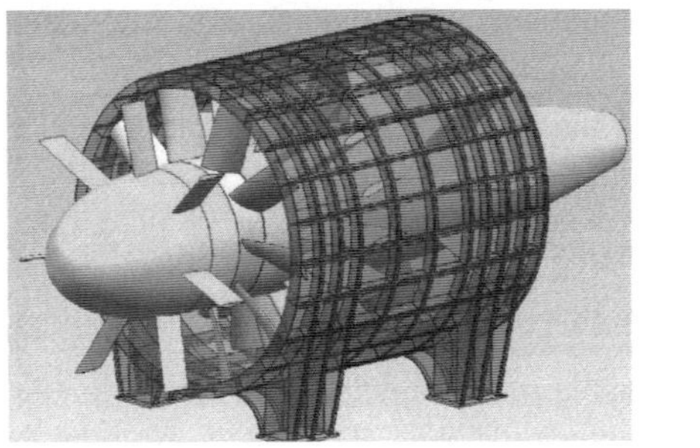

Figure 2. The axial fan in the SJTU wind tunnel

2. Fan design

The Axial fan is used in the SJTU wind tunnel, which is mainly used for the test and research of wind engineering. The test section of the wind tunnel has a section size of 3m (width)×2.5m (height) and a length of 10m. The maximum wind speed of the test section is 60m/s.

Due to the limitation of the construction site, the fan is set between the first corner and the second corner as the long leg of the wind tunnel can be reduced, which will cause ununiform airflow at the inlet of the fan. The ununiform airflow at the inlet of the fan blades will result in a decrease in the aerodynamic efficiency of the fan. Therefore, in the design progress of the fan, the vortex distribution of the fan blades must be optimized to improve the fan performance.

2.1. Design method

In the aerodynamic design of axial fan, the free vortex design method is widely used, and it requires the assumption of two dimensional flow. This vortex method is widely used in the design of axial fan which satisfactorily meets the design requirements of low speed wind tunnel, mine ventilation system, and cooling system, and the details of this method have been described by Collar [27], Pope [28] and Wallis [29]. This method also assumes that the flow velocity has no radial component, and the theoretical axial velocity at the fan inlet and the total pressure rise of the fan are both constant along the fan radius. As mentioned above, the inlet velocity of the fan has an ununiform distribution, if the free vortex design method is employed, the fan efficiency will be low, and there is a risk of flow separation at the hub or tip of the blade. However, another fan design method named arbitrary vortex method does not require radial equilibrium and a real velocity distribution can be applied in the fan design procedure [30]. Moreover, fan design with higher aerodynamic performance can be achieved with this method, so, the arbitrary vortex method was employed in the design of the axial fan in SJTU wind tunnel.

2.2. Fan Blade

The velocity upstream of the fan blades in the SJTU wind tunnel is obtained by the numeral simulation of the wind tunnel circuit, the inlet velocity profile is expressed as follows.

$$u1 = (-x + 1.75)U \tag{1}$$

Where $u1$ is the axial velocity at the fan inlet, U is the mean velocity at the fan cross-section, x is the dimensionless radius.

With the arbitrary vortex method, the fan system can transform the oncoming skewed velocity distribution into a more uniform distribution at the rotor and stator exit by optimizing the swirl velocity component with radius. Thus, the flow leaving the stator has a sufficiently uniform profile of velocity before entering the second corner of the wind tunnel, and the aerodynamic performance of the second corner will be improved. Moreover, adjusting the load distribution on the blade leads to larger output work at the tip of the blade, so the fan can achieve higher efficiency at lower rotational speed. The swirl coefficient ε_s can define as a function of the radius:

$$\varepsilon_s = f(x) \tag{2}$$

Once the swirl velocity and the axial velocity at the fan inlet are known, the axial velocity at the fan rotor outlet $u2$ and the axial velocity at the stator outlet $u3$ can be obtained from the following two equations.

$$u2 = U\sqrt{\left(\frac{U_1}{U}\right)^2 + \left(\frac{2\varepsilon_s}{\lambda_s} - (\varepsilon_s)^2\right) - \left(\frac{2\varepsilon_s}{\lambda_s} - (\varepsilon_s)^2\right)_{x_U} - 2\int_{x_U}^{x} (\varepsilon_s)^2 dx/x} \tag{3}$$

$$u3 = U\sqrt{\left(\frac{U_2}{U}\right)^2 + (\varepsilon_s)^2 - ((\varepsilon_s)^2)_{x_U} + +2\int_{x_U}^{x} (\varepsilon_s)^2 dx/x} \tag{4}$$

Where λ_s is the advance ratio, x_U is the dimensionless radius which the local axial velocity is equal to the mean velocity.

After the axial velocities at the fan rotor outlet and stator outlet have been derived, the chord lengths and setting angles of the rotor and stator blades can be determined according to the detailed design procedure introduced in reference [31].

Figure 3 shows the designed profiles of axial velocity at different fan locations. By adjusting the blade loading, more uniform velocity distributions can be obtained at the rotor outlet and stator outlet. Consequently, a significant redistribution of axial energy occurs in the fan plane and the momentum of the flow near the fan tip is significantly improved and the fan efficiency can be increased. Blade loading coefficient provides a correlation between fan blade lift coefficient and solidity. Figure 4 show blade loading coefficient and blade efficiency with free vortex design method and arbitrary vortex design method. With arbitrary design method, more blade loading and efficiency can be obtained at the fan tip, it is help to improve the hole blade aerodynamic efficiency.

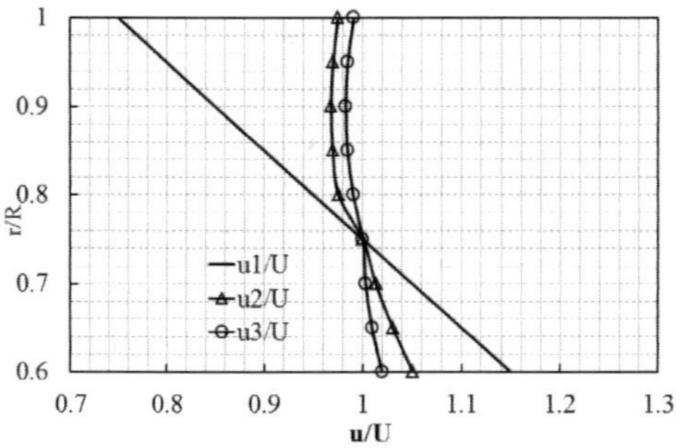

Figure 3. The velocity distribution at different location of the designed fan

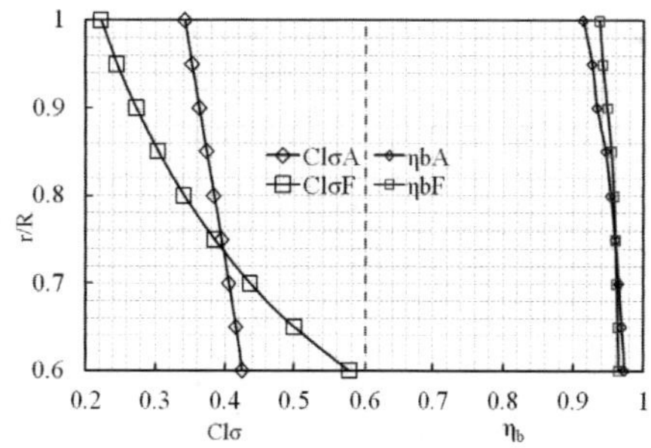

Figure 4. The blade loading coefficient and blade efficiency with different design method (A-arbitrary vortex design method, F-free vortex design method)

The key physical parameters and performance estimates are shown in table 1. The length of the fan is 12m, the maximum fan rotating speed is 600rpm.The fan airline and rotor blade stacking plan are shown in figure 5 and figure 6. This demonstrates the blade twist. The installation angles at the blade root, mid-span and tip are 44.5 degrees, 30.5 degrees and 22.7 degrees respectively. The values of the attack angle alone the whole blade span is not large than 0.4 degrees at the design point, which should give the fan a reasonable stall tolerance at low flow rates.

Table 1. FAN Characteristics

Parameter	Value
Pressure rise	1667Pa
Flow rate	450m^3/s
Diameter	3.8m
Hub-tip ratio	0.5
Estimated whole fan efficiency	0.83
Rotor blade count	10
Stator blade count	17

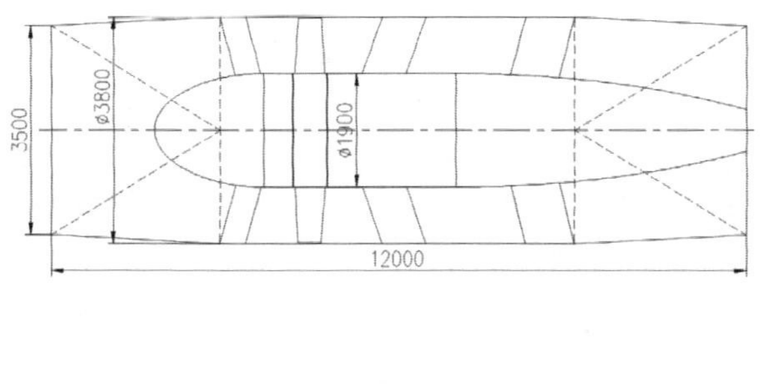

Figure 5. The airline of the fan

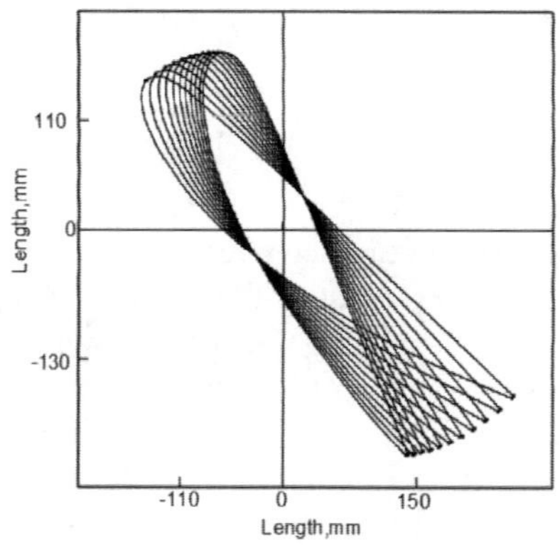

Figure 6. Stacking profile of the fan blade

3. The commissioning test of the fan

Preliminary tests were carried out on the performance of the SJTU wind tunnel and fan, at the initial stage after the wind tunnel assembled. At the design point of the fan, the wind speed test of the test section corresponding to different fan rotation speeds was carried out. The test result tells that the test section wind speed corresponding to the maximum fan rotation speed of 600rpm is 57.8m/s, the corresponding volume flow is 433.5m^3/s, and the fan power is 629kW, which is not satisfied wind tunnel's maximum wind speed requirement. Combined with the comprehensive commissioning plan of the wind tunnel, the installation angle of the fan needs to be adjusted. Under the condition that all wind tunnel sections are not modified, the installation angle adjustment target is that when the fan rotation speed is 575rpm, the wind speed of the test section reaches 60m/. Meanwhile, to save the wind tunnel commissioning test period, the adjustment of the fan installation angle is hoped to be as soon as possible, and a fast prediction method of the fan flow rate under with different installation angle was developed.

3.1. Theoretical derivation

3.1.1. The relationship between the swirl coefficient and blade installation angle. For the axial fan system in the low speed wind tunnel, in order to improve the fan efficiency, the isolated airfoil design method is generally used in the fan design to avoid the efficiency loss caused by the interference between the blades. The fan tip speed is strictly controlled by choosing a reasonable fan diameter, hub ratio, rotation speed and other parameters. To avoid the influence of the three-dimensional effect on the aerodynamic performance of the fan during operation, the solidity of blade at each radial position is optimized by choosing the suitable fan blade number and the chord length [12], [26].

The speed composition of a axial fan without preswirl velocity component is shown in Figure 7.

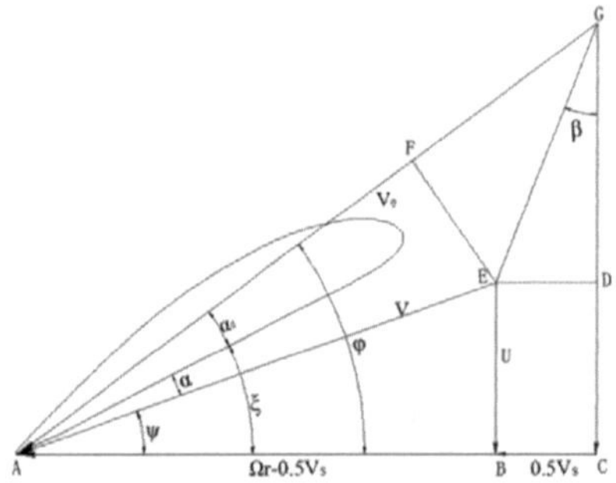

Figure 7. The speed composition of fan blade

Taking the coordinate system of the radial position of the fan blade as a reference, the velocity triangle formed by the axial and circumferential airflow in the blade section is the AEB in figure 7. The axial velocity of the airflow$U$=EB. After the airflow does work through the blade, it will generate a circumferential rotation speed $V_S$behind the blade profile, and the direction of V_S is the same as the blade rotation direction. In fan engineering design, the local airflow rotation speed of the blade profile is generally taken as $0.5V_S$, then relative speed at the circumferential direction speed can be calculated as $\Omega r-0.5V_S$=BC. For the combined speed $V=AE$. In the figure 7, ψ is the advancing angle, which is the angle between the relative combined speed of the blade and the direction of rotation. ξ defines as the installation angle, the angle of attack of the blade can be express as $\alpha=\xi-\psi$.

When the fan runs at different conditions, the local speed triangle of the blade will be different, and the two parameters of the forward angle ψ and the angle of attack α will be changed, which will make it hard to the estimate the fan performance. Since the zero-lift line of the airfoil does not change with the changing incoming flow, the angle between the zero-lift line of the airfoil and the direction of rotation, is considered as one of the parameters for fan performance estimation.

In Figure 7, the zero-lift line of the blade profile is represented by AG, and α_0 is the zero-lift angle of attack. When the local resultant speed of the blade coincides with the zero lift line, the resultant speed V_0=GA ,at this condition, the fan does not put power into the airflow, the airflow downstream of the blade will not get a circumferential induced speed. The point B in the speed triangle moves to a point C, the circumferential relative airflow speed will increase to Ωr, where Ωr=CA, the corresponding axial speed is U_0, U_0=GC.

The axial velocity U can be expressed by the following equation.

$$U = EB = GC - GD = \Omega r \cos\varphi - \frac{0.5\text{Vs}}{\tan\beta} \tag{5}$$

Divided by Ωr, λ can be obtained as follows.

$$\lambda = \cos\varphi - \frac{0.5\varepsilon_S\lambda}{\tan\beta} \tag{6}$$

Where λ is the advance ratio, $\lambda = \text{U}/\Omega r$. ε_S is swirl factor, $\varepsilon_S = \text{Vs/U}$.

The relationship between the swirl factor, the lift coefficient and the advance angle can be obtained using the following equation as derived by Wallis [13]:

$$C_L = 2\sigma\varepsilon_S \sin\psi \tag{7}$$

Where σ is solidity, which represents the ratio between the chord length of a single blade of the rotor and the distance between two adjacent blades of the rotor.

For the axial fan inside the wind tunnel, its operating point should avoid the blade in a stalled state, so the blade lift coefficient has a linear relationship with the angle of attack:

$$C_L = h\sin(\alpha - \alpha_0) = h\sin(\varphi - \psi) \tag{8}$$

Where, h is the linear coefficient of the lift coefficient of the blade airfoil varying with the angle of attack. As the value of the selected blade solidity is small, the airfoil can be regarded as an isolated airfoil, and its aerodynamic performance can be determined according to the test data of the corresponding two-dimensional airfoil. If the value of the selected blade solidity is large, the flow interference of adjacent blades cannot be ignored, the airfoil cannot be regarded as isolated, the lift provided by the blades will be reduced due to the mutual interference effect, in this case, h can be corrected according to the relationship curve with the airfoil advancing angle [13].

From figure 7, two equations can be given:

$$FE = V\sin(\varphi - \psi) = GE\cos(\varphi + \beta) = \frac{0.5V_S\cos(\varphi+\beta)}{\sin\beta} \tag{9}$$

$$\frac{\sin(\varphi-\psi)}{\sin\psi} = \frac{0.5\varepsilon_S\cos(\varphi+\beta)}{\sin\beta} \tag{10}$$

The relation between swirl factor and φ can be calculated by combining eqns. 6, 7, 8 and 10.

$$\varepsilon_S = \frac{2\cos\varphi(\tan\varphi/\lambda - 1)}{4/h\sigma + \sin\varphi} \tag{11}$$

Therefore, the relationship between the installation angle of the fan blade and the swirl factor can be obtained as follows:

$$\varepsilon_S = \frac{2\cos(\xi-\alpha_0)(\tan(\xi-\alpha_0)/\lambda - 1)}{4/h\sigma + \sin(\xi-\alpha_0)} \tag{12}$$

The relationship between the fan pressure, flow rate and blade installation angle. In the process of fan design, the pressure rise provided by the fan can be expressed in a dimensionless form, and the relationship between its pressure rise coefficient, pressure rise, axial speed, air flow density and fan aerodynamic efficiency is defined as:

$$\boldsymbol{K} = \frac{\boldsymbol{\Delta P}}{0.5\boldsymbol{\rho U^2 \eta}} \tag{13}$$

The axial speed can be calculated by the air volume flow rate, the fan radius and the tip-hub ratio

$$U = \frac{Q}{\pi R^2(1-x_b^2)} \tag{14}$$

The relationship between the pressure rise coefficient, swirl factor and the advance ratio can be obtained using the following equation as derived by Wallis:

$$K = \frac{2\varepsilon_S}{\lambda} \tag{15}$$

Combining eqns. 12, 13, 14 and 15, the relation between pressure, air flow rate and the rotor blade installation angle can be obtained as flows:

$$Q = \frac{\Delta P \pi R^2(1-x_b^2)[4/h\sigma + \sin(\xi-\alpha_0)]}{2\Omega r \rho \eta \cos(\xi-\alpha_0)\left[\tan(\xi-\alpha_0)/\frac{Q}{\Omega r \pi R^2(1-x_b^2)} - 1\right]} \tag{16}$$

The energy provided by the fan is used to overcome the pressure loss of the wind tunnel circuit. The pressure loss of the wind tunnel can be calculated by the whole wind tunnel pressure loss coefficient ε and the wind speed U_t of the test section.

$$\Delta P = 0.5\varepsilon\rho U_t^2 \tag{17}$$

The compressibility dose not considered when the air running in a low speed wind tunnel, so the volume flow rate of the test section is considered to be consistent with the volume flow rate of the fan channel. The relationship between wind tunnel circuit loss coefficient, fan efficiency, fan flow rate and installation angle can be given by combining eqns. 16 and 17.

$$Q = \frac{2A_t^2 \Omega r \cos(\xi-\alpha_0)\left[\tan(\xi-\alpha_0)/\frac{Q}{\Omega r \pi R^2(1-x_b^2)} - 1\right]}{0.5\pi R^2(1-x_b^2)[4/h\sigma + \sin(\xi-\alpha_0)]} \frac{\eta}{\varepsilon} \tag{18}$$

The relationship between fan flow rate and installation angle is presented by Equation 18, where A_t is the area of the test section.

Follow Equation 18, if the flow rate of the wind tunnel, that is, the flow rate of the fan does not reach the design value at the fan maximum rotation speed, the installation angle of the fan needs to be adjusted. The adjustment of the wind tunnel fan installation angle is generally to change the maximum wind speed in the test section at the fan's maximum rotation speed. With many wind tunnels and fans commissioning experiences, it is recommended that the wind speed changing value of the wind tunnel test section is generally within 15% of the maximum wind speed and the air Reynolds number will not be changed much. If the structure of the wind tunnel does not change, the pressure loss coefficient of the wind tunnel circuit is almost the same before and after the fan installation angle adjusted. For an axial fan with good performance in the wind tunnel, its aerodynamic efficiency in the design point and the operating conditions near the design point can be higher, if the installation angle of the fan is adjusted to a small extent, the corresponding fan aerodynamic efficiency before and after the adjustment is regarded as a constant. So, $\frac{\eta}{\varepsilon}$ can be regarded as unchanged before and after the fan installation angle adjusted.

According to equation 18, combining the design parameters (fan radius, solidity, hub-tip ratio, zero-lift angle of attack and linear coefficient) of the fan with the initial installation angle and the wind tunnel flow rate with the maximum fan rotating speed, the ratio of the wind tunnel loss coefficient to the fan efficiency, $\frac{\eta}{\varepsilon}$ can be calculated before the fan installation angle is adjusted. With the constant $\frac{\eta}{\varepsilon}$ and the fan design parameters, substituting the flow corresponding to the required wind speed in the test section of the wind tunnel into equation 18, a new fan installation angle ξ_1 can be obtained. So, the adjustment value of the fan installation angle $\Delta\xi$ can be calculated by $\Delta\xi = \xi_1 - \xi_0$. Equation 18 can calculate the fan flow rate without the fan pressure rise test data. Using this equation, when the fan is debugged, the total pressure test of multiple sections of the fan inlet and outlet can be avoided, and the use of test equipment such as the total pressure rack, total pressure tube and electronic pressure scanning valve required for the test can also be avoided, which can save the wind tunnel commissioning test period.

3.1.2. The calculation of the fan blade installation angle adjustment. Using the fan design method described in detail by Wallis [27], the average position of the blade in the span direction can be chosen as the position to predict the fan performance after the fan installation angle adjusted. For the fan in this paper, the average position of the blade in the span is $x/R = 0.75$. In this position, the blade installation angle is 30.5 degrees, and the ratio of the wind tunnel loss coefficient to the fan efficiency, $\frac{\eta}{\varepsilon}$ can be calculated out to be 1.13 after substituting the parameters such as hup-tip ratio, the aero of the wind tunnel test section, fan rotation speed, fan radius, solidity, hub-tip ratio, zero-lift angle of attack and linear coefficient into Equation 18. Combined with $\frac{\eta}{\varepsilon}$ calculated from the initial conditions of fan operation, the target flow rate 450m3/s, speed 575rpm and other parameters are substituted into equation 14 again, and the new installation angle of the average position of the fan blades in the span direction can be calculated as 35 degrees, increased by 4.5 degrees on the original basis.

The angle of attack will increase as the installation angle of the fan increases, so the lift coefficient of the blade will increase. To determine whether the fan has the risk of stalling after the installation angle is adjusted, the lift coefficient of the blade airfoil needs to be checked. Combining Figure 7 and Equations 7, 8 and 12, the lift coefficient can be calculated as follows:

$$C_L = \frac{4\cos(\xi-\alpha_0)(\tan(\xi-\alpha_0)/\lambda-1)\sin\psi}{4/h+\sigma\sin(\xi-\alpha_0)} \tag{19}$$

In the fan design process described before, the maximum value of the lift coefficient in the radial direction of the blade will be at the tip of the blade, so after the installation angle is adjusted, the maximum value of the lift coefficient in the radial direction of the blade will also be at the tip location. The lift coefficient at the tip position of the blade is calculated to be 1.0 after the installation angle is increased by 4.5 degrees by equations 19, which is less than the stall lift coefficient 1.35 of the selected airfoil of the blade. So, after increasing the installation angle of the blades, the fan has enough stall margin, and it is safe for the operation of the fan to increase the installation angle by 4.5 degrees.

3.2. Test verification

Based on the calculation result above, the installation angles of the 10 blades of the fan are all increased by 4.5 degrees, and the installation angle of the average position of the fan blades in the span direction after adjustment is 35 degrees. When the fan speed is 570rpm (0.9% smaller than the theoretically predicted value of 575rpm), the fan power is 674kw, the wind speed in the wind tunnel test section reaches 60m/s, the commissioning target is satisfied. Figure 6 shows the relationship between the wind speed in the wind tunnel test section of the fan speed and fan power before and after the adjustment of the fan installation angle. Before and after the adjustment of the fan installation angle, the linearity between the wind speed in the test section and the fan speed is maintained well, and the fan power and the wind speed in the test section have a cubic relationship. With the same wind speed, the fan operating power of the before and after the installation angle adjustment is basically the same, which shows that the pressure loss of the wind tunnel circuit and the efficiency of the fan hardly change before and after the adjustment of the fan installation angle. The hypothesis that the fan aerodynamic efficiency and wind tunnel loss coefficient do not change due to a small change of the fan installation angle is further verified.

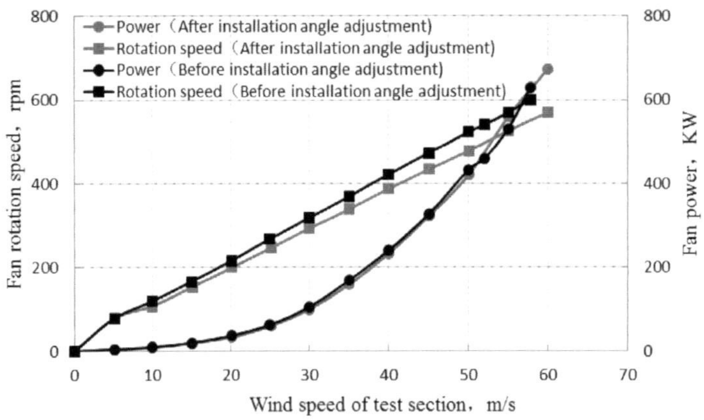

Figure 8. The relationship between fan speed and power and wind speed in test section

4. Conclusion

An axial fan system is designed for SJTU wind tunnel with the arbitrary vortex method, a relationship between the fan pressure, flow rate and blade installation angle is given by this paper, The following lessons are learned in the design and commissioning test of the fan:

(1) The more uniform speed distribution at the outlet of the rotor and stator with a nonuniform inlet flow can be obtain by using the arbitrary vortex method.

(2) According to the particularity of the operation of the axial flow fan in the wind tunnel, combined with the two-dimensional fan element theory, the fan efficiency and the wind tunnel loss coefficient are used as important parameters to deduce the inherent relationship between the fan flow rate and the blade installation angle. Without the fan pressure rise test, based on the fan speed and flow rate, theoretically predict the basic operating parameters of the fan after the installation angle is adjusted.

(3) Aiming at the requirement that the wind speed of the test section of a low speed SJTNU wind tunnel needs to reach 60m/s when the fan speed is 575rpm, the method described in this article is used to calculate the fan installation angle. The theoretically predicted adjustment value is 4.5 degrees. The actual fan speed at 570rpm has achieved the test section wind speed of 60m/s. Although the actual fan speed is 5rpm lower than the set speed of 575rpm, the deviation of 0.9% is acceptable in engineering. The fan installation angle described in this article The adjustment method can fully meet the needs of adjusting the installation angle of the fan during the commissioning of the low speed wind tunnel.

References

[1] Antoine Moreau, Sebastien Guerin, Lars Enghard. Axial-Flow Ventilation Fan Design Through Multi-Objective Optimization to Enhance Aerodynamic Performance [J]. Journal of Fluids Engineering, 2011, Vol. 133./ 101101-3

[2] C. Sarraf, H. Nouri, F. Ravelet, F. Bakir. Experimental study of blade thickness effects on the overall and local performances of a Controlled Vortex Designed axial-flow fan [J]. Experimental Thermal and Fluid Science, 2011, Vol. 35, 684-693.

[3] Zivan Spasic, Milos Jovanovic and Jasmina Bogdanovic-Jovanovic. Design and performance of low-pressure reversible axial fan with doubly curved profiles of blades [J]. Journal of Mechanical Science and Technology, 2018, Vol. 32, 3707~3712.

[4] Lee, C., and Kil, H. G. A Computerized Design System of the Axial Fan Considering Performance and Noise Characteristics [J]. Fluid Mach. (in Korea), 2010,13(2), 48–53.

[5] Lee, C., and Kil, H. G. A Preliminary Axial Fan Design Method With the Consideration of Performance and Noise Characteristics[C]. 2009, The 10th Asian International Conference on Fluid Machinery, Kuala Lumpur, Malaysia, AICFM0095.

[6] Lotfi, O., Teixeira, J. A., Ivey, P. C., Sheard, G., and Kinghorn, I. R. Aerodynamic Optimization

of Industrial Fan Blade Cascades [C]. 2005, ASME Turbo Expo 2005, American Society of Mechanical Engineers , GT2005-68385, New York.
[7] Krishna, S. R., Krishna, R. A., and Ramji, K. Reduction of Motor Fan Noise Using CFD and CAA simulations [J]. Applied Acoustics, 2011,Vol. 72, No. 12, 982–992.
[8] Gue, F., Cheong, C., and Kim, T. Development of Low-Noise Axial Cooling Fans in a Household Refrigerator [J]. Journal of Mechanical Science and Technology, 2011,Vol. 25, No. 12, 2995–3004.
[9] Lee, C, and Kil, H. G. A Preliminary Axial Fan Design Method with the Consideration of Performance and Noise Characteristics [C]. 2009, The 10th Asian International Conference on Fluid Machinery, Kuala Lumpur, Malaysia.
[10] Lee, C., and Kil, H. G. An Aero-Acoustic Performance Prediction Method of Axial Flow Fan [C]. 2009, The 3rd International Conference on Integrity, Porto, Portugal.
[11] Antoine Moreau, Sebastien Guerin, Lars Enghardt. The new NWB ventilator: a practical case of design-to-noise [C]. 2012, 18th AIAA/CEAS Aeroacoustics Conference, Colorado Springs.
[12] Andreas Bergmann. The Aeroacoustic Wind Tunnel DNW-NWB [C]. 2012, 18th AIAA/CEAS Aeroacoustics Conference, Colorado Springs.
[13] Michael Pott-Pollenske, Wilhelm von Heesen, Andreas Bergmann. Acoustical Preexamination Work and Characterization of the Low Noise Wind Tunnel DNW-NWB [C]. 2012, 18th AIAA/CEAS Aeroacoustics Conference, Colorado Springs.
[14] Alexej Pogorelov, Matthias Meinke, Wolfgang Schroder. Effects of tip-gap width on the flow field in an axial fan [J]. International Journal of Heat and Fluid Flow, 2016, VOL 61 466–481.
[15] Michael B. Wilkinson, Johan van der Spuy, Theodor W. von Backstrom. Performance Testing of an Axial Flow Fan Designed for Air-Cooled Heat Exchanger Applications [J]. Journal of Engineering for Gas Turbines and Power, 2018.
[16] Michael B. Wilkinson, Johan van der Spuy, Theodor W. von Backstrom. Performance Testing of an Axial Flow Fan Designed for Air-Cooled Heat Exchanger Applications [J]. Journal of Engineering for Gas Turbines and Power, 2018.
[17] Michael B. Wilkinson, Johan van der Spuy, Theodor W. von Backstrom. THE DESIGN OF A LARGE DIAMETER AXIAL FLOW FAN FOR AIR-COOLED HEATEXCHANGER APPLICATIONS [C]. ASME 2017 GT2017-63331, June 26-30, 2017, Charlotte, NC, USA.
[18] Weifeng He, Yiping Dai, Shi Zhu, Dong Han, Chen Yue, Wenhao Pu. Influence from the blade installation angle of the windward axial fans on the performance of an air-cooled power plant [J]. Energy, 2013, VOL 60, 416–425.
[19] Lei Zhang, Ruiyang He*, Xin Wang, Qian Zhang, Songling Wang. Study on static and dynamic characteristics of an axial fan with abnormal blade under rotating stall conditions [J]. Energy, 2019, VOL 170 305–325.
[20] Jiadong Tang, Yiping Lu, Xuemei Sun, Wanquan Zhang. CFD Simulation of Installation Angle for Axial Flow Fan of a Motor [J]. Applied Mechanics and Materials, 2019, VOL 274, 212–215.
[21] Yan Xiaokang, Wang Lijun, Zhang Jingsong, Wang Xinyong. Numerical and experimental investigation on effect of installation angle of rotor blade on axial flow fan [C] ICMET2011, 2011.
[22] YE Xuemin, LI Xinying, LI Chunxi. Effect of the First-stage Impeller with Single Abnormal Blade on the Performance of a Variable Pitch Axial Fan [J]. Proceedings of the CSEE, 2014, VOL 34, 2297–2306.
[23] YE Xuemin, LI Chunxi, YIN Pan. Effect of Abnormal Blade Reverse Deviation on Performance of Axial Fan[J]. Journal of Chinese Society of Power Engineering, 2013, VOL 33, 450–458.
[24] YE Xuemin, LI Jun, WANG Songling, LI Chunxi, QIAN Hongwei. Aerodynamics of Adjustable Blade Axial Fan Under Abnormal Installation Angles [J]. Proceedings of the CSEE, 2009, VOL 26, 79–84.
[25] Chunxi Li, Qing Lin, Xueliang Ding, Xuemin Ye. Performance, aeroacoustics and feature

extraction of an axial flow fan with abnormal blade angle [J]. Energy, 2016, VOL 103, 322–339.

[26] Francois G.Louw, Theodor W.von Backstrom, Phillippe R.P.Bruneau, Sybrand J.van der Spuy. THE DESIGN OF AN AXIAL FLOW FAN FOR APPLICATION IN LARGE AIR-COOLED HEAT EXCHANGERS [C]. Proceedings of ASME Turbo Expo 2012, June 11-15, 2012, Copenhagen, Denmark, GT2012-69733.

[27] Collar, AR. The Design of Wind Tunnel Fans. 1940.8.

[28] Pope, A. Wind-Tunnel Testing, Second Edition. John Wiley and Sons, New York.

[29] Wallis, RA. Axial flow fans and ducts. John Wiley and Sons Inc. 1983.

[30] Wallis, RA. Axial Flow Fans, Design and Practice. George Newnes Ltd, London 1961.

[31] Qu Xiaoli, Yu Yongsheng, Liao Daxiong, Lv jinlei. The design of the high performance low noise fan of the acoustic pilot wind tunnel, Journal of Experiments in Fluid Mechanics. Vol. 27 (3): 103-108.

Numerical Investigation of Round-to-Slot Holes on Film-Cooling Performance in a Turbine Vane

Ying Huang[1,] *, Can Peng[1], Jingzhou Zhang[2], Xingdan Zhu[2]

[1]Shanghai Aircraft Design and Research Institute, Shanghai, China
[2]College of Energy and Power Engineering, Nanjing University of Aeronautics and Astronautics, Nanjing, China

*Corresponding author email: huangying_0021@163.com

Abstract. The numerical investigation of single row holes on suction surface and pressure surface of a scaled turbine guide vane is carried out. In particular, the effects of round-to-slot holes (RTSHs) on film cooling performance are illustrated. Four RTSHs with different exit-to-inlet area ratios are considered under different blowing ratios. The results show that RTSHs can suppress the normal penetration of cooling jet and broadening the lateral coverage of film layer. With the increase of slot aspect ratio, the anti-kidney vortices pair is generally strengthened, making the film cooling improvement more significantly. The film cooling improving effects of shaped holes on the pressure surface is better than that on the suction surface, particularly under high blowing ratios. A key factor affecting the bleed pressure of coolant flow is exit-inlet area ratio of RTSH. The equivalent-area RTSH achieves nearly the same discharge coefficient (Cd) as the cylindrical hole. Viewed from the comprehensive effects of the Cd and adiabatic film cooling effectiveness on shaped hole, the RTSH-3 with a slot aspect of 9 and an exit-inlet area ratio of 1.27 is suggested to be the most favorable film-hole geometry among the current shaped-holes, either on the pressure surface or on the suction surface.

Keywords: round-to-slot hole; film cooling; turbine guide vane; pressure surface; suction surface.

1. Introduction

Film cooling is an external cooling technology used in the advanced composite cooling configuration to protect the turbine hot section components from thermal damage [1]. Due to the film cooling gas provided by the compressor of engine, the large consumption of cooling gas will undoubtedly induce engine performance penalties. Hence, the proper design and optimization of the film cooling configuration will produce higher film cooling effectiveness while using less cooling air.

During the last decades, the enhanced film cooling strategies had received considerable attention, both in passive and active [2]. The innovation and application of shaped holes was one of the most remarkable technologies of high-efficiency film cooling [3, 4]. The shaped film cooling holes could passively control the mutual interaction of jet-in-crossflow and consequently alleviate the negative effects of kidney vortices in the jet on film cooling performance. As far as we know, the early exploration of shaped holes was mainly concerned on the fan-shaped holes [5-9]. As a result of the influence of the anti-kidney vortices, compared with the conventional film hole, the fan-shaped hole can produce less normal penetration and better span-wise film coverage. As the fan-shaped holes contain many geometric parameters, the influence of multi-parameters about film cooling performance is extremely complicated and still attractive to the researchers so far [10-14].

In recent years, many scholars were devoted to the exploration of newly innovative shaped film cooling holes, and the ultimate goal is to achieve the perfect state of the film cooling art. Of particular interest was the converging slot-hole (namely, console) proposed by Sargison et al. [15, 16]. By using a round-to-rectangular transition section, featured by a divergence in spanwise direction and a convergence in axial direction, the console was constructed. Different from two-dimensional slot, the console remained the discrete-hole geometric feature. Therefore, in real engineering applications, it could overcome the drawback of two-dimensional slot in the structural strength and become practical. Sargison et al. [15-17] carried out the preliminary experiment, and proved that the console could have better spanwise film coverage, close to the two-dimensional slot film cooling. Compared to the fan-shaped or cylindrical holes, its aerodynamic loss was also found to be reduced. They suggested that the console effectively suppresses the normal penetration of coolant injection, thus significantly weakening the influence of coolant outflow on the downstream boundary layer and reducing the mainstream aerodynamic loss. The following investigations on the detailed flow fields (Azzi [18], Yao [19], Wang et al. [20]) revealed that a pair of counter-rotating vortices was formed as a result of mutual interaction of jet-in-crossflow, and the rotation direction of counter-rotating vortices is opposite to the conventional kidney vortices. This specific flow structure has excellent potential in the enhancement of film cooling, because it could suppress the normal penetration of cooling flow into the primary flow and prevent the lateral invasion of the primary flow into the film layer. Liu et al. [21, 22] experimentally studied the influence of console exit-entry area ratio and lateral expansion angle on the thermal and film cooling performances. They pointed out that the consoles with smaller area ratio of exit and entry have smaller discharge coefficient and larger film cooling effectiveness. Pu et al. [23, 24] experimentally examined the use of consoles on turbine endwall. They compared three types of converging-slot configurations by varying exit-entry area ratio and inclination angle. It was suggested that a 30 degree inclined console with a smaller area ratio of entry and exit is a more optimal configuration, in the viewing of film cooling enhancement and mitigation of secondary vortex.

As mentioned above, the consoles are demonstrated to be capable of producing notable benefits in enhancing film cooling effectiveness, but it is at the cost of larger pressure drop inside the console. For this cause, Zhang et al. [25] proposed an evolutionary concept of RTSH from console. When the outlet area of the film hole is larger than the inlet area, the film hole will be the diffusing slot-hole. The console and diffusing slot-hole are collectively referred to as RTSH. Huang et al. [26] conducted the optimization study of film cooling effectiveness to RTSHs on the flat surface. The results show that the optimized RTSH has converging characteristic under low blowing ratio, while the optimized RTSH has diffusing characteristic under high blowing ratio. For addressing the insight into the film cooling performance of the RTSH configuration on the film-cooled turbine vane, a numerical simulation is performed in the present study. Four RTSHs with different area ratios of exit and inlet are adopted. Film cooling from fan-shaped holes and the cylindrical holes are also used for the comparison with the RTSHs.

2. Computational Scheme

2.1. Computational model

Figure 1 displays schematically the computational model of a film-cooled turbine guide vane. Due to the periodic characteristics of turbine cascade flow, the calculation domain only includes one cascade pitch. In the spanwise direction, two film holes are selected in the calculation domain. As seen in Figure 1(a), the coolant channel adopts air intake from both end-side entrances. Figure 1(b) presents the sectional profile of the turbine guide vane. For this turbine guide vane, the axial chord length (C_x) and the chord length (C) are 63.5 mm and 111.6 mm, respectively. The mainstream inflow angle (β) is 90°

and the blade stagger angle (α) is 35.7°. For the cascade channel, its cascade pitch (P) is 90 mm. In the present investigation, a single row of holes at the pressure side and suction surfaces is considered. This film holes is located at a streamwise position of 30.5% C_x at the pressure side, or 59.1% C_x at the suction side apart from the leading edge. The coordinate origin is located at the exit of the film hole on its respective side, as shown in Figure 1(c). The coordinates z, y, s represent the spanwise, normal and streamwise directions of the vane surface, respectively. In the current study, three kinds of film cooling holes are taken into considerations, including cylindrical hole (CH), fan-shaped hole (FSH) and RTSH, as schematically shown in Figure 2. The inclination angle of film holes is 35° (θ) without spanwise deflection. The diameter of CH (d) is 3 mm and the hole-to-hole pitch (P_{hole}) of adjacent holes is fixed as $4d$. The height of the film-hole (t) is $1.5d$. All shaped holes have the same circular inlet as that of the cylindrical hole. For FSH, the expansion part starts from the middle of film hole, and the lateral expansion angle is set to 15°. With regards to the RTSHs, four RTSHs with different area ratios of exit-inlet are designed by varying the width (w) and slot length (l), including an equivalent-area slot (RTSH-2), one converging slot with an area ratio of 0.85 (RTSH-1), and two diffusing slots with the same exit-inlet area ratio of 1.27 (RTSH-3 and RTSH-4). Although RTSH-3 and RTSH-4 have the same slot-exit area, the former has a larger slot aspect ratio of 9 while the latter has an aspect ratio of 4. The main slot-exit geometries of RTSHs are summarized in Table 1.

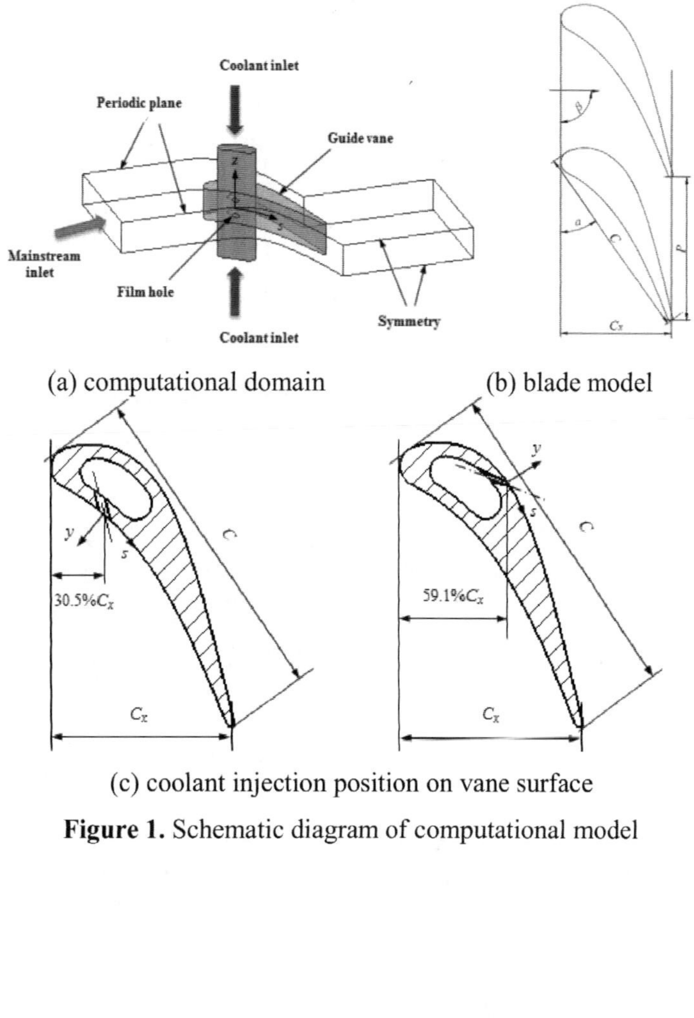

(a) computational domain (b) blade model

(c) coolant injection position on vane surface

Figure 1. Schematic diagram of computational model

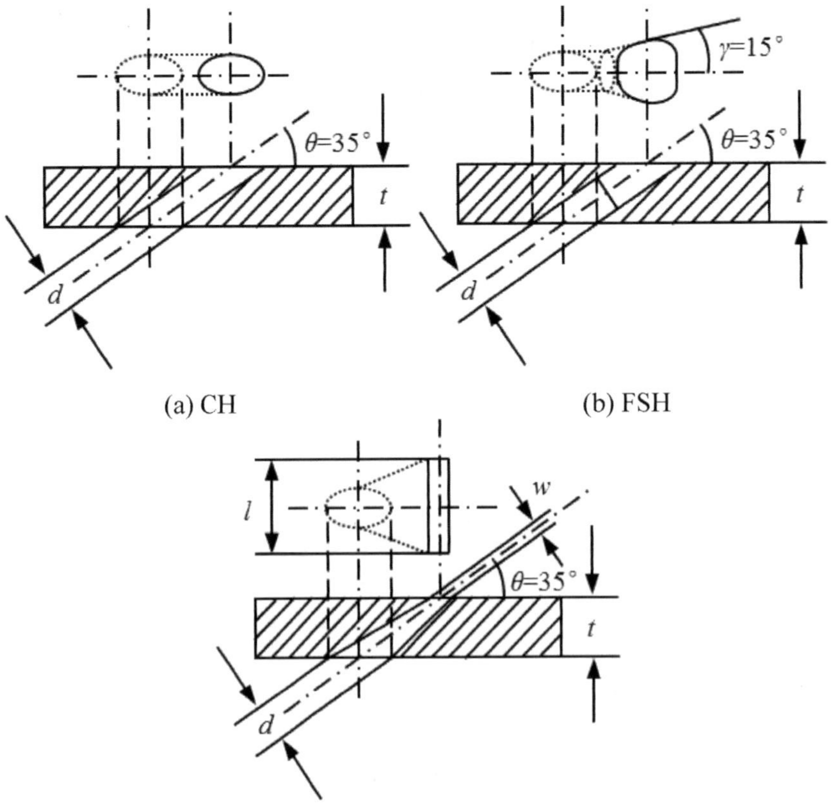

(a) CH (b) FSH

(c) RTSH

Figure 2. Geometric configurations of film holes

Table 1. Geometric dimensions of RTSH exit-slots

Hole	Slot length (*l*)	Slot width (*w*)	Exit-inlet area ratio ($4lw/(\pi d^2)$)
RTSH-1	2*d*	1/3*d*	0.85
RTSH-2	2.37*d*	1/3*d*	1
RTSH-3	3*d*	1/3*d*	1.27
RTSH-4	2*d*	1/2*d*	1.27

According to Figure 1(a), the corresponding boundary conditions are specified as the follows.

At cascade entrance, the temperature of mainstream and the velocity are set as 85°C and 25 m/s, respectively. Correspondingly, the mainstream Reynolds number is approximately 134000. The turbulence length scale is set to 3% and the turbulence intensity is set to 2% at the cascade inlet. A pressure-outlet boundary is applied by setting a standard atmospheric pressure at cascade outlet. Periodic conditions are used for the lateral boundaries of the cascade passage. In the spanwise direction, the segment planes of cascade passage adopt the symmetry boundary condition.

The plenum entrances located at both end-sides are set to the mass flow inlets. The mass flow rate of the coolant depends on the blowing ratio (*M*). The coolant temperature is set as 35°C.

$$M = \frac{(\rho_c u_c)_{\text{inlet}}}{(\rho_\infty u_\infty)_{\text{entrance}}} \quad (1)$$

Where u_∞ and ρ_∞ are the velocity and density of the mainstream at the inlet of cascade; u_c and ρ_c are the velocity and density of the coolant at the entrance of the film hole. Five nominal blowing ratios are considered in the range between 0.4 and 2.0 in the current study.

Adiabatic no-slip condition is applied for the solid wall surfaces, including the outer surface of blade, inner surfaces of coolant channel and film holes.

2.2. Characterization of film cooling performance

Two main characteristic parameters discharge coefficient (C_d) and adiabatic film cooling effectiveness (η) are selected to evaluate the effects of film-hole geometry on film cooling performance.

$$\eta = \frac{T_\infty - T_{\rm aw}}{T_\infty - T_{\rm c}} \tag{2}$$

$$C_{\rm d} = \frac{m_{\rm c}}{A_{\rm c,inlet}\sqrt{2\rho_{\rm c}\left(p_{\rm c}^* - p_{\rm c}\right)}} \tag{3}$$

where the temperature of the adiabatic wall on the outer surface of the guide vane is represented by T_{aw}; the coolant temperature and mainstream temperature are represented by T_c and T_∞; the static pressure of coolant at the film hole exit is represented by p_c and the total pressure of coolant at the coolant channel entrance is represented by ρ_c^*; the actual mass flow rate of coolant ejecting from individual film-cooling hole is represented by m_c; the cross sectional area of film-hole at inlet is represented by $A_{c,inlet}$.

2.3. Computational method

Computational meshes are generated with the aid of commercial software GAMBIT. For the mainstream channel, multi-block structured grids are generated, while, for the irregular RTSHs, the tetrahedral grids are applied. The near wall cell y+ is close to the magnitude of one. The growth ratio of grid is less than 1.2 from the cooling wall. Figure 3 shows detailed grids on the central plane of a film hole.

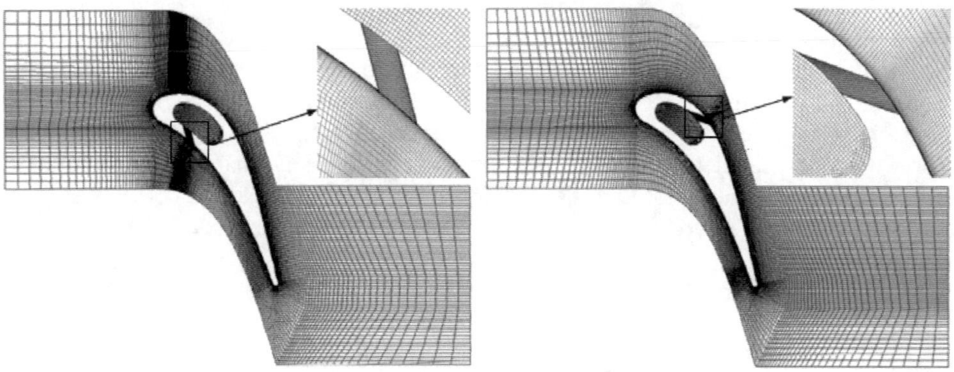

(a) film cooling on pressure surface (b) film cooling on suction surface

Figure 3. Computational grids for blade surface film cooling

The Reynolds-averaged Navier-Stokes equations of three dimensions are solved by applying the commercial software ANSYS 14.5. In the conservation equations, second-order upwind scheme is used for the spatial discretization of convective terms. And SIMPLEC algorithm treats for the pressure-velocity coupling. Based on the previous research [27-29], the k-ε model coupled with the enhanced wall function is used in the present study.

Figure 4 presents two CFD experimental validation examples provided by Arts [30] and Yao et al. [31], respectively. The Mach number distribution obtained by numerical simulation in the cascade channel is presented in Figure 4(a). The mainstream shows significant flow acceleration on the suction side of the vane, due to the flow bend angle of the cascade channel. While, on the pressure side, the mainstream only forms significant flow acceleration near the vane trailing edge. From the comparison of the η of the console (shown in Figure 4(b)) on the surface of guide vane, the numerical simulation results of reliable k-ε turbulence model is in good agreement with the experimental results. The maximum error is within 10%.

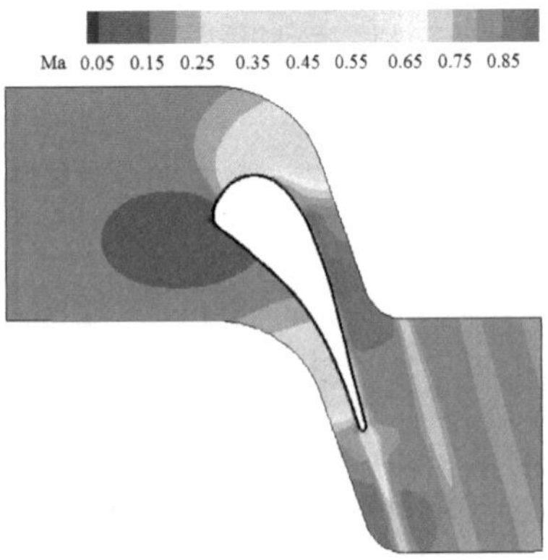

(a) Mach number distribution

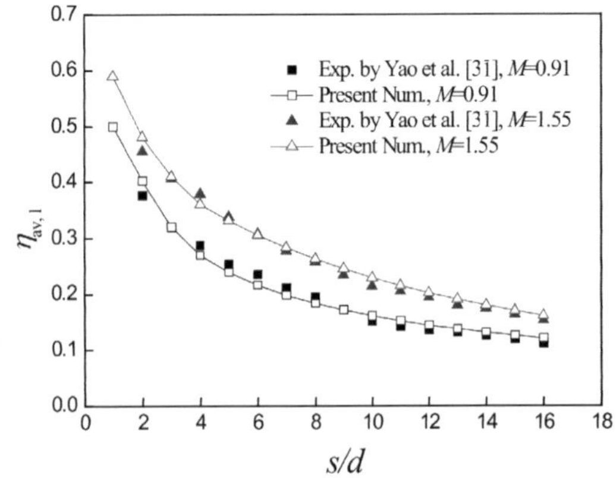

(b) Comparison of current simulation and experimental data [13]

Figure 4. CFD validation

3. Results and Discussions

3.1. Flow fields and discharge coefficients

As illustrated by much research on the cascade flow dynamics, near the blade pressure side, the primary flow field is distinctly different from that near the blade suction side. The mainstream acceleration at the suction side is significantly obvious over the entire suction surface in the cascade channel, because of the flow turning of mainstream. While the mainstream acceleration mainly occurs near the blade trailing edge at the suction side. As the near-wall flow field of primary cascade flow affects the mutual interaction with the cooling jets dominantly, the influence of film-hole shape on film cooling behavior is certainly associated with the location of film holes on the vane. In order to illustrate the mutual interaction between the shaped-hole jet and the near-wall mainstream, some local flow fields are presented for discussion.

Figure 5 and Figure 6 shows static pressure contours and the local streamlines on the hole-centerline plane at the pressure side and suction side respectively under M=2.0. For the cylindrical holes (CH), before the coolant jet is bent over, it presents strong normal penetration to the mainstream, as seen in Figure 5(a) and Figure 6(a). Either at the suction side or pressure side, the lift-off effect from the film-cooled surface can be clearly observed immediately downstream of the film hole. The surface is conjectured to be poorly protected by the film layer coverage, due to this serious coolant jet detachment. As seen in Figure 5(b) and Figure 6(b), for FSH, due to the anti-vortex mechanism of FSH, the normal penetration of coolant jet is effectively restrained. For the RTSH, as seen in Figure 5(c)-(f) and Figure 6(c)-(f), the cooling jet is closer to the cooling wall, either in diffusing type or converging type, when compared to the FSH.

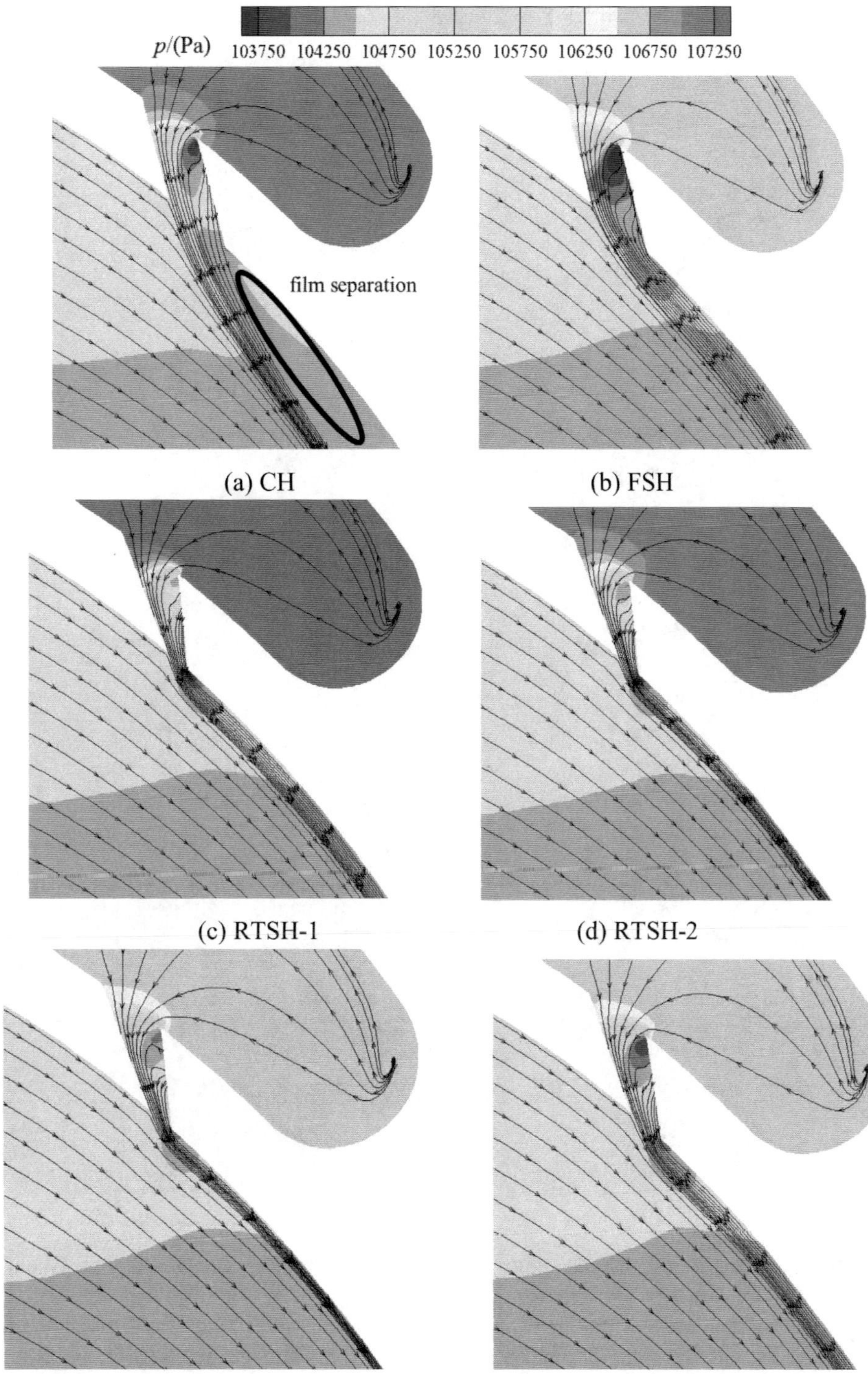

(e) RTSH-3 (f) RTSH-4

Figure 5. Local streamlines and pressure distribution on pressure-side hole-centerline plane under M=2.0

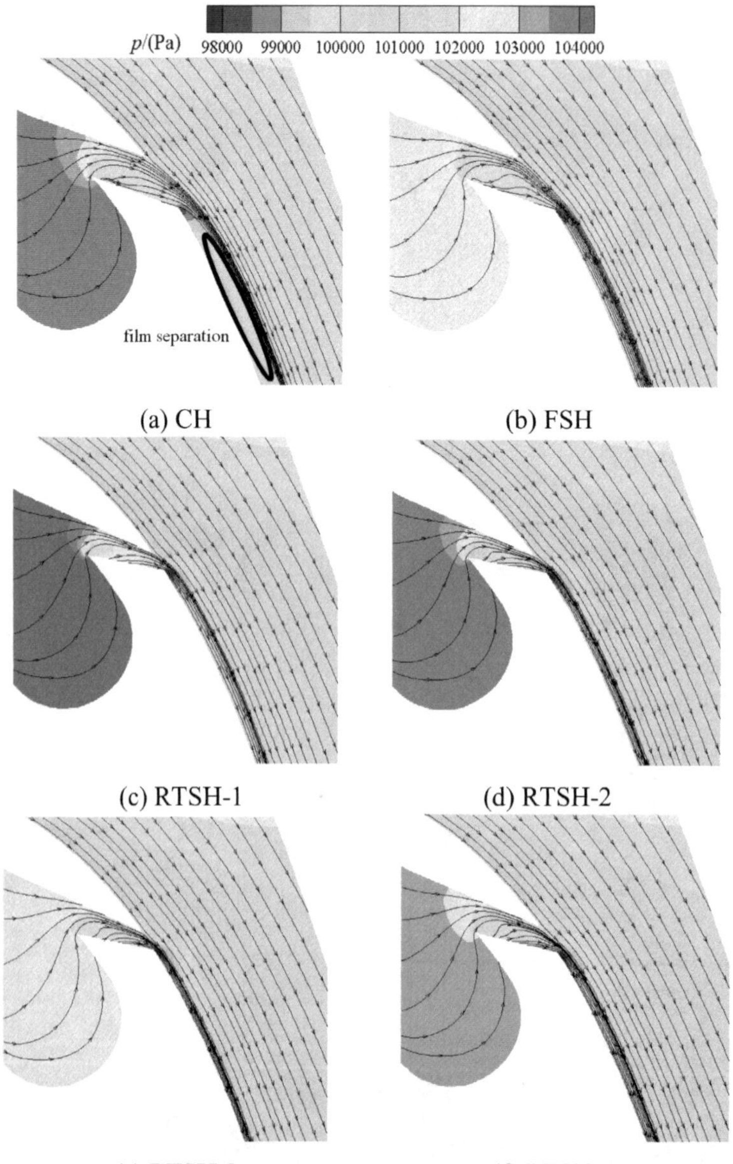

(a) CH (b) FSH

(c) RTSH-1 (d) RTSH-2

(e) RTSH-3 (f) RTSH-4

Figure 6. Local streamlines and pressure distribution on suction-side hole-centerline plane under M=2.0

With regard to the effect of film-hole geometry on coolant plenum pressure, it is found that area ratio of film cooling hole exit-inlet is a key factor affecting the bleed pressure of coolant flow. With respect to the CH, given the same coolant mass-flow rate or nominal M, the RTSH-2 with the equivalent-area transition produces nearly the same plenum pressure. The area-diffusing holes (such as FSH, RTSH-3 and RTSH-4) make the plenum pressure reduce but the area-converging hole (RTSH-1) makes the plenum pressure increase in related to the CH. Evidently, as console is transitioned with a gradually reduced cross-sectional area, a greater flow loss would occur inside the film hole. For this reason, with the same cooling air mass-flow rate as the baseline CH, a greater pressure drop of cooling air is required through the film hole.

It is also found that the plenum pressure is less when the cooling jet is ejected on suction surface, compared Figure 5 and Figure 6. Due to the significant flow acceleration at the suction side, near the suction-side film-hole, the cascade pressure is smaller than that near pressure-side film-hole. Therefore, to achieve the same cooling air mass-flow rate or nominal M for a specific film-hole on both sides of blade, a higher bleed pressure of coolant flow is needed for the pressure-side jet injection. However, in the viewing of discharge coefficient (C_d), as displayed in Figure 7, the C_d of film hole at the pressure side is bigger than the corresponding one at suction side, for every film cooling shape. Referring the definition of C_d (Eq.3), at a fixed coolant mass flow-rate, the C_d is inversely proportional to the pressure drop through the film-hole, it is inferred that the coolant jet ejecting at suction side experiences a larger pressure drop, although a less pressure rise occurs inside the coolant plenum. With regard to the effect of film-hole geometry on C_d, it is certainly that the RTSH-2 has almost the same C_d with the CH, as this RTSH has equivalent exit area with CH. The area-converging RTSH-1 produces an obviously negative influence on the C_d. The area-diffusing holes (such as FSH, RTSH-3 and RTSH-4) are beneficial to reduce the pressure drop across the film-hole and thus result in higher C_d when compared to CH. For these three film cooling holes having the same exit-inlet area ratio, the difference in the C_d is still reflected, especially at suction side. Generally, FSH achieves the largest C_d and RTSH-3 is the next.

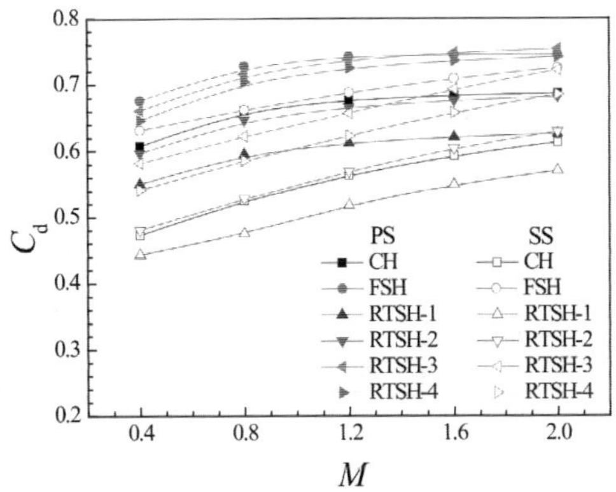

Figure 7. Comparison of discharge coefficient for different film cooling holes

Figure 8 and Figure 9 shows the local streamlines and dimensionless temperature contours originating at s/d=2 downstream of the film-hole on the normal plane at the suction side and pressure side, respectively, under M=2.0. Viewed from these two figures, the different features in flow dynamics of jet-in-crossflow are demonstrated more distinctly. For CH, presented in Figure 8(a) and Figure 9(a), the interaction between the mainstream and the cooling jet leads to the strong kidney vortices originated from the center of the film-hole, lifting the cooling air obviously off the cooling wall and the hot mainstream is easily to be entrained beneath the wake of cooing air from hole-sides. Because the near field of mainstream at pressure side is different from that at suction side, the vortex structure induced by the mutual interaction between the mainstream and the cooling air takes on distinct features. For the cascade flow, on the entire suction surface, the flow acceleration at the suction side is significantly obvious. Upon the impact of accelerating primary flow, compared with the coolant jet penetration at the pressure side, at the suction side, the coolant jet penetration is weakened. Therefore, the kidney vortex at pressure side has a larger size than that at suction side. In addition, the primary flow induces a vertical velocity component normally inward the surface at pressure side, making the mainstream enters beneath the wake of cooling air more easily, because the flow turning inside cascade channel. For FSH, the normal penetration of cooling jet is effectively restrained, because of its diffused exit. When compared to CH, the vortex pair stays closer to the surface. Meanwhile, as presented in Figure 8(b) and Figure 9(b), the cores of the kidney vortices are pushed toward both sides of the film-hole. With regards to RTSHs, as the coolant jet is ejected from a narrow slot, it covers well the surface immediately downstream of film-hole outlet. Interestingly, the vortex structures behave distinct features at suction side and pressure side. The hot primary flow has a vertical velocity component normally inward the concave surface at the pressure side. As seen in Figure 8(c)-(f), the hot primary flow above the coolant layer is pushed toward the concave surface, invading the film layer from the film-hole central zone. The mainstream still invades the film layer through the renal vortex by the kidney vortices near the slot edges. With respect to FSH, the cores of kidney vortices are pushed far away from film-hole center, especially for RTSH-3. In this situation of RTSH-3, the slot is diffused with a largest slot length. The coolant injection has a strongest lateral transition to force the coolant jet flow laterally. As a consequence, a pair of anti-kidney vortices is formed dominantly, as seen in Figure 8(e). The kidney vortices are significantly weakened in the scale size and pushed to the slot edges. However, at suction side, because the primary flow induces a vertical velocity component normally outward the convex surface, the streamlines distribution near the wall behaves distinctly from that at pressure side. As presented in Figure 9(c)-(f), the streamlines near the wall are separated by two parts, especially obvious in Figure 9(d) and 10(e). The flow is directed outward surface, up the separation line, but the flow is directed inward surface from the film-hole central zone, below the separation line. By comparing Figure 9 with Figure 8, for shaped film cooling holes, the coolant coverage in the lateral direction at the suction side is obviously smaller than the respective one at the pressure side. The coolant jet is ejected into an accelerated primary flow at the suction side. Therefore, its lateral flow would be restricted by the accelerated primary flow.

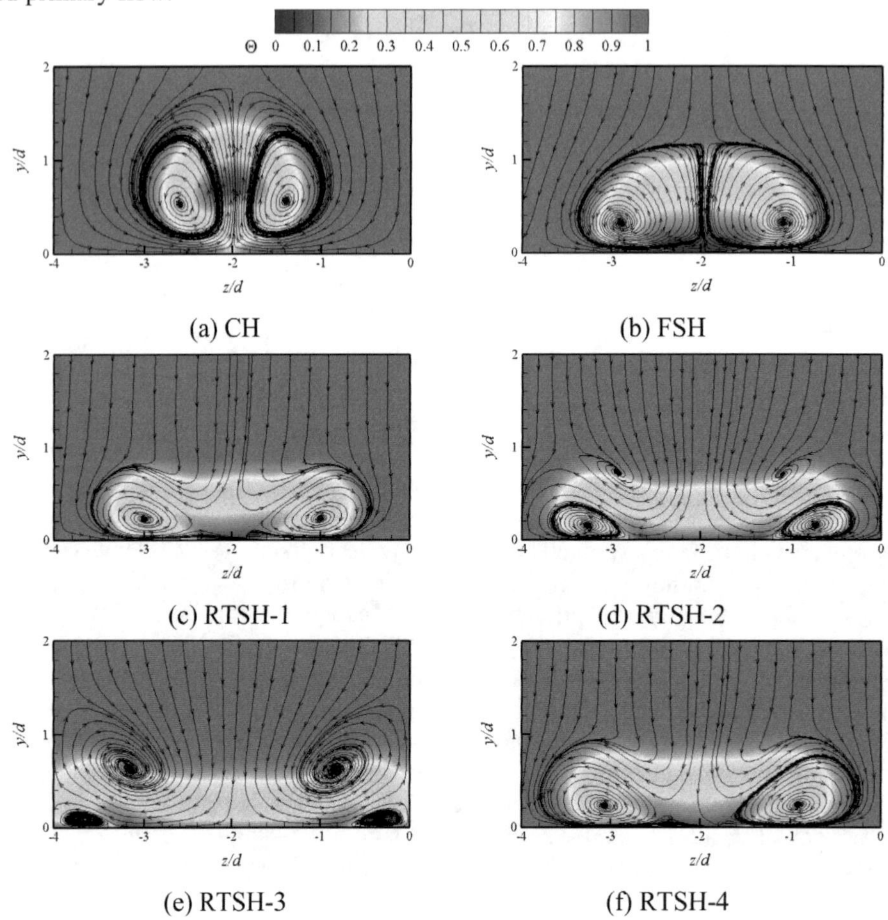

Figure 8. Local streamlines and dimensionless temperature contours on pressure-side normal section at s/d=2 under M=2.0

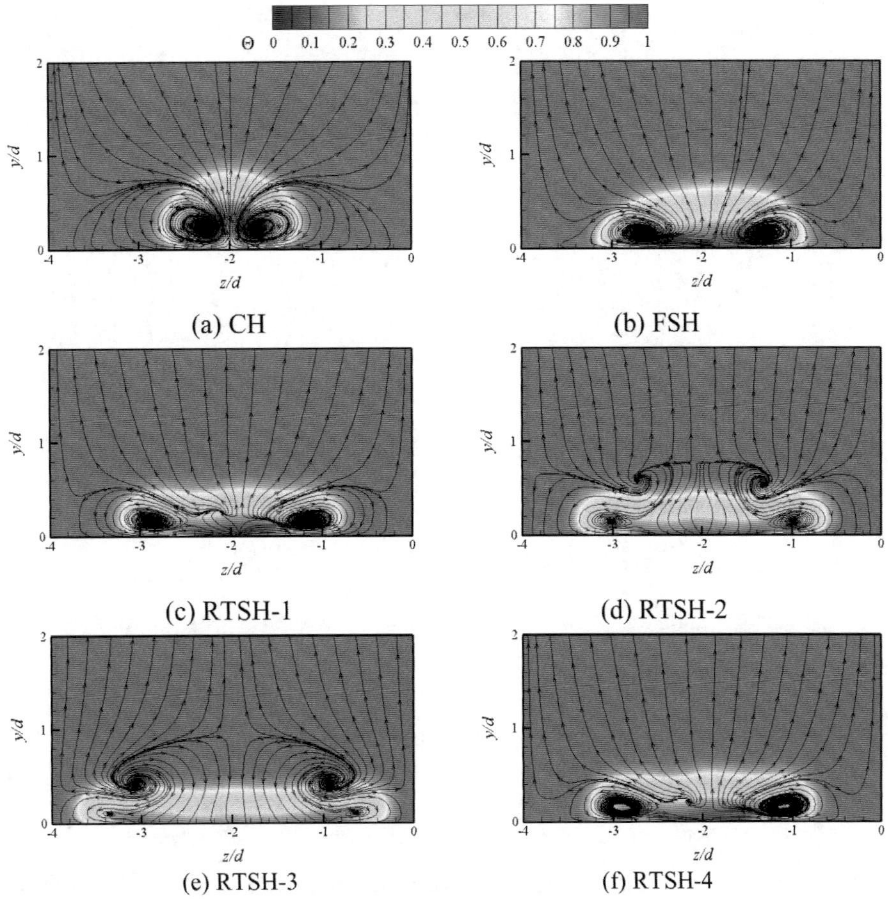

Figure 9. Local streamlines and dimensionless temperature contours on suction-side normal section at s/d=2 under M=2.0

3.2. *Adiabatic film cooling effectiveness* (η)

Figure 10 presents the laterally-averaged adiabatic film cooling effectiveness ($\eta_{\mathrm{av,\,l}}$) at pressure side and suction side. It is clearly illustrated that the effect roles of shaped hole on film cooling are intensively associated with the M and the protected surface curvature. As seen in Figure 10(a), under M=0.4, along the streamwise direction, the $\eta_{\mathrm{av,\,l}}$ is decreased monotonously at both the pressure and suction sides. At pressure side, all the shaped holes produce prominent improvement of $\eta_{\mathrm{av,\,l}}$ downstream of the film-hole exit. Comparing fan-shaped hole, RTSHs are surely more favorable for improving film cooling, especially for RTSH-3. While at suction side, the $\eta_{\mathrm{av,\,l}}$ is generally less than the corresponding one at pressure side for the shaped holes. All the shaped holes achieve nearly the same $\eta_{\mathrm{av,\,l}}$ distribution. A little increase of the $\eta_{\mathrm{av,\,l}}$ related to the baseline hole is vaguely demonstrated immediately downstream of the film-hole exit. Under M= 1.2, the advantage of the shaped-hole on improving film cooling is more obviously demonstrated on the pressure surface, as seen in Figure 10(b). Related to the FSH, all the RTSHs perform more favorably, especially for the RTSH-3 and RTSH-2, which have bigger length-to-width aspect ratio. At suction side, the advantage of the shaped-hole on improving the film cooling related to the cylindrical hole is distinctly displayed within s/d=10. It is found that the difference of the $\eta_{\mathrm{av,\,l}}$ distribution between the RTSHs and the FSH is still small except for the RTSH-3. As seen in Figure 10(c), under M=2.0, it is confirmed that the benefits of the shaped-holes are tremendously featured, either at suction side or at pressure side. In general, in view of film cooling enhancement, the RTSH-3 is likely to be the most favorable film-hole geometry, and the RTSH-2 is the next.

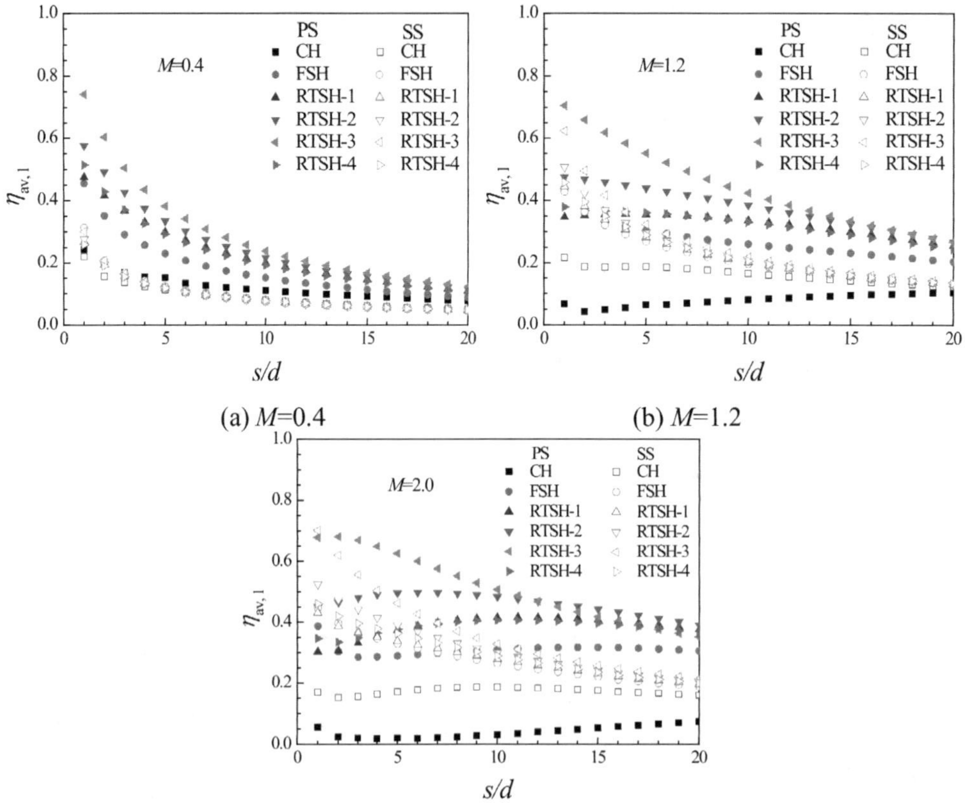

(a) M=0.4 (b) M=1.2

(c) M=2.0

Figure 10. Comparison of laterally-averaged adiabatic film cooling effectiveness

To further evaluate the influence roles of shaped hole on the film cooling, a comparison of spatially-averaged adiabatic film cooling effectiveness ($\eta_{av,s}$) between different film cooling holes under five M is made, as displayed in Figure 11. Here, the specified area is sized as 20d in streamwise direction originated from film-hole exit and two pitches of holes in spanwise direction. For the cylindrical hole, the $\eta_{av,s}$ decreases monotonously with the increase of M at pressure side, but increases a little from M=0.4 to M=1.2 and then tends to unvaried with the increase of M at suction side. Except under M=0.4, at suction side, the $\eta_{av,s}$ is higher than that at pressure side. However, for shaped holes, with the increase of M, the $\eta_{av,s}$ is found to increases monotonously, either at suction side or at pressure side under the current study conditions. At suction side, the $\eta_{av,s}$ is less than the corresponding at pressure side. All the round-to-slot holes are confirmed to be capable of producing distinctly higher $\eta_{av,s}$ related to the FSH among the researched M at pressure side. However, at suction side, the round-to-slot holes show vague beneficial roles on improving $\eta_{av,s}$ under low M. For instances, at pressure side, the RTSH-3 achieves approximately 60% increase of the $\eta_{av,s}$ related to the fan-shaped hole under M=0.8 or M=2.0. While at suction side, only 8% increase of the $\eta_{av,s}$ is obtained by the RTSH-3 under M=0.8 and approximately 30% increase is achieved by the RTSH-3 under M=2.0 in comparison with the FSH.

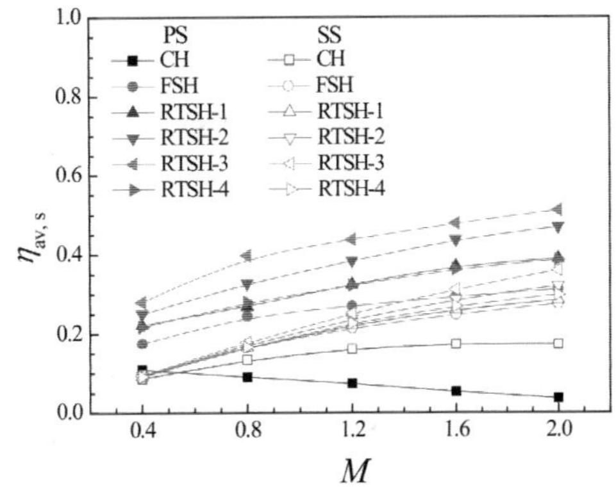

Figure 11. Comparison of spatially-averaged film cooling effectiveness

4. Conclusions

In the present study, the film cooling performance of single row holes on the pressure surface and the suction surface of a scaled turbine guide vane is numerically investigated. The main purpose is to study to study the effects of RTSHs on the film cooling performance under typical M ranging from 0.4 to 2.0. Four RTSHs with different area ratios of exit-to-inlet are adopted, including a converging RTSH-1, a constant-area RTSH-2 and two diffusing RTSHs (RTSH-3 and RTSH-4). A CH and a FSH are also adopted for the comparison. The main conclusions are as follows:

(1) Due to the specific geometric characteristics of RTSH, the normal penetration of cooling jet is effectively restrained and a wider lateral film coverage is produced downstream of the film-hole exit. With the increase of slot aspect ratio, the anti-kidney vortices pair is generally strengthened, making the film cooling improvement more significantly.

(2) The exit-inlet area ratio of RTSH is a key factor affecting the bleed pressure of coolant flow. The equivalent-area RTSH achieves nearly the same discharge coefficient as cylindrical hole. The area-diffusing shaped-hole results in higher discharge coefficient but the area-converging shaped-hole produces a negative influence on the C_d related to the CH. The C_d of film cooling holes is generally larger at pressure side than at suction side.

(3) The effect of film-hole shape on the film cooling behavior is tightly associated with the location of film cooling holes on the vane. Under the same nominal blowing ratio or coolant usage, the coolant jet penetration is relatively weaker at suction side than at pressure side due to the significantly accelerating primary flow. The effects of shaped holes on film cooling improvement behave more obviously at pressure side, especially under high blowing ratios.

(4) Viewed from the comprehensive effects of the shaped holes on the adiabatic film cooling effectiveness and the discharger coefficient, it is suggested that RTSH-3 is the most favorable film-hole configuration, if the mechanical-thermal characteristics are satisfied in practical applications

Acknowledgements

The authors gratefully acknowledge the financial supports for this project from the National Natural Science Foundation of China (grant No: U1508212).

References

[1] Bunker, R. S., "Gas turbine heat transfer: ten remaining hot gas path challenges," ASME Journal of Turbomachinery, Vol. 129, No. 1, 2007, pp. 193-201.

[2] Acharya, S., and Kanani, Y., "Advances in film cooling heat transfer," Advances in Heat Transfer, Vol. 49, 2017, pp. 91-156.

[3] Bell, C. M., Hamakawa, H., and Ligrani, P. M., "Film cooling from shaped holes," ASME Journal of Heat Transfer, Vol. 122, No. 2, 2000, pp. 224-232.

[4] Bunker, R. S., "A review of turbine shaped film cooling technology," ASME Journal of Heat

Transfer, Vol. 127, No. 4, 2005, pp. 441-453.
[5] Goldstein, R. J., Eckert, E. R. G., and Burggraf, F., "Effects of hole geometry and density on three-dimensional film cooling," International Journal of Heat and Mass Transfer, Vol. 17, No.3-4, 1974, pp. 595-607.
[6] Haven, B. A., and Kurosaka, M., "Improved jet coverage through vortex cancellation," AIAA Journal, Vol. 34, No. 11, 1996, pp. 2443-2444.
[7] Thole, K. A., Gritsch, M., Schulz, A., and Wittig S., "Flowfield measurements for film-cooling holes with expanded exits," ASME Journal of Turbomachinery, Vol. 120, No. 2, 1998, pp. 327-336.
[8] Gritsch, M., Schulz, A., and Wittig, S., "Adiabatic wall effectiveness measurements of film-cooling holes with expanded exits," ASME Journal of Turbomachinery, Vol. 120, No. 3, 1998, pp. 549-556.
[9] Yu, Y., Yen, C. H., Shih, T. I-P., Chyu M.K., and Gogineni S., "Film cooling effectiveness and heat transfer coefficient distributions around shaped holes," ASME Journal of Heat Transfer, Vol. 124, No. 5, 2002, pp. 820-827.
[10] Colban, W. F., Thole, K. A., and Bogard, D. A., "A film-cooling correlation for shaped holes on a flat-plate surface," ASME Journal of Turbomachinery, Vol. 133, No. 1, 2011, pp. 011002.
[11] Lee, K. D., and Kim, K. Y., "Surrogate based optimization of a laidback fan-shaped hole for film-cooling," International Journal of Heat and Fluid Flow, Vol. 32, No. 1, 2011, pp. 226-238.
[12] Saumweber, C., and Schulz, A., "Effect of geometry variations on the cooling performance of fan-shaped cooling holes," ASME Journal of Turbomachinery, Vol. 134, No. 6, 2012, pp. 061008.
[13] Huang, Y, Zhang, J. Z., Wang, C. H., and Zhu, X. D., "Multi-objective optimization of laidback fan-shaped film cooling hole on turbine vane suction surface," Heat and Mass Transfer, Vol. 55, No. 4, 2018, pp. 1181-1194.
[14] Wang, C. H., Zhang, J. Z., Feng H.K., and Huang Y., "Large eddy simulation of film cooling flow from a fan-shaped hole," Applied Thermal Engineering, Vol. 129, 2018, pp. 855-870.
[15] Sargison, J. E., Guo, S. M., Oldfield, M. L. G., Lock, G. D., and Rawlinson, A. J., "A converging slot-hole film-cooling geometry - part 1: low-speed flat-plate heat transfer and loss," ASME Journal of Turbomachinery, Vol. 124, No. 3, 2002, pp. 453-460.
[16] Sargison, J. E., Guo, S. M., Oldfield, M. L. G., Lock, G. D., and Rawlinson, A. J., "A converging slot-hole film-cooling geometry - part 2: transonic nozzle guide vane heat transfer and loss," ASME Journal of Turbomachinery, Vol. 124, No. 3, 2002, pp. 461-471.
[17] Sargison, J. E., Oldfield, M. L. G., Guo, S. M., Lock, G. D., and Rawlinson, A.J., "Flow visualization of the external flow from a converging slot-hole film-cooling geometry," Experiments in Fluids, Vol. 38, No. 3, 2005, pp. 304-318.
[18] Azzi, A., and Jubran, B. A., "Numerical modelling of film cooling from converging slot hole," Heat and Mass Transfer, Vol. 43, No. 4, 2007, pp. 381-388.
[19] Yao, Y., and Zhang, J. Z., "Investigation on film cooling characteristics from a row of converging slot-holes on flat plate," Science China Technological Science, Vol. 54, No. 7, 2011, pp. 1793-1800.
[20] Wang, C. H., Zhang, J. Z., and Huang, Y., "Large eddy simulation of film cooling flow from converging slot-holes," International Journal of Thermal Sciences, Vol. 126, 2018, pp. 238-251.
[21] Liu, C. L., Zhu, H. R., Bai, J. T., and Xu, D. C., "Experimental research on the thermal performance of converging slot holes with different divergence angles," Experimental Thermal and Fluid Sciences, Vol. 33, No. 5, 2009, pp. 808-817.
[22] Liu, C. L., Zhu, H. R., Bai, J. T., and Xu, D. C., "Film cooling performance of converging-slot holes with different exit-entry area ratios," ASME Journal of Turbomachinery, Vol. 133, No. 1, 2011, pp. 011020.
[23] Pu, J., Yu, J., Wang, J. H., Yang W. S., Zhang Z. Q., and Wang L., "An experimental investigation

of secondary flow characteristics in a linear turbine cascade with upstream converging slot-holes using TR-PIV," Experimental Thermal and Fluid Sciences, Vol. 59, 2014, pp. 56-71.

[24] Pu J., Wang J. H., Ma S. Y., and Wu X. Y., "An experimental investigation of geometric effect of upstream converging slot-hole on end-wall film cooling and secondary vortex characteristics," Experimental Thermal and Fluid Sciences, 2015, Vol. 69, pp. 58-72.

[25] Zhang, J. Z., Zhu, X. D, Huang, Y., and Wang, C. H., "Investigation on film cooling performance from a row of round-to-slot holes on flat plate," International Journal of Thermal Sciences, Vol. 118, 2017, pp. 207-225.

[26] Huang, Y., Zhang, J. Z., and Wang, C. H., "Shape optimization of round-to-slot holes for improving film cooling effectiveness on a flat plate," Heat and Mass Transfer, Vol. 54, No. 6, 2018, pp. 1741-1754.

[27] Harrison, K., and Bogard, D., "Comparison of RANS turbulence models for prediction of film cooling performance," in Proceedings of the ASME Turbo Expo: Power for Land, Sea, and Air, ASME Paper GT2008-50366, Berlin, Germany, June 2008.

[28] Silieti, M., Kassab, A. J., and Divo, E., "Film cooling effectiveness: comparison of adiabatic and conjugate heat transfer CFD models," International Journal of Thermal Sciences, Vol. 48, No. 12, 2009, pp. 2237-2248.

[29] Ely, M. J., and Jubran, B. J., "A numerical evaluation on the effect of sister holes on film cooling effectiveness and the surrounding flow filed," Heat and Mass Transfer, Vol. 45, No. 11, 2009, pp. 1435-1446.

[30] Arts, T., De Rouvroit, M. L., and Rutherford, A. W., "Aero-thermal investigation of a highly loaded transonic linear turbine guide vane cascade," VKI Technical Note, No. 174, 1990.

[31] Yao, Y., Zhang, J. Z., and Wang, L. P., " Film cooling on a gas turbine blade suction side with converging slot-hole," International Journal of Thermal Sciences, Vol. 65, 2013, pp. 267-279.

Study on optical scattering detection method of bubbly wake of targets in water

Yun Wang, Wei Gu, Zhiben Shen, Weihua Shi, Jianwei Wu, and Liang Dong*

Wuhan Second Ship Design and Research Institute, Wuhan, China

*Corresponding author e-mail: yunwxidian@126.com

Abstract. A method for the detection of the bubbly wake of targets in water is presented, which is based on the features of optical frequency domain of the scattering optical signal. A simulation on the optical frequency domain characteristics of the bubbly wake scattering signal is performed by statistical test method. The influence of the bubble population and the velocity distribution on the optical frequency domain characteristics is analyzed. On this basis, a test platform for the characteristics of optical frequency domain of the bubbly wake scattering signal is designed and established. The characteristics of the bubbly wake scattering signal are studied experimentally. Experimental results agree with the theoretical analysis. The results show that the amplitude and the cut-off frequency of the shift spectra will increase with the increasing bubble population and range of velocity distribution. It indicates the effectiveness of the detection method.

1. Introduction

When the target is navigating in the water, due to gas emission and cavitation of the propeller, a large number of tiny bubbles will be left in the wake area, forming the so-called bubbly wake. By detecting the bubbly wake, long distance indirect detection of underwater targets can be achieved. The bubbly wake is different from the physical properties of the hydrostatic region, which provides the theoretical basis for bubbly wake detection. The optical detection of bubbly wakes is an important direction in the research of wake detection, and the features of optical backscattering are the current research hotspots. Since the scattering and absorption of light waves by water is stronger than the scattering and absorption of light waves by bubbles, the signal-to-noise of the detection of the intensity of the wake optical backscattering is usually very low [1]. The tiny bubbles in the wake of the target in the water are constantly moving under the action of the turbulence [2]. With the time goes, the bubble density and speed distribution of the bubble group will change. Based on the laser Doppler effect [3], the interaction of the moving bubble in the wake and the laser will cause the frequency shift of the scattering light, and the size of the frequency shift is proportional to the speed of the bubbles.

In this paper, a detection method of underwater target wakes based on the frequency features of scattering light is proposed. And theoretical and experimental researches have been carried out. A simulation analysis model of the frequency of the scattering light of the bubbles based on the statistical test method is established. The influence of bubble density and speed distribution on the frequency shift spectrum is analyzed.

2. Theoretical analysis

2.1. Principle of the laser Doppler detection

Optical heterodyne technology is widely used in laser Doppler speedometer systems to detect optical frequency shift [4]. On the one hand, it can better suppress the scattering of the water. On the other hand, it has high optical gain, so it can improve the signal-to-noise ratio of detection. Among the existing Doppler heterodyne detection modes, the magnitude of the differential mode Doppler frequency shift has nothing to do with the direction of the scattering light [5], so the research in this paper adopts the differential mode. The optical path of the differential laser Doppler detection method is shown in Fig. 1.

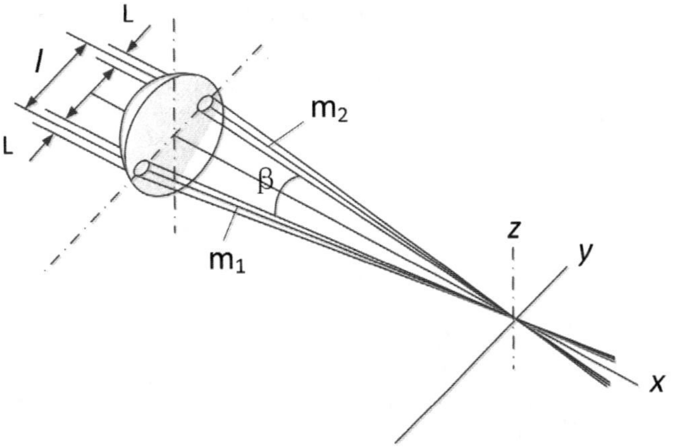

Figure 1. Optical path of the differential Doppler detection.

Two parallel laser beams with the same frequency and intensity have the spot diameter of L, and the space of l. The two laser beams are incident on the moving particles in different directions after being focused by the lens. Since the relative directions of the moving velocity $\mathbf{v}$ of the particle with the wave vectors $\boldsymbol{m}_1$ and $\boldsymbol{m}_2$ of the two beams are different, the optical frequencies f_1 and f_2 emitted by the two beams at the scattering point are also different, respectively are [6]:

$$f_1 = f_0\left(1 - \frac{\boldsymbol{v}(\boldsymbol{m}_1 - \boldsymbol{m}_s)}{c}\right) \tag{1}$$

$$f_2 = f_0\left(1 - \frac{\boldsymbol{v}(\boldsymbol{m}_2 - \boldsymbol{m}_s)}{c}\right) \tag{2}$$

In the formulas (1) and (2), f_0 is the frequency of incident light, c is the speed of light, and $\boldsymbol{m}_s$ is the vector of the direction of movement of the scattering particle.

The scattering light with frequencies f_1 and f_1 interfere on the photo-detector to obtain an intermediate frequency signal. The intermediate frequency is:

$$f_d = |f_1 - f_2| = \frac{f_0}{c}|\boldsymbol{v}(\boldsymbol{m}_1 - \boldsymbol{m}_2)| = \frac{2|v_y|}{\lambda_0}\sin\frac{\beta}{2} \tag{3}$$

In the formula (3), v_y is the component of the scattering particle in the direction of the wave vector difference $(\boldsymbol{m}_1 - \boldsymbol{m}_2)$ between the two incident laser beams, λ_0 is the wavelength of the incident light in vacuum, and β is the angle between the two incident laser beams.

2.2. *Calculation of frequency shift of scattering light by a single particle*

The laser beam propagates in space with Gaussian intensity distribution. Two laser beams with the same polarization direction, intensity and frequency form a certain light intensity distribution in the intersection area. The intersection is called the control body. When a particle is passing through the control body at a certain speed, the photocurrent generated in the photo-detector is

$$i(\mathrm{t}) = i_d \exp\left\{-\left[\frac{2\sqrt{2}(t-t_0)}{T}\right]^2\right\} + i_a \exp\left\{-\left[\frac{2\sqrt{2}(t-t_0)}{T}\right]^2\right\}\cos[2\pi f_d(t-t_0)] \quad (4)$$

In the formula (4), $T = N_f / f_d$ is the finite transit time for the particle to pass through the control body, $N_f = 1.27l / L$ is the number of interference fringes in the control body, t_0 is the moment when the particle reaches the control body and f_d is the Doppler frequency shift, which is proportional to the velocity. i_a and i_d are the photocurrent signal amplitudes, which are proportional to the intensity of the incident light generated by interference modulation. The Fourier transform can obtain the Doppler frequency shift of a single particle.

$$i(f) = \left(\frac{i_d T}{2}\right)\left(\frac{\pi}{2}\right)^{1/2} \exp\left[-\left(\frac{\pi f T}{2\sqrt{2}}\right)^2\right] + \left(\frac{i_a T}{4}\right)\left(\frac{\pi}{4}\right)^{1/2}\left(\exp\left\{-\left[\frac{\pi(f+f_d)T}{2\sqrt{2}}\right]^2\right\} + \exp\left\{-\left[\frac{\pi(f-f_d)T}{2\sqrt{2}}\right]^2\right\}\right) \quad (5)$$

2.3. *Simulation bubble group scattering optical*

On the basis of scattering optical shift of the single particle, in order to set up the statistical test model of the scattering of the bubble group, the following assumptions are made:

Since there is low interference efficiency between scattering light waves of different bubbles, the interference effect between scattering light waves of different bubbles is ignored, that is, the signal received by the detector is linearly generated by all bubbles in the control body in the interference stripe direction component.

$$I_N(f) = \sum_j^N i_j(f) \quad (6)$$

For a single bubble, the amplitude of the scattering signal is proportional to the incident light intensity, and the ratio i_a/i_d of them is related to the scattering characteristic, the polarization characteristic of the incident light, the parameters of the photo-detector, and the intensity ratio of the two beam incident light. In the differential system, the above-mentioned parameters of each bubble in the control body are basically the same, and it can be considered that $i_a/i_d = 1$ for each bubble.

The number of bubbles in the control body is denoted as N, and the velocity in the interference fringe direction is taken as a Gaussian distribution with a mean value of 0. Then, from Equation 3, the corresponding Doppler frequency shift also obeys a Gaussian distribution with a mean value of 0. Assuming the distribution is $M \sim (0, \sigma^2)$. Then, by formula 6, the Doppler frequency shift spectrum of the scattered light signal of the bubble group is as follows.

$$I_N(f)=\sum_j^N i_j(f)=\sum_j^N\left(\frac{i_{d,j}T}{2}\right)\left(\frac{\pi}{2}\right)^{1/2}\exp\left[-\left(\frac{\pi f T_j}{2\sqrt{2}}\right)^2\right]$$
$$+\left(\frac{i_{a,j}T}{4}\right)\left(\frac{\pi}{4}\right)^{1/2}\left(\exp\left\{-\left[\frac{\pi\left(f+f_{d,j}\right)T_j}{2\sqrt{2}}\right]^2\right\}+\exp\left\{-\left[\frac{\pi\left(f-f_{d,j}\right)T_j}{2\sqrt{2}}\right]^2\right\}\right) \quad (7)$$

Due to the large amplitude range of the signal spectrum, the spectrum analyzer displays the value in the form as follow.

$$H_N(f) = 10\, lg(f) \quad (8)$$

Considering the influence of the noise current of the photo-detector, a suitable threshold is set here. The signal amplitude that is smaller than the threshold is taken as the threshold. The frequency value corresponding to the first drop of the amplitude to the threshold is called the cut-off frequency. Finally the frequency shift spectrum of the scattered light signal can be obtained according to the formula (8).

2.4. Analysis of the influence of wake features on the frequency shift spectrum of scattering light

2.4.1. The influence of bubble density on the frequency shift spectrum of scattering light. Fig. 2 shows the frequency shift spectra of scattering light of the bubbly wakes with different bubble density. In the simulation, the number of interference fringes is $N_f = 8$, the current amplitude is $i_a/i_d = 1$mA, the speed distribution is $M \sim (0, 5^2)$ and the signal threshold is -10dBm, and the number of bubbles in the control body is 400, 800, and 1200, respectively.

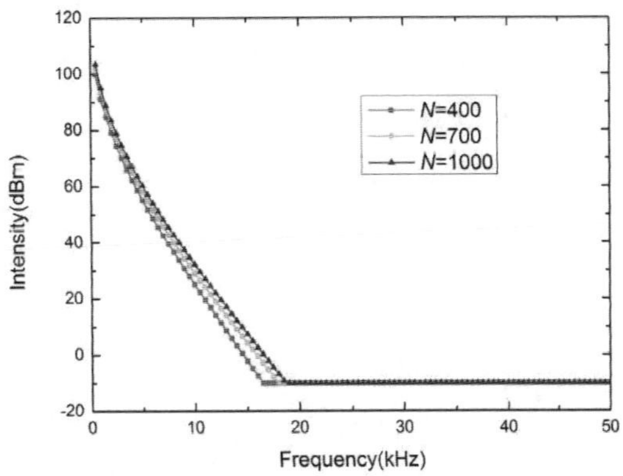

Figure 2. Frequency shift spectra of scattering light of bubbly wakes with different bubble density.

It can be seen from Fig. 2 that under the same condition of the bubble group speed distribution, as the bubble density increases, the amplitude of the frequency shift components of the scattering light increases, and the corresponding cut-off frequency also increases. Specifically, when the number of bubbles in the control body increases from 400 to 1400, the cut-off frequency shift of the scattering light by the wake bubble group is increases from 16.8 kHz to 19.0 kHz.

2.4.2. The influence of bubble velocity distribution on the frequency shift spectrum of scattered light. Fig. 3 shows the frequency shift spectra of the wake scattering light under different bubble speed distribution. In the simulation, the number of interference fringes is $N_f = 8$, the current amplitude is $i_a/i_d = 1$mA, the number of bubbles in the control body is 1000 and the signal threshold is -10dBm, and the standard deviations of the bubble speed distribution in the control body are 4, 8, and 12, respectively.

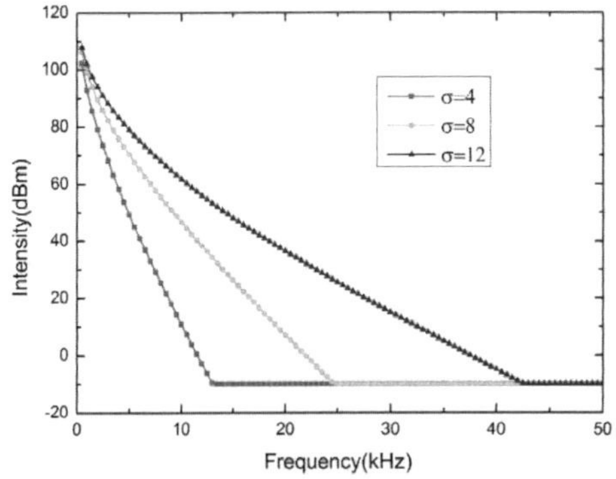

Figure 3. Frequency shift spectra of scattering light of bubbly wakes with different bubble speed distribution.

Fig. 3 shows that under the condition of a certain bubble density, as the range of the particle speed distribution increases, the cut-off frequency of the scattering light frequency shift spectrum gradually increases. Specifically, as the standard deviation of the speed distribution of the bubbly wake increases from 4 to 12, the cut-off frequency shift spectrum of the scattering light of the wake increases from 14.3kHz to 42.2 kHz.

2.4.3. When the bubble density and velocity distribution change at the same time. As time goes by, the bubble density and speed distribution range in the wake will gradually decrease. In order to simulation the evolution of the features of the bubble group over time, the frequency shift spectra of the wake scattering light are simulated. In the simulation, the number of interference fringes is $N_f = 8$, the current amplitude is $i_a/i_d = 1\text{mA}$, the signal threshold is -10dBm, the number of bubbles in the control body is 600, 800, and1000, respectively. And the corresponding standard deviation of the bubble velocity distribution in the control body is 4, 8, and 12.

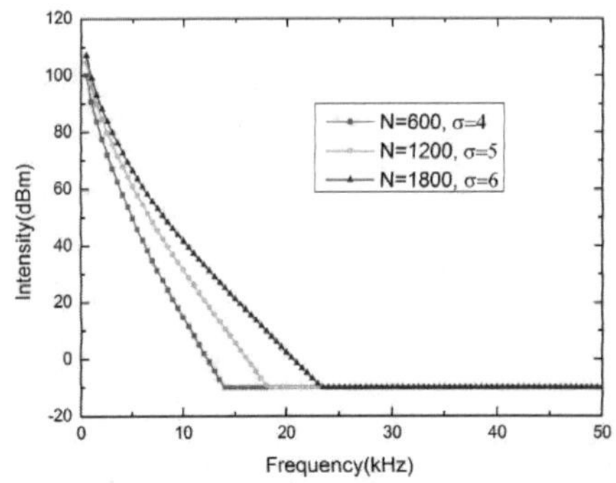

Figure 4. Scattering light shift spectra under different bubble population and velocity distribution.

It can be seen from Fig. 4 that as the bubble density and velocity distribution range increase simultaneously, the cut-off frequency of each frequency shift component and the amplitude of the scattering light frequency shift spectrum also increase. Specifically, under the three conditions, the cut-off frequency shift of the frequency shift spectrum of the scattering light are 13.5 kHz, 16.8 kHz, and 23.7 kHz, respectively. The above simulation result shows that with the prolongation of the wake's existence time, the bubble density and velocity distribution range of the wake bubble group will decrease, the amplitude of each frequency shift component of the scattered will decrease, and the cut-off of the frequency spectrum will also reduce accordingly.

Theoretical analysis results show that the frequency shift spectrum features of the scattering light of the bubbly wake are closely related to the bubble density and velocity distribution range. As the bubble density and velocity distribution range increase, the amplitude of the frequency shift spectrum increases, and the cut-off frequency shift also increases. This indicates that the frequency shift spectrum features of scattering light of the wake can be used as the basis for the identification of the bubble wake.

3. Experimental study

3.1. Experiment platform

The Experimental platform designed and built for the detection of the frequency shift spectrum of the scattering light of the bubbly wake is shown in Fig.5.

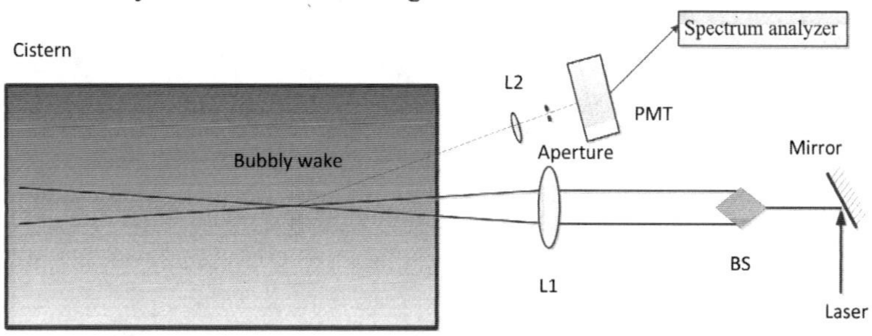

Figure 5. Schematic diagram of experimental platform.

In the experiment, a continuous green laser with a wavelength of 532nm is used, which is divided by the beam splitting prism BS into two parallel beams of equal intensity. After being focused by the lens L1 with a focal length of 400mm, they are incident on the bubbly wake to form an interference field, which is the control body. The scattering light of bubbles in the control body is collected by the lens L2 with a focal length of 120mm, and received by a photomultiplier tube PMT, and is put in a spectrum analyzer for analysis finally. The bubbly wake is produced by the micro-porous ceramic array connected to the air pump. The density and velocity distribution of bubbles can be adjusted through the air pressure of the air pump.

3.2. Experimental results and analysis

3.2.1. Comparison of the frequency shift spectra of scattering light before and after the bubbly wake is generated.

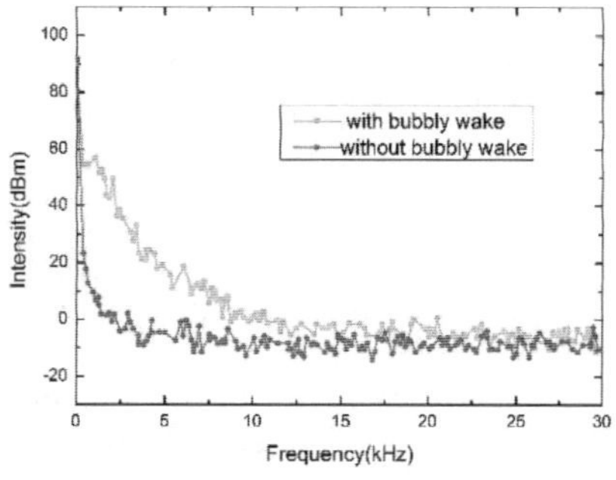

Figure 6. Frequency shift spectra of scattering light without and with bubbles.

Fig. 6 shows the frequency shift spectra without and with bubble wake obtained by the analysis of the spectrometer. The cut-off frequency the frequency shift spectrum of scattering light without bubble wake is 2.6 kHz. And the cut-off frequency the frequency shift spectrum of scattering light with bubble wake is 17.5 kHz. It can be seen from Figure 6 that in the static water state without bubble wakes, the scattering light has almost no Doppler frequency shift. While in the presence of bubble wakes, the scattering light has an obvious frequency shift spectrum.

3.2.2. Frequency shift spectra of scattering light under different wake conditions. Under the premise of keeping other system parameters unchanged, the air pressure applied to the micro-porous ceramic array is set as 0.007MPa, 0.010MPa, and 0.013MPa respectively. The obtained scattering light frequency shift spectra are shown in Fig. 7.

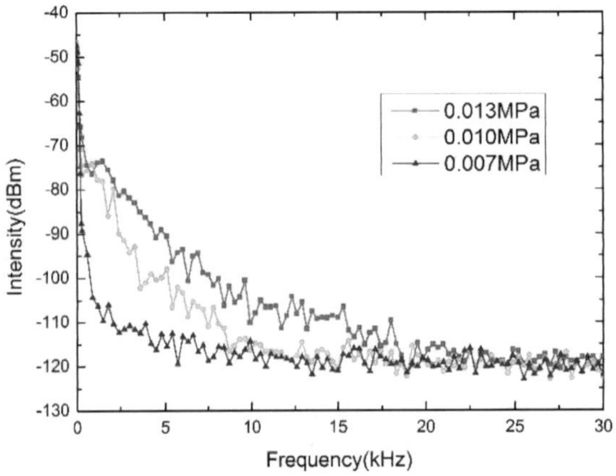

Figure 7. Frequency shift spectra of scattering light under different wake conditions.

From the frequency shift spectra of the scattering light of the bubble wake shown in figure 7, the cut-off frequency shifts under three different conditions can be obtained. According to formula 3, the corresponding speed range of the cut-off frequency shift can be calculated. The obtained speed data are shown in Table 1.

Table 1. Data of different wake conditions

Numble	Pressure(MPa)	Cut-off frequency(kHz)	Speed(mm/s)
1	0.007	10.69	49.23
2	0.011	14.98	69.98
3	0.014	18.72	87.01

According to the data in Table 1, as the pressure applied to the micro-porous ceramic array increases, the bubble density and the velocity distribution range in the wake also increase, and the amplitude of each frequency shift component of the scattering light frequency shift spectrum will increase accordingly, so is the cut-off frequency shift. The experimental results are consistent with the simulation results given in figure 4.

For the target bubbly wake in the water, the bubble density and velocity distribution in the water area will change over time. The closer to the target, the larger the bubble density and movement velocity distribution range, the amplitude and cut-off frequency of the frequency shift spectrum are also larger. Therefore, the method based on the frequency shift spectrum of the scattered light can be used to identify the bubbly wake of the target in the water.

4. Conclusion

In this paper, the features of the frequency shift spectrum of the scattering light by the bubble wake of a moving target in the water are simulated and analyzed. The theoretical analysis results are verified experimentally. According to the research results, a bubbly wake detection method based on frequency shift spectrum characteristics of scattering light is proposed. Studies have shown that when the laser passes through the bubble wake, the scattering light has a significant Doppler frequency shift. With the increase of bubble density and bubble movement speed range, the amplitude of each frequency shift component of the scattering light frequency shift spectrum increases, and the cut-off frequency also increases. The differential mode laser Doppler method is used to measure the frequency shift spectrum of the scattered light of the bubbly wake. According to the characteristics of the amplitude and cut-off frequency of the frequency shift spectrum, the bubble wake of the target in water can be identified and its time-space evolution characteristics can be studied.

References

[1] Yuanyuan Yi, Kecheng Yang, Xiaohui Zhang, Jionghui Rao, Xia Min, Yi Zheng, Xuanhua Lei, Qizhong Liu, Laser backscattering characteristics of bubbles in water, J.Opt. & Optoelectron. Technol. 4 (2006) 38-41.

[2] Jing Cao, Jing Kang, Xiaoqin Jiang, Mairong Xu, An overview in the research of the optical properties of air bubbles in the wake, J.Ship Science and Technology. 27 (2005):5-8.

[3] B.M.Watrasiewicy , M.J.Rudd, Laser Doppler Measurements[M]. Xu Fangtong et al, Transl. Beijing: Water Conservancy Press, 1980.

[4] Donghua Zhu, Xiaohui Zhang, Ship wake detection based on laser Doppler effect, J.Laser Technology. 30 (2006) 298-300.

[5] L. E. Drain. The Laser Doppler Technique, New York: John Wiley, 1980.

[6] Weizhou Gu. Measurement of the liquid current velocity based on the laser Doppler frequency shift technique. J.Laser Journal. 24 (2003) 81-82.

Influence of fan-shaped holes with different transition surfaces on film cooling performance

Hansong Liu*

Shanghai Aircraft Design and Research Institute, Shanghai, China

*Corresponding author email: liuhansong@comac.cc

Abstract. A series of numerical simulations were conducted to investigate the fan-shaped film holes with different transition surfaces on film cooling performance. In the present study, four fan-shaped holes different with the conventional fan-shaped hole were designed by changing the transition surfaces of fan-shaped holes, and their film cooling effectiveness and discharge coefficient were studied under the same boundary conditions. The results show that the anti-s-shaped fan-shaped hole has the best film cooling performance. Under the blowing ratio of M=1 and M=2, compared with the reference fan-shaped hole, the spatially-averaged cooling effectiveness is improved by 8.7% and 23.8% respectively, and the discharge coefficient is increased by 3.9% and 6.4% respectively.

Keywords: film cooling; fan-shaped hole; transition surfaces; film cooling effectiveness; discharge coefficient.

1. Introduction

In the cooling structure design of aircraft engine hot end components, film cooling is an extremely important technical measure. Its technical principle is that the cooling air in the cooling channel is discharged and injected through the film hole to form a cooling protective layer film on the wall. Because the film jet can not only cool the high-temperature wall on the gas side, but also form a cooling protective layer film between the high-temperature gas and the wall, the improvement of the film jet cooling capacity plays a leading role in improving the overall cooling effect of the cooling structure [1].

In the past decades of research, the important development sign of film cooling strengthening technology is the application of shaped holes. Goldstein et al. [2] carried out earlier research on film cooling with shaped holes, and found that the film cooling effectiveness can be significantly improved by using film holes with the expanded exit. Gritsch et al. [3-6] system measured the flow field, the adiabatic film cooling effectiveness, the heat transfer coefficient and the discharge coefficient of the cylindrical hole, the fan-shaped hole and the laidback fan-shaped hole. Compared with the cylindrical holes, the fan-shaped holes and laidback fan-shaped holes can effectively inhibit the penetration of cooling jet into the mainstream and the shear between primary and secondary flows. They can also reduce the heat transfer coefficient downstream, and improve the film cooling effectiveness and discharge coefficient. Lee et al. [7, 8] and Saumweber et al. [9] respectively optimized the inclination angle, the lateral expansion angle and the forward expansion angle of the fan-shaped holes. The results show that the lateral expansion angle of fan-shaped holes has an important influence on the film cooling effectiveness. The fan-shaped holes with reasonable lateral expansion angle can form separation bubbles in the expansion section of the holes, improve the lateral diffusion of film jet and effectively suppress the film separation. Wright et al. [10], Saumweber et al. [11] and Schroeder et al. [12] studied in detail the effects of different mainstream turbulence intensity, Mach number, blowing ratio and density ratio on the flow and cooling characteristics of the fan-shaped holes and the laidback fan-shaped holes. Bell et al. [13] carried out an experimental study on the cooling performance of the fan-shaped hole and the laidback fan-shaped hole, taking into account the effect of the lateral expansion angle, and found that the cooling effect of the fan-shaped hole with the lateral expansion angle was the best. Nguyen et al. [14] studied the cooling performance of a conical film hole. It was considered that the conical hole could improve the lateral spread of the film, and the cooling effectiveness was much higher than that of the cylindrical hole under different blowing ratios.

In summary, as a typical shaped hole, fan-shaped holes have good film cooling performance. However, few scholars have devoted themselves to the study of the effect of the fan-shaped holes with different transition surfaces on film cooling performance. In the present study, five fan-shaped holes with different transition surfaces are selected and designed. The film cooling effectiveness and the discharge coefficient are taken as the evaluation index of film cooling performance. Detailed numerical simulation and comparative analysis are carried out under different blowing ratios.

2. Configuration of fan-shaped holes

2.1. Physical model of computational domain

Due to the periodical distribution feature of fan-shaped holes, the computational domain only includes one pitch of fan-shaped hole. Figure 1 presents the schematic of computational model, the computational domain mainly consists of three parts: the main flow channel, the coolant flow channel and the fan-shaped film hole. The directions of x, y and z are stream-wise direction, vertical to the cooling surface direction and span-wise direction respectively. As shown in Figure 1 (b), the circular entry diameter of the fan-shaped hole (d) is 5 mm, the inclination angle (α) is 35°, and the height of the fan-shaped hole (t) is 6 mm. The coordinate origin is located at the center of the film hole. The length and height of main channel are 31 d and 8 d, respectively. The center of the film hole is 15 d away from the mainstream inlet and 16 d away from the pressure outlet. The length and height of the secondary flow channel are 20 d and 4 d, respectively. As shown in figure 1 (c), the lateral expansion angle (β) of the fan-shaped hole is 6° and the pitch of fan-shaped hole is 2.4 d.

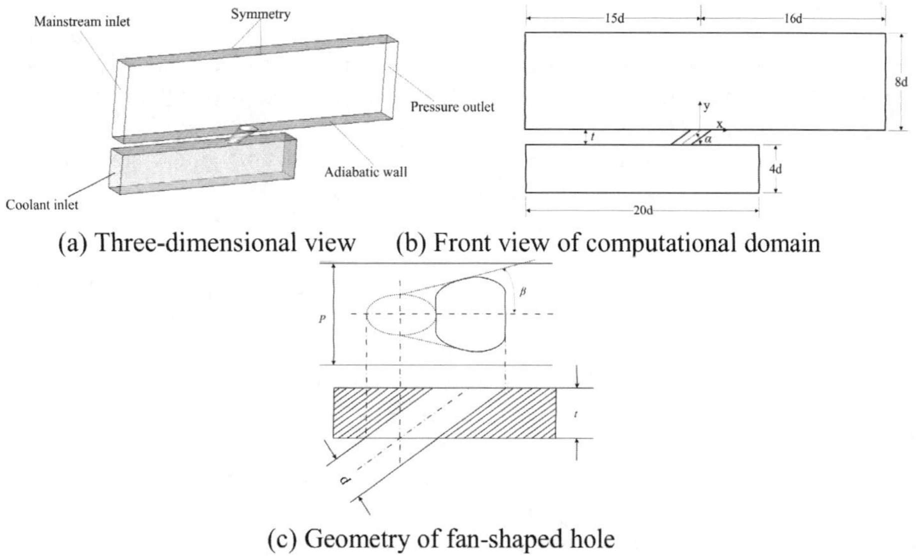

(a) Three-dimensional view (b) Front view of computational domain

(c) Geometry of fan-shaped hole

Figure 1. Schematic of computational model

In order to investigate the influence of the fan-shaped holes with different transition surfaces on film cooling performance, four fan-shaped holes different from traditional fan-shaped hole are designed by changing the lateral transition surfaces. Figure 2 presents the central sections of the reference and four designed fan-shaped holes. The fan-shaped hole with traditional transition surface is selected as the reference film hole. Four fan-shaped holes designed with different lateral transition surfaces are the outer-arc fan-shaped hole, the inner-arc fan-shaped hole, the s-shaped fan-shaped hole and anti-s-shaped fan-shaped hole, respectively.

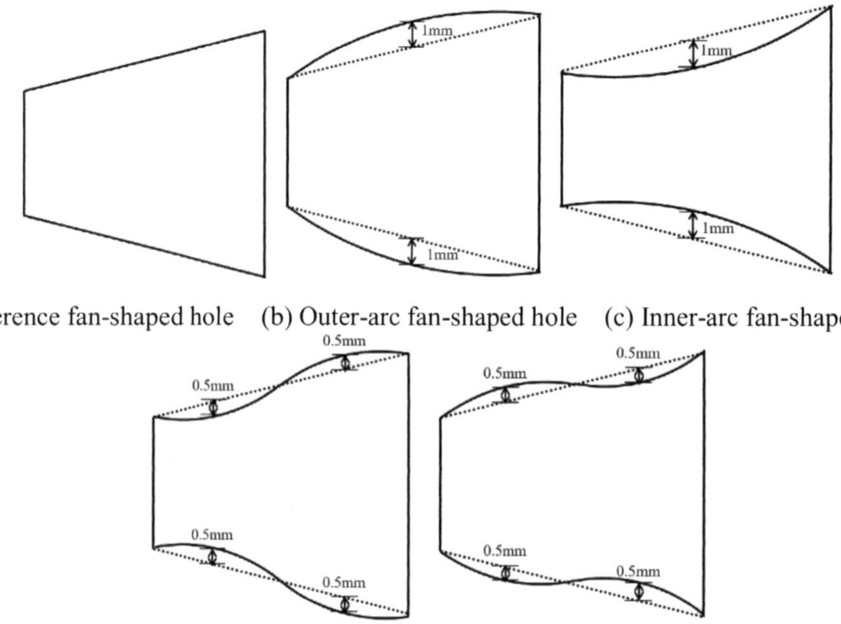

(a) Reference fan-shaped hole (b) Outer-arc fan-shaped hole (c) Inner-arc fan-shaped hole

(d) S-shaped fan-shaped hole (e) Anti-s-shaped fan-shaped hole

Figure 2. The central sections of the reference and four designed fan-shaped holes

2.2. Boundary condition and computational scheme

Two main characteristic parameters discharge coefficient (C_d) and adiabatic film cooling effectiveness (η) are selected to evaluate the effects of film-hole geometry on film cooling performance.

The mainstream channel and the coolant channel inlet are the velocity inlet. The inlet temperature and velocity of the main channel are 25 m/s and 335 K, respectively. The inlet temperature of coolant channel is 300 K. The ideal gas is used for the mainstream and coolant flow. The outlet of mainstream channel is pressure outlet and the pressure is 1 atm. The symmetry boundary condition is adopted on both sides of the mainstream channel. The adiabatic and non-slip velocity conditions are adopted on the other walls. The turbulence intensity of free-stream in the main channel is 2%.

Since the cross-sectional area of the fan-shaped hole varies continuously along the passage, the velocity of the airflow at the inlet and outlet of the film hole is inconsistent. For the convenience of comparison, the inlet parameters of fan-shaped hole are used in the definition of the blowing ratio. In this way, the same mass flow rate can be obtained for all fan-shaped holes as long as the cross-sectional area of the entrance of the film holes is the same under the same blowing ratio.

$$M = \frac{(\rho_c u_c)_{\text{inlet}}}{\rho_\infty u_\infty} \tag{1}$$

Where ρ_c and u_c represent the density and velocity of the secondary flow at the entrance of the film hole, respectively. ρ_∞ and u_∞ represent the density and velocity of the mainstream at the primary flow entrance, respectively.

The adiabatic film cooling effectiveness is an important index for evaluating film cooling performance, which is defined as:

$$\eta_{ad} = \frac{T_{\infty} - T_{aw}}{T_{\infty} - T_{c}} \tag{2}$$

$$\eta_{ad,av} = \frac{1}{11d}\int_{d}^{12d} \eta_{ad,avx} dx \tag{3}$$

$$\eta_{ad,avx} = \frac{1}{2.4d}\int_{-1.2d}^{1.2d} \eta_{ad} dz \tag{4}$$

Where η_{ad} presents the local adiabatic film cooling effectiveness. $\eta_{ad,av}$ presents the spatially-averaged film cooling effectiveness along the streamwise direction between the region of x/d=2 and x/d=12. $\eta_{ad,avx}$ presents the laterally-averaged film cooling effectiveness between the region of z/d=-1.2 and z/d=1.2.

There are various losses when cooling air flows through the film channel, the actual air mass flow rate through the film hole is smaller than the theoretical mass flow rate under the same flow condition, thus the concept of the discharge coefficient is introduced. The magnitude of the discharge coefficient can directly reflect the aerodynamic loss of the fan-shaped holes. A larger discharge coefficient represents a smaller aerodynamic loss, which is defined as:

$$C_d = \frac{m_c}{A_{c,inlet}\sqrt{2\rho_c(p_c^* - p_c)}} \tag{5}$$

Where m_c is the actual mass flow rate of coolant. $A_{c,inlet}$ is the entrance area of the fan-shaped hole. p_c^* is the total pressure at the entrance of the fan-shped hole. p_c is static pressure at the outlet of the fan-shaped hole.

3. CFD solver and results validation

3.1. CFD solver and meshes

In the present study, ANSYS-Fluent 14.5 is used for three dimensional steady-state solution. According to the research of Harrison [15] and Ely [16], the reliable k-ε turbulence model with the enhanced wall function is used for numerical simulation. The second order upwind discrete schemes are adopted in the density, momentum, turbulent kinetic energy, dissipation rate and energy. The simplec algorithm is adopted in the pressure-velocity coupling. The criterion of convergence is that the residual accuracy is less than 10^{-5}, the residual curve tends to be flat and the average temperature change on the cooling wall does not does not exceed±0.1 K.

The multi-block grids meshed by GAMBIT are adopted in the whole computational domain which is shown in Figure 3. The hexahedral structured grids are used for the primary flow and coolant flow channels, and the tetrahedral unstructured grids are used for fan-shaped holes. The mesh density greatly affects the accuracy of the calculation results, it is necessary to study the influence of the number of grids on the calculation results before the formal numerical simulation. The number of grids can be adjusted by adjusting the height of the first grid nodes adjacent the cooling wall. When the number of grids increases to 1.4 million, the influence of the number of grids on the calculation results disappears. At this time, the corresponding maximum y^+ is about 1.

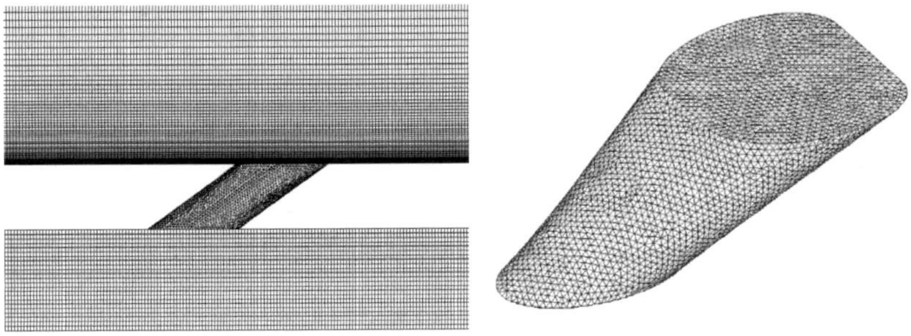

(a) front view of the local grids (b) fan-shaped hole grids

Figure 3. The schematic of the computational domain grid

3.2. CFD calculation results validation

In order to verify the calculation method, the adiabatic film cooling effectiveness of the fan-shaped hole from the research of Saumweber and Schulz [9] et al. is selected to compare with the CFD calculation results. As is shown in Figure 4(a), the $\eta_{ad,avx}$ calculated by CFD is a little higher than the experimental data at M=0.5 and M=2.5. To further verify the calculation method, as is shown in Figure 4(b), the comparison between the $\eta_{ad,av}$ calculated by CFD and the experimental data is also presented. The $\eta_{ad,av}$ calculated by CFD is a little higher than the experimental data under different blowing ratio. However, as a whole, the CFD calculation results of reliable k-ε turbulence model is in good agreement with the experimental results.

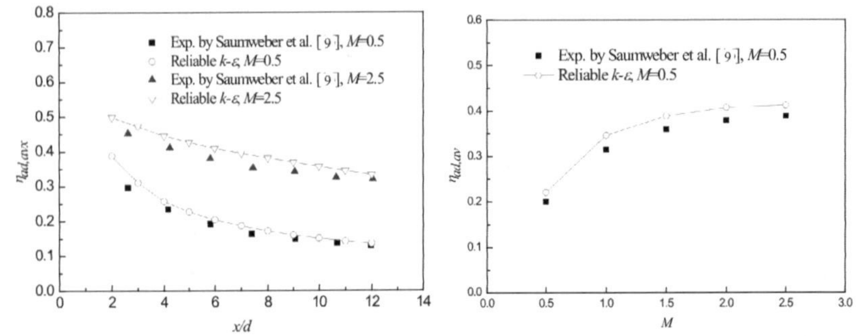

(a) laterally-averaged film cooling effectiveness (b) spatially-averaged film cooling effectiveness

Figure 4. The comparison between CFD calculation results and experimental results [9]

4. Results and discussion

Table 1 shows the comparison of film cooling effectiveness and discharge coefficient for fan-shaped holes with different transition surfaces. For the film cooling effectiveness, compared with the reference film hole, under different blowing ratios, only the film cooling effectiveness with anti-s-shaped transition surface has been improved. And at high blowing ratio, the film cooling effectiveness has been improved more obviously. At M=1 and M=2, the film cooling effectiveness of fan-shaped hole with anti-s-shaped transition surface increases by 8.7% and 23.8% respectively. However, the film cooling effectiveness of the fan-shaped holes with outer-arc and s-shaped transition surfaces deteriorates obviously. At M=1 and M=2, the effectiveness of fan-shaped hole with outer-arc transition surface deteriorates by 19.6% and 24.5%, respectively, and that of the fan-shaped hole with s-shaped transition surface deteriorates by 15.5% and 32.0%, respectively. For the discharge coefficient, compared with the reference film hole, the discharge coefficient of fan-shaped hole with outer-arc and anti-s-shaped transition surfaces increases to a certain extent under different blowing ratios. At M=1 and M=2, the discharge coefficient of fan-shaped hole with outer-arc transition surface increases by 9.7% and 12.1%, respectively, and that of fan-shaped hole with anti-s-shaped transition surface increase by 3.9% and 6.4%, respectively. While, the discharge coefficient of fan-shaped hole with inner-arc transition surface decreases obviously under the different blowing ratio. At M=1 and M=2, the discharge coefficient of fan-shaped hole with inner-arc transition surface decreases by 21.3% and 19.5%, respectively.

Table 1. Comparison of film cooling performance of fan-shaped holes with different transition surfaces

Fan-shaped holes with different transition surfaces	$\eta_{ad,av}$ (M=1)	C_d (M=1)	$\eta_{ad,av}$ (M=2)	C_d (M=2)
Reference	0.515	0.836	0.537	0.844
Outer-arc	0.414	0.917	0.403	0.946
Inner-arc	0.495	0.658	0.522	0.679
S-shaped	0.435	0.748	0.365	0.762
Anti-S-shaped	0.560	0.869	0.665	0.898

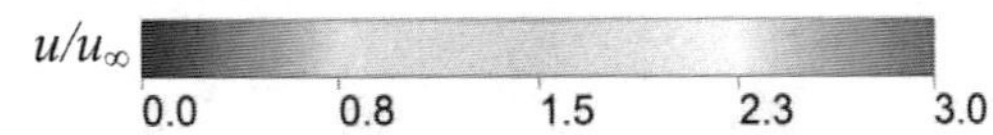

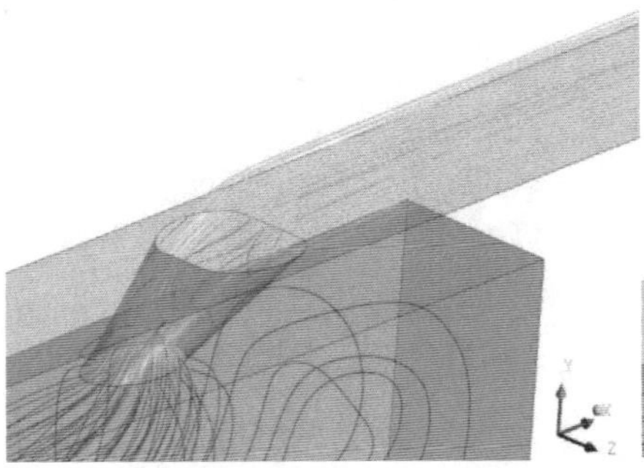

(a) Reference fan-shaped hole

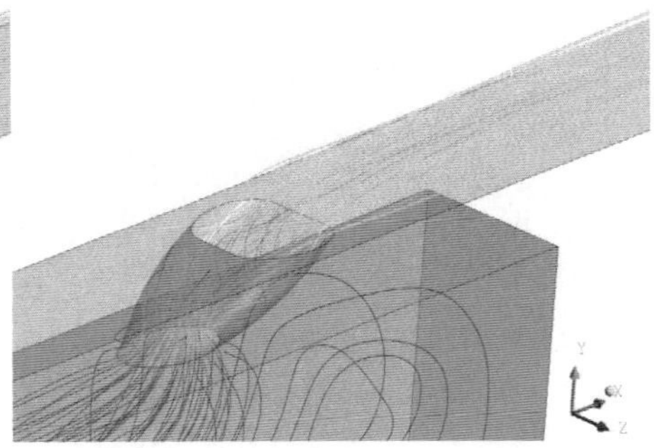

(b) Outer-arc fan-shaped hole

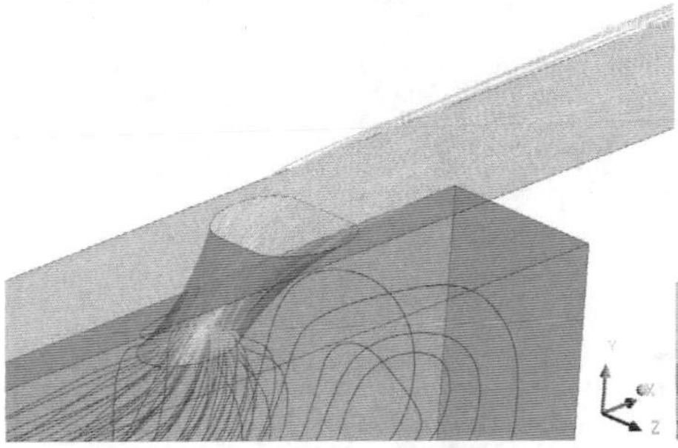

(c) Inner-arc fan-shaped hole

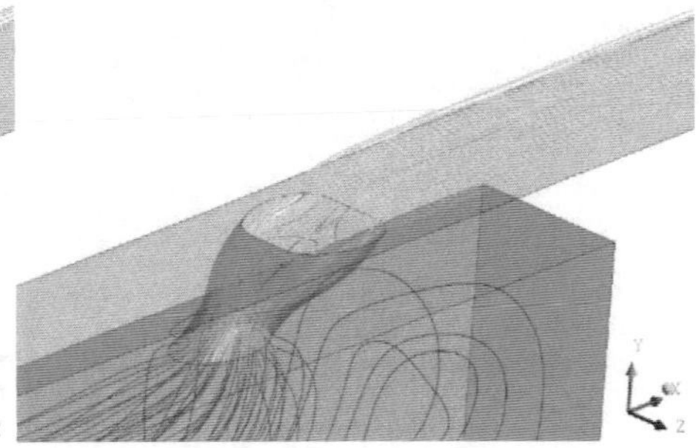

(d) S-shaped fan-shaped hole

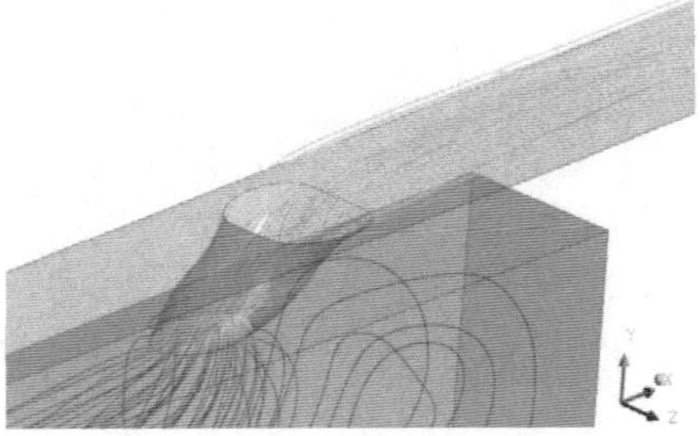

(e) Anti-S-shaped fan-shaped hole

Figure 5. Computed path-lines of cooling air inside film cooling hole at M=1

Figure 5 and Figure 6 present the computed path-lines of cooling air inside film cooling hole at M=1 and M=2. Compared with the reference case, the fan-shaped holes with the outer-arc and the anti-s-shaped transition surfaces guide the secondary flow to the center of the film hole near the entrance of the film hole, so there is a larger film jet at the exit of the film hole, which can be seen intuitively from Figures 6 (b), (e). As a whole, the volume of the film holes with inner-arc and s-shaped transition surfaces shrinks, so the film velocity inside the film holes is faster, which can be observed in Figures 5 (c), (d) and Figures 6 (c), (d). However, near the entrance of the film holes, the inner-arc and s-shaped transition surfaces guide the secondary flow to the two ends of the film hole, so there is a smaller film jet at the exit of film holes. The film jet can better adhere to the wall, which can be seen intuitively from Figures 6 (c), (d).

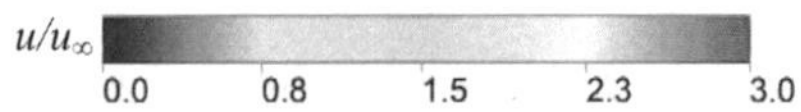

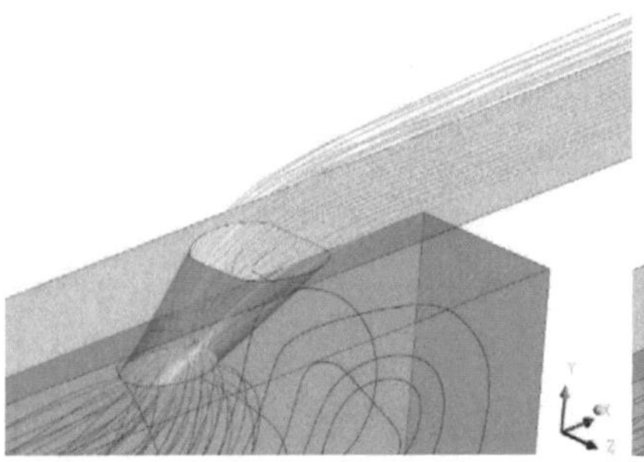

(a) Reference fan-shaped hole

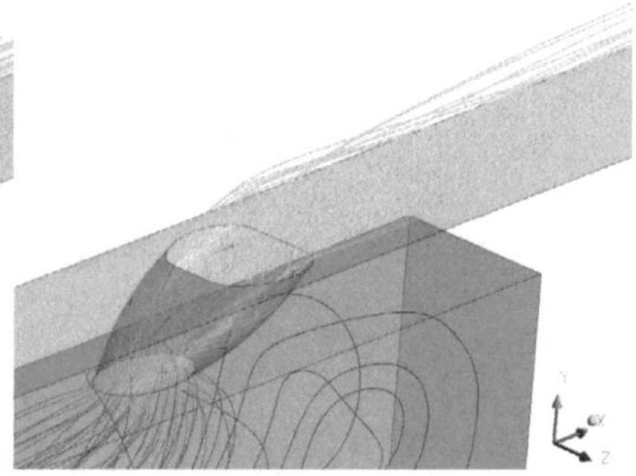

(b) Outer-arc fan-shaped hole

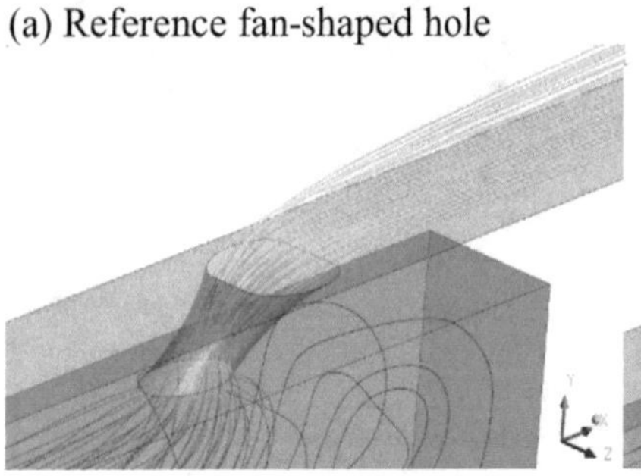

(c) Inner-arc fan-shaped hole

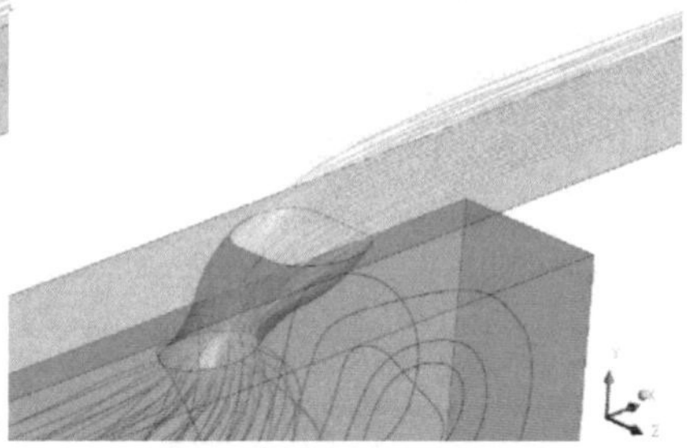

(d) S-shaped fan-shaped hole

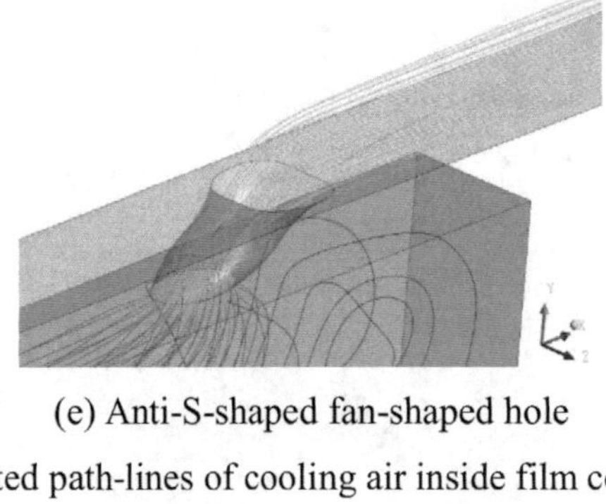

(e) Anti-S-shaped fan-shaped hole

Figure 6. Computed path-lines of cooling air inside film cooling hole at M=2

Figure 7 and Figure 8 present the streamline and dimensionless temperature distribution on the normal plane at x/d=2 at M=1 and M=2. Due to the interaction between the mainstream and the secondary flow, a pair of kidney vortices is formed downstream of the film hole outlet. The high-temperature mainstream is sucked by the kidney vortices from both sides to the lower wall of the cooling jet, resulting in the forced upward lifting of the cold jet and its separation from the wall, which reduces the cooling effect. Compared with the reference case, only the kidney vortices of the anti-s-shaped fan-shaped hole are suppressed significantly, as is seen in Figure 7(e) and Figure 8(e), so the film cooling performance could be improved significantly. However, the kidney vortices of other designed fan-shaped holes are obviously enhanced, the film cooling performance could be significantly reduced. Under the high blowing ratio, a counter rotating pairs of vortices which have the reverse rotating pairs with the traditional kidney vortices are generated by the anti-s-shaped fan-shaped hole, as is seen in Figure 8(e), which can effectively avoid the invasion of the high temperature mainstream and enhance the spanwise ability of the film jet.

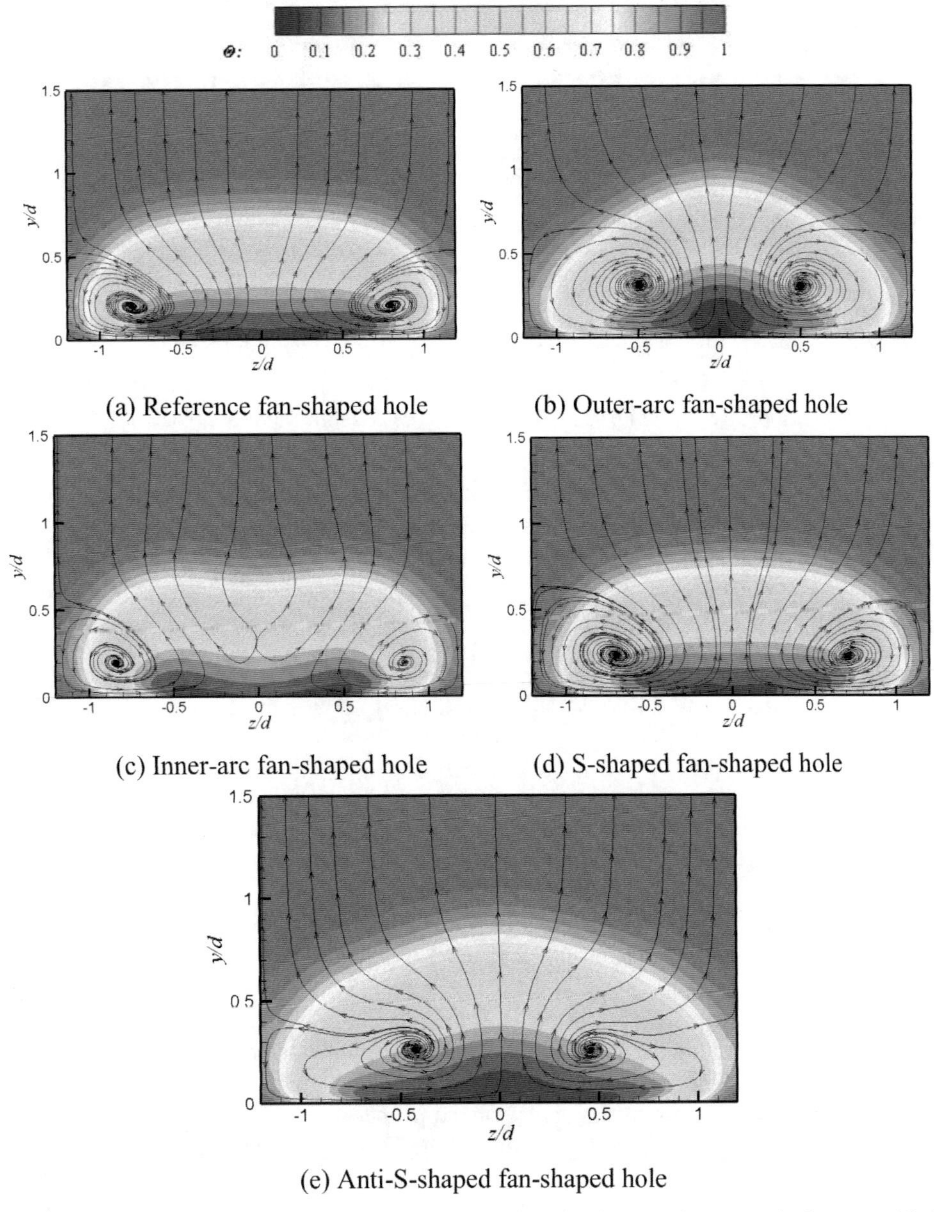

Figure 7. Streamline and dimensionless temperature distribution on the normal plane at x/d=2 at M=1

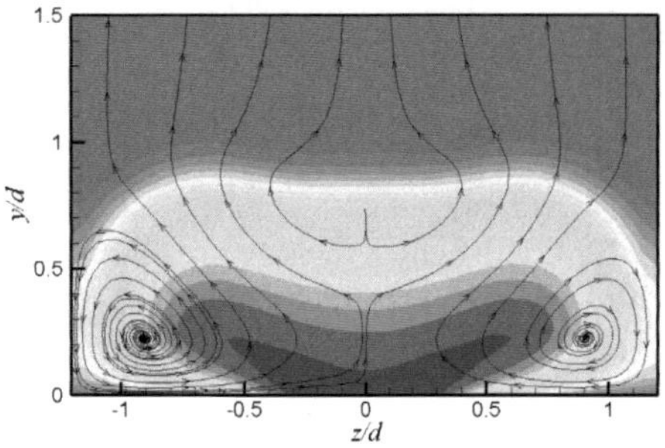

(a) Reference fan-shaped hole

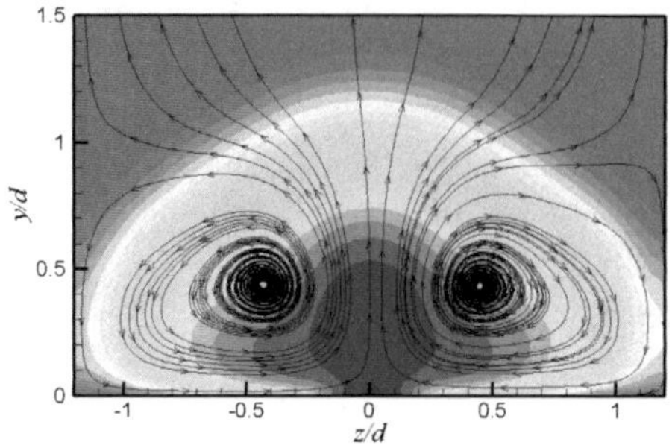

(b) Outer-arc fan-shaped hole

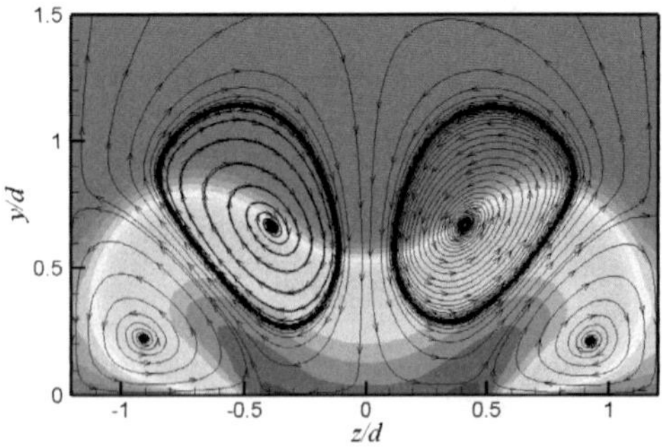

(c) Inner-arc fan-shaped hole

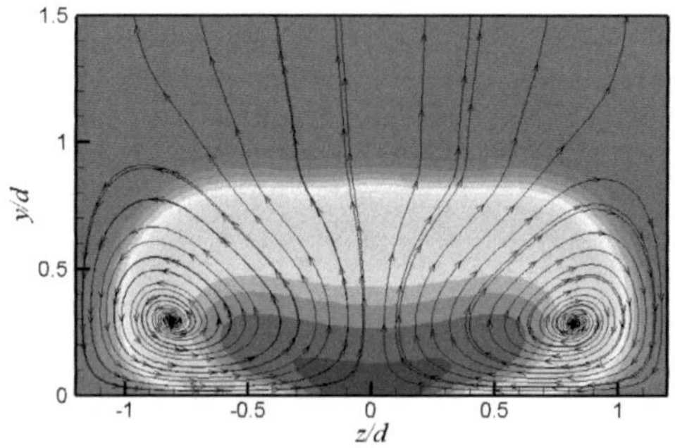

(d) S-shaped fan-shaped hole

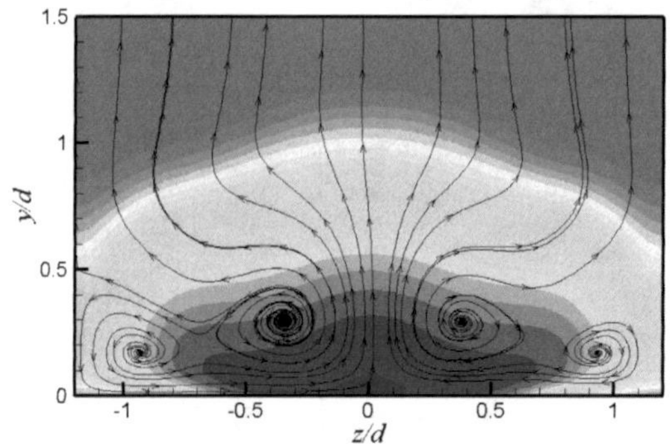

(e) Anti-S-shaped fan-shaped hole

Figure 8. Streamline and dimensionless temperature distribution on the normal plane at x/d=2 at M=2

Figure 9 and Figure 10, respectively, present the dimensionless temperature contours on some cross-sectional planes along x direction at M=1 and M=2. Figure 11 and Figure 12 present the distribution of local adiabatic film cooling effectiveness at M=1 and M=2. As is shown in Figure 9 and Figure 10, compared with the reference case, only the film layer distribution of different cross-sectional planes along the streamwise direction on the anti-s-shaped fan-shaped hole is improved under different blowing ratios, which can also be reflected in Figure 11 and Figure 12. The outer-arc and s-shaped transition surfaces guide the secondary flow to the center of the primary flow near the outlet of the film hole, which is not conducive to the spanwise distribution of the film layer. The reference, inner-arc and anti-s-shaped transition surfaces guide the secondary flow to both ends of primary channel near the outlet of the film hole, while the anti-s-shaped transition surface is obviously more conducive to the spanwise coverage of the film layer.

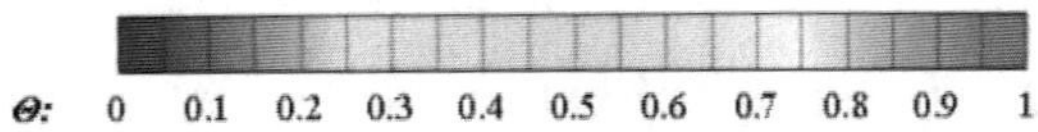

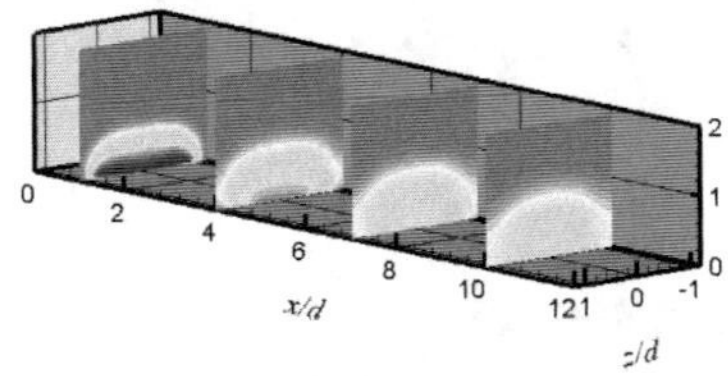

(a) Reference fan-shaped hole

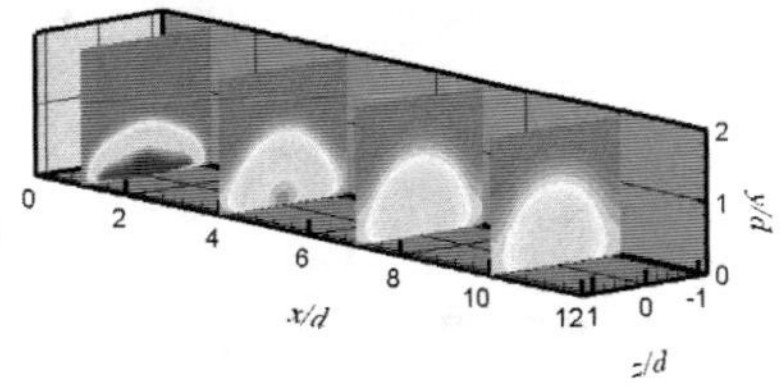

(b) Outer-arc fan-shaped hole

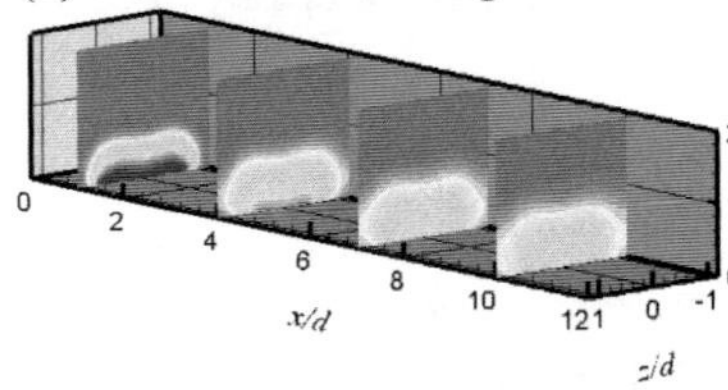

(c) Inner-arc fan-shaped hole

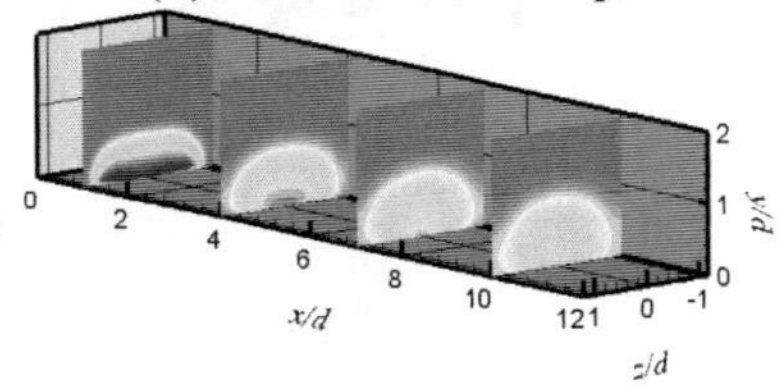

(d) S-shaped fan-shaped hole

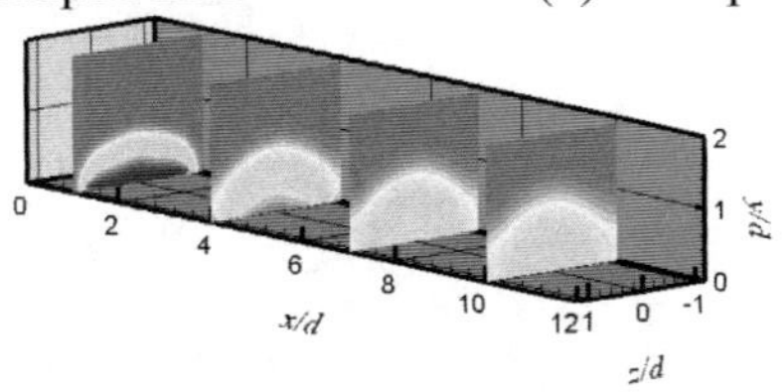

(e) Anti-S-shaped fan-shaped hole

Figure 9. Dimensionless temperature contours on some cross-sectional planes along x direction at M=1

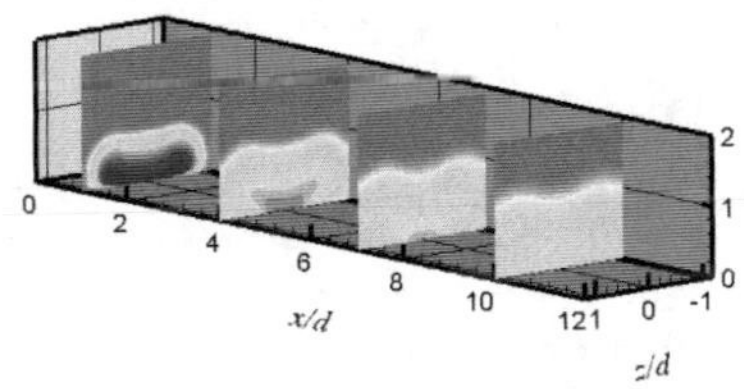

(a) Reference fan-shaped hole

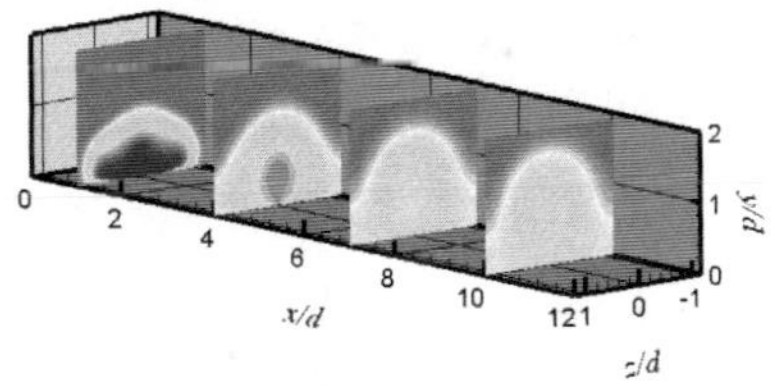

(b) Outer-arc fan-shaped hole

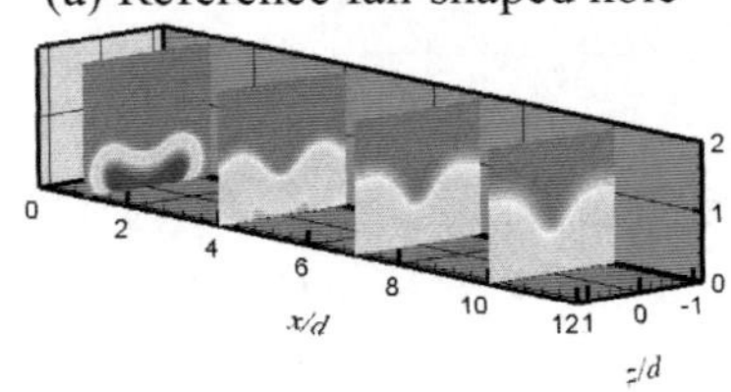

(c) Inner-arc fan-shaped hole

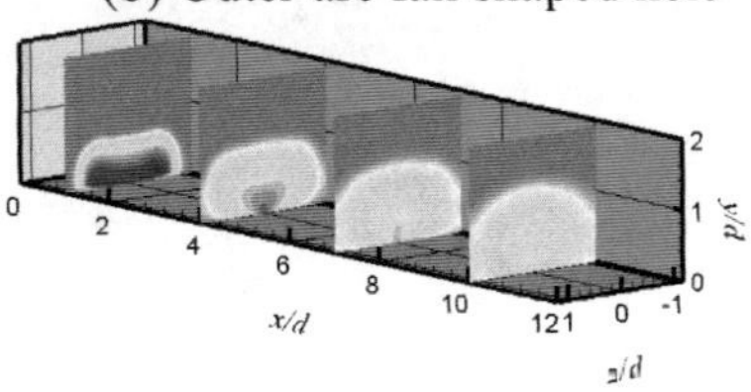

(d) S-shaped fan-shaped hole

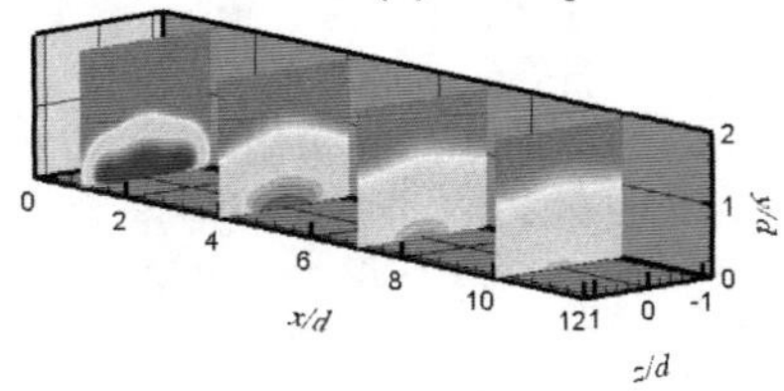

(e) Anti-S-shaped fan-shaped hole

Figure 10. Dimensionless temperature contours on some cross-sectional planes along x direction at M=2

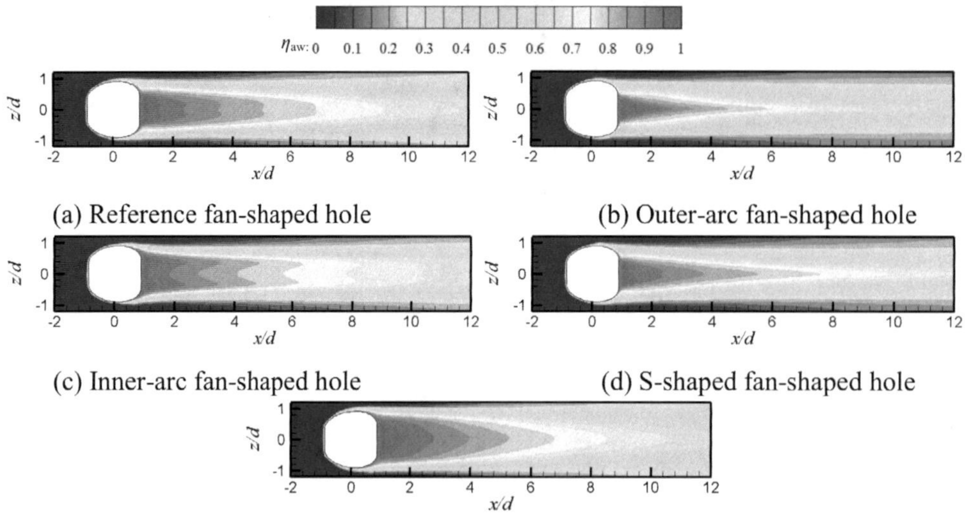

(a) Reference fan-shaped hole (b) Outer-arc fan-shaped hole

(c) Inner-arc fan-shaped hole (d) S-shaped fan-shaped hole

(e) Anti-S-shaped fan-shaped hole

Figure 11. Distribution of local adiabatic film cooling effectiveness at M=1

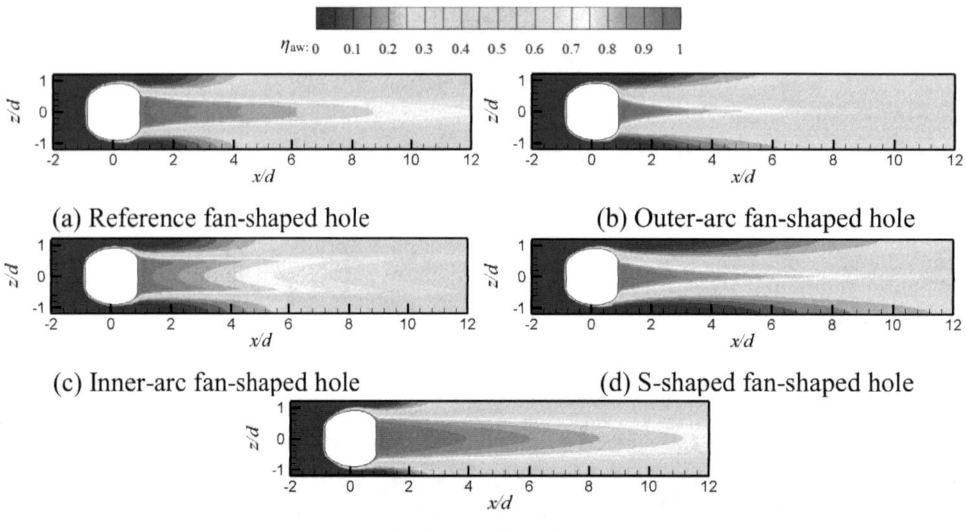

(a) Reference fan-shaped hole (b) Outer-arc fan-shaped hole

(c) Inner-arc fan-shaped hole (d) S-shaped fan-shaped hole

(e) Anti-S-shaped fan-shaped hole

Figure 12. Distribution of local adiabatic film cooling effectiveness at M=2

Figure 13 and Figure 14 show the laterally-averaged adiabatic film cooling effectiveness at M=1 and M=2. Compared with the reference film hole, only the laterally-averaged adiabatic film cooling effectiveness of the anti-s-shaped fan-shaped hole has been significantly improved. At M=1, at the position of x/d=4, the improvement effect is the greatest, which increases by 12.6%. And at M=2, at the position of x/d=3, the improvement effect is the greatest, which increases by 36.1%. Under the low blowing ratio, with the increase of s/d value, the improvement effect of the laterally-averaged adiabatic film cooling effectiveness decreases gradually, due to the continuous mixing of primary flow and film jet. However, under the high blowing ratio, the the laterally-averaged adiabatic film cooling effectiveness decreases first and then increases along the streamwise direction, except for the anti-s-shaped fan-shaped hole. This is due to the detachment-reattachment phenomenon of film jet occurs downstream of the film hole outlet under the high blowing ratio.

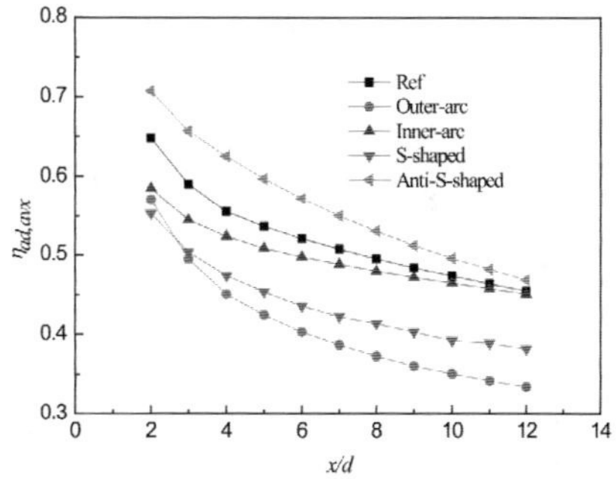

Figure 13. Laterally-averaged adiabatic film cooling effectiveness at M=1

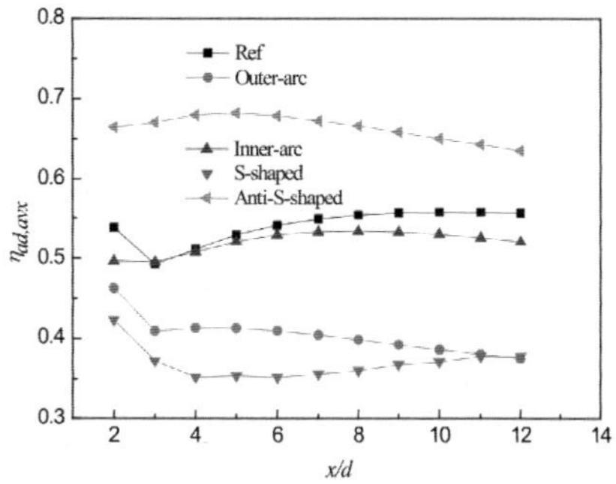

Figure 14. Laterally-averaged adiabatic film cooling effectiveness at M=2

In order to compare the pressure drop in the film hole more intuitively, the pressure drop ratio is introduced, which is defined as:

$$\psi = \frac{\left(p_c^* - p_c\right)}{\left(p_c^* - p_c\right)_{\text{ref}}} \tag{6}$$

The pressure drop ratios for fan-shaped holes with different transition surfaces related to the reference case is shown in Table 2. Figure 15 and Figure 16, respectively, show the streamlines and local static pressure distribution at the y-direction cross section through the console center at M=1 and M=2. As is shown in Table 2, compared with the reference case, the pressure drop ratios of the outer-arc and anti-s-shaped fan-shaped holes are smaller, while the pressure drop ratios of the inner-arc and the s-shaped fan-shaped holes are larger under the different blowing ratios. And these can also be reflected in Figure 15 and Figure 16. This means that the outer-arc and the anti-s-shaped fan-shaped only need overcome less pressure drop to provide the same mass flow rate of secondary flow with the same blowing ratio. This is due to the larger volume of the outer-arc and the anti-s-shaped fan-shaped holes, and the lower average velocity inside the holes, resulting in a smaller flow loss.

Table 2. The pressure drop ratios for fan-shaped holes with different transition surfaces related to the reference case

	Outer-arc	Inner-arc	S-shaped	Anti-s-shaped
$\Psi(M=1)$	0.832	1.517	1.204	0.926
$\Psi(M=2)$	0.795	1.644	1.273	0.882

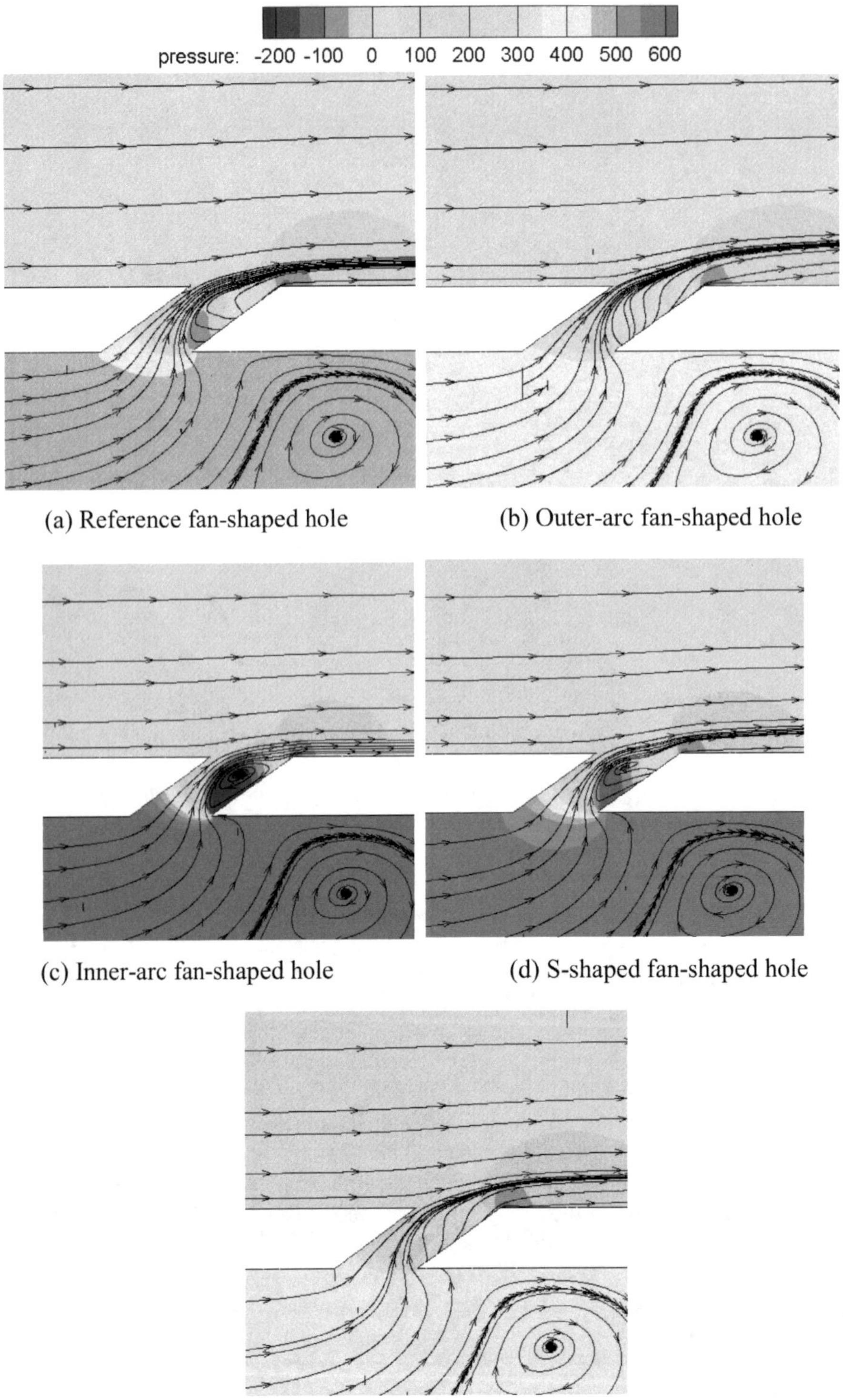

(a) Reference fan-shaped hole (b) Outer-arc fan-shaped hole

(c) Inner-arc fan-shaped hole (d) S-shaped fan-shaped hole

(a) Anti-S-shaped fan-shaped hole

Figure 15. The streamlines and local static pressure distribution at the *y*-direction cross section through the console center at M=1

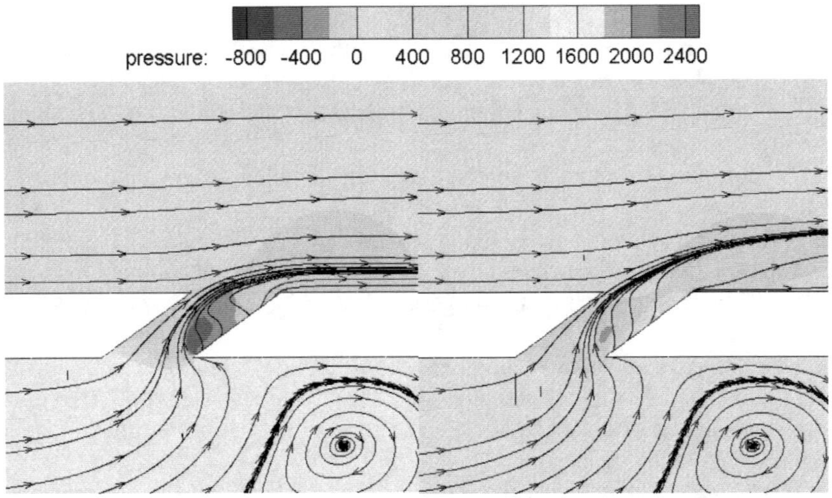

(a) Reference fan-shaped hole (b) Outer-arc fan-shaped hole

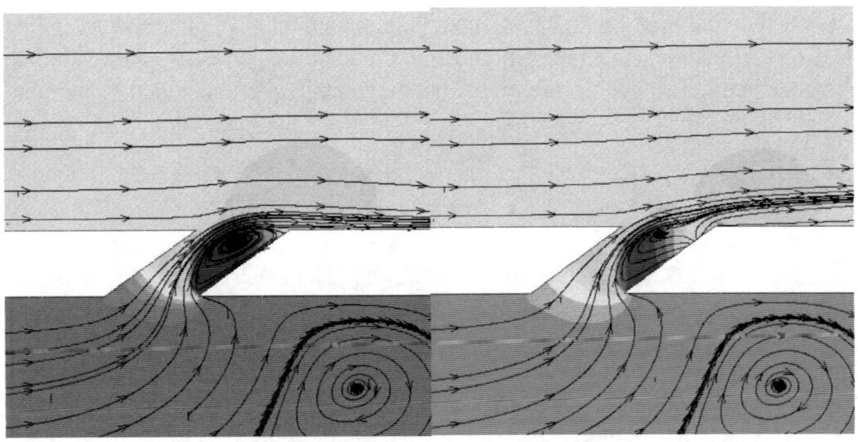

(c) Inner-arc fan-shaped hole (d) S-shaped fan-shaped hole

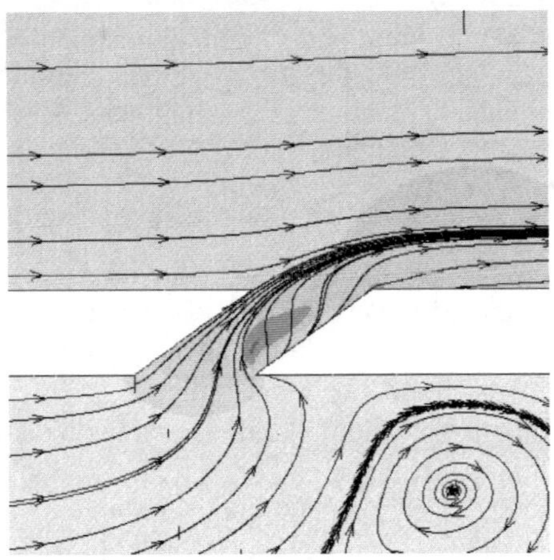

(b) Anti-S-shaped fan-shaped hole

Figure 16. The streamlines and local static pressure distribution at the y-direction cross section through the console center at M=2

5. Conclusions

In order to investigate the influence of fan-shaped holes with different transition surfaces on film cooling performance, four kinds of fan-shaped holes with different transition surfaces from conventional fan-shaped hole are designed and a series of numerical simulations are conducted. The spatially-averaged film cooling effectiveness and discharge coefficient are selected as the index to evaluate the film cooling performance. The optimum structure is selected and the detailed analysis is conducted. The main conclusions are as follows:

(1) Compared with the reference case, the anti-s-shaped fan-shaped hole has better film cooling effectiveness and larger discharge coefficient. At M=1 and M=2, the film cooling effectiveness has been increased by 8.7% and 23.8%, and the discharge coefficient has been increased by 3.9% and 6.4%, respectively.

(2) Compared with the reference case, the discharge coefficient of the outer-arc fan-shaped hole has been increased, but the film cooling effectiveness has been decreased sharply. At M=1 and M=2, the discharge coefficient has been increased by 9.7% and 12.1%, and the film cooling effectiveness has been decreased by 19.6% and 24.5%, respectively.

(3) Compared with the reference case, the inner-arc and s-shaped fan-shaped holes have lower film cooling effectiveness and discharge coefficient. At M=1 and M=2, for the inner-arc fan-shaped hole, the film cooling effectiveness has been decreased by 3.9% and 2.8% respectively, and the discharge coefficient has been decreased by 23.1% and 19.5% respectively. At M=1 and M=2, for the s-shaped fan-shaped hole, the film cooling effectiveness has been decreased by 15.5% and 32% respectively, and the discharge coefficient has been decreased by 10.5% and 9.7% respectively.

Acknowledgments

Special thanks to Ying Huang for his suggestions on this article.

References

[1] Bogard D G, Thole K A. Gas turbine film cooling [J]. AIAA Journal of Propulsion and Power, 2006, 22: 249-269.
[2] Goldstein R J, Eckert E R G, Burggraf F. Effects of hole geometry and density on three-dimensional film cooling [J]. International Journal of heat and mass transfer, 1974, 17(5): 595-607.
[3] Thole K, Gritsch M, Schulz A, et al. Flowfield measurements for film-cooling holes with expanded exits [C]//ASME 1996 International Gas Turbine and Aeroengine Congress and Exhibition. American Society of Mechanical Engineers, 1996: V004T09A010-V004T09A010.
[4] Gritsch M, Schulz A, Wittig S. Discharge coefficient measurements of film-cooling holes with expanded exits [C]//ASME 1997 International Gas Turbine and Aeroengine Congress and Exhibition. American Society of Mechanical Engineers, 1997: V003T09A030-V003T09A030.
[5] Gritsch M, Schulz A, Wittig S. Adiabatic wall effectiveness measurements of film-cooling holes with expanded exits [J]. TRANSACTIONS-AMERICAN SOCIETY OF MECHANICAL ENGINEERS JOURNAL OF TURBOMACHINERY, 1998, 120: 549-556.
[6] Gritsch M, Schulz A, Wittig S. Film-cooling holes with expanded exits: near-hole heat transfer coefficients [J]. International Journal of Heat and Fluid Flow, 2000, 21(2): 146-155.
[7] Lee K D, Kim K Y. Shape optimization of a fan-shaped hole to enhance film-cooling effectiveness [J]. International Journal of Heat and Mass Transfer, 2010, 53(15-16): 2996-3005.
[8] Lee K D, Kim K Y. Surrogate based optimization of a laidback fan-shaped hole for film-cooling [J]. International Journal of Heat and Fluid Flow, 2011, 32(1): 226-238.
[9] Saumweber C, Schulz A. Effect of geometry variations on the cooling performance of fan-shaped cooling holes [J]. Journal of Turbomachinery, 2012, 134(6): 061008.
[10] Wright L M, McClain S T, Clemenson M D. Effect of density ratio on flat plate film cooling with shaped holes using PSP [J]. Journal of turbomachinery, 2011, 133(4): 041011.
[11] Saumweber C, Schulz A. Free-stream effects on the cooling performance of cylindrical and fan-

shaped cooling holes [J]. Journal of Turbomachinery, 2012, 134(6): 061007.
[12] Schroeder R P, Thole K A. Effect of high freestream turbulence on flowfields of shaped film cooling holes [J]. Journal of Turbomachinery, 2016, 138(9): 091001.
[13] Bell C M, Hamakawa H, Ligrani P M. Film cooling from shaped holes [J]. Journal of heat transfer, 2000, 122(2): 224-232.
[14] Nguyen C Q, Johnson P L, Bernier B C, et al. Comparison of film effectiveness and cooling uniformity of conical and cylindrical-shaped film hole with coolant-exit temperature correction [J]. Journal of Thermal Science and Engineering Applications, 2011, 3(3): 031011.
[15] Harrison K L, Bogard D G. Comparison of RANS turbulence models for prediction of film cooling performance [C]//ASME Turbo Expo 2008: Power for Land, Sea, and Air. American Society of Mechanical Engineers, 2008: 1187-1196.
[16] Ely M J, Jubran B A. A numerical evaluation on the effect of sister holes on film cooling effectiveness and the surrounding flow field [J]. Heat and Mass Transfer, 2009, 45(11): 1435-1446.

Effects of blade curvature on flow and heat transfer characteristics of internal cooling at trailing edge

Yan Yan[1], Jianqin Zhu[2] and Zeyuan Cheng[*]

National Key Laboratory of Science and Technology on Aero-Engine Aero-thermodynamics, School of Energy and Power Engineering, Beihang University, Beijing, China

*Corresponding author e-mail: chengzeyuan@buaa.edu.cn

Abstract. In modern turbine blade designs, a large number of internal cooling technologies have been developed and blade curvature is one of the most important factors. This paper numerically simulates the heat transfer process of the internal cooling channel with flat plate and wall curvature at the trailing edge and analyses Nu distribution and deviation between these two structures. The results show that due to the disruptive and reorganising effects of blade curvature on the vortex structure, blade curvature strengthens the heat transfer. There are significant heat transfer differences at the individual surface of each path.

1. Introduction

In order to promote the efficiency and thrust output, gas turbine needs to operate in high inlet temperature conditions, results in the challenge to the thermal protection of the blade. The internal cooling technologies such as integrated impingement and pin fin cooling for turbine blade have been developed to solve this problem.

Integrated impingement and pin fin cooling has been studied on flat plate for several decades. Most scholars investigated the section geometry and parameters of the impingement and pin fin holes [1-3], and concluded that the specific arrangements can achieve the best cooling performance [4]. In pin fin cooling, the heat transfer coefficient of staggered arrangement is higher than that of in-line arrangement for most circumstances. However, blade curvature can cause the change of the flow and heat transfer characteristics in the internal cooling channel, thus influencing these conclusions above.

In summary, there has been a great deal of researches on the effects of internal cooling on flat plate, but less open literature has been done to investigate the blade curvature effects of internal cooling at trailing edge. This paper studies the similarities and differences of heat transfer between these two blade structures—flat plate and wall curvature, and explore the mechanism of blade curvature on flow characteristics.

2. Numerical method

2.1. Geometry and boundary conditions

Figure 1 shows the top view of the trailing edge of a gas turbine blade and the internal cooling structure. The blade consists of 1#, 2# and 3# paths, whose coolant inlet is at 1# path and each path is divided into pressure surface, suction surface, front surface and rear surface. Model 1 and 2 represent the flat plate and blade curvature respectively.

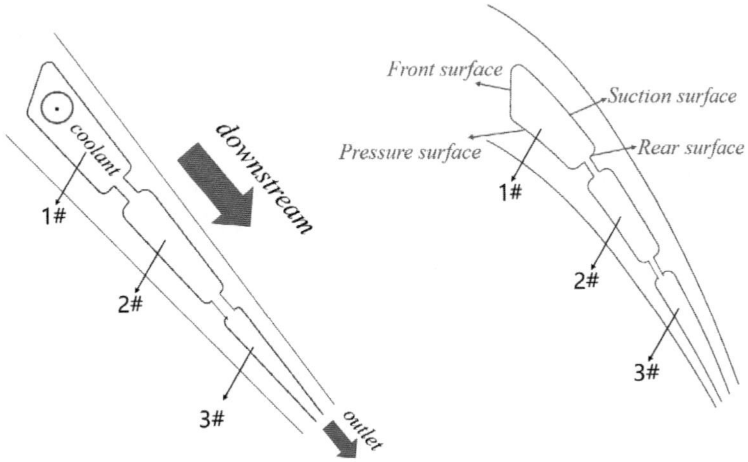

(a) Model 1 **(b)** Model 2

Figure 1. Top view of the blade geometry

The numerical simulation is carried out with CFX software, using a pressure solver and high resolution order difference format. The converged standard is that the residuals are less than 10^{-6}. In this paper, ideal gas is used and the boundary conditions are set as shown in Table 1.

Table 1. Boundary conditions

Boundary	Condition settings
Inlet	Mass flow rate 14.654 g/s, average static temperature 298.15 K
Outlet	Average static pressure 1 atm
Wall	Constant wall temperature 360 K

2.2. Data processing

The convective heat transfer coefficient h is defined as follows:

$$h = \frac{q_w}{(T_w - T_c)} \tag{1}$$

Where q_w is the wall heat flux and T_w is the wall temperature. The average fluid temperature T_c is defined as follows:

$$T_c = \frac{1}{2}(T_{in} + T_{out}) \tag{2}$$

Where T_{in} is the average inlet temperature and T_{out} is the average outlet temperature, T_{out} is defined as follows:

$$T_{\mathrm{out}}=\frac{q_{\mathrm{w}}A_{\mathrm{w}}}{\dot{m}c_{\mathrm{p}}}+T_{\mathrm{in}} \tag{3}$$

Where A_w is the wall area, c_p is the constant pressure specific heat capacity and $\dot{m}$ is the inlet mass flow rate.

The Nussle number is obtained as follows：

$$Nu=\frac{hD_{\mathrm{h}}}{\lambda} \tag{4}$$

Where λ is the thermal conductivity of the fluid and D_h is the channel hydraulic diameter.

2.3. Grid independence and Numerical method validation

The mesh grid is generated by ICEM software, as shown in Figure 2, which produces a tetrahedral mesh over the majority of the channel and prism mesh near the walls and holes. Y plus is less than 1.0, which meets the SST *k-ω* model's requirement.

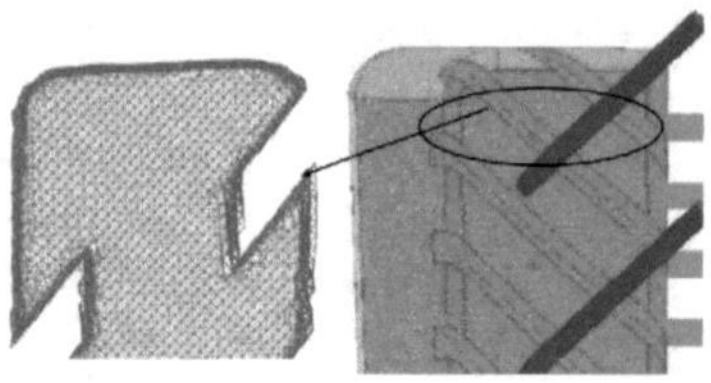

Figure 2. Partial mesh grid of the blade

Six grid scales are selected to obtain the grid independence, as shown in Figure 3. It is seen that when the number of grids is greater than 24 million, the increase in heat transfer coefficient at three surfaces is less than 1%. The mesh of 24 million is finally chosen as the grid independence solution.

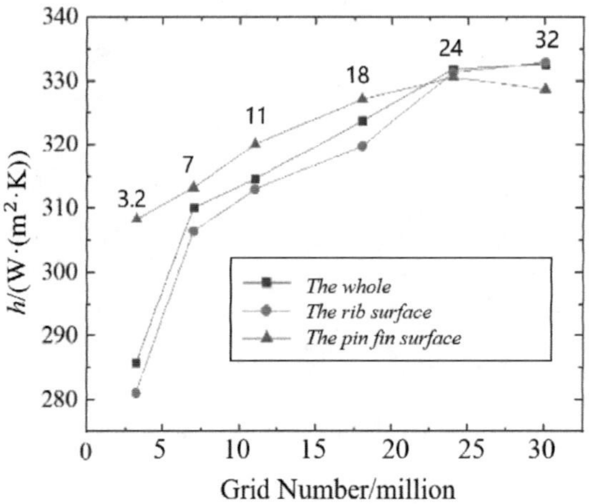

Figure 3. Grid independence solution

In order to accurately predict the flow and heat transfer of the integrated cooling configuration, the integrated impingement and pin fin cooling configuration commonly used for turbine blades was selected for turbulence model validation [5]. Figure 4 shows the comparison between the numerical simulation results and the experimental data, where the film cooling efficiency η is tested. The SST k-ω model is found to be the best agreement with the experimental results within 5% error, which is finally used as the turbulence model for the following work.

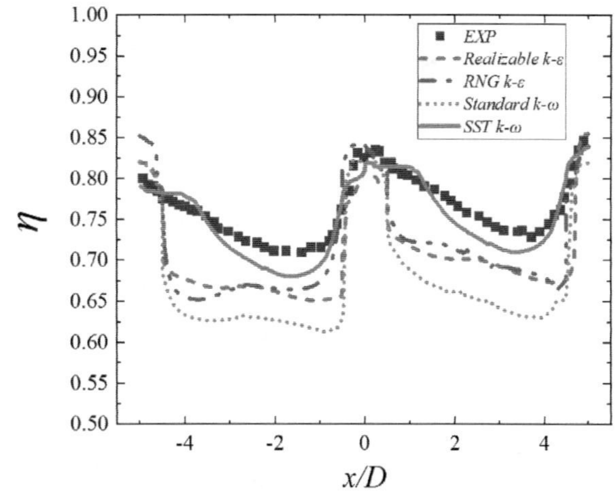

Figure 4. Comparison of numerical results and experimental data of integrated cooling configuration

3. Results and Discussions

3.1. Analysis of heat transfer characteristics

Figure 5 shows a comparison of the Nu differences between model 1 and model 2, focusing mainly on the overall and individual surface of each path. Compared with flat plate, wall curvature strengthens the heat transfer, with the overall heat transfer deviation at around 8% and the local deviation up to 35%. The larger areas of difference are concentrated in the 2# path and the rear surface. Model 2 has a stronger heat transfer effect on the pressure surface than the suction surface under the effect of curvature, while model 1 has the opposite effect.

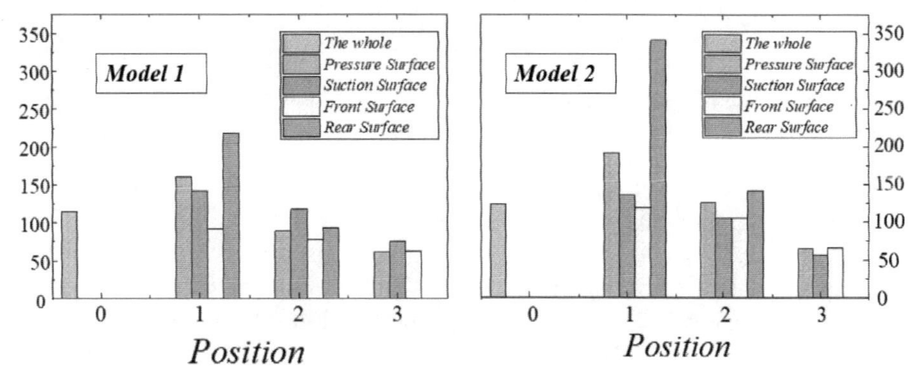

(a) Nu distribution at individual surface of each path

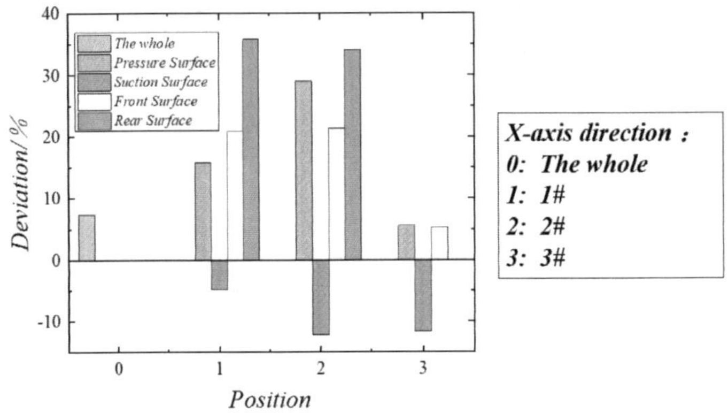

(b) Deviation distribution at individual surface of each path

Figure 5. Comparison of Nu and deviation at individual surface of each path between model 1 and model 2

Figure 6 shows the circumferential Nu for each path along radial direction. The local heat transfer deviation is concentrated at the low radius, which is due to the strong heat transfer between the cooling air and the wall on account of the temperature differences at the low radius. Meanwhile the low radius flow is unstable and significantly influenced by the geometry.

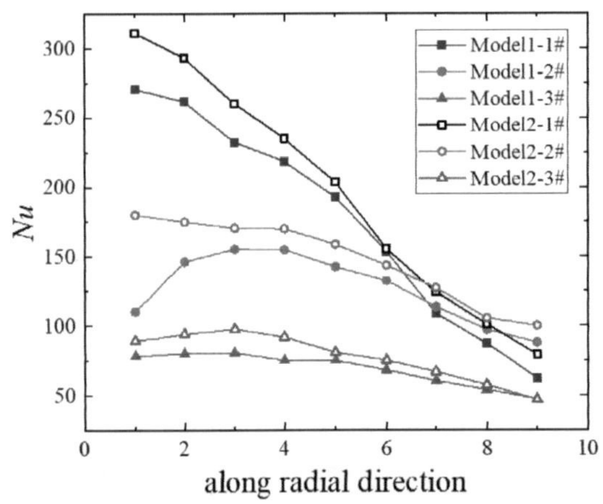

Figure 6. radially Nu at each path between model 1 and model 2

3.2. Analysis of flow characteristics

Figure 7 shows the turbulence kinetic energy and streamline at the low radius of 1# and 2# path crossing section. At the upstream of the impingement cooling (1#①), cooling air in model 2 has a large vortex due to the curvature of the blade. There is a tendency for the high heat transfer zone of the impinging jet towards the pressure surface. There is a small vortex area near the pressure surface in the impingement cooling zone (2# ②) in model 1, pushing the flow towards the suction surface, which also affects cooling air mixed at the rear of the column (2#③). These result in a large difference in heat transfer between the pressure surface and the suction surface.

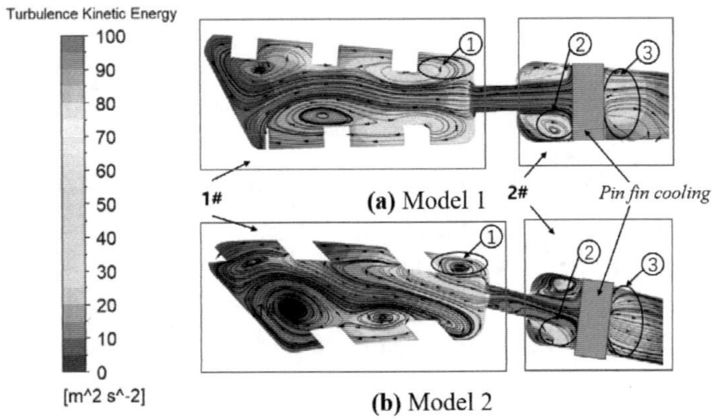

Figure 7. Distribution of turbulence kinetic energy and streamline at low radius of 1# and 2# path crossing section

Figure 8 shows the turbulence kinetic energy and streamline at the inlet of 1# path crossing section. Cooling air in model 2 is prone to collide with the wall surface due to the curvature, which makes it easier to form a vortex structure under the viscous effect of the wall surface, causing the high heat transfer zone move to downstream so that enhancing the rear surface heat transfer.

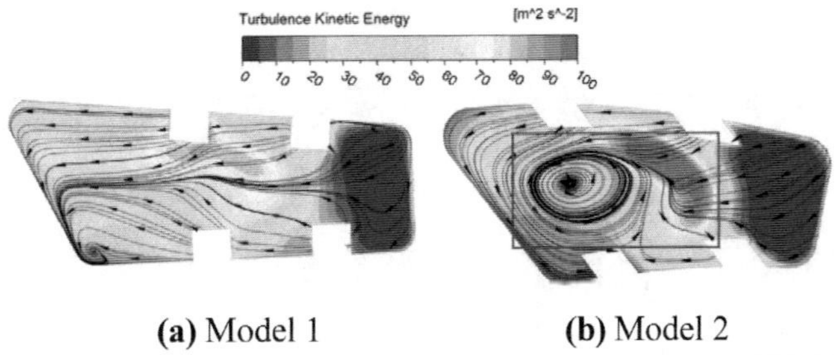

Figure 8. Distribution of turbulence kinetic energy and streamline at the inlet of 1# path section

4. Conclusion

This paper analyses the flow and heat transfer characteristics of the internal cooling channel at the trailing edge of a turbine blade with flat plate and wall curvature. Wall curvature can increase the overall heat transfer deviation of 8% compared with flat plate, where the maximum local heat transfer deviation can be up to 35%. Wall curvature has the tendency for generating vortex at the front and suction surface near low radius, pushing the high heat transfer zone towards the opposite sides, where results in the increase of Nu.

References

[1] Yu Rao. Multiple-jet impingement heat transfer in double-wall cooling structures with pin fins and effusion holes[J]. International Journal of Thermal Sciences, 2018:pp106-119.

[2] Jiaxu Yao. Experimental and numerical investigations on double-jet film-cooling with different mainstream incidence angles[J]. Applied Thermal Engineering, 2020:pp1-10.

[3] Gang Xie. Effects of impingement gap and hole arrangement on overall cooling effectiveness for impingement/effusion cooling[J]. Journal of Heat and Mass Transfer, 2020:pp 1-16.

[4] Maurer M, Ruedel U GritschM, et al. Experimental study of advanced convective cooling techniques for combustor liners[R]. ASME Paper GT2008-51026, 2008.

[5] Nakamata C, Mimura F, et al. Local cooling effectiveness distribution of an integrated impingement and pin fin cooling configuration[R]. ASME Paper GT2007-27020, 2007.

Experimental study of small-scale thin-layer boilover suppression using salt additive

Miao Chen[1], Xuehui Wang[1], Qinpei Chen[2] and Jian Wang[1,]*

[1] University of Science and Technology, Anhui, China
[2] Tianjin Fire Science and Technology Research Institute of MEM, Tianjin, China

*Corresponding author e-mail: wangj@ustc.edu.cn

Abstract. The boilover is a well-known disaster for tank fires. Previous study has shown that changing the boiling point of sub-layer can suppress the small-scale thin-layer boilover, but the detail of suppression mechanism is not claimed. Thus, this study conducted experiments using salt additive with various initial fuel thickness (10mm, 15mm, 20mm).The boilover process and temperature characteristics in liquid pool is discussed. The thermal structure within the liquid pool is figure out in this paper, and the temperature evolution in burning process is analyzed. The results show that the salt additive can change the boiling point of sub-layer and the thermal structure within the liquid pool, which is the main mechanism to suppress the boilover.

Keyword: Thin-layer boilover suppression; Thermal structure; Salt additive.

1. Introduction

The boilover is a well-known disaster for tank fires, which is caused by a rapid phase change of the water sub-layer in fuel tank after a long period heated by the feedback from flame. There are two typical types of boilover which called hot-zone boilover and thin-layer boilover. The thin-layer boilover is usually occurred on the condition that a thin fuel layer burns floating on a water sub-layer. It's characterized by the sharply enlarged flame and a large amount of splashed burning fuel droplets. Therefore, the thin-layer boilover has draw the attentions by many researchers [1, 2].

Arai et al. [3] proposed that the boilover only occurred when fuels whose boiling point was higher than the water boiling point. Garo et al. [4] conducted the boilover experiments, and they obtained that the superheat and nuclear boiling at the interface have significantly influence on the thin-layer boilover. Henceforth researches on the initial parameters like the diameter of fuel pan, initial fuel thickness and so on had been conducted. And researches [5, 6] found that the boilover intensity depended on the burning diameter and initial fuel thickness.

The previous studies about boilover were mainly focused on the hazard of the boilover, such as thermal radiation, flame behaviors. The suppression method of boilover has been rarely concerned. Tseng et al. [7, 8] undertook the small-scale and large-scale experiments using floating perlites, their experiments results revealed perlites can delay and prevent the thin-layer boilover. Ma et al. [9] added sodium chloride to the water sub-layer and found that it effectively suppressed thin-layer boilover by changing the boiling point of water sub-layer. It's shown that the addition of salts into sub-layer is an effect method to prevent the boilover. The occurrence of boilover has the close relationship with the heat transfer between the fuel layer and sub-layer, and the saline solution can change the temperature distribution and heat transfer. But the detail of salt additive suppression mechanism is still not cleared.

Therefore, a series of small-scale experiments using salt additive with various initial fuel thickness (10mm, 15mm, 20mm) were conducted in this paper. The thermal structure and parameters are discussed, and it's significant to have a deep understanding of suppression mechanism of boilover by using salt additive. It's useful for improving the suppression performance for boilover.

2. Experiment setup

The experimental process was conducted on a room with the calm ambient. The ambient temperature is about 10 ± 2 °C, and the relative humidity was about 46 ± 5% during the experimental process.

The fuel pan in experiment was a circular glass pan with diameter of 15 cm and inner depth of 8 cm. The 0# diesel was used as fuel in the experiment, and the physical properties of fuel are shown in Table 1. Three different initial fuel layer thicknesses (10 mm, 15 mm, 20 mm) were considered, and magnesium chloride was used as suppression salt additive. A sub-layer with the thickness of 15 mm is provided below the fuel layer to simulate the conditions of boilover occurrence. The experimental conditions in this paper are shown in Table 2.

Table 1. Properties of 0# diesel.

Fuel	Density (10°C) (kg/m^3)	Boiling point (°C)	Flash point (°C)
0# diesel	835	180~370	78

Table 2. Experimental conditions.

Test label	Fuel thickness (mm)	Type of sub-layer
1	20	water
2	15	water
3	10	water
4	20	$MgCl_2$ solution
5	15	$MgCl_2$ solution
6	10	$MgCl_2$ solution

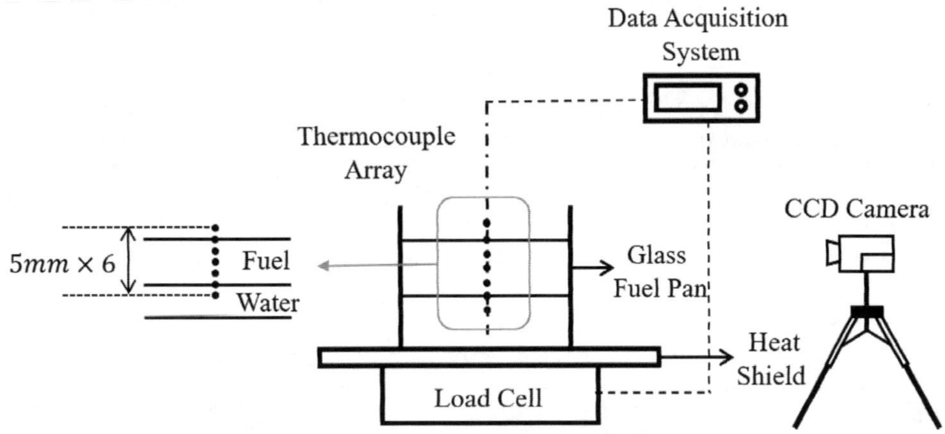

Figure 1. The layout of experimental setup.

The overall experimental setup arrangement is shown in Fig. 1. Seven K-type thermocouples with a diameter of 1 mm were used to measure temperature. The thermocouple array was fixed at the centerline of the fuel pool to record temperature evolution of the liquid layer. The thermocouple spacing is 5 mm, and the lowest thermocouple is 5 mm under the fuel-sub-layer interface. The load cell with accuracy of

0.01 g and measurement interval of 1 s was used to record the mass loss during the combustion, and a heat shield was added between the load cell and the fuel pan to prevent the damage from boilover. A charge coupled device (CCD) camera was placed toward the fuel pan with 25 frames per second to record the flame characteristics.

3. Results and discussion

3.1. The suppression performance of salt additive

The typical burning process of the boilover is shown in Fig. 2 (a). The whole burning process can be divided into three stages based on the flame height and the mass loss rate (MLR): (I) quasi-steady burning stage, the flame height and the MLR nearly remain unchanged in this stage, (II) boilover stage, the flame height and MLR are obviously increased in this stage, (III) Decay stage, the flame height decreases as a third part of the quasi-steady flame height and the flame color becomes pale until fire extinguishment in this stage. And the burning process using salt additive is shown in Fig. 2 (b). Notably, boilover stage disappeared in these conditions. There are only quasi-steady burning stage and fire decay stage.

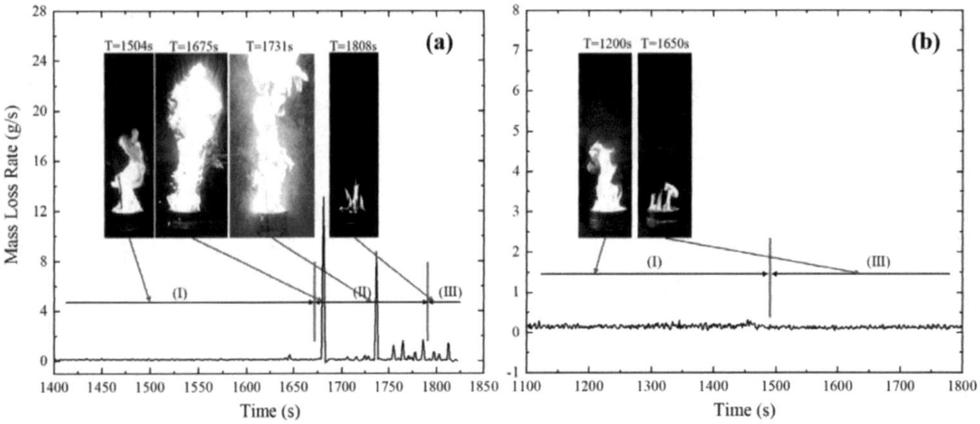

Figure 2. Burning process with 20mm initial fuel thickness versus time: (a) water sub-layer, (b) $MgCl_2$ solution sub-layer

For quantify the boilover phenomenon, the boilover intensity (I_b) is defined as the ratio of the maximum MLR at the boilover stage over the average MLR at the quasi-steady burning stage based on Koseki's work [10], and it can be expressed as:

$$I_b = MLR_{b\text{-}max} / MLR_{qs\text{-}ave} \quad (1)$$

The results of I_b are shown in Table 3. It's shown that all of the I_b with salt additive are zero, which means that the salt addictive has the effective suppression performance on the boilover.

Table 3. The boilover intensity of experiments

Test	Sub-layer	Boilover intensity
1	water	85.25
2	water	61.14
3	water	44.35
4	$MgCl_2$ solution	0
5	$MgCl_2$ solution	0
6	$MgCl_2$ solution	0

3.2. The thermal structure within the liquid pool

The interface temperature of fuel and sub-layer when burning stage just changed were shown in Fig. 3. Fig. 3 clearly shows that the interface temperature (T_{inter}) is independent of the initial fuel thickness. The T_{inter} of sub-layer with salt additive is nearly 30°C higher than the T_{inter} of water sub-layer.

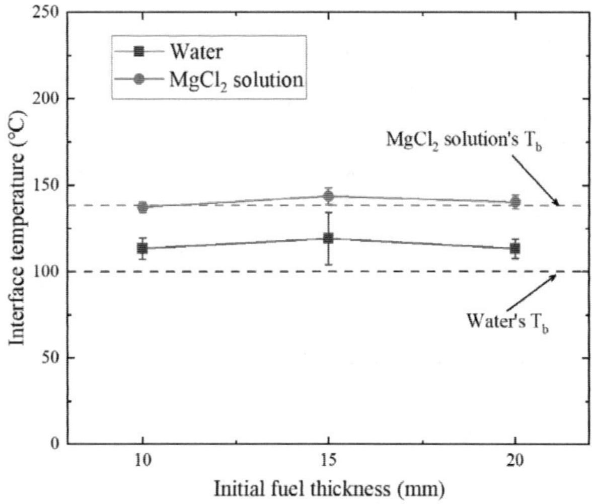

Figure 3. The interface temperature of fuel and sub-layer when burning stage just changed with different initial fuel thickness.

The thermal structure within the liquid pool needs to be analyzed to figure out the influence of salt additive on occurrence of boilover. According to the previous studies by Hamins et al. [11] and Vali's work [12], an adapted heat transfer mechanics of boilover is shown in Fig. 4. There are two layers inside the liquid pool, thus the liquid pool can be divided into three parts: boiling layer (BL), temperature gradient layer inside fuel layer (TGLF) and temperature gradient layer inside sub-layer (TGLS).

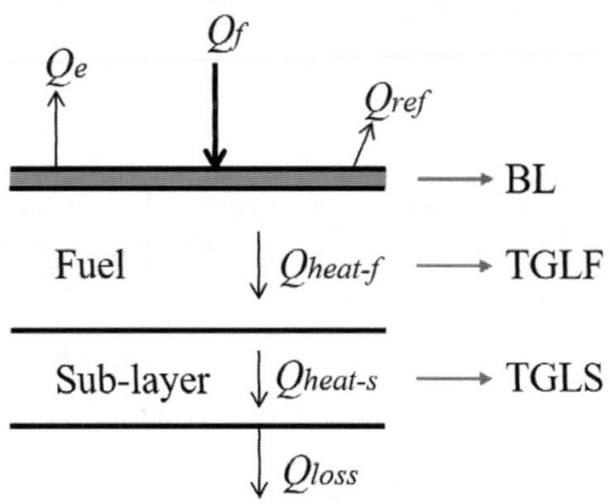

Figure 4. The thermal structure within the liquid pool.

The total heat is supplied by the heat feedback from flame, and it can be expressed as:

$$Q_f = Q_{heat\text{-}f} + Q_{heat\text{-}s} + Q_e + Q_{loss} + Q_{ref} \tag{2}$$

Where Q_f is the heat feedback from flame to the fuel surface; $Q_{heat\text{-}f}$ is the conductive heat from the BL to the TGLF; $Q_{heat\text{-}s}$ is the conductive heat from the TGLF to the TGLS; Q_{loss} is the heat loss to the environment; Q_{ref} is the heat reflection of the fuel surface; Q_e is the total net heat directly used to evaporate.

According to the previous study, in the thin-layer boilover process, Q_{ref} and Q_{loss} can be negligible, so the energy balance equation can be rewritten as:

$$Qf = Q_{heat\text{-}f} + Q_{heat\text{-}s} + Q_e \tag{3}$$

And Qheat-f and $Q_{heat\text{-}s}$ can be expressed respectively as:

$$Q_{heat\text{-}f} = \rho_f c_{pf} V_f (T_b - T_{inter}) \tag{4}$$

$$Q_{heat\text{-}s} = \rho_s c_{ps} V_s (T_{inter} - T_a) \tag{5}$$

Where ρ_f and ρ_s are the fuel and sub-layer densities, kg/m^3; c_p and c_{ps} are the fuel and sub-layer heat capacities at constant pressure, kJ/(kg K); V_f and V_s are the fuel and sub-layer volume, m^3; T_b is the boiling point of fuel, °C; T_a is the ambient temperature, °C.

Table 4. Properties of sub-layer used in the experiments.

Sub-layer	Density (10°C) (kg/m^3)	Boiling point (°C)	Specific heat capacity (kJ/(kg K))	Volume heat capacity (kJ/(m^3 K)
Water	1000	100	4.19	4190
$MgCl_2$ solution	1379	138	4.16	5736.64

According to the Fig. 3 and Table 4, both of the T_{inter} and the volume heat capacity of sub-layer with salt additive are higher than the water sub-layer's, which means that $Q_{heat\text{-}s}$ part increases based on eq. (5), and the sub-layer with salt additive absorbs more heat from the TGLF. In addition, salt additive changes the boiling point of sub-layer, which exceeds the superheat temperature (~113°C in this study) of boilover occurrence's demand. So it effectively demonstrates that salt additive can suppress the boilover.

4. Conclusion

A series of experiments were conducted to investigate the suppression mechanism of thin-layer boilover by using salt additive. The major conclusions are: (1) salt additive changes the boiling point of sub-layer, which exceeds the superheat temperature of boilover occurrence's requirement, (2) the thermal structure changes due to the change of sub-layer, so the sub-layer can absorb more heat from the TGLF. Both of the changes positively suppress the boilover.

Acknowledgments

This work was financially supported by National Science and Technology Major Project (J2019-VIII-0010-0171), Fundamental Research Funds for the Central Universities under Grant (No. WK2320000052).

References

[1] D. Kong, X. Zhao, J. Chen, H. Yang, J. Du, Study on hazard characteristics and safety distance of small-scale boilover fire, Int. J. Ther. Sci. 164 (2021).
[2] J. Zhao, J. Zhang, C. Chen, H. Huang, R. Yang, Experimental investigation on the burning behaviors of thin-layer transformer oil on a water layer, Process Saf. Environ. Protect. 139 (2020) 89-97.
[3] M. Arai, K. Saito, R.A. Altenkirch, A Study of Boilover in Liquid Pool Fires Supported on Water Part I: Effects of a Water Sublayer on Pool Fires, Combust. Sci. Technol. 71 (1990) 25-40.
[4] J.P. Garo, J.P. Vantelon, A.C. Fernandez-Pello, Boilover burning of oil spilled on water, Symp. Combust. 25 (1994) 1481-1488.
[5] J.P. Garo, H. Koseki, J.P. Vantelon, C. Fernandez-Pello, Combustion of liquid fuels floating on water, Therm. Sci. 11 (2007) 119-140.
[6] D. Kong, P. Liu, J. Zhang, M. Fan, C. Tao, Small scale experiment study on the characteristics of

boilover, J. Loss Prev. Process Ind. 48 (2017) 101-110.
[7] T.Y. Tseng, K.C. Tsai. Effect of floating perlites on the occurrence of thin-layer boilover, 11th Asia-Pacific Conference on Combustion, ASPACC 2017, 2 (2017) 1425-1428.
[8] T.Y. Tseng, K.C. Tsai, Hot-zone boilover suppression using floating objects in crude oil tank fires, Fire Saf. J. 118 (2020).
[9] P. Ma, C. Ding, X. Lin, J. Wang, Experimental research on thin-layer boilover by changing the boiling point of water layer, Fire Saf. Sci. 30 (2021) 16-23.
[10] H. Koseki, Y. Natsume, Y. Iwata, T. Takahashi, T. Hirano, A study on large-scale boilover using crude oil containing emulsified water, Fire Safety Journal 39 (2004) 143-155.
[11] A. Hamins, T. Kashiwagi, R. Burch, Characteristics of pool fire burning, Fire Resistance of Industrial Fluids, ASTM International, 1995.
[12] A. Vali, Investigation of the transport phenomena within the liquid phase of a methanol pool fire Ph.D. thesis, University of Alberta, Canada, 2014.

The experimental study of converging slot-holes (consoles) with different transition surfaces

Ying Huang[1,]*, Hao Wang[1], Can Peng[1], Jingzhou Zhang[2]

[1] Shanghai Aircraft Design and Research Institute, Shanghai, China
[2] College of Energy and Power Engineering, Nanjing University of Aeronautics and Astronautics, Nanjing, China

*Corresponding author email: huangying_0021@163.com

Abstract. In order to study the effect of consoles with different transition surfaces on film cooling performance, four kinds of consoles with different transition surfaces were designed. Under the same boundary conditions, a series of experimental studies were carried out under two typical blowing ratios. The experimental results show that whether under M=0.5 or M=1.5, compared with the reference case, the console with anti-S-shaped transition surface has the best spatially-averaged adiabatic film cooling effectiveness ($\eta_{ad,av}$), and the $\eta_{ad,av}$ is improved by 7.6% and 6.6% respectively. Meanwhile, the discharge coefficient (C_d) is decreased by 2.6% and 2.2% respectively.

Keywords: film cooling; converging slot-holes; transition surfaces; film cooling effectiveness; discharge coefficient.

1. Introduction

According to the development trend of advanced aero-engine in the future, the inlet temperature of turbine needs to be further increased. The increase of inlet temperature of turbine engine will inevitably lead to a sharp increase of heat load on the hot end parts. Film cooling has been widely used in turbine blades because of its inherent high efficiency cooling characteristics, and has become one of the main cooling measures for high temperature components of aero-gas turbine engines [1-3]. Coolant flow is introduced from the hole on the cooling wall to the primary stream. This coolant flow bends downstream under the action of primary flow pressure and friction force and adheres to a certain area of the cooling wall. The film with lower temperature is formed to isolate the cooling wall from the high temperature gas, thus playing a good cooling protection role on the wall [4].

Although a lot of research work has been done on the flow and heat transfer characteristics of film cooling, how to further improve the cooling effectiveness is still a challenging and innovative scientific problem. The conventional cylindrical hole jet enters the main stream and has strong entrainment and mixing with the main stream. The interaction between the mainstream and the film outflow will induce a variety of different scale vortices [5]. Among the different scale vortices, the largest scale vortices are the kidney-shaped vortices which rotate reversely downstream of the film holes, and play a dominant role in film cooling. A large number of studies have shown that kidney-shaped vortices easily entrap the high temperature mainstream from both sides to the wall below the cooling jet, resulting in the forced upward lifting of the cold jet and its separation from the wall, and the cooling effect is significantly reduced [6-8]. Therefore, the key point to improve film cooling effectiveness is to restrain the development of kidney-shaped vortex pairs and reduce the penetration of cooling jet to the mainstream. Changing the exit profile of the film hole is a common means to suppress the development of the kidney-shaped vortices. By modifying and optimizing the structure of the shaped hole, the strength of the kidney-shaped vortices can be reduced or the anti-kidney-shaped vortices can be induced to improve the film cooling effect [2]. Film cooling with shaped holes has become a research hotspot of many scholars.

Two dimensional continuous slot film cooling [9] is considered to be an ideal cooling scheme, which can obtain the best film cooling effectiveness. However, due to many conflicts in practical applications, it is impractical to apply two-dimensional continuous slot directly to high-temperature turbine components. Sargisson et al. [10, 11] proposed a new shaped-hole, namely converging slot-hole, which transits from the initial circular inlet to the final slotted outlet. Their experiments preliminarily show that the converging slot-holes make the adiabatic film cooling effectiveness have better uniformity. Compared with fan-shaped holes, converging slot-holes have been proved to have greater transverse average adiabatic film cooling effectiveness and lower aerodynamic loss. Further research shows that converging slot-holes change the coherent kidney vortex structure in film cooling of cylindrical holes [12-14]. Yao et al. [15] conducted a series of numerical simulations to explore the film cooling performance of a single row of converging slot-holes on the suction surface of turbine blades under the real working conditions of the engine. The results show that the film cooling effectiveness of converging slot-hole is significantly higher than that of cylindrical hole.

To our knowledge, the console has excellent film cooling performance. However, few researchers devote to reveal the effect of console transition surfaces on film cooling performance. In the current research, to study the influence of different transition surfaces of film holes on film cooling performance, five consoles with different transition surfaces are selected and designed for numerical and experimental investigation under two types of blowing ratio.

2. Research Procedures

2.1. Computational scheme

In order to further analyze the mechanism of the experimental research, the experimental consoles models with different transition surfaces were numerically simulated. The computational model, meshing, computational method and the definition of film cooling effectiveness and discharge coefficient have been introduced in detail in our previous studies [16], and will not be described in detail here. Figure 1 presents the schematic of computational model.

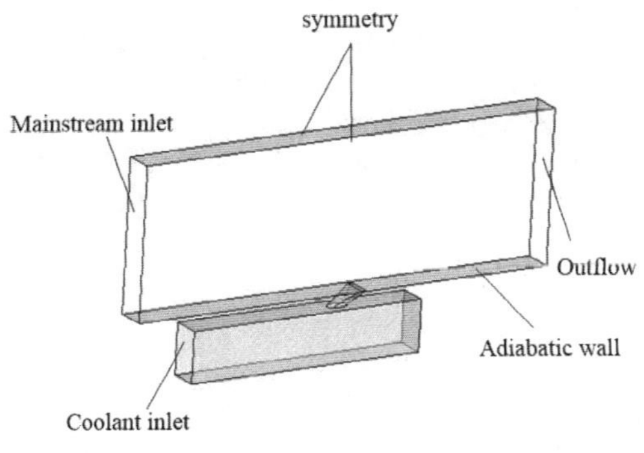

(a) Three-dimensional view

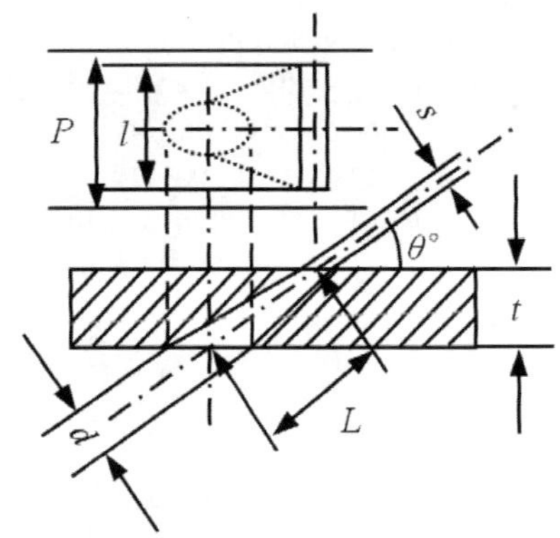

(b) Geometry of console

Figure 1. Schematic of computational model

To study the effect of different transition surfaces of the consoles on film cooling performance, four consoles with different transition surfaces are designed by changing the lateral transition surfaces, which are the outer-arc console, the inner-arc console, the s-shaped console, and the anti-s-shaped console respectively. The central sections of the reference and four designed consoles are shown in Figure 2.

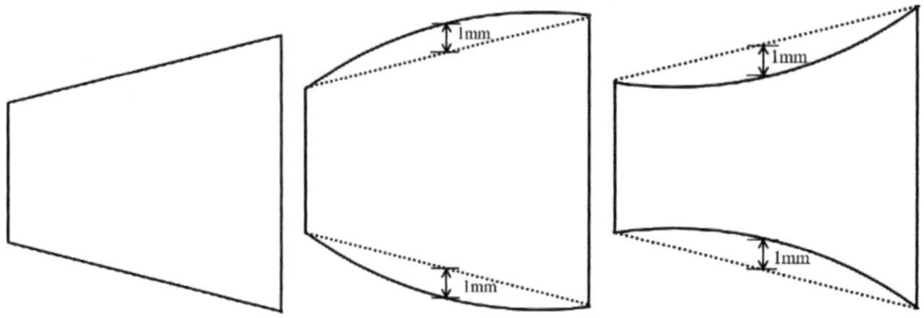

(a) Reference console (b) Outer-arc console (c) Inner-arc console

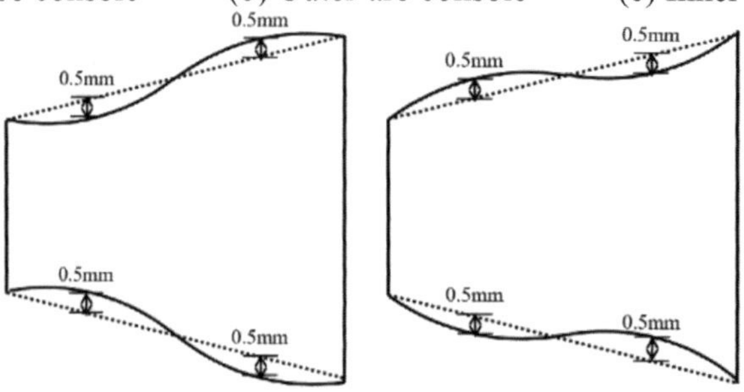

(d) S-shaped console (e) Anti-s-shaped console

Figure 2. The central sections of the reference and four designed consoles

2.2. Experimental scheme

The experiment was carried out in a low speed wind tunnel, shown in Figure 3. The main flow and secondary flow were both air. The primary flow was provided by single screw compressor. After stabled by the surge tank, the primary flow was controlled by the adjusting valve until the vortex flowmeter displayed the specified mass flow. The mainstream could flow smoothly to the test section after passing through baffles section. The secondary flow is provided by the air compressor. After stabled by the surge tank, the secondary flow could reach the target mass flow by adjusting the mass-flow controller, and finally enter the coolant plenum of the test section. In order to avoid water vapor damage to the mass-flow controller, a dryer was installed in front of the mass-flow controller. The temperature of film cooling plate was measured by infrared camera. The total pressure of film hole inlet and static pressure of film hole exit in the test section were obtained by the pressure scanning valve. The total pressure probe and static pressure probe were located 10 mm upstream of the film hole inlet and film hole outlet, respectively. The matt black paint with high emissivity was sprayed on the surface of the film cooling plate, and the emissivity of the black paint was 0.96.

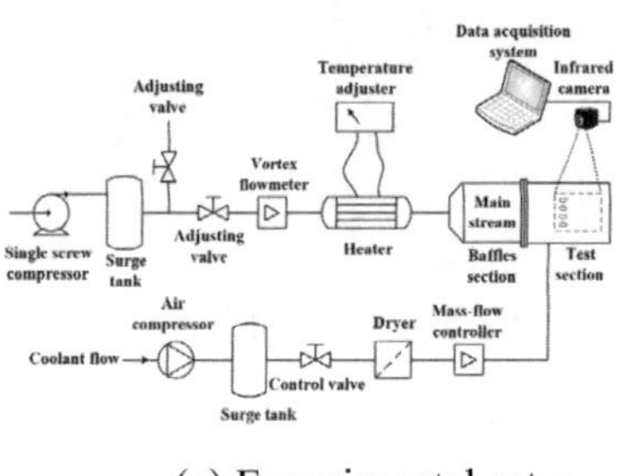

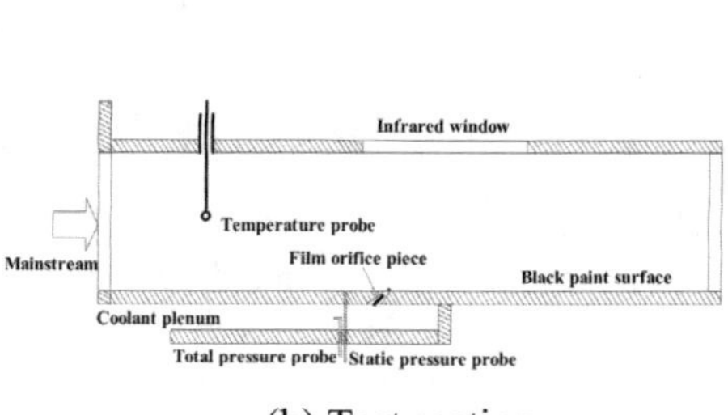

(a) Experimental setup (b) Test section

Figure 3. Experimental system diagram of film cooling

Five film cooling plates were selected to carry out the experimental research, including one reference console and four new designed variable transition surface consoles. The processing material of the film cooling plate was polymethyl methacrylate. Its thermal conductivity was 0.2 W/ (m. K), and its size was 150 mm*80 mm*6 mm.

Adjust the primary flow rate to 0.337kg/s, and the primary flow speed of the test section was 25m/s. At the same time, control the temperature adjuster to make the primary flow temperature reach 80°C. The temperature of the secondary flow was the ambient temperature, which was 30°C.

In the process of experimental measurement, the uncertainty of mainstream temperature, secondary flow temperature and wall temperature were ±0.5°C, ±0.5°C and ±1°C, respectively. The uncertainty of independent variable pressure was ± 0.05%. According to the theory of error transfer, the maximum uncertainty of film cooling efficiency measurement and discharge coefficient is ± 6% and ± 4%, respectively.

3. Results and Discussions

3.1. Flow fields and dimensionless temperature distribution

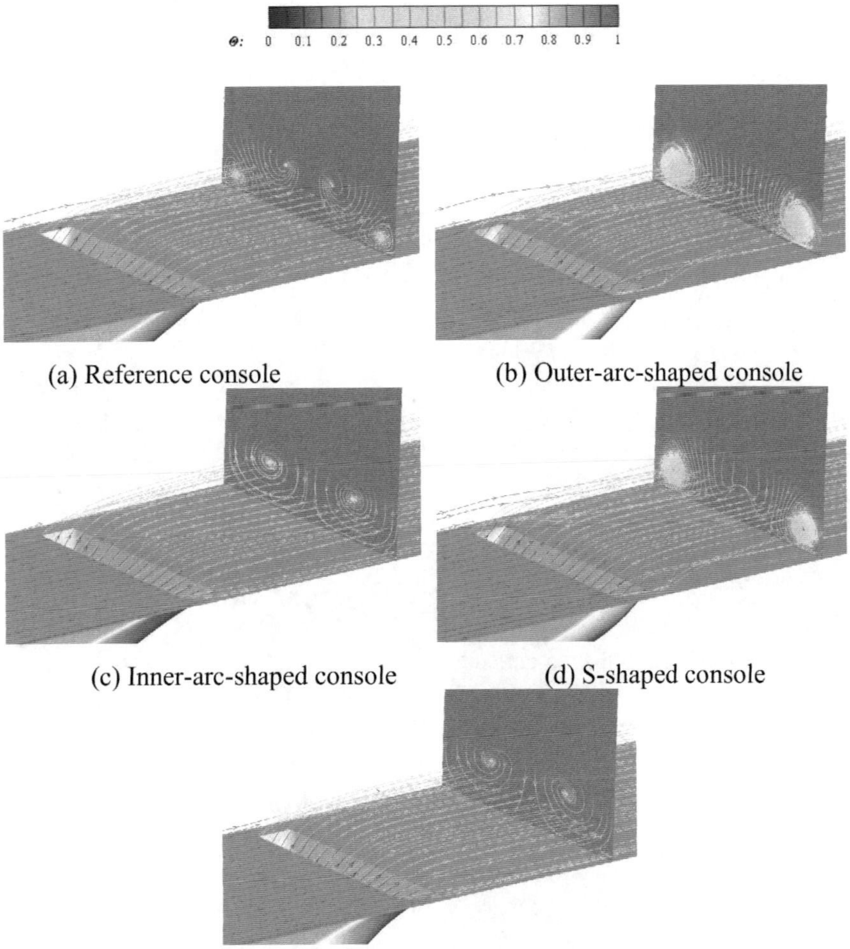

(a) Reference console (b) Outer-arc-shaped console

(c) Inner-arc-shaped console (d) S-shaped console

(e) Anti-S-shaped console

Figure 4. Streamlines in the mainstream channel and on the y-direction cross section of x/d=2 under M=0.5

Figure 4 and Figure 5 present streamlines in the mainstream channel and on the y-direction cross section of x/d=2 under M=0.5 and M=1.5. For the reference console, a pair of anti-kidney vortex is formed in the downstream of the console outlet, which is opposite to the rotating direction of the traditional cylindrical holes. This anti-kidney vortex pairs can effectively prevent the high-temperature primary flow from being sucked into the cooling wall below the cooling jet. In addition, a pair of small-scale corner vortices with the same rotation direction as the kidney vortex is formed on both sides of the outlet downstream. The formation of these unique vortex pairs is closely related to the unique structure of console. In addition to the reference film hole, only the inner-arc-shaped and anti-S-shaped consoles form a pair of anti-kidney vortices with vortex cores far away from the cooling wall surface downstream of the film hole outlet. The formation of the anti-kidney vortex is more conducive to the lateral expansion of the cooling jet. While, a pair of kidney vortices with vortex cores close to the cooling wall surface are formed in the outer-arc-shaped and S-shaped consoles downstream of the film hole outlet, which lead to some extent inhibits the lateral expansion of the cooling jet.

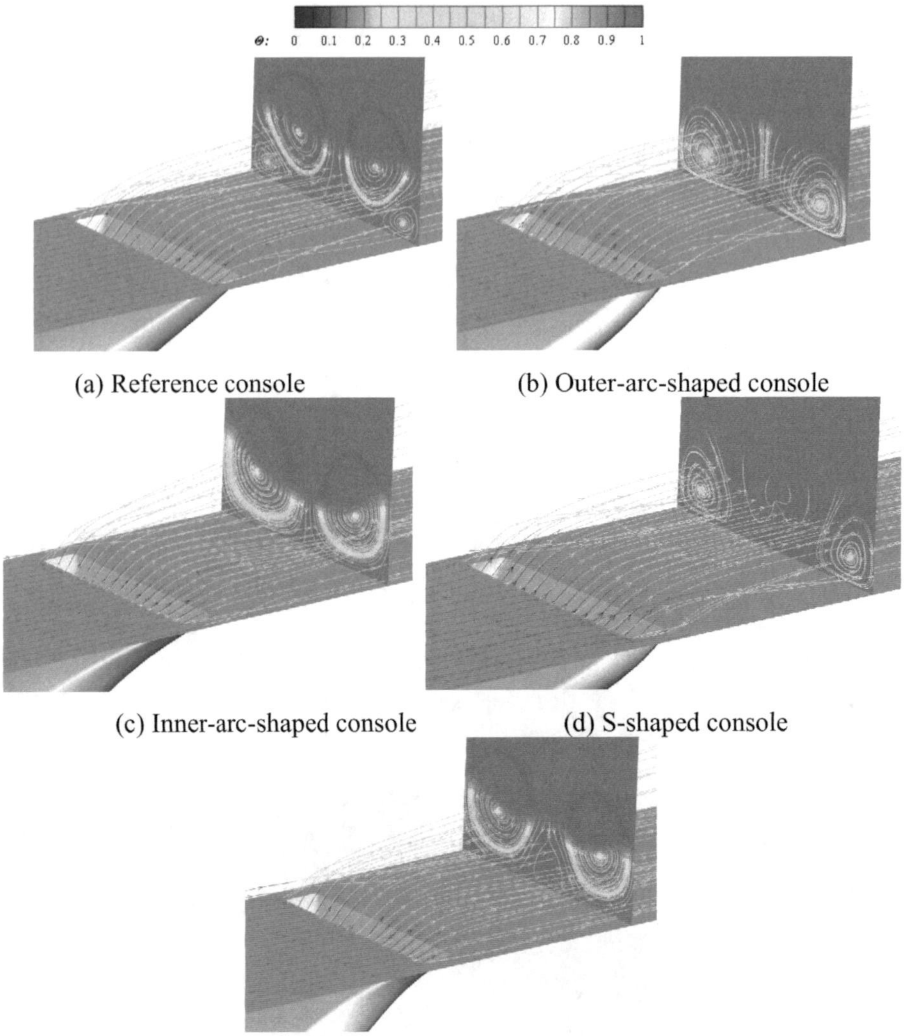

(a) Reference console (b) Outer-arc-shaped console

(c) Inner-arc-shaped console (d) S-shaped console

(e) Anti-S-shaped console

Figure 5. Streamlines in the mainstream channel and on the y-direction cross section of x/d=2 under M=1.5

3.2. Film cooling effectiveness

Figure 6 and Figure 7, respectively, present the experimental results of local η_{ad} distribution under M=0.5 and M=1.5. Under M=0.5, compared with the reference case, the local η_{ad} of inner-arc-shaped console and anti-S-shaped console is obviously improved, while the local η_{ad} of outer-arc-shaped console and S-shaped console is becoming worse. Under M=1.5, compared with the reference case, similarly, the local η_{ad} of anti-S-shaped is obviously improved, the local η_{ad} of outer-arc-shaped console and S-shaped console is becoming worse. While the local η_{ad} of inner-arc-shaped console downstream near the console outlet is improved, and the local η_{ad} becomes worse downstream far away from the film hole exit.

Table 1 presents the $\eta_{ad,av}$ of different transition surface consoles under M=0.5 and M=1.5. The results show that the transition surfaces on different film holes have a significant influence on the $\eta_{ad,av}$ downstream of the console exit between the ranges of s/d=2 and s/d=12. Under M=0.5, compared with the reference case, the $\eta_{ad,av}$ of inner-arc-shaped and anti-S-shaped consoles is improved, while that of outer-arc-shaped and S-shaped consoles is sharply deteriorated. Under M=1.5, compared with the reference case, only the $\eta_{ad,av}$ of anti-S-shaped console is improved, while that of other designed consoles is sharply deteriorated. Under M=0.5, the $\eta_{ad,av}$ of inner-arc-shaped and anti-S-shaped consoles is improved by 6.9% and 7.6%, respectively, while that of outer-arc-shaped and S-shaped consoles is deteriorated by 14.9% and 8.3%, respectively. Under M=1.5, the $\eta_{ad,av}$ of anti-S-shaped console is improved by 6.6%, while that of outer-arc-shaped, inner-arc-shaped, and S-shaped consoles is deteriorated by 34.5%, 5.1% and 15.7%, respectively.

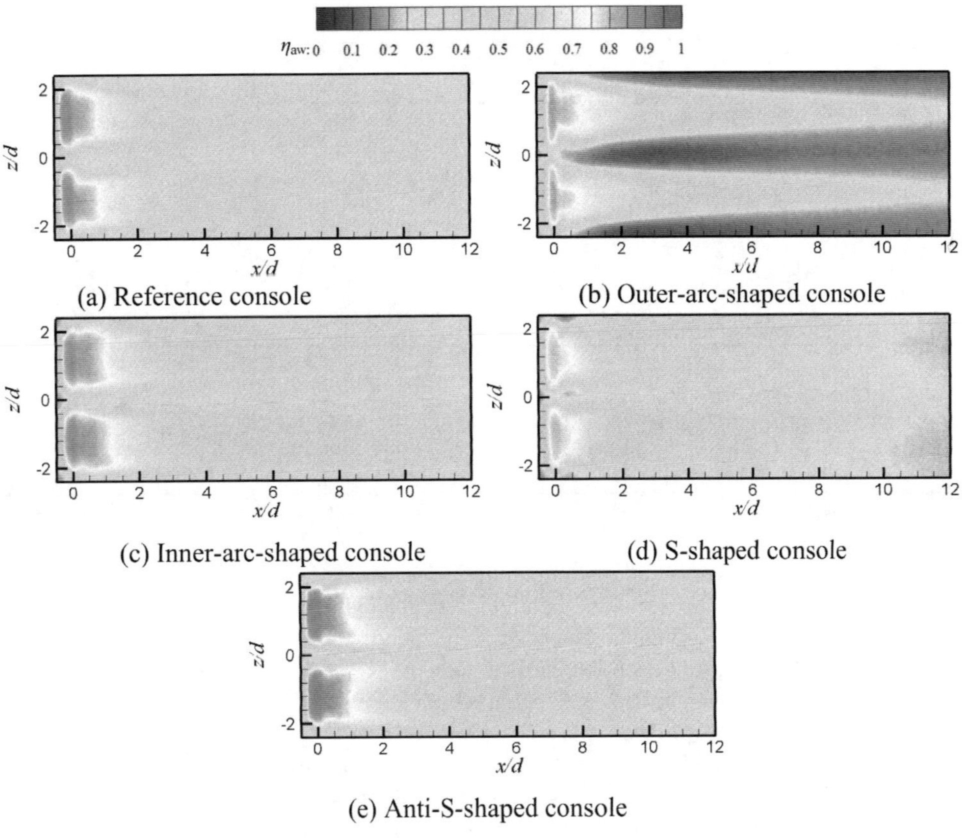

Figure 6. Distribution of local η_{ad} under M=0.5 (Exp)

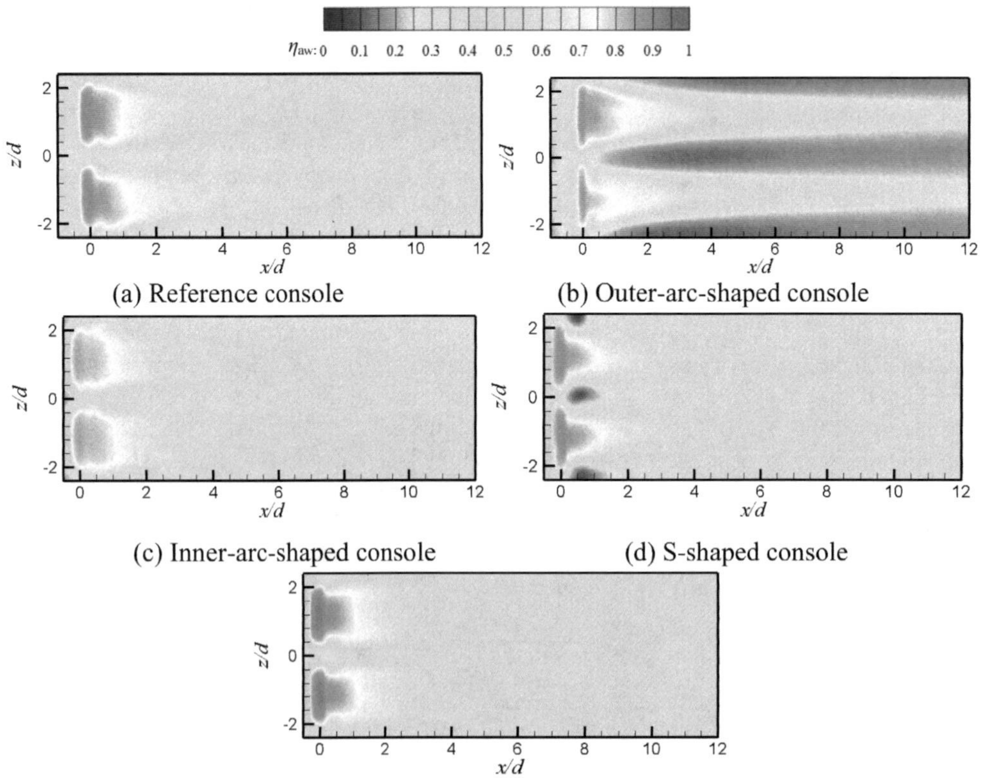

(a) Reference console

(b) Outer-arc-shaped console

(c) Inner-arc-shaped console

(d) S-shaped console

(e) Anti-S-shaped console

Figure 7. Distribution of local η_{ad} under M=1.5 (Exp)

Table 1. The $\eta_{ad,av}$ of different transition surface consoles under M=0.5 and M=1.5

Different transition surface consoles	$\eta_{ad,av}$ (M=0.5)	$\eta_{ad,av}$ (M=1.5)
Reference	0.436	0.594
Outer-arc-shaped	0.371	0.389
Inner-arc-shaped	0.466	0.564
S-shaped	0.400	0.501
Anti-S-shaped	0.469	0.633

3.3. Discharge coefficient

Figure 8 presents the comparison of experimental and computed C_d. Compared with the reference console, the inner-arc-shaped and anti-S-shaped consoles need larger bleeding pressure to overcome the pressure drop inside film holes, while the outer-arc-shaped and S-shaped consoles need smaller bleeding pressure to overcome the pressure drop inside film holes under M=0.5 and M=1.5. Compared with the reference console, the outer-arc-shaped and S-shaped consoles have larger discharge coefficients, i.e. smaller aerodynamic losses. Their discharge coefficients respectively increase by 5.8% and 2.9% under M=0.5 and increase by 4.9% and 2.7% under M=1.5. The inner-arc-shaped and anti-S-shaped consoles have smaller discharge coefficients, i.e., more greater aerodynamic losses. Their discharge coefficients respectively decrease by 4.7% and 2.6% under M=0.5 and decrease by 4.6% and 2.2% under M=1.5.

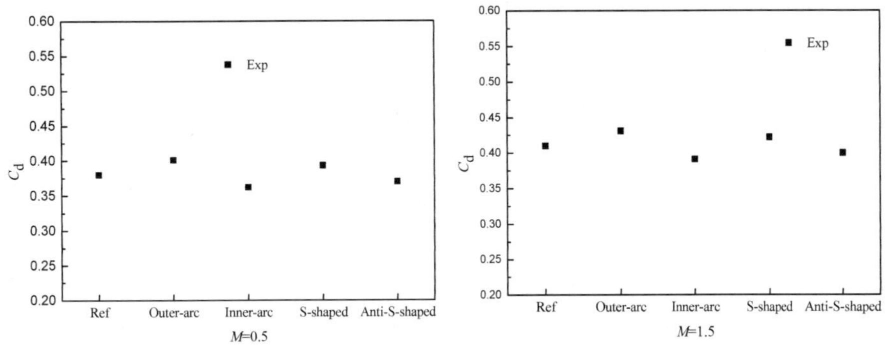

(a) M=0.5 (b) M=1.5

Figure 8. Comparison of experimental and computed C_d

4. Conclusions

To study the influence of transition surfaces for the console on film cooling performance, except the transition surface of reference console, four kinds of transition surfaces were designed, including outer-arc-shaped, inner-arc-shaped, S-shaped, and anti-S-shaped transition surfaces. A series of experiments were carried out under two typical blowing ratios of M=0.5 and M=1.5. The main conclusions are as follows:

1. The anti-S-shaped console has the best $\eta_{ad,av}$ under two typical blowing ratios. Compared with the reference console, the $\eta_{ad,av}$ is improved by 7.6% and 6.6%, respectively, while the discharge coefficient is decreased by 2.6% and 2.2%, respectively, under two typical blowing ratios.

2. The outer-arc-shaped console has the worst $\eta_{ad,av}$ under two typical blowing ratios. Compared with the reference console, the $\eta_{ad,av}$ is decreased by 14.9% and 34.5%, respectively. Simultaneously, it has the best dynamic performance, and the discharge coefficient is increased by 5.8% and 4.9%, respectively, under two typical blowing ratios.

3. Compared with the reference case, the inner arc-shaped console has a better $\eta_{ad,av}$ and its performance is improved by 6.9% under M=0.5. While it has a worse $\eta_{ad,av}$ and its performance is decreased by 5.1% under M=1.5. It also has the worst dynamic performance, and the discharge coefficient is decreased by 4.7% and 4.6%, respectively, under two typical blowing ratios.

4. Compared with the reference case, the S-shaped console has a worse $\eta_{ad,av}$ under two typical blowing ratios, and the $\eta_{ad,av}$ is decreased by 8.3% and 15.7%, respectively. While it has better dynamic performance and its discharge coefficient is increased by 2.9% and 2.7%, respectively, under two typical blowing ratios.

References

[1] Bogard D G, Thole K A. Gas turbine film cooling [J]. Journal of propulsion and power, 2006, 22(2): 249-270.
[2] Bunker R S. A review of shaped hole turbine film-cooling technology [J]. Journal of heat transfer, 2005, 127(4): 441-453.
[3] Wen Z X, Liang J W, Liu C Y, et al. Prediction method for creep life of thin-wall specimen with film cooling holes in Ni-based single-crystal superalloy [J]. International Journal of Mechanical Sciences, 2018, 141: 276-289.
[4] Wang C H, Sun X K, Zhang J Z. Uncertainty analysis of trench film cooling on flat plate [J]. Applied Thermal Engineering, 2019, 156: 562-575.
[5] Fric TF, Roshko A. Vortical structure in the wake of a transverse jet [J]. Journal of Fluid Mechanics, 1994, 279: 1-47.
[6] Burdet A, Abhari R S, Rose M G. Modeling of Film Cooling—Part II: Model for Use in Three-Dimensional Computational Fluid Dynamics [J]. Journal of turbomachinery, 2007, 129(2),

221-231.
[7] Wang C, Zhang J, Feng H, et al. Large eddy simulation of film cooling flow from a fanshaped hole [J]. Applied Thermal Engineering, 2018, 129: 855-870.
[8] Wang C, Fan F, Zhang J, Huang Y, et al. Large eddy simulation of film cooling flow from converging slot-holes [J]. International Journal of Thermal Sciences, 2018, 126: 238-251.
[9] Bunker R S. A review of shaped hole turbine film-cooling technology [J]. ASME Journal of Heat Transfer, 2005, 127: 441-453.
[10] Sargison J E, Guo S M, Oldfield M L, et al. A converging slot-hole film-cooling geometry-part 1: low-speed flat-plate heat transfer and loss [J]. Journal of Turbomachinery, 2002, 124: 453-460.
[11] Sargison J E, Guo S M, Oldfield M L, et al. A converging slot-hole film-cooling geometry: part 2: transonic nozzle guide vane heat transfer and loss [J]. ASME Journal of Turbomachinery, 2002, 124: 461-471.
[12] Azzi A, Jubran B A. Numerical modelling of film cooling from converging slot-hole [J]. Heat and Mass Transfer, 2006, 43: 381-388.
[13] Yao Y, Zhang J Z. Investigation on film cooling characteristics from a row of converging slot-holes on flat plate [J]. Science China-Technological Sciences, 2011, 54: 1793-1800.
[14] Yao Y, Zhang J Z, Wang L P, et al. Film cooling on a gas turbine blade suction side with converging slot-hole [J]. International Journal of Thermal Sciences, 2013, 65: 267-279.
[15] Yao Y, Zhang J Z, Tan X M, et al. Numerical study of film cooling from converging slot-hole on a gas turbine blade suction side [J]. International Communications in Heat and Mass Transfer, 2014, 52: 61-72.
[16] Huang Y, Zhang J Z, Wang C H. Shape-optimization of round-to-slot holes for improving film cooling effectiveness on a flat surface [J]. Heat and Mass Transfer, 2018, 54(6): 1741-1754.

Influence of front wheel air dam and diversion channel on heat dissipation performance of brake disc

Qingyun Guan*, Yang Zhou, Yizheng Liang, Ling Xiao, Xiayi Yuan

Guangzhou Automobile Group Co, Ltd Automotive Engineering Institute, Guangzhou, China

*Corresponding author e-mail: guanqingyun@gacrnd.com

Abstract. In order to improve the heat dissipation performance of the brake disc and reduce the air drag coefficient of the vehicle, the influence mechanism of the front wheel air dam and brake cooling guide structure is studied. Firstly, according to the flow field around the front wheel, six feasible air dam and guide flow optimization solutions are designed; then, CFD simulation method is used to simulate the flow rate, heat transfer coefficient and cooling time; secondly, for the air drag coefficient and brake cooling time of the whole vehicle, aerodynamic-acoustic wind tunnel and road test are used to test and verify respectively; finally, we get the solution of small front wheel air dam and open a guide flow channel around the plate (solution 4) ,the air drag coefficient is only increased by 3ct, and the brake cooling performance is much better.

1. Introduction

Automobile braking is the process of converting the kinetic energy of the car movement into heat energy through the friction of the brake and exhausting the heat to the atmosphere. A large amount of heat is generated during the braking process, especially when braking down long slopes or continuously, the temperature of the brake disc will rise continuously, up to 500-600°C, which can easily lead to the occurrence of heat recession of the brake disc [1-2]. Due to the thermal recession of the brake disc or high temperature damage to the brake components, resulting in brake failure, resulting in countless traffic accidents [3-5].

High-temperature brake disc cooling process mainly through convection and radiation heat dissipation, and more than 90% of the heat through convective heat transfer away [6,7], and convective heat transfer is inevitably inseparable from the flow field around the brake disc. The current domestic brake heat dissipation performance research mainly focuses on the temperature field finite element analysis, and relatively little CFD method to analyze the impact of the flow field around the brake disc on heat dissipation. Liu Changhong et al [8] pointed out that based on the CFD analysis results of different working conditions, a reasonably designed ventilation orifice can guide the high-speed airflow through the brake disc, thus taking away the heat from the brake disc. Ding Qun et al [9] and Lu Kelong [10] simulated and analyzed the air flow field characteristics inside and around the brake disc, and optimized the shape design of the brake disc to improve the heat dissipation performance. Huang Huan [11] simulated and analyzed the flow field around the brake disc and the brake disc cover, and the heat dissipation performance was significantly improved by improving the brake disc cover to increase the air flow through the brake disc.

In summary, it can be found that improving the flow field around the brake disc is crucial to its heat dissipation performance. In this paper, the effect of different schemes on the flow rate, heat transfer coefficient and cooling time is evaluated by simulation means through the method of opening the flow channel on the wheel cover and combining with the air dam structure optimization scheme to explore the flow field distribution law around the brake disc and the direction of air dam design in high-speed motion.

2. Analysis of the factors affecting the heat dissipation performance of brake discs

The heat dissipation of the brake disc mainly depends on the convective heat exchange between the surface of the brake disc and the air, according to the Newtonian cooling formula.

$$Q=hA\Delta T \tag{1}$$

Where Q is the heat transfer power; h is the convective heat transfer coefficient between fluid and solid, the unit is W/(m2k); A is the surface area of the brake disc; ΔT is the temperature difference between fluid and solid surface. Set the starting temperature of the brake disc temperature is 500 ℃, the air temperature is 20 ℃, the study of the cooling time of the brake disc temperature from 500 ℃ down to 100 ℃, through the equation (1) can be seen in the cooling process △ T the same, while in the brake disc structure remains unchanged, the surface area of the brake disc A is also the same, at this time the main factor affecting the heat dissipation performance of the brake disc is the convective heat transfer coefficient h. Research shows that h is mainly depends on the airflow around the brake disc, and improving the flow rate, increasing the turbulence intensity and improving the geometry of the contact body can improve the convective heat transfer coefficient of the brake disc [12].

3. Analysis of the flow field around the brake disc

During the driving process of the car, a part of the airflow flows into the bottom of the chassis, and after flowing through the guard, it flows into the complex wheel cavity, forming a larger vortex, as shown in Fig. 1. The airflow formed around the brake disc not only has an impact on the heat dissipation performance of the brake disc, but also has a great impact on the wind resistance. There have been tests show that when a car to 80km / h forward, 60% of the fuel consumption is used to overcome the wind resistance. In recent years, the shortage of oil resources and rising prices to the transportation industry has caused tremendous pressure to improve vehicle fuel economy has become extremely severe, and reducing wind resistance is one of the important ways to improve fuel economy. One of the effective measures to reduce wind resistance is to install the front wheel air dam on the wheel cover, which can reduce the airflow into the wheel cavity and reduce the flow rate of airflow around the brake disc, but this is very detrimental to the heat dissipation performance of the brake disc. How to balance the relationship between the two, on the basis of improving the brake disc cooling performance, while keeping the wind resistance basically unchanged, is the focus of this paper. Based on the above considerations, different front wheel air dams and deflector channels were designed to verify their effects on brake disc cooling performance and wind resistance, which are shown in Table 1 below. Option 2~5 are the design optimization options, and Option 6 is the front wheel air dam option without considering the wind resistance design.

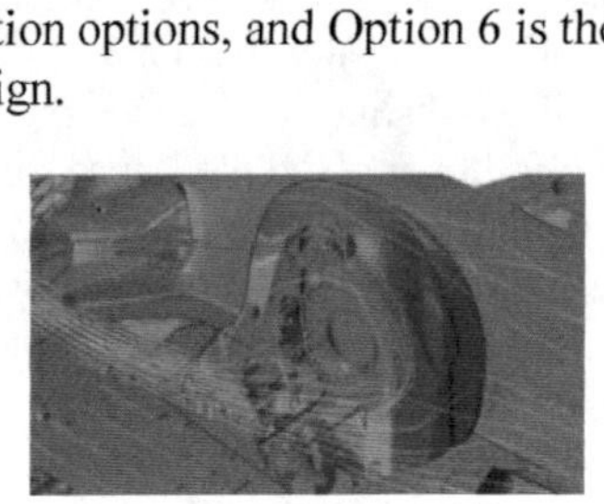

Figure 1. Schematic diagram of the flow field in the wheel cavity

Table 1. Front wheel air dam and guide channel solution

Scheme	Illustration	Notes
Scheme 1		Front wheel air dam solution without considering the heat dissipation performance of the brake disc
Scheme 2		Front wheel air dam and deflector channel designed by combining brake disc cooling performance and wind resistance
Scheme 3		
Scheme 4		
Scheme 5		
Scheme 6		Without wind resistance design without front wheel air dam solution

4. Modeling and numerical calculation of the whole vehicle and brake disc

4.1. Vehicle and brake disc model

A SUV is selected as the object of study, and the model of the whole vehicle contains the real cabin and lower body, the schematic diagram of the whole vehicle model is shown in Fig. 2. The size of the whole vehicle is 2900mm×4950mm×1940mm. There are 43 ventilation channels in the front brake disc, and the ventilation channels are constructed by the straight square ribbed plate, the 3D model of the brake disc and the section of the ventilation channels are shown in Fig. 3.

Figure 2. Whole vehicle model

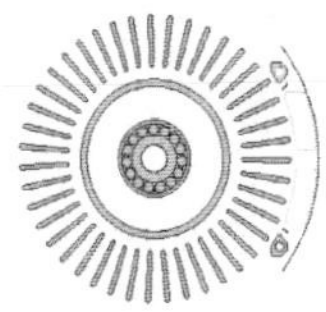

Figure 3. Sectional view of brake disc

4.2. Model meshing

The calculation domain is a rectangular body, with the body as the reference, the entrance X coordinate is -5m; the exit X coordinate is 15m; the left and right wall Y coordinates are -7m and 7m respectively; the height coordinate is the horizon translation 10m. The brake disc and caliper grid size is 2mm, the rim grid size is 8mm, the wheel cavity and the front wheel air dam are set around 4mm encrypted body, the whole calculation domain body grid is cut body grid, the total number of grid is about 66 million.

4.3. Calculation of boundary conditions

4.3.1. Definition of physical parameters. The air in the calculation domain is set as an ideal incompressible gas, the brake disc material is cast iron, the density is 7200kg/m^3, the specific heat capacity is 470J/(kg • K), and the thermal conductivity is 36W/(m • K).

4.3.2. Definition of motion area. In the braking process, the wheel and the brake disc are rotating, and the ventilation channel of the brake disc and the wheel spokes are equivalent to the impeller, which makes a large amount of air pass through and take away heat [13], setting the calculation domain within the ventilation channel of each brake disc and the cavity area surrounded by each wheel spoke as the MRF area; each wall surface on the wheel and each surface of the brake disc adopt the rotating wall method, and the rotation speed of both is the same. The ground is set to move in the X-direction with the same speed as the forward speed of the car.

4.3.3. Simulation scheme boundary parameters. Define the initial temperature of the brake disc 500 ℃, the air temperature is 20 ℃, the speed of the car is 100km / h, calculate the time required to cool the temperature of the brake disc from 500 ℃ to 100 ℃.

4.4. Calculation method

The process of brake disc cooling involves heat transfer between fluids, heat transfer within a solid, and heat transfer between a fluid and a solid. The coupled heat transfer between fluid and adjacent solid becomes conjugate heat transfer, and for coupled heat transfer analysis, the energy equation is solved for effective implicit thermal coupling at the intersection of fluid and solid.

The brake disc heat dissipation is calculated using the fluid-solid coupling method, where the convective heat transfer coefficient and fluid temperature are solved at the velocity and solid surface temperature. This is then transferred to the heat transfer solver, where the material properties are added to solve for the solid surface temperature, and then cycled again, solving for one transient temperature of the brake disc per cycle.

5. Numerical calculation results and analysis

5.1. Analysis of the flow field around the brake disc for each scheme

The air flow rate is one of the key indicators to evaluate the heat dissipation performance. The air mass flow rate from the inner side of the wheel cavity to the outer side of the car for each scheme is shown in Fig.4. From the figure, it can be seen that the front wheel air dam without inflow channel in scheme 1 has the least air mass flow rate from the inner side of the wheel cavity to the outer side, and scheme 6 without front wheel air dam has the most air mass flow rate from the inner side of the wheel cavity to the outer side. Scenarios 2-5, where the front wheel air dam is installed and the inflow channel is started at the same time, intervene between the first two in terms of air mass flow from the inside of the wheel cavity to the outside. The air flow from the wheel cavity was 0.44 kg/s, which was about 33% higher than that of the basic scheme 1.

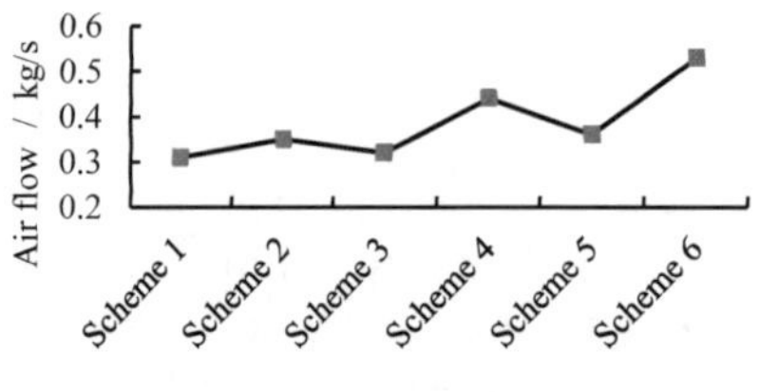

Figure 4. Air flow rate out of the wheel cavity

5.2. Convective heat transfer coefficient of brake disc for each scheme

The convective heat transfer coefficient is the most direct parameter to evaluate the heat dissipation performance, which characterizes the size of the heat transfer capacity of the brake disc. The convective heat transfer coefficients on the surface of the brake disc and in the ventilation channel for each scheme are shown in Fig. 5. It can be seen that the convective heat transfer coefficient increases when the air flow into the wheel cavity increases. The convective heat transfer coefficient of the front wheel air dam without the deflector channel in scheme 1 is the smallest, and the convective heat transfer coefficient of scheme 6 without the front wheel air dam is the largest. It can be seen from Fig. 5 that the trend of convective heat transfer coefficient is basically the same as the trend of air flow, which is due to the positive correlation between heat transfer and flow.

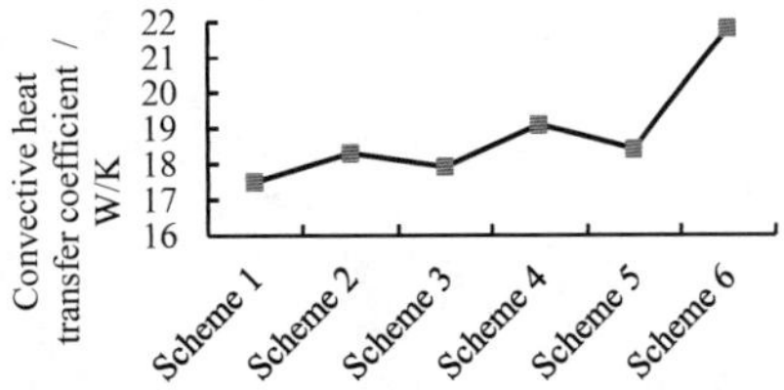

Figure 5. Convection heat transfer coefficient of each solution

5.3. Analysis of braking cooling time for each scheme

Combined with the analysis results of 5.1 and 5.2, it can be seen that under the same heat dissipation, when the air flow into the wheel cavity increases, the convective heat transfer coefficient increases, and the brake disc cooling time is the shortest. Fig. 6 shows the brake cooling process, the brake disc temperature change curve with time, Table 2 shows the time used to reduce the brake disc from 500 ℃ to 100 ℃, consistent with the previous conclusions, option 6 without front wheel air dam when the brake cooling time is the shortest, single to heat dissipation performance as the evaluation goal, option 6 is undoubtedly the best choice, but this paper is to wind resistance and brake disc cooling performance as a comprehensive goal assessment, so the final program The choice of the final solution, but also need to study the impact of each program on wind resistance.

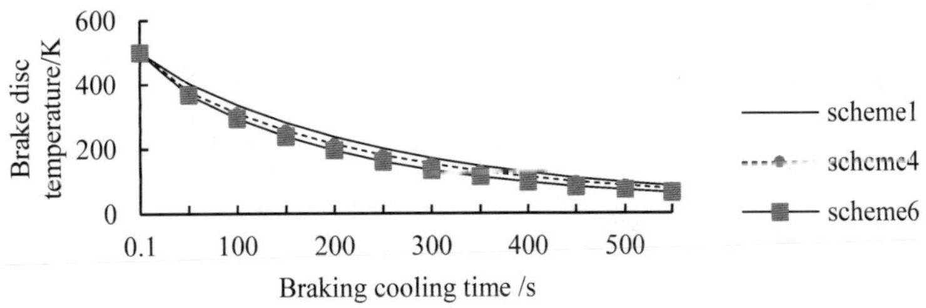

Figure 6. Variation curve of brake disc temperature with time

Table 2. Simulation value of brake cooling time of each solution

Scheme	Brake cooling time /s
Scheme 1	483
Scheme 2	462
Scheme 3	471
Scheme 4	442
Scheme 5	459
Scheme 6	387

6. Analysis of test results

6.1. The brake disc cooling time of each scheme

This experiment was tested on the runway high loop section, and the brake disc temperature measurement was done with a contact thermocouple, as shown in Fig.7. The test method is to carry out at least 11 consecutive braking, the brake disc temperature heated to more than 550 ℃, and then driving at a constant speed of 100km / h until the brake disc temperature drops below 100 ℃, monitor the brake disc from 500 ℃ → 100 ℃ used time t.

Figure 7. Schematic diagram of contact thermocouple

The test results are shown in Table 3 below, from the table we can see that: the shortest brake cooling time for scheme 6 without front wheel air dam, the second shortest brake cooling time for scheme 4 with small front wheel air dam and brake inflow channel around the guard, which are basically consistent with the simulation conclusions; the longest brake cooling time for scheme 3 in the test, which is somewhat different from the simulation, the analysis may be because the test is outdoors, the test process will be affected by wind speed However, overall, the error between the simulation and the test is within 10%, which meets the expected target, so the brake disc cooling performance can be evaluated by simulation.

Table 3. Comparison of brake cooling time test simulation of each solution

Scheme	Test/s	Simulation /s	Error
Scheme 1	508	483	4.9%
Scheme 2	508	462	9.1%
Scheme 3	513	471	8.2%
Scheme 4	456	442	3.1%
Scheme 5	492	459	6.7%
Scheme 6	416	387	7.0%

6.2. Wind resistance coefficient of each scheme

The wind resistance test of each scheme was conducted in the wind tunnel laboratory of Tongji University, using the front wheel air dam of scheme 1 without inflow channel as the reference benchmark, and the wind resistance change of the remaining schemes relative to the benchmark, as shown in Table 4. the test results show that: the wind resistance of the front wheel air dam without inflow channel is optimal; scheme 6 without front wheel air dam is the worst for wind resistance; the other schemes with inflow channel, relative to base, the wind resistance deteriorates about 1%.

Table 4. Air drag coefficient of each solution

Scheme	Effect on wind resistance
Scheme 1 (base)	-
Scheme 2	+4ct
Scheme 3	+2ct
Scheme 4	+3ct
Scheme 5	+3ct
Scheme 6	+12ct

When the wind resistance and brake disc heat dissipation performance as a comprehensive target assessment, neither the front wheel air dam without deflector channel with optimal wind resistance (Scheme 1) nor the scheme without air dam with the shortest braking time (Scheme 6) can be selected, while the small front wheel air dam and the opening of a deflector channel around the guard (Scheme 4) is the scheme with relatively short braking time and less wind resistance deterioration, and after comprehensive consideration, Scheme 4 is finally selected.

7. Conclusion

1) Under the same heat dissipation, when the air flow into the wheel cavity increases, the convective heat transfer coefficient increases and the brake disc cooling time is the shortest, but the larger flow rate will inevitably have a greater impact on the wind resistance.

2) The results of the analysis using CFD simulation are within 10% of the experimental error, so it is reasonable to evaluate the heat dissipation performance of the brake disc by CFD method.

3) The wind resistance of the front wheel air dam without inflow channel is optimal, and the braking cooling time of the scheme without air dam is the shortest, so the small front wheel air dam and the opening of the inflow channel around the guard is the best choice when the wind resistance and brake disc cooling performance are evaluated as a comprehensive goal.

References

[1] Chen Zhaohui. Effect of Automobile Brake on Safety Driving [J]. Engineering of Tropical Agriculture, 2003(2):31-32.
[2] Yang Zhigang, Zou Yuqi. Analysis of cooling performance of brake disc on FLUNT and Workbench [J]. Jiamusi Univ: Nat Sci,, 2013, 31(2): 161-164
[3] Jin Xiaoxing. Research of Temperature Field for Disc Brake [D]. Hefei:Hefei University of Technolgy, 2007.
[4] Ding Qun, Xie Jilong. The Temperature Field and Stress Field Calculation of Brake Disc Based on 3-Dimension Model [J]. Journal of the China Railway Society,2002,24(6):34-38.
[5] Liu Ying,Hu Yuyong,Ren Qifeng,et al. Simulation of the Temperature Field and Thermal Stress for Automotive Disc Brake Based on ANSYS [J]. Nanchang University Journal (Engineering), 2013, 35(3): 276-280, 290.
[6] Wang H, Al-Qadi I L. The combined effect of moving wheel loading and three-dimensional contact stresses on perpetual pavement responses [C] Transportation Reaserch Board 2009 Annual Meeting. Washington D C: TRB, 2009:1-14.
[7] Du Xuzhi, Yang Zhigang, Li Qiliang, et al. Brake disc cooling characteristics of a passenger car[J]. Journal of Tongji University (Natural Science), 2016,44(5):787-793.
[8] Liu Changhong, Hou Sichao. Thermal analysis in car disk brake [J]. Tractor & Farm Transporter, 2008,35(1):67-69.
[9] Ding Qun, Xie Jilong. The temperature field and stress field calculation of brake disc based on 3-dimension model [J]. Journal of The China Railway Society, 2002, 24(6): 35-37.
[10] Lu Kelong. Numerical Calculation and Optimization of the Automobile Brake Disc Heat Dissipation Based on CFD [D]. Hunan: Hunan University, 2011:5-19.
[11] Huang huan. The influence of the structure of the brake disc cover on the heat dissipation of the brake disc. Scientific and technological information, 2009, (34): 311-312.
[12] Wu Jiawei, Yang Zhigang. Influence of Airflow Direction on Cooling Performance of Vented Brake Discs [J]. Chinese Journal of Automotive Engineering, 2014, 4(6): 1-6.
[13] Yu wuxian. The simulation and analysis of the transient temperature field of automotive disk brake [D]. Hefei: Hefei University of Technology, 2009.

Study on boiling heat transfer mechanism based on microlayer evaporation theory

Wu-han Dong, Ming Gao *, Zhong-xiang Shen and Li-xin Zhang

Shanghai Key Laboratory of Multiphase Flow and Heat Transfer in Power Engineering, School of Energy and Power Engineering, University of Shanghai for Science and Technology, Shanghai 200093, China

*Corresponding author e-mail: gaoming@usst.edu.cn

Abstract. Based on dynamic microlayer evaporation model. Heat transfer mechanism of nuclear boiling and microlayer thickness under single bubble on heated surface. It has a lot to do with the temperature of the wall. Therefore, this paper studies the mechanism of bubble microlayer and makes comparative analysis with its experiments. It is helpful to master nuclear boiling heat transfer more comprehensively. In the study of individual bubble microlayer thickness. The effects of time and radius on wall temperature are considered in this paper. The theoretical formula of initial thickness of microlayer is deduced. The theoretical formula is compared with the experimental data of microlayer evaporation under the nucleation bubble. It was found that the experimental data fell within ±25% of the formula range. It is consistent with the previous experimental research.

Keywords: Bubble; Microlayer; Initial microlayer thickness; Evaporation

1. Introduction

As an important and efficient heat transfer method in energy conversion and transfer, nuclear boiling has been widely used in life and industry. Boiling heat transfer has such a high heat transfer intensity mainly due to the formation of bubbles, growth and separation from the heating wall caused by various disturbances.

Between the expanding bubble and the heat transfer surface there is a microlayer with a thickness of several microns (Fig. 2 (b)). Microlayer play an important role in the heat transfer of nuclear boiling because the evaporation process of microlayer can remove a large amount of heat from the heated surface. In the past few decades, a large number of studies have been carried out. Snyder and Edwards[1] modeled the growth of bubble in uniformly superheated liquid and first proposed the existence of microlayer. It is pointed out that the micro-evaporation of the microlayer provides a lot of energy for the growth of the bubble, and the thermal resistance is very small. During the growth of the bubble, the vapor in the bubble is mainly generated by its evaporation. Moore and Mesler[2] indirectly confirmed the existence of microlayer in the boiling process for the first time by measuring the unsteady temperature change of the heat transfer surface, and measured the temperature change of the heat transfer surface by using a specific micro thermocouple. Cooper and Lloyd[3] were the first to prove the existence of the microlayer at the bottom of the bubble in the experiment and put forward a theoretical model for the change of the microlayer thickness. It was found that the growth rate of the bubble showed the same order of magnitude as the evaporation rate of microlayer. The shape of the bubble and the formation of microlayer were discussed. MacGregor and Jawurek[4] first measured the microlayer thickness with the interference method, and also observed the ethanol bubble microlayer with the interference method, and analyzed the error. Koffman and Preesset[5] experimentally studied the formation process of microlayer at the vaporization core of water and ethanol and the change of microlayer thickness with time. Voutsinos and Judd[6] mainly studied the contribution of microlayer evaporative heat transfer and the thickness of microlayer in boiling heat transfer. Utaka et al. [7] used laser extinction method to study the initial microlayer thickness in water and ethanol. Zhao Yaohua et al. [8] proposed a microlayer model considering both microlayer and thick layer to predict the critical heat flux of subcooled boiling.

Most of the previous studies on the mechanism of microlayer evaporation are mainly experiments and simulations. The theoretical study on the mechanism of microlayer evaporation is still not perfect. In order to reduce the complexity of the model, most researchers consider the linear increase of wall temperature with time. The influence of wall temperature with time and radius is not considered. But nucleated boiling heat flow is mainly caused by evaporation of microlayer. The initial thickness of the microlayer decreases rapidly with the increase of overheating. The influence of time and radius on wall temperature is particularly important. Therefore, this paper considers the influence of wall temperature with time and radius distance, and makes a more detailed work on this problem.

2. Evaporation model

2.1. Dynamic microlayer evaporation model

In the microlayer model, the process of bubble growth can be divided into two stages: the initial microlayer growth stage and later rewetting stage. In the initial growth stage of microlayer, it is assumed that bubble grows in a hemispherical shape, the bubble diameter and bubble contact radius increase with time. A microlayer is formed under the bubble and the heating wall. The microlayer expands continuously at the bottom of the bubble until the microlayer disappears and the initial growth stage of the microlayer ends. In the later rewetting stage, the microlayer did not expand, and the bubble diameter remained almost unchanged. Bubbles continue to grow in the vertical direction, and the volume of bubbles still increases, but the growth rate of bubbles is less than that of the growth period of microlayer. At this time, there is a shrinkage phenomenon at the bottom of the bubble, the liquid flows to the bottom center, rewetting phenomenon occurs.

Fig. 1 (a) is shown below. Clear interference fringe images are obtained for experimental processing. According to the principle of interferometry, the radius of fringe is obtained . Then the thickness of microlayer corresponding to different interference fringe can be calculated. Fig. 1 (b) below shows the drying area (dry spots) at the later rewetting stage, and the dividing line between the two stages is the time t_g when the microlayer disappears.

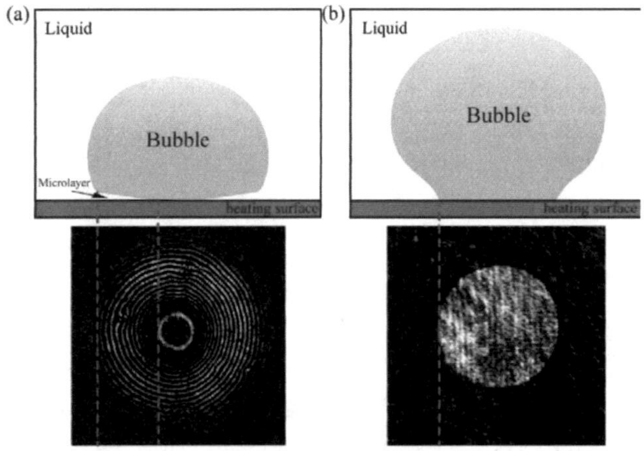

Fig. 1. The schematic diagram of bubble growth process in different stages :(a) initial microlayer growth stage ($t \leq t_g$). (b) the later rewetting stage ($t > t_g$).

Cooper and Lloyd[3] studied the mechanism of microlayer thickness change theoretically and experimentally, and obtained the expression of initial microlayer thickness as follows:

$$\delta_m = C\sqrt{\nu \cdot t} = \sqrt{c\alpha t} \tag{1}$$

In the formula δ_m is the maximum initial microlayer thickness at time t ,ν is the kinematic viscosity of the liquid, C is the empirical constant (Cooper and Lloyd[3] the empirical constant C is defined as 0.8).α is the thermal diffusion coefficient of liquid working fluid. At this time $C = 0.64P_r$ (P_r is the Prandtl number) , because the bubble growth time is very short, although the surface tension is not considered in the above equation, the equation still holds.

2.2. *Bubble growth at the initial microlayer growth stage*

In the process of bubble growth during the initial microlayer growth period, bubbles grow in a hemispherical shape (Fig. 2 (a)). The evaporation of microlayer provides a lot of latent heat of evaporation for the growth of bubbles, Therefore, in this stage, the heat balance equation can be established by the latent heat required for the bubble growth and the heat transmitted by microlayer (that is, the evaporation heat transfer on the surface of microlayer).

$$\frac{1}{2} \cdot \frac{4}{3}\pi R^3 \rho_v h_{fg} = \int_0^{t_g} \int_0^R 2\pi r dr k_l \frac{\Delta T(r,t)_s}{\delta_m} dt \quad 0 \leq t \leq t_g \tag{2}$$

Where R is the bubble radius, ρ_v is the density of steam, h_{fg} is the latent heat of evaporation, and k_l is the thermal conductivity of liquid.

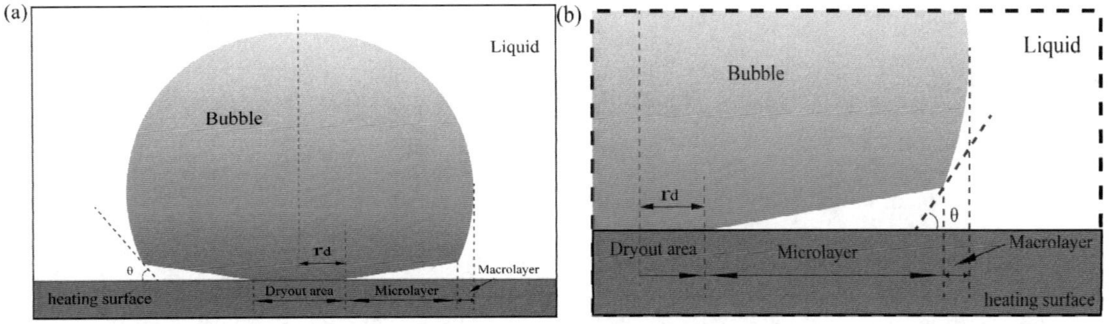

Fig. 2. Schematic diagram of the microlayer below the nucleating bubble at a certain time :(a) the microlayer below the nucleating bubble, (b) the extension of the microlayer.

2.2.1. Initial thickness of microlayer (considering the influence of radius and time). In this study, not only the time effect on wall temperature, but also the radius effect will be considered. Therefore, it is assumed that the function of the wall superheat along with time and the distance from the center of the bubble is:

$$\Delta T_s = T_0 + c_1 t + c_2 r^2 \tag{3}$$

Where ΔT_s is the wall overheating, T_0 is the initial boiling wall temperature, t is the growth time of microlayer, c_1 and c_2 are constants. In combination with Eq. (1), (2) and (3), the instantaneous radius r expression of the bubble can be obtained:

$$r = \frac{\rho_v h_{fg}\sqrt{c\alpha} - \sqrt{\rho_v{}^2 h_{fg}{}^2 c\alpha - 18k_l{}^2 c_2 T_0 t - 6c_1 c_2 k_l{}^2 t^2}}{3k_l c_2 t^{\frac{1}{2}}} \quad 0 \le t \le t_g \tag{4}$$

When the interference fringes of microlayer disappear, the growth process of initial microlayer ($t = t_g$) ends, and the bubble diameter d can be expressed as:

$$d = \frac{2(\rho_v h_{fg}\sqrt{c\alpha} - \sqrt{\rho_v{}^2 h_{fg}{}^2 c\alpha - 18k_l{}^2 c_2 T_0 t - 6c_1 c_2 k_l{}^2 t^2})}{3k_l c_2 t^{\frac{1}{2}}} \tag{5}$$

It can be deduced that the time required for the microlayer to reach position r is :

$$t = f(r) \quad 0 \le t \le t_g \tag{6}$$

At the position of radius r , the initial thickness of the microlayer is expressed as

$$\delta_0 = \sqrt{c\alpha t} = \sqrt{c\alpha f(r)} \tag{7}$$

2.2.2. Initial thickness of microlayer (only considering time effect). For simplicity, only consider when the wall temperature increases linearly with time (when $c_2 = 0$), namely :

$$\Delta T_s{}' = T_0 + c_1{}' t' \tag{8}$$

At this point, the instantaneous radius of the bubble r expression is as follows:

$$r' = \frac{3k_l T_0 t'^{\frac{1}{2}} + k_l c_1{}' t'^{\frac{3}{2}}}{\rho_v h_{fg}\sqrt{c\alpha}} \tag{9}$$

When the growth process of the initial microlayer ($t' = t_g$) ends, the bubble diameter d can be expressed as:

$$d' = \frac{2(3k_l T_0 t'^{\frac{1}{2}} + k_l c_1' t'^{\frac{3}{2}})}{\rho_v h_{fg}\sqrt{c\alpha}} \tag{10}$$

At the position of radius r, the initial thickness of the microlayer is expressed as:

$$\delta_0' = \sqrt{c\alpha t'} = \sqrt{c\alpha f(r')} \tag{11}$$

The thickness of the microlayer affects the evaporation heat flux, which is transmitted through the bubble substrate at the wall. However, the radial range of each direction of the bubble is about 10 times of its thickness, and there are reasons such as short time, difficulty to observe and small size, which the measurement of microlayer very difficult. Therefore, the microlayer parameters derived from experimental observations are usually prone to uncertainties, so the theoretical calculation is more accurate.

3. Results and analysis

To date, a large number of experiments have investigated reliable qualitative and quantitative work on microlayer thickness under individual bubbles. Liu et al.[9] used optical methods to directly measure the depleted microlayer thickness. For example, laser interferometry was used to explore the microlayer evaporation mechanism under the nucleation bubble, which was usually combined with high-speed camera. Using the interference measurement method, clear interference fringe images are obtained through simple image processing such as contrast change and background correction. As shown in Fig. 3, Interference fringe image (left) and synchronous bubble growth and departure photo (right).The radial range and thickness of the microlayer change with time and radius. This is the same as the experiment we conducted to explore the evaporation mechanism of the microlayer at the bottom of boiling bubble (Fig. 1 (a) below).

As shown in Fig. 4, the initial microlayer thickness δ_0 is in the range of about 0-5μm, and the general trend of the change of the initial microlayer thickness can be observed to increase with the increase of time and distance from the initial position of the bubble, which is consistent with the trend of the experimental data.

In the initial microlayer growth stage, this study uses the heat balance equation (Eq. (2)) between the latent heat required by bubble growth and the evaporation heat transfer on the microlayer surface, combined with the assumed function of wall superheat changing with time and radius (Eq.(3)). The relation of initial thickness of microlayer can be derived (Eq.(7)). According to our experimental research, we can get the microlayer thickness corresponding to the different interference fringes we need. By fitting the experimental data with the theoretical formula calculated by us, the values of c_1 and c_2 constant consistent with the fitting are determined. The constant values of c_1=5000 and c_2=-0.002 given in this paper.

Compared with previous studies on Jung and Kim[2019][10], Jung and Kim[2014][11], Utaka[12], Yabuki and Nakabeppu[13]. The experimental data of initial microlayer thickness with time and radius fall within ±25 % of the formula range, which is found to be in good agreement with previous experimental studies.

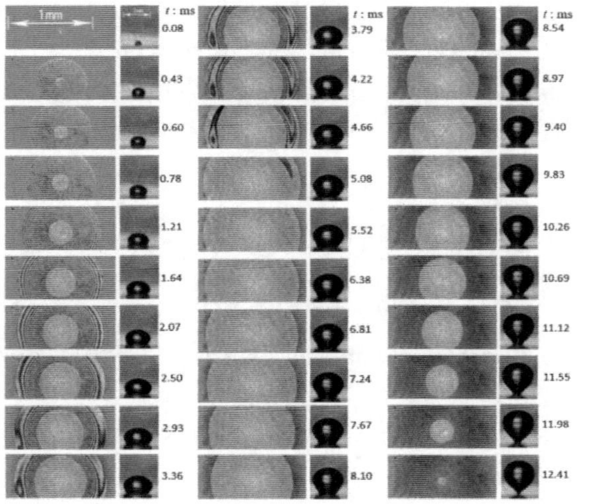

Fig. 3. Interference fringe image (left) and synchronous bubble growth (from 0.08 ms to 4.66 ms) and departure (from 4.66 ms to 12.41 ms) photo (right).

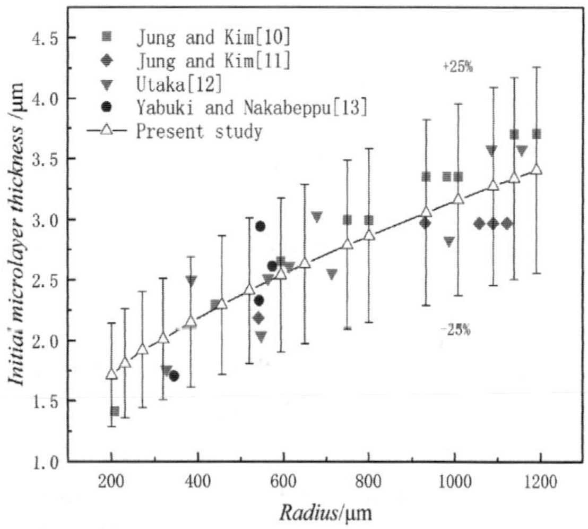

Fig. 4. Comparison of the formula of initial microlayer thickness variation with experimental data under the influence of time and radius

4. Conclusions

Study on boiling heat transfer mechanism of microlayer evaporation theory. The main conclusions are as follows:

(1) Based on the latent heat required for bubble growth and the evaporation heat transfer on the microlayer surface. Combine with assumed wall temperature as a function of time and radius. The initial thickness of microlayer at radius R is calculated Eq. (7).

(2) In this paper, the theoretical formula of microlayer thickness under a single bubble is fitted and analyzed with the experimental data of microlayer evaporation mechanism at the bottom of the boiling bubble explored by the laser interferometry and high-speed camera. Define c_1 and c_2 with constant values of 5000 and -0.002.

(3) We substitute the constant values c_1 and c_2 into the initial thickness Eq. (7) of microlayer under a single bubble to obtain the empirical formula. Compared with the experimental data of previous researchers (Fig. (4)), it was found that the experimental data fell within ±25% of the formula range, which was in good agreement with the data of previous experimental studies.

Acknowledgments
This work was supported by National Natural Science Foundation of China through Grant No. 51976127.

Reference

[1] Snyder N, Edwards D. Summary of conference on bubble dynamics and boiling heat transfer[J]. JPL Memo. 1956,20:20-137.

[2] Moore FD, Mesler RB. The measurement of rapid surface temperature fluctuations during nucleate boiling of water[J]. AIChE Journal. 1961,7(4):620-624.

[3] Cooper M, Lloyd A. The microlayer in nucleate pool boiling[J]. International Journal of Heat and Mass Transfer. 1969,12(8):895-913.

[4] MacGregor H, Jawurek H. High speed ciné laser interferometry technique for microlayer studies in boiling [J]. N&O Joernal April. 1992:1-7.

[5] Koffman LD, Plesset MS. Experimental Observations of the Microlayer in Vapor Bubble Growth on a Heated Solid[J]. Journal of Heat Transfer. 1983,105(3):625-32.

[6] Voutsinos CM, Judd RL. Laser Interferometric Investigation of the Microlayer Evaporation Phenomenon[J]. Journal of Heat Transfer. 1975.97(1):88-92.

[7] Utaka Y, Kashiwabara Y. Ozaki M. Microlaver structure in nucleate boiling of water and ethanol at atmospheric pressure[J]. International Journal of Heat and Mass Transfer.2013,57(1):222-230.

[8] Zhao Y, Masuoka T, Tsuruta T. Theoretical studies on transient pool boiling based on microlayer model[J]. International journal of heat and mass transfer. 2002, 45(21): 4325-4331.

[9] Liu J, Gao M, Zhang L, et al. A laser interference/high-speed photography method for the study of triple phase contact-line movements and lateral rewetting flow during single bubble growth on a small hydrophilic heated surface[J]. International Communications in Heat and Mass Transfer. 2019, 100: 111-117.

[10] Jung S, Kim H. Hydrodynamic formation of a microlayer underneath a boiling bubble[J]. International Journal of Heat and Mass Transfer. 2018, 120: 1229-1240.

[11] Jung S, Kim H. An experimental method to simultaneously measure the dynamics and heat transfer associated with a single bubble during nucleate boiling on a horizontal surface[J]. International journal of heat and mass transfer. 2014, 73: 365-375.

[12] Utaka Y, Kashiwabara Y, Ozaki M. Microlayer structure in nucleate boiling of water and ethanol at atmospheric pressure[J]. International Journal of Heat and Mass Transfer. 2013, 57(1): 222-230.

[13] T. Yabuki, O. Nakabeppu, On heat transfer mechanism of nucleate boiling withMEMS senors (3rd Report, evaluation of the approach method and heattransfer characteristics of isolated boiling bubble), Trans. Jpn. Soc. Mech. Eng. (Ser. B) 76 (771) (2010) 1932–1941.

Causes and troubleshooting of high bearing bush temperature of a nuclear power steam turbine

Shuaishuai Li [a], Tianzuo Qu [b]

Shandong Nuclear Power Company Ltd, Yantai, 265100, China

[a] lishuaishuai@spic.com.cn, [b] qutianzuo@spic.com.cn

Abstract. As an important part of the steam turbine unit, once the temperature of the bearing bush rises abnormally, it will not only reduce the application performance of the bearing itself, but also affect the normal operation of the whole unit, resulting in shutdown accidents. Aiming at the problem of abnormal temperature rise of bearing bush of nuclear power steam turbine, combined with the operation and maintenance of the unit, the real cause of high bearing bush temperature is clarified, and then targeted treatment measures are taken to solve the problem, which is beneficial to ensure the safety of the unit.

Key words: bearing bush; temperature; reason.

1. Introduction

Bearing is an important part of steam turbine unit, and its working condition directly affects the safe and economic operation of the unit. Turbine bearings are divided into radial support bearings and axial thrust bearings. The radial support bearing is used to bear the weight of the rotor and the centrifugal force caused by the unbalance of the rotor mass, and determine the radial position of the rotor to keep the rotor rotation center consistent with the cylinder center, so as to ensure that the radial clearance between the rotor and static parts such as cylinder, steam seal and diaphragm is within the standard range, Thrust bearing is used to bear the axial thrust of steam and generator magnetic field on the rotor, and determine the axial position of the rotor to ensure the correct position of axial clearance between the dynamic and static parts of the flow passage. Therefore, during the operation of the steam turbine unit, it is necessary to strengthen the control of various unit operation parameters to make them within the range of meeting the operation standards, especially the bearing bush temperature, which is an important parameter for the operation of the steam turbine unit. Once it rises abnormally, it will affect the stable and efficient operation of the whole steam turbine unit. By analyzing the causes of abnormal temperature rise of radial support bearing of nuclear power steam turbine, the solutions are found out, which provides a reference for the design, construction and operation of steam turbine in the future.

2. Bearing overview

All the supporting bearings of the steam turbine are cylindrical self-aligning bearings. The bearing is divided into upper and lower halve. The outer surface of the bearing pad iron and the bearing seat pits in contact with it are spherical. When the rotor journal is inclined, the bearing can rotate and adjust automatically, so that the gap between the Journal and the bearing remains unchanged within the whole bearing length. The fit between the bearing and the bearing seat shall not only ensure the free sliding of the bearing during operation, but also be too loose to prevent bearing vibration. Jacking oil hole is set for bearing bush of low pressure part, and jacking oil is not set for bearing bush 1 and 2 due to relatively low load. The position of No. 2 bearing is shown in Figure 1.

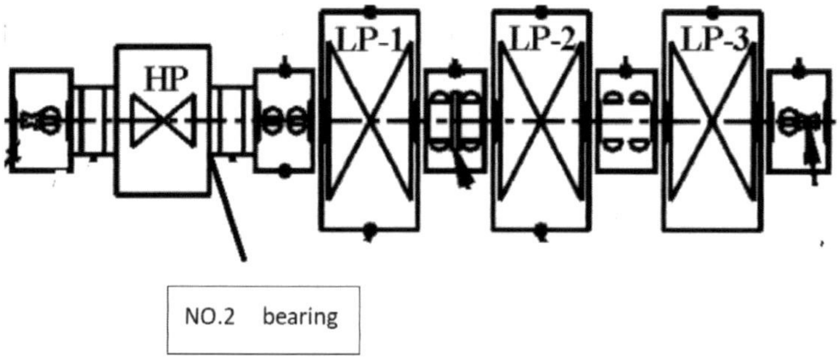

Fig. 1 No. 2 bearing position

3. Cause analysis of high bearing temperature of steam turbine

3.1. High temperature of No. 2 bearing bush of steam turbine

Since the commercial operation of unit 2 in 2019, the temperature of No. 2 bearing bush of steam turbine has always been higher than that of other bearing bushes. According to statistics, before the first overhaul, the temperature of No. 2 bearing bush was maintained in the range of 75-80 °C, about 7 °C higher than the maximum value of other bearing bushes; During the first overhaul, the load distribution of bearing bush was adjusted, but the temperature of No. 2 bearing bush increased slightly again, up to 87.5 °C, as shown in Figure 2.

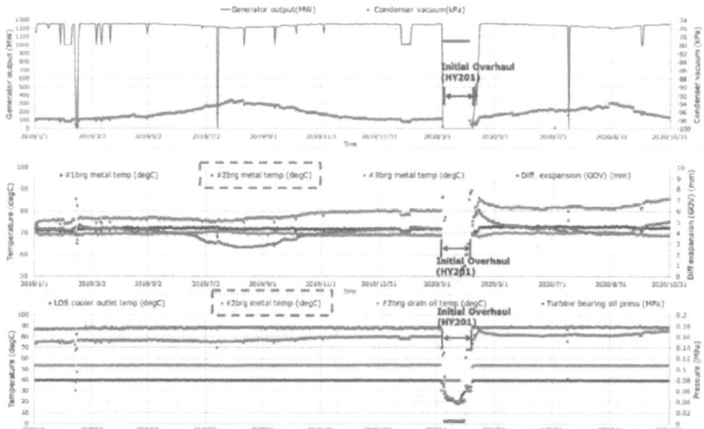

Fig. 2 temperature change trend of No. 1 bearing and No. 2 bearing bush

It can be further found from Figure 2:

(1) The temperature of No. 2 bearing bush changes with time and increases with the decrease of ambient temperature.

(2) The general trend of No. 2 bearing bush temperature is to increase, especially after September, the temperature rise of No. 2 bearing bush is relatively large compared with that in the early stage.

(3) The temperature of No. 2 bearing bush of steam turbine decreases with the decrease of load and increases with the increase of load.

(4) No. 2 bearing bush is sensitive to the change of unit load.

3.2. Cause Analysis

3.2.1. influence of turbine lubricating oil system on temperature rise of No. 2 bearing bush. (1) Check the oil pressure and lubricating oil temperature of turbine lubricating oil system. The oil pressure of lubricating oil main pipe at No. 2 bearing bush of lubricating oil system is 0.12Mpa, the lubricating oil supply temperature is 40 °C, and the return oil temperature of each bearing is normal.

(2) There is no obvious change in the oil return flow of each bearing, and there is no water drop in the oil return window of No. 2 bearing.

(3) Moisture content of lubricating oil of the unit is 12.7%; The viscosity of lubricating oil is 3.7mm^2/s and the particle size is nas4, meeting the requirements.

It can be judged from the above points that the lubricating oil system of steam turbine operates normally and will not cause the temperature rise of No. 2 bearing bush.

3.2.2. influence of unit vibration. Check that the shaft vibration and bearing vibration of the unit are stable without increasing trend. See Table 1 for vibration conditions; Eliminate the bearing temperature rise caused by abnormal vibration of the unit.

Table 1. data sheet of No. 2 shaft vibration and pad vibration

Bearing number	X-direction Shaft vibration	Y-direction Shaft vibration	bearing vibration
1	0.0143	0.0115	0.007
2	0.0102	0.0051	0.0086
3	0.0164	0.0013	0.0103
4	0.0190	0.0115	0.0091
5	0.0134	0.0083	0.0081
6	0.0186	0.0118	0.0086
7	0.0154	0.0071	0.0120
8	0.0255	0.0071	0.0120

3.2.3. troubleshooting of external heat sources. Check the heat source near No. 2 bearing housing on site. The insulation of high-pressure cylinder is complete, and the high-pressure cylinder does not constitute a heat source affecting the temperature rise of turbine bearing.

The shaft seal was inspected on site and no steam leakage was found; The infrared thermometer is used to measure the temperature of the high-pressure cylinder shaft seal body and the outer oil baffle of the bearing on site. The measurement results are 77.2 °C for the shaft seal body and 71.6 °C for the outer oil baffle.

Based on the above points, it can be judged that these factors may contribute to the higher temperature of No. 2 bearing bush than other temperatures, but there is no external heat source, which directly causes the high temperature of No. 2 bearing bush of steam turbine. Therefore, it can be judged that the temperature rise of No. 2 bearing bush has nothing to do with the external heat source.

3.2.4. the temperature of No. 2 bearing bush increases due to the change of bearing load. With the change of ambient temperature, the temperature of circulating water decreases and the condenser vacuum increases. The condenser vacuum of unit 2 increases from -94.197kpa on September 14 to -97.082kpa on November 15; When the unit load remains unchanged, the vacuum increases, the downward tension of the low-pressure rotor increases under the influence of vacuum, the load distribution of the rotor shafting of the steam turbine generator unit changes, and the load of the No. 2 bearing of the high-pressure rotor increases, which shows that the temperature of the No. 2 bearing increases. The variation curve of No. 2 bearing temperature and condenser vacuum is shown in Figure 3.

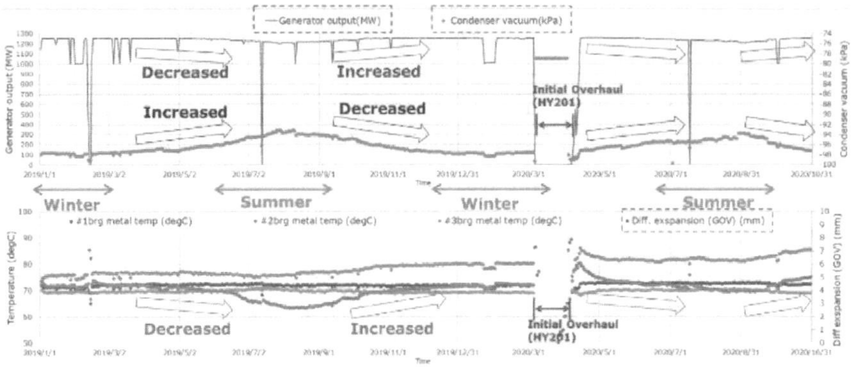

Fig. 3 Variation Trend of No. 2 bearing bush with vacuum

3.2.5. problems found in the first disassembly of No. 2 bearing. (1) Before the first overhaul, the high-pressure rotor was 0.205mm higher than the low-pressure rotor, which greatly increased the load of No. 2 bearing bush. 201 overhaul reduced the deviation to 0.0025mm.

(2) There is a certain degree of wear on the black gold surface of No. 2 bearing, and there is a certain degree of eccentric wear, and the contact with the journal is uneven. See Figure 4 for contact conditions.

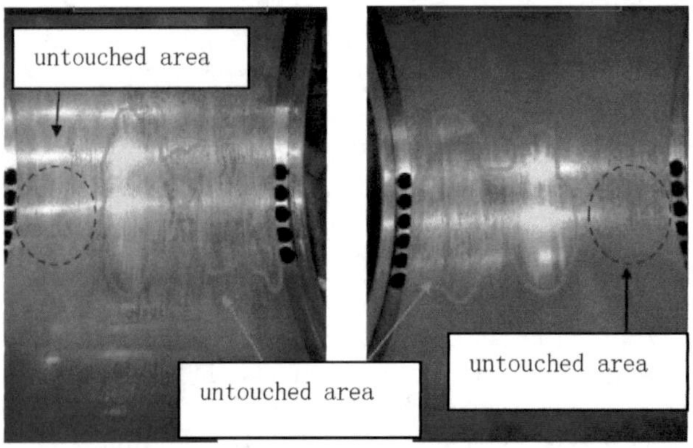

Fig. 4 contact of No. 2 bearing bush

(3) The clearance of No. 2 bearing pillow exceeds the standard

The structural diagram of the bearing is shown in Figure 5, and the specific data are shown in Table 2.

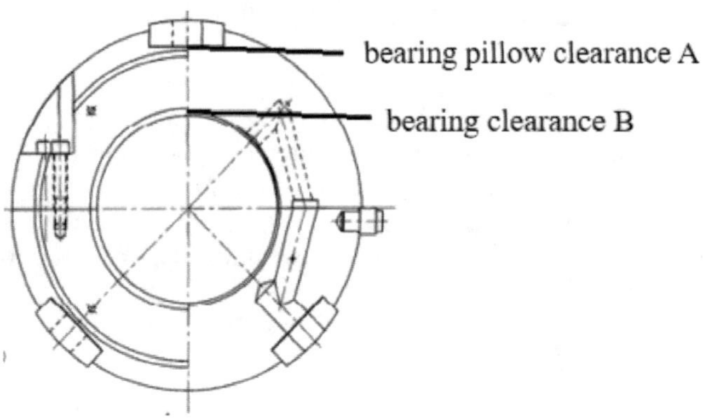

Figure 5. Structural drawing of bearing

Table 2. data sheet of bearing bush clearance

	item	Standard: 0.05- 0.18mm (bearing pillow clearance), 0.98-1.08 (bearing clearance)
Data before the first disassembly	bearing pillow clearance (A)	-0.06
	bearing clearance (B)	1.05
Data after the first disassembly	bearing pillow clearance (A)	0.09mm
	bearing clearance (B)	1.05

3.3. problem solving

(1) Through the above analysis, it can be seen that the excessive clearance between the bearing pillow causes problems in the bearing bush adaptation, which biases the contact between the shaft diameter and the bearing bush to one side, resulting in collision and wear at the contact place due to the increased load, resulting in the high temperature of the bearing Bush. With the improvement of the vacuum, the temperature gradually increases, which is caused by the load distribution, which is a normal phenomenon. After an overhaul of the steam turbine, the clearance of the bearing pillow was adjusted, but the problem was not solved. The bearing temperature of No. 2 bearing was still 7 °C higher than that of other bearings, up to 87 °C, which had a certain impact on the safety of the unit.

(2) During the second unit overhaul, the No. 2 bearing was disassembled again, and it was found that there was still a certain degree of eccentric wear, as shown in the figure below:

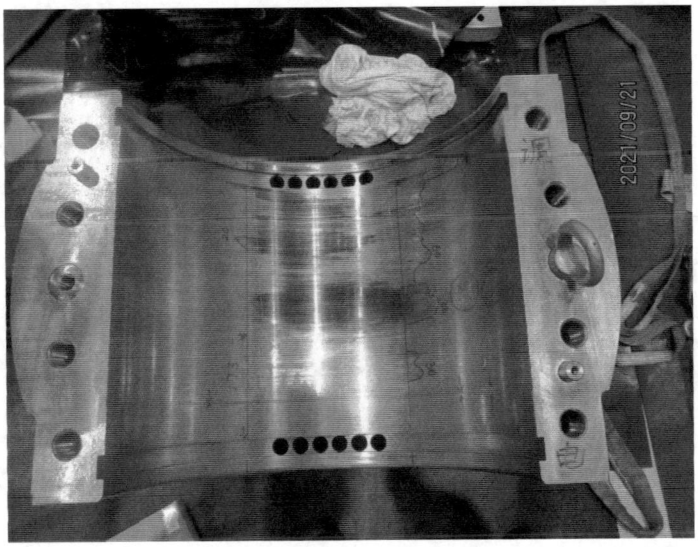

Fig. 6 contact diagram of No. 2 bearing bush during the second disassembly

(3) According to the two disassembly of No. 2 bearing, eccentric wear occurred. After the first installation of No. 2 bearing in the construction and installation period, the bearing bush was fixed to one position due to the excessive clearance of bearing pillow. In the process of adaptation, the bearing bush was used to the original position and could not adapt to the appropriate position. Therefore, a new bearing needs to be replaced.

(4) After the new bearing bush is replaced, the clearance between bearing pillow and bearing bush is adjusted again. The data are as follows:

Table 3. disassembly clearance data of the second bearing

	item	Standard: 0.05- 0.18mm (bearing pillow clearance), 0.98-1.08 (bearing clearance)
data before bearing bush replacement	bearing pillow clearance (A)	0.10
	bearing clearance (B)	1.05
data after bearing bush replacement	bearing pillow clearance (A)	0.15
	bearing clearance (B)	1.01

(5) On January 13, 2021, the temperature of bearing bush was 86.3 °C, which was replaced after the new bearing bush, the temperature decreased by about 5 °C on January 13, 2022, and the problem was solved.

4. Conclusion

As an important part of the steam turbine unit, once the temperature of the bearing bush rises abnormally, it will not only reduce the application performance of the bearing itself, but also affect the normal operation of the whole unit, resulting in shutdown accidents. Aiming at the problem of abnormal temperature rise of No. 2 bearing bush, combined with the operation and maintenance of the unit, the real cause of the abnormality was clarified, and then targeted treatment measures were taken to solve the problem. It provides a new analysis idea for solving similar problems in the future.

References

[1] Analysis and the Adopted Measures of Nuclear Turbine Bearing High Temperature [J]. Turbine Technology, 2013.4.
[2] Chinese Society of Power Engineering. Technical manual of thermal power equipments: The 2nd volume, steam Turbines [M]. Beijing: China Machine Press, 2002(in china).
[3] The Analysis of the Cause of 1000 MW Steam Turbine Bearing Metal Temperature on the High Side and Countermeasures [J]. Turbine Technology, 2016.2.
[4] High temperature of turbine bearing in nuclear power plants: reason analysis and countermeasures [J]. Thermal Power Generation, 2013.8.
[5] Countermeasures for high temperatures of bearing pad of 328.5 MW steam [J]. Turbine Technology, 2005.5.
[6] Cause Analysis of High Temperature of Bearing for 350MW Turbine and Disposal [J]. Turbine Technology, 2006.12.

Comparative experimental study on dynamic characteristics of bubble microlayers in small channel flow boiling and pool boiling

Zhongxiang Shen, Ming Gao*, Wuhan Dong and Lixin Zhang

Shanghai Key Laboratory of Multiphase Flow and Heat Transfer in Power Engineering, School of Energy and Power Engineering, University of Shanghai for Science and Technology, Shanghai 200093, China

*Corresponding author e-mail: gaoming@usst.edu.cn

Abstract. The microlayer present at the bottom of the bubble plays a very important role in heat transfer during nucleation boiling. In this paper, the dynamic characteristics of microlayer at the bottom of boiling bubble in a small channel were studied by laser interferometry method and high speed camera, and the results were compared with pool boiling experiments. The results show that the bubbles have an obvious tendency to slip in the flow boiling. The microlayer interference fringe is deformed and no longer a complete Newtonian ring. By analyzing the thickness distribution of microlayer in different directions in flow boiling and the change of micro-contact angle in the flow direction, it was found that the micro-contact angle in the flow direction became larger with the growth of bubbles.

Keywords: Laser interference; Small channel; Flow boiling; Bubble; Microlayer.

1. Introduction

With the goal of "carbon peak and neutrality", the improvement of energy utilization will be more urgent, in which boiling heat transfer, as a very important heat transfer method, has a vital role in improving energy utilization. Snyder et al [1] first introduced the concept of microlayer, which is a thin layer of liquid existing between the bottom of the bubble and the heating surface. The microlayer of several microns in thickness at the bottom of the bubble in nucleation boiling will transfer a large amount of heat to the bubble during evaporation [2], and therefore plays an important role in boiling heat transfer. The experimental methods for studying the microlayer can be broadly classified into three types: 1) temperature distribution method; 2) laser extinction method; and 3) laser interferometry. In terms of temperature distribution method, some experts have measured the change of temperature at the heat transfer surface using special thermocouples [3] or precision thermometers [4], which indirectly confirmed the existence of microlayer and analyzed that the microlayer is a "wedge-shaped" structure. Gerardi et al [5] observed the temperature distribution of the heat transfer surface and analyzed that the influence of microlayer evaporation on bubble growth could not be ignored. In laser extinction method, Utaka et al [6] measured the initial microlayer thickness distribution of water and ethanol by laser extinction method. In laser interferometry, Sharp et al [2] measured the thickness of microlayer by interferometry using a mercury lamp as the light source and also obtained a "wedge-shaped" microlayer structure. Voutsinos and Judd [7]measured the microlayer thickness of methylene chloride using a He-Ne laser with good coherence and a high-speed camera, and the measurements showed that the microlayer still appeared as a " wedge-shaped" structure. MacGregor and Jawurek [8] measured the thickness change of methanol microlayer and obtained the shape of bubble simultaneously. It was concluded that evaporation of the microlayer contributed significantly to the volume of bubble and verified the accuracy of laser interferometry in measuring the microlayer. Gao et al [9]and Liu et al [10] measured the microlayer thickness distribution of alcohol by combining laser interferometry and high-speed camera, and summarized the dynamic characteristics of the microlayer. Chen [11, 12]et al. measured the microlayer thickness distribution of water at different heat fluxes and found a special microlayer structure near the edge of the microlayer.

In previous experimental studies, researchers have mostly studied the structural changes of microlayer in pool boiling, and comparative studies of pool boiling and flow boiling are relatively rare. In this paper, the structural changes of microlayer at the bottom of flow boiling bubbles are studied by combining laser interference and high-speed camera, and the comparative study with pool boiling microlayer [9, 10, 12] is important to elucidate the boiling heat transfer mechanism.

2. Experimental setup

The schematic diagram of experimental system s is shown in Fig.1. The experimental system can be roughly divided into four parts: the fluid flow part; optical system; data, Image Acquisition and Power Supply. Driven by the injection pump, the working medium ethanol passes through the preheater (adjusting the inlet temperature of the working fluid), and the working fluid then enters the test section and then into the reagent bottle placed on the balance. In the optical section, the laser emitted by the laser first passes through the laser beam expander into a large diameter parallel laser beam, When the laser beam irradiates on the beam splitter, and part of the laser beam is vertically reflected onto the test section, When this part of the laser passes through the bubble, as shown in Fig.2, part of the laser will reflect on the upper and lower surfaces of the microlayer at the bottom of the bubble, and the two reflected laser beams will interfere. The interference fringes are magnified by the microscope after passing through the beam splitter, and then reflected by the all-mirror onto the CCD and record. There is also an auxiliary light source LED above the test section, which is used to photograph the contour of the bottom of the bubble and recorded by CCD. The temperature acquisition instrument collects the temperature of the inlet and outlet of the test section. The power supply is used to control the voltage input to the test section and LED, thereby controlling the heat flux and LED brightness.

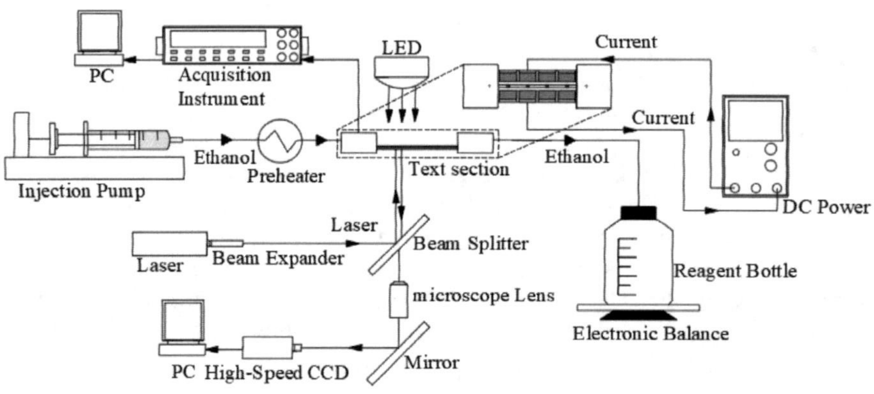

Fig. 1 Principle of experimental system

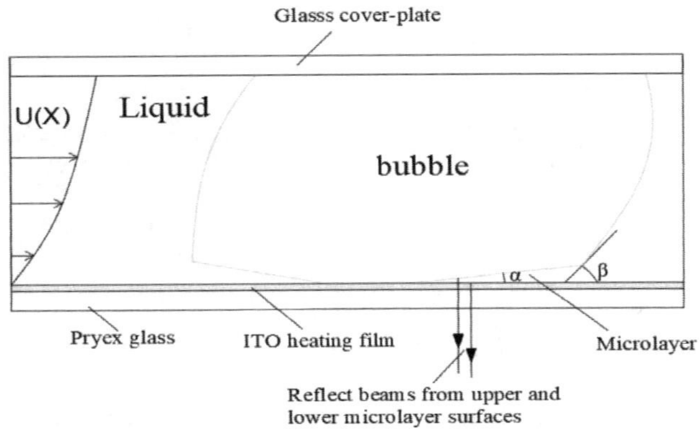

Fig. 2 Schematic of microlayer beneath a nucleation bubble in flow boiling

3. Comparison of interference fringe and bubble contour between small channel flow boiling and pool boiling

In this experiment, ethanol was used as the experimental working medium, and the interference fringes of the microlayer at the bottom of the bubble and the contour of the bubble bottom were recorded. In this paper, the height of all experimental channels is 1 mm.

When the heat flux is 854kW /m2 at saturation state, the pool boiling bubble microlayer with water as the working medium is shown in Fig 3(a). It can be seen that the microlayer is deformed, and there is a long stripe, and the dry spot is still roughly round. As shown in Fig. 3(b), it can be seen that the interference fringes are deformed, interference fringes show different characteristics in different directions. As shown in Fig. 3(c), a long interference fringe is found downstream of the bubble. The dry spot at the bottom of the bubble (highlighted in red in Fig 3 (c)) is no longer circular due to the flow of fluid. The reason for the difference between flow boiling in small channel and pool boiling interference fringe is that fluid flow exerts shear force on bubbles.

The side photos of pool boiling bubbles is shown in Fig. 4(a). It can be clearly seen from the figure that bubbles have relatively symmetrical shapes. Bottom shot of flow boiling bubbles in small channels is shown in Fig. 4(b). It can be seen that due to fluid flow, bubbles have an obvious tendency to slip in the flow direction.

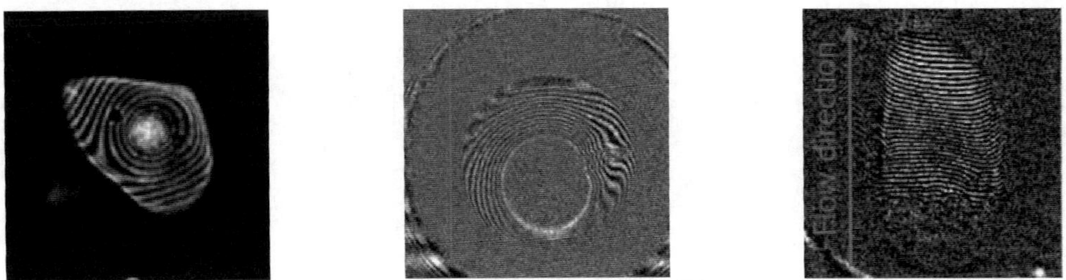

(c)Pool boiling Q=854 kW/m^2 [12]

(b) Q=850 kW /m^2 G=74 kg/(m^2·s) ΔT= 12°C

a) Q=860 kW/m^2 G=162 kg/(m^2·s) ΔT= 4°C

Fig. 3 Comparison o Interference fringe images

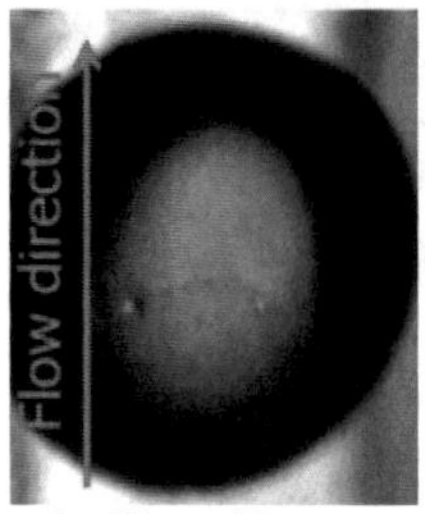

(a) Pool boiling (Side photos) [10] (b) Flow boiling (Bottom photos)

Fig. 4 Comparison of bubble contour

4. Dynamic characteristics of small channel flow boiling interference fringe

Fig. 5 shows the interference fringes of the microlayer at the bottom of the bubble with a heat flux of 645 kW/m^2, a mass flow velocity of 10 kg/ (m^2·s) and a subcooling degree of 12 °C. It can also be seen that the interference fringe has been deformed and is no longer a complete Newtonian ring. Different from the heat flux of 850 kW/m^2, mass flow velocity of 74 kg/ (m^2·s) and subcooling degree of 12 °C, the microlayer has a larger radius under this condition. In order to facilitate the analysis of the characteristics of the microlayer at the bottom of the flow boiling bubble, this paper adopts Fig 5 as the research object for data analysis.

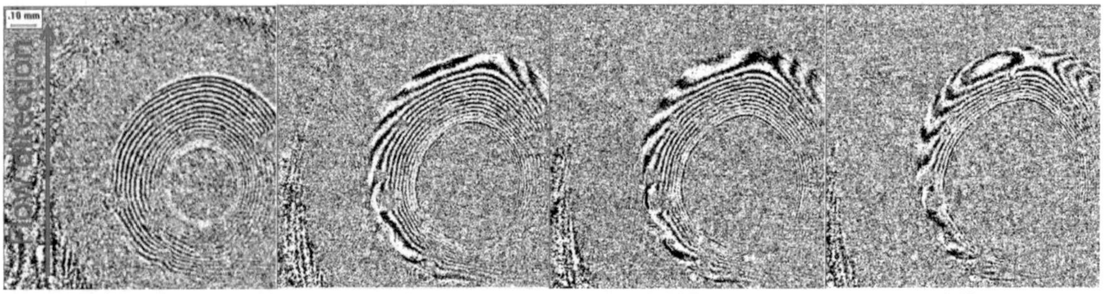

t=0.347ms t=0.694ms t=0.799ms t=1.042ms

Fig. 5 Fringe patterns of an ethanol vapor bubble in flow boiling
(Q=645 kW/m^2 G=10 kg/(m^2·s) ΔT=4 °C)

4.1. Measurement of microlayer thickness

According to the conditions for interference light to form light and dark fringes, the optical thickness of the two adjacent light rings is:

$$2n\delta_m = m\lambda, \quad m = 1,\ 2 \tag{1}$$

and the optical thickness of the two adjacent dark rings:

$$2n\delta_m = (2m+1)\frac{\lambda}{2}, \quad m = 0,\ 1,\ 2 \tag{2}$$

Where δ_m is the thickness of microlayer; n is the refractive index of ethanol; λ is the wavelength of laser, and m is the fringe order. The thickness distribution of microlayer is shown in Fig. 6.

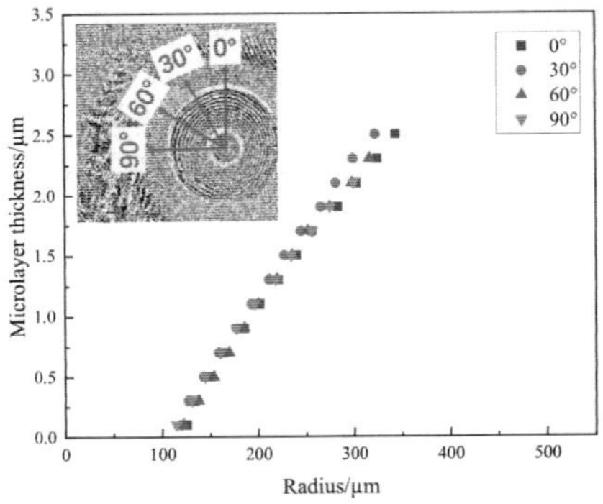

(a) t=0.174 ms

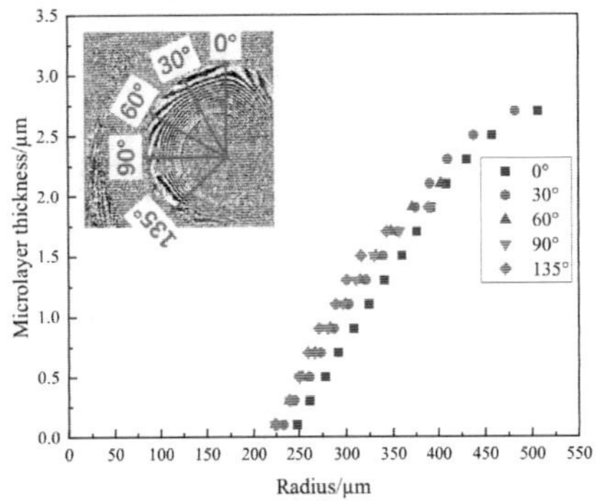

(b) t=0.694 ms

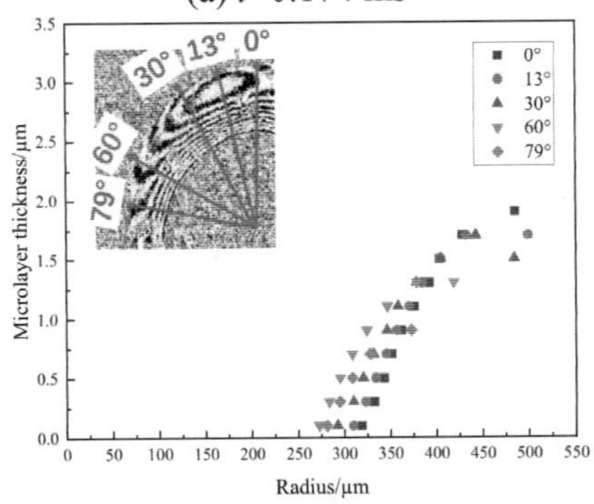

(e) t=1.042 ms

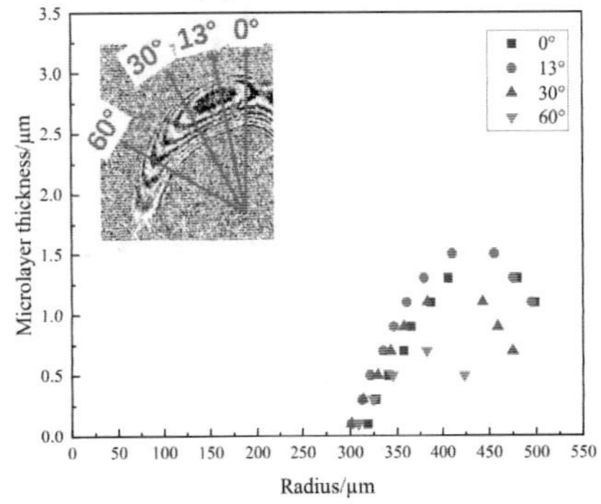

(f) t=1.215 ms

Fig. 6 Thickness changes of microlayer in different directions

4.2. *Characteristics of microlayer thickness*

It can be seen from Fig.5 that the characteristics of the microlayer are different in different directions, so we take the flow direction as the positive direction of the X-axis (i.e., 0 ° in the same direction as the flow), and the angle increases counterclockwise. The thickness distribution of the microlayer in different directions at different times is shown in Fig.6. It can be seen that in the early stage of bubble growth, the thickness of the microlayer is roughly linear with the radius, and the microlayer is still "wedge-shaped". With the growth of bubbles, part of the small channel is occupied by bubbles, the flow area of the fluid becomes smaller, and the microlayer is subjected to more and more shearing force, which eventually leads to the deformation of the microlayer, and the microlayer is no longer a centrosymmetric structure. It can be clearly seen from Fig.6 that the maximum thickness of microlayer is different in different directions. In most of the time, the thickness of microlayer in the 0 ° direction is greater than that in other directions. This is because the microlayer in other directions is greatly affected by the direct disturbance of the fluid flow, and the microlayer is broken, resulting in the thinning of the microlayer. Moreover, the bubble is "tilted forward" due to the fluid flow, which also makes the microlayer in the 0 ° direction thicker. Fig.6 shows that at the late stage of bubble growth, a special "bending structure" appears in the microlayer. Special microlayer bending structure has been found in pool boiling, but the difference is that the bending microlayer structure appears in the same direction as the flow direction in flow boiling.

4.3. *Change of micro contact angle*

As shown in Fig.2, in addition to a macroscopic contact angle β, there is also a micro-contact angle α in the microlayer region. Since the thickness distribution of the micro-liquid layer is linear with the radius, the micro-contact angle can be calculated by using the thickness of the microlayer and the corresponding radius. The change of micro-contact angle is shown in Fig.7. It can be clearly seen from the figure that the micro-contact angle in the flow direction is different at different times. In general, the micro-contact angle increases with the growth of bubbles. This is contrary to the variation rule of micro-contact angle studied by Gao et al [9]. In the early stage of bubble growth (t<0.799 ms) the change of micro-contact angle is relatively mild, in the late growth stage of bubble (t>0.799 ms) the micro-contact angle increases rapidly. It is worth noting that when the time is 0.799ms, the interference fringes appear obvious deformation.

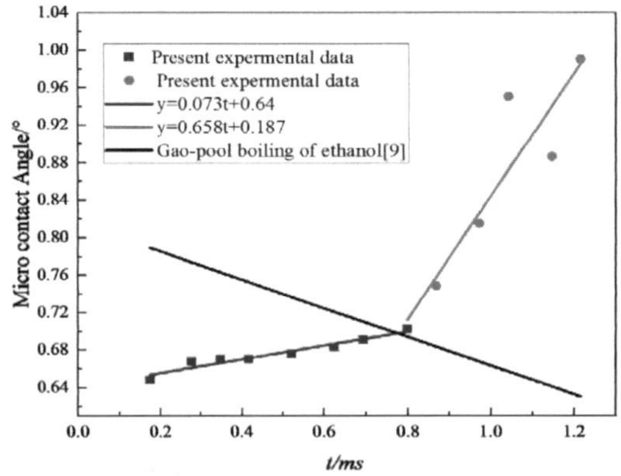

Fig.7 Change of micro-contact angle in flow direction

5. Conclusion

The dynamic characteristics of microlayer at the bottom of boiling bubble in a small channel were studied by laser interferometry method and high speed camera, and the results were compared with pool boiling experiments to reach the following conclusions:

1. The growth of bubbles is limited by the glass cover-plate of the small channel, and the shape of bubbles shows an obvious trend of slipping towards the flow direction due to the fluid flow.

2. In small channel flow, there are two different structures in the microlayer at the bottom of bubble under different working conditions: the combination of a long interference fringe downstream of the bubble and the combination of interference fringe downstream of the bubbles. The microlayer interference fringe is deformed and no longer a complete Newtonian ring.

3. Under the working conditions of heat flow density of 645 kW/m^2, mass flow velocity of 10 kg/(m^2·s) and subcooling degree of 12 °C, the structure of the microlayer is different in every direction, and the micro-contact angle in the flow direction increases gradually with the growth of the bubbles.

Acknowledgments
This work was supported by National Natural Science Foundation of China through Grant No. 51976127.

References

[1] Snyder N W. Summary of conference on bubble dynamics and boiling heat transfer [J]. JPL Memo, 1956, 20.
[2] Sharp R R. The Nature of Liquid Film Evaporation during Nucleate Boiling [M]. NASA TN D-1997, Lewis Research Center, Cleveland, Ohio, 1964.
[3] Moore F D, Mesler R B. The measurement of rapid surface temperature fluctuations during nucleate boiling of water [J]. AIChE Journal, 1961, 7(4): 620-4.
[4] Cooper M G, Lloyd A J P. The microlayer in nucleate pool boiling [J]. International Journal of Heat and Mass Transfer, 1969, 12(8): 895-913.
[5] Gerardi C, Buongiorno J, Hu L W, et al. Study of bubble growth in water pool boiling through synchronized, infrared thermometry and high-speed video [J]. International Journal of Heat and Mass Transfer, 2010, 53(19-20): 4185-4192.
[6] Utaka Y, Kashiwabara Y, Ozaki M. Microlayer structure in nucleate boiling of water and ethanol at atmospheric pressure [J]. International Journal of Heat and Mass Transfer, 2013, 57(1): 222-

230.
[7] Voutsinos C M, Judd R L. Laser Interferometric Investigation of the Microlayer Evaporation Phenomenon [J]. Journal of Heat Transfer, 1975, 97(1): 88-92.
[8] MacGregor H G, Jawurek H H. High speed ciné laser interferometry technique for microlayer studies in boiling[J]. N&O JOERNAAL, 1992, 4: 1-7.
[9] Gao M, Zhang L, Cheng P, et al. An investigation of microlayer beneath nucleation bubble by laser interferometric method[J]. International Journal of Heat and Mass Transfer, 2013, 57(1): 183-189.
[10] Liu J, Gao M, Zhang L, et al. A laser interference/high-speed photography method for the study of triple phase contact-line movements and lateral rewetting flow during single bubble growth on a small hydrophilic heated surface[J]. International Communications in Heat and Mass Transfer, 2019, 100: 111-117.
[11] Chen Z, Haginiwa A, Utaka Y. Detailed structure of microlayer in nucleate pool boiling for water measured by laser interferometric method [J]. International Journal of Heat and Mass Transfer, 2017, 108: 1285-1291.
[12] Chen Z, Hu X, Hu K, et al. Measurement of the microlayer characteristics in the whole range of nucleate boiling for water by laser interferometry[J]. International Journal of Heat and Mass Transfer, 2020, 146: 118856.

Analysis on Welding Temperature Field and Stress Field of Boring Bar of New Boring Machine

Liang Han, Yuning Song*

Yingkou Institute of Technology, Department of Mechanical and Power Engineering, Yingkou, Liaoning, China

*Corresponding author e-mail: songyuning@yku.edu.cn

Abstract. Ansys software is used to simulate the welding process of the boring bar of the new cylinder head boring machine. The indirect coupling method is used to analyze the temperature field, then the stress field and deformation. It is found that for the double support structure, the highest temperature occurs at the outer stiffened plate of the small diameter end, and the lowest temperature occurs at both ends of the inner support structure. The deformation of the whole structure mainly comes from the axial deformation, and the maximum deformation occurs at the large diameter end of the outer rib plate. Because the role of the supporting web is to limit the large deformation of the whole structure, the structure at the small diameter end is firmer and has the ability of anti-deformation. In terms of technology, the influence caused by deformation can be compensated by the compensation design of machining amount.

1. Introduction

Two symmetrical step holes in the Cummins QSK60 engine cylinder head are processed using a conventional boring machine. Traditional boring processing mainly adopts two forms: processing boring machine spindle moving, bench grip parts movement or spindle expansion movement, bench grip parts, no matter which case workpiece need to be secondary clamp or boring cutter secondary cutting, easy to cause the processing accuracy deviation of accumulated and waste of time [1].

A boring machine is currently designed to process the cylinder head, with fixed boring rod length and two-hole integral boring form. The processing process can be processed at one time, ensuring the processing accuracy, while saving a lot of time.

In view of the small boring rod section size and the boring rod stiffness, the large section boring rod is designed within the allowable range to transform the stiffness boring process to equal stiffness boring [2]. Because the back end of the boring bar is connected to the deceleration box, which is expected to be fixed, and the gravity of the boring head and the cutting force at the processing work on the other end, the greater the closer the boring bar is to the deceleration box [3], Considering the smoothness of the welding process, the boring rod is designed into a hollow octahedral with a large diameter near the spindle box and a small size close to the boring head. In the process of the boring rod diameter changes, it is difficult to make a whole connection, and it is also connected to the belly, so it is designed to weld multiple cut steel plates together. However, in the actual trial production process of many products, because the outer wall of the boring rod is a positive octagonal welding splicing structure with varying diameter, and a great bending deformation is produced in the production process, and the boring rod itself is long, it is easy to produce bending and droop, resulting in the minimum bending of the boring rod is about 32mm, which affects the processing accuracy of the middle part of the hole depth. Also for the welding structure for the boring rod

Once there is a harmful welding deformation, it will greatly reduce the actual carrying capacity of the components, and it takes a lot of time to correct them, consuming a lot of manpower, material resources and financial resources. Therefore, it is necessary to control the welding deformation of the boring rod.

2. Experimental Materials and Methods

Figure 1. Boring sample machine

The outer wall of the boring rod is a positive octagonal welded splicing structure with variable diameter, with a circular pipe inside, and the boring head drive rod is placed inside the pipe. At the same time, in order to increase the strength of the boring rod and prevent a large deformation during the welding process, the pipe and the outer wall are connected together by two webs. Its external shape is as shown in Figure 1.

Because the shape is a splicing structure, the welding process adopts manual arc welding. The welding power supply adopts ZX7-400 inverter welding machine and DC reverse polarity wiring method. The electrode model is E5015, and the welding current is 100~110A [10], Welding speed is 9 to 11 c m / min.

3. Establishment of a finite-element model

The geometric model was simulated using a holistic model, with two layers of abdominal structures inside the structure. The weld area is independent body, and the weld split type modeling is convenient for subsequent selection, grid division, loading and other operations.

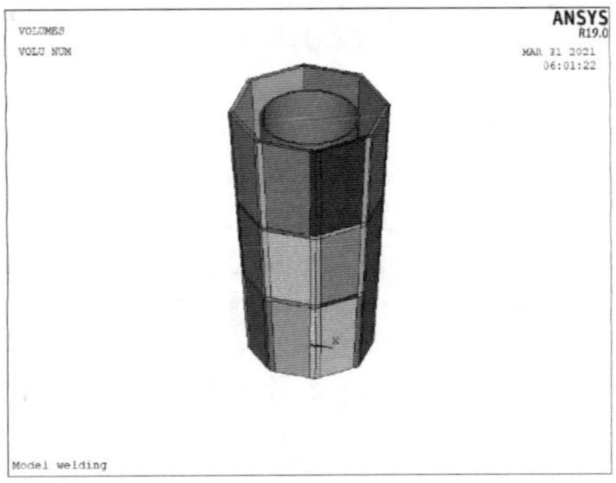

Figure 2. Overall modeling

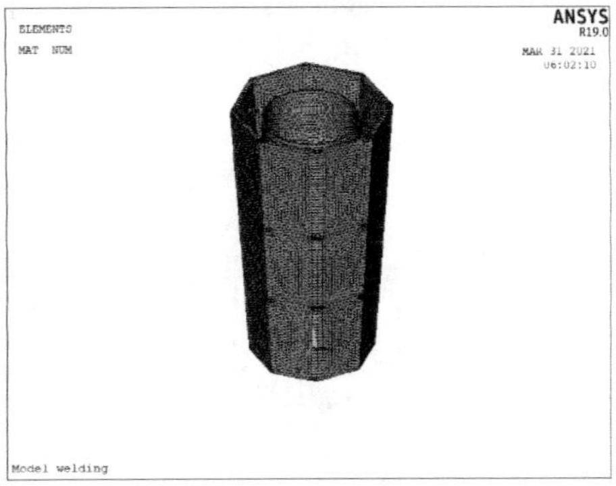

Figure 3. Model meshing

4. Calculation results and analysis

4.1. Welding Temperature Feld Analysis

Temperature field distribution of the first, fourth, eighth weld and 46 minutes after welding are shown in the figure below.

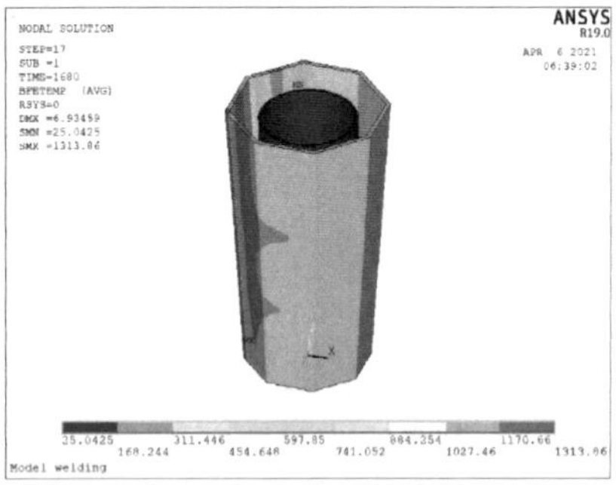

Figure 4. End of first welding

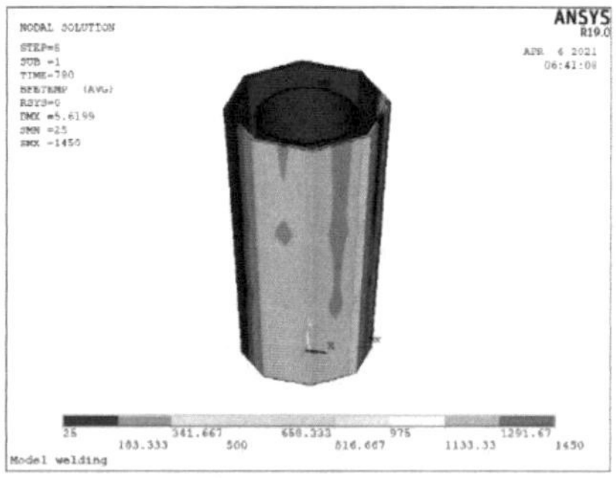

Figure 5. End of the fourth welding

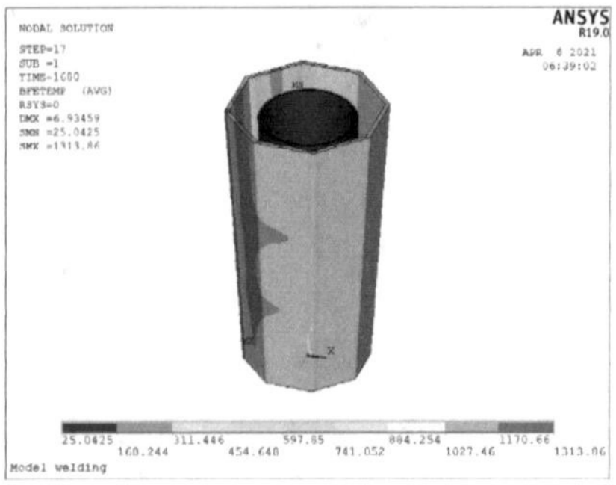

Figure 6. End of the eighth welding

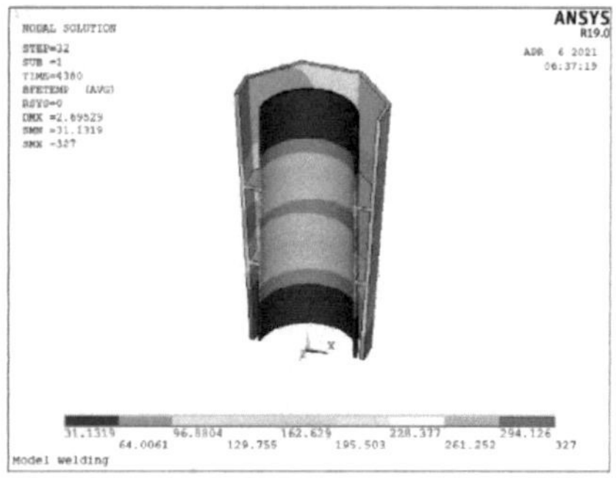

Figure 7. Cooling for 46 minutes after split welding

The results show: (1) At the end of the first weld welding, the welding temperature is 1450℃, exceeding the material melting point, indicating that a good starting welding fusion can be achieved under the welding process conditions.(2) The highest temperature difference after the fourth and eighth welding is not big, which is produced in the weld and near weld area, which is due to the large size of the overall component and the long weld path distance, and the thermal input in the welding process is concentrated in the weld area. However, it can also be seen from the overall temperature cloud map that after the sequential welding, the temperature of the overall parts, especially the outer plate, is gradually increasing, and the heat is gradually transmitted from the weld area to the surrounding area of the weld.(3) After the cooling of the overall structure welding, the maximum temperature is 327℃ and the minimum temperature is 31.12℃.The maximum temperature generation site is in the outer laminate at the small caliber end, and the minimum temperature is generated at both ends of the internal support structure. This is due to the concentration of heat at the small caliber end during transmission, and the heat of the internal tendon plate needs to be transferred through the belly of the two-layer structure, while the inner tendon plate temperature is lower at both ends under air convection heat exchange conditions.

After the overall structural cooling calculation, the solid state phase transition at the temperature and the temperature of the welding repair area has been basically consistent, where the residual stress calculation can be performed.

4.2. Stress Field and Analysis

In this paper, the thermal structure indirect coupling calculation method is adopted to realize the transformation from the thermal analysis Solid70 unit to the structural division Solid185 unit by using the Ansys indirect coupling unit type automatic conversion function. The welding process deformation and residual stress of each load step are calculated by reading the temperature field results and combined with the mechanical related parameters.

The stress field distribution of the first, fourth, eighth weld and 46 minutes after welding is shown in the figure below.

The result is shown: (1) At the end of the first weld welding, the maximum Vonmiss stress is 367Mpa.At the end of the fourth weld weld, the maximum Vonmiss stress is 418Mpa.

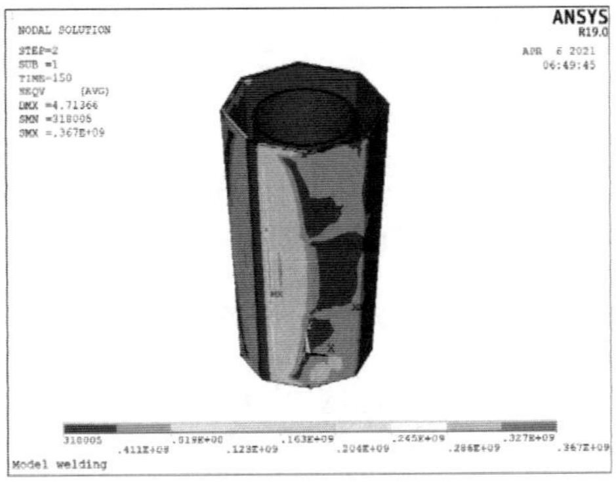

Figure 8. Stress field distribution at the end of the first welding

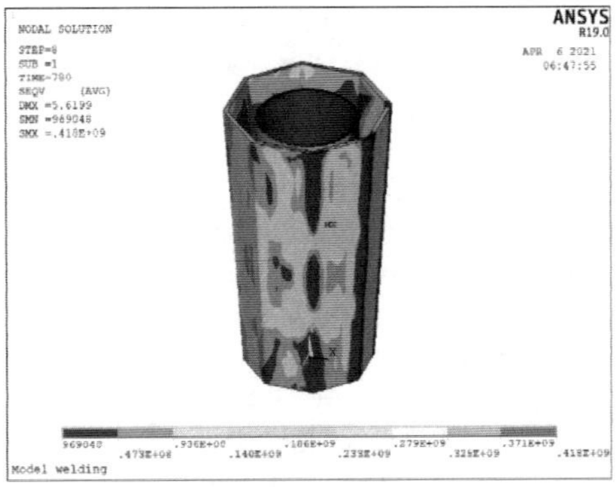

Figure 9. Stress field distribution at the end of the fourth welding

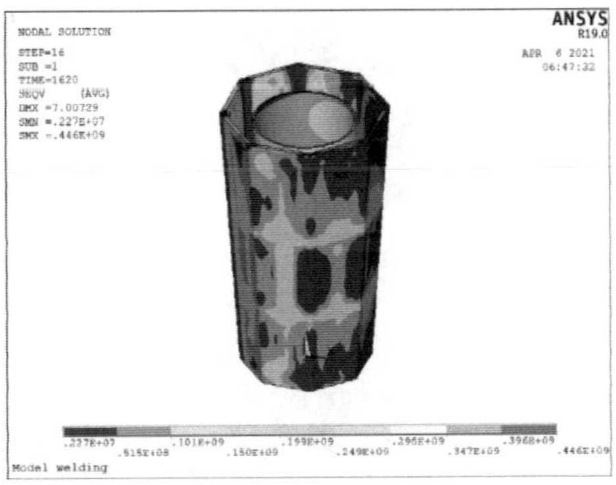

Figure 10. Stress field distribution at the end of the eighth welding

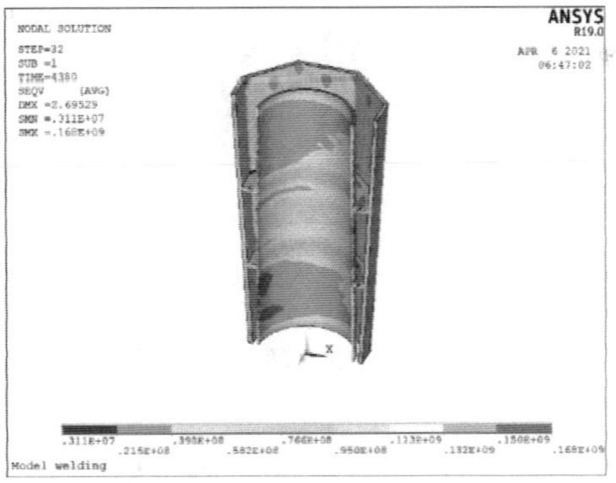

Figure 11. Distribution of stress field 46 minutes after split weldingg

(2) In terms of stress distribution, maximum stress occurs at the end of the eighth weld weld with a maximum v onmiss stress of 446 Mpa. It shows that the dynamic stress of the welding process increases with the deformation and gradually approaches the yield strength of the material. Welds and nearby areas are due to the largest thermal gradient of the weld during welding. In terms of process angle, in order to avoid dynamic thermal cracks, the weld and thermal influence area shall be cooled, that is, insulation cotton shall be covered during interlayer cooling. (3) The maximum value of the residual vonmiss stress is 168Mpa, and the main area is distributed between the inner layer structure and the peritoneal plate structure. Analysis of the reasons, the outer structure deformation is large, then the reaction to the abdomen, the deformation of the overall outer structure and produced a large pull pressure,

resulting in a large residual stress.(4) From the perspective of process optimization, the residual stress is not overconcentrated, and its maximum value is also within the acceptable range, indicating that the overall welding process is reasonable.

4.3. Deformation Analysis

The total residual deformation and the deformation in all directions are shown in the figure below.

The result shows: (1) The maximum residual total deformation is 2.69mm, produced at the large diameter end of the outer tendon plate. (2) Radial deformation X and Z direction are consistent. Although the maximum deformation value is slightly different, namely 1.37 mm and 2.08 mm, respectively, the overall trend depends on the external expansion of the outer tendon plate structure. This is related to the plastic deformation of the thermal expansion during welding.(3) Axial deformation The maximum deformation amount of the axial Y direction is 2.47mm, produced at the large diameter end of the outer tendon plate, which shows that the deformation of the overall structure mainly depends on the axial deformation, that is, the Y direction deformation. Numerically, during the whole welding process, the structural deformation is gradually increasing, which is due to the increase of the amount of welding, heat accumulation, the material under thermal expansion and cold contraction deformation, accompanied by elastic and plastic deformation, the more heat accumulation of the overall structure, the greater the internal deformation and stress.(4) From the perspective of the cloud map distribution, the dynamic deformation process of the welding is more complex, which is also determined by the thermodynamic theory of the welding process. The deformation here is minimum due to fixed constraints at both ends of the internal support structure. As the welding process goes on, the maximum deformation is produced at the large diameter end of the lateral tendon plate, which is consistent with the theories related to structural mechanics, because the role of supporting the abdomen is to limit the large deformation of the overall structure, and the trail end structure is more firm and has anti-deformation ability.

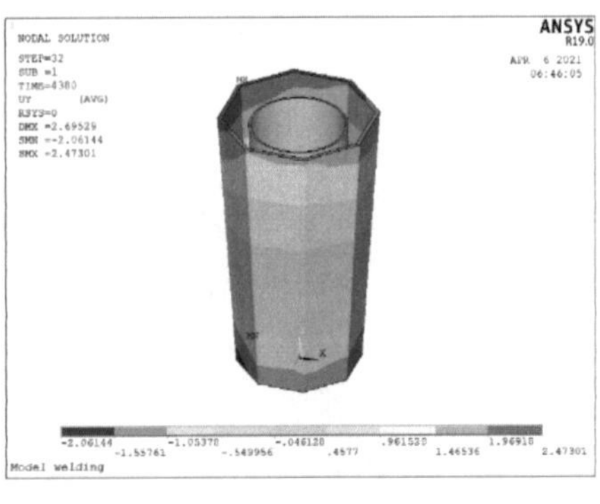

Figure 12. Total residual deformation

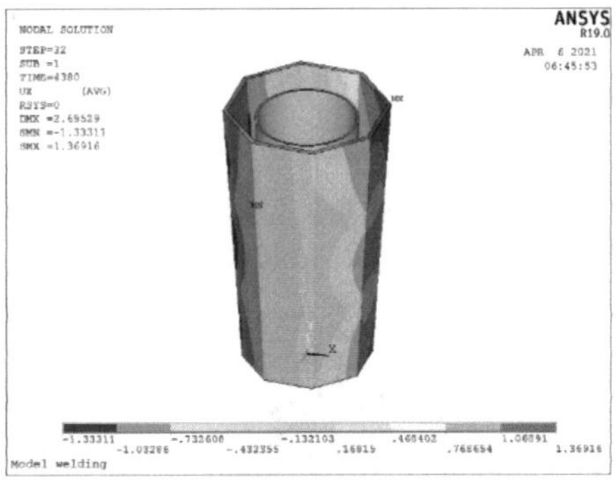

Figure 13. X-direction deformation

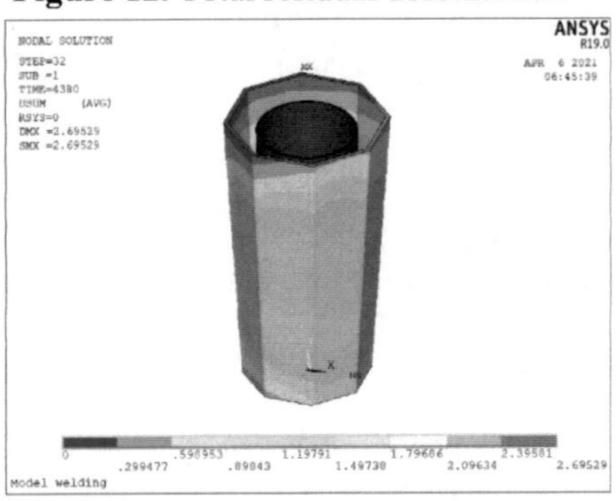

Figure 14. Y -direction deformation

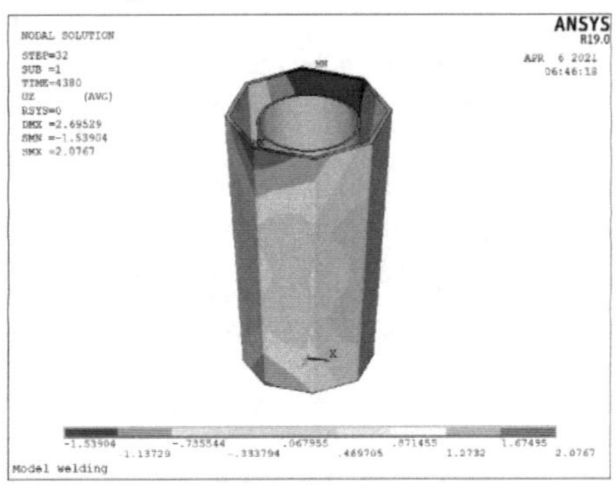

Figure 15. Z -direction deformation

5. Conclusion

Through numerical simulation and analysis of the temperature field and stress field of the head structure:

Accumul of heat during welding. From the process point of view, the heat input should be appropriately increased at the initial welding to obtain a good fusion. The heat input can be appropriately reduced in the later welding repair period to prevent the melting pool temperature from forming a welding hot crack. It can also effectively reduce the composition analysis between welds, and improve the overall strength and tissue uniformity of welds.

For double support structures with interlayer, the maximum temperature generation during welding is at the outer laminate plate at the small caliber end and the minimum temperature at both ends of the internal support structure. Therefore, in the welding process, accelerate the welding speed of small diameter to reduce heat aggregation.

During the whole welding process, the structure deformation and stress are gradually increased, until the cooling process after welding, deformation and stress began to gradually reduce, due to the material high temperature plastic deformation, the residual deformation, mainly from the whole structure of axial deformation, the maximum deformation of the lateral plate, because support web is to limit the overall structure deformation, the path end structure is stronger, with deformation resistance, process through the processing quantity compensation design.

During the cooling process after welding, the initial cooling rate is larger and slower in the later stage. Therefore, in order to avoid thermal cracks after welding, effective measures such as insulation cotton should be covered quickly after welding. In late cooling, hammering can be performed appropriately to release the residual stress within the weld. The overall welding residual stress is small, and the current welding process can be appropriately optimized, but it is already feasible, and the welding process design is reasonable.

Acknowledgments

This work was financially supported by Scientific Research Funding Project of Liaoning Provincial Department of Education- "Study on Welding Deformation of Boring Bar for Serving the Local Economy Talents" (L2020012).

References

[1] Zhang Wenwu, Qian Lin, Han Liang, et al. Structure design of boring head of boring machine of twenty-roller mill rack [J]. Manufacturing Technology and Machine Tool, 2020, 12 (3): 54-57

[2] Qian Lin, Zhang Wenwu. Development of boring machine of twenty roller mill [J]. Science and Technology Innovation, 2017, 36 (28): 23-25.

[3] Chen Xiao. Frame plum blossom hole finishing process scheme [J]. China Heavy Equipment, 2015, 4 (1): 27-29.

[4] Li Chaowen, Wang Yong, Li Liying, et al. Three-dimensional dynamic FEM simulation of temperature distribution of T-joint [J]. Transactions of the China Welding Institution, 2011, 32(8): 33-36.

[5] Brickstad B, Josefson B L.A rarametric study of residual stresses in multi-pass butt-welded stainless steel pipes [J]. International Journal of Pressure Vessels Piping, 1998,75(1): 11-25.

[6] Xie yuanfeng. Numerical simulation study of the welding temperature field and the stress field based on Ansys [D]. Wuhan: Wuhan University of Technology, 2006.

[7] You Min, Zheng Xiaoling. Connection structure analysis [M]. Hubei: Huazhong University of Science and Technology Press, 2004.

[8] Chen Xiangrong, Zhao Zhiyun, Jiang Yuanyuan, et al. Study on residual stress test of box section welding with excessive width ratio [J]. Civil Engineering Journal, 2010,43 (2): 281-285.

[9] Zhang Zhichao, Zhang Guofu, Su Wei, et al. Influence of thermal welding process on welding residual stress of Q235 steel [J]. Metal casting and forging welding technology, 2010, 39 (11): 156-157.

[10] QuXinFang. Study on the Welding Performance of Q235 Steel Plate for Construction [J].Casting technology, 2014, 35 (8): 1825-1827.

Numerical Simulation of Robustness to Large Displacement of Optical Flow Based on Deep Learning

Chao Wang[1], Xiucheng Dong[1,*], Shifu Gu[1], and Zhengyu Zhang[2]

[1]School of Electrical Engineering and Electronic Information, Xihua University, Chengdu, China
[2] High Speed Aerodynamics Institute, China Aerodynamics Research and Development Center, Mianyang, China

*Corresponding author e-mail: dxc136@163.com

Abstract. Traditional optical flow algorithm based on prior is sensitive to large displacement, in order to solve the problems, one optical flow algorithm based on deep learning is introduced to measure the global velocity of fluorescent oil film. The test results of numerical simulation show that the Average Endpoint Error (AEE) of HS, LK and Flownet2 are 2.4604 pixels, 2.3898 pixels and 0.9916 pixels in terms of warping 0.1s respectively, and the error of FlowNet2 is always the lowest under the conditions of large displacement. Therefore, the algorithm is more robust to large displacement than the others that is currently being used, which has certain exploration significance and engineering reference value.

1. Introduction

Reducing the skin friction of aircraft is of great significance for reducing fuel consumption and improving performance [1-3]. When the thickness of oil film is extremely thin, the simplified measurement model for skin friction of fluorescent oil film is related to the velocity, dynamic viscosity and thickness of oil film [4]. The global velocity is usually calculated by the optical flow algorithms.

The optical flow algorithms can be mainly divided into two categories [5]: the traditional optical flow algorithm based on prior knowledge and the optical flow algorithm based on deep learning. The traditional optical flow algorithm mainly establishes a variational model based on some assumptions [6-7], but the assumptions are too ideal to be fully satisfied in real scenes, which makes it more sensitive to environmental variation. Large displacement and illumination variation have been studied by some researchers [8-9]. In recent years, Dosovitskiy et al [10] first introduced the optical flow algorithm based on deep learning, called FlowNet, which used convolutional neural networks to learn the features of adjacent frames. Subsequently, Ilg et al [11] stacked multiple FlowNet networks and used different training strategies to create a new network called FlowNet2, which increased the model size but effectively improved the performance compared to FlowNet.

The oil film moves fast in transonic wind tunnel, and the pixel displacement is large, which has a significant impact on the optical flow algorithms. In this paper, the global velocity measurement of oil film by optical flow algorithm based on deep learning is proposed, and the results of numerical simulation verify the robustness of the algorithm in terms of large displacement, which has certain exploration significance and engineering application value.

2. Optical Flow Algorithm

2.1. HS

Horn and Schunck et al.[6] proposed the traditional optical flow variational model based on three assumptions, called HS optical flow algorithm. First, based on the assumption of constant brightness, that is, the brightness of the same pixel in adjacent frames remains unchanged.

$$I(x, y, t) = I(x+dx, y+dy, t+dt) \tag{1}$$

According to the Taylor series, we have:

$$I(x, y, t) = I(x, y, t) + \frac{\partial I}{\partial x}dx + \frac{\partial I}{\partial y}dy + \frac{\partial I}{\partial t}dt + \sigma \tag{2}$$

Where σ represents second-order infinitesimal that can be neglected under the condition of small displacement. When the time interval approaches 0, we can obtain Eq. (3).

$$\frac{\partial I}{\partial x}\frac{dx}{dt} + \frac{\partial I}{\partial y}\frac{dy}{dt} + \frac{\partial I}{\partial t}\frac{dt}{dt} = 0 \tag{3}$$

Let $u = dx / dt$ and $v = dy / dt$ be the mean velocity in the direction x and the direction y, respectively, we see that,

$$I_x u + I_y v + I_t = 0 \tag{4}$$

Where abbreviations I_x, I_y and I_t are partial derivatives of image brightness with respect to x, y and t, respectively. Let E_{rg} be the gradient smooth energy function.

$$E_{rg}(u, v) = \iint_\Omega (|\nabla u|^2 + |\nabla v|^2)dxdy \tag{5}$$

Then, we can get the energy function of the variational model.

$$E(u, v) = \iint_\Omega [(I_x u + I_y v + I_t)^2 + \lambda(|\nabla u|^2 + |\nabla v|^2)]dxdy \tag{6}$$

Where λ is weight coefficient. Eq. (6) can be solved by using the Euler-Lagrange equation, and the corresponding Euler-Lagrange equations are as follows.

$$\frac{\partial L}{\partial u} - \frac{\partial}{\partial x}\frac{\partial L}{\partial u_x} - \frac{\partial}{\partial y}\frac{\partial L}{\partial u_y} = 0, \frac{\partial L}{\partial v} - \frac{\partial}{\partial x}\frac{\partial L}{\partial v_x} - \frac{\partial}{\partial y}\frac{\partial L}{\partial v_y} = 0 \tag{7}$$

Where L is defined as $L = (I_x u + I_y v + I_t)^2 + \lambda(|\nabla u|^2 + |\nabla v|^2)$. The derivation of Eq. (7) yields,

$$\begin{aligned} I_x(I_x u + I_y v + I_t) - \lambda\Delta u = 0 \\ I_y(I_x u + I_y v + I_t) - \lambda\Delta v = 0 \end{aligned} \tag{8}$$

The Laplacian operator is usually approximated by the difference between the speed of a point and the mean of its surrounding speed, namely,

$$\begin{aligned} \Delta u(x, y) = \bar{u}(x, y) - u(x, y) \\ \Lambda v(x, y) = \bar{v}(x, y) - v(x, y) \end{aligned} \tag{9}$$

Substitute Eq. (9) into Eq. (8),

$$\begin{aligned} (I_x{}^2 + \partial^2)u + I_x I_y v = \lambda\bar{u} - I_x I_t \\ I_x I_y u + (I_x{}^2 + \partial^2)v = \lambda\bar{v} - I_y I_t \end{aligned} \tag{10}$$

Finally, the iterative equations for optical flow velocity are as follows:

$$\begin{aligned} u^{k+1} = \bar{u}^k - I_x(I_x\bar{u}^k + I_y\bar{v}^k + I_t) / (\lambda + I_x{}^2 + I_y{}^2) \\ v^{k+1} = \bar{v}^k - I_y(I_x\bar{u}^k + I_y\bar{v}^k + I_t) / (\lambda + I_x{}^2 + I_y{}^2) \end{aligned} \tag{11}$$

2.2. LK

Lucas and Kanade [7] introduce sparse consistency to estimate flow field. It is the same as HS, Eq. (4) can be obtained based on the constant brightness and small displacement, then if we assume a small neighborhood with the same motion, such as 3×3pixels, yields:

$$\begin{bmatrix} I_{x1} & I_{y1} \\ I_{x2} & I_{y2} \\ \vdots & \vdots \\ I_{x9} & I_{y9} \end{bmatrix} \begin{bmatrix} u \\ v \end{bmatrix} = \begin{bmatrix} -I_{t1} \\ -I_{t2} \\ \vdots \\ -I_{t9} \end{bmatrix} \tag{12}$$

Let $P = \begin{bmatrix} I_{x1} & I_{x2} & \cdots & I_{x9} \\ I_{y1} & I_{y2} & \cdots & I_{y9} \end{bmatrix}^T$, $V = \begin{bmatrix} u & v \end{bmatrix}^T$, and $Q = \begin{bmatrix} -I_{t1} & -I_{t2} & \cdots & -I_{t9} \end{bmatrix}^T$, according to least square method, we have:

$$P^T PV = P^T Q \tag{13}$$

Finally:

$$\begin{bmatrix} u \\ v \end{bmatrix} = \begin{bmatrix} \sum I_{xi}^2 & \sum I_{xi} I_{yi} \\ \sum I_{xi} I_{yi} & \sum I_{yi}^2 \end{bmatrix}^{-1} \begin{bmatrix} \sum I_{xi} I_{ti} \\ \sum I_{yi} I_{ti} \end{bmatrix} \tag{14}$$

Actually, the assumptions are too ideal to be satisfied in real scenes, but pyramid module [12-13] can overcomes the problem to a certain extent. The main idea of that is to reduce the image resolution to obtain a multilayer image pyramid by downsampling, and to correct rough value solved by the traditional optical flow algorithm from top to bottom to obtain the fine flow. By the way, the module is widely used in velocity measurement of fluorescent oil film now. Flowchart of pyramid module is shown in Fig.1.

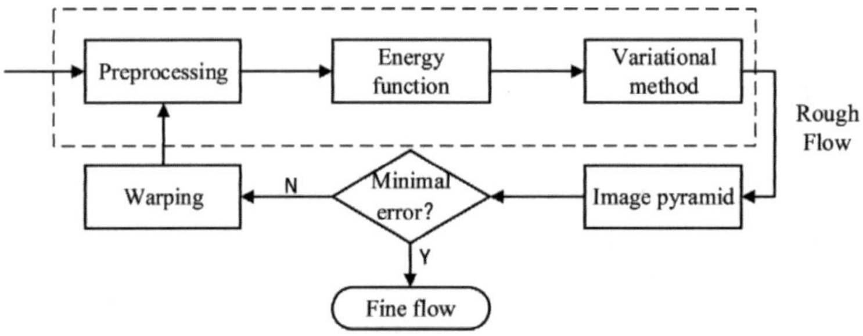

Figure 1. Flowchart of pyramid module

2.3. FlowNet2

In 2015, Dosovitskiy et al. first adopt CNNs to learn optical flow, they design two networks based on encoder-decoder architecture, called FlowNetS and FlowNetC. In the encoder part, FlowNetS stacks adjacent frames images into the network, and directly obtains the features map through six convolution layers. But the input of FlowNetC is divided into two branches, and each branch is composed of three convolution layers that used to extract features map of the corresponding image. Then, the outputs of the two branches are fed into a correlation layer that used to calculate the matching cost of features map. Besides, there are four deconvolution layers in the decoder part, which can expand the scale of features map to a quarter of the original image scale. Finally, a bilinear interpolation method is used to restore the scale to the same as original image size.

As the first-generation, FlowNet has a simple architecture, but single dataset and poor generalization ability, its comprehensive performance cannot reach state-of-the-art of traditional optical flow algorithm. Then, Ilg et al. obtain many variants networks with different scale by stacking

FlowNet. To deal with different displacement, they also introduce a sub-network by using different dataset, named FlowNetSD. Finally, they denote the final network with the best performance as FlowNet2, which marginally runs slower than the original network, but increases the estimation precision by more than 50%. The schematic diagram of FlowNet2 is shown in Fig.2.

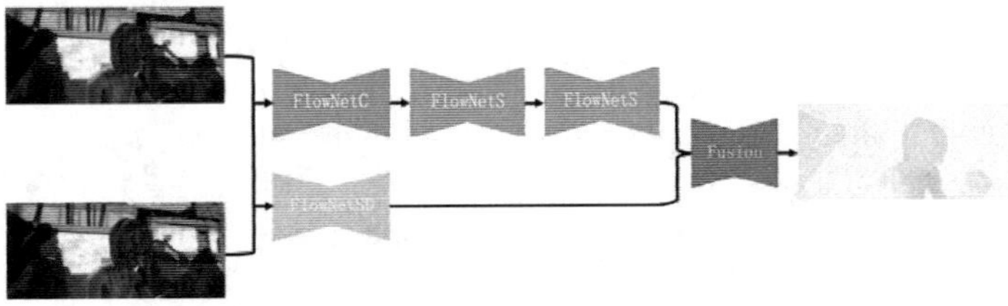

Figure 2. The schematic diagram of FlowNet2.

Flownet2 was trained on Chairs and fine-tuned on Things3D [4], which include a large number of images with different scale displacement. Characteristic of the dataset is shown in Table 1.

Table 1. Characteristic of the dataset.

	Chairs	Things3D
Training frames	22232	21818
Test frames	640	4248
Resolution	512×384	960×540

3. Numerical simulation

Since the ground truth cannot be accurately measured by experimental equipment, the warping method [14-15] is used to compare and analyze the algorithm performance. According to the Oseen vortex velocity field formula:

$$U_r = [1-\exp(-r^2/r_0^2)]\frac{\Gamma}{2\pi r} \quad (12)$$

If we let Γ=3000, $r_0 = 10$, and superimpose it on a uniform velocity field with a size of 10, we can get the simulated velocity streamlines as shown in Fig.3(a), which is used as the ground truth. In order to be as close as possible to the fluorescent oil film image in the real scene, the Jupiter surface image [4] shown in Fig.3(b) is used as the first frame simulation image, with a size of 320×192 pixels.

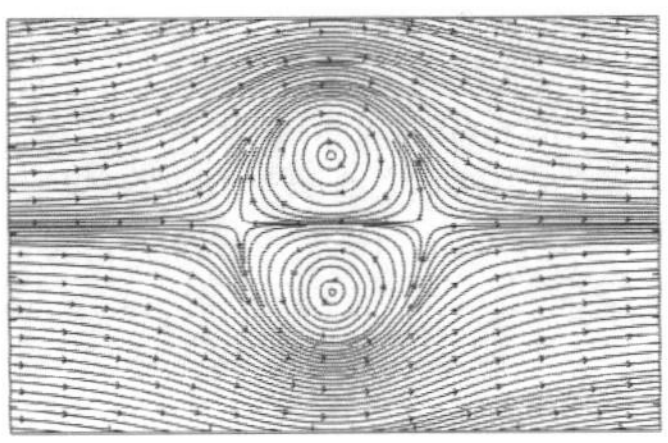

(a)

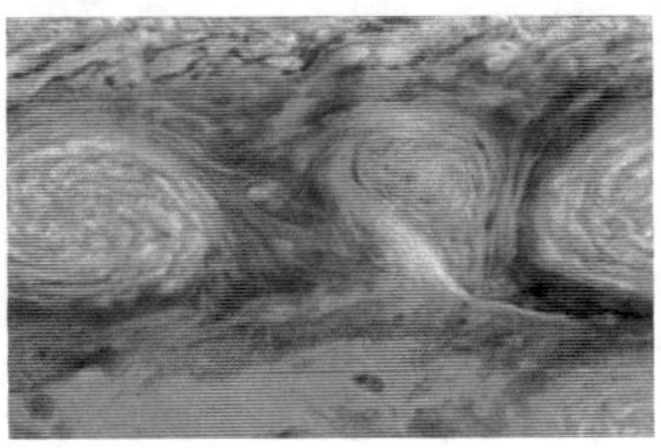

(b)

Figure 3. Simulation images

EPE (Endpoint Error) and AEE (Average Endpoint Error) are used as evaluation indicators [5]. The EPE represents the Euclidean distance between the calculated velocity vector and the ground truth on one pixel, and the AEE is the mean of the EPE for all pixels of the image.

$$EPE_{(i,j)} = \sqrt{(u^t_{(i,j)} - u_{(i,j)})^2 + (v^t_{(i,j)} - v_{(i,j)})^2}$$
$$AEE = \frac{1}{m \times n}\sum_{i=1}^{m}\sum_{j=1}^{n} EPE_{(i,j)} \quad (13)$$

Where (u,v) is velocity vector estimated by optical flow algorithm, and (u^t,v^t) is the ground truth, m and n are scale of the images. Large displacement is a longstanding problem in the optical flow field [15]. On the one hand, there will be a lot of outliers on the image edge because of large displacement, then the image details will be lost. On the other hand, it contradicts the small displacement precondition of the optical flow algorithm, which will directly lead to a significant decrease in the accuracy of the algorithm. In view of that, the above warping method is also used, the time interval is set to 0.01s, then we can obtain a total of 10 groups simulation images between 0.01s and 0.1s, the longer the time, the larger the displacement. The images at 0.01s and 0.1s are shown in Fig.4.

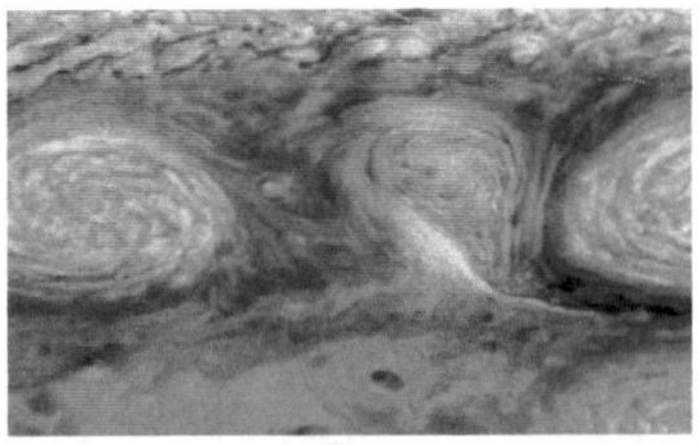
(a)0.01s

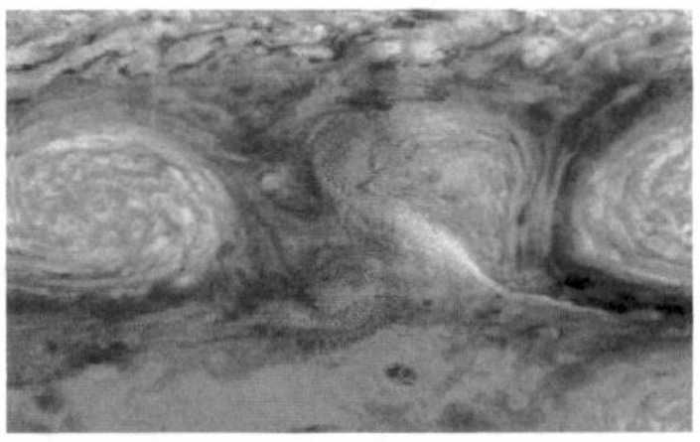
(b)0.1s

Figure 4. Simulation images at 0.01s and 0.1s

Tab 2 shows the AEE of the three algorithms at different warping time, and Fig.5 shows the error curve of them. It is clear that the errors of HS and LK are always higher than that of FlowNet2, and the upward trend is obviously stronger than FlowNet2 when the warping time is long, the maximum can be up to 2.4604 pixels, 2.3898 pixels and 0.9916 pixels respectively at 0.10s. So, the performance for large displacement of FlowNet2 is the best of these algorithms.

Table 2. The AEE at different warping time

Warping time/s	AEE/pixels		
	HS	LK	FlowNet2
0.01	0.5373	0.3953	0.3627
0.02	0.8958	0.5433	0.3926
0.03	1.1635	0.7379	0.4417
0.04	1.3951	0.9912	0.5122
0.05	1.4570	1.1554	0.5467
0.06	1.7104	1.4436	0.6333
0.07	1.7878	1.6739	0.7389
0.08	1.9461	1.8971	0.7887
0.09	2.1896	2.1726	0.9206
0.10	2.4604	2.3898	0.9916

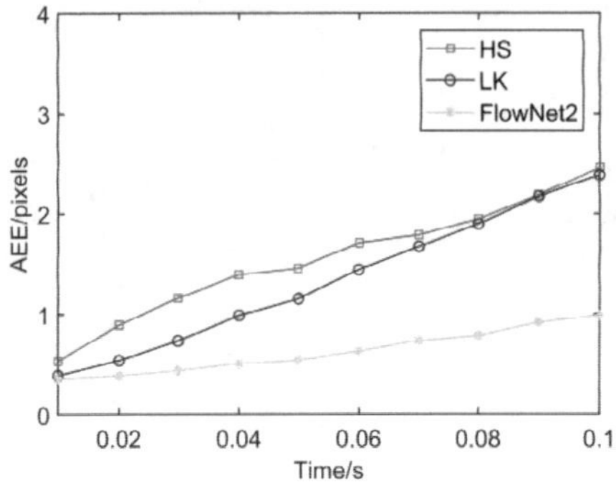

Figure 5. AEE curve at different warping time

4. Conclusion

This paper introduces the algorithm principles of HS, LK and FlowNet2, and compares their robustness to large displacement by numerical simulation. The results show that AEE of the FlowNet2 is always the lowest of these algorithms with large displacement. Therefore, FlowNet2 based on deep learning is more robust to large displacement than the algorithms based on priori, which is of great significance to the research on the skin friction measurement of aircraft. In the future work, we will continue to dig deeper to improve its comprehensive performance and engineering applications.

Acknowledgments

The research was supported by the National Natural Science Foundation (11872069), and Central Government Funds of Guiding Local Scientific and Technological Development for Sichuan Province (2021ZYD0034).

References

[1] XING Yu, LUO Dongming, YU Xiongqing. Optimization strategy of supercritical laminar flow airfoil design[J]. Journal of Beijing University of Aeronautics and Astronautics,2017, 43(8):1616-1624.

[2] WANG Shuai, HE Guoqiang, QIN Fei, et al. Research on skin-friction drag and drag reduction technics in a supersonic inner flow path[J]. Journal of Aerospace Power,2019,34(4):908-919.

[3] QIAN Hongjiang, DONG Xiucheng, ZHANG Zhengyu. Improved prediction model of gray and thickness of fluorescent oil film based on Hankel matrix[J]. Journal of Aerospace Power. 2021,36(10):2061-2071.

[4] LIU T, SULLIVAN J P. Luminescent oilfilm skin-friction meter[J]. AIAA Journal, 2015, 36(8):1460-1465.

[5] ZHAI M, XIANG X, LV N, et al. Optical flow and scene flow estimation: a survey[J]. Pattern Recognition, 2021, 114(3):107861.

[6] HORN B, SCHUNCK B G. Determining optical flow[J]. Artificial Intelligence, 1981, 17(1-3):185-203.

[7] Lucas B, Kanade T. An iterative technique of image registration and its application to stereo. The 7th International Joint Conference on Artificial Intelligence, IJCAI, 674–679.

[8] Qian H, X Dong, Zhang Z, et al. Research on Stability of Gray Value of Excited-State Fluorescent Oil Film Based on Variable Light Vector Angle[J]. Mathematical Problems in Engineering, 2021.

[9] WEINZAEPFEL P, REVAUD J, HARCHAOUI Z, et al. DeepFlow: Large displacement optical flow with deep matching[C]. IEEE International Conference on Computer Vision. IEEE, 2014.

[10] FISCHER P, DOSOVITSKIY A, et al. FlowNet: learning optical flow with convolutional networks[C]. IEEE International Conference on Computer Vision(ICCV). IEEE, 2015.

[11] ILG E, MAYER N, SAIKIA T, et al. FlowNet 2.0: evolution of optical flow estimation with deep networks[J]. IEEE, 2017.

[12] BROX T, BRUHN A, PAPENBERG N, et al. High accuracy optical flow estimation based on a theory for warping[J]. ECCV, 2004.

[13] A. Ranjan, M.J. Black. Optical flow estimation using a spatial pyramid network[C]. IEEE Conference on Computer Vision and Pattern Recognition (CVPR), 2017, pp. 2720–2729

[14] LI Peng. Studies of direct measurement techniques about global skin friction[D]. Nanjing: Nanjing University of Aeronautics and Astronautics, 2012.

[15] ZOU Yifeng, ZHANG Zhengyu, WANG Xueyuan, et al. Velocity measurement of fluorescent oil film path movement on wind tunnel testing model surface[J]. Acta Aeronautica et Astronautica Sinica, 2019(6):14.

CFD analysis of flow field and hydrodynamic compensation of poppet valve

Ruichuan Li, Jilu Liu*, Xinkai Ding, Yi Cheng and Qi Liu

College of Mechanical and Automotive Engineering, Qilu University of technology, Jinan, China

*Corresponding author e-mail: liugiroud@163.com

Abstract. The steady-state hydrodynamic force of cartridge valve poppet valve core reduces the stability of cartridge valve. A circular arc structure is added to the main valve core of the poppet valve to compensate the steady-state flow force. Referring to the structure and parameters of the actual cartridge poppet valve, the three-dimensional geometric model of the flow channel in the valve is established with UG software. CFD method is used to calculate and analyze the flow state of fluid in the poppet valve under the working state of the valve core, and the influence of the arc size and position of the valve core on the steady-state hydrodynamic compensation of the main valve core is studied. The results show that the circular arc structure can realize the hydrodynamic compensation of the valve core and increase the stability of the valve core. When the opening amount is small, properly increasing the height h of the circular arc structure can increase the static hydrodynamic compensation of the valve core; Keep the circular arc structure size L less than 1mm, and the static hydrodynamic compensation effect of the valve core is better.

1. Introduction

As an important part of the hydraulic system, the static and dynamic characteristics of the hydraulic poppet valve have an important impact on the stability of the hydraulic system [1]. The hydraulic force has a great relationship with the dynamic and static characteristics of the valve, which is an important factor to be considered in the design of the hydraulic valve. Flow force is divided into transient flow force and steady-state flow force, in which steady-state flow force has a great influence on the static characteristics of hydraulic valve. Poppet valve is one of the main structural forms of hydraulic valve, especially in cartridge valve technology. Most of the poppet valves widely used in cartridge valves are flat bottom poppet valves [2]. Gao used the finite element method to simulate the internal flow field of the external flow poppet valve, approximated it with two-dimensional flow action, and studied the hydrodynamic conditions under different opening degrees, different valve core structures and different valve seat sizes [5]; M. Borghi [3] used CFD software to analyze the pressure distribution diagram, velocity distribution diagram and the curve diagram of flow, pressure and differential pressure of the poppet valve when the fixed opening is 0.4, 0.7, 1.0, 1.1mm. The steady-state flow force of the valve is calculated through the pressure integral on the valve core surface, and the ratio of transient flow force and steady-state flow force is estimated according to the formula. Si studied the influence of valve core diameter on the flow field characteristics and steady-state hydrodynamic characteristics of poppet valve by numerical simulation technology. The results show that the change of valve core diameter has little effect on the flow and hydrodynamic characteristics of poppet valve [8-9].

Aiming at the outflow condition of the poppet valve, taking the main valve core of a cartridge valve as the object, adding an arc structure on the main valve core, the three-dimensional flow field model of the poppet valve core is established, and the grid is divided in ICEM CFD. CFD is used to simulate the flow field of the original structure and the main valve with compensation structure, and the stress of the valve core is analyzed.

2. Geometric model of poppet valve

The geometric model of the flow channel in the valve can be established by using the internal program of FLUENT software, but it is not suitable to construct complex geometry due to functional limitations. The advantage of using UG software to model the research object is that it can draw very complex geometry and extract the runner model in fluent. Figure 1 is the geometric model of the poppet valve required for calculation. The model size adopts the standard cartridge valve with 18 mm diameter.

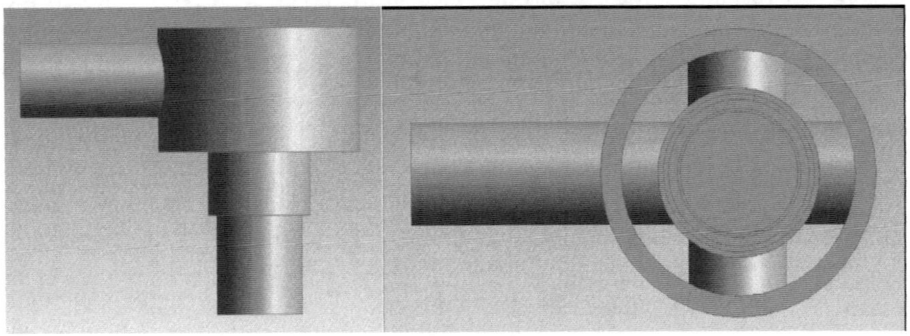

Figure 1. Geometric model of poppet valve

3. Spool geometry

The improved structure of poppet valve core with arc is shown in Figure 2.

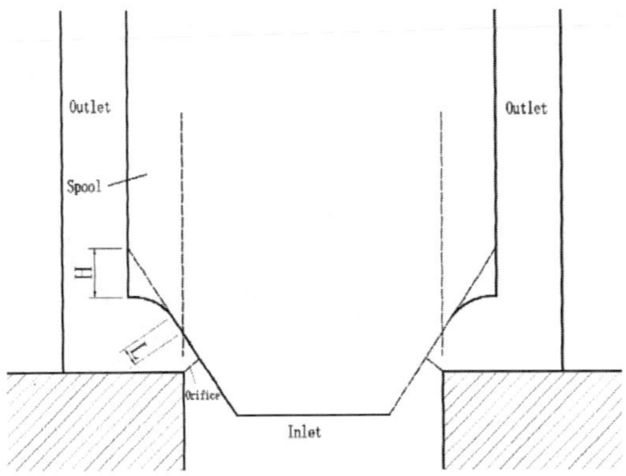

Figure 2. Arc structure of main valve core

According to the structural parameters of the poppet valve, through the analysis of the flow field in the valve, the velocity and pressure distribution of the flow field in the valve can be obtained, and then the flow of the poppet valve under different structural parameters and different opening can be obtained by integration; The flow coefficient reflects the flow capacity of the valve, which is expressed by the following formula:

$$C_d = \frac{Q}{A(x)\sqrt{\frac{2\Delta P}{\rho}}} \tag{1}$$

$A(x)$ is the flow area at the throttle of the valve port, Q is for flow, ρ is density of fluid medium in poppet valve, ΔP is the pressure difference between inlet and outlet, x is the valve core opening, α is the half angle of the top angle of the poppet valve core, d is the diameter at the inlet.

$$C_d = \frac{Q}{\pi dx\left[1 - \frac{x\sin(2\alpha)}{2d}\sin\alpha\sqrt{\frac{2\Delta P}{\rho}}\right]} \tag{2}$$

When the opening of the valve core is fixed, the force acting on the valve core due to the change of momentum when the hydraulic oil passes through the throttle is the steady-state flow force, which can be decomposed into axial and radial components in the treatment of steady-state flow force. Because the oil hole set on the valve sleeve will offset the radial force in the circumferential direction of the valve core, the force affecting the performance of the poppet valve is axial. Through FLUENT software, we can get the pressure distribution of the internal flow field of the poppet valve, and then get the force on the surface of the valve core through the integration method, so as to get a more accurate flow force on the valve.

The steady-state flow force of the internal flow of the poppet valve is:

$$Fs = \rho Q\left(v_{2x} - v_{1x}\right) = \rho Q\left(v_2 \cos\alpha - v_1\right) = \Delta P\frac{\pi}{4}d^2 - F \tag{3}$$

$$Fs = \rho Q\left(v_2 \cos\alpha - v_1\right) = C_d \pi dx\left(1 - \frac{x\sin(2\alpha)}{2d}\right)\sin\alpha\sqrt{\frac{2\Delta P}{\rho}} \tag{4}$$

Fs is steady-state flow force; F is the axial force of the liquid on the valve core.

According to formula (3), increasing F can reduce the steady-state flow force of the valve core; According to formula (4), reducing the half angle of the top angle α of the poppet valve core can reduce the flow force of the valve core.

4. Computational modeling

4.1. Meshing

The three-dimensional model of the flow field of the main valve core is established in UG, and then imported into ICEM CFD to divide the tetrahedral mesh. Due to the throttling effect, the liquid pressure and velocity gradient at the valve port are large, so the grid is locally densified here. The grid of the flow field of the main valve core is shown in Figure 3. In order to compare with the main valve with circular arc structure, the original model of the flow field of the main valve is established, the grid is divided, and CFD simulation is carried out under the same conditions.

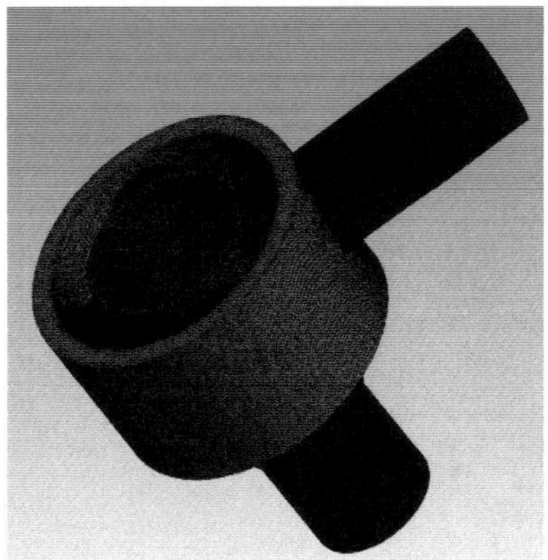

Figure 3. Tetrahedral grid of main valve flow field with arc structure

4.2. Simulation conditions and parameter settings

The following assumptions are made for the numerical simulation:

1. The fluid is a constant Newtonian fluid, and the fluid is incompressible.
2. Ignore the gravity of the fluid itself and the influence of heat transfer inside the valve cavity.
3. Assuming that the fluid flow state is turbulent, the standard k—ε turbulence calculation model is adopted in fluent numerical simulation.
4. The density of fluid medium in the valve chamber is $870\,kg/m^3$ and the dynamic viscosity is $0.04\,Pa\cdot s$.

5. Analysis of simulation results

5.1. Influence of arc parameter h on steady-state hydrodynamic compensation

Keep the valve core opening of 0.4mm and the pressure difference between inlet and outlet of 1MPa. CFD simulation is carried out for the original valve core structure and the valve core flow field when h = 0.7mm and L= 0.3mm. As shown in Figure 4, the pressure of the valve core channel with circular arc structure is significantly higher than the original structure, so the axial force increases and increases. According to equation (3), the flow force decreases.

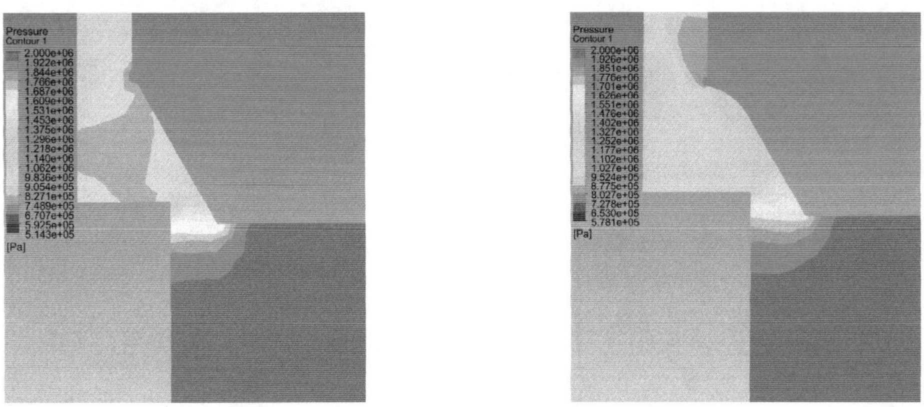

(a) (b)

(a) Original structure; (b) H=0.7mm, L=0.3mm

Figure 4. Pressure nephogram

Keep the opening of the valve core unchanged at 0.4mm, the arc size parameters h are 0.4mm, 0.7mm and 1.0mm respectively, and the arc size L is 0.3mm. Simulate the steady-state flow force of the valve core under different inlet and outlet pressure differences, and the results are shown in Fig. 5.

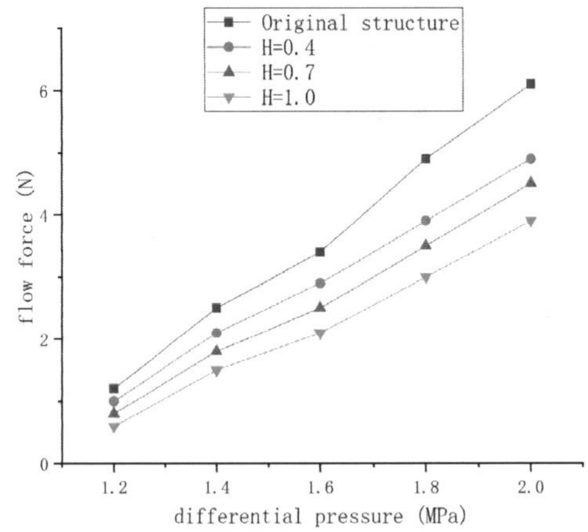

Figure 5. Steady state flow force of valve core under different differential pressure

It can be seen from Fig. 5 that the steady-state flow force of the poppet valve core with the original structure increases linearly with the increase of the pressure difference, which is consistent with the theory; The conical valve core with circular arc structure has a radial guiding blocking effect on the liquid flow direction. The angle between the speed of oil flowing out of the valve port and the axial direction of the valve core increases, which greatly reduces the momentum change of the liquid axial direction. The steady-state flow force is less than the steady-state flow force of the conical valve core with original structure, and the decrease gradually increases with the increase of differential pressure; The larger the arc size h is, the greater the radial guiding effect on the liquid flow direction is, the smaller the steady-state hydrodynamic value is, and the more sufficient the compensation is.

The pressure difference between the inlet and outlet of the main valve is 1MPa, the arc size L is 0.3mm, and the arc size h is 0.3mm respectively 4 mm, 0. 7 mm, 1.0 mm. Simulate the changes of the steady-state flow force of the original structure of the main valve and the structure with circular arc under different opening quantities. The results are shown in Fig. 6.

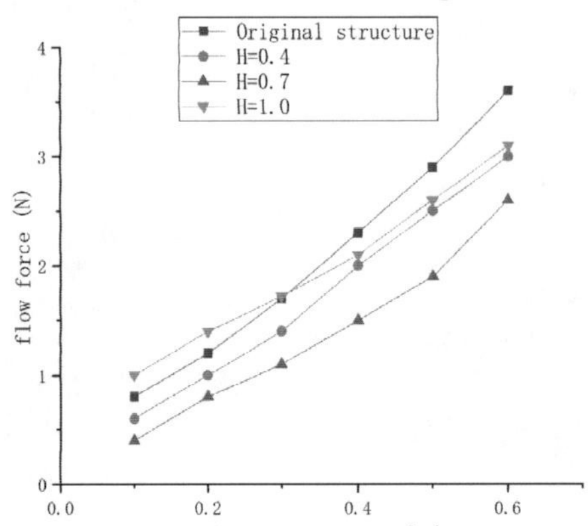

Figure 6. Steady state flow force of valve core with different arc size H

It can be seen from Figure 6 that under each opening amount, when the arc size h of the valve core is less than 0.7mm, the steady-state flow force of the valve core with arc compensation under different opening amounts is significantly less than that of the original structure; H is 1mm. When the opening amount is small, the flow force is greater than the original structure, resulting in over compensation. However, with the increase of the opening amount, the steady-state flow force of the circular arc structure spool gradually becomes positive compensation. As shown in the velocity cloud diagram in Figure 5, because h is too large, the arc structure has too much effect on the reduction of the cross-sectional area of the flow channel. When the opening amount is small, the flow of the flow field is too small, the velocity increases significantly, and the axial force F on the valve core decreases. According to formula (3), the flow force increases, resulting in over compensation. For a single poppet valve, hydrodynamic over compensation has an impact on the stability of the valve. Therefore, the smaller the over compensation, the better. The circular arc structure H also has a great influence on the steady-state hydrodynamic compensation effect. When the opening is small, the longer the length of arc dimension h, the greater the over compensation. When the flange length is less than 0.7 mm, there is no over compensation, and the compensation effect is good.

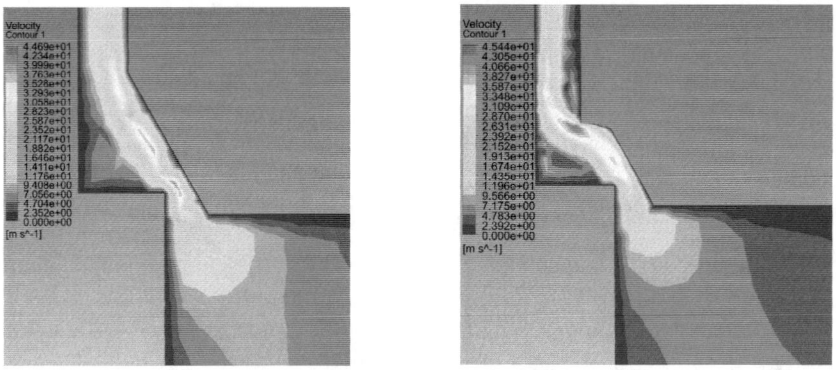

(a) (b)

(a) Original structure; (b) H=1.1mm, L=0.3mm

Figure 7. Velocity cloud chart

5.2. *Influence of arc dimension L on steady-state hydrodynamic compensation*

The pressure difference between the inlet and outlet of the main valve is 1MPa, the arc height h is kept unchanged at 0.7mm, and the arc size distance L is changed to 0.3mm, 0.7mm and 1.1mm respectively. CFD simulation is carried out to observe the hydrodynamic compensation effect of the valve core under different opening quantities. The results are shown in Fig. 8.

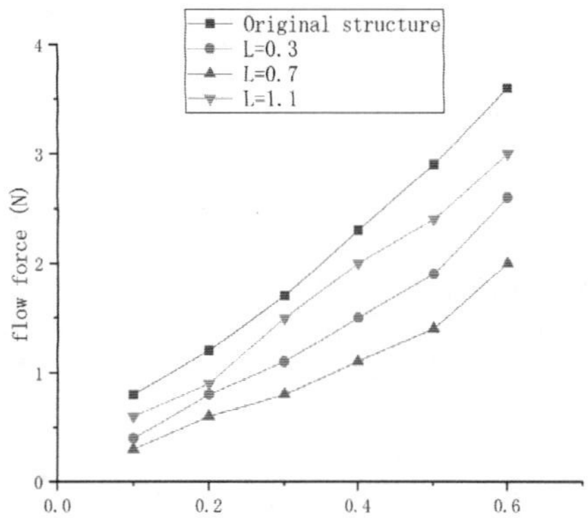

Figure 8. Steady state flow force of L valve core with different arc sizes

It can be seen from Fig. 8 that the steady-state flow force of each valve core with arc size L increases with the increase of opening degree; For the valve core with circular arc, when l is less than 0.7mm, the steady-state flow force decreases with the increase of L, and the compensation effect is better. This is because the larger L is, the greater the distance between the circular arc structure and the valve seat is, the larger the flow area of the channel increases and the liquid velocity decreases. According to Bernoulli equation, the velocity decreases, the pressure increases and f increases. According to formula (3), the steady-state flow force decreases; When the arc size L increases, the height of the flange structure from the valve seat is too large, the diversion and blocking effect of the arc structure on the liquid is greatly reduced, and the hydrodynamic compensation effect becomes worse.

6. Conclusion

This paper presents an improved scheme to reduce the flow force of the poppet valve. By increasing the circular arc structure to block and guide the liquid in the channel, increasing the jet angle at the outlet, reducing the momentum of the oil in the axial direction and reducing the hydraulic force. Through the analysis of the structural parameters and flow conditions of the valve core, we come to the following conclusion:

(1) The poppet valve core with circular arc structure can realize steady-state hydrodynamic compensation, and the compensation amount increases gradually with the increase of differential pressure;

(2) When l remains unchanged and H is small, the hydrodynamic compensation of circular arc structure increases with the increase of valve core opening; When h reaches 1mm and the opening is small, there will be less over compensation;

(3) Keeping h unchanged, when l increases to 1mm, the hydrodynamic compensation effect of circular arc structure will decrease.

Acknowledgments

This work was financially supported by Development of high pressure common rail fuel injection system for diesel engine fund. The project code is 2021CXGC010207.

References

[1] Yan. H, Li. J, Cai. C, Numerical investigation of erosion wear in the hydraulic amplifier of the deflector jet servo valve, Appl. Sci, 2020,pp.1299.

[2] Qu. D.H, Zhou. Y .S, Luo. W, Influences of fit clearances on steady-state fluid forces of relief valves, China Mech.Eng, 2018, pp. 893–899.

[3] M. Borghi, M. Milani, R. Paoluzzi, Transient Flow Force Estimation on the Pilot Stage of a Hydraulic Valve, fluid Power systems and teehnology, 1998, pp. 15–162.

[4] Lu. Q.Q, Tiainen. J, Kiani-Oshtorjani. M, Wu. Y.F, Lateral force acting on the sliding spool of control valve due to radial flow force and static pressure, IEEE Access, 2021, pp.126658–126669.

[5] Gao. D,R, Wang, Y,Q, Numerical simulation and experimental visualization of the flow field inside hydraulic control poppet valve, Chinese journal of mechanical engineering, 2002,pp. 66-70.

[6] Hailing. An, Jungsoo. Suh, Michael. W. Plesniak, 2003 ASME International Mechanical Engineering Congress, 2003, pp. 457-465.

[7] Qian. Long, Wang. S.J, Numerical Simulation of Steady State Hydrodynamic Force and Surface Pressure Distribution of Poppet Valve, Journal of Gansu Sciences, 2017, pp. 69-72.

[8] Si. J.Y, Li. H, Xie. Y.H, The Effect of Valve Core Diameter on the Performance of Hydraulic Poppet Valve, Machine Tool & Hydraulics, 2010, pp. 66-68.

[9] Si. J.Y, Study on the Effect of Cone Diameter to the Cone Applied Force for Hydraulic Poppet Valve, Hydraulics Pneumatics & Seals, 2013, pp. 22-23.

[10] Sheng. J.C, Discussion about Direction of Flow-force on Poppet Valves under Inlet Flow with

Authors of Some Textbooks on Hydraulic Transmission, Chinese Hydraulics & Pneumatics, 2006, pp. 26-28.
[11] O. Shigeru, I. Tsuneo, Cavitation phenomena and performance of oil hydraulic poppet valve: 2nd report, influence of the chamfer length of the seat and theflow performance, Bull. JSME 28 , 1985, pp. 2272–2279.
[12] G. Hong, F. Xin, Y. Huayong, T. Tetsuhiro, Numerical investigation of cavitating flow behind the cone of a poppet valve in water hydraulic system, J. Zhejiang.

Design and analysis of a new heavy oil pump based on CFD

Yi Sun [a], Fanxiang Meng [b], Lichao Pan [c], Xu Wang [d]

Department of Mechanical Science and Engineering, Northeast Petroleum University, Daqing, China

[a] 332667988@163.com, [b] 2261849770@163.com, [c]panlichao1126@163.com, [d]514521086@163.com.

Abstract. At present, submersible electric pump is the main oil production equipment in the world's oil fields. Because the pump is tested with clean water as the medium when it leaves the factory, it can not be widely used in all oil fields. In this paper, a prototype submersible electric pump is improved and designed according to the actual working conditions, and a new heavy oil pump is obtained, which can be widely used in heavy oil Wells. The static pressure distribution and velocity vector distribution under the actual working conditions were simulated by CFD, and the feasibility was verified. It provides theoretical basis and technical guidance for solving the adverse effect of high viscosity produced liquid on pump, and has certain guiding significance for oilfield pump selection. So as to effectively save electricity and rationally exploit energy.

Keywords: viscosity, CFD, electric submersible pump, heavy oil, electricity.

1. Introduction

In this era of energy shortage, esp production has become the main means of stable production in the world's oil fields. The performance and working stability of esp are directly related to the production efficiency of oil well [1-2]. To achieve efficient and green oil recovery, you need to select the right pump for different well conditions and reduce pump failures, which can effectively avoid the waste of electricity and energy. There are considerable heavy oil reserves in the world, but heavy oil contains a large amount of asphalt and gum and has a high viscosity, so it is relatively difficult to exploit [3-5].

2. Simulation analysis of impeller performance of new heavy oil pump

The traditional submersible electric pump unit is composed of multistage centrifugal pump, submersible motor, protector, separator and tubing. In order to make it more suitable for lifting high viscosity produced liquid, we widen the flow channel of the multistage centrifugal pump and reduce the surface roughness of the flow channel, so that the multistage centrifugal pump has better hydraulic performance [6] (Fig. 1).

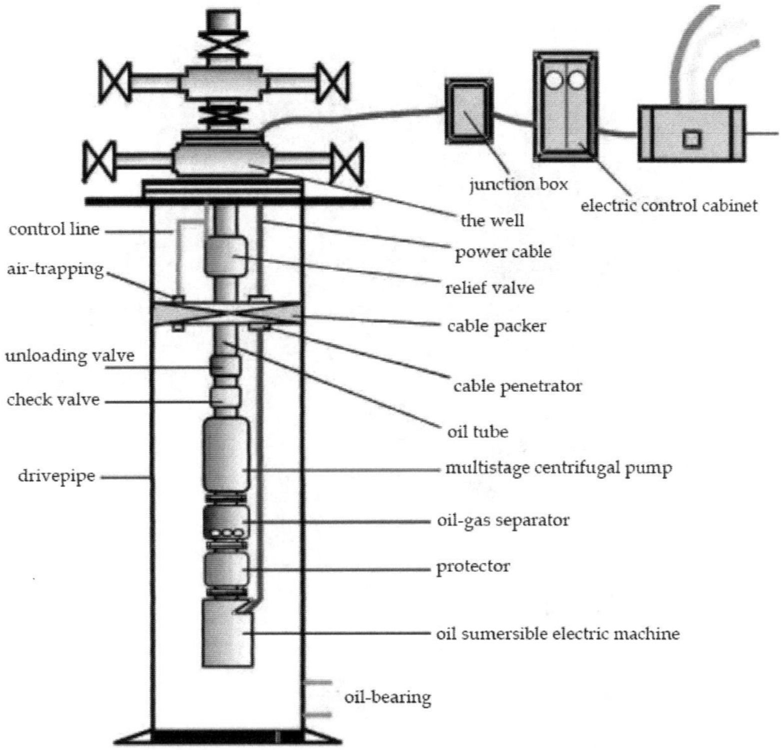

Figure 1. Schematic diagram of submersible electric pump unit

2.1. 3D modeling of a new heavy oil pump

According to the design parameters and structural parameters of the impeller of the new heavy oil pump, such as the rated speed of the pump is 2900r/min, the rated flow is 160m^3/d, and the single-stage head is 8.55m, the three-dimensional model of the vane vane is established by Solid Works (Figs. 2-4).

Figure 2. Impeller

Figure 3. Guide vane

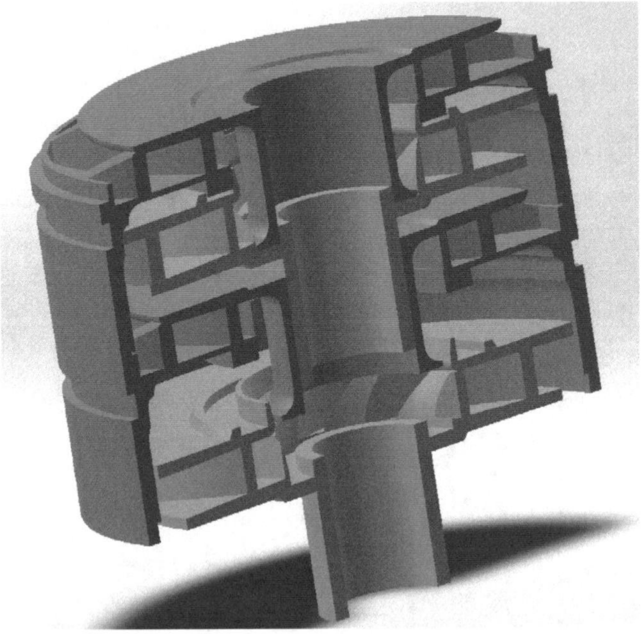

Figure 4. Impeller guide vane fit (two-stage)

2.2. Mesh

This paper uses Autogrid togrid the flow channel. The total number of grids is 445609, impeller passage is 213661, regulating wheel passage is 231948. After inspection, the grid indicators all meet the calculation requirements (Fig. 5).

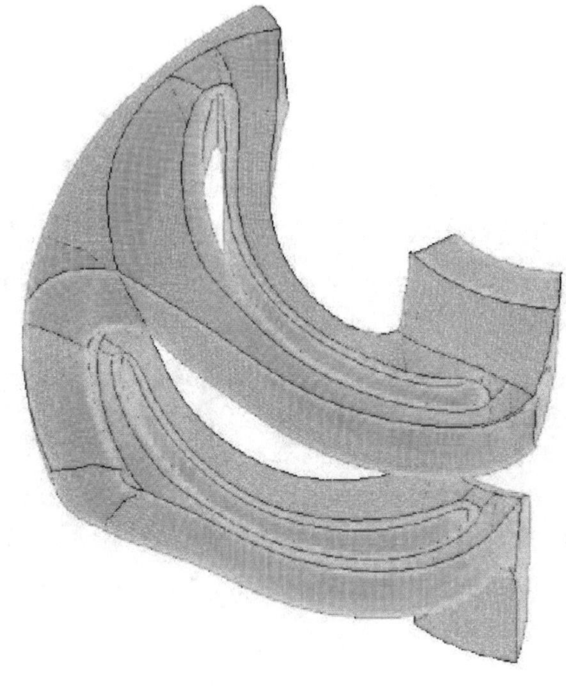

Figure 5. Flow channel three-dimensional grid diagram

3. CFD simulation calculation

(1) With clean water as the medium, the displacement of 100m3/d, the prototype submersible electric pump simulation calculation, other working conditions remain unchanged (Fig. 6).

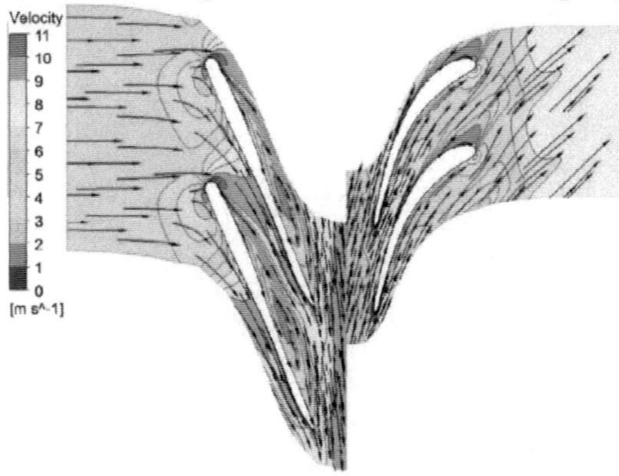

Figure 6. Cloud diagram of velocity vector distribution with viscosity μ=1mPa·s and displacement of $100m^3/d$

(2) Taking the heavy oil of B oilfield as the medium and the displacement of $100m^3/d$, the simulation calculation of the prototype esp was carried out, while other working conditions remained unchanged (Figs. 7-8).

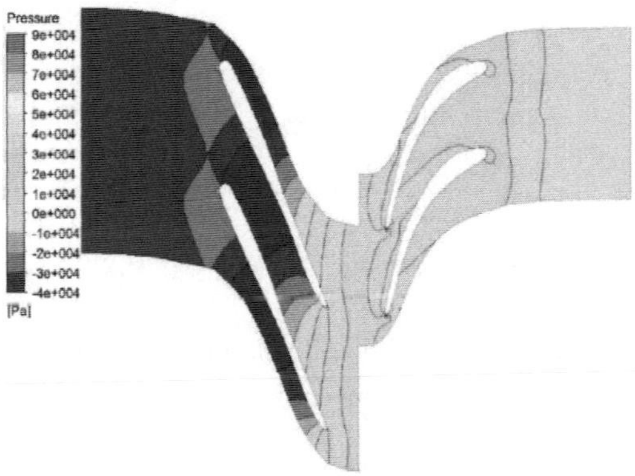

Figure 7. Cloud diagram of static pressure distribution at viscosity μ=320mPa·s and displacement of $100m^3/d$

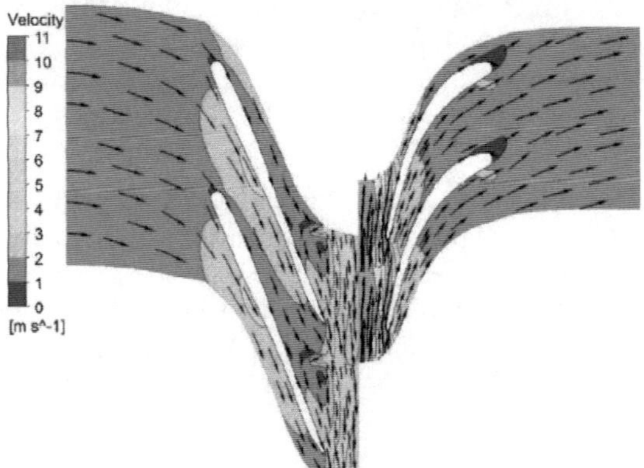

Figure 8. Cloud diagram of velocity vector distribution with viscosity μ=320mPa·s and displacement of $100m^3/d$

(3) B oilfield heavy oil as the medium, displacement 160m^3/d, flow channel surface roughness reduced, the new heavy oil pump simulation calculation, other conditions remain unchanged (Fig. 9).

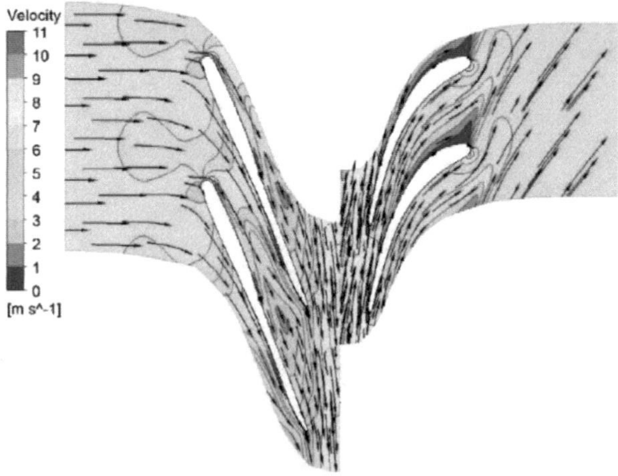

Figure 9. Cloud image of velocity vector distribution when viscosity μ=320mPa·s and displacement 160m^3/d

4. Conclusion

It is proved that the model can effectively deal with the production problems of B oilfield in theory, and can play a guiding role in selecting pump in the future. It is of great significance to select suitable pumps for different types of crude oil with different viscosity for efficient operation of pumps and efficient production of oil fields. Then can save a lot of manpower and electricity, more reasonable and efficient green mining energy.

References

[1] Xu Jianjun, Huang Lida, Yan Limei, Yi Na. Insulator Self-Explosion Defect Detection Based on Hierarchical Multi-Task Deep Learning. Transactions of China Electrotechnical Society, 2021, 36(07): 1407-1415.

[2] Yan Limei, LIU Yongqiang, XU Jianjun, et al. Broken string diagnosis of composite insulator based on Grabcut segmentation and filler area discrimination. Power System Protection and Control, 2021, 49(22): 114-119.

[3] Yi, Q. Wang, L. Yan, et al., A multi-stage game model for the false data injection attack from attacker's perspective. Sustainable Energy Grids & Networks 28 (2021).

[4] Na Yi, Jianjun Xu, Limei Yan, Lin Huang. Task Optimization and Scheduling of Distributed Cyber-physical System Based on Improved Ant Colony Algorithm. Future Generation Computer Systems, 109(Aug. 2020), 134-148.

[5] Yang Zhao, Jianjun Xu, Jingchun Wu. A New Method for Bad Data Identification of Oilfield Power System Based on Enhanced Gravitational Search-Fuzzy C-Means Algorithm. IEEE Transactions on Industrial Informatics. VOL. 15, NO. 11, NOVEMBER 2019, pp. 5963-5970.

[6] Jing Han, Xi Wang, LiMei Yan, Aida Dahlak, et al. Modelling the performance of an SOEC by optimization of neural network with MPSO algorithm. International Journal of Hydrogen Energy, Volume 44, Issue 51, 22 October 2019, pp. 27947-27957.

CFD-DEM based wind-deposited sand high concentration slurry pipe transport characteristics and erosion study

Boqiang Wu*

Xi'an Research Institute Co. Ltd., China Coal Technology and Engineering Group Corp, Xi'an 710077, China

*Corresponding author e-mail: wuboqiang@cctegxian.com

Abstract. In order to study the conveying process of wind-accumulated sand slurry pipeline, this study simulates the flow process of wind-accumulated sand slurry at 1m/s inlet flow rate with straight pipe bend and uphill by coupled CFD-DEM method. The fluid flow, particle motion and position distribution, pipe erosion and force during the whole process are analyzed based on the selection of a suitable traction model. The results show that the flow velocity of highly viscous slurry at the wall of the pipe is small which can easily lead to the deposition of sand particles, which is proved by the particle velocity cloud diagram. The extent of erosion and the degree of erosion increase with time, and the peak of erosion occurs at the bend. The force changes along the Y-direction in different parts of the pipe show that the extreme values of force and erosion are located at the bend. This part is the weak point of the whole pipeline slurry flow and erosion project should pay attention to reinforcement or monitoring.

Keywords: wind-bound sand slurry; pipeline transport; CFD-DEM; flow and erosion.

1. Introduction

The problem of particle deposition and erosion and wear of pumping pipes at bends as well as uphill in coal mine projects conveying wind-accumulated sand slurry filling projects is one of the big problems often encountered in pumping gangue slurry projects containing two-phase flow. In the study of concrete fluidity, Wang Z et al. found that the hydration of cement will produce a large number of hydration products such as Ca(OH) 2 and C-S-H, which will increase the viscosity of the flowing slurry system, thus increasing the slump of concrete through time loss, so the essence of the study of concrete slump loss lies in the study of concrete Cui W et al. in the study of concrete rheology process concluded that viscosity has a serious effect on particle deposition during slurry pumping and particle to pipe bend collision efficiency. The rheology was also measured by slump experiments. The concrete behavior in the pipe was directly investigated using computational fluid dynamics (CFD) simulations, which are based on the Eulerian method and the dense discrete phase model (DDPM). The concrete behavior, including flow profiles, aggregate distribution and migration, was analyzed and discussed. It is obtained that the flow characteristics change from shear flow to plug flow with increasing viscosity, with a more uniform aggregate distribution along the central axis. The radial migration of aggregates becomes more pronounced with increasing shear time, decreasing viscosity, and increasing aggregate size. Concrete with viscosities between 12 and 22 Pa-s is more suitable for pumping. In a similar pipeline erosion numerical simulation study, Shuai M used pipeline transport to transfer the gangue slurry to the user from the perspective of comprehensive utilization of coal gangue for brick making [3]. Through the basic characteristic parameters and rheological parameters of the gangue slurry prepared in the laboratory, the data of the pipeline transport characteristics of the slurry and its relationship with the particle size distribution and weight concentration of the slurry were determined and analyzed. At present, most of the major studies in China and abroad focus on the impact of particles in the multiphase flow in oil and gas pipelines on the bend leading to wear and damage, the main reason is the material particles in the process of pressurized pumping on the steel pipe erosion at the bend. you-nan and Vieira R E et al [4-6] found that different sand content and the type of sand on the bend erosion and wear situation are different. Chen X et al [7] found that the bend at the gas and solid two-phase flow pipeline is the most serious part of the erosion damage. Based on the above related research [8], this paper takes the bend pipe and uphill pipe sand slurry pumping project as the background, and studies the dynamics of slurry and fine pipe internal performance by CFD-DEM method, and analyzes the relevant parameters to understand the whole process of granular flow pumping in bend pipe with slope.

2. Numerical simulation theory

Fluid-particle flow has a wide range of applications in coal mine engineering such as slurry pumping, coal mine particle sieving, and particle mixing. The main challenge facing the modeling of such complex multiphase multi-scale problems is to accurately describe the nature of the problem particle-scale interaction forces between the fluid and particles. The discrete unit method is a popular and preferred computational method to simulate granular materials and particle motion and interactions. Coupled CFD-DEM coarse mesh-based models of fluid-particle flow are gaining importance in approximation. The fluid motion is solved by the locally averaged Navier-Stokes equation, which allows the transfer of local porosity and momentum (including fluid-particle interactions) to the DEM model.

CFD theory

For the pure fluid part (including air and water), computational fluid dynamics is relied upon. A finite volume method (FVM) is used for the computation of fluid flow, where the particle fraction is introduced into the N-S equation for the fluid through the volume fraction.

$$\frac{\partial}{\partial t}(\alpha\rho) + \nabla\cdot(\alpha\rho\mathbf{U}) = 0, \tag{1}$$

$$\frac{\partial}{\partial t}(\alpha\rho) + \nabla\cdot(\alpha\rho\mathbf{U}\otimes\mathbf{U}) = -\alpha\nabla p - \mathbf{F}_p + \nabla\cdot[\alpha\mu(\nabla\mathbf{U} + \mathbf{U}\Delta)] + \alpha\rho\mathbf{g}, \tag{2}$$

t is the time, α is the fluid volume fraction, ρ is fluid density, $[\alpha\mu(\nabla\mathbf{U} + \mathbf{U}\Delta)]$ is the viscous stress, $\mathbf{F}_p$ is fluid force on the particle, and g is the acceleration of gravity.

Equation (2) is an equation of velocity that can be used to solve the velocity problem. Unlike compressible flow, equation (1) is also a velocity equation, which does not contain pressure. On closer inspection, it looks more like a constraint: the divergence of the velocity field must be zero. The second term of equation (2) is a nonlinear term; the third term is a pressure term, the fourth term is a fluid action term, the fifth term is a viscous stress term, and the sixth term is a gravity term.

In an incompressible fluid the equation becomes equation (3):

$$\frac{\partial}{\partial t}(\alpha) + \nabla\cdot(\alpha\mathbf{U}\otimes\mathbf{U}) = -\frac{\alpha\nabla p}{\rho} - \frac{\mathbf{F}_p}{\rho} + \frac{\nabla\cdot[\alpha\mu(\nabla\mathbf{U}+\mathbf{U}\Delta)]}{\rho} + \alpha\mathbf{g}, \tag{3}$$

In incompressible flows, the speed of sound tends to infinity, which causes the time step of the stable solution to go to zero. For these characteristics, the incompressible solution is usually used by the projection method. Currently, the implicit projection method is usually used in CFD to remove the time step limit. The PISO algorithm embedded in OpenFOAM is an implicit projection method. In the projection method, there are two unknown quantities in this set of equations: velocity U and pressure P. Equation (1) cannot be solved for velocity but can only be transformed into a limiting condition, and equation (2) is also an equation about velocity. In order to solve the coupled solution of pressure and velocity, i.e., to construct the pressure Poisson equation.

The momentum prediction is first performed. The momentum direction is written as follows [9]:

$$\mathbf{CU} = -\frac{\alpha \nabla p}{\rho} - \frac{K U_S}{\rho} + \alpha g, \tag{4}$$

C is a coefficient matrix, *K* is the momentum exchange field, and Us is the particle velocity.

In cfdem *C*U is calculated by the fvVector UEqn code.

The following treatment is used for the explicit part $-\frac{\alpha \nabla p}{\rho} - \frac{K U_S}{\rho} + \alpha g$. To facilitate the calculation, the *C* matrix is discretized into a diagonal array consisting of diagonal elements and a matrix consisting of non-diagonal elements, denoted by A for the diagonal array and H for the non-diagonal matrix representation.

$$H = A\mathbf{U} - C\mathbf{U}, \tag{5}$$

Equation (4) can be written as:

$$A\mathbf{U} - H = -\frac{\alpha \nabla p}{\rho} - \frac{K \mathbf{U_S}}{\rho} + \alpha g, \tag{6}$$

H can be obtained explicitly from the non-diagonal matrix term and the velocity term of the previous iteration step, so it is also known, and the term is part of the source term of the pressure Poisson equation.

The velocity field can be derived from the above equation that:

$$\mathbf{U} = A^{-1}H - A^{-1}\left(\frac{\alpha \nabla p}{\rho} + \frac{K \mathbf{U_S}}{\rho} - \alpha g\right), \tag{7}$$

Substituting the obtained velocity equation into the continuity equation yields the Poisson equation for the pressure of the granular flow.

$$\nabla\left(A^{-1}\left(\frac{\alpha \nabla p}{\rho} + \frac{K \mathbf{U_S}}{\rho} - \alpha g\right)\right) = \nabla(A^{-1}H), \tag{8}$$

The N-S equation of equation (1)(2) can be transformed into the solution of equation (4)(8).

In the PISO method the velocity prediction is followed by a velocity correction, and the corrected velocity U** is calculated from the predicted velocity U* as well as the corrected pressure P*. The specific iterative process can be described as follows.

1. The predicted velocity is obtained by first solving equation (4).
2. Pressure is obtained by solving equation (8) and the predicted velocity.
3. The velocity is corrected by the pressure, which is equation (7).

Return to step 2, first solve equation (5) and loop until convergence.

2.1. DEM theory

The contact interaction between the colliding particles is governed by the contact between the spheres ik and jl. Having all the required variables (position and velocity) in a spatially fixed frame, the contact force between these two spheres can be calculated, similarly to the interpretation for spherical particles. The net contact force acting on this particle. $\mathbf{f}_{\text{cont}}$ is the sum of the contact forces acting on all spheres on this particle.

$$\mathbf{f}_{cont} = \sum \mathbf{f}_{cont,ik,jl}, \tag{9}$$

where $\mathbf{f}_{cont\ ik\ jl}$,is the contact force between the sphere ik in particle i and the sphere jl in particle j. Since the contact force acts on the surface and it is considered to be at the center of mass of the particle. So the torque around this point should be considered. The net contact moment on the particle is the net contact moment of the particle in a spatially fixed frame, obtained by

$$M_{cont,i} = \sum_{CL}(R_{C,ij} \times f_{cont,ij}), \tag{10}$$

The motion of the particle is controlled by Newton's equations of motion, for particle i to achieve motion in space by translational and rotational, the controlling equation for translational motion can be expressed as

$$m_i \frac{d\mathbf{v}_i}{dt} = m_i \mathbf{g} - V_i(\overline{\nabla p} + \overline{\nabla \cdot \tau}) + \vec{f}_{pf} + \sum(\mathbf{f}_{i,j} + \mathbf{f}_{i,w}), \tag{11}$$

m_i denotes the mass of particle i , Vi denotes the volume of particle i. According to Gidaspow, two formulations to describe the fluid motion are proposed, namely model A and model B. The difference between these two models lies in the pressure source term in the control equation, model A assumes that the pressure drop is shared between the gas and solid phases, while model B considers the pressure drop only in the gas phase, where the viscous effect of the fluid generated at the particle walls is considered Choosing model A, the stress tensor can be expressed as: $\tau = -\mu(\nabla \mathbf{U} + \mathbf{U}\nabla)$, $\mathbf{f}_{i,j}$, $\mathbf{f}_{i,w}$ are particle-particle and particle-wall collisions.

On the other hand, the rotational motion of the particles is controlled by

$$\mathbf{I}_i \frac{d\omega_i}{dt} = \sum(\mathbf{T}_r + \mathbf{T}_t), \tag{12}$$

Where $\mathbf{I}_i$ is the rotational inertia matrix, $\mathbf{T}_r$ and Tt are the rolling friction moments and moments caused by other particles or wall tangential forces, and ω_i is the angular velocity.

The above is the whole theoretical process of CFD-DEM coupling, and the whole calculation flow is shown in Fig 1.

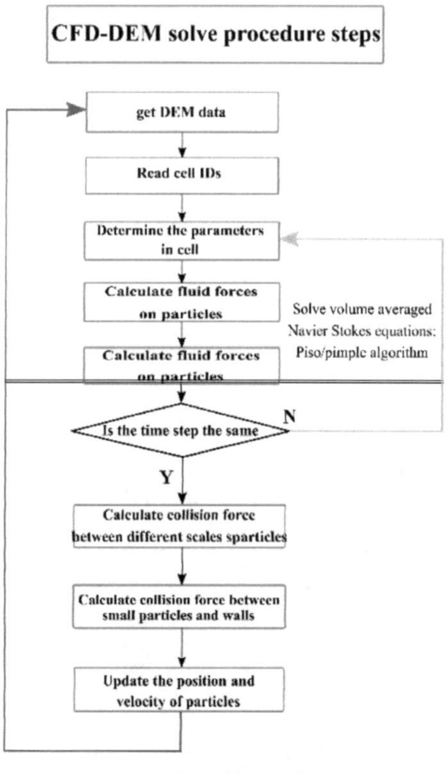

Figure 1. CFD-DEM coupling calculation flow chart

3. Numerical calculation model

In order to realize the analysis of pipe conveying with bends and slopes, it is necessary to establish the corresponding numerical model. For CFD-DEM calculation, it is mainly for model establishment and numerical calculation. In this paper, three analyses of physical model establishment, material parameter setting, and numerical calculation convergence verification are carried out here.

3.1. *Physical Model Establishment*

The physical model of this paper is an 80 mm diameter circular pipe with bend and 40 degree slope, wall thickness h = 8 mm, bend section radius R= 120 mm, geometric dimensions L1 = 880 mm, L2 = 760 mm, L3 = 880 mm pipe material is carbon steel, its elastic modulus is 200 GPa, Poisson's ratio is 0. 31, density is 7850 kg/m3. As shown in Figure 2.

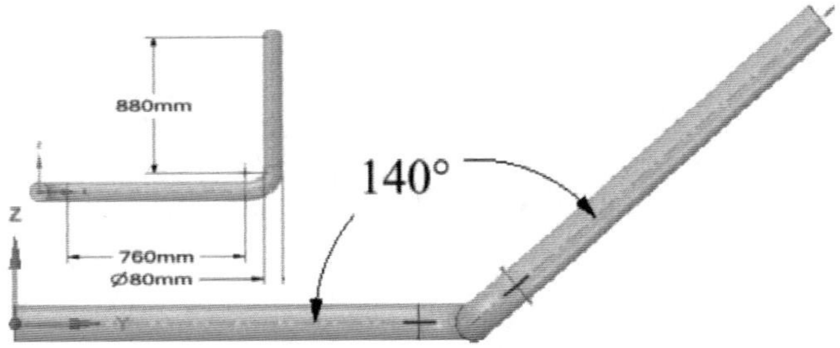

Figure 2. CFD fluid domain geometry model and parameters

The flow medium inside the bend is a high concentration slurry of anemophilous sand, consisting of a slurry phase and anemophilous sand, with a dynamic viscosity determined by the flow viscosity and material parameters. This is described in the next subsection.

3.2. *Flowing slurry material parameters*

For the CFD-DEM method: determining the fluid properties and particle material characteristics parameters is a prerequisite to ensure accurate calculations. Here we refer to the relevant literature [8] to obtain the particle and fluid related parameters. As shown in Table 1.

Table 1. Fluid material and particle material parameters

Material properties	Density (kg/m^3)	Viscosity (Pa·s)
Slurry	1300	15
Wind Accumulated Sand Particles	2660	1000

Table 2. Particle material size gradation

Particles (mm)	0.075	0.1	0.2	0.3	0.4	0.5	0.6	0.7	0.8	0.9
Content %	3.83	5.82	20.75	40.20	64.87	84.74	95.39	98.84	99.83	100

3.3. *Mesh convergence verification*

The verification of mesh irrelevance is extremely important in CFD, and mesh convergence can ensure the validity of numerical results. In this study, 1cm and 0.5cm meshes are used for meshing the computational model.

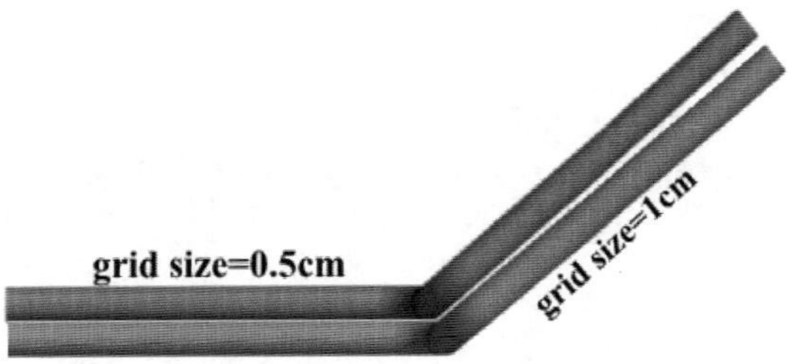

Figure 3. The grid situation at different sizes

The monitoring of the same points A and B in the computational domain for different grid sizes has been marked in the figure, and the monitoring results are shown in Table 3.

Table 3. Grid convergence verification

Monitoring points	Velocity (m/s)	Error (%)
A	1.08	0
B	1.12	3.7

Comparative analysis of the flow velocities of A and B shows that the error is only 3.7% in both grids, which is lower than the more accepted error of 5%, and the simulation results show that the grid has converged in the case of 1cm, so the 1cm fluid grid fully meets the requirements.

4. Analysis of numerical results

Based on the coupled simulation method of CFD and DEM, a Double double precision dense particle flow model is used in CFD, and a discrete element model is used in DEM to track the motion process of solid particle population. In order to investigate the law of motion evolution of the discrete solid particles in this coupled model and their erosion of the pipe wall inside the pipe as well as the motion process, this paper analyzes the flow velocity changes at different locations of the pipe, the motion state of the particles inside the pipe, and the erosion and force of the pipe in conveying slurry.

4.1. Analysis of the internal flow state of the pipeline

The fluid state inside the pipeline mainly contains the movement of the slurry and the movement of the particles, as the slurry is a high viscosity fluid in the flow process belongs to the higher viscous flow, the flow velocity inside the pipeline also has an important impact on the transport here we refer to the flow velocity at the pumping inlet of the pipeline set at 1m/s. The turbulence model uses the k-model. The flow velocity clouds inside the pipeline are shown in Figure 4.

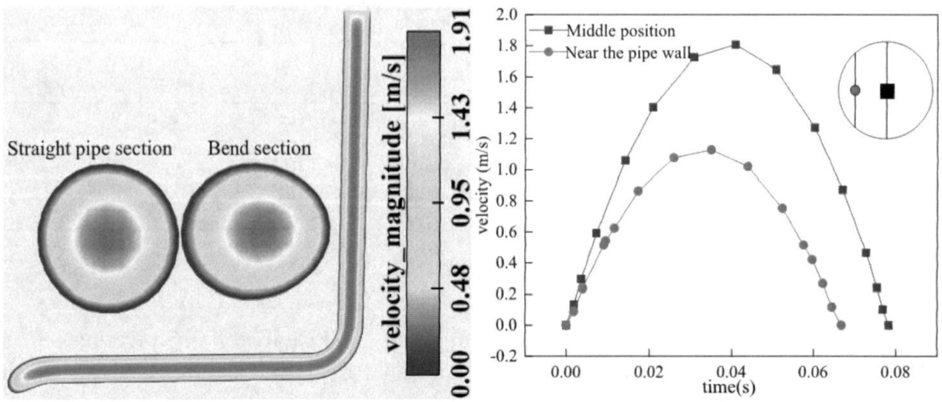

Figure 4. Flow cloud diagram of the slurry inside the pipe with the flow velocity values at different locations in the pipe

From the changes of the cloud diagram in Fig. 4 and the flow velocities at different monitoring points in the pipe, it can be found that the flow velocity is maximum at the center of the pipe and decreases to both sides of the pipe. The peak velocity in the pipe is about 1.9m/s, but the velocity near the wall is zero, which also indicates that the viscous boundary layer of the highly viscous fluid is relatively thick and the flow resistance at the boundary level is very high. On the whole, the pipeline shows that the velocity of motion varies with the curvature of the pipeline, the smaller the curvature the faster the flow velocity; on the same section, the velocity near the wall is relatively low, while the velocity of motion of the particles in the middle of the pipeline is relatively large.

4.2. The movement state of particles inside the pipe
For the flow process of highly viscous slurry can be further understood by the flow state of wind accumulation sand particles. As shown in Figures 5 and 6.

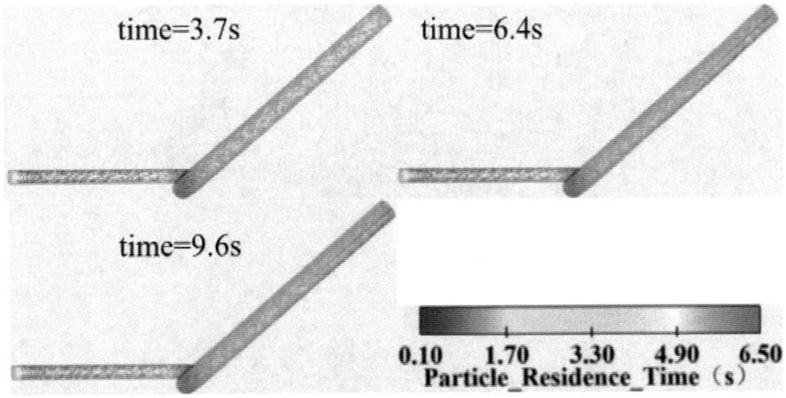

Figure 5. Residence time of wind-deposited sand in the pipeline at different moments

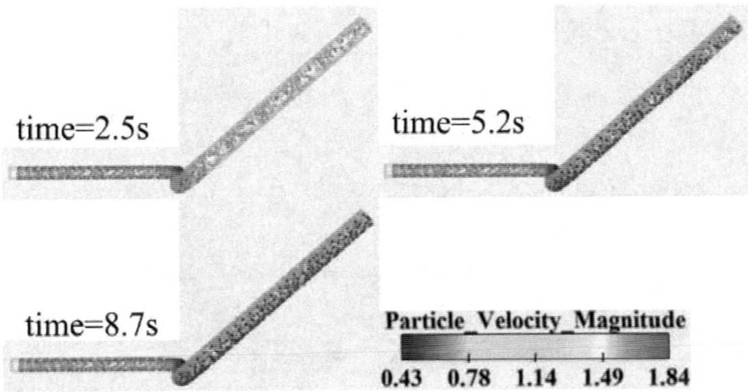

Figure 6. Velocity variation of wind-bound sand inside the pipe at different times

It can be seen that at the beginning of the time the particles stay in the interior of the pipe for a small period of time and move faster, with the extension of time at the near wall surface to produce a highly viscous boundary layer particle movement in the viscous boundary layer makes the movement of particles slow down and may be deposited at the wall surface of the pipe. The particle motion pattern can be further understood by the particle motion state of the particle pipe cross section. As shown in Figure 7.

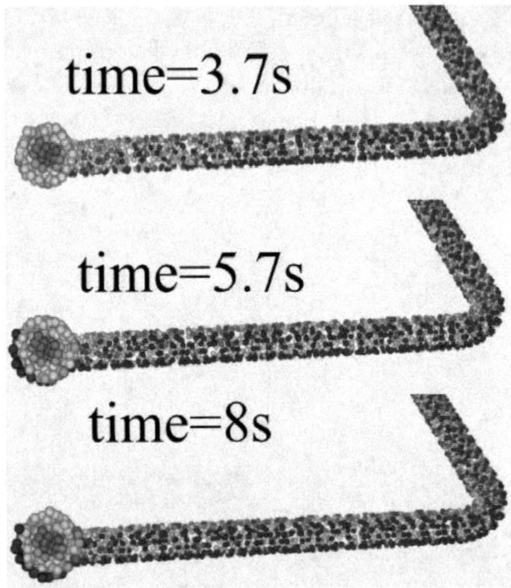

Figure 7. Velocity of particle motion at the pipe cross-section at different moments

The variation of particle velocity at the pipe cross-section at different time points clearly shows that the velocity of particles near the wall decreases significantly with time. This is the main factor that may lead to pipe erosion and long particle residence time and deposition.

4.3. Punching and force of pipeline transport slurry

The movement and deposition of particles near the wall surface inevitably lead to the erosion and force of the particles on the pipe, as shown in the erosion clouds of the pipe at different moments in Figure 8.

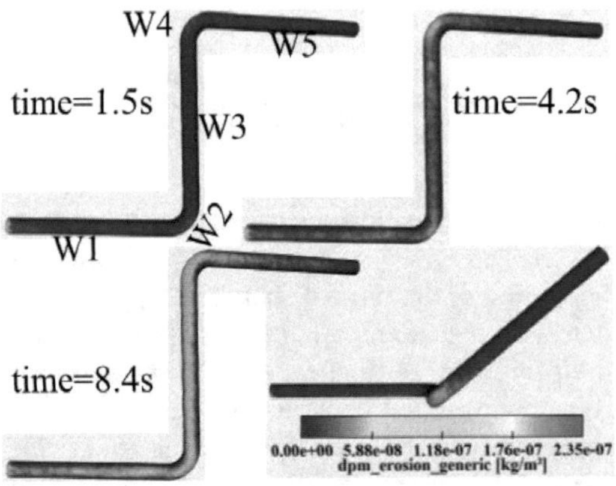

Figure 8. Erosion cloud of the pipe at different times

The erosion cloud can be sehen that with the extension of time, the erosion area and erosion volume inside the pipe show an increase in the law with the extension of time which will, for example, lead to the damage of the pipe, obviously showing the most serious erosion at the bends w2 and w4.

On this basis combined with the internal Y-directional force state of the pipe, as shown in Figure 9.

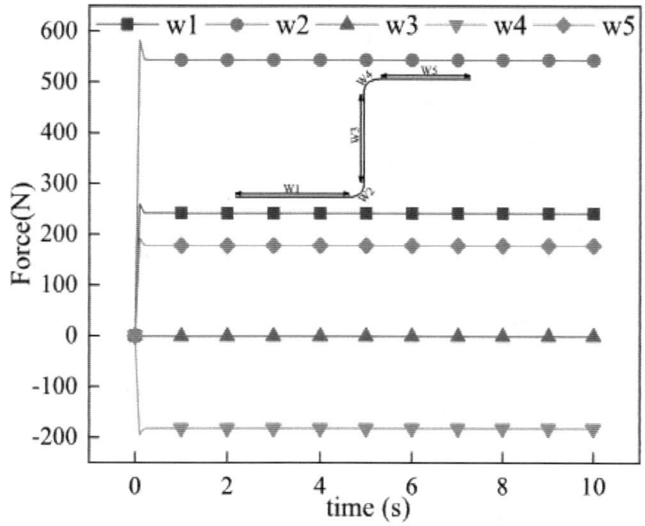

Figure 9. Force on different parts of the pipe

The clouds and data in Figs. 8 and 9 show that the force and erosion at the bend of the pipe are the largest in the process of transporting slurry, making the bend very vulnerable to breakage. In summary, the numerical simulation of the slurry and the particles has shown that the high viscosity of the slurry and the resistance near the wall lead to the flow erosion and deposition of the plugging pattern inside the pipe.

5. Conclusion

Based on the CFD-DEM discrete phase model to study the mixed multiphase flow problem in the concrete pumping process, the calculation results show that: CFD-DEM can more accurately simulate the flow changes of high viscosity multiphase fluid inside the pipe at the straight pipe and bend. The particles tend to impact the outside of the pipe at the bend leading to pipe breakage and solidify and accumulate at the inside leading to pipe blockage, which is consistent with the breakage of the pipe in the field. Subsequent research will further explore in reducing the surface wetting angle, reducing frictional resistance to prevent particle deposition and erosion on the inner wall of the pipe, and improving the wear resistance of the pipe by incorporating new materials.

Acknowledgments

This work is supported by Science and Technology Innovation Fund of Xi'an Research Institute of CCTEG (2020XAYJS02).

References

[1] Wang Z, Wang Z, Xia H, et al. Analysis of hydration mechanism and microstructure of composite cementitious materials for filling mining [J]. Journal of Wuhan University of Technology-Mater. Sci. Ed., 2017, 32(4): 910-913.

[2] Cui W, Tang Q W, Song H F. Effects of Viscosity on Pumping Concrete Behavior Using Computational Fluid Dynamics Method [J]. ACI Materials Journal, 2021, 118(2): 117-126.

[3] Shuai M, Xiaodong W, Jiande W, et al. Study on resistance loss characteristics of iron concentrate pipeline transportation [C]//The 27th Chinese Control and Decision Conference (2015 CCDC). IEEE, 2015: 2037-2041.

[4] You-nan P Y L U, Xi-yun Y I N. The Influence of Sediment Concentration and Particle Size on the Erosion Wearing Characteristics of QT500 [J]. Materials for Mechanical Engineering, 2006, 4.

[5] Sinkov K F, Spesivtsev P E, Osiptsov A A. Simulation of particles transport in multiphase pipe flow for cleanup of oil and gas wells [C]// BHR Group-19th International Conference on Hydrotransport 2014. 2014.
[6] Vieira R E, Parsi M, Zahedi P, et al. Sand erosion measurements under multiphase annular flow conditions in a horizontal-horizontal elbow [J]. Powder technology, 2017, 320: 625-636.
[7] Chen X, McLaury B S, Shirazi S A. Application and experimental validation of a computational fluid dynamics (CFD)-based erosion prediction model in elbows and plugged tees [J]. Computers & Fluids, 2004, 33(10): 1251-1272.
[8] Ma H , Sun J , Wu C , et al. Study on the Pore and Microstructure Fractal Characteristics of Alkali-Activated Coal Gangue-Slag Mortars [J]. Materials, 2020, 13(11): 2442.
[9] Nan X, Hou J, Shen Z, et al. CFD-DEM coupling with multi-sphere particles and application in predicting dynamic behaviors of drifting boats[J]. Ocean Engineering, 2022, 247: 110368.

Optimization of Dual Unloading Grooved Piston Pump Distributor Based on Penalty Function Method

Yuning Song, Liang Han*

Yingkou Institute of Technology, Department of Mechanical and Power Engineering, Yingkou, LiaoNing, China.

*Corresponding author e-mail: hanliang@yku.edu.cn

Abstract. According to the problem of noise caused by fluid pulsation in the piston pump, and related to the opening form, position and size of damping slot on the flow plate, a new type of distribution disc with double V discharge slot is designed again. Then CFD simulation software is used to simulate the design of the flow distribution disc into the piston pump. Finally, through simulation, it is concluded that the width of the triangle slot of the distribution plate is about 10°, the slope angle is about 12°, and the diameter of the unloading hole is about 0.8mm. When the diameter of the hole in the unloading hole is about 1.5mm, the flow curve is the smoother, and the pressure is the smoothest. Force shock and noise are also minimized at this time.

1. Introduction

According to the working principle of the axial piston pump, we know that the axial piston pump plays an important role in the interactive oil distribution plate, and the working reliability and life of the hydraulic pump have a great impact on the [1-3]. When the axial plunger pump conducts the interactive flow distribution, the hydraulic oil constantly rises and drops the pressure drop, and the hydraulic impact thus forms the [4-6]. Pressure drop is too sudden, the common method is to open small grooves or holes to solve this impact phenomenon, so that the oil pressure effect will not be too intense, so as to resolve this sudden impact. The flow state of the fluid and the overall high efficiency of the axial plunger pump are [7-9].

Use the 3 D modeling software SolidWorks to conduct 3 D modeling of the distribution disk, and then use the analysis software ANSYS to obtain the stress and strain curve and deformation cloud map, and use the simulation results obtained from the finite element software analysis to verify whether the distribution plate can meet the requirements under the actual working conditions.

2. Wear mechanism of the flow distribution plate

Plunger pump in the process of actual work, workers usually unconsciously ignore the process of small hole phenomenon, familiar do not know the small hole is caused by [10-12], under the study of the scholar, he found that small bubbles in the burst will not only cause great noise, but also in the disc or even cylinder plunger cavity caused pit [13-15]. And the impulse borne per unit area when the bubble burst is also inferred, and the specific algorithm can be calculated with the following impulse formula.

$$I = \sqrt{2dl\rho(p - p_0)} \tag{1}$$

In the formula:

p_0—Pressure of air in the hole;

l—Length of the column in the flush oil;

d—Backstroke bubble diameter;

It is concluded that when the bubble burst or annihilate, air erosion or pit phenomenon will form on the wall of the cylinder body or the distribution plate. The production of this phenomenon and the [16-18] is closely related to the diameter of the bubble and the pressure of the bubble is the length of the plunger. Therefore, high pressure small diameter vacuoles can not be ignored.

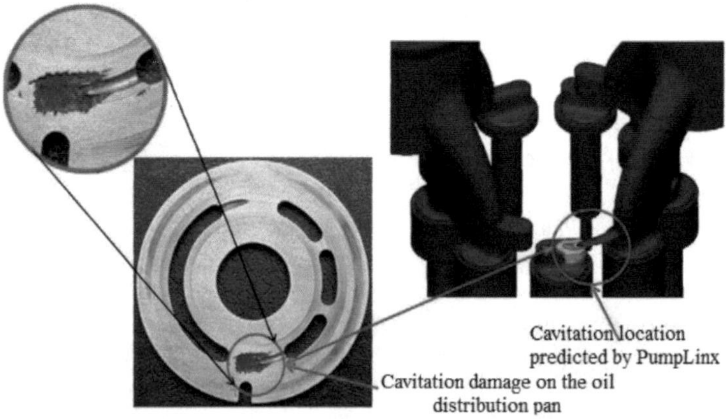

Fig. 1 Validation of cavitation on the distribution plate.

3. Analysis of different types of buffer unloading tank

At present, although there are many types of distribution plate usually used at home and abroad, there are about three basically. As summarized as shown in Figure 2, its common feature is to open a certain buffer structure above the waist groove. When People first designed the distribution plate, they didn't think of opening any shape on the waist slot, one at the left in Figure 2, which people often call them the traditional [19]. However, when this flow distribution plate is installed into the plunger pump, the plunger pump will produce a lot of noise, which is that the flow of the oil outlet shows a pulsating phenomenon. So, people began to analyze what is the cause of noise and pulse, after analysis found that in the process of transition from low pressure waist trough to high pressure waist trough, because there is no buffer groove, buffer has no room but directly into the high pressure chamber, high pressure cavity oil pressure, directly into will cause pulse and noise, so to try to improve, so that the birth of Figure 2 of the other two types of buffer groove, triangle groove or U groove [20].

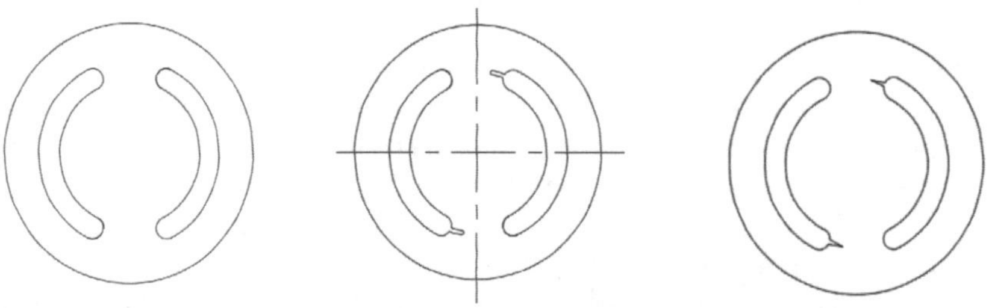

(a) No damping groove (b) U-shaped damping groove (c) Triangle damping groove

Fig. 2 The structure type of the buffer tank.

The two buffer tanks largely deflect to the use of the triangular trough over years. However, it is found that although the opening of the triangle groove can play a certain buffer role, the tip of the triangle groove is prone to form stress concentration. In order to solve the problem of stress concentration, small round holes are opened at the top of the triangle groove to solve the problem of stress concentration [21]. This again forms a new type of buffer groove, hole groove binding type. Although this distribution plate can play a certain buffer role, it can not ensure that the pre-rise pressure when the piston working chamber turns to the high pressure trough and the predischarge pressure when turning to the low pressure trough are equal, which is easy to cause cavitation phenomenon.

However, for the distribution plate in the triangle tank is prone to air erosion and noise phenomenon, after reading a large number of information on the distribution plate direction, [22], it is decided to change the opening of the single triangle channel to a double triangle channel under the damper hole.

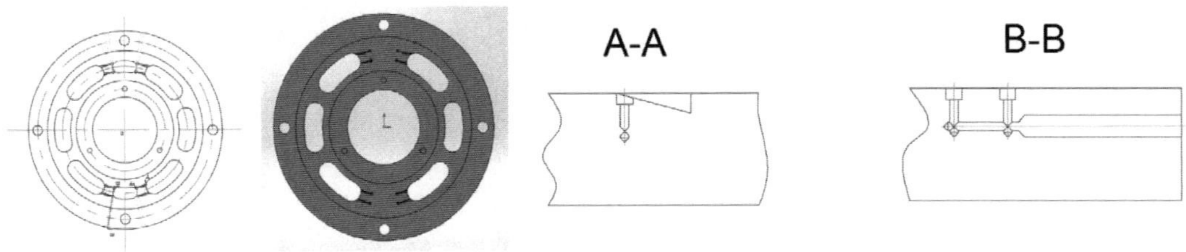

Fig. 3 Double buffered damping tank.

4. Structural design of the triangular damping groove

Generally speaking, there must be a group or several sets of combined data, so that the inclined disc axial cylinder pump in the flow distribution process of minimal pressure impact, the smallest flow pulsation, noise is also the smallest combined data. However, due to such complex engineering problems, detailed and precise expression functions cannot be given in practice, so we optimize the design again on the basis of the original design. The specific process is shown in Figure 4.

4.1. Constraints of the buffer slot and the optimization target function

When designing the simulation of the buffer tank on the distribution plate, we first should choose the simulation factor, and after the simulation factor is selected, we can determine the right and error of our simulation, the simulation purpose [23]. Select the buffer tank, we specify the triangle tank span a and width b as the design factors, record the span of the high pressure side buffer tank of the distribution plate as a_1 width as b_1, record the span of the low pressure side buffer tank as a_2 width as b_2, and use the full factor method to obtain some arrangement combinations of the two factors. The so-called all factor analysis method is actually the process of each factor experiment in accordance with the permutation combination results minus 1, combination of all of them are participated in no less than two independent repeated experiment, we just after the full permutation combination of the new experimental factor method is called full factor method. For example, the full arrangement combination of factors m_1, m_2, m_3, m_4 is 2k-1 (k =4), then we need to do 15 trials, in which the main effect is m_1, m_2, m_3, m_4, m_1m_2, m_1m_3, m_1m_4, m_2m_3, m_2m_4, $m_1m_2m_3$, $m_1m_2m_4$, $m_2m_3m_4$, $m_1m_2m_3m_4$, a total of 15 experiments.

In the test process, in order to avoid the problem of large calculation quantity and low efficiency in the redesign process, the second-order response surface function is generally selected as the target performance approximate model of the inclined disc axial plunger pump, as shown in equation 2.

$$F = a_0 + \sum_{i=1}^{n} a_i X_i + \sum_{i=1}^{n}\sum_{j=1}^{n} b_{ij} X_i X_j \tag{2}$$

According to the structural characteristics of the pump and the installation requirements, the width a1, b_1, a_2, a_{11}, a_{12}, a_{13}, and a_{14}, and the width of b_{11}, a_{12}, b_{22}, a_{23}, a_{24}, a_{21}, widths b_{21}, b_{21}, b_{22}, b_{23}, b_{24}, with the design range as shown in formula 3

To use the constraint function expression [24], the optimization variable low pressure side damping groove cross angle a_1, low pressure side damping groove width b_1, high pressure side damping groove cross angle a_2, high pressure side damping groove width b_2 are expressed as constraint functions, as shown in Equation 4.

The obtained theoretical expression of the objective performance is shown in Equations 5, 6, and 7.

$$\begin{cases} 0 \le a_{11} \le 20° \\ 0 \le b_{11} \le 5 \\ 0 \le a_{21} \le 20° \\ 0 \le b_{21} \le 5 \end{cases} \begin{cases} 0 \le a_{12} \le 20° \\ 0 \le b_{12} \le 5 \\ 0 \le a_{22} \le 20° \\ 0 \le b_{22} \le 5 \end{cases} \begin{cases} 0 \le a_{13} \le 20° \\ 0 \le b_{13} \le 5 \\ 0 \le a_{23} \le 20° \\ 0 \le b_{23} \le 5 \end{cases} \begin{cases} 0 \le a_{14} \le 20° \\ 0 \le b_{14} \le 5 \\ 0 \le a_{24} \le 20° \\ 0 \le b_{24} \le 5 \end{cases} \tag{3}$$

$$\begin{cases} g_1(X) = -x_1 \le 0, \quad g_2(X) = 20 - x_1 \le 0 \\ g_3(X) = -x_2 \le 0, \quad g_4(X) = 5.0 - x_2 \le 0 \\ g_5(X) = -x_1' \le 0, \quad g_6(X) = 20 - x_1' \le 0 \\ g_7(X) = -x_2' \le 0, \quad g_8(X) = 5.0 - x_2' \le 0 \end{cases} \tag{4}$$

$$\begin{cases} \Delta P_H = 10^{-3} \cdot \left[-0.108x_1^2 - 0.814x_2^2 - 12.455x_1 \cdot x_2 + 148.800x_1 + 5.914x_2 - 876.252\right] \\ Q_h = 10^{-3} \cdot \left[0.0896x_1^2 + 2.132x_2^2 - 13.81x_1x_2 - 9.633x_1 + 55.039x_2 + 197.98\right] \end{cases} \tag{5}$$

$$\begin{cases} \Delta P_L = 10^{-3} \cdot \left[-2.2x_1'^2 - 1.8x_2'^2 + 1.8x_1' \cdot x_2'^2 + 73.2x_1' - 66.7x_2' - 618.8\right] \\ Q_l = 10^{-3} \cdot \left[20.0x_1'^2 - 24.1x_2'^2 + 10.4x_1' \cdot x_2' - 130.3x_1' + 188.3x_2' - 372.1\right] \end{cases} \tag{6}$$

$$R^2 = 1 - \frac{\sum_{j=1}^{N}\left[y_{rsm}(j) - y(j)\right]^2}{\sum_{j=1}^{N}\left[y(j) - \bar{y}\right]^2} \tag{7}$$

Formula: ΔP_H is the difference between the maximum pressure and the minimum pressure when the buffer tank reaches the cylinder chamber with high pressure; Q_h is the instantaneous inversion of flow in the plunger chamber, $y(j)$ is representing the plane response value of the planning size; $y_{rsm}(j)$ is the simulation mean of each point in the design space; N represents the number of test points in the design space. R_2 represents the determination coefficient, and the size of the value represents the difference between the reaction plane and the simulation plane value, which is taken within 0-1. When the value is 1, they are basically the same.

Tab. 1 Optimized results for manifolds.

Optimize the variable	Structure of the low-pressure zone buffer groove								Structure of the high-pressure zone buffer groove							
	a_{11}	a_{12}	a_{13}	a_{14}	b_{11}	b_{12}	b_{13}	b_{14}	a_{21}	a_{22}	a_{23}	a_{24}	b_{21}	b_{22}	b_{23}	b_{24}
Before optimization	10				3		3		10				3		3	
postoptimality	9.884				3.532		3.332		11.365				2.561		2.31	

The function expression can basically reflect the performance of the objective function on the response [25] of local structural changes, which can further replace the CFD simulation, and finally optimize the redesigned objective function, which can improve the integrated design efficiency. Then through the construction of the model and the CFD simulation test, the target performance function can be obtained, because of the maximum order of magnitude difference between the inversion and the pressure difference, so we introduce a new coefficient and, so that the optimization values F_1 and F_2 about the overall structural design scheme can be obtained.

$$F_1 = \Delta P_H + \lambda_1 Q_H \tag{8}$$

$$F_2 = \Delta P_L + \lambda_2 Q_l \tag{9}$$

$$\lambda_1 = \frac{(\Delta P_H)_{max} - (\Delta P_H)_{min}}{(Q_H)_{max} - (Q_H)_{min}} \tag{10}$$

4.2. The procedure and results optimized using the penalty function method

Here to optimize the algorithm, we choose the penalty function algorithm [26], and adopt the sorting method [27], which is suitable for solving the minimum value in Matlab, and the design results are shown in Table 1.

The optimized triangular groove is modeled with SolidWorks and uses Fluent software to simulate the relationship between pressure and flow in the plunger chamber. The simulation results are shown in Figure 4.

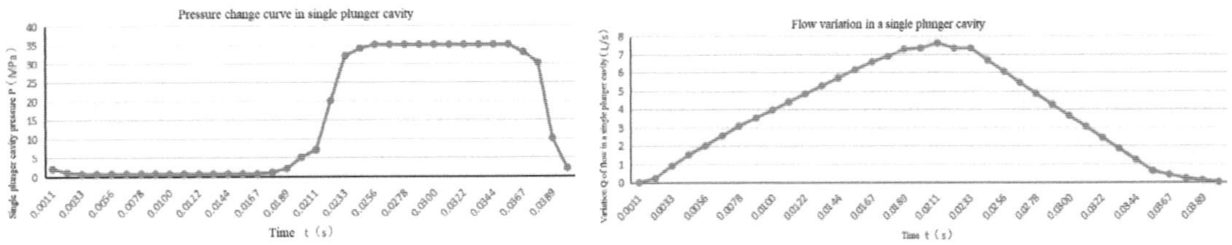

(a) (b)

Fig. 4 Damping groove optimization design cavity after the curves of pressure flow curve.

5. Static and mode analysis of the new double unloading tank distribution plate

Because the project actually needs the distribution plate to rotate between the cylinder block. Since there is mutual rotation, there must be a certain friction between them, which requires that the materials used by the distribution plate must have good wear resistance.

Since the waist groove of the distribution disc is located on both sides of the axial symmetric centerline of the distribution disc, on one side, the area with high pressure is connected to the drainage window, on the other side is the area with lower pressure, which is connected with the suction window and the pressure is different on both sides, so the first thing we need to do in the simulation experiment is to divide the two areas with different pressure above the distribution disc, and then divide the grid above the already divided area.

According to the service conditions of the distribution plate, a load pressure of 35Mpa is applied to the oil outlet of the distribution plate and applied to the inner surface of the oil outlet of the distribution plate. The force caused by friction is ignored when the pressure is applied, and then the constraints imposed on the distribution disc include the constraints preventing the axial movement of the distribution disc and the radial movement of the distribution disc.

Through the analysis, we can draw a basic conclusion that the distribution plate is installed at one end of the cylinder body, but it is not directly connected to the cylinder body, but the load it receives is still related to the angle of the cylinder body rotation. Starting from the center of a cylinder plug just coincides with the point on the segment line on the waist groove above the distribution plate, the phase angle of the cylinder body is 0 degrees, and then as the cylinder body rotates, the angle of the cylinder body, as shown in Figure 5.

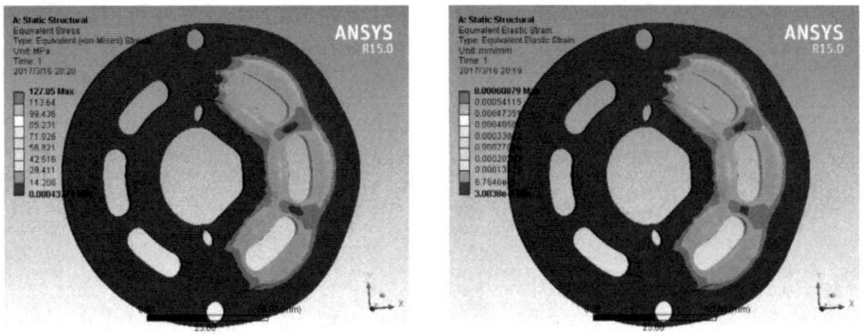

(a) (b)

Fig. 5 Stress and Strain distribution of flow plate deformation.

By analyzing the images of the stress and the strain, we can draw the following conclusions.

(1) The simulation results are the same as we have previously guessed. The maximum stress is concentrated in the buffer groove on the waist groove with high pressure. Although the pressure distribution is different, the stress value of the maximum distribution plate is basically the same, and the stress distribution diagram is basically the same.

(2) According to the simulation curve on the figure, the overall deformation of the distribution plate obtained at different angles is basically similar, with the largest deformation at the buffer groove in the waist window of the high pressure area.

To make the results more clear, the change curve of stress and change of the cylinder cylinder within 1 / 9 cycles. As shown in Figures 6 and 7.

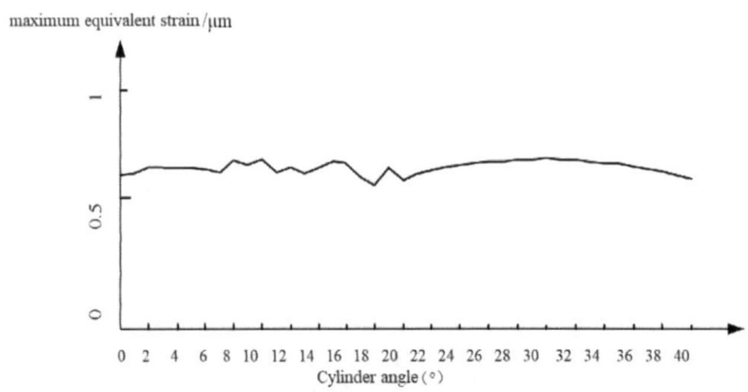

Fig. 6 Relationship between Maximum Equivalent Strain and Cylinder Corner

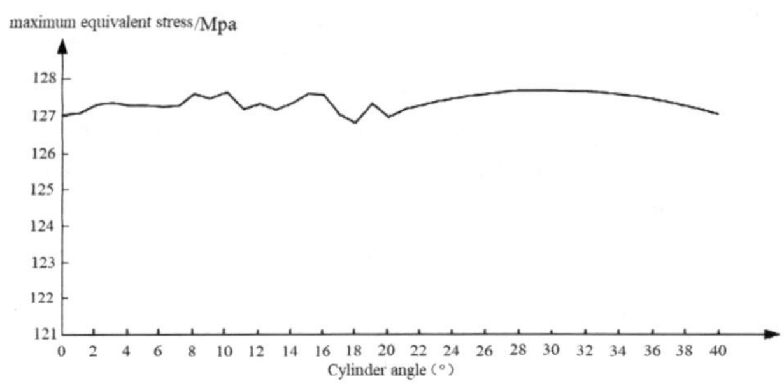

Fig. 7 Relationship between Maximum Equivalent Stress and Cylinder Corner

The relationship between stress, strain and the cylinder body angle can draw the following conclusions:

(1) The stress and strain on the distribution plate, including the buffer tank, are not constant, but change with the rotation angle of the cylinder body during the cylinder body rotation. It can be clearly seen in the figure that the maximum stress and strain value basically appear around 10 degrees, because in the previous optimization design, the buffer tank is opened around 10 degrees from the waist tank, which is in line with the conclusion obtained under actual conditions. The maximum stress value is 127.85Mpa and the maximum strain value is 0.00060879mm.Then its design material is nitride 38GrMoAl, compared with its performance index, to meet the needs of the work.

(2) Through the analysis of the above figure above, it can also be concluded that the maximum deformation of the cylinder body reaches about 10 degrees when the cylinder body rotates. You can know that the distance of the maximum deformation is about 0.6 m. At this time, the distribution plate deformation along the axis, once it appears deformed, then the slit width between it and the cylinder block will increase, the oil will be instantly filled into the gap, that is to say, the instantaneous oil film thickness will also increase several times. Generally, the gap width is ideal between 0.3 μm and 12 μm.

6. Conclusion

The 3 D modeling software SolidWorks uses 3 D modeling of the distribution plate, and then the analytical software ANSYS to obtain the stress and strain curve and deformation cloud map. Finally, the width of the distribution plate with width is about 10, the slope angle is about 12, the hole diameter of the discharge hole is about 0.8mm, the flow curve of the discharge hole is about 1.5mm is smooth, the pressure impact and noise are also minimum.

Acknowledgments

This work was financially supported by Liaoning Natural Science Foundation Guidance Plan "Research on Key Technology of Anti-impact Protection Safety Large Flow Valve for Hydraulic Support" (20180550584), Scientific Research Funding Project of Liaoning Provincial Department of Education-"Optimization Research of Double Pger Pump for Young Science and Technology Talents" (L2020011).

References

[1] YU Yongping, YIN Xia. Design and Simulation Research of Dual Piston Pump Flow Distribution Disk [J]. Machine Tool & Hydraulics, 2015, 43(21): 163-167.

[2] SHI Xiang, ZHAO Dongbiao. Structure Optimization of the Valve Plate of Aviation Piston Pump[J]. Machine Tool & Hydraulics, 2015, 43(22): 57-60.

[3] ZHANG Y, GAO You-shan, QUAN Long, JING Zhen-hua, LLANG Yi-nan. Simulation and Analysis of Axial Piston Pump with Four Distribition Windows [J]. Hydraulics Pneumatics & Seals, 2016, 36(03): 18-21.

[4] Pan Yang, Li Yibo, Huang Minghui, Ma Jun, Liang Dedong. Valve Plate Improvement and Flow Ripple Characteristic A nalysis for Double Compound A xial Piston Pump [J]. Transactions of the Chinese Society for Agricultural Machinery,2016, 47(04): 391-398.

[5] SHI Ying-xun, MENG Guang-yao, HUANG Ju-xin, MENG Hang, SHENG Kai-ming. Simulation Study on Effects of the Shape of Damping Groove on Cavitating Flow. CHINAESE HYDRAULIC & PNEUMATICS, 2016, 0(8): 108-111.

[6] Wang Yixiang. Study on Damping Groove Characteristics and Optimization of Axial Piston Pump [D]. Zhejiang University, 2014.

[7] TIAN Zhendong, HUANG Jiahai, WANG Hao, HUANG Yue. Research Based on Full Cavitation Model for Cavitation of High Pressure Axial Piston Pump Valve Plate [J]. Chinese Hydraulics & Pneumatics, 2018(01): 17-22.

[8] Huang Huangnan, He Xueming, Chen Shuangcheng, Zhang Rong. Structural optimization of the distribution plate of the axial plunger pump for desalination [J]. Pump Technology, 2018 (01):

10-15.

[9] KONG Xiang -chun, NIE Song -lin, YIN Fang -long . Study on the Vibration Characteristic of Port Plate for Water Hydraulic Axial Piston Pump [J]. Hydraulics Pneumatics & Seals, 2018, 38(04): 28-32.

[10] Zhang Tingbiao, Dai Zhengqiang, Xiao Xiao, Wang Junxiong, Deng Pengzhi, Zeng Gushan. Experimental study on the influence of low noise Axial Cylinger Pump [J]. Times Agricultural Machinery, 2018, 45(01): 110-111+113.

[11] GENG Fushuai, WEI Xiuye. Simulation Study of Valve Plate Structure Based on Bee Colony Algorithm [J]. Machine Tool & Hydraulics, 2018, 46(09): 151-155.

[12] Chi Yurong, Yang Junru, Li Ruichuan, Gongxubo, Guo Kun. Simulation and Research on Flow Pulsation of A xial Piston Pump [J]. Machine Tool & Hydraulics, 2018, 46(13): 138-143.

[13]]Hao Xiyang, W ang Yulin, Geng Chao, Li Zhiming, Zhang Jian. Study on Damping Groove Characteristics of Axial Plunger Pump [J]. Agricultural Equipment & Vehicle Engineering, 2018, 56(07): 48-51.

[14] Li Peiyong. Analysis of the inclined disc torque and instantaneous flow output characteristics of the pump with the axial plunger variable [D]. Lanzhou University of Technology, 2018.

[15] LIU Cheng-qiang, LIU Yin-shui, YANG Guan- zhong. New Type of Swash Plate Piston Hydraulic Transformer' s Noise [J]. Chinese Hydraulics & Pneumatics, 2017(02): 1-4.

Study on the flow around a cylinder by candle flame

Junwei Huang, Pengyu Zhang, Zhuoxuan Xie, Haojie Zhao, Chenkun Xu, Jing Chi and Huiping Lu*

School of Physical Science and Technology, TianGong University, Tianjin, 300387, China

*Corresponding author e-mail: luhuiping@tiangong.edu.cn

Abstract. Interferometric measurement, schlieren method and shadow graph are commonly used to observe the low-speed steady flow. It is difficult to directly observe the flow direction in the experiment of flow around a cylinder at medium and high Reynolds number. On this premise, this paper mainly studies the relationship between the movement state of candle flame and the direction of candle airflow. This paper studies the influence of different wind speeds on the flow around a circular cylinder at medium and high Reynolds numbers. The innovation of this paper is to propose a new idea to directly observe the complex air flow, and to compare the candle flame state under different experimental conditions and the calculation images of different turbulence models.

1. Introduction

The flow around a circular cylinder has been studied for a long time. Reynolds number is a dimensionless number of the ratio of inertial force to viscous force, which is usually used to characterize fluid flow characteristics. By calculating the Reynolds number of the flow around a cylinder, the flow state can be divided into the following regions [1].

Table 1. Reynolds number ranges for different regions.

region	Laminar flow region	Subcritical region	Supercritical region	Ultra supercritical region
Reynolds number	$1\times10^0 \sim 3\times10^2$	$3\times10^2 \sim 3\times10^5$	$3\times10^5 \sim 3.5\times10^6$	$\geq 3.5\times10^6$

For the calculation of Reynolds number, we can establish the Coanda effect model [2] to get the simplified formula of Reynolds number:

$$\mathrm{Re} = \frac{U_j \cdot R}{v} \tag{1}$$

Where U_j is the speed of the jet leaving the Coanda nozzle, and R is the cylinder radius and v is the coefficient of air viscosity.

In the experiment of flow around a cylinder, a stable jet is needed, which also meets the applicable conditions of Coanda effect.

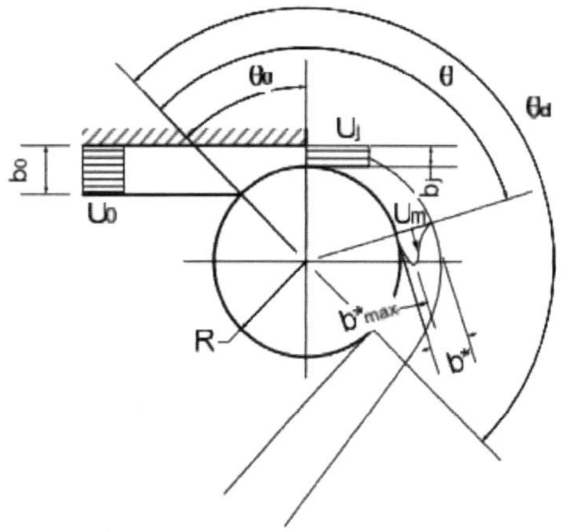

Figure 1. Coanda effect model

Previously, people have solved the flow around a cylinder at lower Reynolds number (Re < 300), and obtained the flow without separation or stable and regular vortex. However, laminar transition makes the whole process of air flow very complicated at medium and high Reynolds numbers. Even if the computer with strong computing power is used, the exact solution can not be obtained. So people usually use different turbulence models of RANS or LES to describe the turbulent motion, and the accuracy of the final solution depends on the accuracy of the simulation method. Different simulation methods have their own advantages and disadvantages in dealing with different problems.

In this paper, we try to visualize the wake with Reynolds numbers from 1.5×10^4 to 2.7×10^4. Several wake research methods are compared. This paper mainly studies the influence of different wind speed and cylinder radius on the flow around the cylinder, and compares the simulation results of different candle flame states under different models.

2. Experimental details

2.1. Verify the significant influence of related variables on the experiment

Due to the disorder of turbulence, it is necessary to use a numerical value to measure whether the change of variables in formula (1) has a significant impact on the flow state of fluid

In a closed laboratory, we put the anemometer on the Coanda nozzle and measured the average velocity of five incoming streams. After that, we put the candle 12 cm away from the cylinder to keep the convex part of the center of the candle level with the position of the vent. Providing a steady jet until the candle flame goes out. Repeat the experiment five times, record the time, get the average time t, then change the radius and speed of the cylinder, and do another experiment. We took the duration t of the whole experiment as the quantitative standard and recorded the results in Table 2 and table 3.

Table 2. Effect of wind speed on experimental time

V (m/s)	4.280	5.230	6.270	7.450	8.310	9.110
T(s)	2.702	3.885	4.021	4.425	0.338	0.068

Table 3. Effect of cylinder radius on experimental time

R(cm)	4.500	3.500	3.000
T(s)	4.021	0.608	0.236

According to the data in Table 2 and Table 3, draw the following figure.

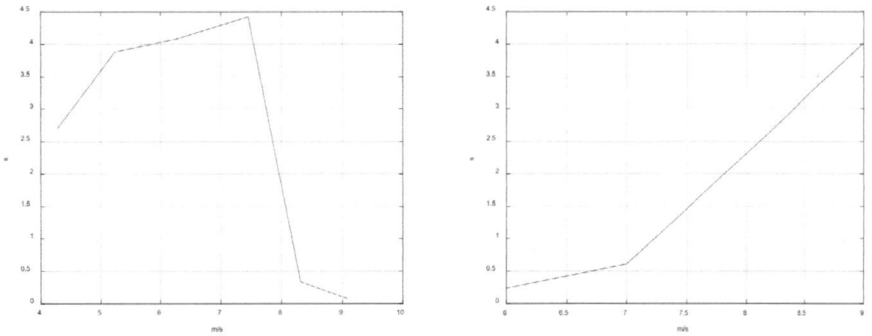

Figure 2. Experimental time corresponding to six kinds of wind speed and three kinds of cylinder radius

When we observe the values of Figure 2. We can see the obvious change. This represents that the corresponding variable (R, V) we changed has a significant effect on the observed flame state of the candle.

2.2. Effect of wind speed on flow around a circular cylinder

In the later experiments, we first make sure that the fan and the cylinder are coaxial with the candle, and the state of each candle is consistent. Let's take a cylinder with a radius of 4.5 cm and put it on the abscissa of 10 cm. When the candle is placed 16cm away from the cylinder, the wind speed is 5.23cm/s, 7.45cm/s and 9.11cm/s respectively. Compare the observed flame state of the candle with the images calculated by the two models, and mark the black spot on the image as the coordinate position of the candle, as shown in the figure below:

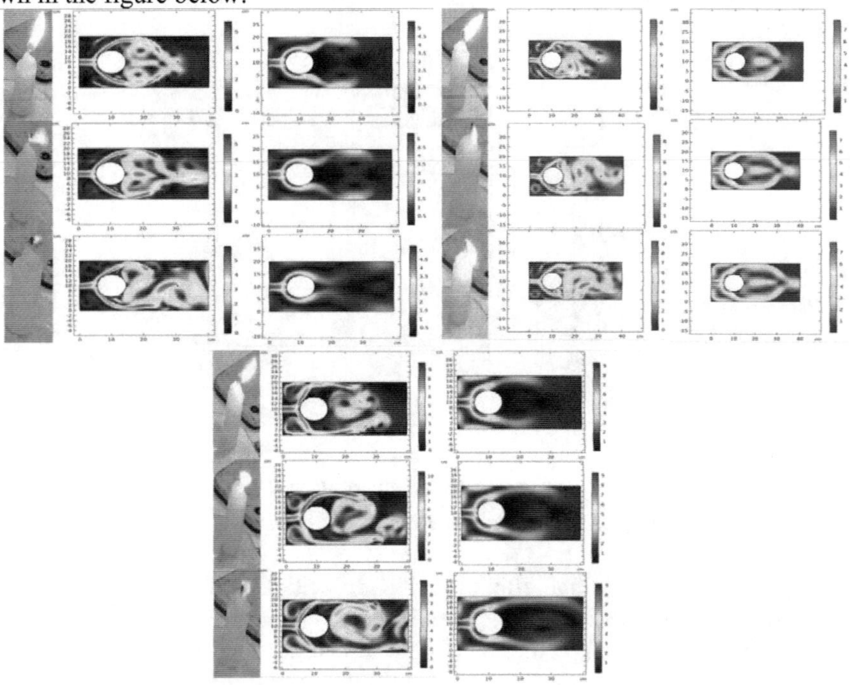

Figure 3. At the speed of 5.23m/s, 7.45m/s and 9.11m/s, the image of candle flame changing with time when the candle is placed at the abscissa of 30.5cm, as well as the wind speed distribution calculated by LES model and k - ω model.

Comparing the experimental results with the model, it can be seen from the above figure that the wind speed field calculated by LES model is more consistent with the observed phenomenon, and with the increase of wind speed, the wind speed distribution map calculated by LES model gradually changes from a pair of stable and symmetrical vortexes to random shedding vortexes. However, there is no obvious change in the calculation results of $k - \omega$ model, which can not solve the irregular shedding of small-scale vortexes. It should be noted that when analyzing the simulation image, we should not only analyze the wind speed distribution near the candle coordinates, but also consider the current airflow situation.

2.3. Compare the flame state of the candle at different positions with the images calculated by the two models

We keep using a cylinder with a radius of 4.5 cm and placing it at the abscissa of 10 cm. Then we put the candle 4cm, 12cm and 20cm away from the bottle. Set the speed to 7.45m/s, as shown in the following figure:

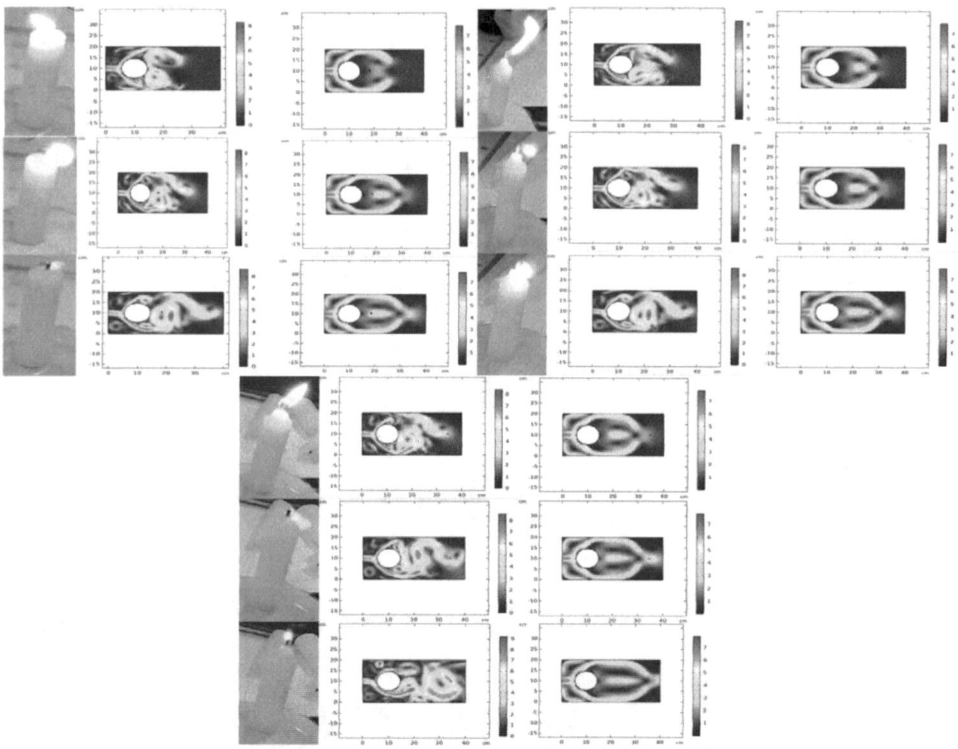

Figure 4. When the wind speed is 7.45m/s, the abscissa of the candle is set to 26.5cm, 30.5cm and 34.5cm respectively. The flame image of the candle changes with time. The LES model and $k - \omega$ model are used to calculate the wind speed distribution.

It can be seen from the above figure that the fluid velocity field calculated by LES model is more consistent with the results of theoretical calculation. Under the same grid density, RANS model is more difficult to obtain more detailed fluid motion at high Reynolds number. According to the candle flame state, this is a method that can roughly observe the fluid movement in three-dimensional space at high Reynolds number.

3. Conclusion

In this paper, we propose a method to observe the complex air flow roughly, and verify the visualization results with the turbulence model. Due to the limitation of experimental conditions, the limited experimental results are compared with the simulation results of two different turbulence models, and the wind speed distribution map obtained by LES model is compared, which is well verified.

It is planned to use more stable and observable tools instead of candles in the future, and the use of more powerful computers will also have an impact on the simulation results. This provides a new way to directly observe the air flow at high Reynolds number

Acknowledgments

This work was supported by the National Science Foundation of China (No. 61904123), the Natural Science Foundation of Tianjin (No. 18JCQNJC71800), Scientific Research Project of Tianjin Educational Committee (No. 2018KJ220).

References

[1] ZHAN Hao, LI Wan-ping, FANG Qin-han, LI Long-an. (2008) Numerical Simulation of the Flow Around a Circular Cylinder at Varies Reynolds Number. Journal of wuhan university of technology, 1671-4431(2008)12-0129-04.

[2] Trancossi, Dumas, Vucinic, Mathematical Modeling of Coanda Effect, SAE Technical Paper n. 2013-01-2195, 2013.

[3] Franke J, Frank W. Large Eddy Simulation of the Flow Past a Circular Cylinder at Re =3900 [J]. Journal of Wind Engineering and Industrial Aerodynamics, 2002, 90: 1191-1206.

[4] Pietro Catalano. Numerical Ismulation of the Flow Around a Circular Cylinder at High Reynold Numbers [J], International Journal of Heat and Fluid Flow, 2003, 24: 463-469

[5] Anagnostopoulos, P., 1994.Numerical investigation of response and wake characteristics of a vortex-excited cylinder in a uniform stream. Journal of Fluids and Structures 8, 367-390

[6] Anagnostopoulos, P., Iliadis, G., Richardson, S., 1996.Numercial study of the blockage effects on viscous flow past a circular cylinder. International Journal of Numerical Methods in Fluids 22, 1061-1074

Flow Field Analysis of Gas Injection and Brine Discharge in Natural Gas Storage Salt Cavern

Yi Zhang *, Wei Huang, Xueqi Cen, Dandi Zhao, Guoxing Chai

Petroleum Exploration and Production Research Institute of SINOPEC, Beijing, China

*Corresponding author e-mail: zhangyi5801@yeah.net

Abstract. Gas injection and brine discharge is the key link in the construction and safe operation of the salt cavern gas storage. Accurate description of the bottom flow field of the salt cavern in the process of gas injection and brine extraction is the basis and premise for determining the depth of the brine drainage string. According to the characteristics of gas injection and brine discharge, a numerical simulation calculation model of the brine flow field near the inlet of the brine discharge tubing was established. The brine flow field under different working conditions is simulated, and the safe brine drainage rate at different residual brine depths in the later stage of gas injection and brine drainage is determined, which provides technical support for the adjustment of gas injection brine drainage rate in the later stage of construction.

Keyword: natural gas storage salt cavern, gas injection and brine discharge, flow field.

1. Introduction

Salt cavern gas storage is an artificial cave built in the underground salt layer by means of water-soluble mining, and its function is to inject, store and produce natural gas [1-3]. Salt rock has good creep and damage self-healing functions, and the permeability of salt rock is very low, and the sealing of salt cavity is good [4-6]. The safety of salt cavern gas storage is much higher than that of ground storage facilities, and it is known as a highly strategically safe storage [7]. Gas injection and brine discharge are the key links in the construction and safe operation of salt cavern underground gas storage, and the smooth discharge of brine is the key to forming an effective storage capacity [8-10]. Figure 1 is a schematic diagram of this process. In this process, the brine in the solution cavity is discharged from the brine discharge inner tubing by injecting natural gas into the annulus between the inner tubing and the outer pipe.

In the process of natural gas replacing brine, when the replacement is in the later stage, since the brine submerged the drainage pipe is very shallow, a large amount of natural gas may be discharged along the pipe wall, causing unnecessary safety accidents. Properly controlling the brine discharge rate in the later stage and reducing the flowing rate at the discharge pipe inlet can effectively reduce the risk of natural gas being brought out during the brine discharge process [11, 12]. Based on the actual working conditions of brine discharge in the underground storage, the flow field analysis theory was used to simulate the flow field at the bottom of the salt rock storage cavity in the later stage of process. According to the safety critical distance between the inlet of the brine discharge pipe and the gas-water interface, the safe brine drainage rate at different depths of residual brine in the later stage of gas injection and brine drainage was analyzed and determined.

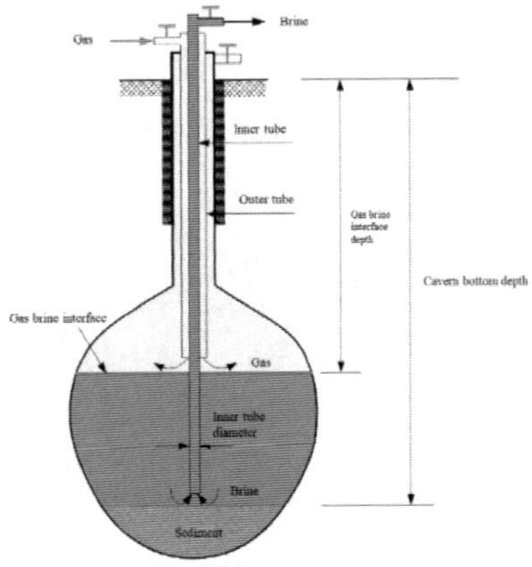

Figure 1. Schematic diagram of gas injection and brine discharge in salt cavern.

2. Distribution characteristics of flow field around the discharge tubing inlet

In the process of gas injection and brine discharge, due to the large diameter of the cavity, which can generally reach several tens of meters, and the diameter of the brine discharge tubing is generally 0.114 m, the liquid level in the solution cavity falls slowly, and the gas-liquid interface decline is basically laminar flow. With the progress of the brine discharge, the distance between the gas-liquid interface and the discharge tubing inlet is gradually reduced, and the diameter of the cavity bottom is also smaller, which may form a vortex at the discharge tubing inlet, causing the natural gas to be carried out with the brine. Figure 2 shows the schematic diagram of the flow field distribution near the inlet of the brine discharge pipe. The potential vortex field is a non-swirling flow field, the Bernoulli equation is established in the whole flow field, and the total energy of the fluid micro-element is conserved at two points at any height. Let the asymptote of the horizontal boundary of the free surface be $z=z_0$, and the infinity point of the asymptote has a relational expression with any point on the free surface

$$\frac{v^2}{2} + gz_s + \frac{p_s}{\rho} = \frac{v_0^2}{2} + gz_0 + \frac{p_0}{\rho} \tag{1}$$

In the formula, v is the fluid flow velocity at the calculated position; zs is the liquid column depth at the calculated position; ps is the pressure on the fluid; v0 is the fluid flow velocity at the position of the gas-liquid interface; z0 is the depth of the gas-liquid interface which is taken as 0; p0 is the pressure of the fluid at the gas-liquid interface position.

On the horizontal boundary, $r_0 \to \infty$, $v_0 = c/r_0 \to 0$; and on the free surface, $p_s = p_0$, it can be obtained from the above formula

$$\frac{v^2}{2} + gz_s = gz_0 \tag{2}$$

Substituting $v = C/r$ into the above formula, the free-surface equation can be obtained as the rotating hyperbola equation as

$$z_0 - z_s = \frac{C^2}{2gr^2} \tag{3}$$

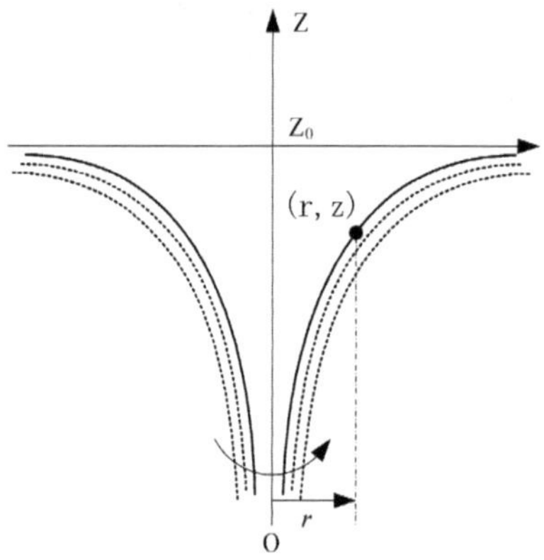

Figure 2. Schematic diagram of flow field near the brine discharge pipe inlet.

Considering that the cross-sectional area of the solution cavity is much larger than the cross-sectional area of the brine discharge tubing, the influence of the falling velocity of the gas-liquid interface can be ignored. Using the above calculation formula, the expression of the critical distance between the discharge tubing inlet and the gas-liquid interface can be obtained as follows:

$$\Delta H = \frac{1}{4g} v^2 \tag{4}$$

In the formula: ΔH is the distance between the discharge pipe inlet and the gas-liquid interface, $\Delta H = z_s - z_0$.

Considering that the diameter of the discharge pipe used in the field is 114mm, the critical distance between the discharge pipe inlet and the gas-liquid interface can be calculated under the conditions of different brine discharge rates, as shown in Figure 3. It can be seen from the figure that with the increase of the brine discharge rate, the critical distance keeps increasing, so as to prevent the formation of vortices on the gas-brine interface during the process of brine discharge, resulting in the natural gas being carried out. The calculation results show that when the 114mm pipe string is used for brine discharge, when the brine discharge rate reaches 100 m^3/h, the depth of the vortex formed is about 0.19 m, 0.27m at 120m^3/h, and 0.37m at 140m^3/h, 0.48m at 160m^3/h.

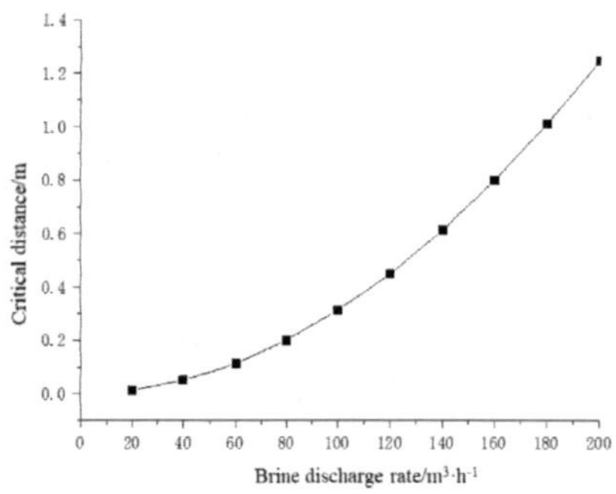

Figure 3. The relationship between the discharge critical distance and the brine discharge rate when the tubing diameter is 114mm.

3. Flow field analysis in the process of gas injection and brine discharge

Considering that the diameter of the cavity is much larger than the diameter of the brine discharge tubing, the brine discharge process only has a greater impact on the brine flow field near the pipe inlet, while the effect on the brine flow field at the cavity wall is basically negligible. Therefore, the influence of the morphological differences of different cavities on the flow field in the cavity during the process of brine discharge is not considered when establishing the numerical simulation model. Figure 4 shows the mesh of the flow field in the solution cavity during the gas injection and brine discharge process when the distance between the gas-brine interface and the discharge tubing inlet is 4m. When the distance between the gas-brine interface and the discharge pipe inlet is 1m, 2m, and 3m respectively, the model grids are the same, but the distances are different. The cavity is axisymmetric, and the mesh is half of the cavity in order to save computation.

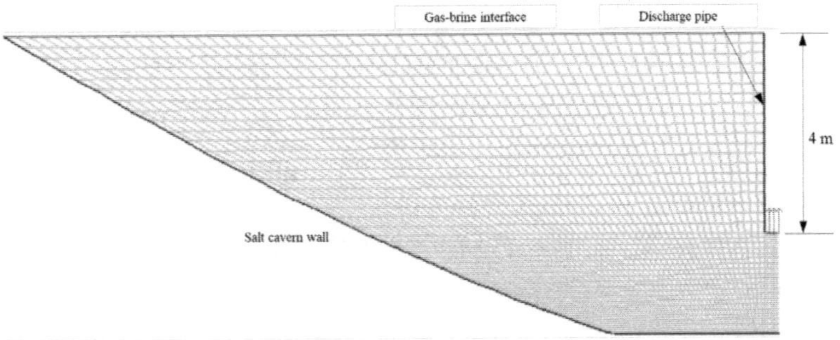

Figure 4. Mesh in the solution cavity when the distance between the gas-brine interface and the discharge pipe inlet is 4m.

Using the above mesh, the simulation of the flow field in the dissolution cavity was carried out when the brine discharge rate was 100m^3/h, 80m^3/h and 60m^3/h respectively. The simulation results are shown in Figure 5 - Figure 8. It can be seen from the figure that the area where the flow field inside the cavity changes during the process of gas injection and brine discharge is mainly concentrated near the brine discharge pipe, generally ranging from a few meters to a dozen meters. The area of the region where the flow field changed significantly decreased with the decrease of the brine discharge rate, and increased with the decrease of the distance between the gas-brine interface and the pipe inlet.

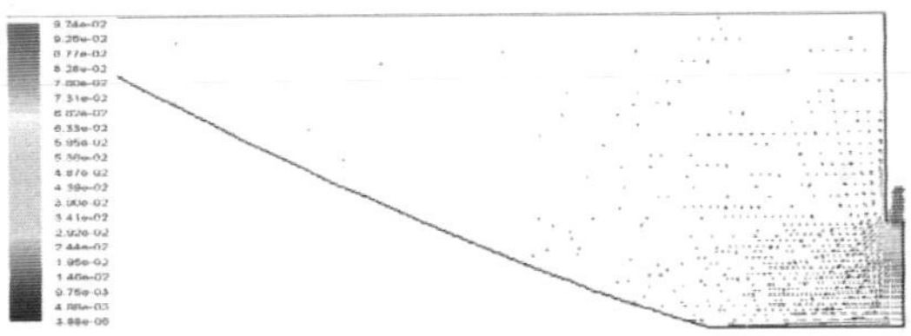

(a) 100 m^3/h Velocity Vector

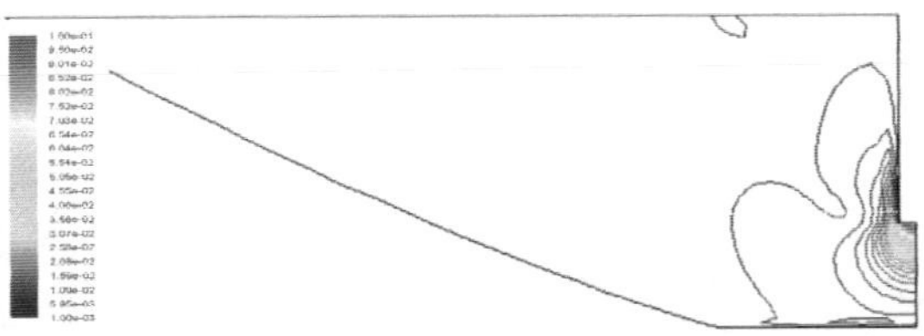

(b) 100 m^3/h Velocity Contour

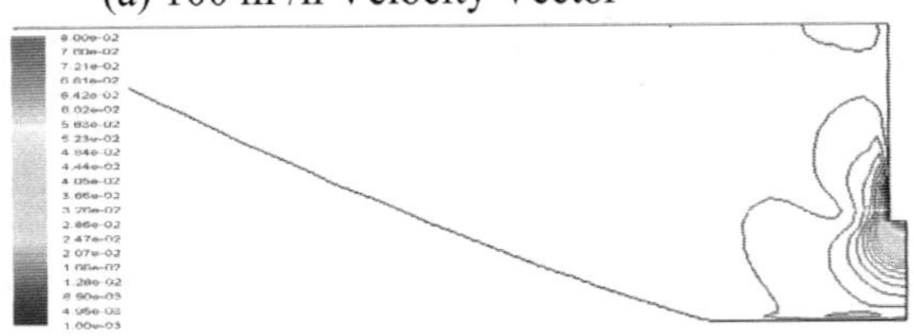

(c) 80 m^3/h Velocity Contour

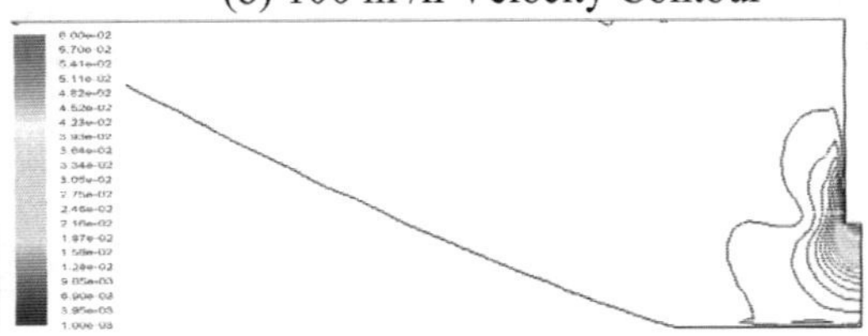

(d) 60 m^3/h Velocity Contour

Figure 5. Flow field distribution at different brine discharge rates under the condition that the distance between gas-brine interface and discharge pipe inlet is 4m.

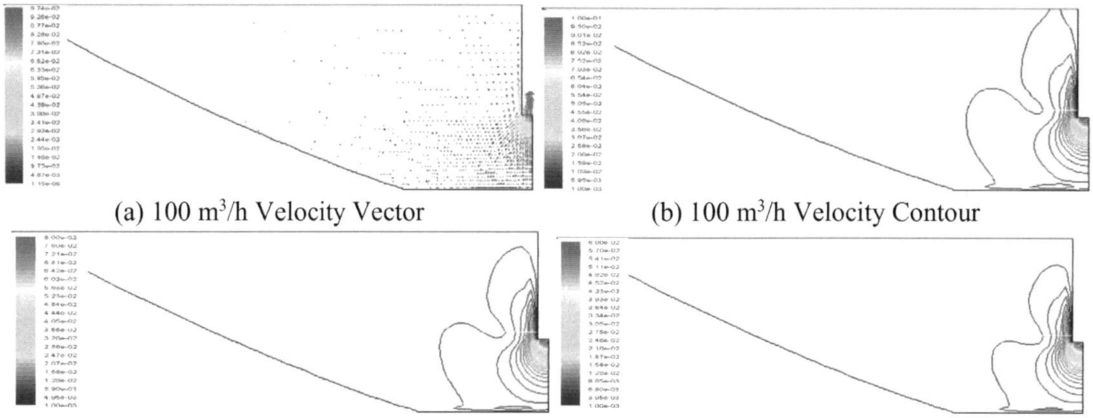

(a) 100 m^3/h Velocity Vector (b) 100 m^3/h Velocity Contour

(c) 80 m^3/h Velocity Contour (d) 60 m^3/h Velocity Contour

Figure 6. Flow field distribution at different brine discharge rates under the condition that the distance between gas-brine interface and discharge pipe inlet is 3m.

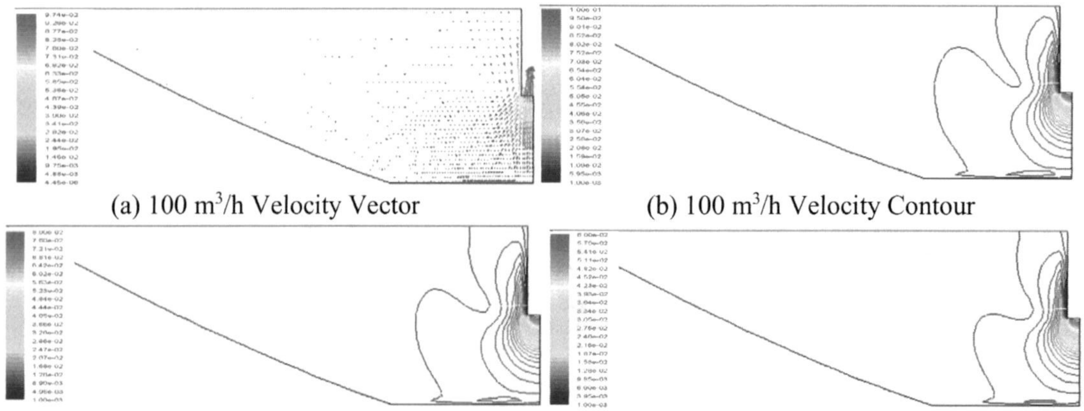

(a) 100 m^3/h Velocity Vector (b) 100 m^3/h Velocity Contour

(c) 80 m^3/h Velocity Contour (d) 60 m^3/h Velocity Contour

Figure 7. Flow field distribution at different brine discharge rates under the condition that the distance between gas-brine interface and discharge pipe inlet is 2m.

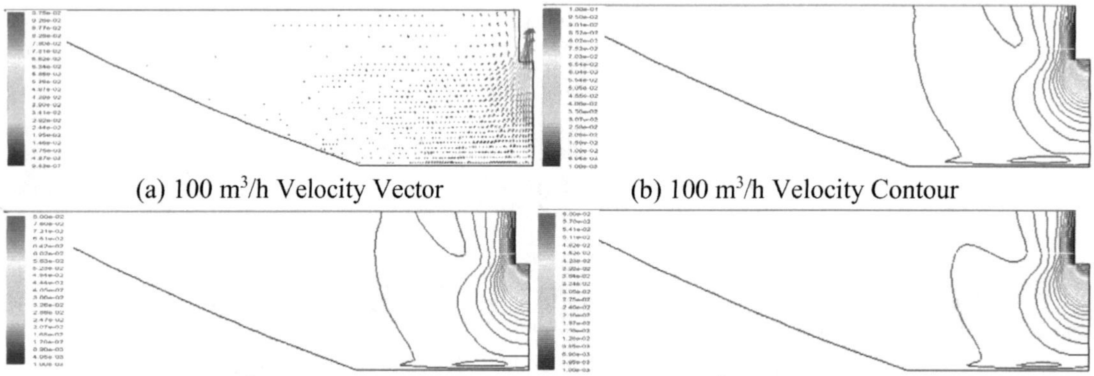

(a) 100 m^3/h Velocity Vector (b) 100 m^3/h Velocity Contour

(c) 80 m^3/h Velocity Contour (d) 60 m^3/h Velocity Contour

Figure 8. Flow field distribution at different brine discharge rates under the condition that the distance between gas-brine interface and discharge pipe inlet is 1m.

From the Figure 5 - Figure 8, it can also be seen that as the distance between the gas-brine interface and the discharge pipe inlet decreases, the flow velocity around the discharge pipe inlet and gas-brine interface changes drastically, especially when the brine discharging rate is large. This shows that in the later stage of brine drainage, the rate of brine drainage needs to be reduced in time to prevent a large amount of natural gas from being carried to the ground by brine which is a potential safety hazard.

4. Conclusion

In the process of gas injection and brine drainage in underground salt caverns, natural gas and brine coexist. The depth of the brine discharge pipe entering the brine and the rate of brine drainage determine whether natural gas will be brought out to the surface by brine.

How to adjust the rate of brine drainage in the later stage and reduce the risk of bringing out the natural gas is a technical problem in the process of gas injection and brine removal. Aiming at the research on this technical problem, the conclusions are as follows:

In the later stage of gas injection and brine drainage process, when the remaining brine in the cavity is 4m deep, the brine drainage rate of $100m^3/h$ will not carry out natural gas. When the brine depth of 1 m remains in the cavity, the area around the brine discharge pipe inlet is a rapidly changing area of flow field. When the brine is discharged at the rate of $40\text{-}100m^3/h$, the velocity gradient at the gas-brine interface near the discharge pipe is large, easy to result in gas carried out. When the remaining brine in the cavity is 1m deep, the brine should be drained slowly at a rate of less than $40m^3/h$.

References

[1] C. Yang, T. Wang, J. Li, et al. Feasibility analysis of using closely spaced caverns in bedded rock salt for underground gas storage: a case study, Environ. Earth Sci. 75(15) (2016) 1-15.

[2] F. Bayram, I. Bektasoglu, Determination of actual dissolution rates from some rock properties in construction of deep salt cavern for natural gas storage, Intern. J. of Rock Mecha. & Mining Sci. 126 (2020) 104183.

[3] H. Schweinsberg, Etzel Cavern Facility–An Energy Hub for Crude Oil and Gas Storage in Europe, Solution Mining Research Institute Fall 2012 Technical Conference, Bremen, Germany, 2012, pp. 1-8.

[4] P. Berest, B. Brouard, J. G. Durup, Tightness tests in salt-cavern wells, Oil & Gas Sci. & Tech. 56(5) (2001) 451-469

[5] W. Liang, C. Yang, Y. Zhao, et al. Experimental investigation of mechanical properties of bedded salt rock, Inter. J. of Rock Mecha. & Mining Sci. 44(3) (2007) 400-411.

[6] J. Li, X. Shi, C. Yang, et al. Mathematical model of salt cavern leaching for gas storage in high-insoluble salt formations, Sci. Reports, 8(1) (2018) 1-12.

[7] D. G. Caglayan, N. Weber, H. U. Heinrichs, et al. Technical potential of salt caverns for hydrogen storage in Europe, Inter. J. of Hydrogen Energy, 45(11) (2020) 6793-6805.

[8] T. Wang, S. Ding, H. Wang, et al. Mathematic modelling of the debrining for a salt cavern gas storage, J. of Natural Gas Sci. & Eng. 50 (2018) 205-214.

[9] H. Bernhardt, S. Boor, W. Micus. New Developments in the Gas First Fill Process of Natural Gas Storage Caverns, SMRI Fall 2011 Technical Conference, York, Unite d Kingdom, 2011, pp. 3-4.

[10] T. Wang, C. Yang, H. Wang, et al. Debrining prediction of a salt cavern used for compressed air energy storage, Energy, 147 (2018) 464-476.

[11] Z. Ren, J. Li, J. Zhu, et al. Numerical simulation on technological parameters of gas injection and brine discharge in salt-cavern gas storage, Oil&Gas Storage & Trans. 37(4) 2018 403-406.

[12] T. Wang, G. Chai, X. Cen, et al. Safe distance between debrining tubing inlet and sediment in a gas storage salt cavern, J. of Petro. Sci. & Eng. 196 (2021) 107707.

A new unstructured hybrid mesh generation method based on BP-ANN

Peng Lu[1,2,3], Nianhua Wang[1], Yi Lin[1,2], Xiaorong Zhang[2], Yadong Wu[4] and Hongying Zhang[2,*]

[1]Stake Key Laboratory of Aerodynamics, China Aerodynamics Research and Development Center, Mianyang, China
[2]Southwest University of Science and Technology, Mianyang, China
[3]School of Intelligent Manufacturing Engineering, Chongqing University of Arts and Science, Yongchuan, China
[4]School of Computer Science and Technology, Sichuan University of Science & Engineering, Yibin, China

*Corresponding author e-mail: zhywyd@163.com

Abstract. The automation and intelligence of grid generation are still bottleneck problems in computational fluid dynamics. This problem can be solved by generating anisotropic quadrilateral by advancing layer method and isotropic triangular mesh by advancing front method. However, the advancing layer method needs to calculate the advancing direction, step size, a correction factor of the concave and convex area, so the automation level and efficiency of grid generation need to be improved. To solve the above problems, a hybrid grid generation method based on BP-ANN is proposed. The grid data generated by traditional methods are learned by a neural network, and the neural network can predict the above key parameters. Finally, a mesh generation example of 30P30N multi-element airfoil shows that the proposed algorithm can effectively improve the adaptability, efficiency and quality of hybrid mesh generation.

1. Introduction

Unstructured mesh generation technology has been developed vigorously because of its adaptability and high efficiency. However, such isotropic unstructured grids are not suitable for the simulation of viscous flows because there exists a very thin boundary layer near the object surface. To capture viscous flow characteristics, a denser grid should be generated in the boundary layer. Because of the strong anisotropy of the flow in the boundary layer, it is more reasonable to use structured or semi-structured anisotropic grids in this region. In the far-field of the object surface, to improve the grid discrete efficiency, the isotropic triangular mesh should be generated, it is necessary to generate isotropic and anisotropic hybrid mesh to simulate viscous flow [1].

The advancing layer method (ALM) proposed by Pirzadeh is the most influential method for generating quadrilateral grids for boundary layer anisotropy [2]. Its basic idea is as follows: After the object plane mesh is determined, the advancing direction and distance are determined at the vertices of each layer to advance to the computational domain layer by layer with recursive layer idea.

With the rapid development of science and technology, artificial intelligence (AI) technology has been preliminarily applied in grid generation. Reichert [3] introduced an expert system in two-dimensional (2D) grid generation. Yao [4] proposed an element extraction method based on an artificial neural network (ANN) for automatic grid generation and adopted an artificial neural network to replace expert databases for pattern recognition. Yilmaz [5] proposed particle swarm optimization to optimize the quality of hexahedral grids, which solved the problem of the poor effect of the Laplacian mesh smoothing algorithm in concave regions. In recent years, more and more researchers applied machine learning in grid generation research [6-9]. The above AI method improves the automation and intelligence level of grid generation. However, these studies do not solve the problem of isotropic and anisotropic hybrid mesh generation. Therefore, this paper tries to improve the adaptability and efficiency of the hybrid mesh generation method based on the backpropagation artificial neural network (BP-ANN).

2. Hybrid mesh generation method based on BP-ANN

This paper introduces the anisotropic mesh generation method based on BP-ANN as follows:

(1) The training data of the neural network are automatically extracted from the boundary layer grid generated by traditional methods, and the data are normalized preprocessed.

(2) The neural network structure is designed and trained to obtain the neural network model that can predict the key parameters in the process of layer propulsion.

(3) The environment around the front in the actual layer propulsion process was collected and used as the neural network input. The advancing direction, step size adjustment factor and multi-direction factor are predicted through the neural network forward path.

(4) The growth height of each layer is controlled. Since the physical quantity gradient inside the boundary layer is very large and then gradually decreases, the step size increases exponentially.

$$\delta_l = \delta_0 (1+\beta)^{l-1} k \tag{1}$$

Where δ_l and δ_0 represent the advance distances of layer 1 and layer 0 (object surface) respectively. The growth height of layer 0 depends on the Reynolds number of the incoming flow. β is the rate at which the grid size grows, depending on flow states such as laminar or turbulent flow.

(5) After calculating the direction and advance distance of a point in the current layer, the position of the new point can be determined by Formula (2).

$$\boldsymbol{r}_{l+1} = \boldsymbol{r}_l + \boldsymbol{n}_p \delta_{l+1} \tag{2}$$

(6) If the quadrilateral element formed by the newly generated points intersects other existing edges that are not directly connected, the advance will stop. To prevent the occurrence of very close proximity without intersection, resulting in poor quality of triangular mesh generated in the later period, the circle with the new point as the center of the circle and 0.5 times the advance step as the radius shall not contain other edges that are not directly connected.

(7) Repeat steps (2) ~ (7), until the aspect ratio of all quadrilateral reaches the set threshold. To make the transition from quadrilateral element to triangular element natural, the threshold is usually set as 1.

2.1. Neural network structure design

The neural network has two main functions: prediction and classification. The backpropagation (BP) learning algorithm is used to optimize the weights between nodes in the neural network. The algorithm is divided into two parts. The first part (forward propagation process) calculates the output of each unit layer by layer from the input layer through the hidden layer. The output error of the second part (backpropagation process) is calculated layer by layer, and the error is used to correct the weight of the previous layer until the actual neural network output and the expected output tend to be minimum.

(1) Determine the number of nodes in the input layer

As shown in Figure 1, the traditional method to determine the direction of a point is firstly to find the normal direction of the front, and then to find the growth direction of the point through the smooth average iterative algorithm. Theoretically, the environment of a propulsion point is composed of many adjacent points, but only the nearby points can affect the direction of the point. To extract training data quickly and conveniently, the experiment found that the best comprehensive effect was obtained when the four adjacent points of the layer where the advancing point was selected as the input. As shown in Figure 1(b), it is assumed that point P_{mid} is the advancing point, and its four neighboring points are the neighboring points P_{l1} and P_{l2} on the left and P_{r1} and P_{r2} on the right. The coordinates of five points are selected as the input of the ANN, so the input node number of the ANN is 10.

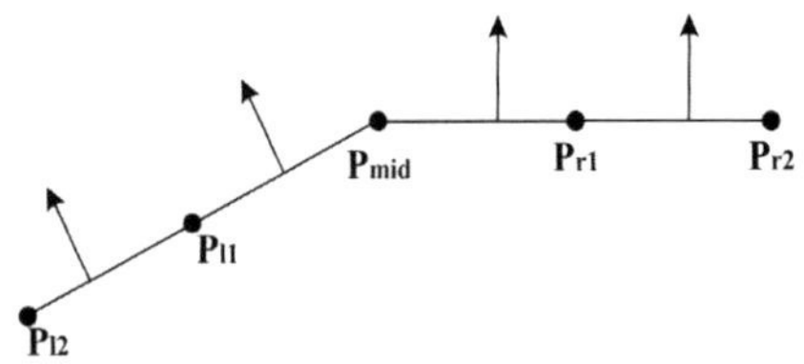

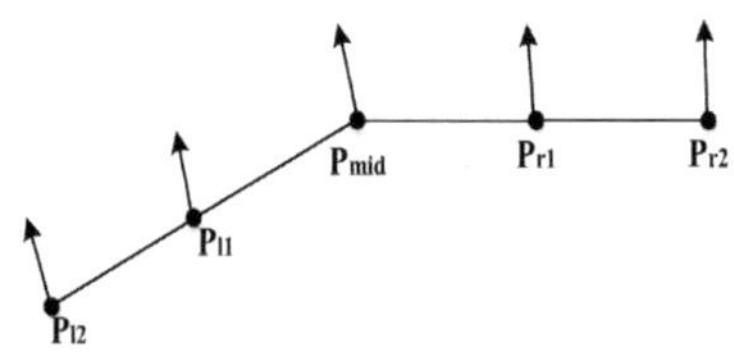

(a) The normal vector to the edge (b) The normal vector to the point

Figure 1. 2D grid edge normal and point normal in the traditional method.

(2) Determination of the number of nodes in the output layer

The traditional ALM firstly needs to determine two key parameters of advance direction and advance step size of each layer. To improve the mesh quality of the concave and convex region, it also needs to calculate the smooth factor of concave and convex region step size and the number of multi-direction advances. Due to the step is associated with Reynolds number, which is easy to obtain, and the smooth factor is a concave and convex area of the step length and level of the ratio of the step length, has been included in the neural network training samples of the step length in the relationship. Finally, the output of the neural network is determined to promote the coordinate values, perspective prediction, and calculate the angle sizes. Therefore, the number of output nodes of the ANN is 3.

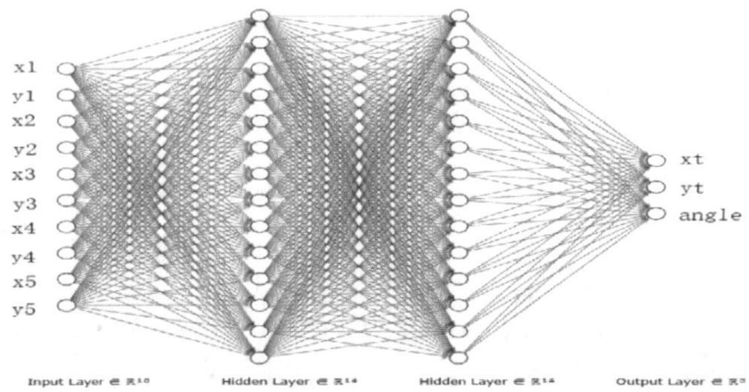

Figure 2. Neural network forward pathway structure.

2.2. Extraction and pretreatment of neural network training samples

As shown in Figure 3, there are convex and concave angles similar to many airfoils. The adjacent points on the left of P_{mid} are P_{l1} and P_{l2}, and the adjacent points on the right are P_{r1} and P_{r2}. Taking point P_t as an example, the coordinates of P_{l2}, P_{l1}, P_{mid}, P_{r1} and P_{r2} are used as the training sample data set of the ANN, and the input data labels are point P_t coordinates and the included angle related to multi-direction propulsion $\angle P_{l1}P_{mid}P_{r1}$.

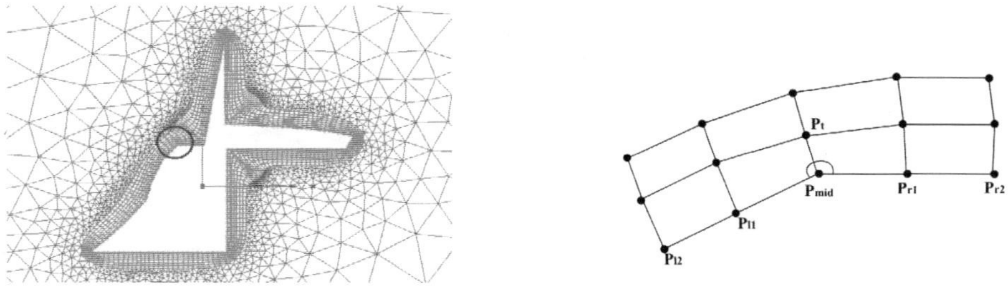

(a) training sample grid (b) Single training sample

Figure 3. Grid of training neural network and extraction of training data.

After feature extraction, it is the processing of feature data, among which the key step is whether to normalize the feature data, the unified standard of input points can be obtained by rotating, translating and scaling the front as shown in Figure 4.

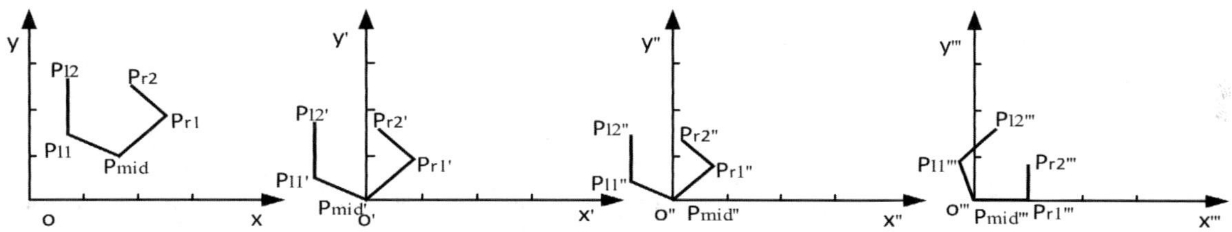

(a) original location (b) Translation (c) scaling (d) rotation

Figure 4. Normalization of training sample data.

$$X_{norm} = \frac{X - X_{\min}}{X_{\max} - X_{\min}} \tag{3}$$

Where X is the original data, $X_{\max}$ and $X_{\min}$ are the maximum and minimum values of sample data respectively.

2.3. *The neural network training process*

The weight of the neural network can be optimized layer by layer through the BP algorithm, and the goal is to minimize the loss function. The adam optimization algorithm is adopted to optimize the weight coefficients of neurons layer by layer, as shown in Formula (4).

$$\Delta w_{ij} = -\eta \frac{\partial loss_{MSE}}{\partial w_{ij}} \tag{4}$$

3. Experimental examples and analysis

Figure 5 shows the actual output of the neural network in some common cases. The neural network predicted that the advance step of the new point was longer at the concave, but shorter at the convex, so the concave-convex adjustment factor could be derived from the ratio of the length of the new point at the concave-convex angle to the horizontal position. However, the new point here is not the position of the last point. In the case of no multi-direction growth, the predicted new point here is only the final direction of the new point, and the specific position shall be determined through step (4) in Section 1. For the place with a large convex Angle, the number of multi-direction propulsion should be calculated by the Angle predicted by the neural network, and the coordinates of the new generation points after the optimization of multi-direction propulsion should be recalculated.

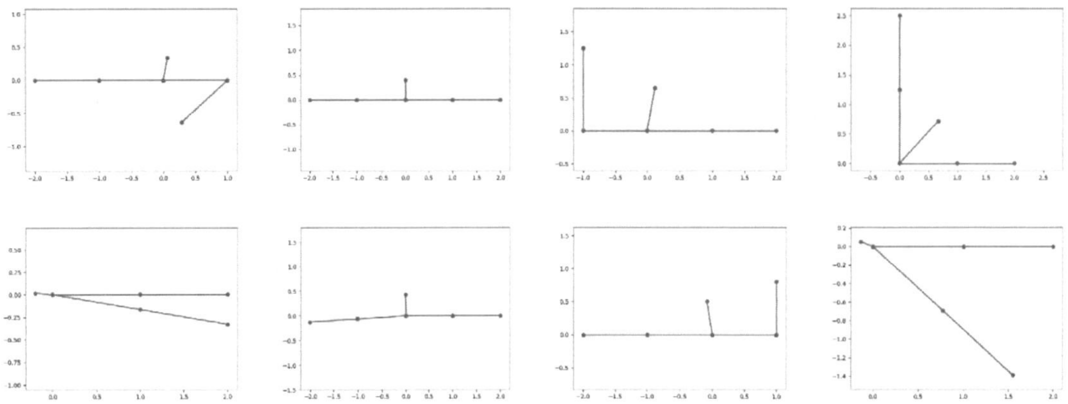

Figure 5. Prediction of the new point direction in different states.

As shown in Figure 6, a 30P30N airfoil grid contains 24,798 cells (quadrilateral cells account for 75% of the total), 37,030 edges, and 12,619 points. In the three sections of the wing, mesh refinement is carried out near the area with large curvature, while multi-direction propulsion technology is adopted in the area with a particularly large convex angle of the airfoil, as shown in Figure 6(c-d). The ANN can predict the generating direction of anisotropic grid points and the number of multiple directions. The grid generation efficiency is improved by 18% by eliminating the steps of finding surface normal, point normal, smoothing and the adjusting factors in the concave and convex area. For isotropic grids, the efficiency of grid generation can be increased by more than 30% due to the reduction of a large number of intersecting judgments by machine learning.

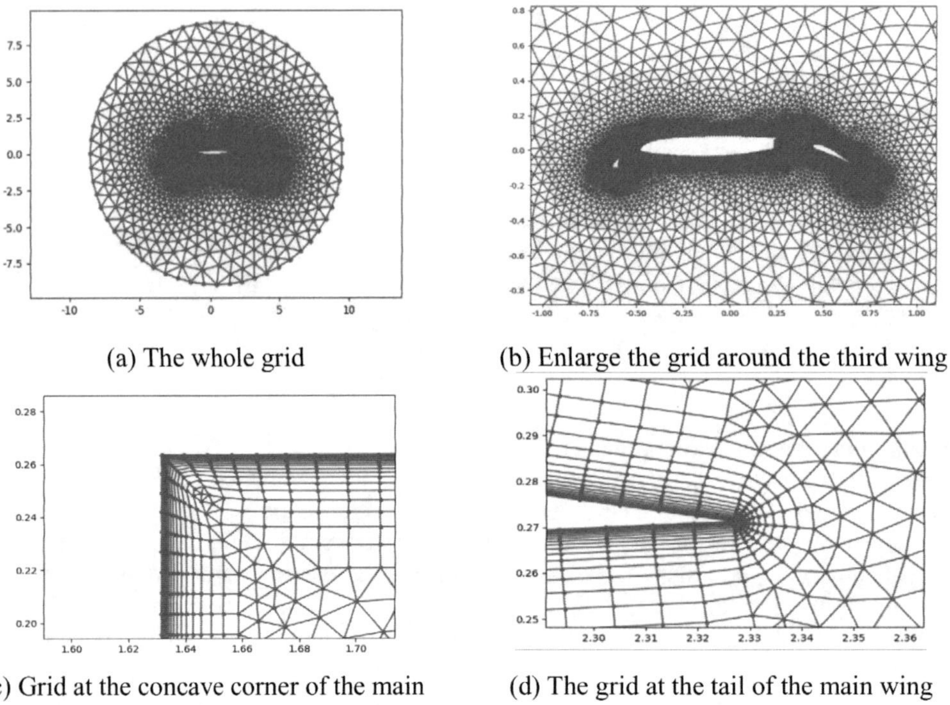

(a) The whole grid

(b) Enlarge the grid around the third wing

(c) Grid at the concave corner of the main wing

(d) The grid at the tail of the main wing

Figure 6. Grid around 30P30N three wings.

4. Conclusion

In this paper, a 2D hybrid mesh generation technique based on machine learning is proposed. First, the grid data of training samples generated by traditional methods are automatically extracted and pre-processed. Second, the key parameters in the process of generating anisotropic meshes by the neural network prediction ALM are designed. Finally, the performance of the mesh generation method is tested by using the 30P30N multi-element airfoil. The results show that the new method can improve the adaptability and efficiency of mesh generation, and the quality of the mesh is improved obviously, which shows that the machine learning method has a good application prospect in the unstructured hybrid mesh.

Acknowledgments

This work was supported by the Open Research Program of State Key Laboratory of Aerodynamics (20200203), and the Technology Innovation and Application Development Special Project of Chongqing (cstc2020jscx-msxm0434).

References

[1] Slotnick J, Khodadoust A, Alonso J, Darmofal D, Gropp W, Lurie E, Mavriplis D. CFD vision 2030 study: A path to revolutionary computational aero sciences. NASA/CR–2014-218178, 2014.

[2] Pirzadeh S. Three-dimensional unstructured viscous grids by the advancing-layers method. AIAA Journal, 1996; 34(1): 43-49.

[3] Reichert K, Skoczylas J, Tarnhuvud T. Automatic mesh generation based on expert-system-methods. IEEE Transactions on Magnetics, 1991; 27(5): 4197-4200.

[4] Yao S, Yan B, Chen B, et al. An ANN-based element extraction method for automatic mesh generation. Expert Systems with Applications. 2005; 29: 193-206.

[5] Yilmaz AE, Kuzuoglu M. A particle swarm optimization approach for hexahedral mesh smoothing. International Journal for Numerical Methods in Fluids, 2010;60(1): 55-78.

[6] Papagiannopoulos A, Clausen P, Avellan F.How to teach neural networks to mesh: Application on 2-D simplicial contours. Neural Network. 2021; 136: 152-179.

[7] Guo Y, Wang C, Ma Z, et al. A new mesh smoothing method based on a neural network. Computational Mechanics, 2021.

[8] Zhang Z, Jimack P K, Wang H.MeshingNet3D: Efficient generation of adapted tetrahedral meshes for computational mechanics. Advances in Engineering Software, 2021; 157-158.

[9] Mitusch S K, Funke S W, Kuchta M.Hybrid FEM-NN models: Combining artificial neural networks with the finite element method. Journal of Computational Physics, 2021; 446.

Rotordynamic characteristic of water lubricated plain journal bearing under transient load

Pingwei Chen[1, 2], Tong Wang[1, *],Wensheng Ma[2], Jie Chen[3], Rui Cao[2]

[1] School of Mechanical and Powder Engineering, Harbin University of Science and Technology, Harbin 150080, China
[2] Chongqing pump industry, Chongqing 400033, China
[3] CSSC Chongqing Hydraulic Mechanical-electronical CO.LTD, Chongqing 400038, China

*Correspondence to: wangtong@hrbust.edu.cn

Abstract. Transient startup operating parameter has a serious influence on the stability of water-lubricated bearings. In this study, CFD method was used to predict the rotordynamic coefficients of water lubricated bearings under transient loads. The stiffness and damping coefficients versus transient time for water-lubricated bearings are given. The stiffness and damping coefficients versus different transient loads for water-lubricated bearings are also given. The results show that the stiffness and damping coefficients will increase rapidly in a very short time. With the increase of impact time, both of them come to a steady value. With the increase of impact load, the direct stiffness and the direct damping coefficients of water lubricated bearing will increase. The value of cross-coupling stiffness and the cross-coupling damping coefficients will decrease.

Keywords: Water lubricated; plain journal bearing; transient load; rotordynamics.

1. Introduction

It is becoming a trend to use water-lubricated bearing due to its environmental protection and many other benefits. Compared to the oil-lubricated bearing, the water-lubricated bearing has the advantages of simplified design, low friction, low cost because of its low viscosity. The water-lubricated bearings are used in many areas like ships, pumps and turbines [1-3]. Many Researches on water-lubricated bearings have done on the friction [4–5] and pressure distribution in the water film [6-7]. Water lubricated bearings have good performance, but as the core part of the rotor-bearing system, it also has an important role in rotor dynamics. It is need to study the dynamic coefficients of water-lubricated bearings. Xiuli Zhang [8] studied the relationships between the stiffness coefficients and the load for bearings with different relative clearances, different length-to- diameter ratios and different rotational speeds are obtained. The linear & nonlinear stiffness and damping coefficients are estimated from medium to large journal orbit by V. Meruane [9]. However, there are few studies on the stiffness and damping changes in the process of instant start-up.

In this paper, CFD (Computational Fluid Dynamics) method is used to analyze the rotordynamic coefficients of water-lubricated bearings under transient loads. The stiffness and damping coefficients versus different transient loads for water-lubricated bearings are given.

2. Nomenclature:

Fx,Fy: components of the load in the X direction and Y direction
Kij: stiffness coefficients
Cij: damping coefficients
ΔFdij: the change of water film force caused by disturbed displacement
ΔFvij: the change of water film force caused by disturbed velocity
Δx,Δy: disturbed displacement
Δx',Δy': disturbed velocity

3. Computational model

3.1. Plain journal bearing model

Fig.1 shows a schematic of the plain journal bearing. The bearing is subjected to a constant vertical load. When the rotating journal deviates from the bearing center, dynamic pressure is generated in the water film. The distance between the journal center and the bearing center is the journal eccentricity (e).

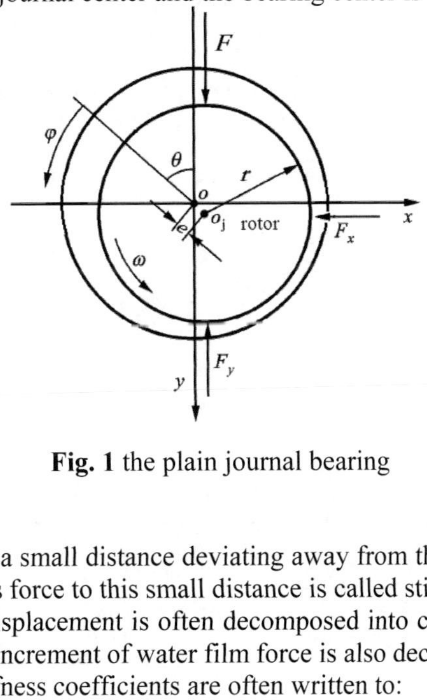

Fig. 1 the plain journal bearing

3.2. Stiffness and damping

When the rotor is disturbed by a small distance deviating away from the equilibrium position, the ratio of the increment of water film's force to this small distance is called stiffness.

In the analysis, the small displacement is often decomposed into components along the coordinate axis x and y direction, and the increment of water film force is also decomposed into components along the x and y direction. Four stiffness coefficients are often written to:

$$K_{xx} = \frac{\Delta F_{dxx}}{\Delta x} \qquad K_{yy} = \frac{\Delta F_{dyy}}{\Delta y}$$

$$K_{yx} = \frac{\Delta F_{dyx}}{\Delta x} \qquad K_{xy} = \frac{\Delta F_{dxy}}{\Delta y}$$

When the rotor is disturbed by a small velocity deviating away from the equilibrium position, the ratio of the increment of water film's force to this small velocity is called damping.

The small velocity is often decomposed into components along the coordinate axis x and y direction, and the increment of water film force is also decomposed into components along the x and y direction. Four damping coefficients are often expressed:

$$C_{xx}=\frac{\Delta F_{vxx}}{\Delta x'} \quad C_{yx}=\frac{\Delta F_{vyx}}{\Delta x'}$$
$$C_{xy}=\frac{\Delta F_{vxy}}{\Delta y'} \quad C_{yy}=\frac{\Delta F_{vyy}}{\Delta y'}$$

4. Finite element model and calculation results

4.1. CFD model

In CFD software, the mesh of the model is hexahedron grid. The double precision calculations are employed. The model of water lubricated bearing is shown in Figure 2. The fluid is 25°C water. The bearing's radius gap c=0.07mm, the length L=120mm, Rotor radius R=40mm, the rotor speed n= 3000r/min. Since the bearing is fully submerged in the water, both the inlet and outlet boundary conditions are 0 Pa (gauge pressure). According to actual working conditions, the outer surface of the water film model is "stationary wall" and the inner surface is "moving wall". No-slip condition is used on the two walls in Fluent [10] software. In fluent software, *k-ε* turbulence model is used.

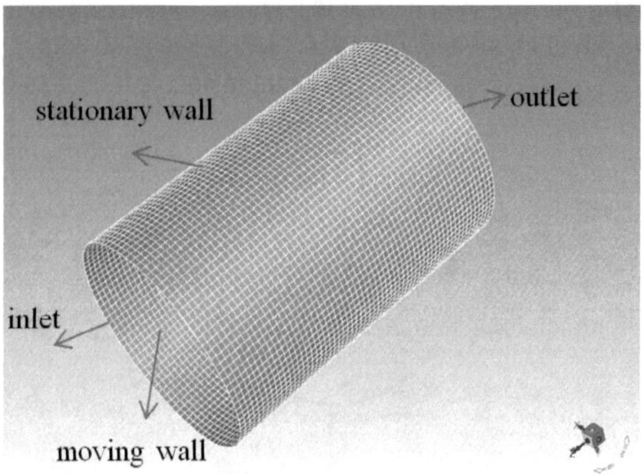

Fig. 2 CFD model

Fig.3 shows the water film pressure distribution at transient time 0.05s when the load is 1000N. We can see the effective bearing area of the dynamic water pressure film.

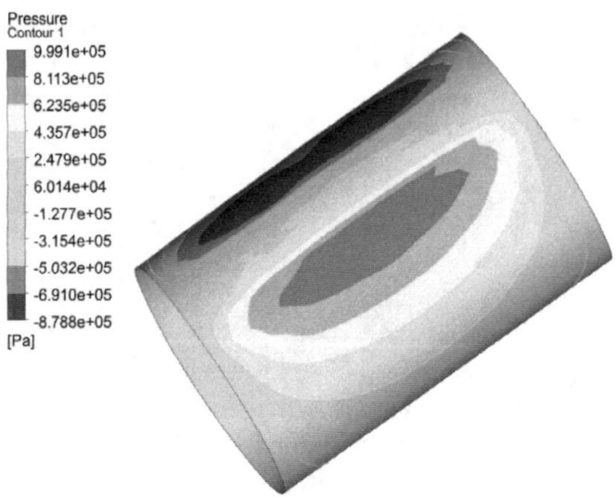

Fig. 3 Pressure contour distribution at 0.05s

Fig.4 illustrates the axis trajectory within transient time 0.1s, with the increase of transient time, the amplitude of its axis trajectory becomes smaller, and gradually tends to be stable.

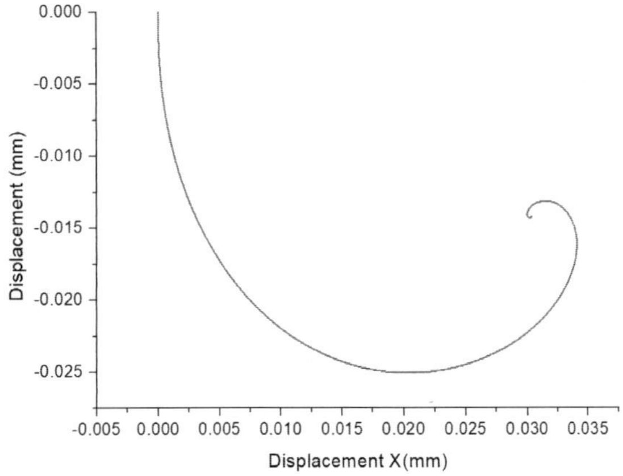

Fig. 4 Axis trajectory

4.2. Stiffness and Damping

Fig5 gives the time-varying stiffness and damping coefficient when the water lubricated bearing is subjected to transient load of 1000N.The stiffness Coefficient and Damping Coefficient also have a great change obviously under the transient load. The maximum Kxx is 4.64e7N/m and the stabled value is 3.52e7N/m, the maximum Kyy is 6.23e7N/m and the stabled value is 4.12e7N/m, the maximum Kxy is 3.4e7N/m and the stabled value is 2.13e7N/m, the maximum Kyx is -7.2e7N/m and the stabled value is-5.62e7N/m. the maximum Cxx is 2.21e5N·S/m and the stabled value is 1.93 e5 ,the maximum Cyy is 4.0e5 N·S/m and the stabled value is 3.1e5 N·S/m, the maximum Cxy (Cyx) is -1.34e5N·S/m and the stabled value is -1.0e5N·S/m. The transient impact coefficient of stiffness and damping is1.1-1.6.

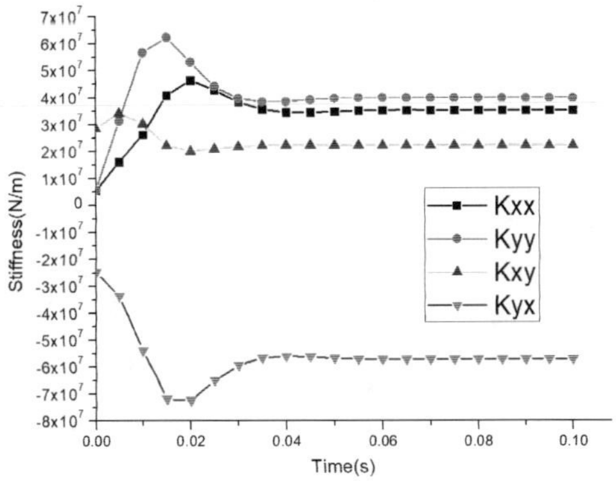

(a) the time-varying change of stiffness

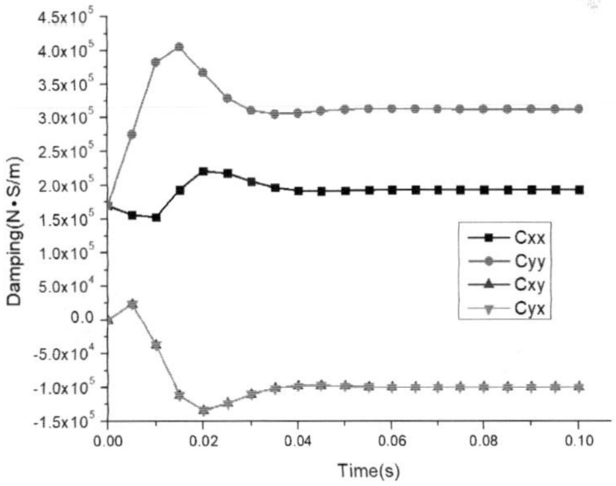

(b) the time-varying change of damping

Fig. 5 Rotordanamic coefficients versus time

In order to compare the stiffness and damping coefficients of water-lubricated bearings under different transient loads (500N, 1000N, 2000N, 3000N), we choose the transient time at 0.025s, 0.05s, 0.075s, 0.1s.

Figure 5 shows 4 types of stiffness versus transient time with different transient load. Figure 6 gives 4 types of damping versus transient time with different transient load. With the transient load increases, the direct stiffness coefficient increases and the value of the cross-coupling stiffness decreases. The direct damping coefficient increases, while the cross-coupling damping coefficient decreases.

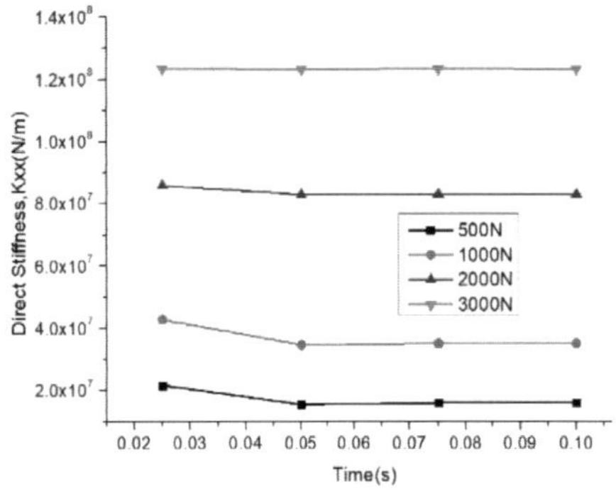

(a) direct stiffness Kxx

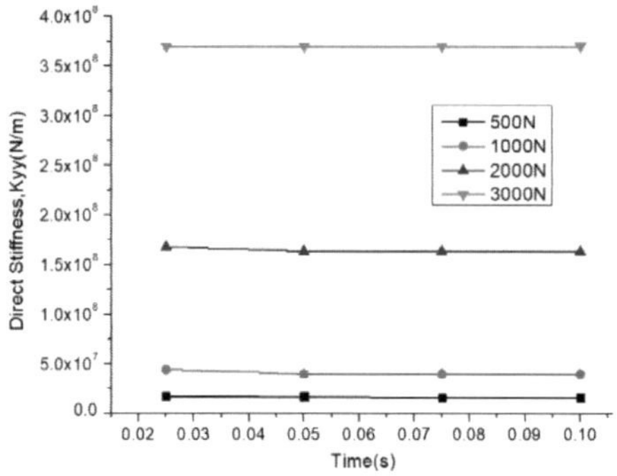

(b) direct stiffness Kyy

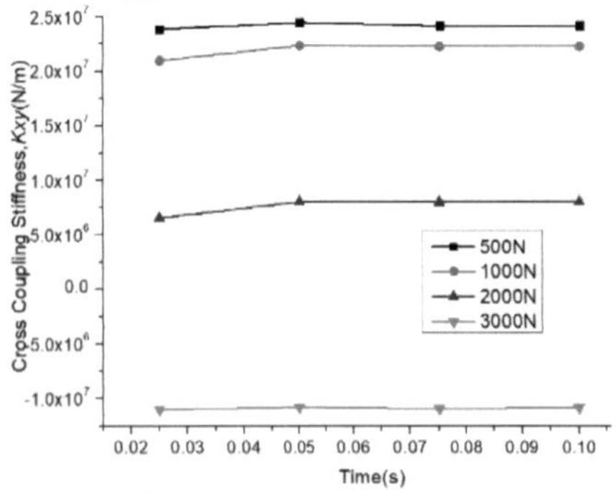

(c) cross-coupling stiffness Kxy

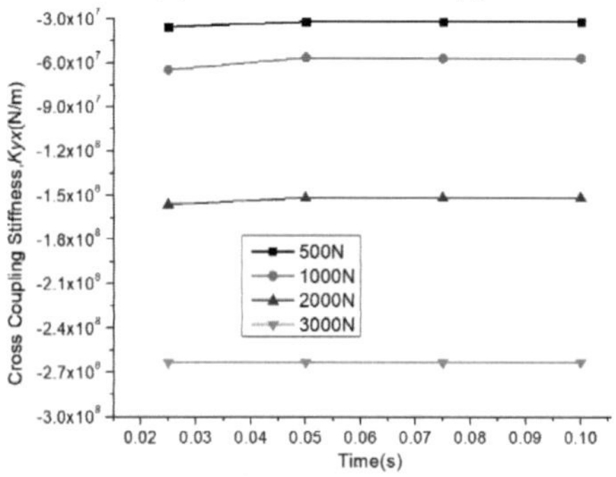

(d) cross-coupling stiffness Kyx

Fig. 6 stiffness coefficients versus transient time

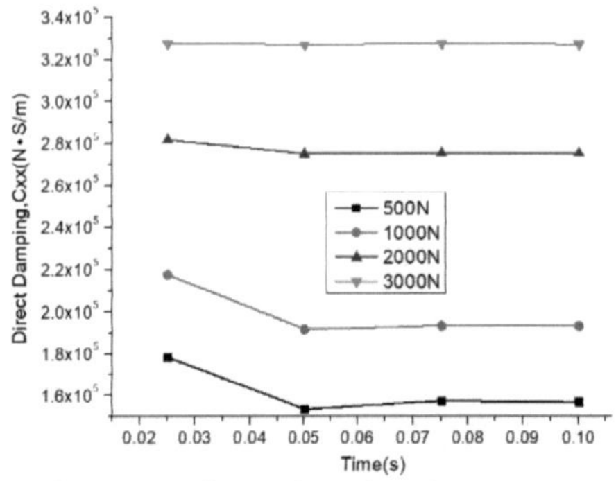

(a) direct damping Cxx

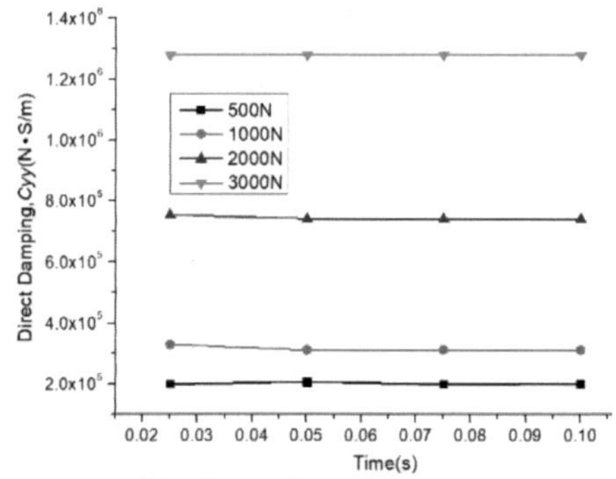

(b) direct damping Cyy

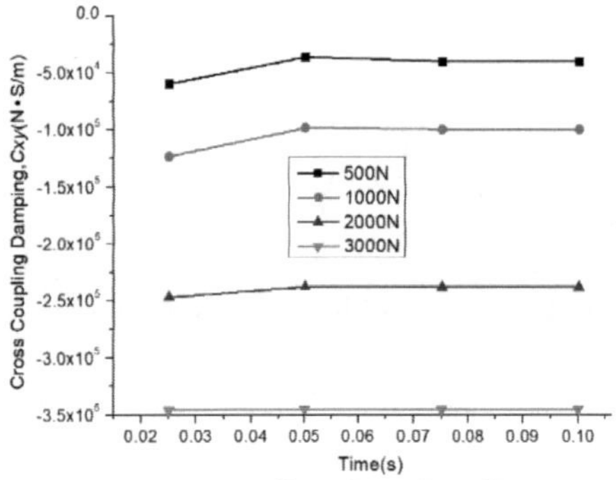

(c) cross-coupling damping Cxy

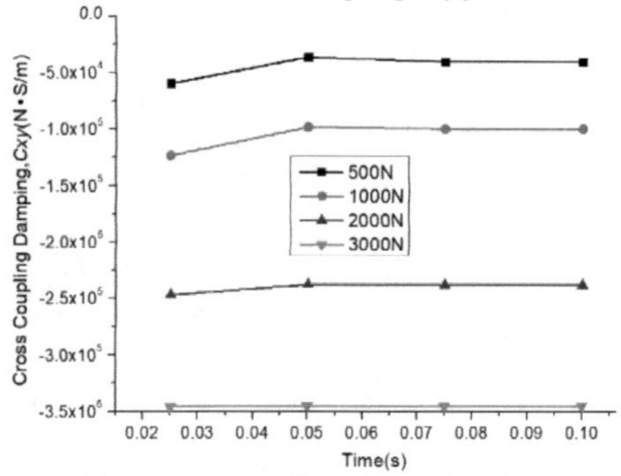

(d) cross-coupling damping Cyx

Fig. 7 damping coefficients versus transient time

5. Conclusions

In this paper, CFD method is used to calculate the rotordynamic characteristic of water lubricated bearing. We found that the transient load has a great influence on the water lubricated bearing. The stiffness and damping coefficients will increase rapidly in a very short time. With the increase of impact time, both of them come to a steady value. With the increase of impact load, the direct stiffness and the direct damping coefficients of water lubricated bearing will increase. The value of cross-coupling stiffness and the cross-coupling damping coefficients will decrease.

The stiffness Coefficient and Damping Coefficient also change obviously under the transient load. The impact factor of stiffness and damping is generally 1.1-1.6.

Acknowledgments

This study has been funded by Key projects of technological innovation and application development of Chongqing, China (Grant NO.cstc2020jszx-msxm0411).

References

[1] Pai R,Hargreaves DJ. Water lubricated bearings. In: Nosonovsky M ,Bhushan B, editors. Green tribology.Berlin–Heidelberg: Springer; 2012.p. 347–91.

[2] Lingwall BA,SextonTN,Cooley CH.Polycrystalline diamond bearing testing for marine hydrokinetic application.Wear 2013; 302: 1514–9.

[3] Heberley BD. Advances in hybrid water-lubricated journal bearings for use in ocean vessels. ([Ph.D.thesis]). AnnArbor, USA: Massachusetts Institute of Technology; 2013.

[4] UnalH,MimarogluA. Friction and wear characteristics of PEEK and its composite under water lubrication.J Reinf Plast Compos 2006; 25:1659–67.

[5] JiaJ-h, ChenJ-m, ZhouH-d, WangJ-b, ZhouH. Friction and wear properties of bronze–graphite composite under water lubrication. TribolInt2004; 37: 423–9.

[6] Zhong-liang Xie, Na Ta, et al. The lubrication performance of water lubricated bearing with consideration of wall slip and inertial force. [J] Journal of Hydrodynamics 2017, 29(1):52-60.

[7] Pingwei Chen et al. Characteristic Analysis of water lubricated Plain Journal Bearing under Transient Load.2022 J.Phys.: Conf.Sel. 2160 012073.

[8] Xiuli Zhang, Zhongwei Yin, et al. Determination of stiffness coefficients of hydrodynamic water-lubricated plain journal bearings [J] Tribology International 85(2015)37–47.

[9] V. Meruane, R. Pascual. Identification of nonlinear dynamic coefficients in plain journal bearings [J] Tribology International 41(2008)734–754.

[10] ANSYS fluent 20.0 Theory Guide, ANSYS Inc., USA, 2019.

FMIA-2022 IOP Publishing
Journal of Physics: Conference Series 2280 (2022) 012047 doi:10.1088/1742-6596/2280/1/012047

Research on Influence Factors of Fluid-induced Force in labyrinth liquid Seals for pump

Pingwei Chen [1, 2], Tong Wang[1,*], Wensheng Ma[2], Tianzhou Wang[1], Rui Cao[1]

[1] School of Mechanical and Powder Engineering, Harbin University of Science and Technology, Harbin 150080, China
[2] Chongqing pump industry, Chongqing 400033, China

*Corresponding author e-mail: wangtong@hrbust.edu.cn

Abstract. A 3D model of straight-through labyrinth liquid seal for water pump was built to analyse the fluid-induced forces. We found that the eccentricity and pressure difference have great impact on both the tangential force and the radial force. The tangential force increases along with the rotating speed. When tooth hight is T_h=2.4mm, the tangential force comes to the minimum. When tooth width Tw =2.0mm, the radial force comes to the maximum. The tooth numbers have a great influence on the fluid-induced force. The more number of sealing teeth will cause the instability the sea.

1. Introduction

The labyrinth seal is commonly seen in turbo-machinery and has been widely used [1].With the development of high performance, high temperature and high pressure, the exciting force generated by seal is becoming more and more important to the stability of rotor system [2-5].The fluid excitation forces are non-negligible especially for the liquid turbo-machinery, which often works against the dense process fluid with the large pressure differences at high rotational speed. The typical liquid annular seals include the smooth plain seal, labyrinth seal [6], and the helical groove seal [7].

The scholars have done a lot of research on the labyrinth seal, the main methods are the bulk flow model, CFD-perturbation method and three dimensional (3-D) CFD method. The Bulk Flow has advantages of simple program, high calculation speed and high calculation precision, so it is widely used in engineering field. In 1980, a single control system of Bulk Flow model for predicting the dynamic characteristics of labyrinth seal rotors was first proposed by Iwataubo [8].It ignores the radial gradients of flow variables and is actually a two dimensional (2-D) flow model. The CFD perturbation method was firstly presented by Dietzen and Nordmann [9], and was further developed by Feng, Arghir and Kim [9]. Generally, the more accurate method may be the 3-D CFD method; it directly solves the Reynolds Averaged Navier-Stokes equations without special simplifications.

In this paper, we studied the fluid-induced force of center seal. The different structure parameters and operating conditions were considered into account.

2. Nomenclature:

T_s: Groove width, *mm*
T_w: Tooth width, *mm*
T_h: Tooth hight, *mm*

D_s: Seal diameter, *mm*
C_r: Radial clearance, *mm*
e: Eccentricity
N: Rotor speed, r/min
Ω*:* Whirling speed, r/min
F_τ : The tangential force, N
F_r : The radial force, N
Fx: force in X direction, N
Fy: force in Y direction, N
Δp*:* Pressure difference, Mpa

3. CFD Models

In the rotating coordinate system, as shown in Fig.1.The eccentric flow field is divided into inner and outer layers by using the median plane of the seal clearance as the interface, The inner layer fluid near the rotor is greatly affected by the Rotor whirling, and is solved in the rotating reference frame, this transforms the transient flow solver into the steady solver.

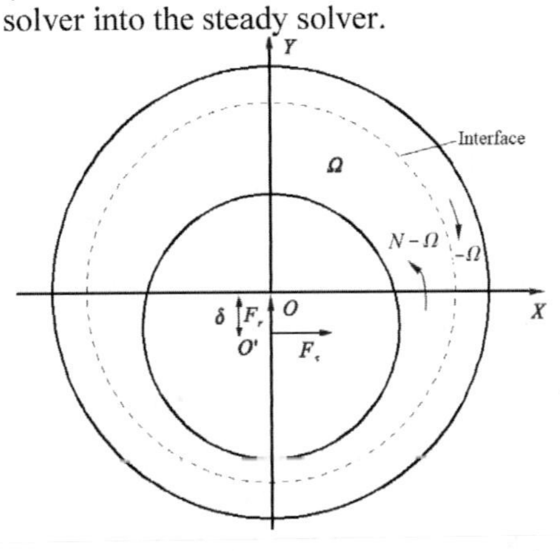

Fig. 1 Multiple reference system model

When the rotor has a circular whirling, the tangential force and the radial force can be written:

$$\begin{pmatrix} F_r \\ F_\tau \end{pmatrix} = \begin{bmatrix} \cos\theta & \sin\theta \\ -\sin\theta & \cos\theta \end{bmatrix} \begin{pmatrix} F_x \\ F_y \end{pmatrix} \tag{1}$$

Assuming Ω is the rotation velocity of the inner layer fluid, the relative angular velocity of the rotating wall around its axis is $N-\Omega$, the relative angular velocity of the interface around origin is $-\Omega$, Ω is the whirling velocity, N is the rotor speed.

4. Computational model and Results

4.1. Computational model

The CFD model of a grooved-stator/smooth-rotor is calculated using FLUENT 19.2 package. All grids are hexahedron grids. Fig. 2 shows computational fluid model.

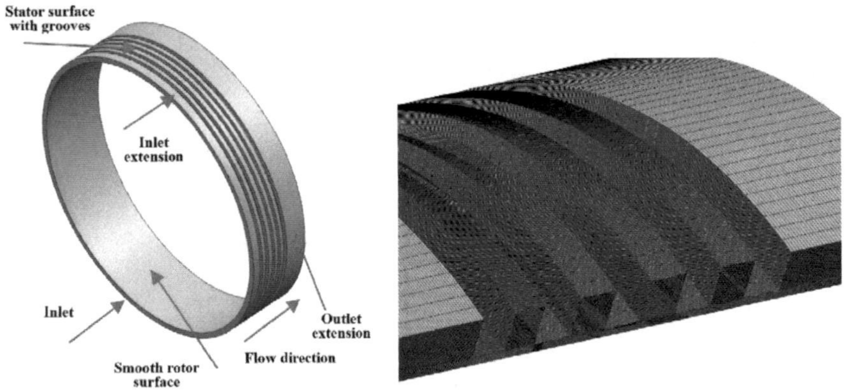

Fig. 2 CFD model

The related geometric dimensions and parameters can be seen in Table 1, and the fluid is 25oC water. A pressure-based solver and k- ε viscous model are chosen for this analysis.

Table 1. Related geometric dimensions and parameters

Descriptions	Value
Diameter, D_s (mm)	100
Radial Clearance, C_r (mm)	0.25
Eccentricity, e	0.1,0.2,0.3
Tooth Height, T_h (mm)	2.0,2.2,2.4,2.6,2.8
Tooth width, T_w (mm)	1.6,1.8,2.0,2.2,2.4
Groove width, T_s (mm)	2.0
Number of Teeth	5,10,15
Rotor speed, N (r/min)	0,3000,6000,9000
Pressure difference, ΔP (Mpa)	1,5,10

4.2. Influence factors on fluid-induced force

Fig.3 shows the fluid-induced force versus eccentricity and pressure drop when the rotor speed is 6000r/min. The tangential force and radial force increase with the increasing of rotor eccentricity and pressure drop. The tangential force is generally larger than the radial force (absolute value). With the increase of eccentricity, the fluid bearing effect of seal is enhanced, which causes the increase of fluid force.

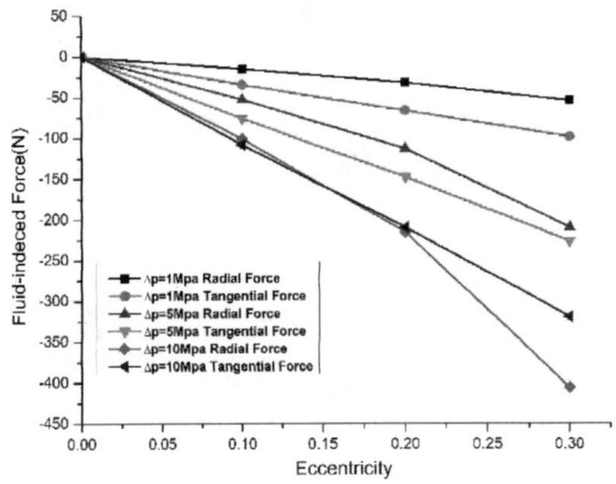

Fig. 3 The fluid-induced force versus eccentricity

Fig.4 illustrates the fluid-induced force versus rotor speed and pressure drop when the rotor eccentricity ratio is 0.1. With the increase of rotor speed (from 0 to 9000r/min), the radial force has little change but the tangential force increases greatly. This is caused by the increase of the tangential velocity around the rotor when the rotor speed increases.

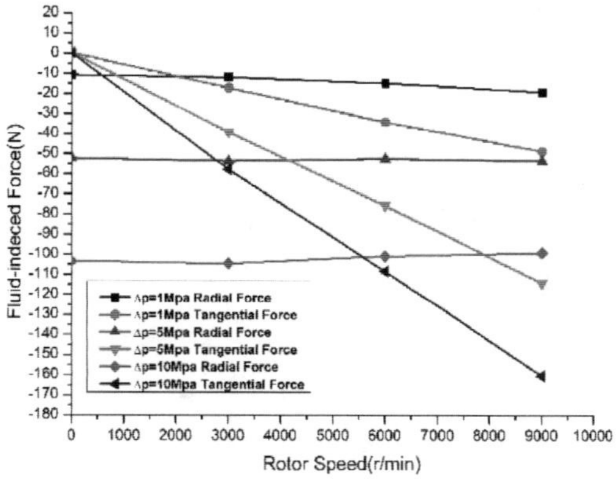

Fig. 4 The fluid-induced force versus rotor speed

Fig.5 shows the fluid-induced force versus tooth height and pressure drop when the rotor speed is 6000r/min and eccentricity ratio is 0.1. It can be seen that the radial force increases (absolute value) with the increase of tooth height, and the tangential force is generally larger than the radial force (absolute value). The tangential force increases at first and decreases secondly.

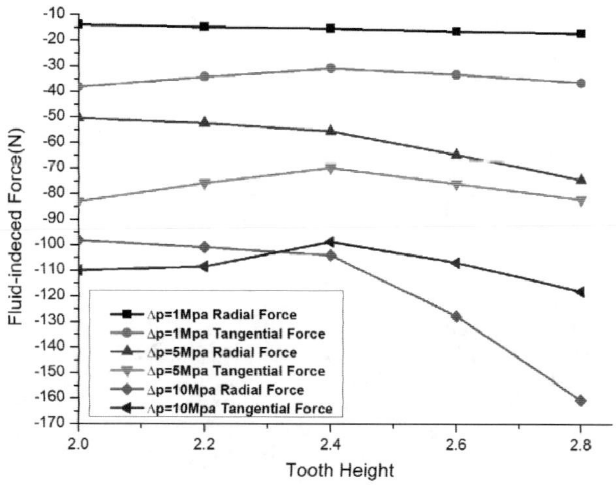

Fig. 5 The fluid-induced force versus tooth height

Fig.6 gives the fluid-induced force versus tooth width and pressure drop when the rotor speed is 6000r/min and eccentricity ratio is 0.1. It can be seen that the tangential force increases (absolute value) with the increase of tooth width, and the tangential force is generally larger than the radial force (absolute value). The radial force increases at first and decreases secondly (absolute value).

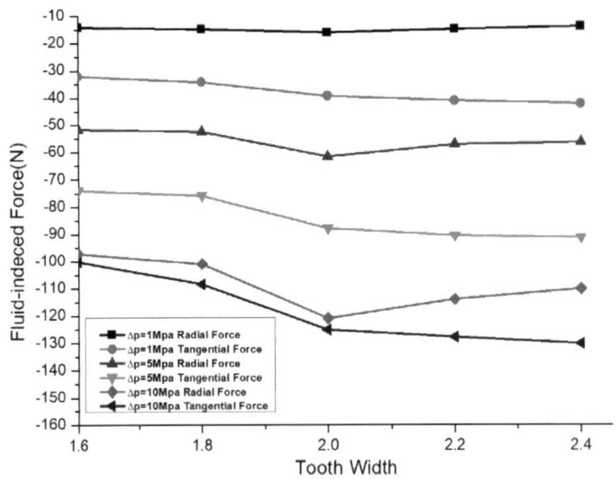

Fig. 6 The fluid-induced force versus tooth width

Fig.7 shows the fluid-induced force versus tooth number and pressure drop when the rotor speed is 6000r/min and eccentricity ratio is 0.1. It can be seen that the tangential force increases (absolute value) with the increase of tooth number (from 5 to 15). The tangential force is generally larger than the radial force (absolute value). It shows that the more number of sealing teeth will cause the instability the seal.

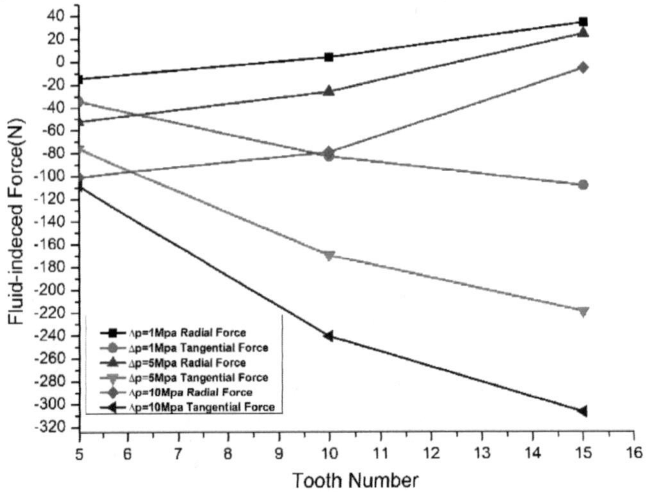

Fig. 7 The fluid-induced force versus tooth number

5. Conclusions

Multiple reference system of 3D model is used to calculate the fluid-induced forces. The results show that the tangential force and radial force increase with the increase of rotor eccentricity. The results also show that when tooth hight is T_h=2.4mm, the tangential force comes to the minimum. When tooth width Tw =2.0mm, the radial force comes to the maximum. The tooth numbers have a great influence on the fluid-induced force. The tangential force increases (absolute value) with the increase of tooth number. The more number of sealing teeth will cause the instability the seal.

Acknowledgments

This study has been funded by Key projects of technological innovation and application development of Chongqing, China (Grant NO.cstc2020jszx-msxm0411).

References

[1] HE Li-dong, YUAN Xin, YIN Xin. The advance in the investigation of brush seal. Proceedings of the CSEE, 2001, 21(12):28-32, 53.

[2] Untaroiu, A., Hayrapetian, V., Untaroiu, C. D., Wood, H. G., Schiavello, B.,and McGuire, J., 2013, "On the Dynamic Properties of Pump Liquid Seals,"ASME J. Fluids Eng., 135(5), p. 051104.

[3] LANG Ji, YANG Jiangang,CAO Hao.Calculation and Experimental Comparison of Fluid Induced Force Between the Titling Pad Seal and the Fixed Pad Seal [J].Journal of mechanical engineering, 2013,49(3): 101-105.

[4] Leng xiaolei, Meng guang.The Bifurcation Phenomena of Rubbing Rotor under Flow Induced Force [J].Turbine technology, 2006, 48(1):1-6.

[5] TIWA RIR, MANIKANDAN S, DUIVEDY S K. A review of the experimental estimation of the rotor dynamics parameters of seals [J].The Shock and Vibration Digest, 2005, 37(4): 261-284

[6] San Andres, L., Wu, T., Maeda, H., and Tomoki, O., 2018, "A Computational Fluid Dynamics Modified Bulk Flow Analysis for Circumferentially Shallow Grooved Liquid Seals," ASME J. Eng. Gas Turbines Power, 140(1), p. 012504.

[7] Kanki, H., and Kawakami, T., 1988, "Experimental Study on the Static and Dynamic Characteristics of Screw Grooved Seals," ASME J. Vib. Acoust, 110(3), pp. 326–331.

[8] Iwatsubo T. Evaluation of Instability of Forces of Labyrinth Seals in Turbines or Compressors [R]. Workshop on Rotor dynamic Instability Problems in High-Performance Turbo machinery, Texas A&M University, NASA Conference Publication No. 2133, 1980: 139–67.

[9] F. J. Dietzen and R. Nordmann, Calculating rotordy namic coefficients of seals by finite-difference techniques, ASME Journal of Tribology, 109 (3) (1987) 388-394.

Lattice Boltzmann simulations of gas bubble transport in proton exchange membrane water electrolysis cells

Gang Li*

School of Aeronautics and Astronautics, Zhejiang University, Hangzhou 310027, China

*Corresponding author e-mail: ligang0701@zju.edu.cn

Abstract. Gas/liquid two-phase transport is the key fundamental scientific issue in the proton exchange membrane (PEM) water electrolysis cell, which has important effects on the overall performance of water electrolysis and microstructure optimization. In this study, a combination of the lattice Boltzmann two-phase model and the QSGS numerical method was used to numerically model the porous diffusion layer and the micro-channel structure, and investigate the effects of porosity and wettability properties on the gas/liquid two-phase transport process. Simulation results show that the value of porosity will obviously affect the association and distribution pattern of solid particles inside the porous diffusion layer, and the reduction of porosity will cause the association part of solid particles to be narrower and longer, resulting in complex and variable pore channels. Some closed pore channels will appear inside the diffusion layer, which is not conducive to the smooth transmission of gas bubbles. In addition, the increase of contact angle will enhance the interaction force between the bubble and the solid wall, making it difficult for the bubble to fall off from the solid surface. Moreover, the increased interaction force will slow down the sliding speed of gas bubbles, which will result in the fusion of more gas bubbles and increase the risk of blocking the flow micro-channel. This study has initially grasped the mechanisms of micro/meso scale gas bubble transport in the PEM water electrolysis cell, which will provide theoretical basis for the optimized development of high-performance water electrolysis system for hydrogen production.

1. Introduction

The proton exchange membrane (PEM) water electrolysis technology is recognized as one of the most promising technologies in hydrogen production because of its excellent start-stop performance, high gas purity, and high energy efficiency[1, 2].

During the operation process of the PEM water electrolysis cell, oxygen bubbles generated at the anode need to be transferred through the porous layer and discharged from the micro-channel in time. Excessive gas bubbles will occupy the reaction sites in the catalytic layer and block the micro-channel, thus hindering the timely supply of reactant water and affecting the overall electrochemical performance. Therefore, gas/liquid two-phase transport in a PEM water electrolysis cell, particularly gas bubble evolution mechanisms, play an important role in water electrolysis performance.

The existing researches in this area are mainly focusing on preliminary experimental observations and macroscopic modeling. For example, Lim et al [3] investigated the effects of the content of hydrophobic polymers in the carbon paper diffusion layer on the dynamic performance of a hydrogen/air PEM water electrolysis cell. The measurements results of contact angle showed similar hydrophobicity between porous diffusion layers impregnated with different amounts of fluorinated ethylene propylene (FEP), with a hydrophobicity range of 10 ~ 40%. Grigoriev et al [4] investigated the porous diffusion layer of sintered titanium particles and found that the average pore size of the particles and the thickness of the titanium plate have a significant effect on the performance of water electrolysis cell. They observed that the optimum spherical particle size is between 50 - 75 mm and the optimum pore size is between 12-13 mm, and found that varying the porosity between 0.35 and 0.40 has no significant effect on the electrolysis efficiency. Through mathematical modeling, there are some macroscopic models for the preliminary study of mass transport in the PEM water electrolysis cell [5, 6]. In addition, Arbabi et al [7] used the volume of fluid (VOF) method to simulate the transport process of gas bubbles inside a porous transport layer. Lafmejani et al [8] used the VOF method to predict Taylor flow within a micro-channel and validated their simulation results through experiments.Although some progress has been made in the transport phenomena of PEM water electrolysis cell, the existing experimental and macro modeling methods are unable to effectively study the transport mechanisms inside the porous layer and micro-channel at micro/meso scale.

In this study, the lattice Boltzmann two-phase flow model based on mesoscopic theory is used to numerically simulate and investigate the gas bubble transport process and the effects of some physical parameters on the evolution and distribution of gas bubbles inside the porous diffusion layer and micro-channel of the PEM water electrolysis cell.

2. Numerical method

2.1. The lattice Boltzmann model

In this study, the Shan-Chen lattice Boltzmann two-phase flow model is used to simulate the gas/liquid two-phase transport process in a single PEM water electrolysis cell. The full discrete lattice Boltzmann equation can be expressed as:

$$f_\alpha^k(\boldsymbol{x}+\boldsymbol{e}_\alpha\delta_t, t+\delta_t) - f_\alpha^k(\boldsymbol{x},t) = -\frac{1}{\tau_k}\left[f_\alpha^k(\boldsymbol{x},t) - f_\alpha^{k(eq)}(\boldsymbol{x},t)\right] \tag{1}$$

Where $f_\alpha^k(\boldsymbol{x},t)$ is the distribution function of k components at the location of point $\boldsymbol{x}$ at the moment, k=1,2... .S denotes the component; i=1,2... .q denotes the discrete velocity direction; τ_k denotes the relaxation time of the k phase, which is related to the viscosity of the k fluid phase; $f_\alpha^{k(eq)}(\boldsymbol{x},t)$ is the corresponding equilibrium distribution function, which takes the form of the D2Q9 model proposed by Qian et al [9], the discrete velocity model is shown in Figure 1.

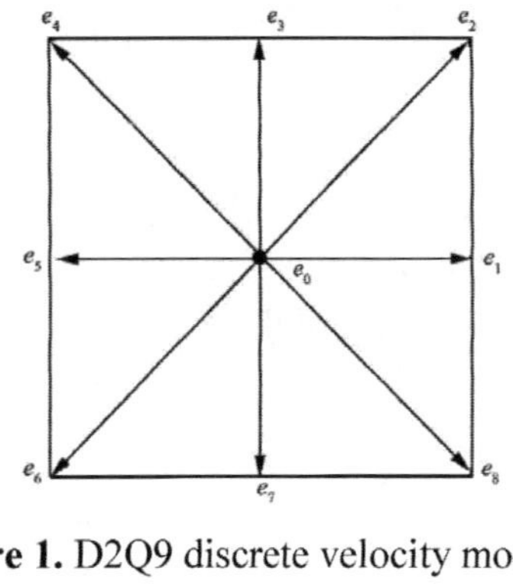

Figure 1. D2Q9 discrete velocity model

The velocity configuration of the D2Q9 model is given as:

$$\boldsymbol{e}_\alpha = \begin{cases} (0,0) & \alpha = 0 \\ c(cos[(\alpha-1)\frac{\pi}{2}], sin[(\alpha-1)\frac{\pi}{2}]) & \alpha = 1,2,3,4 \\ \sqrt{2}c(cos[(2\alpha-1)\frac{\pi}{4}], sin[(2\alpha-1)\frac{\pi}{4}]) & \alpha = 5,6,7,8 \end{cases} \tag{2}$$

The equilibrium distribution function for the k fluid phase particle is:

$$f_\alpha^{k(eq)} = w_\alpha \rho_k \left[1 + \frac{\boldsymbol{e}_\alpha \cdot \boldsymbol{u}_k}{c_s^2} + \frac{(\boldsymbol{e}_\alpha \cdot \boldsymbol{u}_k)^2}{2c_s^4} - \frac{\boldsymbol{u}_k^2}{2c_s^2}\right] \tag{3}$$

The values of the density ρ_k and velocity $\boldsymbol{u}_k$ of the k particle in the model can be calculated by equation (4) and (5), and the conservation of mass and momentum are ensured by this formal representation.

$$\rho_k = \sum_i f_\alpha{}^k \tag{4}$$

$$\boldsymbol{u}_k = \frac{1}{\rho_k} \sum_\alpha \boldsymbol{e}_\alpha f_\alpha{}^k \tag{5}$$

The following equations (6)-(8) are used to calculate the total forces $\boldsymbol{F}_k$ in the pseudo-potential model, which consists of the interaction force between adjacent fluid particles $\boldsymbol{F}_f$, the forces between the fluid and the solid wall $\boldsymbol{F}_w$, and other external forces (gravity) $\boldsymbol{F}_g$.

$$\boldsymbol{F}_f(\boldsymbol{x}) = -G_{kk'}(\boldsymbol{x}, \boldsymbol{x}')\psi_k(\boldsymbol{x}) \sum_\alpha w_\alpha \psi_{k'}(\boldsymbol{x} + \boldsymbol{e}_\alpha \delta_t)\, \boldsymbol{e}_\alpha \tag{6}$$

$$\boldsymbol{F}_w(\boldsymbol{x}) = -\psi_k(\boldsymbol{x}) \sum_\alpha G_w w_\alpha s(\boldsymbol{x} + \boldsymbol{e}_\alpha \delta_t)\, \boldsymbol{e}_\alpha \tag{7}$$

$$\boldsymbol{F}_g(\boldsymbol{x}) = -\rho_k \cdot \mathbf{g} \tag{8}$$

In equations (6)-(8), $\boldsymbol{x}'$ denotes the position of the proximity of the lattice position $\boldsymbol{x}$. $\psi_k(\boldsymbol{x})$ is the pseudo-potential function of the k phase particle, which is related to the density, $G_{kk'}(\boldsymbol{x}, \boldsymbol{x}')$ is the Green's function, which denotes the strength of the interaction between fluid particles, and $\mathbf{g}$ is the acceleration of gravity.

In order to satisfy the conservation of total momentum of the system, the velocity in the equilibrium distribution function needs to be redefined as shown in equation (9):

$$\boldsymbol{u}_k^{eq} = \boldsymbol{u}' + \frac{\tau_k}{\rho_k} \boldsymbol{F}_k \tag{9}$$

where $\boldsymbol{u}'$ is the modified macroscopic velocity of the mixed fluid obtained by satisfying the conservation of total momentum, and it can be expressed as:

$$\boldsymbol{u}' = \frac{\sum_k \frac{\rho_k \boldsymbol{u}_k}{\tau_k}}{\sum_k \frac{\rho_k}{\tau_k}} \tag{10}$$

In this way, the total momentum of the entire fluid system can be conserved if the interaction matrix is a symmetric matrix [10] even if the local momentum of the system is not conserved.

The macroscopic density ρ and pressure p of the mixed fluid can be expressed as:

$$\rho = \sum_k \rho_k \tag{11}$$

$$p(\boldsymbol{x}) = c_s^2 [\sum_k \rho_k(\boldsymbol{x}) + \psi_k(\boldsymbol{x})\psi_{k'}(\boldsymbol{x})] \tag{12}$$

2.2. Laplace validations

According to the Laplace law, the pressure difference ∇P inside and outside the gas bubble should be proportional to the surface tension σ acting on the gas/liquid interface and inversely proportional to the radius R of the bubble [11], which satisfies the following equation (13):

$$\nabla P = p_{in} - p_{out} = \frac{\sigma}{R} \tag{13}$$

Through relevant validations, the results of the collated data are shown in Figure 2, and it can be seen that it is obvious that the pressure difference ∇P inside and outside the bubble is linearly related to the inverse of its radius.

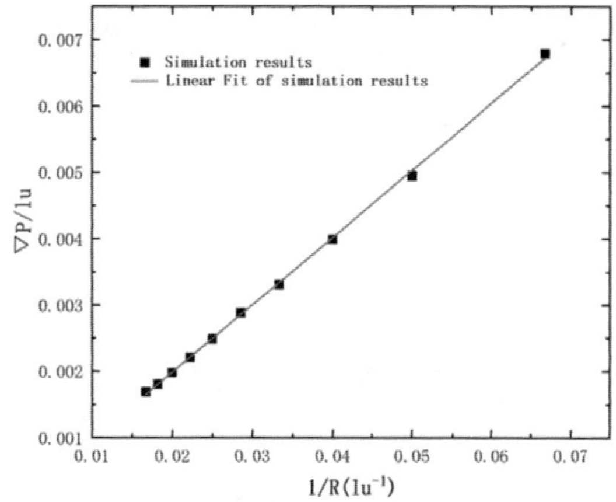

Figure 2. Laplace validation results

2.3. Wettability validations

Using equation (7), we can simulate different contact angles by varying the mutual coefficients G_{w1} between the fluid and the solid wall with different wettability characteristics. Figure 3 shows the relationship between contact angle θ and fluid-solid coefficient G_{w1}.

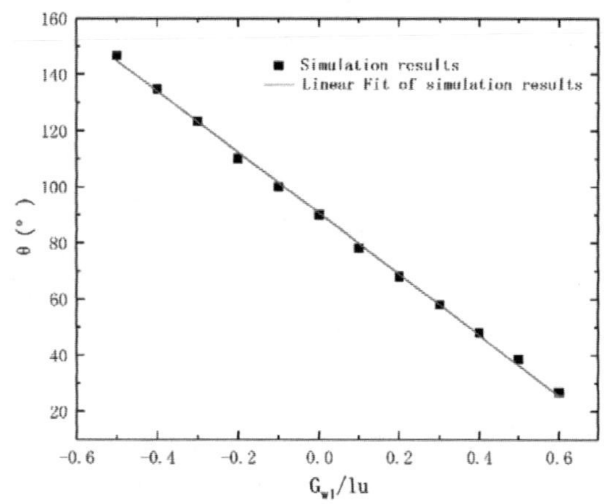

Figure 3. Relationship between contact angle θ and fluid-solid coefficient G_{w1}

3. Results and Discussions

3.1. Effects of porosity in the porous diffusion layer

In the simulation, 200×50 (lu×lu) grids was used to model the computational region. Since this research subject focuses on the gas/liquid two-phase transport in the porous diffusion layer, the structure of the catalytic layer was simplified as a boundary layer, and the bottom surface had been set as a layer of gas initially. Under specific boundary conditions, the bottom layer of gas is subject to fluctuation and the bubble clusters will be formed, which can simulate the phenomenon of generated bubbles in the PEM water electrolysis cell. At the same time, in the Y=2-32lu region, the porous diffusion layer was constructed based on the QSGS method [12], the rest of boundaries were set as non-equilibrium extrapolation boundary, and the porosity of the diffusion layer was set as the control variable. Simulation results of gas bubble distribution with different porosities are shown in Figure 4.

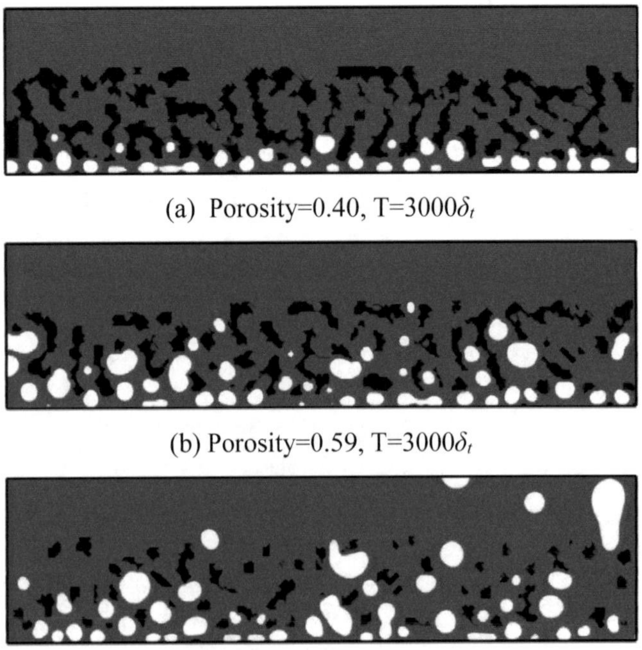

(a) Porosity=0.40, T=3000δ_t

(b) Porosity=0.59, T=3000δ_t

(c) Porosity=0.78, T=3000δ_t

Figure 4. The distribution of gas bubbles inside the porous diffusion layer with different porosities (White: gas phase, Gray: liquid phase, Black: solid phase)

As shown in Figure 4, we can see that more gas bubbles can pass smoothly in the porous diffusion layer with larger porosity in a given time. This indicates that the larger porosity will make a relatively small and dispersed volume of solid particles within the diffusion layer, and provide a greater number of pore channels and larger pore size, which enable bubbles to smoothly find effective transport paths to pass through the gas diffusion layer. Besides, it can be found that when the porosity is low, the interconnected parts of the solid particles of the porous layer tend to be longer, which causes the pore size to be reduced. More than that, some closed channels will appear, which is not conducive to the smooth transmission of gas bubbles.

In order to further quantify and compare the simulation results of bubble transport process in porous diffusion layer under different porosities, the value of the vertical coordinate Y was used as the independent variable, and the percentage of the number of gas phase distribution lattices in the same vertical coordinate (Gas phase percentage) was chosen as the dependent variable. The transport process and distribution of gas bubbles inside and outside the porous diffusion layer can be compared quantitatively through statistical analysis. Considering the simplicity and accuracy of data processing, the interpolation statistics were carried out with the values of the five grid vertical coordinates as the benchmark, and the distribution of the gas phase percentage along the Y-value at the above three moments were shown in Figure 5.

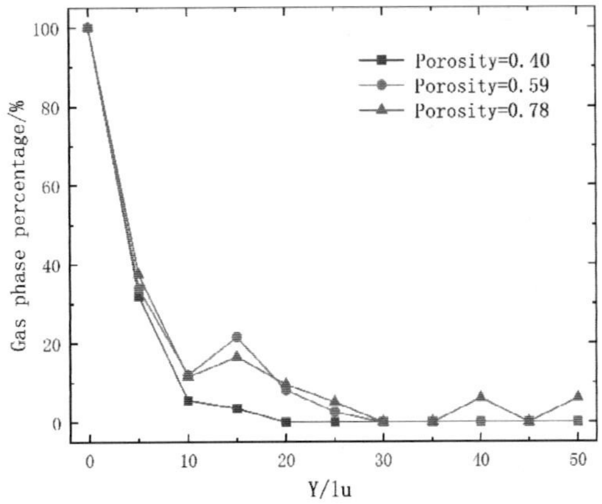

(a) T=500δ_t

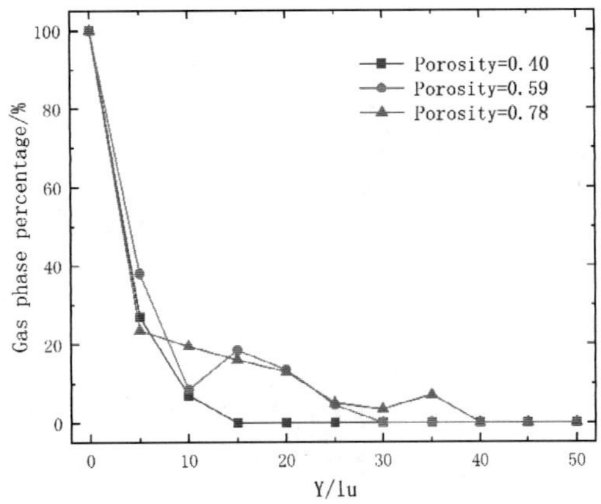

(b) T=1500δ_t

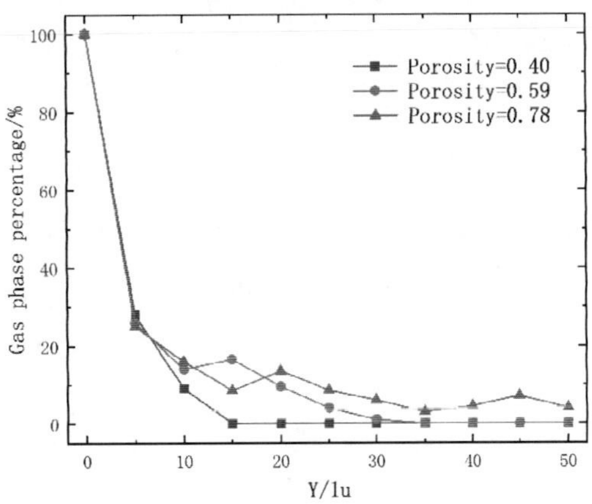

(c) T=3000δ_t

Figure 5. Distribution of gas phase percentage along Y-value with different porosities

In Figure 5, we can see that when the porosity is 0.40, the distribution of gas phase percentage along Y hardly changes significantly with time. Gas bubbles will suffer great transport resistance when they enter the porous diffusion layer with small porosity and accumulation occurs in the bottom region of the diffusion layer (Y=5-10lu). When the porosity is 0.59, we find that a certain number of bubbles can pass through the diffusion layer. Combining with the results in Figure 4, it can be seen that some bubbles with an early shedding time and a very small average diameter can pass through a few pores in the diffusion layer. However, due to the interlacing complexity of the pores, the tiny bubbles that cannot pass smoothly in a very short time will occur in the fusion process inside the diffusion layer channels, so most of the bubbles finally accumulate in the middle region of the diffusion layer (Y=15lu). It can be seen from Figure 5 that the percentage of gas phase in this region is higher than ones in the other two sets of cases. Further, when the porosity is 0.78, the average pore size of the diffusion layer channel will be larger and the number of connected pores increases, making it easier for the gas bubbles to find an effective path in a shorter time. More bubbles can eventually pass through the diffusion layer into the flow channel region (Y>30lu) smoothly eventually.

3.2. Effects of wettability in the micro-channel

In Figure 6, based on the reconstructed porous structure from QSGS method, the evolution behaviors of multiple gas bubbles on the micro-channel surface were investigated with different wettability properties (hydrophilic and hydrophobic properties). The 200×50 (lu×lu) grids was used in the simulation, and the contact angle between the bubble and the solid wall was set as the control variable.

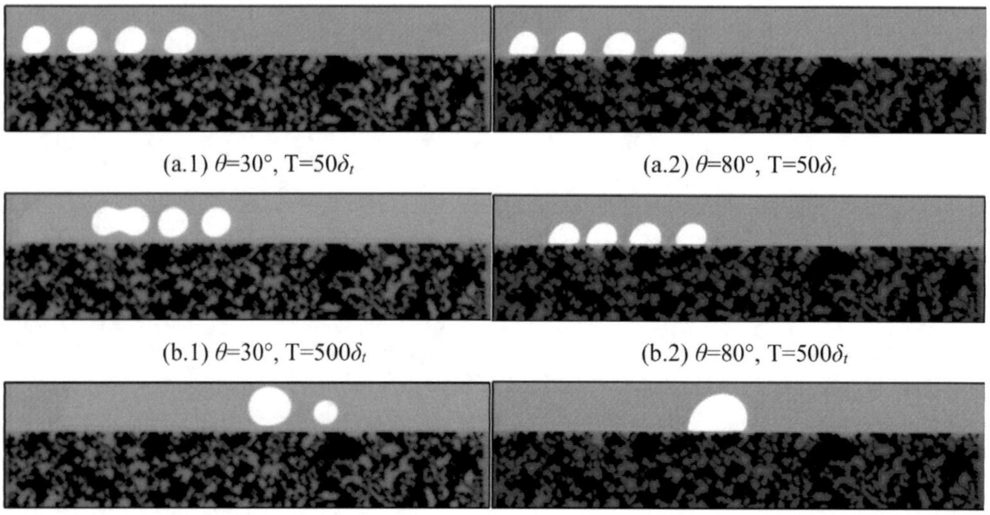

(a.1) θ=30°, T=50δ_t (a.2) θ=80°, T=50δ_t

(b.1) θ=30°, T=500δ_t (b.2) θ=80°, T=500δ_t

(c.1) θ=30°, T=2000δ_t (c.2) θ=80°, T=2000δ_t

Figure 6. Gas bubbles evolution on the micro-channel surface with different contact angles

(White: gas phase, Gray: liquid phase, Black: solid phase)

As shown in Figure 6, we can see that contact angle has a significant effect on the gas bubble movement in the flow micro-channel. When the contact angle θ is 30°, the bubbles are already in the shedding state in the initial stage, and then they fuse one by one as they transmit in the flow micro-channel under the interaction of incoming flow. However, due to the fact that the fusion process and the transport process are synchronized, there will be some bubbles discharged from the flow micro-channel before the fusion occurs. When the contact angle θ is 80°, we can see that four bubbles eventually fuse into one large bubble as time goes by,. This is because the increase in contact angle will lead to the augment of interaction force between the gas bubble and the solid wall, which will not only make it difficult for the gas bubble to fall off into a free form, but also cause resistance to the gas bubble when it slides along the solid wall surface. In addition, as the gas bubbles move slowly, the size of the converged bubbles after continuous fusion will also increase, which may eventually cause the problem of gas bubbles accumulation in the flow micro-channel.

4. Conclusion

In this study, a two-phase lattice Boltzmann model was used to simulate the gas bubble transport phenomena inside the reconstructed porous diffusion layer and micro-channel. The effects of some physical parameters such as porosity and contact angle on the gas bubbles evolution and distribution were analyzed in detail. Simulation results indicate that gas bubble transport process is very sensitive to the porosity of porous diffusion layer. It is found that more gas bubbles can move through the diffusion layer as the porosity increases. In addition, the gas bubbles movement on the micro-channel surface is significantly different with different contact angles. The value of the contact angle determines the strength of the interaction force between gas bubble and solid wall surface. When the contact angle increases, the interaction force will consequently increase, making it difficult for the gas bubbles to detach from the surface. Results obtained in this study can provide theoretical understanding of gas/liquid two-phase transport in the PEM water electrolysis cell.

Acknowledgments

This work was financially supported by the National Natural Science Foundation of China (No.11702245).

References

[1] Wang Xunying, Zhang Linsong, Li Guangfu, Zhang Geng, Shao Zhigang, Yi Baolian. The influence of Ferric ion contamination on the solid polymer electrolyte water electrolysis performance[J]. Electrochimica Acta, 2015, 158: 253-257.

[2] Wang Xunying, Shao Zhigang, Li Guangfu, Zhang Linsong, Zhao Yun, Lu Wangting, Yi Baolian. A cocrystallized catalyst-coated membrane with high performance for solid polymer electrolyte water electrolysis[J]. Journal of Power Sources, 2013, 240: 525-529.

[3] LIM C, WANG C Y. Effects of hydrophobic polymer content in GDL on power performance of a PEM fuel cell[J]. Electrochimica Acta, 2004, 49(24): 4149-4156.

[4] GRIGORIEV S A, MILLET P, VOLOBUEV S A, et al. Optimization of porous current collectors for PEM water electrolysers[J]. International Journal of Hydrogen Energy, 2009, 34(11): 4968-4973.

[5] Han Bo, Mo Jingke, Kang Zhenye, Zhang Fengyuan. Effects of membrane electrode assembly properties on two-phase transport and performance in proton exchange membrane electrolyzer cells[J]. Electrochimica Acta, 2016, 188: 317-326.

[6] Han Bo, Mo Jingke, Kang Zhenye, Yang Gaoqiang, BARNHILL W, Zhang Fengyuan. Modeling of two-phase transport in proton exchange membrane electrolyzer cells for hydrogen energy[J]. International Journal of Hydrogen Energy, 2017, 42(7): 4478-4489.

[7] ARBABI F, MONTAZERI H, ABOUATALLAH R, et al. Three-Dimensional Computational Fluid Dynamics Modelling of Oxygen Bubble Transport in Polymer Electrolyte Membrane Electrolyzer Porous Transport Layers[J]. Journal of The Electrochemical Society, 2016, 163(11): F3062-F3069.

[8] LAFMEJANI S S, OLESEN A C, KæR S K. VOF modelling of gas–liquid flow in PEM water electrolysis cell micro-channels[J]. International Journal of Hydrogen Encrgy, 2017, 42(26): 16333-16344.

[9] QIAN Y H, D'HUMIèRES D, LALLEMAND P. Lattice BGK Models for Navier-Stokes Equation[J]. Europhysics letters, 1992, 17(6): 479-484.

[10] SHAN X, DOOLEN G. Multicomponent lattice-Boltzmann model with interparticle interaction[J]. Journal of Statistical Physics, 1995, 81(1-2): 379-393.

[11] Yang Shiming, Tao Wenquan. Heat transfer[M]. Beijing: Higher Education Press, 1998.

[12] WANG M, WANG J, PAN N, CHEN S. Mesoscopic predictions of the effective thermal conductivity for microscale random porous media[J]. Physical Review. E, Statistical, Nonlinear, and Soft Matter Physics, 2007, 75(3): 036702.

Development and application of drilling and expanding integrated bit based on high pressure water jet coal breaking mechanism

Fan Wang*

China Coal Science and Industry Group Co., Ltd, Beijing, China

*Corresponding author e-mail: wangfan@ccteg.cn

Abstract. Outburst coal seam and high gas coal seam and using borehole to extract coal seam gas are the main measures to prevent coal and gas outburst, reduce gas emission and reduce air flow gas overrun in working face. Increasing borehole diameter and improving coal seam permeability are one of the common methods to improve the effect of this measure. In this paper, the drilling and expanding integrated bit is developed and designed to improve the construction efficiency. Firstly, the mechanical analysis of high-pressure water jet combined with coal breaking of drilling teeth is carried out to provide the basis for the optimization of drilling and expanding integrated bit. The optimization principles of bit and nozzle are put forward respectively. The bit material is optimized. The hardness is increased by 20 ~ 40 times by using diamond composite chip (PDC) bit. Through numerical simulation analysis, it is determined that when the nozzle l/d value is 2, the jet development can maintain good stability. The nozzle is arranged with 2 ~ 3 nozzles in radial direction and 1 ~ 2 nozzles in axial direction, and the best contraction angle is 13 ° ~ 15 °. The field test shows that the average equivalent diameter of coal 6 after reaming is 720mm, which is 9.6 times that before reaming, and the effect of reaming and pressure relief is remarkable.

Key words: Drilling and expanding integrated bit; Nozzles; Numerical simulation; The field test.

With the reduction of shallow coal resources, the mine gradually enters deep mining. However, with the continuous increase of mining depth, many fundamental changes have taken place in the mechanical properties and in-situ stress characteristics of the mine rock mass compared with the shallow part, resulting in the continuous transformation of the mine type from low gas mine to high gas mine and non outburst mine to outburst mine, which makes the number and degree of outburst of high gas mine and outburst mine increase day by day [1-5].

For outburst coal seam and high gas coal seam, using borehole to extract coal seam gas is the main measure to prevent coal and gas outburst, reduce gas emission and reduce air flow gas overrun in working face. Increasing borehole diameter and improving coal seam permeability is one of the common methods to improve the effect of this measure. There are some technical problems in large-diameter drilling with drilling rig, such as increased risk of drilling outburst, serious hole collapse, difficult slag removal, short hole length and geometric multiple increase of drilling rig load; Therefore, it is very necessary to find a technical way to effectively expand the borehole diameter and improve the influence range of borehole pressure relief [6-7].

In recent years, with the development of high-pressure water jet technology, the use of high-pressure water has achieved certain results in coal seam cutting, drilling and reaming in the constructed boreholes, resulting in large borehole diameter and increasing the pumping effect. But, there are obvious deficiencies in studying and solving the problem of drilling slag removal, and the slag removal effect directly affects the hole length, pumping efficiency and the popularization and application of the process. Therefore, it is necessary to develop and design drilling expansion integrated bit to improve construction efficiency and promote the application of high-pressure water jet technology.

1. Mechanical analysis of coal breaking by high pressure water jet combined with drill teeth

1.1. Mechanical analysis of coal breaking by water jet

The damage process of jet to coal and rock mass materials is very complex, which is closely related not only to the working condition of jet, but also to the heterogeneity of materials. The effects of jet on coal and rock mass damage can be summarized as follows: ① Jet impact force; ② Water wedge action; ③ Stress wave produced by jet; ④ Cavitation damage.

When the compression wave of jet impacting rock mass propagates to the free surface of rock mass, the stress of rock mass changes from the compressive stress at incident to the tensile stress at total reflection. When the tensile stress exceeds the tensile strength of some low-strength rock mass, tensile fracture occurs there and cracks are formed. The water jet crushing mechanism of heterogeneous materials is shown in Fig 1.

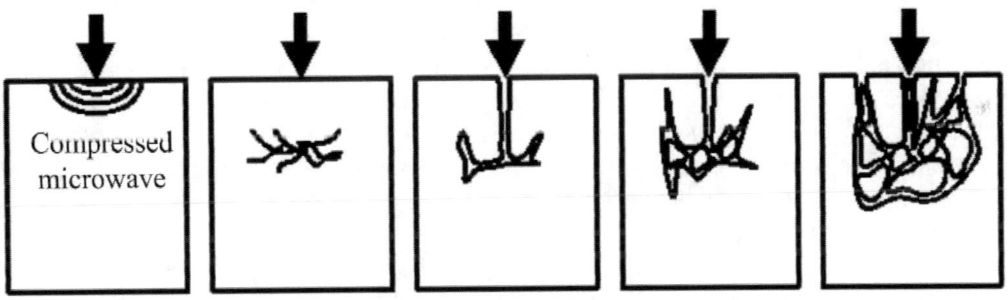

Fig. 1 water jet crushing mechanism of heterogeneous materials

1.2. Mechanical analysis of coal breaking by high pressure water jet combined with drilling teeth

The erosion of water jet on coal body and cutting groove on coal body mainly occur in the deformation stage of drilling teeth cutting coal body. When the drill teeth open cracks on the coal under the auxiliary action of water jet, the water jet immediately enters the cracks, and the auxiliary drill teeth expand the cracks until the coal is broken into blocks. This stage is the maximum stress stage of the drill teeth. When the water jet assists the drill teeth to break coal, the stress of the drill teeth is greatly reduced. The stress of the rock when the water jet enters the coal rock fissure opened by the drilling teeth is shown in Fig 2.

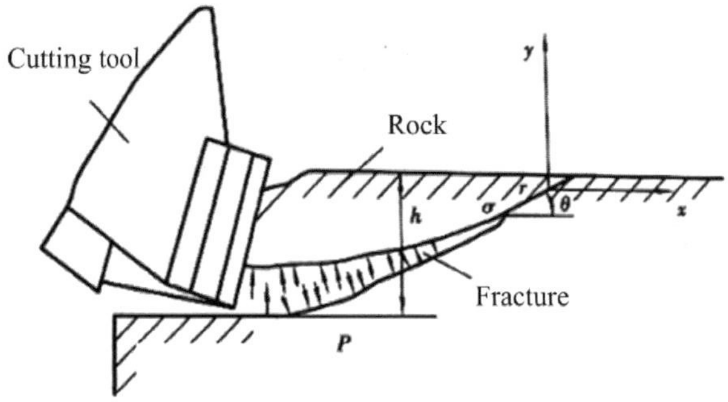

Fig. 2 stress state of coal breaking by water jet combined with drill teeth

r - distance from the crack tip to any point; θ— The included angle between the line from the crack tip to this point and the horizontal direction; - Rock fracture opening; P - pressure of flow entering the crack tip Critical stress of coal.

In order to ensure that the jet can slot the drill teeth on the coal body during the crack generation process, the jet pressure must be greater than the critical stress of the coal body, or the jet pressure must be reduced in order to greatly reduce the stress of the drill teeth during the crack expansion process, $P_0 > 1.38\sigma e^{(\frac{1.1\beta h}{D})}$.

1.3. Coal breaking mechanism of high pressure water jet combined with drilling teeth

In the process of high-pressure water jet combined with drilling tooth joint coal breaking, the high-pressure rotating water jet impacts the dynamic load of the coal body, produces impact load, forms stress wave on the contact surface between water and coal wall, the jet pressure outside the contact point decreases sharply, the coal body produces tensile damage, and the rotating water jet produces shear damage to the coal body and forms microcracks. Under the continuous action of high-pressure rotating water jet, the coal is gradually peeled off. In the axial direction of the borehole, the drill teeth further mechanically cut the coal under the driving action of the drilling rig, expand the high-pressure rotating water jet, form cracks on the coal wall, peel off the coal and break the coal block, which is conducive to slag removal. The axial and radial high-pressure water jet further expands the cracks formed by the drill teeth cutting the coal body. Under the joint action of the two, the borehole diameter and fracture area are increased, and the permeability of the coal seam around the borehole is improved.In addition, the mechanical cutting of drilling teeth can further break the large coal stripped by water jet, which is conducive to the transportation of coal cuttings to the orifice and slag discharge in the process of high-pressure water jet drilling and expansion.

2. Optimization principle

Different bit structures form different jets and have different rock breaking effects. The performance of drill bit determines the flow pattern and working performance of jet. Due to the short service life of the bit in the field test, the analysis shows that the reason is that the material of the bit cutting part has poor wear resistance, and the nozzle parameters affect the jet effect. Therefore, it is necessary to improve the material of the cutting part of the bit to improve the service life of the bit and optimize the nozzle parameters to improve the reaming efficiency.

2.1. Bit optimization principle

The design of drilling and expanding integrated bit mainly considers two uses, that is, to realize both drilling and reaming processes at the same time. Compared with the traditional reaming equipment, its feature is that it can realize rotary drilling driven by the mining drill and use the nozzle to spray high-pressure water jet to cut the coal radially and axially, so it can better adapt to the bad working conditions such as hole collapse and hole plugging, and ensure the success and efficiency of hydraulic reaming.

2.2. Nozzle optimization principle

The parameters of bit nozzle determine the flow pattern and working performance of jet. On the contrary, different application requirements and different working conditions require various bit designs. Therefore, the study of the relationship between the parameters of the three can improve the working efficiency and quality of the jet. It is necessary to determine reasonable parameters such as nozzle installation angle, nozzle diameter, nozzle quantity and jet pressure.

3. Optimization and improvement of 3 drilling expansion integrated bit

3.1. Bit optimization

According to the field test, in view of the coal seam conditions in some mining areas, the traditional bit adopts cemented carbide material, which has the disadvantages of poor wear resistance and short service life. In view of the poor wear resistance of the product, it is decided to upgrade the bit material and adopt diamond composite chip (PDC) bit. Diamond composite chip (PDC) bit takes diamond composite chip as the main cutting element and uses the scraping shear principle to break the rock stratum. It is suitable for drilling in soft medium hard stratum, and has better advantages in overcoming incomplete stratum drilling. By making full use of the high wear resistance and high impact resistance of diamond composite, the hardness of diamond composite bit is increased by 20 ~ 40 times and the service life is increased by 3 ~ 5 times. It has the advantages of high construction efficiency and high cost performance. See Fig. 3.1 and Fig. 3.2 for conventional bit and PDC bit respectively.

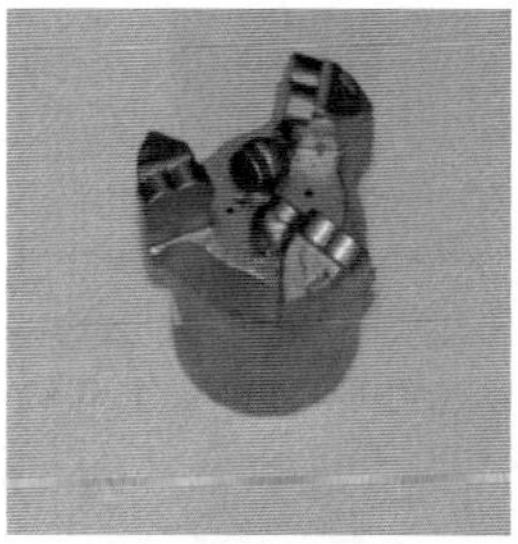

Fig. 3 (1) physical drawing of ordinary reaming bit, (2) diamond composite bit

3.2. Nozzles optimization

The geometric parameters of the nozzle mainly include the contraction angle of the nozzle α, The transition shape of the inlet and outlet, the radius of curvature R1 and R2 of the chamfer, the outlet diameter D, the ratio of the length L of the cylindrical section to the diameter D, the length of the nozzle and the surface roughness of the inner wall, etc. The numerical analysis model of mechanical characteristics of high-pressure water jet is established, as shown in Fig. 4; The performance characteristics of different jets under different nozzle structures are numerically simulated, as shown in Fig. 5. It lays a theoretical foundation for the structural design of drilling teeth and nozzles.

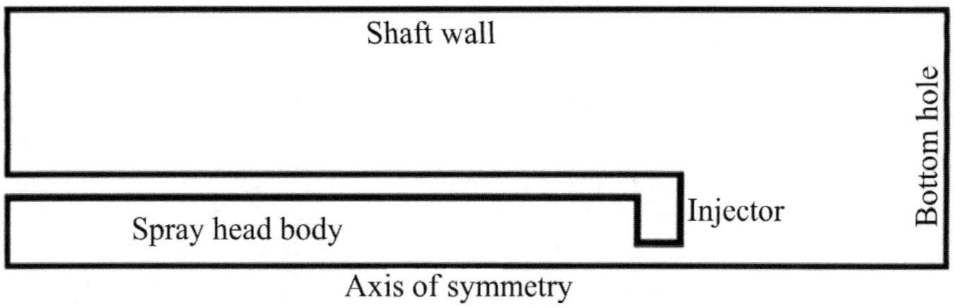

Fig. 4 Model diagram of jet drilling hole bottom

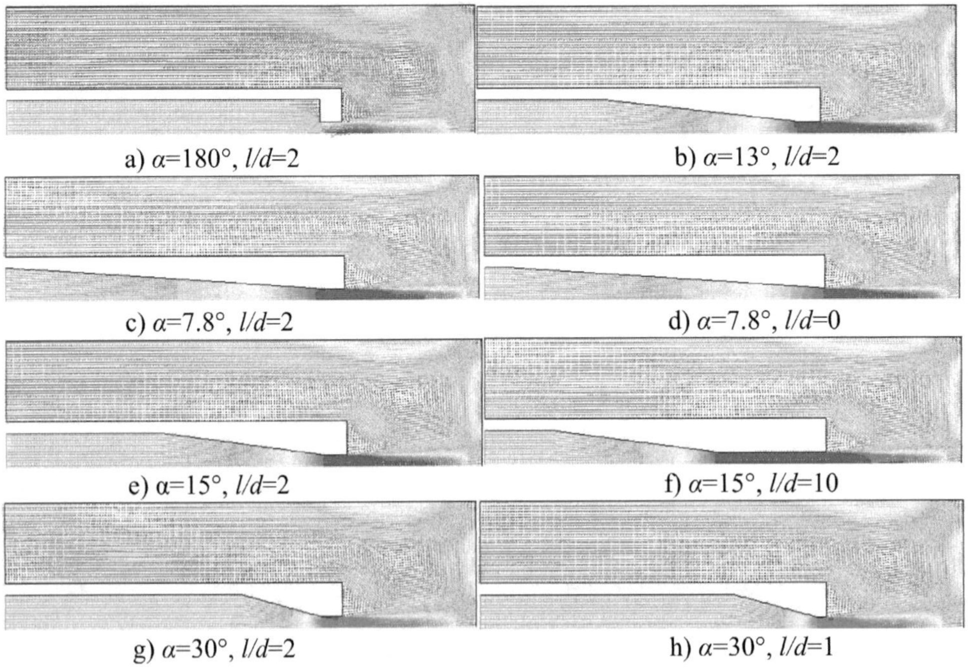

Fig. 5 jet velocity vector diagram of each scheme

As shown in Fig. 5, the energy of the jet is concentrated at the axis of the jet. With the continuous advancement of the jet, the jet gradually diffuses, resulting in the increase of the impact area. The axial jet flow pressure and jet velocity at the same jet distance are different for jets with different contraction angles and outlet length, but they all decay gradually with the increase of jet distance.

The flow pressure of symmetrical axis is shown in Fig. 6.

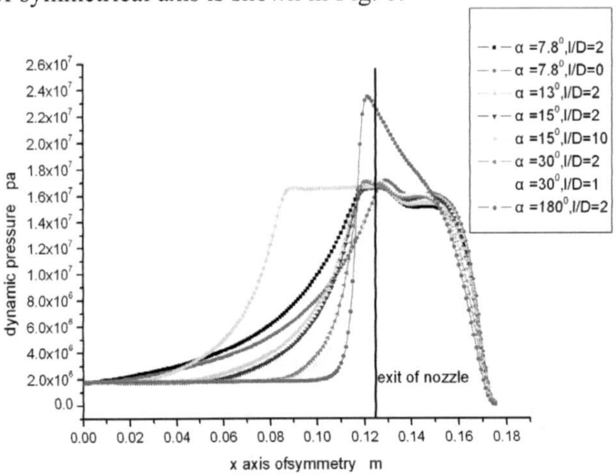

Fig. 6 Axisymmetric jet flow pressure

It can be seen from Fig. 5 and 6 that the contraction angle should be appropriately reduced if conditions permit. When the contraction angle is reduced to 30 °, the dynamic pressure attenuation does not change significantly with the contraction angle. It can also be seen from Fig. 5 that under the same conditions, the influence of L / D value at the nozzle outlet relative to the contraction angle on the axial dynamic pressure is not very obvious. When the L / D value is 2, the jet development can maintain good stability.

At the same spray distance, the total pressure distribution at the hole bottom (1 / 2 hole bottom range) is shown in Fig 7.

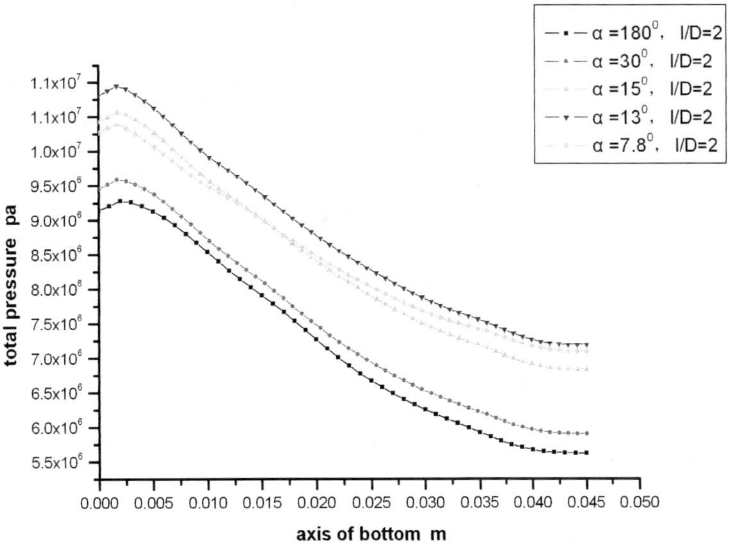

Fig. 7 total pressure distribution at hole bottom

As can be seen from Fig. 7, the total pressure at the hole bottom gradually decreases as it is away from the jet axis. When the length of the nozzle outlet section is the same, the attenuation trend of the total pressure distribution at the bottom of the jet hole with different jet contraction angles is similar, but there are differences in the maximum pressure at the bottom of the jet hole under the same conditions: for the conical convergent nozzle, the best contraction angle is 13 ° ~ 15 °.

With the extension of striking time, the borehole diameter increases continuously, but with the extension of time, the change of borehole diameter tends to be flat. The length of reaming time is mainly determined by the amount of reaming coal and reaming diameter to achieve the optimization of safety and benefit. Due to the differences in the physical structure and properties of each coal and rock stratum and the different drilling inclination, the flow and velocity of high-pressure water jet required in the process of drilling and reaming are different. After comprehensive analysis, it is determined that 2 ~ 3 nozzles are arranged radially and 1 ~ 2 nozzles are arranged axially.

4. The field test

4.1. The situation of working face

The central pedestrian concealed inclined shaft of Xinji No. 1 coal mine was selected as the test site to expose 6 coal seams. The coal seam permeability was poor. The high-pressure water jet was used to expand the hole to increase the coal seam permeability. Coal seam 6 contains coal seams 6-1 and 6-2, with an average dip angle of 33 ° and an average thickness of 3.36M and 0.87m. The coal seam gas pressure and content are P_6 = 0.82MPa and X_6 = 5.81m^3/t respectively. The firmness coefficient of 6 coal seam is f = 0.5. The water jet drilling expansion integrated device mainly includes: reaming bit, high-pressure reaming drill pipe, high-pressure dynamic seal rotating water tail, high-pressure rubber hose, water tank, high-pressure water pump, etc. before the downhole test of the high-pressure water jet drilling expansion integrated device, it is generally necessary to carry out commissioning test in the well, as shown in Fig. 8.

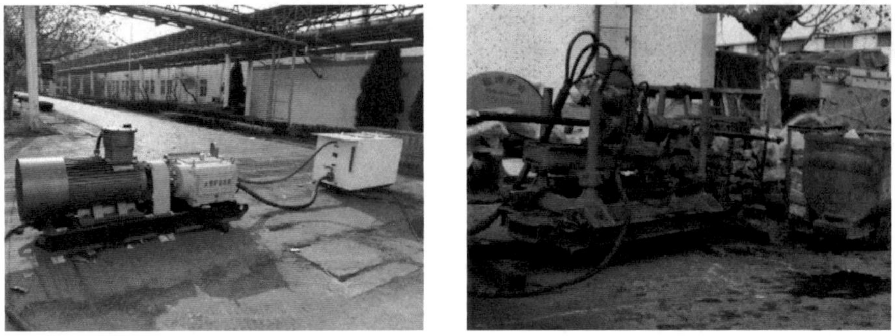

Fig. 8 Ground commissioning of high pressure water jet device

4.2. Test scheme and effect analysis

A total of 92 regional outburst prevention boreholes with a hole diameter of 75mm were constructed in coal seam 6 in the coal uncovering area, of which 35 were reamed by high-pressure water jet- See 4.2 for borehole layout of outburst prevention measures in the area where 700 central pedestrian concealed inclined shaft exposes coal seam 6.

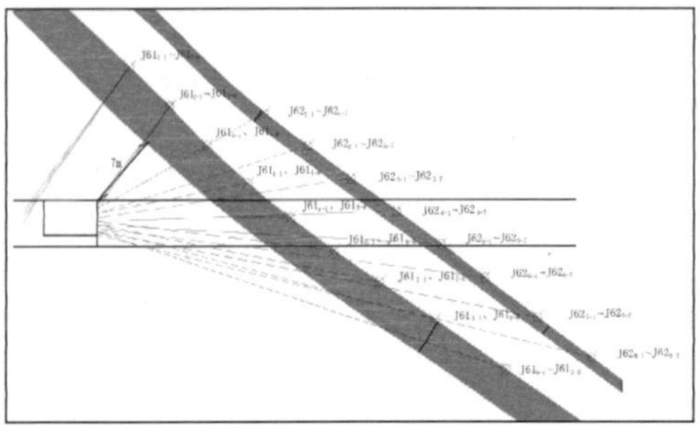

Fig. 9 - 700 borehole layout profile of outburst prevention measures in the area where the central pedestrian concealed inclined shaft exposes coal seam 6

The statistical results of some coal drills in 6 coal seams are shown in table 1.

Tab. 1 single hole expanded coal cuttings of coal seam 6 exposed

Hole number	Dip angle/°	Hole depth/m	Length of coal hole/m	Nozzle parameters/mm	Drilling diameter before reaming/mm	Expanded coal content/m^3	Drilling diameter after reaming/mm	Borehole diameter ratio before and after reaming
J62$_{4-2}$	1.7	29.1	6.8	2.5	75	3.3	790	10.5
J62$_{4-4}$	1.8	28.2	6.5	3.0	75	3.0	770	10.3
J62$_{4-6}$	1.7	29.1	6.8	2.0	75	3.1	766	10.2
J62$_{5-1}$	-2.9	34.2	8.1	3.0	75	3.5	746	9.9
J62$_{5-3}$	-3.1	32.7	7.5	3.0	75	2.7	681	9.1
J62$_{5-5}$	-3.1	32.7	7.5	3.0	75	2.5	656	8.7
J62$_{5-7}$	-2.9	34.3	8.1	2.5	75	2.8	668	8.9
J61$_{6-8}$	-5.7	26.3	6.8	3.5	75	2.5	688	9.2
Average						3.1	740	9.9

-700 central pedestrian concealed inclined shaft exposes 6 coal seams and uses high-pressure water jet reaming equipment for hydraulic reaming. The statistical distribution of equivalent borehole diameter after reaming is shown in Fig 10.

Fig. 10 statistical distribution map of equivalent borehole diameter after high pressure water jet enlargement of coal seam 6

After the 75mm drill bit is used to form the hole, the high-pressure water jet is used for reaming. After reaming, the drilling diameter is more than 650mm, of which the drilling diameter of 650 ~ 700mm is the most, accounting for 57%. The coal output of single hole expansion is 2.5 ~ 3.5m3, with an average of 2.9 m3, and the effect is very remarkable.

Through the analysis of borehole diameter after reaming, the equivalent borehole diameter of coal 6 after reaming is 656 ~ 790mm, 8.7 ~ 10.5 times that before reaming, and the average equivalent diameter is 720mm, 9.6 times that before reaming. See Fig. 11 for the relationship between drilling diameter and nozzle diameter after reaming with high-pressure water.

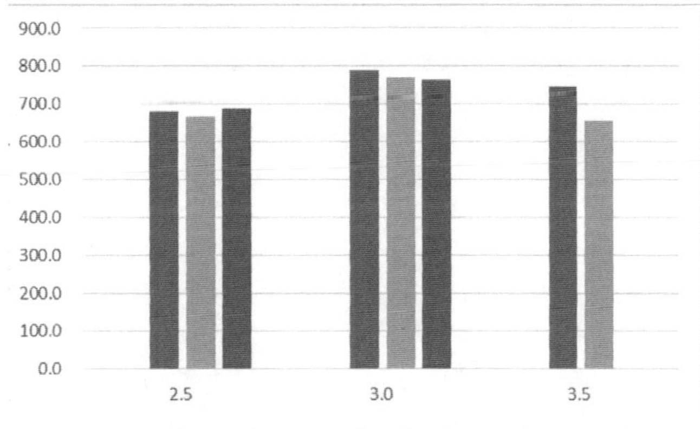

Fig. 11 Relationship between borehole diameter distribution and nozzle diameter after coal flat hole reaming

The reaming bit with nozzle diameter of 3.0mm is used to ream coal seam 6, and the drilling diameter increases obviously. Compared with other types of reaming bits, the bit with nozzle diameter of 3.0mm has better reaming effect under the conditions of firmness coefficient f = 0.5 and flat hole reaming of Xinji No. 1 coal mine.

5. Conclusion

(1) The mechanical analysis of high-pressure water jet combined with drill tooth coal breaking is carried out, including the mechanical analysis of water jet coal breaking, the mechanical analysis of high-pressure water jet combined with drill tooth coal breaking and the mechanism of high-pressure water jet combined with drill tooth coal breaking, which provides a theoretical basis for the optimization of high-pressure water jet drilling and expanding integrated bit.

(2) According to the principle of bit optimization, the bit material and nozzle parameters are optimized respectively. After using diamond composite bit, the hardness is increased by 20 ~ 40 times and the service life is increased by 3 ~ 5 times. Through numerical simulation analysis, it is determined that when the L / D value of the nozzle is 2, the jet development can maintain good stability. 2 ~ 3 nozzles are arranged radially and 1 ~ 2 nozzles are arranged axially, and the best contraction angle is 13 ° ~ 15.

(3) The field test shows that after the reaming of coal 6, the single hole expanded coal volume is 2.5 ~ 3.5m3, with an average of 2.9 m3, the equivalent drilling diameter is 656 ~ 790mm, with an average equivalent diameter of 720mm, which is 9.6 times that before the reaming, and the reaming effect of the drill bit with nozzle diameter of 3.0mm is better.

References

[1] Gao Kui, Liu zegong, Liu Jian. Role of in-situ stress in coal and gas outburst induced by soft coal in Shimen uncovering structure [J]. Journal of rock mechanics and engineering, 2015 (2): 305-312.

[2] Yuan Liang. Strategic thinking on deep coal and gas CO mining in China [J]. Journal of coal, 2016 (1): 1-6.

[3] Liu Qingquan, Cheng Yuanping, Li Wei, et al. Gas solid coupling model of coal and gas in the first mining layer with deep low permeability [J]. Journal of rock mechanics and engineering, 2015 (S1): 2749-2758.

[4] Zhang Yongjiang. Identification of coal, rock and gas dynamic disasters in deep mines [J]. Energy technology and management, 2014: 7-9.

[5] Pan Yishan. Study on the integration of coal and gas outburst and rockburst composite dynamic disasters [J]. Journal of coal, 2016 (1): 105-112.

[6] Li Chengquan, Li Zhonghua, pan Yishan. Experimental study on prevention and control of coal mine rockburst by high pressure rotating water jet [J]. Journal of Liaoning University of engineering and technology, 2003: 75-77.

[7] Zhang Yongjiang, Meng Xianzheng. Investigation and Study on the extraction radius of high pressure water jet hydraulic reaming [J]. Mining safety and environmental protection, 2012 (S1): 8, 53-54.

Flow regulation and bottom hole pressure control under gas influx condition based on RMR system

Wenzhen Zhang[1,2], Jie Zhang[1,2*], Zhiheng Wang[1,2], Ruitao Sun[1,2], Sheng Yu[1,2], Zhijun Yang[1,2]

[1]State Key Laboratory of Oil and Gas Reservoir Geology and Exploitation, Southwest Petroleum University, Sichuan Chengdu, China
[2]Petroleum Engineering School, Southwest Petroleum University, Sichuan Chengdu, China;

*Corresponding author e-mail: 202021000933@stu.swpu.edu.cn

Abstract. During the drilling process of RMR system, when the gas reservoir is drilled, the bottom hole pressure and formation pressure are in an unbalanced state, and the formation fluid invades the bottom hole under the action of pressure difference, further causing complex situation such as overflow and kick. In order to effectively control the bottom hole pressure during drilling and ensure drilling safety, a bottom hole pressure calculation model of RMR system under gas influx condition is established based on the gas-liquid two-phase flow theory. Three flow regulation methods of increasing the outlet flow of drilling pump and subsea pump at the same time, reducing the outlet flow of subsea pump while the outlet flow of drilling pump remains unchanged, and increasing the outlet flow of drilling pump while reducing the outlet flow of subsea pump are selected to dynamically simulate the change process of bottom hole pressure after flow regulation. Through data fitting, The relationship equation between them and bottom hole pressure is obtained, and the flow regulation method of RMR system when controlling bottom hole pressure under gas influx condition is formed.

1. Introduction

In order to solve the problems encountered by conventional offshore drilling technology in deep water drilling, Norway AGR has developed a technology suitable for safe drilling under narrow density windows in deep water - Riserless Mud Recovery System Dual-gradient Drilling, RMR[1,2]. This technology gets rid of a series of problem caused by conventional risers. During the drilling process of the RMR system, the drilling fluid is pressurized and pumped into the drill pipe by the drilling pump, then passes through the drill bit to the bottom of the well, and then returns to the suction module through the wellbore annulus, and finally the subsea pump lifts the drilling fluid to the surface platform through the return line[3,4]. As shown in Figure 1 below.

This paper uses the theory of managed pressure drilling and combines the characteristics of RMR dual gradient drilling. By adjusting the outlet flow of the drilling pump and the subsea pump, and with the aim of controlling the bottom hole pressure, the flow regulation method under the gas influx condition is analyzed in detail.It provides a reference for the control of bottom hole pressure in the practical application of RMR system[5,6]. This research is expected to reduce the probability of complex downhole accidents of the RMR system and improve the safety and stability of the operation of the circulation system.

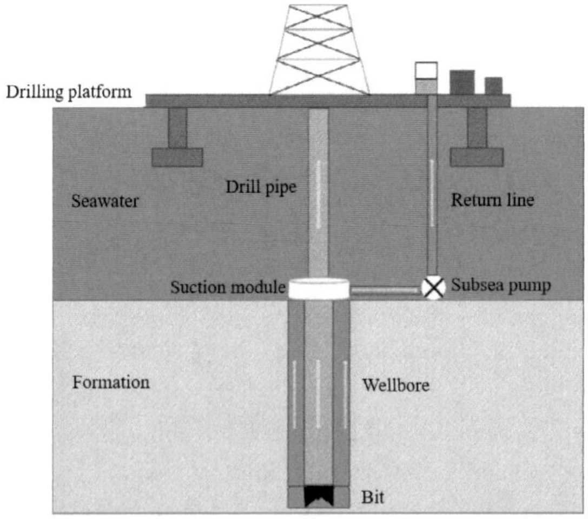

Figure 1. Schematic diagram of RMR dual gradient drilling technology

2. Establishment of main governing equations under gas influx condition

2.1. Gas influx calculation equation

Assuming that the invading gas satisfies the non-Darcy seepage formula, the gas influx volume can be calculated as follows[7,8]:

$$P_e^2 - P_{bt}^2 = Aq_g + Bq_g^2 \tag{1}$$

$$A = \frac{1.291\times10^{-3}\mu Z T_f}{Kh}\ln\frac{r_e}{r_{bt}} \tag{2}$$

$$B = \frac{2.828\times10^{-21}\beta\gamma_g T_f}{\mu r_{bt} h^2} \tag{3}$$

$$\beta = \frac{7.644\times10^{10}}{K^{1.5}} \tag{4}$$

P_e—Formation pressure, MPa; q_g—Gas intrusion in standard state, m^3/s; μ—Gas viscosity, mPa·s; Z—Natural gas compression factor, dimensionless; T_f—Formation temperature, K; K—Formation permeability, mD; h—Gas layer exposed thickness, m; r_e—Gas effective intrusion radius, m; r_{bt}—Bottom hole radius, m; γ_g—Relative density of gas, dimensionless.

2.2. Continuity Equation

Take a micro-element segment dz along the flow direction of the fluid for research, and let the area of its flow cross section be Aann. According to the law of conservation of mass, each micro-element can get the mass change. Take gas phase as an example[9]:

Gas Continuity Equation:

$$\frac{\partial A_{ann}\rho_g E_g}{\partial t} + \frac{\partial A_{ann}\rho_g E_g v_g}{\partial z} = q_g \tag{5}$$

Liquid Continuity Equation:

$$\frac{\partial A_{ann}\rho_d\left(1-E_g\right)}{\partial t} + \frac{\partial A_{ann}\rho_d v_d\left(1-E_g\right)}{\partial z} = 0 \tag{6}$$

ρ_g—gas density，kg/m^3; ρ_d—liquid density，kg/m^3; v_g—gas flow rate，m/s; v_d—liquid flow rate，m/s; E_g—cross-sectional air content; A_{ann}—Annular cross-sectional area，m^2.

2.3. Momentum Equation

According to Reynolds transport theorem, there are the following relation for any selected control body: The amount of momentum accumulated in the controller is equal to the sum of the momentum flowing into the controller and the impulse of the resultant fluid acting on the controller minus the momentum flowing out of the controller

Mixed momentum equation:

$$\begin{aligned}&\frac{\partial}{\partial z}\left[\rho_d\left(1-E_g\right)v_d^2A_{ann}+\rho_gE_gv_g^2A_{ann}\right]+\frac{\partial}{\partial t}\left[\rho_d\left(1-E_g\right)v_dA_{ann}+\rho_gE_gv_gA_{ann}\right]\\&=-A_{ann}\frac{dP}{dz}-A_{ann}\left[\rho_d\left(1-E_g\right)+\rho_gE_g\right]g-\left(\tau_gD_g+\tau_dD_d\right)\end{aligned} \tag{7}$$

The friction resistance between the gas-liquid phase and the well wall is expressed by the friction pressure gradient:

$$\begin{aligned}&\frac{\partial}{\partial z}\left[\rho_d\left(1-E_g\right)v_d^2A_{ann}+\rho_gE_gv_g^2A_{ann}\right]+\frac{\partial}{\partial t}\left[\rho_d\left(1-E_g\right)v_dA_{ann}+\rho_gE_gv_gA_{ann}\right]\\&=-A_{ann}\frac{dP}{dz}-A_{ann}\left[\rho_d\left(1-E_g\right)+\rho_gE_g\right]g-A_{ann}\left(\frac{dP}{dz}\right)_{fr}\end{aligned} \tag{8}$$

$\frac{dP}{dz}$—Annular pressure per unit length，MPa; $\left(\frac{dP}{dz}\right)_{fr}$—Annular friction per unit length，MPa; τ_g—Shear stress of gas phase and tube wall，N/m^2; τ_i—Shear stress between gas and liquid phases，N/m^2; D_g—Gas phase wet cycle，m; D_i—Gas-liquid two-phase internal wet cycle，m; τ_d—Shear stress of liquid phase and tube wall，N/m^2; D_d—Liquid phase wet cycle，m.

2.4. Bottom hole pressure governing equation

The bottom hole pressure control equation under gas intrusion conditions mainly includes four parts: annular hydrostatic column pressure, annular pressure loss, subsea pump inlet pressure and subsea pump inlet boost pressure, and its expression is[10]:

$$P_{bt}=P_{inlet}+P_a+\Delta P_{f,a}+\Delta P \tag{9}$$

According to the above equation, the bottomhole pressure control equation of RMR system under gas influx condition is different from that of conventional land drilling. In addition to changing the annular pressure consumption by increasing the outlet flow of drilling pump, the bottomhole pressure can also be controlled by decreasing the outlet flow of subsea pump to increase the inlet pressure of subsea pump.

3. Flow regulation and bottom hole pressure control under gas influx

The relevant data of a vertical well in deep sea are selected for calculation. The wellbore structure is shown in Figure 2. The basic parameters of the three sections are shown in Table 1, and the basic data under gas influx are shown in Table 2. The formation pore pressure and fracture pressure are 80.16 MPa and 86.56 MPa, respectively ; the mechanical drilling speed is 5 m/h ; the cuttings diameter is 0.4 cm ; the cuttings density is 2.6 g/cm^3 ; the pump pressure is 25 MPa ; the pump power is 1600 KW ; the pump efficiency is 0.8 ; the screw drill tool is 5LZ197×7Y ; the drilling pump is 3NB1600 ; and the punch is 60~80/min.

Table 1. Basic parameter data table of three sections

Parameter	Value	Parameter	Value
The depth of the sea /m	800	Formation depth /m	5800
Drilling fluid density /g·cm^{-3}	1.23	Wellbore size /mm	215.9
Outer diameter of drill pipe /mm	127.0	Drilling pump outlet flow /L·s^{-1}	30
Plastic viscosity of drilling fluid /mPa·s	25	Dynamic shear /Pa	8

Table 2. Basic data for calculation under gas influx condition

Parameter	Value	Parameter	Value
Gas layer exposed thickness /m	10	Rate of penetration /m·h^{-1}	5
Bottom hole temperature /K	393	Permeability /mD	50
Average gas compression coefficient	0.91	Average gas viscosity / mPa·s	0.02
Bottom hole radius /m	0.108	Effective gas influx radius /m	140
Formation pressure /MPa	80.16	Bottom hole pressure /MPa	79.16

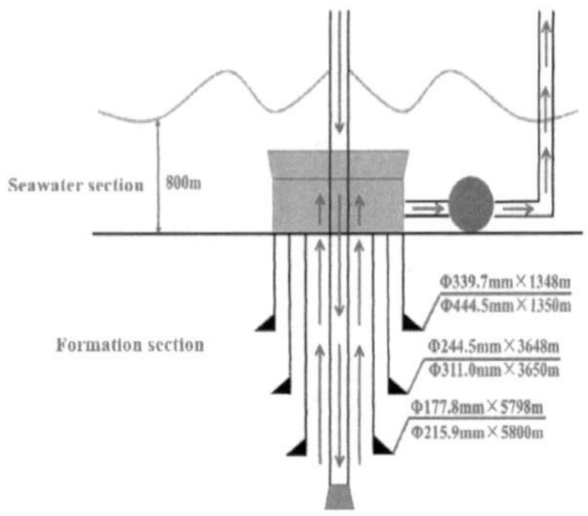

Figure 2. Schematic diagram of well structure

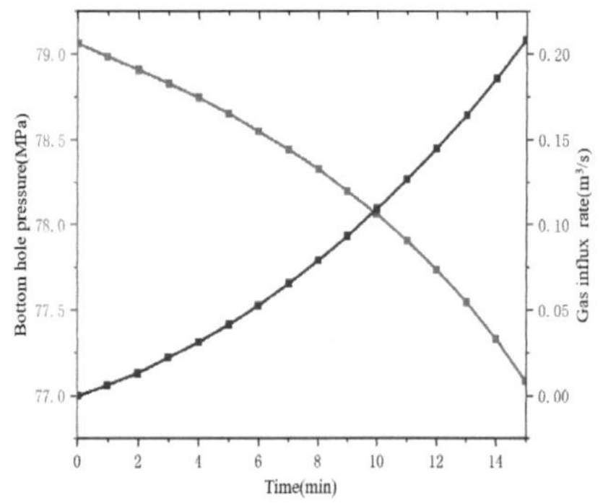

Figure 3. Changes of gas influx and bottom hole pressure at each moment

As shown in Fig. 3, with the increase of time, the reservoir opening thickness increases, and the gas invasion increases gradually, which makes the bottom hole pressure decrease continuously. In order to make the bottomhole pressure equal to the formation pressure all the time, three methods of adjusting the flow rate to control the bottomhole pressure are proposed. They are increasing drilling pump outlet flow and seabed pump outlet flow, drilling pump outlet flow unchanged while reducing seabed pump outlet flow and increasing drilling pump outlet flow while reducing seabed pump outlet flow.

3.1. Increase drilling pump flow and subsea pump flow at the same time

Firstly, the gas influx rate, bottom hole gas volume fraction, bottom hole pressure and mud pool increment under different circulation flow rates are analyzed.

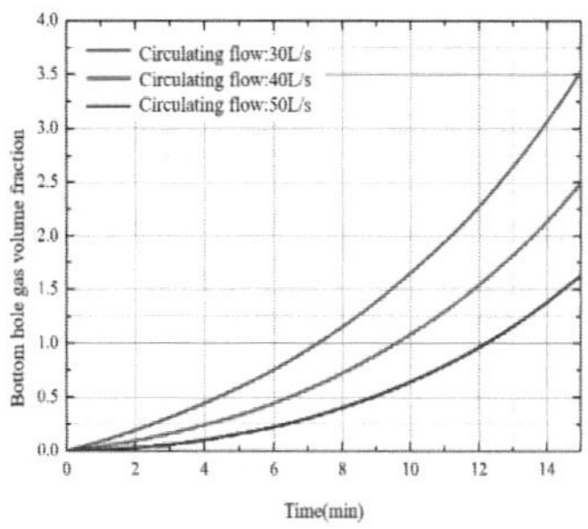

(a)

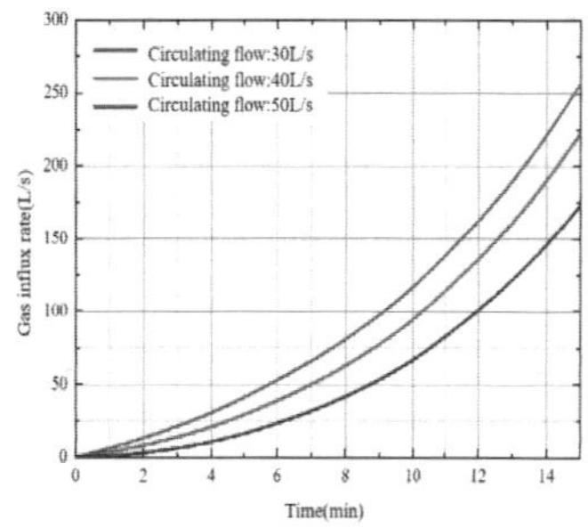

(b)

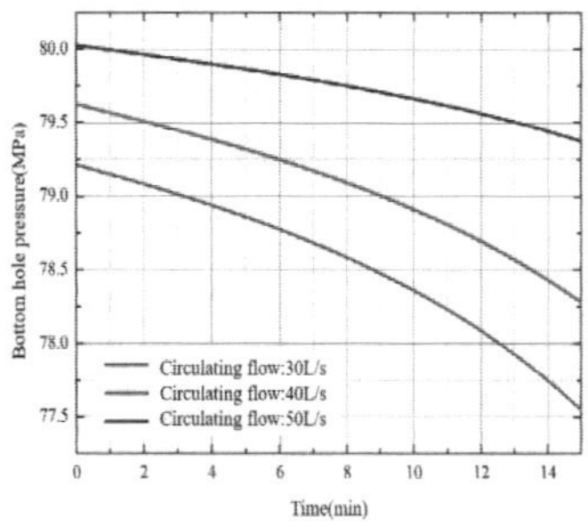

(c)

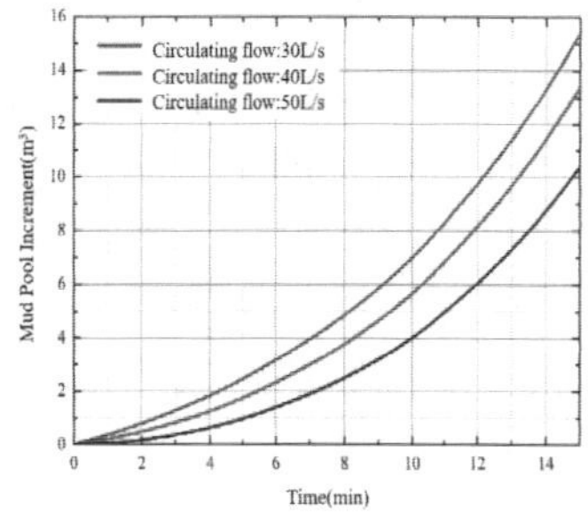

(d)

Figure 4. Analysis of Influencing Factors under Different Circulation Flow Rates

As can be seen from Fig. 4(a), at the initial moment, since the reservoir has not been opened and the gas has not entered the wellbore, the volume fraction of the gas phase at the bottom of the wellbore at the initial moment is 0, and the larger the circulating flow is, the smaller the volume fraction of the gas phase at the same moment. This is because the larger the circulation flow rate, the greater the apparent velocity of the liquid phase, the smaller the relative gas holdup.

It can be seen from Fig. 4(b) that at thc initial time, the reservoir is not opened, and the gas influx rate is 0. With the increase of time, the gas influx rate increases continuously, but the larger the circulation flow is, the smaller the gas influx rate at each time is.

It can be seen in Fig. 4(c) that with the increase of time, the bottom hole pressure under each circulation flow shows a downward trend, but the decline rate is different. The larger the circulating flow rate is, the slower the downward trend is, and the higher the bottom hole pressure at 15 min is.

It can be seen from Fig. 4(d) that at the initial time, there was no gas influx at the bottom of the well, and the increment of the mud pool was 0. With the increase of time, the increment of slurry pond increases continuously, but the larger the circulating flow rate is, the smaller the amount of gas invasion is. At 15 min, the increment of slurry pond with circulating flow rate of 30 L/s is the largest, with a value of 15.402 m3, while the increment of slurry pond with circulating flow rate of 50 L/s is only 10.392 m3.

Based on the above analysis of various influencing factors under different circulation flow, it can be seen that increasing the circulation flow under the condition of gas influx can increase the bottom hole flow pressure, reduce the pressure difference between bottom hole pressure and formation pressure, and effectively inhibit the invasion of gas, which also proves that the circulation flow can effectively control the bottom hole pressure. Therefore, the relationship between flow regulation and bottom hole pressure under different viscosity and dynamic shear force will be further analyzed.

3.2. Constant Drilling Pump Flow and Reduce Submarine Pump Flow

By reducing the outlet flow of submarine pump, the pressure increment at the inlet of submarine pump is increased to suppress gas influx and control bottomhole pressure.

（1）Variation of subsea pump inlet pressurization under each subsea pump outlet flow during pressurization operation

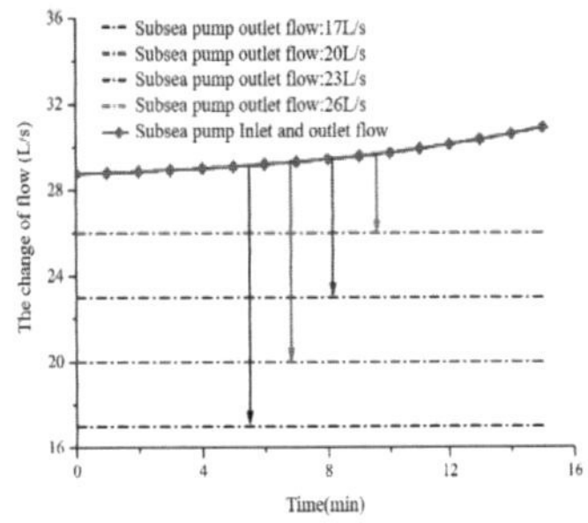

Figure 5. Schematic diagram of flow change of subsea pump

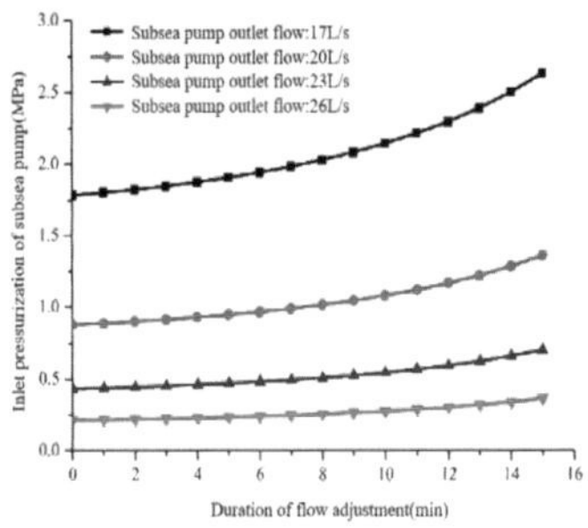

Figure 6. Variation of inlet pressure under the outlet flow of each subsea pump

It can be seen from Figure 5 that with the extension of time, the reservoir opening thickness increases, and the gas invasion rate increases, which makes the mixed flow gradually increase, that is, the flow rate of submarine pump inlet and outlet increases. Before the bottomhole pressure control, the flow rate of submarine pump outlet is equal to that of inlet.

At each moment, the outlet flow of the subsea pump is adjusted to 17 L/s, 20 L/s, 23 L/s and 26 L/s to obtain the corresponding inlet pressurization of the subsea pump, as shown in Fig. 6. It can be seen that, at each time, the smaller the adjusted subsea pump outlet flow value, the larger the subsea pump inlet pressure increment will be. In order to ensure the requirement of carrying rock, the outlet flow rate of submarine pump cannot be lower than 16.4 L/s.

（2）Briefly reduce subsea pump outlet flow

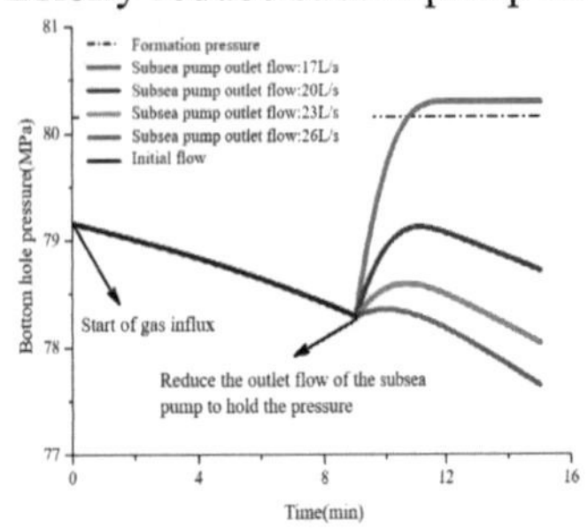

Figure 7. The change of bottom hole pressure when the submarine pump flow is reduced in a short time

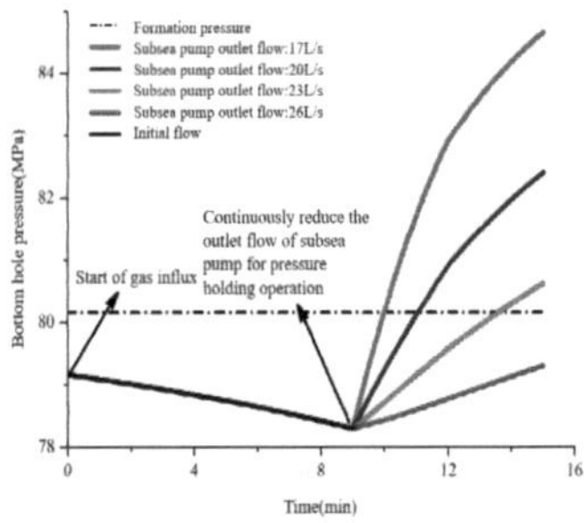

Figure 8. Bottomhole Pressure Change with Duration Reducing Subsea Pump Flow

Fig. 7 shows the change process of bottom hole pressure after applying different flow rates at 9 min. Through a short time flow regulation, it can be found that the bottomhole pressure will increase more or less. Thus, if the bottomhole pump outlet flow remains unchanged, the bottomhole pressure will eventually tend to balance, which also shows that it is feasible to reduce the bottomhole pump outlet flow to suppress pressure.

（3）Duration reduces subsea pump outlet flow

Fig. 8 shows the variation trend of bottom hole pressure in the whole gas influx time when the outlet flow of submarine pump is adjusted to 17 L/s, 20 L/s, 23 L/s and 26 L/s.

By comparing the bottom hole pressure changes under the outlet flow of each submarine pump, it can be seen that the smaller the outlet flow of submarine pump is, the earlier the formation pressure can be balanced. This is because the smaller the outlet flow of submarine pump is and the smaller the opening is, the larger the inlet pressure increment of submarine pump is, and the gas influx can be suppressed earlier and better.

3.3. Increase drilling pump flow rate and decrease subsea pump flow rate

The above analysis shows that increasing the outlet flow of the drilling pump or reducing the outlet flow of the subsea pump can control the bottom hole pressure to a certain extent. Now increase the outlet flow of the drilling pump and reduce the outlet flow of the subsea pump at the same time, and observe the change of the bottom hole pressure. When the seabed pump outlet flow is 26 L/s, the drilling pump outlet flow is 32 L/s, 34 L/s and 36 L/s respectively. When the drilling pump outlet flow is 32 L/s, the seabed pump outlet flow is 26 L/s, 27 L/s and 28 L/s respectively.

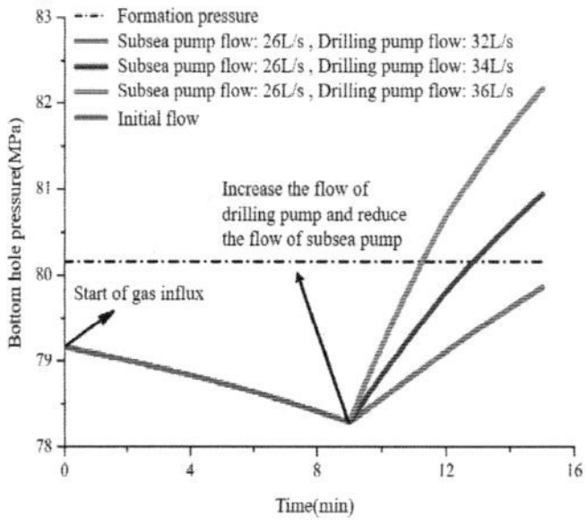

Figure 9. Change of bottom hole pressure when fixing the flow of subsea pump and adjusting the flow of drilling pump

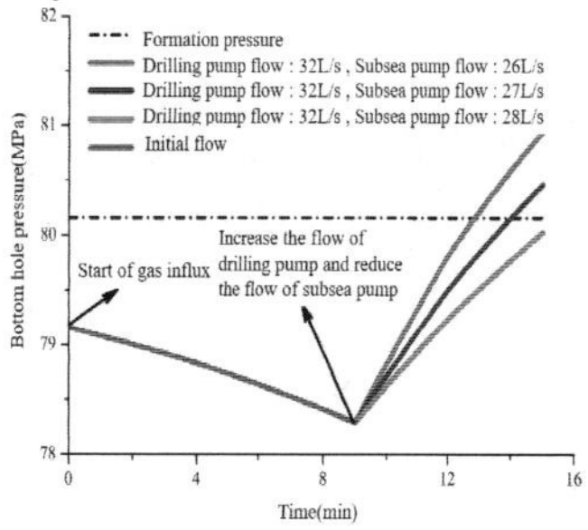

Figure 10. Change of bottom hole pressure when the flow rate of drilling pump is fixed and the flow rate of subsea pump is adjusted

It can be seen from Figure 9 that when the outlet flow of the subsea pump is fixed, the larger the outlet flow of the drilling pump, the larger the ratio of the outlet flow of the subsea pump to the inlet flow, that is, the opening becomes larger, so the inlet pressurization of the subsea pump becomes larger, which makes the bottom hole pressure rise rapidly, and the larger the outlet flow of the drilling pump, the greater the rise of the bottom hole pressure, When the outlet flow of drilling pump is 32 L/s and 34 L/s, the formation pressure and bottom hole pressure are equal at 10.8 min and 13.1 min respectively.

It can be clearly seen from Fig. 10 that when the outlet flow rate of the subsea pump is 26 L/s and 27 L/s, the formation pressure is balanced at 13.1 min and 14 min, respectively. When the outlet flow rate of the drilling pump was slightly increased (the initial flow rate was 28.8 L/s), a small change in the outlet flow rate of the subsea pump also caused a large change in the pressure.

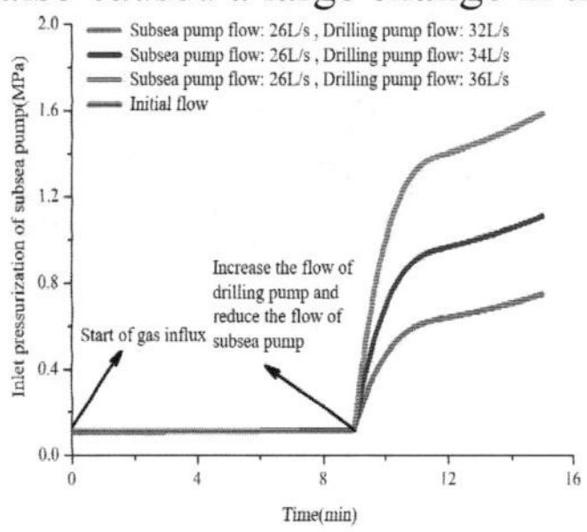

Figure 11. Variation of subsea pump inlet pressure with time during dual-pump adjustment

It can be seen from Fig. 11 that when the outlet flow of the subsea pump decreases to a certain value, a slight increase in the outlet flow of the drilling pump (the initial flow is 28.8L /s) leads to a rapid increase in the pressurization of the subsea pump inlet, and the larger the outlet flow of the drilling pump is, the larger the final inlet pressurization of the subsea pump is. This is due to the dual effect of increasing outlet flow of drilling pump and decreasing outlet flow of subsea pump, while the change of outlet flow of drilling pump is small, and the effect of annular pressure consumption is not obvious, so it is mainly the effect of throttling and pressurization.

It can be concluded that a larger outlet flow rate of the drilling pump and an appropriate reduction of the outlet flow rate of the subsea pump can effectively restrain the gas influx and control the bottom hole pressure. When there is a synergy between the two, whether it is to increase the outlet flow of the drilling pump or reduce the outlet flow of the subsea pump, only changing the smaller flow value can effectively increase the inlet pressure of the subsea pump and balance the formation pressure. Therefore, the method of simultaneously adjusting the flow rate to control the bottom hole pressure can meet the requirements of bottom hole pressure balance during gas influx.

4. Conclusion

From the above analysis, the following conclusions can be drawn:

（1）During the drilling process of the RMR system, when the gas layer is encountered, the bottom hole pressure and the formation pressure are in an unbalanced state, which leads to further problems such as overflow and well kick. Based on the main control equations and auxiliary equations of the RMR system under gas influx condition, combined with the finite difference method, the bottom hole pressure calculation model of the RMR system under gas influx condition is established.

（2）Through the calculation model of bottom hole pressure of RMR system under gas invasion condition, the simulation analysis of various parameters of gas invasion process under different flow rate and the change of bottom hole pressure under different flow rate regulation method is carried out. Three methods of regulating flow to control bottom hole pressure are proposed, which are to increase the outlet flow of drilling pump and the outlet flow of subsea pump, keep the outlet flow of drilling pump unchanged, reduce the outlet flow of subsea pump, and increase the outlet flow of drilling pump and reduce the outlet flow of subsea pump.

Reference

[1] Zhang Xiaodong, Wang Haijuan. Progress and Prospect of deepwater drilling technology [J]. Natural gas industry, 2010, 30 (9): 46-48.
[2] Zhang Jie Mao Gezhen Li Xin Sun Xuefeng.Study on Influencing Factors of Annular Pressure in Riser-free Drilling Fluid Recovery System[J].China Petroleum Machinery,2020,48(12):60-66.
[3] Jing Liu, Chuan Wang, Wen Hou, et al. Effect of Deepwater Drilling Platform Motion on Lateral Vibration of Drill String [J]. Journal of Research in Science and Engineering, 2020, 2(5).
[4] Cayeux E, Daireaux B, Dvergsnes E W W, et al. Principles and Sensitivity Analysis of Automatic Calibration of MPD Methods Based on Dual-Gradient Drilling Solutions [J]. Society of Petroleum Engineers, 2014, 2(1).
[5] PENG Q, FAN H H, JI R Y, et al. Analysis of circulating pressure drop in surface formation wellbore in riseless drilling[J]. China Petroleum Machinery, 2015,43(8): 73-77
[6] Xu Peng. Study on dynamic well killing method in deep water drilling [D]. China University of petroleum, 2008
[7] Cai Wentao, Cai Jinliang. Study on circulating displacement and influencing factors of deep-water dynamic kill method [J]. Guangdong chemical industry, 2016, 43 (22): 38-40.
[8] Cohen, J., Kleppe, J., Martin, T. Gulf of Mexico's First Application of Riserless Mud Recovery for Top-hole Drilling-A Case Study [C]. Offshore Technology Conference, 2010.
[9] Han Tianwang, Jiang Hongwei, Yang Guang. Classification and research progress of deepwater double gradient drilling technology [J]. Petroleum field machinery, 2019, 48 (05): 83-89.
[10] Liu Jie, Chen Guoming, Zhou Changjing, et al. Drilling hydraulics calculation of deepwater riser free drilling fluid recovery [J]. Petroleum drilling and production technology, 2010, 32 (05): 20-23.

Dynamic Simulation Analysis of Aircraft Hydraulic Pipeline System under Different Pressure

Zhenhua Liao[1, a], Guangjie Zhai[2, b] and Jie Tang[1, c]

[1] School of Aeronautical Engineering, Civil Aviation University of China, Tianjin 300300, China;
[2] CNSIG Inner Mongolia Chemical Industry Co., Ltd. Sodium plant production technology room, China.

[a] jasticpin@163.com, [b] gjzh30@163.com, [c] tangjie_cauc@163.com

Abstract. The hydraulic system is very important in the aircraft structural system, but the hydraulic pipeline is prone to cracks or even breaks due to vibration problems during system operation, causing major hidden dangers to flight safety. Using ANSYS Workbench to carry out modal analysis and harmonic response analysis of pipelines, the influence of different inlet pressures on pipeline vibration under pulsation excitation is compared and analyzed, which provides theoretical guidance for pipeline vibration research.

Keywords: Pipeline Vibration; Modal Analysis; Harmonic Response Analysis; Fluid-Structure Coupling.

1. Introduction

In the development of aviation industry, hydraulic pipeline system has been widely used, which provides energy conversion, distribution and control for aircraft fuel system, hydraulic system and landing gear system, and seriously affects the takeoff, cruise, landing and taxiing of aircraft. In the hydraulic pipeline work for a long time, the impact and corrosion of hydraulic oil will affect the inner wall of the hydraulic pipeline, so that the inner surface of the pipeline cracks, these cracks will intensify the vibration of the hydraulic oil at work, is a hidden safety hazard. In addition, the impurities mixed in the hydraulic oil will accumulate on the inner pipe wall over time, forming local bulges. This type of pipe wall defect will also affect the vibration of the pipeline.

As early as the late 19th century, Toshiyuki Sakai studied the vibration of pipelines [1], In 1972, Benson introduced the vibration of pipelines [2], Hatfied et al. established a mathematical model of fluid-structure coupling, calculated the modal of fluids and pipelines and got the natural frequency of pipelines [3]. The hydraulic system reformed by Wang Zhongqiang improved the vibration of the pipeline. In the process of research, he found that the oil supply and pipeline stress had a certain influence on the vibration of the pipeline [4]. Yang Fengyu et al. conducted time-domain analysis of pipelines and found that the coupling effect of hydraulic pipelines and hydraulic oil caused the vibration of pipelines [5].Stangl et al. modeled nonlinear vibration of fluid pipeline system by using extended Lagrange equation and Sandretz method [6]. Tornabene et al. studied the stability of cantilever pipe conveying fluid by constructing and solving algebraic eigenvalue problems using the generalized differential quadrature method [7].Throughout 2011 jia-shun hu for containing the straight crack of pipeline vibration characteristic analysis, to the establishment of the finite element model, crack line type beam crack line are analyzed, and the vibration of the cantilever type crack line features, the final results show that the vibration of the pipeline associated with the depth of the crack and crack position [8].In 2013, Du Jinbo studied the identification of pipeline cracks based on NOFRF algorithm, analyzed it with Matlab software, and obtained the influence of different positions of cracks in pipelines on time-frequency signals of vibration [9].Yang Wei et al. conducted time domain and frequency domain analysis of pipeline cracks by developing the lower and upper computers of the cracked pipeline vibration system and using the METHOD of HHT transformation [10].On the basis of the above research, this paper studies the vibration problem caused by the coupling of hydraulic oil and the pipeline with inner wall defects during the hydraulic pipeline work, and analyzes the influence of the inner wall structure of the pipeline on the vibration of the pipeline, which is of great significance to the fault detection, non-destructive testing and vibration optimization of the hydraulic pipeline.

2. Pipeline Moodel Establishment

2.1. Fluid-Structure Coupling Mathematical Model of Hydraulic Pipelines

In the research process, it is considered that the size of the pipeline and liquid contact surface is much smaller than the length of the pipeline. On this basis, bernoulli-Euler beam model is used to carry out dynamic modeling. First, it is assumed that the tube body is still perpendicular to the centreline profile after deformation. The characteristics of pipeline structure are studied without considering the shear force. Based on the principle of finite element analysis, the pipeline system is modeled by a certain unit of plane pipeline.

The relation between bending deformation curvature K and displacement of pipeline displacement model is:

$$K=\cos\theta=\frac{1}{\left(1+\varepsilon_{x_0}\right)^2}\left[\left(1+\frac{\partial U}{\partial X_0}\right)\frac{\partial^2 W}{\partial X_0{}^2}-\frac{\partial^2 U}{\partial X_0{}^2}\frac{\partial W}{\partial X_0}\right] \tag{1}$$

The hydraulic pipeline movement system consists of two parts, one is fluid movement, and the other is associated movement caused by the interaction between fluid and pipeline. Therefore, the velocity of pipeline is expressed by vector:

According to Hamiltonian principle, the variational equation that meets when the pipeline system is stable is as follows:

$$\delta\left[\iint Xdxdt\right]=0 \tag{2}$$

After simplification of the equation of motion by Hamiltonian principle:

$$m\frac{\partial^2 u}{\partial t^2}+2m_f v_f\frac{\partial^2 u}{\partial x\partial t}-2m_f v_f\frac{\partial^2 u}{\partial x\partial t}-EA_P\frac{\partial^2 u}{\partial x^2}=0 \tag{3}$$

2.2. Hydraulic Pipeline Calculation Model

In this paper, an aviation hydraulic straight pipe with a length of 400mm, an outer diameter of 24mm and a wall thickness of 3mm is selected as the research object. The 3d model of hydraulic pipeline is drawn in SolidWorks. In order to study the influence of the defects on the inner surface of the pipeline on the pipeline vibration.

The Fluent software was used to process the model. Face Mesh in the Mesh was selected and Relevance was set as 100 in the Mesh detail panel. When setting the Relevance, the Mesh quality and computing time were taken into consideration and the roughness of the Mesh was refined on the premise of satisfying the research requirements. The grid model is shown in Fig.1. Set the inlet and outlet of the flow field, and set the relative pressure to 100Pa., other boundary types were defined as Wall, and simulation analysis was conducted at last.

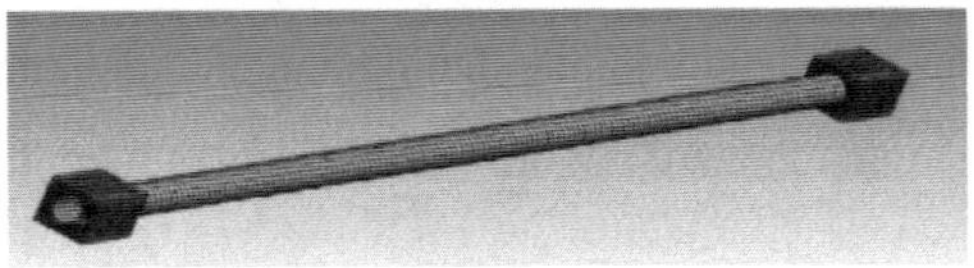

Fig. 1 Hydraulic pipeline grid

3. Results and Discussion

3.1. Modal Analysis under Steady-State Setting
New modules Static Structural and Modal are added to the established steady-state setting module, and geometric structure, basic Engineering Data and Model Model are shared. Under the setup of the steady-state structure module, the pipe and the pipe wall section are constrained, pressure is applied to the pipe wall, and the relative stress distribution of the pipe can be obtained by solving, as shown in Fig.2. It can be seen that the stress is mainly concentrated on the pipe wall, with the maximum stress of 975.17[MPa] and the minimum stress of 225.13 [MPa].

Fig. 2 Distribution of relative stress in pipeline

The first six orders of natural frequency were selected for analysis, and the first six orders of natural frequency of pipeline structure were obtained in ANSYS Workbench, when the pipeline modal order is from 1 to 6, the natural frequencies are 135.05Hz, 135.06Hz, 769.12Hz, 769.13Hz, 1564.2Hz and 2083.8Hz respectively.

In the first-order mode diagram, the deformation in the middle of the pipeline increases mainly along the negative direction of the Y-axis, and the maximum deformation reaches 65.194mm.In the second vibration mode, the deformation of the middle part of the pipe structure also increases along the Y-axis direction, but the deformation intensity is larger, and the maximum deformation reaches 65.195mm.In the third-order mode diagram, the deformation in the middle of the pipeline is mainly in the z-axis direction and the maximum deformation is 69.693mm.In the fourth mode diagram, the middle part of the pipe is deformed towards the X axis, and the maximum deformation is 69.697mm.According to the fifth mode diagram, it is found that the fifth mode diagram is obviously different from other modes. The pipe wall in the fifth mode diagram is obviously deformed. Along the Y-axis, the deformation of the pipe wall gradually increases, and the maximum deformation reaches 51.624mm. In the sixth order mode diagram, the pipe swings and deforms along the YZ direction, with the maximum deformation reaching 67.945mm.

It can be seen from the mode diagram of the pipeline structure that the stress direction of the pipeline structure is different, and the pipeline deformation mostly appears in the weak links such as pipeline bending. Because there is no constraint set in the middle of the pipeline, only the pipeline section is set, so it mainly reflects the deformation swing in the middle of the pipeline.

3.2. Harmonic Response Analysis
In this study, the main input load is frequency, load pressure and displacement can also be input, and the output can be stress and strain caused by deformation. The Harmonic Response analysis module needs to be added in ANSYS Workbench. Its parameter Settings can be directly shared with the data in

previous Modal, and the sharing mode is similar to Modal setting mode. Set the highest frequency to 10000Hz and the lowest frequency to 0Hz. Fig.3 and shows the processing results.

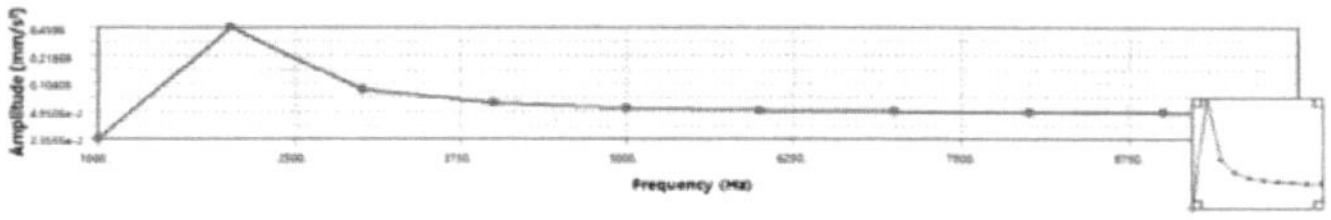

Fig. 3 Frequency amplitude diagram of the pipe

It can be seen from the figure that when the frequency is close to 2000Hz, the maximum amplitude of the pipeline occurs, which is about 0.4596mm. Therefore, when preventing the structural damage caused by resonant deformation of the pipeline, the external frequency of 2000Hz is mainly considered.

3.3. Vibration Response of Pipe Wall under Pulsating Excitation

In order to study the vibration response of pipe wall under pulsating excitation, combined with Transient analysis of pipeline, Fluid Flow CFX module of Fluid Transient analysis in ANSYS Workbench was used for simulation, followed by Transient Structure module and Mesh module. When solving THE FSI problem, the main form is to analyze the vibration response of any point on the tube. Under Analysis Type, attention should be paid to the solution to be selected. Analysis Type Option is Transient solution. Select Expressions under The Outlinie setting to set the function and create a pulsation input that changes with time. The pulsation time function set here is a sine function:, and the function setting and function time of the pulsation input are shown in Fig.4.

Change the import conditions, at this time the import Settings need to be changed to Opening pres. And Dirn, select the function parameters set previously, create the transient file, set the output Settings, change the coupling times, set the calculation parameters of solution in CFX, In Rusult's CFD-POST, solid element stress changes can be generated. Fig.5 is the displacement cloud diagram of solid shell element.

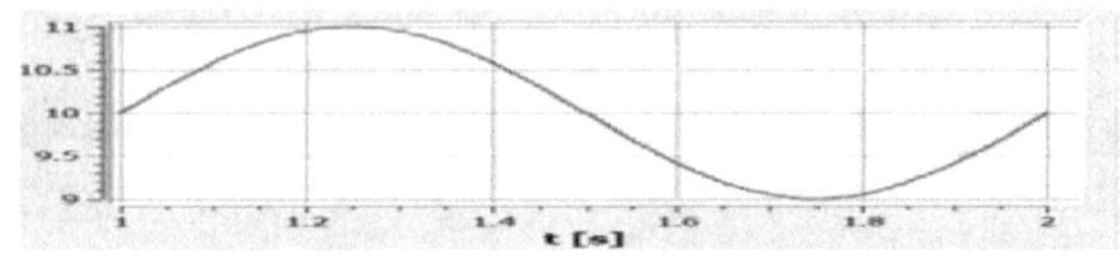

Fig. 4 Function time diagram of pulsating input

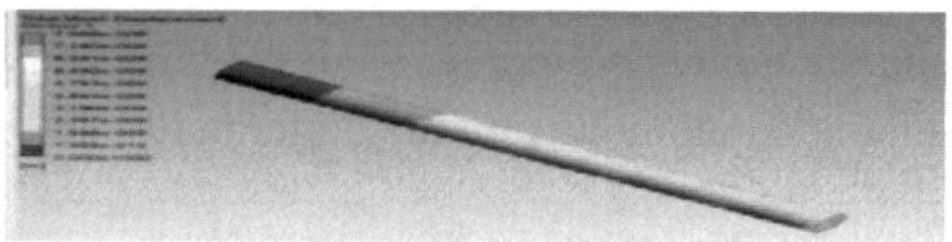

Fig. 5 Displacement diagram of solid shell element

According to the previous time function diagram, some special time values are selected for simulation analysis, and the pressure input values of 9MPa, 10MPa and 11MPa are selected to carry out the corresponding modal harmonious response analysis of the pipeline. The analysis method is the same as that used to solve the modal harmonious response of the pipeline structure in the steady-state setting. Therefore, the sixth-order natural frequencies under the three pressures can be obtained, as shown in Table 1.

Table 1. Natural frequencies of pipes under different input pressures (Hz)

Order Number	1	2	3	4	5	6
9MPa	136.37	136.37	806.92	806.93	1564.8	2129.7
10 MPa	136.49	136.49	806.52	806.53	1564.8	2129.2
11 MPa	136.6	136.6	806.13	806.13	1564.7	2128.7

The harmonic responses of the three input pressures are shown in Fig.6a, b, c.

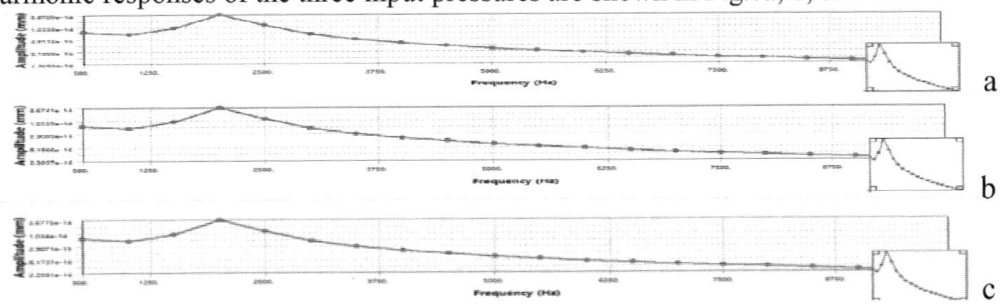

a. (9MPa Frequency amplitude diagram) b. (10MPa Frequency amplitude diagram) c. (11MPa Frequency amplitude diagram)

Fig. 6 Harmonic response diagram of pipeline under different input pressures

According to the figure, the points with the maximum amplitude under the three pressure inputs all appear at about 2000Hz. The amplitudes at the frequency of 2000Hz are selected for comparison, When the input pressure is 9Mpa, 10Mpa and 11Mpa, the amplitudes are 3.6709E-014mm, 3.6741E-014mm and 3.6775E-014mm respectively.

By comparing the data, it is found that at the maximum amplitude, the amplitude is different with the inlet pressure value, and when the inlet pressure value is larger, the amplitude is larger and the vibration is more intense.

Set the inlet pressure of the second group as 10MPa, 20MPa and 30MPa respectively, and the inlet pressure of the third group as 10MPa, 100MPa and 190MPa respectively. Ensure that the increasing amount is the same when setting the inlet pressure. The second and third groups were analyzed by analogy with the first group.

Using the same method, the sixth order natural frequencies are obtained when the inlet pressure values are 10MPa, 20MPa and 30MPa.Table 2.

Table 2. Natural frequencies of pipes under different input pressures (Hz)

Order Number	1	2	3	4	5	6
10 MPa	136.49	136.49	806.52	806.53	1564.8	2129.2
20 MPa	137.5	137.5	802.55	802.55	1564.4	2124.2
30 MPa	138.21	138.21	798.52	798.53	1564.0	2119.2

Harmonic responses of pipe vibration under three different inlet pressures are shown in Fig.7a, b, c.

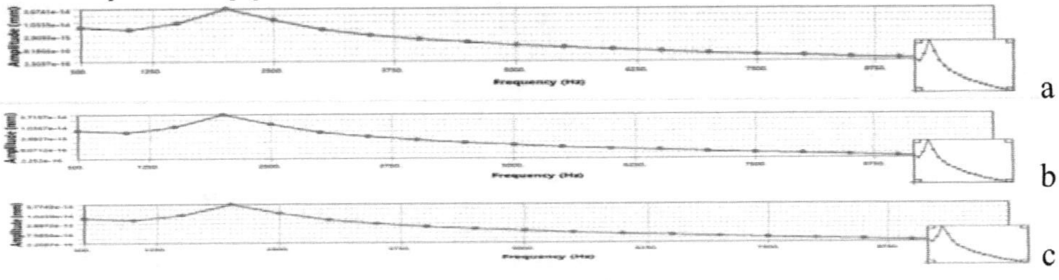

a. (10MPa Frequency amplitude diagram) b. (20MPa Frequency amplitude diagram) c. (30MPa Frequency amplitude diagram)

Fig. 7 Harmonic response diagram of pipeline under different input pressures

According to the figure, the points with the maximum amplitude under the three pressure inputs all appear at about 2000Hz. The amplitudes at the frequency of 2000Hz are selected for comparison, When the input pressure is 10Mpa, 20Mpa and 30Mpa, the amplitudes are 3.6741E-014mm, 3.7157E-014mm and 3.7743E-014mm respectively.

By comparing the data, it is found that with the increase of input pressure, the amplitude of pipeline structure still increases. Simulation verification is carried out for the third group.Using the same method, the sixth order natural frequencies are obtained when the inlet pressure values are 10MPa, 20MPa and 30MPa. Table 3.

Table 3. Natural frequencies of pipes under different input pressures (Hz)

Order Number	1	2	3	4	5	6
10 MPa	136.49	136.49	806.52	806.53	1564.8	2129.2
100 MPa	135.05	135.05	769.12	769.13	1561.2	2083.8
190 MPa	108.26	108.3	727.67	727.68	1557.6	2036.7

The harmonic response of pipe vibration under three different inlet pressures is shown in Fig.8a, b, c.

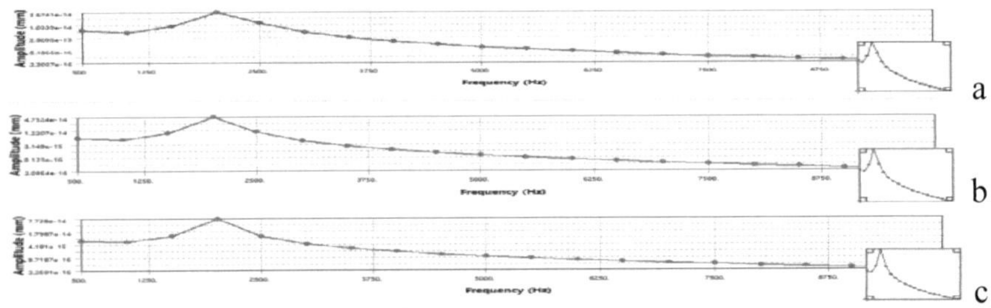

a. (10MPa Frequency amplitude diagram) b. (100MPa Frequency amplitude diagram) c. (190MPa Frequency amplitude diagram)

Fig. 8 Harmonic response diagram of pipeline under different input pressures

In the figure, the point with the maximum amplitude under the three pressure inputs still appears at about 2000Hz. The amplitude at the frequency of 2000Hz is selected for comparison, When the input pressure is 10Mpa, 100Mpa and 190Mpa, the amplitudes are 3.6741E-014mm, 4.7324E-014mm and 7.738E-014mm respectively.

By comparing the data, it is found that with the increase of the input pressure, the amplitude of the pipeline structure still increases. The larger the input pressure value is, the more intense the vibration of the pipeline is.Therefore, in the pulsating input only for pressure change, the change of liquid pressure mainly affects the change of amplitude, and has no effect on the position of pipeline resonance point.

4. Conclusion

This paper mainly studies the vibration of aircraft hydraulic pipeline system, mainly simulates the steady and transient Settings of pipeline system, and analyzes their modal characteristics and harmonious response characteristics:

(1) For the pipeline finite element analysis problem, the Workbench module in ANSYS was used for simulation solution, and the imported model was used for structural meshing, so as to obtain the strain cloud diagram of shell element under steady and transient Settings.

(2) After the pipe wall is constrained, the mode analysis is carried out to obtain the sixth order natural frequency of the pipe vibration and the harmonious response diagram of the pipe vibration pattern under the corresponding order in the pipeline structure set in the steady state. Through the analysis of the diagram, the natural frequency at the maximum amplitude is about 2000Hz.

(3) In the transient setting, the pulsation function is inserted to analyze the vibration response caused by the pipe wall. The results show that when other conditions are constant, the higher the inlet pressure

is, the more intense the vibration will be, while the inlet velocity has little effect on the pipeline vibration and can be ignored in the harmonic response analysis.

References

[1] Toshiyuki Sakai et al. Study on pulsation of pipe system of reciprocating compressor, Bulletin of JSME, 1973, 16-18.
[2] Benson R.S, Veer A S. Some Recent Researched in Gas Dynamic Modeling of Multiple Single Stage Reciprocating Compressor System[C]. Purdue Compressor Technology Conference, 1972, 490-498.
[3] Hatfied F J, Wiggert D C, Qtwell R S. Fluid structure interaction in piping by component synthesis [J]. Fluids Engineering, 1998, 104: 318-325.
[4] Wang Zhongqiang, QIN Satian et al. Vibration and Control of hydraulic Control System of φ140 Pipe Mill [J]. Key Technology, 2002, (4): 63-65.
[5] Yang Fengyu, Yang Rui, et al. Vibration and Noise Reduction of 200MN Hydraulic Press Hydraulic System [N]. Journal of Gansu Sciences, 2006, (4): 95-97.
[6] Stangl M, Gerstmayr J, et al. An alternative approach for the analysis of nonlinear vibrations of pipes conveying fluid. J Sound Vib 2008, 310:493-511.
[7] Tornabene F, Marzani A, Viola E, et al. Critical flow speeds of pipes conveying fluid using the generalized differential quadrature method. Adv Theor Appl Mech. 2010, 3: 121 -- 38.
[8] Hu Jiashun, Feng Xin et al. Vibration Characteristics of Pipe with Non-penetrating Straight Crack [J]. Journal of Vibration and Shock, 2011, 30(04): 21-25.
[9] Du jinbo. Time-frequency analysis of vibration signal of hydraulic pipeline crack fault [D] :(master's thesis). Anshan: University of Science and Technology Liaoning, 2014.
[10] Yang wei. Research on real-time monitoring and analysis of hydraulic pipeline crack fault [D] :(master's thesis). Anshan: University of Science and Technology Liaoning, 2016. Feng Guotai, Huang Jiahua, Li Haibin, et al. Journal of university of Shanghai for science and technology, 2001, 23 (3): 189-192.

Numerical Simulation on Environmental Capacity Assessment in Sanmen Bay, China

Yanming Yao[1], Jiahao Zhu[1], Li Li[1, 2, *], Jiachen Wang[1]

[1] Ocean College, Zhejiang University, Zhoushan 316021, China
[2] The Engineering Research Center of Oceanic Sensing Technology and Equipment, Ministry of Education, China

*Corresponding author e-mail: lilizju@zju.edu.cn

Abstract. Estuarine environmental capacity is the foundation for coastal biological diversity and self-purification capacity. Hence, studies on the marine environmental capacity (MEC) are the foundation for the total discharge control and water quality improvement of land-based pollutants. In the article, A calibrated two-dimensional hydrodynamic model was used to study the environmental characteristics of Sanmen Bay. The model results were used to estimate the environmental capacity of the bay. Taking the pollution problem in Sanmen Bay as an example, the method of response factor, the sub-unit control method, and the phased control method were used to estimate the environmental capacity, pollutant amounts, and the pollutant reduction in the bay. The concentrations of COD, inorganic nitrogen, and acid salt in Sanmen Bay are spatially varied, with higher values occurring in the western part and in the inner bay. The half exchange time of the whole bay is about 23 days, and the exchange time of 95% water body is about 60 days. The evaluation of MEC can not only provide technical support for the offshore aquaculture industries but also provide a scientific basis for the total control of terrigenous pollutants in coastal cities in Southern Zhejiang Province.

Keywords: Marine environmental capacity (MEC), response factor, water exchange, residual, Sanmen Bay.

1. Introduction

With the fast development of the coastal economy and marine aquaculture, estuaries are suffering a lot from terrestrial and oceanic pollutants. Understanding the characteristics of pollutants and estimating the marine environmental capacity (MEC) in estuaries provides a theoretical foundation for coastal management. Therefore, identifying sources and sinks of pollutants, and estimating the estuarine environmental capacity is quite important for coastal development (Yoon et al., 2020; Halpern et al., 2008; Syvitski et al., 2005).

Sanmen Bay is located on the coast of the East China Sea (Figure 1a). It is a macro-tidal turbid estuary, with a maximum tidal range at the bay mouth and suspended sediment concentration at the middle of the bay. It has four tributaries.

In this study, we take Sanmen Bay as an example, to study the characteristics of water exchange ability and marine environmental capacity, using both numerical models and field data.

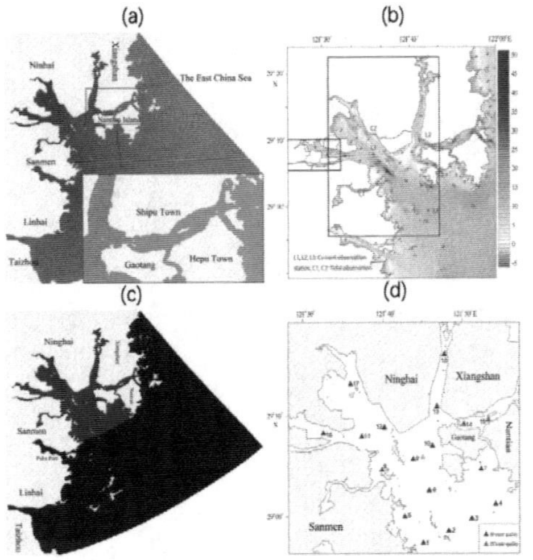

Figure 1. (a) Model domain for Sanmen Bay, (b) Location of observation stations, (c) Initial conditions of water exchange model, (d) Water quality control stations in Sanmen Bay

2. Methodology

2.1. Model descriptions

A calibrated two-dimensional hydrodynamic model was used to reproduce the coastal oceanic and environmental characteristics of Sanmen Bay. Flow, Hydrodynamics, WAQ were used in this study. The governing equations of the Delft3D hydrodynamic model in vertical sigma coordinate system and horizontal curvilinear coordinate system are expressed as follows:

$$\frac{\partial \zeta}{\partial t}+\frac{1}{\sqrt{G_{\xi\xi}}\sqrt{G_{\eta\eta}}}\frac{\partial((d+\zeta)U\sqrt{G_{\eta\eta}})}{\partial \xi}+\frac{1}{\sqrt{G_{\xi\xi}}\sqrt{G_{\eta\eta}}}\frac{\partial((d+\zeta)V\sqrt{G_{\xi\xi}})}{\partial \eta}=(d+\zeta)Q$$

where ξ is the coordinate direction under the Delft3D curvilinear coordinate system corresponds to the X-axis of the rectangular coordinate system, η is the Y-axis, ζ is the height of the water surface above the zero scale line of the Z coordinate, d is the depth from the zero scale line of the Z coordinate to the bottom of the water, U is the velocity for X-axis, V is the velocity for Y-axis, $\sqrt{G_{\eta\eta}}$ is conversion coefficient for X-axis and $\sqrt{G_{\xi\xi}}$ is conversion coefficient for Y-axis.

The three-dimensional convection-diffusion equation in the water quality module is as follows:

$$\frac{Cover\partial C}{\partial t}+v_x\frac{\partial C}{\partial x}-D_x\frac{\partial^2 C}{\partial x^2}+v_y\frac{\partial C}{\partial y}-D_y\frac{\partial^2 C}{\partial y^2}+v_z\frac{\partial C}{\partial z}-D_z\frac{\partial^2 C}{\partial z^2}=S+f_R(C,t),$$

where C is substance concentration, D is diffusion coefficient, S is the inflow term, $f_R(C,t)$ is the reaction term.

2.2. Model configurations

The model domain contains Sanmen Bay and its adjacent seas. The open ocean boundary is from 28°31'N to 29°26'N, the east part can extend to 122°27'E. Jiaojiang River runoff is considered at the open boundary. The measurements were conducted in the bay between the Nantian station and the Linhai station (Figure 2a).

2.3. Model validations

Compared with the measured values, the calculated values of the total tidal mean concentration in the whole sea area are also close, with an error of only 1.96%, indicating that the water quality model has successfully simulated the distribution of inorganic nitrogen concentration in Sanmen Bay.

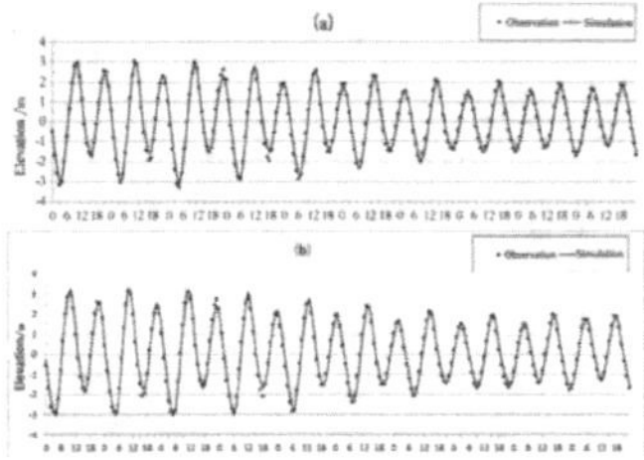

Figure 2. Tidal elevation in (a) Jiantiao station and (b) Gangdi station (2/12/2009 to 12/12/2009)

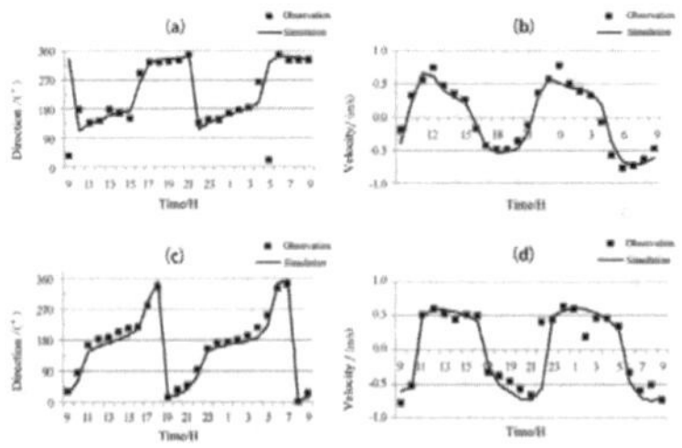

Figure 3. Current velocity and direction during spring tides and (c) stand for direction, (b) and (d) stand for velocity.

Table 1. Validation of COD

Station	Flood(model)	Ebb(medel)	Average(model)	observation	Error/(%)
S1	0.6777	0.7055	0.6916	0.717	-3.60
S2	0.6225	0.6749	0.6487	0.661	-1.91
S3	0.5673	0.6360	0.6016	0.640	-5.93
S4	0.5399	0.5942	0.5671	0.556	2.06
S5	0.5376	0.5483	0.5430	0.535	1.49
S6	0.6963	0.7078	0.7020	0.728	-3.51
S7	0.6539	0.6997	0.6768	0.630	7.43
S8	0.6105	0.6739	0.6422	0.583	10.25
S9	0.6986	0.7020	0.7003	0.660	6.06
S10	0.6865	0.6965	0.6915	0.594	16.51
Average	0.6432	0.6658	0.6545	0.642	1.96

2.4. Method to estimate the MEC

Given that velocity and diffusion coefficient is determined, a convection-diffusion equation can be regarded as a linear equation. As the equation satisfying the principle of superposition, the situation where balanced concentration fields formed together by more than one source of pollution is equivalent to the individually existing one. Formula are shown as follows:

$$C(x,y,z) = \sum_{i=1}^{m} C_i(x,y,z) \quad (1)$$

Where $C(x,y,z)$ means the concentration of each position (mg/L). $C_i(x,y,z)$ means the concentration of the ith position (mg/L).

Concentration field formed separately by each source can also be regarded as some times of the one formed when certain source strongly discharges pollution.

$$C_i(x,y,z) = Q_i \cdot \alpha_i(x,y,z) \quad (2)$$

where Q_i is the ith pollution source's emission; α_i is the response coefficient field of the ith pollution source, referring to one point's concentration under unit source intensity. α_i reflects the ith point's response level to the ith pollution source.

According to the response coefficient and controlling goals of each controlling point, to work out the environment capacity by the linear programming method. The main steps are as follows:

(1) After working out each source's response coefficient field, researchers should extract each controlling point's corresponding response coefficient value.

(2) Pollution source's permitted emission might correspond to the total residual emission maximum worked out by linear programming (LP) under the condition where controlling points meet the demand of water quality.

Taking Sanmen Bay for example, this paper studies total amount control and emission cutdown management. One of the central basis of management of the marine environment is environment capacity. Contaminants picked to be calculated should reflect the water quality, degree of contamination and operability on the management of environment capacity and contamination controlling and other aspects. Considering pollution source, water quality of Sanmen Bay and linking to control index of land-sourced pollutants, Sanmen Bay's prime pollutants are nutrient salt formation such as nitrogen, phosphorus, etc. COD is a comprehensive index to describe the degree of water pollution. According to relevant provision, CODs, (chemical oxygen demand) TN (total nitrogen), TP (total phosphorus) should be taken into consideration of calculation on environment capacity or cutdown. Based on investigations of Sanmen Bay and relevant provisions on seawater quality, this paper will calculate by controlling the content of CODMn, inorganic nitrogen and labile phosphate in seawater then gets a result. Finally, the research will make a conversion from the result to environment capacity or cutdown for COD, TN and TP.

3. Water exchange ability in the bay

Luff et al. (1996) introduced the concept of Half-life period, which is defined as the time required for the concentration of conservative substance to be diluted to half of the initial concentration by convection diffusion. The definition is based on the fact that it is almost impossible for the final concentration of a substance to be zero, and the rate of dilution represents the rate of water quality change, that is, the exchange capacity of the sea area.

Based on the concept of half exchange time, this study calculated the diffusion, transport, and dilution speed of conservative matter in each grid point of Sanmen Bay by using the transport and diffusion model of conservative materials, so as to study the water exchange capacity of Sanmen Bay.

Passive-tracer concentrations:

Based on the previous hydrodynamic model, water exchange model for the transport of regional passive-matter concentration is established (Figure 2c).

In the closed boundary condition, the current is 0 ($\frac{\partial C}{\partial n} = 0$), different treatment methods are adopted for the inflow and outflow of the model, and the inflow material concentration is 0, and the outflow material concentration is calculated by the model. The flow conditions are automatically obtained from the flow model.

Figure 5 shows the distribution of conservative materials at different times in three months. Note that the concentration value of conservative substances on each grid point not only represents its own concentration, but also is an important indicator of local water exchange degree. The period of water exchange rate reaching 95% is about 60 days.

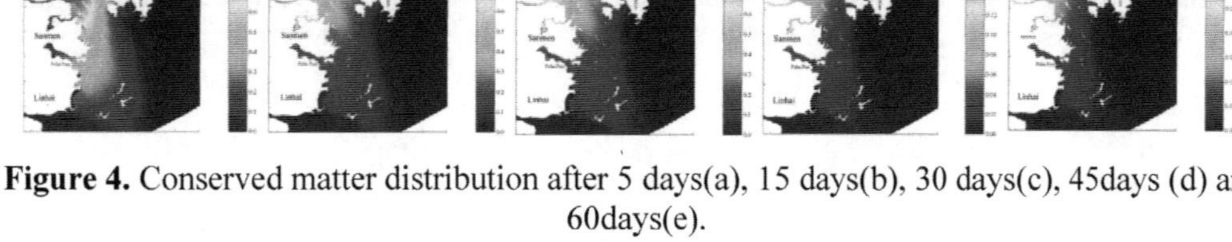

Figure 4. Conserved matter distribution after 5 days(a), 15 days(b), 30 days(c), 45days (d) and 60days(e).

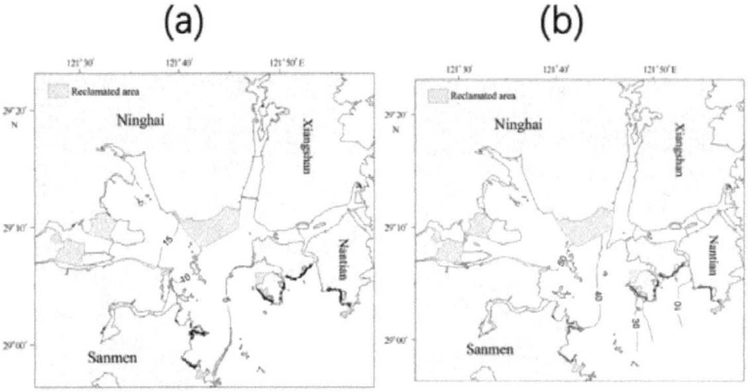

Figure 5. Water semi-exchange time (a) and 95% water exchange time (b).

4. Results

4.1. The response factor field of COD

According to the method of response factor, the response factor field of pollution source in each catchment unit needs to be calculated at first. The response factor field uses the same region and mesh as the pollution diffusion model. Initial condition was set as 0 to exclude other source strength's influence. The concentration field (also known as response factor field) is as the Table 2 shown.

Table 2. COD_{Mn} concentration field (mg/L)

Number	Bapu	Jiantiao	Haiyou	Qimen	Liyang	Baijiao	Shipu
1	0.0043	0.0035	0.0030	0.0028	0.0030	0.0019	0.0010
2	0.0011	0.0026	0.0025	0.0023	0.0025	0.0017	0.0010
3	0.0005	0.0015	0.0015	0.0015	0.0016	0.0012	0.0010
4	0.0002	0.0004	0.0005	0.0005	0.0005	0.0008	0.0012
5	0.0039	0.0046	0.0037	0.0037	0.0036	0.0021	0.0010
6	0.0010	0.0042	0.0053	0.0053	0.0054	0.0032	0.0013
7	0.0003	0.0010	0.0012	0.0012	0.0013	0.0034	0.0037
8	0.0010	0.0115	0.0097	0.0097	0.0096	0.0028	0.0012
9	0.0008	0.0040	0.0059	0.0059	0.0064	0.0053	0.0019
10	0.0007	0.0025	0.0029	0.0029	0.0029	0.0157	0.0039
11	0.0008	0.0057	0.0239	0.0239	0.0150	0.0029	0.0012
12	0.0008	0.0052	0.0108	0.0108	0.0196	0.0037	0.0014
13	0.0005	0.0020	0.0024	0.0024	0.0025	0.0281	0.0039
14	0.0005	0.0020	0.0023	0.0023	0.0024	0.0114	0.0069
15	0.0004	0.0016	0.0019	0.0019	0.0020	0.0077	0.0076
16	0.0007	0.0052	0.0273	0.0273	0.0140	0.0027	0.0011
17	0.0007	0.0051	0.0118	0.0118	0.0198	0.0028	0.0011
18	0.0004	0.0019	0.0022	0.0022	0.0023	0.0380	0.0035

4.2. Environmental capacity of COD_{Mn}

To figure out COD_{Mn} permitted emission at each catchment according to Linear Programming, This study took the maximum concentration during a whole tide as background concentration to estimate the capacity. Parameters about the permitted emission of COD_{Mn} is shown in Table 3.

Table 3. Parameter on COD_{Mn} (mg/L)

Number	C_{0i}	C_{si}	Number	C_{0i}	C_{si}
1	0.952	3	10	0.827	3
2	0.944	3	11	0.852	3
3	0.921	3	12	0.844	3
4	0.882	3	13	0.815	3
5	0.892	3	14	0.820	5
6	0.857	3	15	0.830	5
7	0.852	3	16	0.877	3
8	0.842	5	17	0.849	3
9	0.837	3	18	0.811	3

5. Conclusion

A calibrated two-dimensional hydrodynamic model was built and fully validated to study the environmental characteristics of Sanmen Bay, including the tides, the residual currents, the tidal prism, and water exchange abilities.

The concentrations of COD, inorganic nitrogen, and acid salt in Sanmen Bay showed a trend of being higher in the inner estuary and lower outside of the bay, and was higher in the western part and lower in the eastern part. The concentration of COD was lower than 0.60 mg/L in most areas of the eastern part of the bay, while was higher than 0.65 mg/L in the western part of the bay. The concentration of inorganic nitrogen was more than 0.70 mg/L near the west coast. The concentration of acid salt was lower in the outer bay (less than 0.045 mg/L), while was higher in the inner bay (greater than 0.05 mg/L).

References

[1] Burn D H, Lence B. Comparison of optimization formulations for waste load allocation. Environmental Engineering [J], 1992, 119(4): 597 - 612.
[2] Cardwell H, Eills H. Stochastic dynamic programming models for water quality management. Water Resources Research [J], 1993, 29(40): 803 - 813.
[3] Sasikumar K, Mujumdar PP. Fuzzy optimization model for water quality management of a river system. Journal of Water Resources Planning and Management - ASCE, 1998, 124(2): 79 - 88.
[4] Stebbing, A. R. D. Environmental Capacity and the Precautionary Principle. Marine Pollution Bulletin [J], 1992, 24(6): 287 - 295.
[5] Subhankar Karmakar, P. P. Mujumdar. A two - phase grey fuzzy optimization approach for water quality management of a river system. Advances in Water Resources [J], 2007, (1): 1218 - 1235.
[6] Zhang X, Sun Y. Study on the environmental capacity in Jiaozhou Bay. Marine Environmental Science [J], 2007, 26(4): 347 - 359.
[7] Yu L, Liu S, Wang M. Research on SMS-Based Water Environment Capacity. Henan Science [J], 2006, 24(6): 874 - 876.
[8] Wang H, Pang Y, Ding L. Calculation of the water environment capacity for a waterfront body [J]. Acta Science Circumstantiae [J], 27(12): 2067 - 2073.
[9] Niu Z, Zhang H. Study on estimation of coastal water environmental capacity by geostatistics and GIS. Journal of Tianjin Polytechnic University [J], 2006, 25(1): 74 - 80.
[10] Liu H, Yin B. Numerical Calculation on the Content of Nitrogen, Phosphate and COD in the Liaodong Bay. Marine Science Bulletin [J], 2006, 25(2): 46 - 54.
[11] Su S, Li H, Xia J. Dapeng Bay water environment capacity analysis on the base of Delft 3D model. Research of Environmental Sciences [J], 2005, 18(5): 91 - 95.
[12] Guo L, Jiang W, Li F, Wang X. Environmental capacity calculation of COD and PHs in the Bohai Sea, Periodical of ocean university of China [J], 2007, 37(2): 310 - 316.

Numerical simulation of saturated oil process based on digital pore model

Haibo Liu*

Exploitation and Development Research Institute, PetroChina Daqing Oilfield Company; Heilongjiang Daqing, China.

*Corresponding author e-mail: zhenxinhaibo@126.com

Abstract. In order to study the saturated oil process in the digital pore model, numerical simulations are performed by building a model. The results show that the direction of the capillary force is judged according to whether there is an odd or even number of fluid segments in the throat, and the odd segments cancel each other out, while the even segments have a capillary force; Three cases need to be considered when calculating the drag coefficient: (1) only single-phase fluid exists in the throat; (2) the corner of the throat is in the wetted phase, while the center of the throat is in the non-wetted phase; (3) the corner is in the wetted phase and the center of the throat is a multi-phase fluid.After performing saturated oil simulation calculations, the digital pore model can visualize the saturated oil process and results, which is realistic, fast, accurate and low cost, and is of great significance to the subsequent research on efficient reservoir development methods.

1. Introduction

With the development and progress of the computer industry, numerical simulation technology has become the main tool for solving complex problems in petroleum engineering. This discipline can be applied to various disciplines because it has the advantages of cost saving and short work cycle compared with traditional physical simulation experiments, so research scholars will construct reservoir structure pore models with the help of numerical simulation methods to numerically simulate saturated oil and oil drive processes. In this paper, based on the theory of fluid percolation in porous media, the direction of the capillary force under the condition of odd or even number of fluid segments is considered, the resistance coefficient in the throat is calculated, the oil saturation of the model is calculated, and the saturation process of the digital pore model is visualized, this technique can provide strong support for improving the recovery of crude oil and tapping the micro residual oil[1-3].

2. Digital pore modeling

The process simulation for saturated oil uses a quasi-static model and incorporates percolation theory, with the left side of the pore model in order of repulsion, with the inlet on the left and the outlet on the right.

2.1. Determination of the direction of the capillary force

In Figure 1, the dark blue color in the throat represents water, and the dark yellow color represents oil. In the picture on the left, the throat is hydrophilic in nature, and dividing the fluid in the throat into five sections, there will be four oil-water interfaces, which means there are four capillary forces present in the throat, and since the inner tangent radius characterizing the shape of the throat varies according to the cosine formula, there is not much difference between these four capillary forces in the throat.The direction pointing to the concave surface of the oil-water interface is the direction of the capillary force, so it can be seen in Figure 1 that the four capillary force directions in the throat have two directions pointing to the left and two directions pointing to the right, and can cancel each other out. And in the right figure, there are three capillary forces in total, one pair can cancel each other, and the remaining one is the capillary force that the whole throat channel is subjected to. From this, it is easy to find that in the digital pore model constructed in this paper, the capillary forces in the throat channel are related to the number of segments of the fluid. If there is an odd or even number of fluid segments in the throat, then the former will have an even number of oil-water interfaces and the latter an odd number, and the capillary forces in the throat with the odd number of segments will cancel each other out, while the throat with the even number of segments will have a capillary force that is not canceled out [4-6].

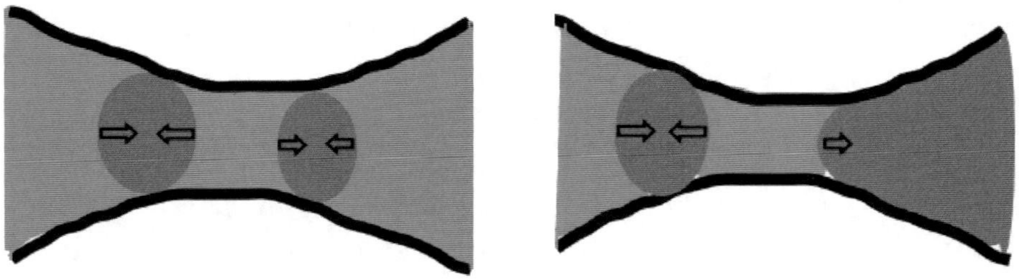

Figure 1. Diagram of the direction of the throat capillary force

2.2. Calculation of resistance coefficient

Darcy's law is that when a fluid in a reservoir is subjected to a certain pressure, the fluid overcomes the resistance to produce flow. Since there are both wetted and non-wetted phase fluids in the throat, the next three cases in this paper are described for the calculation of the drag coefficient:

(1) Only single-phase fluids are present in the throat

If the wetting phase of the fluid in the center of the throat channel is the same as the wetting phase in the corner, only single-phase flow exists throughout the throat channel. The conductivity coefficient of the throat channel can be calculated by the modified Poiseuille's law, as shown in the following equation:

$$g = \frac{\pi}{128}\left(\sqrt{\frac{A_t}{\pi}} + r_t\right)^4 \tag{1}$$

In the formula: g-perveance, μm^4; A_t-Cross sectional area of the whole throat, μm^2; rt-Throat radius, μm.

$$A_t = r_t^{\,2} \sum_{i=1}^{n} \cot\alpha_i = C r_t^{\,2} \tag{2}$$

In the formula: α_i-half angle of the inner angle of the cross-sectional shape of the throat, rad; n-Number of sides of the throat section; C-Area constant of the throat cross section.

Combined equations (1) and (2) calculate the resistance coefficient of the throat:

$$R = \frac{u_f 128}{\pi} \frac{1}{\left(\sqrt{\frac{C}{\pi}} + 1\right)^4} \frac{x}{r_t^{\,4}} \tag{3}$$

In the formula: R-resistance coefficient of the throat channel, μ_f-viscosity of the fluid; x-Length of throat channel.

(2) The corner of the laryngeal tract is in the wetting phase, while the center of the laryngeal tract is in the non-wetting phase

The corner of the throat is in the wetted phase, while the center of the throat is in the non-wetted phase. This situation means that the corner of the throat is filled with fluid, while there is only a section of fluid at the center of the throat that is completely filled by the non-wetted phase, in which case the area of the corner within the cross-section of the throat is obtained in order to obtain the drag coefficient. The area of the corner in the polygon can be calculated by the following equation:

$$A_{ci} = r^2\left[\cos\theta\ (\cot\alpha_i\cos\theta - \sin\theta) + \theta + \alpha_i - \frac{\pi}{2}\right] \tag{4}$$

In the formula: A_{ci}-the area of the corner of an interior angle of a polygon, μm^2; r-the radius of curvature at this inside corner, μm; θ-oil-water contact angle, rad.

Then the corner area and the area of the center of the throat can be found as follows:

$$A_{con} = \sum_{i=1}^{n} A_{ci} \tag{5}$$

In the formula: A_{con}-total area at the corner of the cross section of the throat, μm^2.

$$A_{cen} = A_t - A_{con} \tag{6}$$

In the formula: A_{cen}-area at the center of the cross-section of the throat, μm^2.

The fluid conduction coefficient at the center of the throat is obtained by substituting Acen for At in equation (1):

$$g_{cen} = \frac{\pi}{128}\left(\sqrt{\frac{A_{ten}}{\pi}} + r_t\right)^4 \tag{7}$$

In the formula: g_{cen}-inflow coefficient at the center of the throat channel.

And the inflow coefficient at the corner is related to θ and ai, when $\theta+\alpha_i<\frac{\pi}{2}$:

$$g_{cor} = \sum_{i=1}^{n} \frac{A_{ci}^{\ 2}(1-\sin\alpha_i)^2(\varphi_2\cos\theta-\varphi_1)\varphi_3^{\ 2}}{12\sin^2\alpha(1-\varphi_3)^2(\varphi_2+\varphi_1)^2} \tag{8}$$

$$\varphi_1 = \frac{\pi}{2} - \alpha_i - \theta \tag{9}$$

$$\varphi_2 = \cot\alpha_i\cos\theta - \sin\theta \tag{10}$$

$$\varphi_3 = \left(\frac{\pi}{2} - \alpha_i\right)\tan\alpha_i \tag{11}$$

When $\theta+\alpha_i>\frac{\pi}{2}$:

$$g_{cen} = \frac{\pi}{128}\left(\sqrt{\frac{A_{ten}}{\pi}} + r_t\right)^4 \tag{12}$$

$$g_{cor} = \sum_{i=1}^{n} \frac{A_{ci}^{\ 2}\tan\alpha(1-\sin\alpha_i)^2\varphi_3^{\ 2}}{12\sin^2\alpha(1-\varphi_3)(1+\varphi_3)^2} \tag{13}$$

After obtaining the coefficient of conduction, bring the corresponding drag coefficient into equation (14):

$$R = \frac{\mu_f x}{g} \tag{14}$$

After finding the resistance coefficients for both cases based on the above equation, the resistance at the corner and the fluid resistance through the center of the throat are simplified to parallel resistance based on the hydroelectric similarity principle to calculate the resistance of the entire throat.

(3) Wetting phase in the corner, multiphase fluid in the center of the throat

When the corner of the throat channel is filled with the wetted phase, there are multiple segments of bulk flow present in the throat channel. This situation can be obtained by equation (7) and equation (14) to find the drag coefficient of the fluid present in each segment in the center of the throat channel. If you want to find the resistance of the entire throat, it is simplified to connect multiple sections of fluid in the center of the throat in series, and then in parallel with the fluid at the corner of the throat, as shown in Figure 2.There are two sections of fluid in the throat, the resistance of the first section of fluid is R_{cen1}, the resistance of the second section of fluid is R_{cen2}, and the resistance of fluid in the corner is R_{cor}, and the resistance of the whole throat is calculated by equation (15).

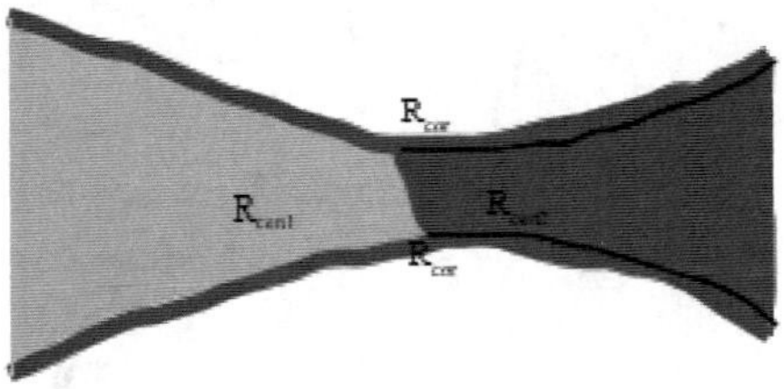

Figure 2. Schematic diagram of a throat with corner and two central fluid sections

$$\frac{1}{R}=\frac{1}{R_{cor}}+\frac{1}{R_{cen1}+R_{cen2}} \tag{15}$$

2.3. *Calculation of oil content saturation*

The saturated oil simulation process in this paper uses static repulsion, and at each time step in the model, the water in the corresponding throat and the pore connected to the throat will be repelled, indicating that the pore model is saturated, and if the throat is oil wet, then:

$$v_{(o,\ i)}=v_{(o,\ i-1)}+A_t\times l_i \tag{16}$$

In the formula: $v\ (0,\ i\text{-}1)$-volume of oil in the model at the previous step length; $v(0,\ i)$-volume of oil contained in this step model; l_i-the length of the saturated throat in the step, if it is water-wet:

$$v_{(o,i)}=v_{(o,\ i-1)}+A_{cen}\times l_i \tag{17}$$

There are also corresponding pore spaces saturated with oil:

$$v_{(o,i)}=v_{(o,\ i-1)}+v_{pi} \tag{18}$$

Formual: v_{pi}-the step is saturated with the pore volume of the pore.

The oil saturation in the pore model is calculated using equation (19):

$$S_o=\frac{v_{(o,i)}}{\sum_{i=1}^{m}V_{pi}+\sum_{j=1}^{k}V_{tj}} \tag{19}$$

2.4. *Model visualization at the end of saturated oil simulation*

After the simulation of the saturated oil process is completed, the oil and water distribution in the pore model is output as text by VB program, and then directed to 3Dmax software for final output imaging, Figure 3 output model in red represents oil and blue represents water, you can observe the front view and side view.

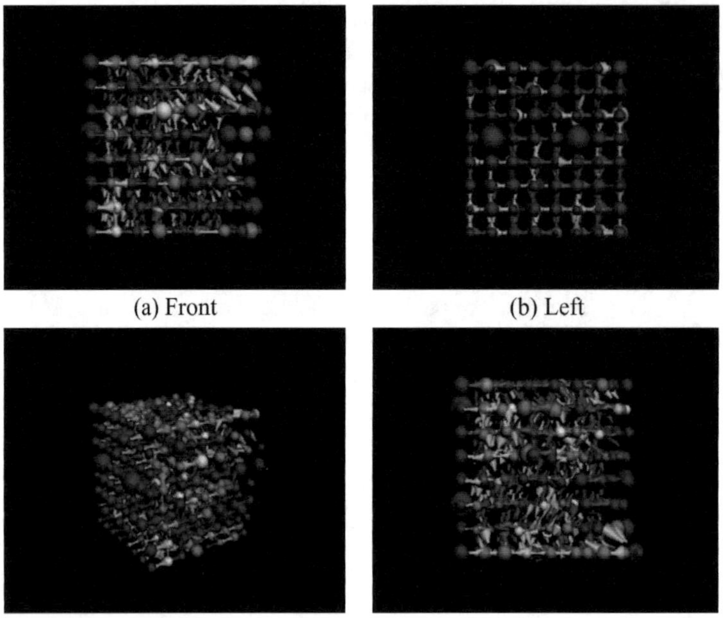

(a) Front (b) Left (c) Side (d) Forward

Figure 3. Digital pore model saturated oil

3. Conclusion

(1) In the process of building the digital pore model, when judging the direction of the capillary force, there exists an odd number of fluid numbers in the throat between the capillary action force to cancel out, even number of fluid numbers in the throat will exist a capillary force, this capillary force is not canceled out.

(2) Calculation of the drag coefficient needs to consider three situations:①only single-phase fluid in the throat; ②throat corner for the wetting phase, while the center of the throat in the non-wetting phase; ③corner for the wetting phase, the center of the throat for multiphase fluid.

(3) Construct a digital pore model, and through simulation calculations, the 3DMax software can be used to visualize the model saturated oil process.

References

[1] Sui Wei, Cheng Si Calculation method of absolute permeability of sandstone digital core based on convolution neural network [J] Oil and gas geology and recovery, 2022,29 (01): 128-136 .

[2] Xu Jing, Wang Pengfei, Li Junfei, LV Zuobin, Gao Zhennan Study on micro seepage mechanism of fractured metamorphic rock reservoir based on digital core-Taking buried hill of JZ oilfield as an example [J] Journal of Shaanxi University of science and technology, 2021,39 (04): 96-102.

[3] Fan Yufei, pan Baozhi, Guo Yuhang, Zhang Lihua Evaluation of conductive model of clay bearing sandstone by digital core technology [J] Journal of Jilin University (Earth Science Edition), 2021,51 (03): 919-926.

[4] Li Qian, Zhang Yuxiang, Li Tao, Yang shenglai, Hu Die, Wu Tingting, Wang Bei, Li Qiu A method for evaluating the complete pore throat structure of carbonate rocks based on digital core-Taking Qixia Formation in Northwest Sichuan as an example [J] Oil and gas geology and recovery, 2021,28 (03): 53-61.

[5] Liu Wenmo, Zhang Wenwei, Liu songmo, Liu Bolai Study on seepage channel of core slice based on lattice Boltzmann method [J] Journal of petrochemical colleges and universities, 2020,33 (06): 43-48.

[6] Li Bo, Han Tongcheng, Fu Liyun Study on dielectric properties of fractured reservoir sandstone based on digital core [J] Journal of Geophysics, 2020,63 (12): 4578-4591.
[7] Zhao Ling, Shi Xue, Xia Huifen Construction method of digital core pore network model [J] Science, technology and engineering, 2018,18 (26): 32-38.

Experimental Study on Fluid Elastic Instability of Concentric Circular Arrangement Heat Exchanger Tubes

Mengmeng Liu [1], Donghui Zhang [2,*], Guangdong Song [1]

[1] China Institute of Atomic Energy, Beijing 102413, China
[2] CNNP Xiapu Nuclear Power Co., Ltd, Fujian, 355199, China

*Corresponding author e-mail: zhangdh03@cnnp.com.cn

Abstract. In the industrial production, there exists the phenomenon of structure vibration caused by flow, which will cause the deformation and failure of the structure and lead to the economic loss, and even cause the industrial accident and great danger. In this paper, the one mechanism of fluid-induced vibration, namely fluid elastic instability, of the concentric circle arrangement tube bundles in the sodium-sodium heat exchanger is studied experimentally. According to different central angle positions, three kinds of tubular plates are designed, which are approximately triangle arrangement, approximately square arrangement and transitional arrangement respectively. In the small-size water tunnel non-contact graphic measurement is adopted to obtain the critical flow rate Ucr when the tube bundle is unstable. The experimental results show that the flow induced vibration resistance of the transitional arrangement is better, and the critical flow velocity is about 3.78.

1. Introduction

Flow-induced vibration is a very complex physical phenomenon, which involves structural mechanics, material mechanics and fluid mechanics. At present, it is generally believed that the mechanisms of fluid-induced bundle vibration mainly include vortex shedding, turbulence buffering, fluid elastic instability and acoustic resonance [1-3]. In heat exchangers, fluid elastic instability is an important cause of fluid-induced vibration. Mitra [4] used fully flexible pipe and single flexible pipe to conduct the fluid-induced vibration experiment of square arranged bundle, and found that the fully flexible pipe is more prone to fluid elastic instability than the single flexible pipe in both single-phase flow and two-phase flow. Ricciardi et al. [5] used air-water to study the elastic instability of two-phase fluid-induced bundle, and found that the equilateral triangle arrangement was more stable than the angular equilateral triangle arrangement. it is very important to obtain the critical velocity for the analysis of fluid elastic instability.

In this paper, the flow induced vibration of the heat exchange tubes and concentric tube bundles of the sodium-sodium heat exchanger in the China prototype fast reactor is analyzed. The heat exchanger tube bundles of sodium-sodium heat exchanger are arranged in concentric circles. Due to the particularity of their arrangement, there is no relevant standard to calculate the critical velocity of the tube bundles in this arrangement. In this experiment, through the water tunnel test of the experimental body in the straight pipe segment, combined with high-speed photography technology measurement, the vibration characteristics of the concentric circular arrangement bundle at different flow rates were obtained, and finally the critical flow rate Urcwere obtained. The experiment is to take apart concentric arrangement of tube bundle, designing and manufacturing three arrangement forms of tube plate, respectively approximate triangle configuration, the approximate square arrangement and transitional arrangement of tube plate, the three tube plate in small size tunnel, non-contact measurement and high-speed photography, to obtain bundle buckling critical velocity.

2. Experiment of the critical flow velocity measurement

2.1. Experimental apparatus and process

The experimental apparatus in this study is shown in Fig. 1. High-speed photography is used to photograph the vibration of the tube bundle in the visual window. The function of the high-frequency lamp is to improve the contrast between the elastic tube and the rigid tube in the shooting image.

The water tunnel test apparatus in this study can be used to study the fluid-induced vibration of the bundle under the condition of single-phase flow. The specific test steps are shown in Fig. 2.

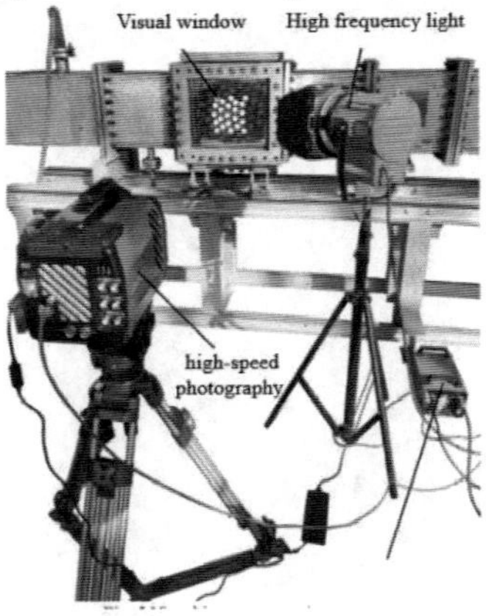

Fig. 1 Small size water tunnel test apparatus.

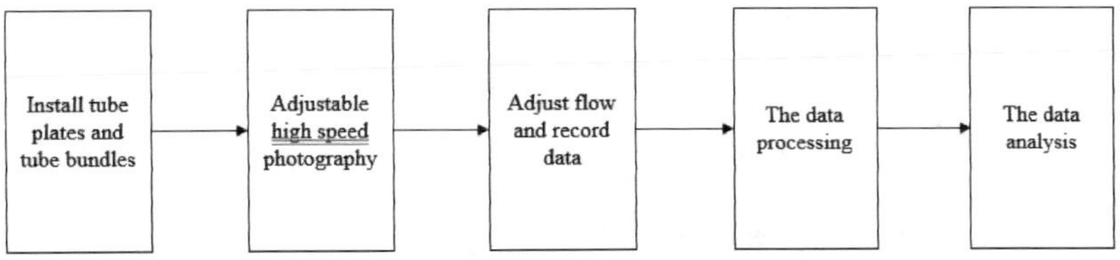

Fig. 2 Flow chart of water tunnel test

Captions should be typed in 9-point Times. They should be centred above the tables and flush left beneath the figures.

The water tunnel test can be conducted to study the fluid-induced vibration of the straight section of the concentric circular tube bank. The whole test device can be divided into three parts: the circulation system, the main platform and the data acquisition system. The circulation system includes the water tank, the loop, the centrifugal pump, the main and bypass regulating valves, the pressure gauge, the flow meter and the main platform. Among them, the flow in the pipeline can be adjusted by adjusting the valve in the main road and the valve in the bypass. The flow regulation range is 0-180m3/h. Water under normal temperature is used as the experimental medium in the circulating pipeline. The main platform consists of three parts: the steady flow section, the test section and the wake section. The overall shape is a rectangle with a width of 0.255m and a height of 0.330m. Among them, the length of the steady flow section is 5m, and 5 orifice plates whose aperture gradually becomes fine are set at equal intervals in the steady flow section, so as to ensure the uniform incoming flow of the test section tube bundle and reduce the irregular pulsation impact on the tube bundle. Formed by high-speed photography and computer data acquisition system, the high-speed photography Camera (PCO. dimax S1) produced by the German PCO company, the maximum resolution is 1008×1000, maximum frame rate is 152811 fps. In this experiment, the frequency of 400 frames/second is used to capture the vibration of the bundle, and software Camera Pco is installed on the computer to make relevant Settings, then the connection between the computer and the high-speed Camera can be completed. The video of the vibration of the bundles captured by the high-speed Camera is recorded by the computer, and the vibration characteristics of the bundles are obtained through data processing.

In order to ensure that the identification of elastic tubes photographed during the experiment is high enough, the top of all elastic tubes is coated with white waterproof paint before filling with water, and the top of rigid tubes is coated with red waterproof paint. The image of tube bundles captured by adjusting high-frequency lights is shown in Fig.3. As can be seen from the figure, the contrast between the elastic tube and the rigid tube is very high in the images and photos taken. The elastic tube is very clear, while the rigid tube is almost invisible. The higher the contrast, the more accurate the experimental data will be.

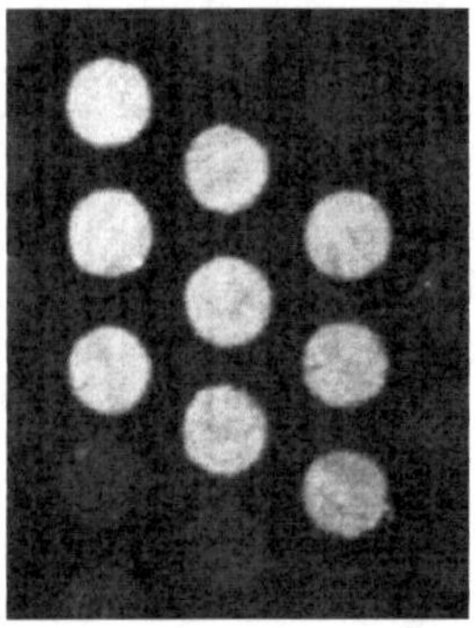

Fig. 3 Elastic tube high-speed photography captures images

After the installation of tube plate and tube bundle as well as the debugging of high-speed photography, water can be opened and filled, and the test data can be recorded. The main steps are as follows:

(1) Open the exhaust valve in the wake section and close the drainage valve. At the same time, use the operating table to open the centrifugal pump, open the main road and bypass valve, exhaust the air in the device.

(2) Close the exhaust valve after exhausting the air in the device.

(3) Adjust the flow rate. After the flowmeter on the operating table is stable, a camera will be used to shoot the bundle for 10s and store it as a video file.

(4) After recording, increase the flow rate and repeat the operation in step (3) until violent vibration and even collision of the bundle can be observed in the images taken.

(5) Close the centrifugal pump, open the drain valve and exhaust valve, and close the high-frequency lamp and camera.

(6) After the water is drained out of the device, repeat the experiment with the tube plate.

2.2. Experimental tube plate and tube bundle design

In this experiment, water tunnel tests were carried out on concentric circular tube bundles at three different positions. The tube plates are made of 304 stainless steel. Fig. 4 is the plan of the water tunnel test tube plate of the concentric tube bundle. The blue part in the center is the location of the elastic tube, and all the tubes are distributed in the small rectangle in the middle. The size of the small rectangle is consistent with the size of the visual window on the runner, so the vibration of the bundle can be clearly observed during the test. The round hole on the bundle is a 10mm diameter through hole.

In the experiment, different positions of concentric circular tube bundles were selected as the research objects. Among them, there are 9 elastic tubes, the circumferential spacing of each layer of tubes is 23.1mm, and the radial spacing of each concentric circle is 22mm. It can be seen from Fig. 4 that the tubes in the tube plate are approximately triangular in arrangement, approximately square in arrangement and transitional in arrangement.

In this experiment, two different kinds of tube bundles were used, rigid tubes and elastic tubes. The material of the tube bundles was 6063 aluminum alloy. The density of the material is 2.69g/cm3 and the elastic modulus is 69GPa. The reason why 6063 aluminum alloy is used as tube bundle material is that the texture of 6063 aluminum alloy is soft, and the low order mode frequency of the tube bundle with the same size is lower than that of stainless steel and cast iron, which is convenient for experiments. At the same time, the density is small, easy to install and fix, and 6063 aluminum alloy has a certain corrosion resistance, can meet the requirements of the experiment.

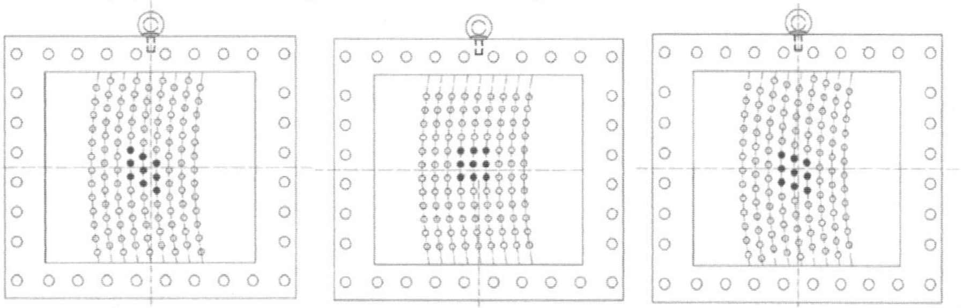

(a) (b) (c)

Fig. 4 Water tunnel test tube bundles with concentric rings: (a) tube bundles No. 1 (Approximately triangular arrangement) (b) tube bundles No. 2 (Approximately square arrangement) (c) tube bundles No. 3 (Transitional arrangement)

Fig. 5 is a schematic diagram of an elastic tube and a rigid tube. As shown in Fig. 5, the radius of the rigid tube along the axial direction is unchanged, while the elastic tube has two sections along the axial direction, namely the thick axis and the thin axis. In this study, the diameter of the elastic pipe is 16mm, which is the same as the diameter of the rigid pipe. The length of the elastic pipe is 235mm, the diameter of the fine pipe is 8mm, and the length is 110mm. The tail is all composed of a 10mm fine rod and M10 thread. The commercial software ANSYS Modal is used for Modal calculation of the elastic tube, as shown in Fig. 6, and the first-order natural frequency of the elastic tube is 24.124Hz.

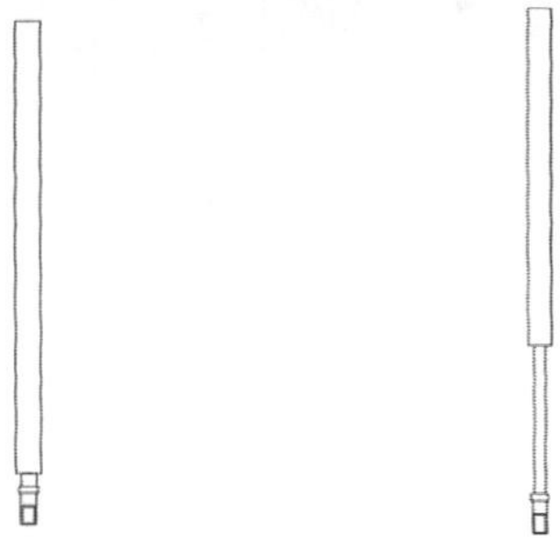

(a) (b)

Fig. 5 Elastic tube and rigid tube: (a) rigid tube and (b) elastic tube

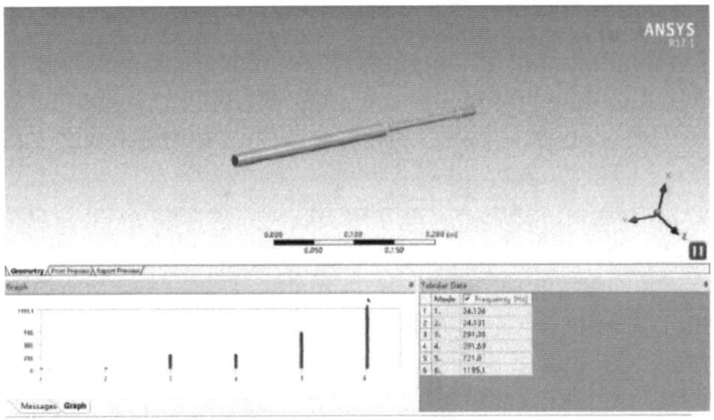

Fig. 6 Modal analysis and calculation of elastic pipe

2.3. Experimental data processing

(1) Video files can be captured by high-speed camera. The picture of bundle vibration can be obtained frame by frame by Camware software.

(2) The self-compiled software was used to process the picture obtained in step (1), identify the position of the bundle in the picture, and thus obtain the bundle displacement data. The picture is processed with the self-made software, as shown in Fig. 7.

(3) Matlab software was used to process the bundle displacement data obtained in step (2) to obtain the bundle's root-mean-square amplitude, trajectory and other data under experimental conditions.

(4) Origin software was used to draw root mean square amplitude-velocity images and spectral maps to obtain the vibration of the bundles in the transverse direction and the clockwise direction.

(5) By further analyzing the data, the law of the influence of nodal diameter ratio, two-direction stiffness ratio and bundle arrangement on the instability of stray bullets was obtained.

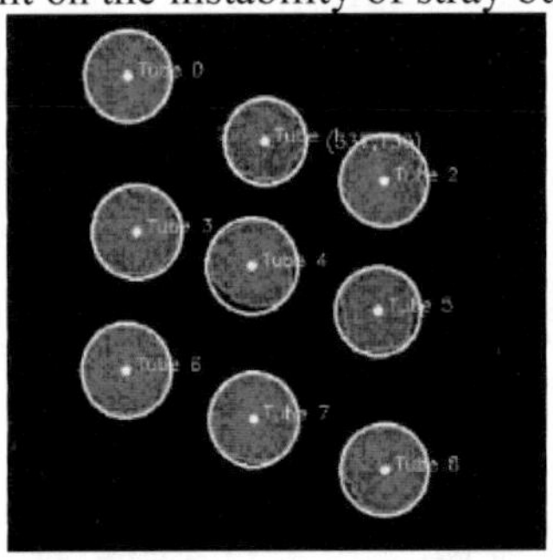

Fig. 7 Image processing of bundle vibration

2.4. Experimental parameter setting

2.4.1. The incoming flow velocity. In the water tunnel experiment, it is not possible to directly measure the flow rate through the tube bundles. We can only get the flow rate through the whole pipe through the flow meter in the pipe. To facilitate the analysis of the relationship between the vibration of the bundle and the flow rate, the flow rate needs to be converted to obtain the flow rate, as shown in Eq. (1).

$$V=\frac{G}{A} \tag{1}$$

Where V¬ is the incoming flow velocity, m/s, G is the flow data recorded in the experiment, m^3/h, A is the cross-sectional area of the test segment, m^2 (In this experiment, the test section is a rectangle with a width of 0.255m and a height of 0.330m. A=0.08415m^2).

2.4.2. Calculation of interstice flow velocity of concentric circle bundles. The common tubular array includes square array, Angle square array, equilateral triangle array, Angle triangle array. The flow velocity through the gap of the tube bundle will change due to the different array arrangement. In the common arrangement of bundles, some scholars have derived the corresponding formula for conversion between incoming flow velocity and clearance velocity. For the concentric tube bundles in this experiment, there is no appropriate formula for the conversion of incoming flow velocity and clearance velocity. Moreover, the arrangement of the tube bundles in concentric rings will change at different central angles, so it is unreasonable to use the same formula to calculate clearance velocity.

For the arrangement of the three bundles in this experiment, the equal flow rate method was used to calculate the clearance velocity. The flow rate is equal when the water does not pass through the bundle area and when it passes through the bundle area. The clearance velocity can be obtained by calculating the cross-sectional area of the bundle area. As shown in Eq. (2):

$$V_p = \frac{G}{A_0} \tag{2}$$

Where V_p is the clearance velocity, m/s, A0 is cross-sectional area of passable bundle area, m^2.
The calculation method of A0 is shown in Eq. (3):

$$A_0 = A - A_T \tag{3}$$

Where A_0 is cross-sectional area of passable bundle area, m^2 (In the experiment, take the sum of the cross-sectional areas of the first column of pipes).

3. Results and discussions

3.1. Analysis of water tunnel test results of No.1 tube bundle

The 9 elastic tubes in Fig. 4(a) were named, as shown in Fig. 8.

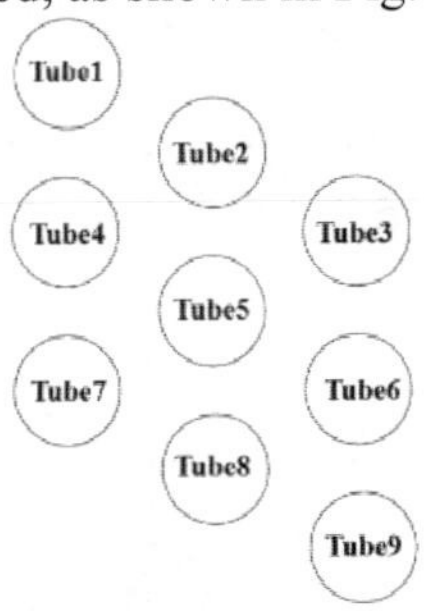

Fig. 8 No.1 Tube bundle naming method

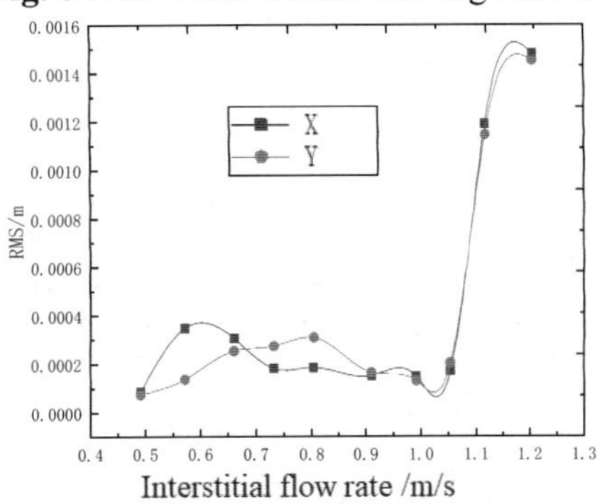

Fig. 9 Diagram of root mean square amplitude of tube bundle of No.1 tube-plate and clearance velocity

3.1.1. Results of bundle root mean square amplitude analysis. Through data processing, the relationship between root mean square amplitude and clearance velocity of 9 elastic tubes in No.1 tube bundle is obtained.

In the experiment, Tube 8 has the largest amplitude and the most obvious vibration. Fig. 9 shows the relationship between root mean square amplitude and clearance velocity of Tube 8 along X (downstream direction) and Y (cross-flow direction) directions. As can be seen from Fig. 9, when the clearance velocity of the tube bundle is 1.05m/s, the root mean square amplitude of the tube along the X and Y directions will change suddenly. Therefore, the critical velocity of the tube bundle in the arrangement mode of tube bundle No.1 is about 1.1m/s. Before the instability of the bundle, the amplitude of the tube first increased and then decreased, which was caused by the vortex shedding between the bundles. With the increase of velocity, vortex shedding disappears and bundle amplitude decreases.

3.1.2. Bundle spectrum analysis. It can be observed from Fig. 9 that the critical velocity of the bundle is about 1.1m/s. In order to further determine the magnitude of the critical velocity, the spectrum analysis of Tube 8 is performed. Take the data at five clearance velocities of 0.91m/s, 0.99m/s, 1.05m/s, 1.11m/s and 1.20m/s to draw the spectrum diagram.

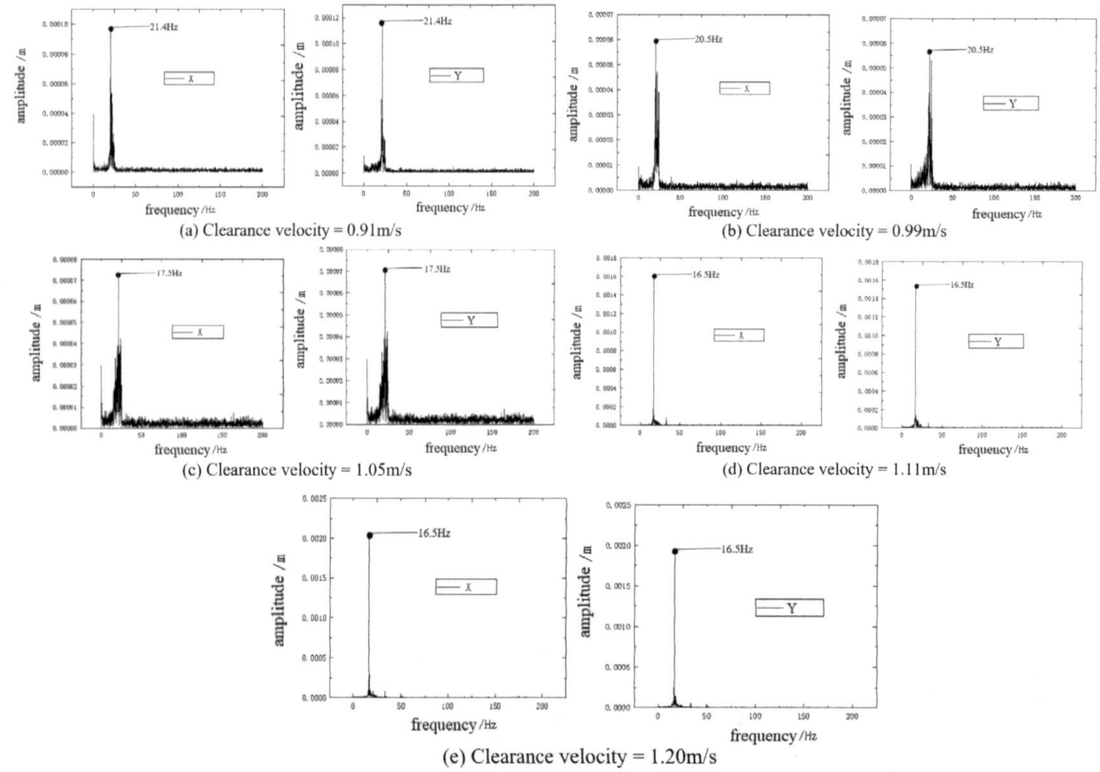

Fig. 10 Tube 8 spectrum diagram of Tube bundle No. 1

As can be seen from Fig. 10, when the clearance flow velocity was 0.91m/s, 0.99m/s and 1.05m/s, the frequency value corresponding to the maximum amplitude in the frequency graph decreased with the increase of the flow velocity, and multiple peaks would appear. When the clearance velocity is greater than 1.11m/s, the frequency value corresponding to the maximum amplitude will not change, and only one peak appears in the spectrum diagram. Therefore, it can be verified that the critical velocity of the tube bundle in tube bundle No. 1 is 1.11m/s. The velocity is dimensionless, that is $U_{cr} = v/Df$ (the same below), and the critical velocity Ucr=2.88 is obtained.

3.2. Analysis of water tunnel test results of No.2 tube bundle
The 9 elastic tubes in Fig. 4 (b) were named, as shown in Fig. 11.

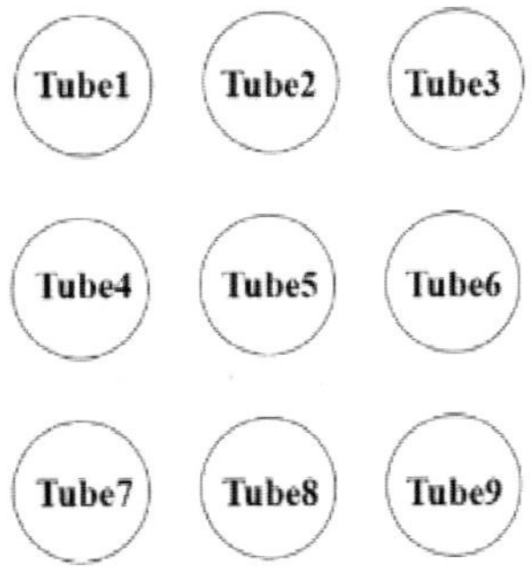

Fig. 11 No.2 Tube bundle naming method

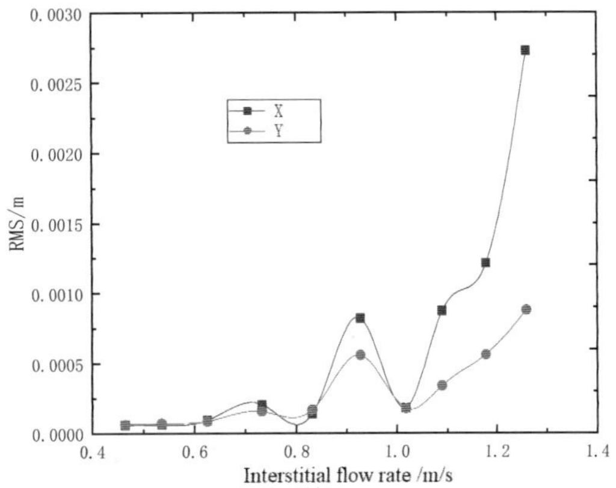

Fig. 12 Diagram of root mean square amplitude of tube bundle of No.2 tube-plate and clearance velocity

3.2.1. Results of bundle root mean square amplitude analysis. Through data processing, the relationship between root mean square amplitude and clearance velocity of 9 elastic tubes in tube bundle No. 2 was obtained.

The data results of Tube 5 are shown in Fig. 12. In the figure, the relationship between root mean square amplitude and clearance velocity of Tube 5 in the direction of X (downstream direction) and Y (transverse direction) is shown. As can be seen from the figure, when the clearance velocity of the tube bundle is between 1.0m/s and 1.1m/s, the root-mean-square amplitude of the tube along the X and Y directions will change suddenly. Therefore, the critical flow rate of the tube bundle in the no.2 tube-plate arrangement is between 1.0m/s and 1.1m/s. When the clearance velocity was 0.8m/s, vortex shedding occurred between the bundles, leading to increased amplitude of the bundles. With the increase of clearance velocity, vortex shedding disappears and bundle amplitude decreases.

3.2.2. Bundle spectrum analysis. It can be observed from Fig. 12 that the critical flow rate of No.2 tube bundle is between 1.0m/s and 1.1m/s. In order to further determine the magnitude of the critical flow rate, Tube 5 was analyzed by spectral analysis. The spectrum diagram was drawn from the data recorded in the experiment at five clearance flow rates of 0.93m/s, 1.02m/s, 1.09m/s, 1.18m/s and 1.26m/s.

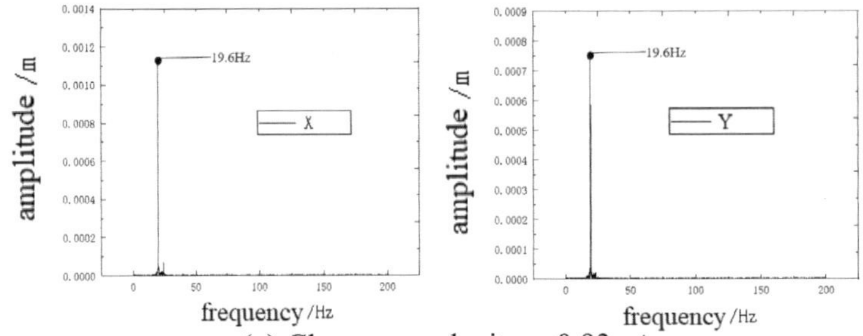

(a) Clearance velocity = 0.93m/s

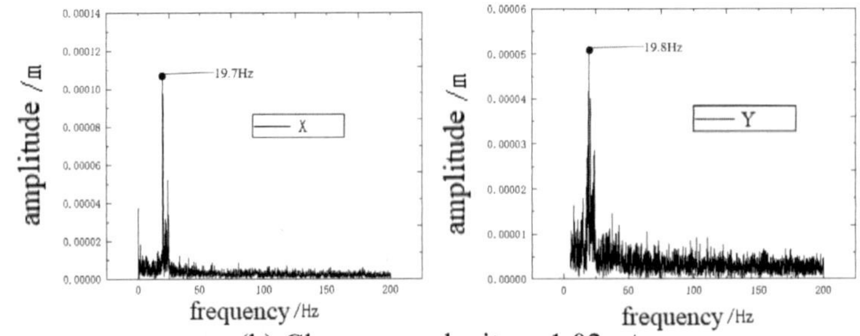

(b) Clearance velocity = 1.02m/s

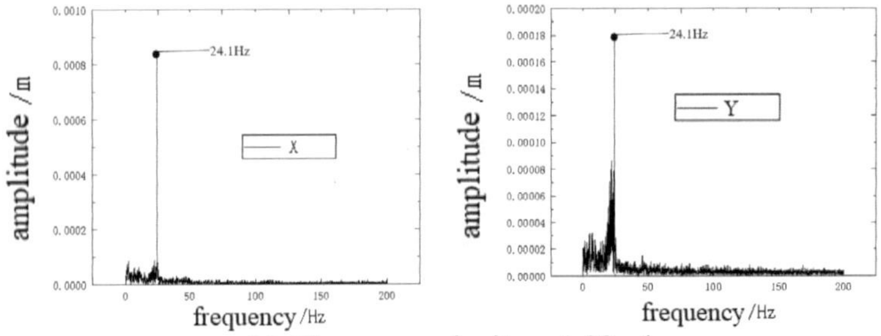

(c) Clearance velocity = 1.09m/s

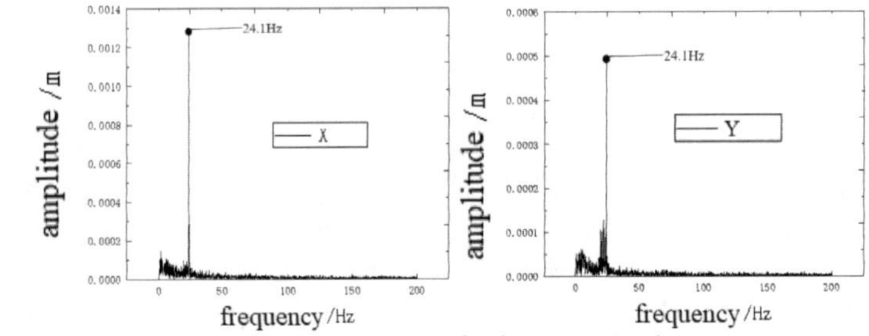

(d) Clearance velocity = 1.18m/s

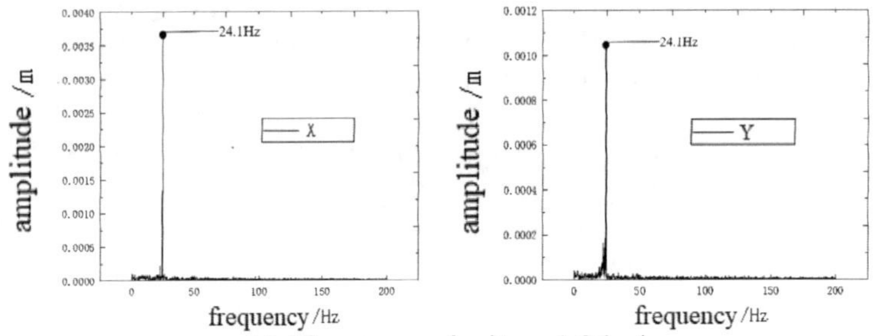

(e) Clearance velocity = 1.26m/s

Fig. 13 Tube 5 spectrum diagram of Tube bundle No. 2

As can be seen from Fig. 13, when the clearance velocity is greater than 1.09m/s, the frequency value corresponding to the maximum amplitude will not change, so it can be verified that the critical velocity of the bundle in tuber 2 is 1.09m/s. The dimensionless critical velocity Ucr = 2.82 is obtained.

3.3. Analysis of water tunnel test results of No.3 tube bundle

The 9 elastic tubes in Fig. 4 (c) were named, as shown in Fig. 14.

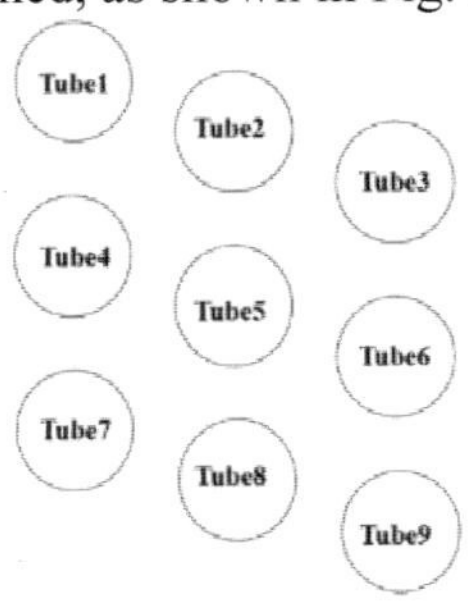

Fig. 14 No.3 Tube bundle naming method

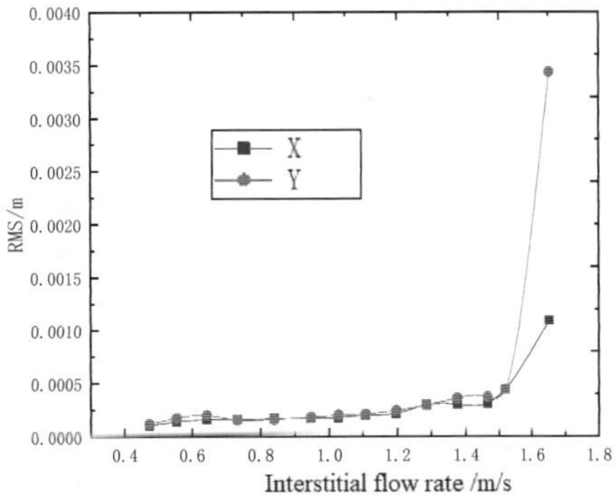

Fig. 15 Diagram of root mean square amplitude of tube bundle of No.3 tube-plate and clearance velocity

3.3.1. Results of bundle root mean square amplitude analysis. Through data processing, the relationship between root mean square amplitude and clearance velocity of 9 elastic tubes in No. 3 tube bundle is obtained. The data results of Tube 5 are shown in Fig. 15, which shows the relationship between root mean square amplitude and clearance velocity of Tube 5 along X (downstream direction) and Y (cross-flow direction) directions. As can be seen from the figure, when the clearance velocity of the tube bundle is between 1.4m/s and 1.6m/s, the root-mean-square amplitude of the tube along the X and Y directions will change suddenly. Therefore, the critical flow rate of tube bundle in the arrangement mode of No.3 tube-plate is between 1.4m/s and 1.6m/s, and the amplitude increment in Y direction is significantly larger than that in X direction. As shown in Fig. 15, no vortex shedding occurs with the increase of clearance velocity.

3.3.2. Bundle spectrum analysis. It can be observed from Fig. 15 that the critical flow rate of No.3 tube bundle is between 1.4m/s and 1.6m/s. In order to further determine the magnitude of the critical flow rate, Tube 5 was analyzed by spectral analysis. The spectrum diagram was drawn from the data recorded in the experiment at five clearance flow rates of 1.29m/s, 1.38m/s, 1.46m/s, 1.52m/s and 1.65m/s.

Journal of Physics: Conference Series **2280** (2022) 012054 doi:10.1088/1742-6596/2280/1/012054

frequency /Hz

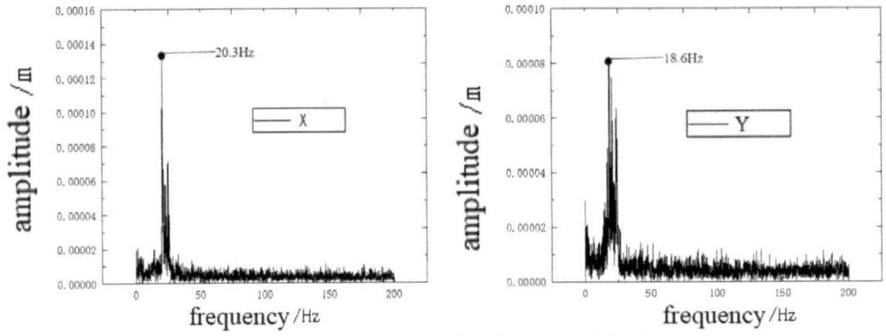

(a) Clearance velocity = 1.29m/s

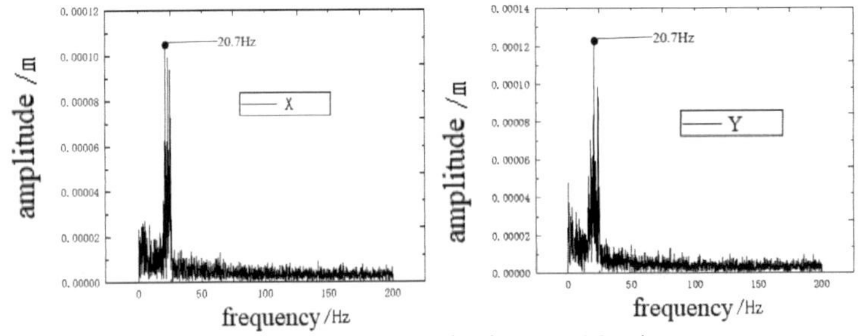

(b) Clearance velocity = 1.38m/s

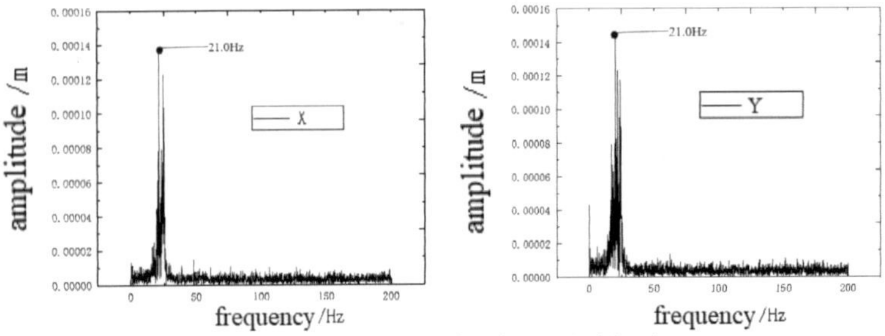

(c) Clearance velocity = 1.46m/s

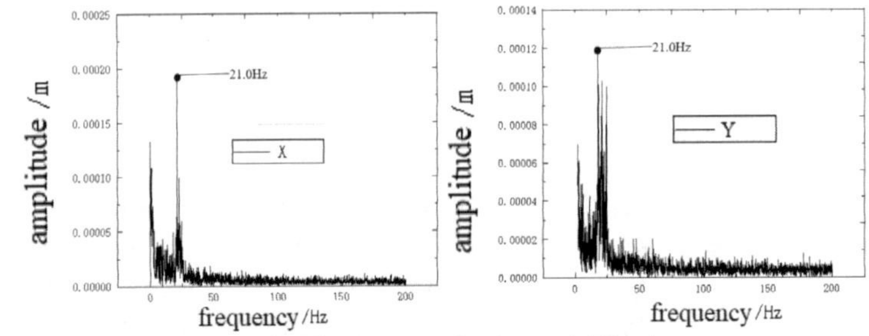

(d) Clearance velocity = 1.52m/s

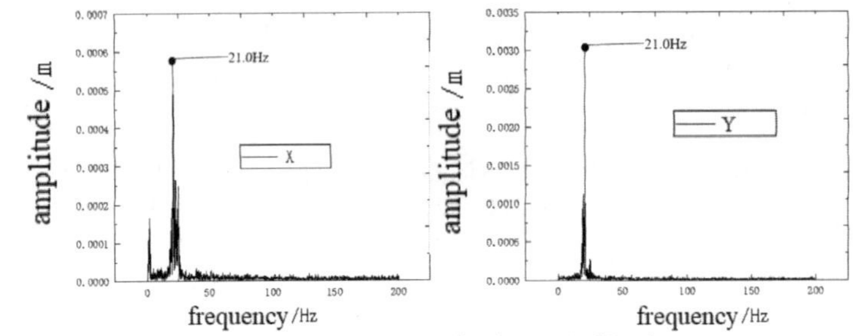

(e) Clearance velocity = 1.65m/s

Fig. 16 Tube 5 spectrum diagram of Tube bundle No. 3

As can be seen from Fig. 16, when the clearance velocity is greater than 1.46m/s, the frequency value corresponding to the maximum amplitude will not change, so it can be verified that the critical velocity of the tube bundle in tube bundle No. 3 is about 1.46m/s. The dimensionless critical velocity Ucr = 3.78 is obtained.

In conclusion, the fluid-induced vibration of three kinds of concentric tube bundles was studied by water tunnel test, and the following conclusions could be obtained through data processing: the critical velocity of the tube bundles in tube plate No. 1 was about 2.88, the critical velocity of the tube bundles in tube plate No. 2 was about 2.82, and the critical velocity of the tube bundles in tube plate No. 3 was about 3.78. It can be observed from tube bundles of No.1 and No.2 tube plates that vortex shedding occurs during the process of clearance velocity increasing, while no vortex shedding occurs in tube bundles of No.3 tube plate.

4. Conclusion

There is no standard to calculate the vibration of the concentric tube bundles of the heat exchanger. Therefore, in this project, according to different central Angle positions, three kinds of tubular plates are designed, which are approximately triangle arrangement, approximately square arrangement and transition arrangement. In the small-size water tunnel experiment, high-speed photography non-contact measurement is adopted to obtain the critical velocity when the tube bundle is unstable. The main conclusions are as follows:

(1) The critical velocity of the bundle in the approximate triangle arrangement tube plate No. 1 is about 2.88, the critical velocity of the bundle in the approximate square arrangement tube plate No. 2 is about 2.82, and the critical velocity of the bundle in the transitional arrangement No. 3 is about 3.78.

(2) It can be observed from tube bundles of No.1 and No.2 tube plates that vortex shedding occurs during the process of clearance velocity increasing, while no vortex shedding occurs in tube bundles of No.3 tube plate.

(3) The results show that the sensitivity of the tube bundles in different positions in the concentric circular tube bundles is different, and the performance of the tube bundles in the transitional arrangement is betterin perspective of thc critical vclocity.

References

[1] Blevins R D. Flow induced vibration. 2nd ed. New York: Van Norstrand Tein hold Company (1977).

[2] Kassera Volker, Kacem Hamouda. Numerical simulation of flow induced vibrations of a tube bundle in uniform cross flow. ASME, Pressure Vessels and Piping Division. New York: ASME, (1995): 37-43.

[3] Lin TK, Yu MH. Experimental Study on the Cross-flow Vibration of a Flexible Cylinder in Cylinder Arrays. Experimental Thermal and Fluid Science, 29 (2005).

[4] Mitra D, Dhir V K, Catton Ⅰ. Fluid-elastic instability in tube arrays subjected to air-water and steam-water cross-flow. Journal of Fluids and Structures, 25(2009) 1213-1235.

[5] Ricciardi G, Pettigrew M J, Mureithi N W. Fluidelastic Instability in a Normal Triangular Tube Bundle Subjected to Air-Water Cross-Flow. Journal of Pressure Vessel Technology 133, 6 (2011).

Analysis of rainwater overflow in underground pipe network

Tian Zhou*, Fan Yang, Wei Liu, Yue Tan

Beijing Institute of Radio Metrology and Measurement, China

*Corresponding author e-mail: zhoutians88@163.com

Abstract. Urban rainwater pipe network is a direct way to resist urban waterlogging, and its safe and stable operation is the premise to ensure the normal operation of the city. However, in China and even around the world, the monitoring means for the operation of urban rainwater pipe network are very limited. For the occurrence and solution of urban waterlogging, it often mainly depends on the means of personnel inspection to monitor where the rainwater pipe network overflows, and then take measures. There is no effective monitoring method for the whole rainwater pipe network. Therefore, an accurate and efficient rainwater pipe network monitoring method is urgently needed. To solve the current problem of real-time monitoring of rainwater pipe network, the core problem is how to monitor the liquid level data of any pipe section of the whole rainwater pipe network in real time, so it is necessary to introduce hydrodynamic model. The traditional hydrodynamic model mainly simulates the hydraulic and hydrological, and does not simulate the rainwater pipe network, but makes a generalized estimation of the drainage of the rainwater pipe network. In addition, some hydrodynamic models for drainage pipe network are not established for the problem of rainwater pipe network monitoring. The ultrasonic liquid level gauge with real-time acquisition function is not introduced into the calculation simulation, and it does not have the ability of adaptive correction. Therefore, the traditional rainwater monitoring methods and hydrodynamic models can not fundamentally solve the problem of real-time monitoring of the whole rainwater pipe network. Using hydrodynamic model calculation, combined with rain gauge and liquid level gauge to monitor and modify model parameters, so as to intelligently monitor the whole rainwater pipe network, which can greatly improve the current monitoring level of urban rainwater pipe network and reduce the work cost of manual patrol monitoring.

1. Introduction

At present, there is no particularly effective monitoring method for the whole rainwater pipe network. The existing monitoring means often monitor the situation of a certain rainwater pipe section through the liquid level gauge, which is unable to monitor the situation of each pipe section of the whole rainwater pipe network in real time. However, some existing rainwater hydrodynamic models remain in the stage of simulation calculation, are not effectively combined with the monitoring means of liquid level gauge, do not have the ability of adaptive adjustment, and are not actually applied to the monitoring of rainwater pipe network. Therefore, when a rainstorm occurs in a city, it is often found and corresponding measures are taken after a large-scale overflow of the rainwater pipe network, resulting in personnel and property losses [1].

SWMM model is used as the prototype basis of rainwater overflow model. SWMM model has been more and more widely used as an urban rainwater and flood management model, but it is not easy to establish a model with good accuracy, and it often needs to spend considerable energy on parameter calibration [2]. The precondition of parameter calibration or parameter adjustment is that the model has been basically built, the basic data of rainfall and pipe network have a certain accuracy, and the generalization processing on this basis is relatively reasonable. Otherwise, the parameter adjustment has little significance.

2. SWMM model

SWMM is a comprehensive mathematical model, which can completely simulate the process of urban rainfall runoff and pollutant transport. The model can be summarized and analyzed according to the input data and relevant data, and output the flow of each section and the change process line of each pollutant. It can be used for planning, design and practical operation. It is an applied model widely used for rainstorm runoff simulation and prediction. Includes the following modules [3]:

2.1. Hydrological module

The hydrological module of SWMM mainly simulates the runoff generation and concentration process of surface runoff. Among them, the runoff production process includes time-varying rainfall, surface water evaporation, snow and its melting, depression storage, etc; The confluence process includes rainfall infiltration into unsaturated soil layer, runoff infiltration into groundwater layer, flow exchange between groundwater and drainage pipeline, slope confluence, and various micro impact processes that reduce or delay precipitation and runoff.

2.2. Hydraulic module

The hydraulic module mainly simulates the flow of runoff and external water flow in pipelines, channels, water storage and treatment units, water diversion structures, etc., including the flow in closed pipelines and open channel pipelines of various shapes, the flow in some special parts (such as water storage and treatment units, diverter valves, water pumps, weirs and drainage orifices, etc.), the input and confluence of external water flow and water quality, And various forms of water flow (such as backwater, countercurrent, overflow, etc.).

2.3. Water quality module

According to the functional area or land cover type, the same drainage area can be divided into different hydrological response units, and the accumulation model and scouring model of various surface pollutants can be defined to simulate the growth, scouring, transportation and treatment process of pollutants in surface runoff.

Generally speaking, for runoff coefficient or yield, impervious rate, watershed area, characteristic width and slope are the most influential parameters, and the influence of other parameters can be almost ignored. Generally speaking, the most influential parameters for peak runoff are watershed area, characteristic width, impervious rate, depression storage in permeable area, Manning coefficient in impervious area and slope. Of course, the sorting of this parameter will be slightly different in different regions and situations, but there will be no big difference.

3. Study on overflow analysis model

Combined with SWMM hydrodynamic model calculation method and liquid level instrument to monitor the liquid level of the whole rainwater pipe network, the monitoring method has a good effect on judging the overflow results. The overflow analysis model uses three-dimensional geographic information system (GIS) to visually display the overflow points of rainwater pipe network, and highlight the pipe sections exceeding the liquid level warning value, which is convenient for the monitoring personnel to view the real-time status of rainwater pipe network. The overflow analysis model has self-adaptive ability, which can be reflected in the comparison between the measured value of the liquid level gauge and the calculated value of the hydrodynamic model. If the difference is large, the program calculation results are consistent with the measured values by modifying the pipe section friction parameters of the calculation model, so as to improve the adaptability of the monitoring method. The overflow analysis model can greatly improve the monitoring level of rainwater pipe network, monitor the liquid level of any pipe section of the whole pipe network in the process of rain, and overcome the change of pipeline drainage friction coefficient over time. When the rainwater pipeline overflows or approaches the warning liquid level, the three-dimensional geographic information system can be used to prompt the monitoring personnel to facilitate the monitoring personnel to locate the location of the overflow point, so as to go to the overflow area for corresponding disposal, so as to reduce the loss of personnel and property caused by urban rainstorm and improve the overall management level of urban rainwater pipe network. The overflow analysis model is not only suitable for the liquid level monitoring of urban rainwater pipe network, but also suitable for the liquid level monitoring of drainage pipe network with rainwater and sewage confluence.

The technical principle of the overflow analysis model is: take the liquid level value of the key points of the pipe network collected by the rainfall and liquid level gauge as the independent variable, transfer the independent variable into the hydrodynamic model of the rainwater pipe network, and calculate the current liquid level value of any pipe section of the whole pipe network. If the liquid level calculated by the hydrodynamic model is quite different from the liquid level measured by the liquid level gauge, the friction coefficient in the model will be automatically corrected to make the calculated value of the calculation model consistent with the measured value. Therefore, the rainwater pipe network monitoring method proposed by the overflow analysis model has adaptive ability. The monitoring method of the constructed overflow analysis model can monitor the liquid level value of any pipe section of the whole rainwater pipe network in real time. On the three-dimensional pipe network geographic information system (GIS) platform, the liquid level of any pipe section of the whole pipe network is displayed through different colors, and when the pipeline points on the rainwater pipe network overflow, The overflow volume of the overflow node can be calculated and the three-dimensional scene of the overflow can be displayed at the same time. The current operation of rainwater pipe network will be intuitively and conveniently displayed to the monitoring personnel. Using this method, the monitoring personnel can monitor the whole operation of the rainwater pipe network in real time, so as to greatly improve the monitoring efficiency of the rainwater pipe network.

The specific steps of the overflow analysis model are as follows:

(1) For the rainwater hydrodynamic calculation model, the model calculation program will be established by using the rainfall intensity formula, runoff generation model formula, confluence and rainwater pipeline transportation formula.

(2) For the establishment of rainfall intensity formula, first count the historical rainfall data of the region or city where the rainwater pipe network to be monitored, and collect the historical rainfall data of the region for 10 ~ 30 years. The selection method of rain samples can adopt the annual maximum value method, that is, take 3 ~ 4 maximum values every year and intercept an average of one every year; The rainfall frequency distribution curve is counted by exponential distribution method. Using historical rainfall data and rainfall frequency distribution curve, the rainfall intensity formula of each season in this area is fitted.

(3) The runoff yield model formula is used to calculate the runoff yield (R), that is, the value of rainfall (P) minus initial loss (s) and infiltration (f), r = p-s-f. The rainfall can be collected in real time by the rain gauge and calculated in combination with the rainfall intensity formula in this area. The initial loss can be determined by estimating evaporation, surface hardening rate, surface plant interception and surface depression storage through empirical formula. The empirical formula for calculating initial loss is:

$$S = S_z + S_j + S_w$$

Where s is the total amount of initial loss, SZ is the rainwater evaporation in the area in the corresponding season, SJ is the loss of plant interception, and SW is the loss of surface filling. The infiltration amount can be calculated by the formula

$$f(t) = fc + (f0 - fc)e^{-kt}$$

Where f (t) is the infiltration rate at time t, F0 is the maximum infiltration rate, FC is the minimum infiltration rate, and K is the attenuation constant.

(4) The confluence formula is used to determine the net flow of surface rainwater and then determine the amount of rainwater flowing into the underground pipe network. The net surface flow is determined by the continuity equation, which is set as follows:

$$\frac{dV}{dt} = A \cdot \frac{dh}{dt} = A \cdot i - Q$$

$$Q = W \cdot \frac{1}{n} \cdot (h - hp)(h - hp)^{\frac{5}{3}} \cdot S_0^{0.5}$$

Where, v = ah is the surface water collection, h is the water depth, t is the time, a is the surface area, I is the net rain intensity, and Q is the flow. W is the basin width, n is the surface roughness coefficient, HP is the surface water storage depth, and S0 is the basin slope. Where S0 is the basin slope and W is the basin width, which can be obtained from the digital elevation model data (DEM) obtained by the surveying and mapping department. The surface net discharge can be obtained by combining the above two formulas.

(5) When the surface net flow flows into the urban rainwater pipe network, the specific liquid level value in each pipe section can be determined by using the rainwater pipeline transportation formula. The rainwater pipeline transportation formula is as follows:

$$\frac{\partial A}{\partial t} + \frac{\partial Q}{\partial x} = q_t$$

$$\frac{v}{g} \cdot \frac{\partial v}{\partial x} + \frac{1}{g} \cdot \frac{\partial v}{\partial t} + \frac{\partial h}{\partial x} = S_0 - S_f$$

Where q is the flow, a is the cross-sectional area, V is the flow velocity, h is the water depth, t is the time, X is the distance, SF is the friction slope, S0 is the bottom slope, and QT is the lateral inflow per unit length. The liquid level and drainage flow rate of each pipe section can be calculated through the transportation formula. Where SF is the friction gradient, which is not a fixed parameter [4]. It is associated with the actual liquid level value collected by the liquid level gauge arranged on the key nodes of the pipe network. If the measured liquid level value is greatly different from the calculated value of the model program, the whole calculation model is modified by modifying the friction slope parameter to make the calculation model conform to the actual situation.

(6) The rainwater pipe network monitoring method of the overflow analysis model has the ability of adaptive correction. Specifically, by comparing the liquid level value collected by the full-scale liquid level monitor installed at the key nodes of each rainwater pipe network with the variance value of the liquid level gauge of the pipe section calculated by the dynamic model. If the square difference is greater than the set domain value, the pipe friction coefficient of the pipe section and the upstream and downstream pipe sections related to the spatial topology tree of the rainwater pipe network shall be corrected. Through the change of friction coefficient, the new friction gradient value is obtained, and then the whole rainwater hydrodynamic calculation model program is modified. After correction, if the newly calculated liquid level value is equivalent to the measured value of the liquid level gauge, the parameter correction will not be carried out. If the difference is still large, it is necessary to further modify the corresponding model parameters, and modify the model parameters in the way of step-by-step correction until the difference between the calculated value and the measured value is below the domain value. In this way, the rainwater pipe network [5] monitoring method has adaptive ability, so as to improve the accuracy and intelligent level of rainwater pipe network monitoring.

4. Summary

This paper mainly studies the rainwater overflow analysis model. Combined with the rainfall and the liquid level gauge arranged according to the geographical elevation data and the distribution of the pipe network, the rainfall information and the actual liquid level information of the key points of the pipe network can be obtained, and the operation of the rainwater pipe network calculated by the hydrodynamic model can be used to test the areas and pipeline locations prone to overflow [6]. Through the historical monitoring data, the liquid level gauge and rain gauge can be modified or added to monitor the liquid level at the position prone to overflow, so as to effectively avoid the blindness of simply relying on the liquid level gauge to monitor the liquid level value of the pipeline, improve the layout scientificity of the liquid level gauge and rain gauge, and more effectively monitor the situation of the rainwater pipe network.

References

[1] Gu Yaming, Yin Hailong, Xie Qingjie Study on dynamic model of wastewater interceptor system around Dianchi Lake [J] Environmental engineering. 2016, 34 (3). 53-57 doi: 10.13205/j.hjgc. two hundred and one million six hundred and three thousand and eleven.

[2] Climax Application of SD model in Overflow Pollution Control of combined drainage network [D] Southwest Jiaotong University, 2014.

[3] Zhao Qinghao Investigation and analysis of water quality and quantity of combined drainage system in rainy days [D] Wuhan University of technology, 2013.

[4] Jorge Gironás,Larry A. Roesner,Jennifer Davis. Storm water management model applications manual [M].USA: National risk management and research laboratory office of research and development U.S.environmental protection agency, 2009.

[5] Lewis A. Rossman. Storm water management model user's manual [M]. USA: National risk management and research laboratory office of research and development U.S. environmental protection agency, 2004.

[6] Haestad Methods, S. Rocky Durrans. Storm water conveyance modeling and design [M].USA: Bentley Institute Press, 2004.

[7] LewisA. Rossman.SWMM5 Interfacing Guide [OL]. [2009]. http://www.doczj.com/doc/fb6b9c 2c1711cc7931b716d7. html /ednnrmrl/models/swmm/

Uncertainty Quantification of Multivariate Correlated Output in the Simulation of Airfoil Flow

Chao Zhang, Jiangtao Chen*, Wei Zhao, Xiaojun Wu

China Aerodynamics Research and Development Centre, Mianyang, Sichuan, China

*Corresponding author e-mail: chenjt@cardc.cn

Abstract. Due to the limitation of human cognition, there are uncertainties in the turbulence model closure coefficients that cannot be ignored. The key to assess model objectively is quantifying the uncertainty of model coefficients scientifically. In the CFD simulation of flow around the airfoil, wall pressure distribution is a multivariate correlated quantity. This paper developed a surrogate model method based on artificial neural network to characterize the mapping relationship between multivariate input and multivariate output. In order to reduce the output dimension and modelling difficulty, this paper use Proper Orthogonal Decomposition (POD) method. An RAE2822 benchmark case under transonic conditions is used to study the uncertainty of Spalart-Allmaras (SA) turbulence model coefficients and verify the method developed by this paper.

1. Introduction

There are various uncertainties in CFD simulation of complex engineering profiles. NASA [1] summarize commonly encountered sources of uncertainty in scientific computing, including model-form, model input parameters, model prediction-bias, numerical solution-bias and simulation results processing and inference. In general, model-form, model input parameters and numerical solution-bias uncertainties are most discussed and studied [2]. In our former study [3,4], the conclusion is that the model parameters' uncertainty is very important.

The main work of parameter uncertainty quantification includes identification and characterization of uncertain parameters, uncertainty propagation, sensitivity analysis and model calibration [1]. CFD simulations have the characteristics of multivariate input, multivariate output and possibly correlated, such as turbulence model coefficients with about 10-dimensional uncertainties [4]. The quantities of interest in CFD are not only the lift coefficients, drag coefficients and other global integral components, but also the time or spatial sequence responses, such as surface pressure distribution of airfoil. All of these multivariate output responses may be correlated. Therefore, it is necessary to develop efficient and credible uncertainty quantification method for CFD.

In this paper, a framework of uncertainty quantification will be presented for multivariate input and multivariate output. Section 2 presents the flowchart of uncertainty propagation, which includes POD reduced-order method and artificial neural networks (ANN) method. Section 3 presents the RAE2822 airfoil case to demonstrate the uncertainty quantification framework proposed worked. Finally, section 4 concludes the paper.

2. Uncertainty propagation method

Consider the quantities of interest in CFD are multivariate and correlated, the surrogate model method based on samples is used to propagate uncertainties. Artificial neural network is established to represent the mapping relationship between multivariate input and multivariate output. In order to reduce the training difficulty of ANN, POD is used firstly to reduce the dimension of the multivariate output. Fig. 1 shows the flowchart of uncertainty propagation.

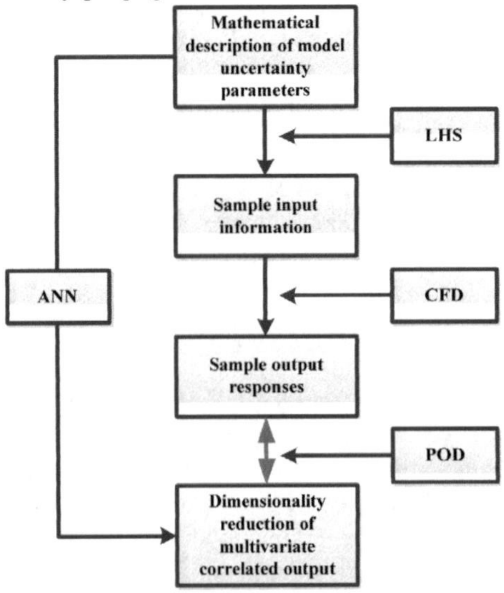

Figure 1. Flowchart of uncertainty propagation

2.1. *POD reduced-order method*

Proper Orthogonal Decomposition (POD) is an efficient and powerful method for data processing [5], and has received a continued attention in the fluid dynamics. The fundamental principle of POD is to find a new vector basis to maximize the projection of the original vector group.

In this paper, the multivariate output $\left\{\mathbf{f}\left(\theta_k\right)\right\}_{k=1}^{M}$ of M samples constitutes the snapshot set, and can be approximated using a linear combination eigenmodes,

$$\mathbf{f}\left(\theta_k\right) \approx \overline{\mathbf{f}} + \sum_{i=1}^{s} \alpha_i^k \boldsymbol{\varphi}_i \tag{1}$$

Where θ_k is the multivariate input, $\overline{\mathbf{f}}$ is the mean value, $\boldsymbol{\varphi}_i$ is the eigenfunction, α_i^k is the coefficients. The detail of solving α_i^k can be found in [6].

2.2. *ANN surrogate model*

Considering the multivariate correlated output of the system, ANN is used in this paper to represent the behavior of the computational model with uncertain inputs. ANN imitates the interactive reflection of biological nervous system to the real world. Neurons in the input layer receive external input, neurons in the hidden layer and neurons in the output layer process signals, and the final results are output by neurons in the output layer. Error backpropagation algorithm is commonly used to backpropagate errors to hidden layer neurons and adjust the connection weights and thresholds between neurons. ANN is a classical machine learning algorithm, and due to the length of this paper will not be introduced in detail.

3. Numerical example

In this section, the methods discussed in section 2 are applied to flow over RAE2822 airfoil with uncertain coefficients of SA turbulence model. The multivariate output is wall pressure at grid points.

In this case, the SA model coefficients are uncertain and are assumed to be uniformly distributed within a certain interval. In the calculation, it is assumed that all flows are turbulent and the trip term in the original model is ignored. The uncertain coefficients are $c_{b1}, \sigma, c_{b2}, \kappa, c_{w2}, c_{w3}, c_{v1}, c_{t3}, c_{t4}$. It is worth noting that the mathematical description of model uncertainty parameters, including value interval, uncertainty distribution and other information, needs in-depth communication with the model developer. This paper focuses on the propagation of uncertainty and does not discuss the mathematical description of uncertainty.

The calculation condition is set as: Ma=0.729, Rec=6.5×106, α=2.31°. The flow solver used is MFlow [7], and the grids and numerical algorithms used can be found in reference [4]. The uncertainty multivariate input is 9 SA model coefficients and the multivariate output is the pressure coefficient of 304 wall grid points. This is a typical numerical example with multivariate input and multivariate correlated output.

3.1. Sampling
The proposed method in this paper is data-driven, and 110 sample points are obtained by Latin Hypercube Sampling (LHS) to ensure the uniformity of input space and projection. Another more 20 sample points are used to test the accuracy of POD and ANN.

3.2. POD
The pressure coefficients of all the wall grid points calculated from each sample points are combined into a 304-dimensional output response vector. POD is applied for 110 multivariate output responses, and the first 2 eigenmodes are selected according to the nominal criterion of 99% energy. Then, the original high-dimensional response is reduced to a vector of eigenfuction coefficients projected in a 2-dimensional eignemode space. The reduction from 304 dimensions to 2 dimensions indicates that the POD method is well applied in this case.

To verify the accuracy of POD, another 20 samples are used. Fig. 2 shows the original wall pressure distribution of 4 samples and the wall pressure distribution restored by POD. The results show the difference is small, the maximum error is only 0.003, determining that POD has excellent generalization ability and can ensure the accuracy of dimensionality reduction.

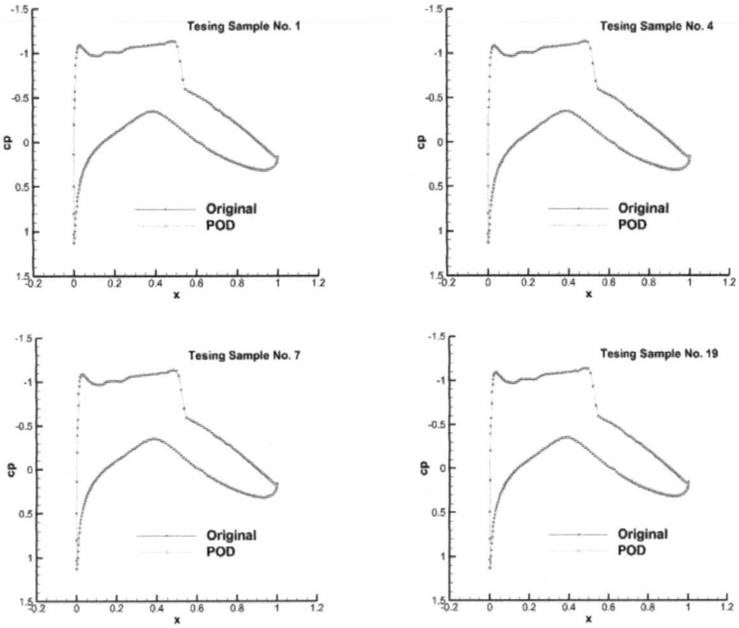

Figure 2. Comparison of original pressure distribution and POD reduced dimensional analysis

3.3. ANN

First, 80% of all 110 samples are used as training samples to train the ANN, and the remaining 20% are used as testing samples to evaluate the model generalization error. Then, stop training when generalization ability is no longer improved. Finally, additional 20 samples are used to test the generalization ability of the model independently.

3.4. Test the accuracy of POD and ANN

In this paper, POD and ANN together constitute a complete set of methods. To verify the accuracy of POD, additional 20 independent samples are used. Fig. 3 shows the original wall pressure distribution of 4 samples and the wall pressure distribution restored by POD. The results show the difference is small, the maximum error is only 0.0043, determining that POD has excellent generalization ability and can replace the original numerical simulation system to predict the multivariate output under new model parameter values. It is a reliable foundation for subsequent uncertainty quantification.

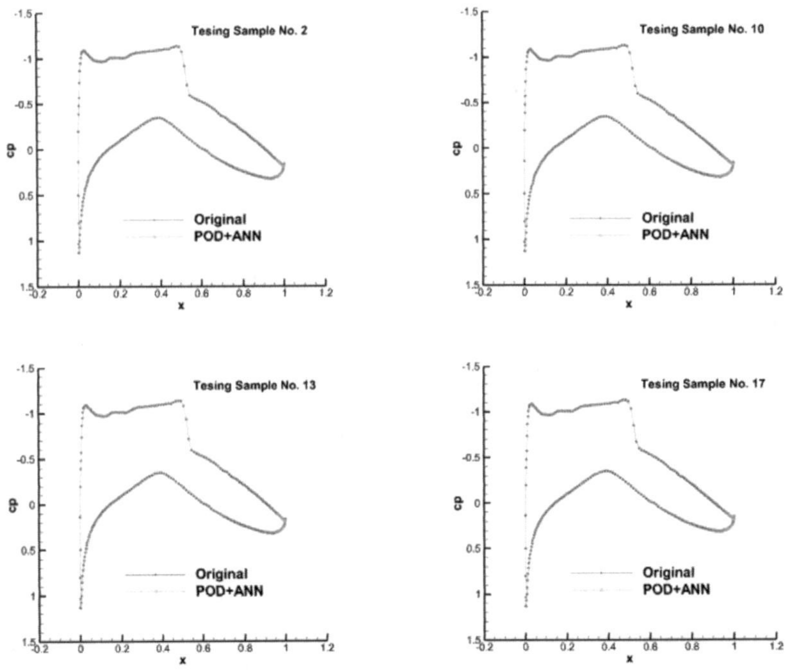

Figure 3. Comparison of original pressure distribution and current analysis in this paper

3.5. Statistics on wall pressure distribution

The ANN coupled POD method effectively reduces the resource consumption of obtaining samples, so that large-scale sampling can be carried out in the value space of uncertain parameters to obtain a large number of output results, and then restore to pressure distribution by POD. The mean and standard deviation of the wall pressure distribution are calculated from a large number of samples generated by the ANN. As shown in Fig. 4 and Fig. 5, the results of 10000 samples and 100000 samples are almost the same, and the time consumed by the ANN model can be ignored. The uncertainty of turbulence model parameters results in the large fluctuation of the pressure coefficient near the shock wave, which fully indicates that the shock wave location is very sensitive to the model coefficient.

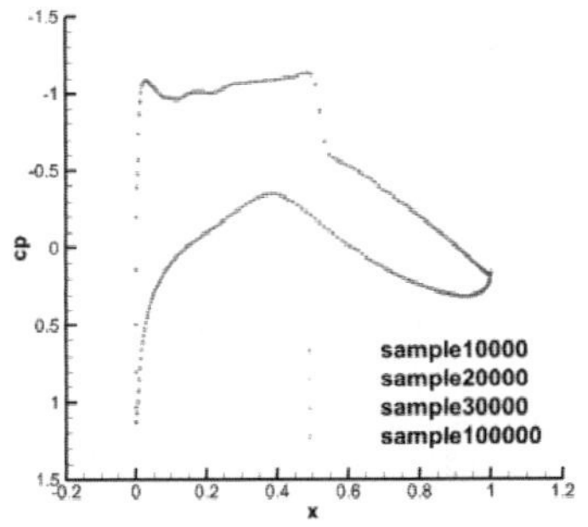

Figure 4. Mean value of pressure distribution

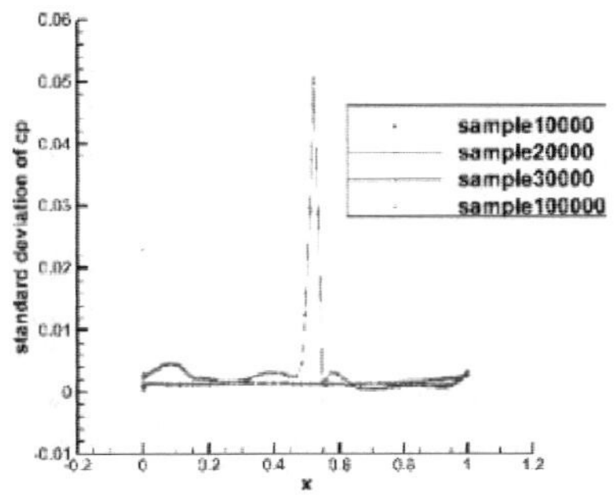

Figure 5. Standard deviation of pressure distribution

4. Conclusion

In order to study the influence of the uncertainty of SA model coefficients on the prediction of RAE2822 airfoil pressure distribution under transonic inflow, a complete set of methods based on Proper Orthogonal Decomposition and Artificial Neural Network. Research findings include:

The use of POD can effectively reduce the dimension of multivariate correlated output. In this paper, 304-dimensional output is successfully reduced to 2-dimensional output. The accuracy of reduced dimension expression is confirmed to meet the requirements through additional independent testing samples. POD reduced-order method greatly reduces the difficulty of subsequent ANN training.

The ANN coupled POD model can replace the original numerical simulation system with high accuracy to predict the airfoil surface pressure distribution under the new model parameters, which is a reliable foundation for using MC sampling to study the propagation of the uncertainty of model parameters in CFD simulation.

It should be noted that the analysis conclusion in this paper is for RAE2822 airfoil, and whether it can be extended to other shapes needs further study. However, the proposed research method is universal.

Acknowledgments

This work is supported by National Numerical Wind tunnel Project.

References

[1] MEHTA U B. Simulation Credibility. Subtitle: Advances in Verification, Validation, and Uncertainty Quantification. NASA/TP, 2016.

[2] Oberkampf W L, Roy C J. Verification and validation in scientific computing. Cambridge: Cambridge university press, 2010.

[3] CHEN J T, ZHAO J, ZHANG C, et al. Effects of numerical simulation approaches on drag prediction of NASA CRM. Acta Aeronautica et Astronautica Sinica, 2020, 41(4): 123383(in Chinese).

[4] ZHAO H, Hu X Z, ZHANG J, CHEN J T, et al. Effects of the uncertainty in turbulence model closure coefficients on the simulation of flow over airfoil. Acta Aeronautica et Astronautica Sinica, 2019, 40(6): 122581(in Chinese).

[5] Barber, T. J., Ahmed M. H. POD Snapshot Data Reduction for Periodic Fluid Flows. AIAA 2005-287, 43rd AIAA Aerospace Sciences Meeting and Exhibit, Reno, NV, January 10-13 2005.

[6] Sirovich, L. Turbulence and the Dynamics of Coherent Structures. I-Coherent Structures. II-Symmetries and Transformations. III-Dynamics and Scaling. Quarterly of Applied Mathematics, 1987, 45, 561-571.

[7] CHEN J T, ZHANG Y B, ZHOU N C, et al. Numerical Investigations of the High-Lift Configuration with MFlow Solver. Journal of Aircraft, 2015, 52(4): 1051-1062.

Structural reliability design of liquid oxygen supply pipeline based on fluid-solid-thermal coupling

Jingying Zhou, Jiawei Ding *

Xi'an Aerospace Propulsion Test Technique Institute, Xi'an 710100, China

*Corresponding author e-mail: dingjw163@163.com

Abstract. Liquid oxygen supply pipeline is a key part of liquid oxygen kerosene engine test system. If leakage occurs, it will cause flow measurement error and affect the accuracy of test data. Therefore, in the design of liquid oxygen supply pipeline, the fluid solid thermal coupling simulation analysis method is used to analyze the stress of "Z" bend, vessel bottom elbow and balance device elbow pipeline of the supply pipeline in the process of ultra-low temperature fluid flow, the optimal laying scheme of the "Z" bend route is obtained, and it is verified that the added elbow pipeline fixing support structure can effectively resist the blind plate force and improve the pipeline reliability, which provides an important reference for the structural design of liquid oxygen supply pipeline.

1. Introduction

In the aerospace industry, liquid oxygen, as a clean and environment-friendly strong oxidant, is usually used with liquid hydrogen or kerosene as rocket fuel. The ultra-low temperature characteristics of liquid oxygen cause large cold stress and deformation in the supply process, which is easy to cause the damage of system structure and interface leakage. At the same time, it will also cause flow measurement error and can not provide accurate performance parameters for engine test. Therefore, it is urgent to study the reliability of low-temperature liquid oxygen system under the action of fluid, solid and thermal physical fields, so as to improve the safety and reliability of liquid oxygen supply system.

The structural design problem of low temperature pipeline generally exists in engineering practice. Ding Chang et al. [1] used the finite element analysis method to analyze the stress of a low-temperature pipeline, and proposed adding two π bends in the straight pipe section of the pipeline to improve the flexibility of the pipeline and reduce the stress. Deng Fang et al. [2] studied the pipeline system of liquid oxygen storage tank in the design range of - 196 ℃ / 65 ℃ by using the finite element analysis method, and came to the conclusion that the temperature difference stress is the maximum load stress of low-temperature pipeline and the maximum stress of pipeline elbow. Zhang Caigong et al. [3] established the thermal structural coupling mathematical model of low-temperature pipeline, solved it by finite element analysis method, and analyzed the variation laws of temperature, stress and deformation of adiabatic support under different apertures. The axial displacement and stress of the low-temperature bellows are strictly controlled by the finite element method [4]. However, the above analysis does not couple multiple physical fields together, and there is a certain gap between the simulation conditions and the actual state of low-temperature pipeline.

In this paper, ANSYS software is used to carry out the fluid solid thermal coupling simulation analysis of liquid oxygen supply pipeline, calculate the temperature field, flow field and structural stress and deformation of different pipeline directions and fixed states of cryogenic system, realize the quantitative analysis of fluid force, fluid heat and structural coupling response, and provide an important reference for the structural design of ultra-low temperature liquid oxygen supply system.

2. Analysis of liquid oxygen supply system

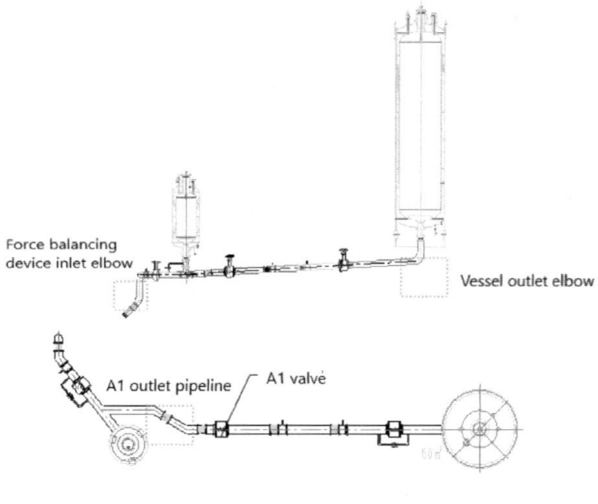

Figure 1. Pipeline layout of liquid oxygen supply system

As shown in Figure 1, there is a blind plate force at the bottom of the container and the inlet of the force balancing device, but there is a lack of foundation support at the bottom of the pipeline, so it is necessary to design and increase the support structure to meet the load-bearing requirements of the low-temperature pipeline; In order to avoid the existing system design, the pipeline behind A1 valve needs to optimize the pipeline direction through simulation calculation to reduce the system stress and the risk of liquid oxygen leakage. Therefore, this paper focuses on the simulation analysis of these three parts, and improves the reliability of liquid oxygen supply system through structural check and optimal design of pipeline support and direction.

3. Analysis ideas

In this paper, the relaxation coupling method is used to simulate the liquid oxygen pipeline and its support structure, that is, the flow field and temperature field are solved first, and then the solution results of fluid pressure and structure temperature field are applied to the solid structure at the same time. The coupling calculation flow chart is shown in Figure2.

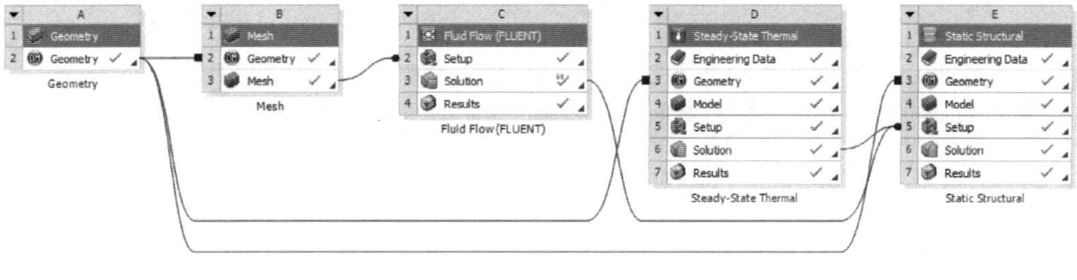

Figure 2. Calculation flow of fluid solid thermal coupling

4. 3D model and boundary conditions

4.1. 3D model

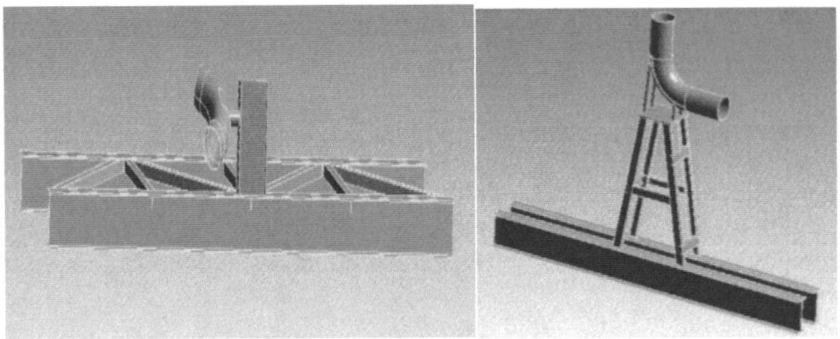

Figure 3. Vessel bottom, inlet pipeline and support structure of force balancing device

The pipeline material of liquid oxygen supply system is stainless steel, and the specification of main pipeline is Φ325×8, pipeline flow rate 305kg/s; The supporting structure of the supply pipeline is made of carbon steel. According to the system pipeline layout, create the three-dimensional model of the container bottom, the inlet pipeline of the force balance device and the support structure (as shown in Figure 3).

4.2. boundary condition

Liquid oxygen pipeline and support structure can be divided into fluid domain and solid domain. The fluid domain is the liquid oxygen fluid in the pipeline, and the liquid oxygen density is 1180kg/m^3. Set the boundary conditions of the fluid domain: inlet velocity: 3.58m/s, inlet fluid temperature: 90K; Pressure outlet: 0.4MPa; The coupling boundary is the inner wall of the pipeline.

The solid domain is liquid oxygen pipeline and support structure. The stress boundary conditions are fixed constraints at the bottom of the support or at both ends of the steel structure, and elastic constraints are applied at both ends of the pipeline to simulate the state with compensator.

5. Simulation results and analysis

5.1. Analysis of "Z" bend at A1 outlet

5.1.1. Pipeline routing analysis and design. According to the site conditions, three schemes of valve A1 outlet pipeline laying are designed. Scheme I is 135 ° pipeline laying, scheme II is 90 ° pipeline laying, and scheme III is straight pipe laying.

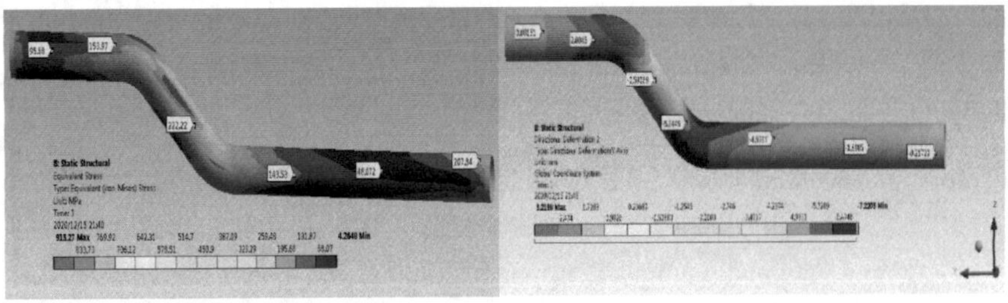

Figure 4. Scheme 1 stress nephogram and axial deformation diagram

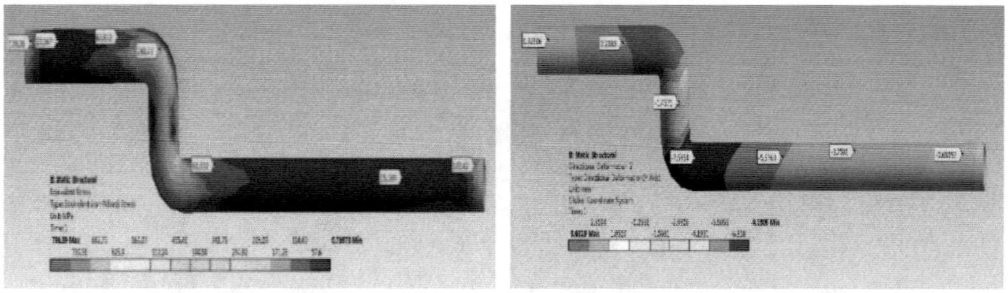

Figure 5. Scheme 2 stress nephogram and axial deformation diagram

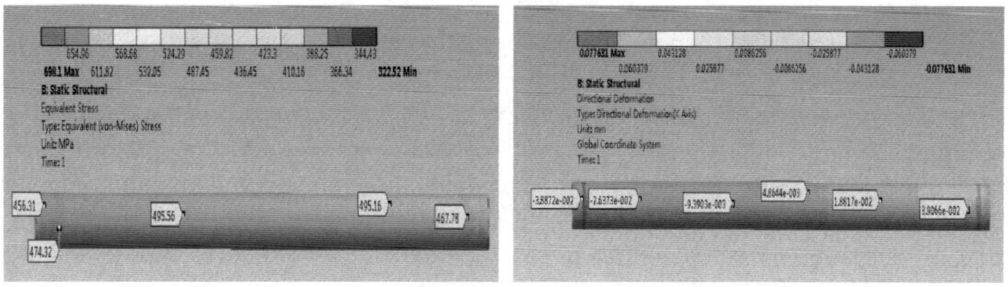

Figure 6. Scheme 3 stress nephogram and axial deformation diagram

The simulation results of the pipeline under the condition of low temperature -183 ℃ and the pipeline without thermal coupling are obtained, as shown in Fig. 4. The stress and deformation results of the three schemes are comprehensively analyzed. Scheme I has less pipeline stress and better pipeline flow resistance. Therefore, scheme I is selected as the laying scheme of "Z" bend road at valve A1 outlet.

5.1.2. Simulation analysis of fluid solid thermal coupling. It can be seen from the observation in Figure 7 that the stress value of the pipeline and support is large, which is caused by the insufficient compensation of the whole pipeline. Therefore, a pipeline compensator is added at the black box, with axial compensation capacity of ± 25mm and radial compensation capacity of ± 5mm. Combined with the natural compensation of "Z" bend, the pipeline between valve A1 and trigeminal pipe is better compensated. In the precooling process of the test run after the transformation, there is little leakage at this position, and the reliability of the pipeline has been improved.

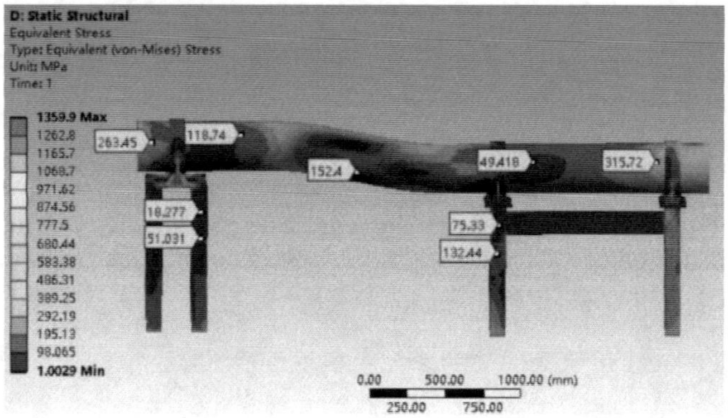

Figure 7. Cloud diagram of stress distribution of "Z" bend road

5.2. Analysis of elbow piping system at the bottom of vessel

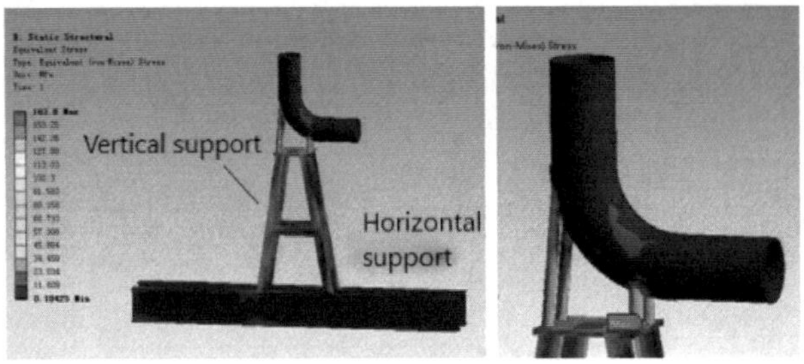

Figure 8. Stress nephogram of pipeline at the bottom of vessel

According to the analysis in Figure 8, except that there is stress concentration at the connection between the pipeline leg at the bottom of the container and the fixed support platform, the rest of the pipeline structure meets the allowable stress of the pipeline material, indicating that setting a compensator at the inlet and outlet of the elbow at the bottom of the container can effectively reduce the low-temperature deformation and low-temperature stress, At the same time, it shows that the elbow fixing method in Figure 8 can effectively resist the action of 3T blind plate force.

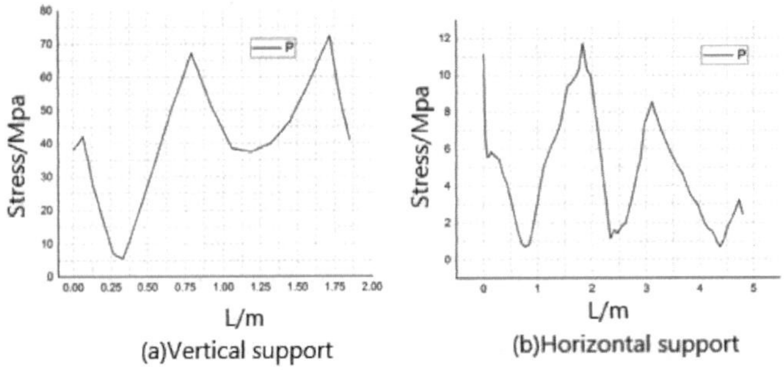

Figure 9. Stress curve of pipeline support at the bottom of vessel

According to the stress curve of the transformation of the stress of the pipeline support with the support length L, the stress value of the vertical support of the pipeline does not exceed 75MPa, and the maximum stress value of the horizontal support is about 12MPa. The high points of the stress values of both appear near the structural connection, but both meet the allowable stress requirements of carbon steel materials.

5.3. Analysis of inlet elbow pipeline system of force balance device

The inlet of the force balance device is a 90 ° elbow of DN300. When the pressure in the liquid oxygen pipeline is 0.4MPa, a blind plate force of 4.2t in horizontal and inclined upward directions of 135 ° will be generated and act on the elbow. Therefore, a fixed support needs to be added at the 90 ° elbow.

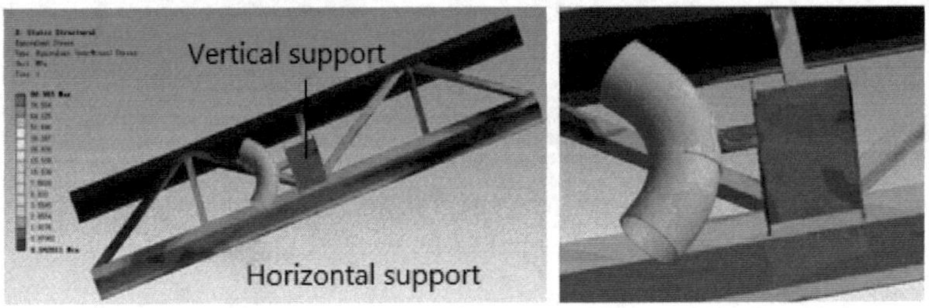

Figure 10. Cloud diagram of inlet pipeline stress balance device

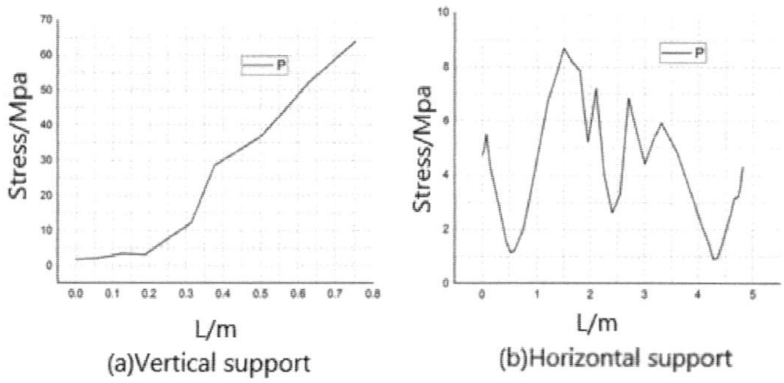

Figure 11. Stress curve of pipeline support of force balancing device

According to the analysis of the pipeline stress nephogram, the maximum stress value of the pipeline structure of the force balance device is located at the connection between the elbow hoop and the vertical support, which is about 89MPa, meeting the material strength requirements; The stress distribution on the elbow is relatively uniform, and the stress value does not exceed 10MPa. According to the analysis of the stress curve of the pipeline support, the closer to the connection point with the horizontal support, the greater the stress value of the vertical support, and the maximum value is about 65MPa, which meets the allowable stress of carbon steel materials.

6. Conclusion

Through the fluid solid thermal coupling simulation calculation and structural optimization design of three key positions of liquid oxygen supply pipeline system, the following conclusions are obtained:

a. The "Z" bend at the outlet of valve A1 of liquid oxygen supply pipeline should choose the 135 ° pipeline laying scheme. By adding a compensator in the middle of the 135 ° pipeline, the compensation capacity of the pipeline is increased, the occurrence of liquid oxygen leakage is greatly reduced, and the reliability of the pipeline is improved;

b. After the fixed support is added to the elbow pipeline at the bottom of the vessel, through the fluid solid thermal coupling simulation analysis, its strength meets the material requirements;

c. The foundation support of parallel I-steel structure connected through channel steel is added in the pipeline design of force balance device, which can effectively resist the blind plate force. The maximum stress of pipeline and support structure is about 89MPa, which meets the material requirements.

References

[1] Chang Ding, Yantang Fu, Zhizhen Zhao, Jinzhong Si. Stress analysis and optimization design of sandwich pipeline of low temperature storage tank [J] Cryogenics and Superconductivity.2016, (08): 27-32.

[2] Fang Deng, Shuiguang Tong, Tao Jin. Pipeline stress analysis of large cryogenic liquid storage tank [J] Low temperature engineering, 2005, (04):37-40.

[3] Caigong Zhang, Changjun Li, Wenlong Jia, Qianwei He. Research on thermal insulation support of low temperature pipeline based on steady-state thermal structural coupling [J] Low temperature engineering, 2018, (06): 47-53.

[4] Shuping Chen, Hongwei Mao, Shuting Yao, Fulu Liu, etc. Nonlinear finite element analysis of stress of bellows in high vacuum multilayer insulation low temperature pipeline [J] Natural gas industry, 2016, 36(04): 84-89.

Numerical Simulation of Noise Characteristics of Open Rotor

Qing Wei [1, 2], Zhengwu Chen[1], Fan Tong[1, *], Guangming Li[2], Panwang Cui[3]

[1]Laboratory of Aerodynamic Noise Control, China Aerodynamics Research and Development Center, Mianyang 621000, China
[2]School of Shanghai University, Shanghai, China
[3]Hunan Key Laboratory of Mechanical Equipment Health Maintenance, Hunan University of Science and Technology, Hunan, 411201, China

*Corresponding author e-mail: tongfan@cardc.cn

Abstract. In order to investigate the noise characteristics of open rotor, numerical simulation was performed for an open rotor which contains 11 blades for the front row and 9 blades for the rear row. The aerodynamic noise of the open rotor was compared with a single propeller, i.e. the front row propeller. Numerical results show that predicted noise field results agree well with the literature. The noise spectrum of the open rotor is obviously more complex and higher than that of the single propeller. In addition, the contribution of different noise generation mechanisms was studied for the interaction noise. Results show that the blade tip vortex interaction noise and the wake interaction noise are the main noise generation mechanisms while the hub vortex interaction noise contributes less to the interaction noise. Blade tip vortex interaction noise and wake interaction noise need to be paid more attention. The investigation result may help to develop relevant noise reduction techniques for the open rotor.

1. Introduction

The significant advantage of the open rotor engine is the high efficiency, low fuel consumption and the relatively high-speed performance. Its cruising Mach number can reach 0.8 [1]. From the 1970s to the 1990s, NASA ran the Advanced Propeller Program [1]. Systematic experimental and numerical studies of a series of high-speed propellers were carried out and funded by the European Union's "Clean Sky" program [1]. Safran has embarked on an eight-year propeller engine project. It began research on open rotor engines in 2013, and the open rotor engine is expected to enter service in the near future. GE's un-ducted fan was designed to consume 30% less fuel than contemporary turbofan engines, but its noise failed to meet the FAR-36 standards [1]. By using URANS numerical simulation method and DPG technology, Schwenbo et al. [1] studied the unsteady aerodynamic interference of the front blade wake and tip vortex with the rear blade. The results show that the interaction of tip vortex is the main cause of interference noise, and the presence of the rear blade reduces the intensity of tip vortex of the front blade. Jin Haibo et al. numerically predicted the noise of an open rotor based on Mohring sound analogy method. Xia Zhenfeng et al. [1] calculated the unsteady flow of counter-rotating propeller by using URANS method and dynamic grid docking technology, and found that the pull coefficient and power coefficient of the front propeller decreased, while the pull coefficient and power coefficient of the rear propeller increased. The rear propeller can recover the swirl energy of the front propeller, therefore improve the total efficiency. Yan Wenhui et al. [1] studied the complex flow field of contra-rotating propeller and the unsteady aerodynamic interference between propellers by using URANS method. Results showed that unsteady wake vortex of the front rotor had a great influence on the rear rotor.

Despite a lot of research was carried out to study the flow field and noise field of open rotors (or contra-rotating propellers), the noise characteristics and the noise generation mechanisms of open rotor still need to be studied. Especially, the contributions of different noise generation mechanisms should be further studied. Therefore, the current paper carried out numerical simulation to study the characteristics of the open rotor and evaluate the different noise generation mechanisms.

2. Numerical Method

2.1. Numerical method and the governing equation

N-S (Navier-Stokes) equation is the most complete equation describing the motion law of viscous fluid, and the governing equation based on the unsteady RANS equation is:

$$\frac{\partial}{\partial t}\iiint_{Q} Q d\Omega + \iint_{S} F \cdot dS = \iint_{S} F^{V} \cdot dS \tag{1}$$

Where $Q=[\rho\ \rho u\ \rho v\ \rho w\ \rho e]^T$ is a conservative vector, $(u\ v\ w)$ is the velocity component in the $(x\ y\ z)$, ρ and e represent density and total energy, respectively. Ω is the control volume, S is the surface of control body, F and F^V are inviscid and viscous fluxes through the surface S.

2.2. Grid and simulation set-up

An open rotor model was studied in this paper. The relevant parameters of the open rotor are shown in Table.1. The front blade row has 11 blades and the rear blade row has 9 blades. The speed of the front blade was -7633RPM and that of the rear blade was 7695RPM. The incoming flow velocity is 67.784m/s, the static pressure is 98595Pa, the temperature is 285.88K, and the angle of attack was 0°. In addition to the open rotor, a single row of propeller was also simulated. The single row of propeller was the front row of the open rotor. The meshes for the open rotor and single propeller are shown if Fig.1 and Fig.2 respectively. Fig.3 shows the mesh around the blade. The computational domain is shown in Fig.4. The number of grids of open rotor is about 760W, and that of single rotor propeller is about 500W. The SA turbulence model was used in the simulation.

Table 1. Parameters of the open rotor

Open rotor parameters	Value
Number of blade (fwd/aft propeller)	11/9
Design cruise Mach number	0.72
Nominal diameter,mm	622/607
Nominal design cruise tip speed,m/s	238
Nominal design advance ratio	2.82
Hub-to-tip ratio	0.42
Geometric tip sweep,deg	34/31
Rotors spacing,mm	150

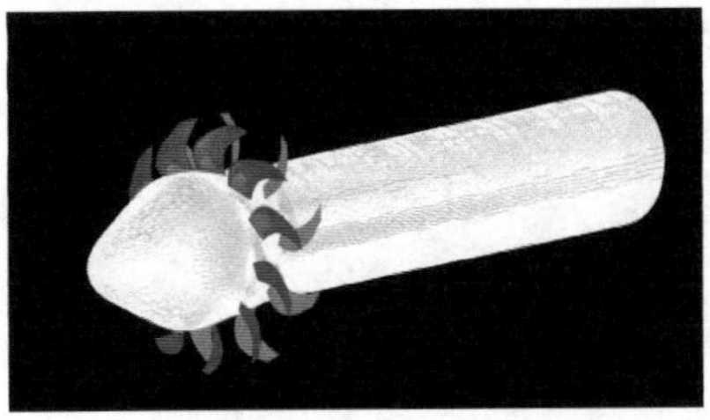

Fig. 1 Open rotor mesh

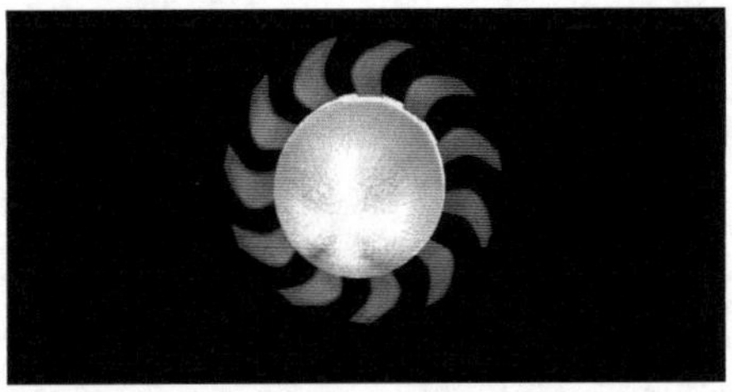

Fig. 2 Single propeller mesh

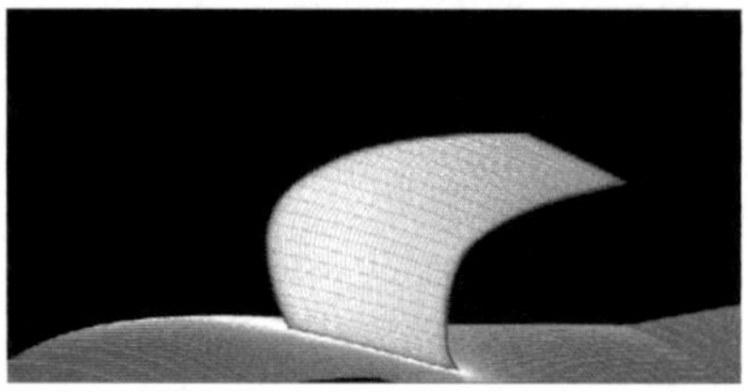

Fig. 3 Mesh around the blade

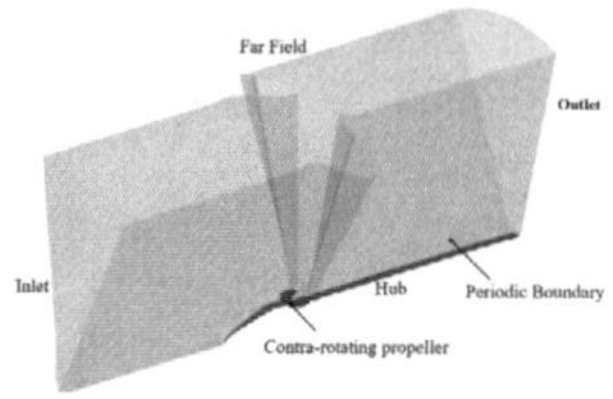

Fig. 4 Computational domain

3. Results and Analysis

The flow field of the open rotor is calculated by the nonlinear harmonic method, and the corresponding aerodynamic noise results are obtained by the FW-H equation. Nine far-field microphones were used to evaluate the noise. The microphone array is located above the open rotor, 0.61 m from the axis of rotation. The angle ranges from 50 degrees to 130 degrees with 10 degrees spacing. The direction of the air flow corresponds to 0 degree.

The sound pressure level spectrum of the open rotor is shown in Fig.5. The sound pressure level spectrum of a single propeller (the front blade row) is shown in Fig. 6. The noise of the open rotor contains not only noise at blade passing frequency (BPF), i.e. BPF1 (1399.38Hz), BPF2 (1154.25Hz), and its harmonics but also the interaction noise, such as BPF1+BPF2 (2553.63 Hz) and 2*BPF1+BPF2 (3953.03 Hz). The noise of the single propeller is mainly composed of BPF1 (1399.38 Hz), 2*BPF1 (2798.76 Hz) and 3*BPF1 (4198.14 Hz). BPF1 is the blade passing frequency for the front rotor, and BPF2 is the blade passing frequency for the rear rotor. Results show that the sound pressure level of the open rotor is obviously more complex and higher than that of the single propeller.

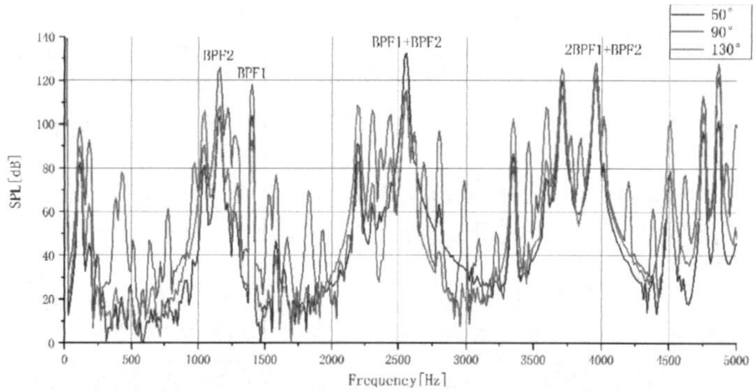

Fig. 5 Numerical predictions of the noise spectra at three different locations (open rotor)

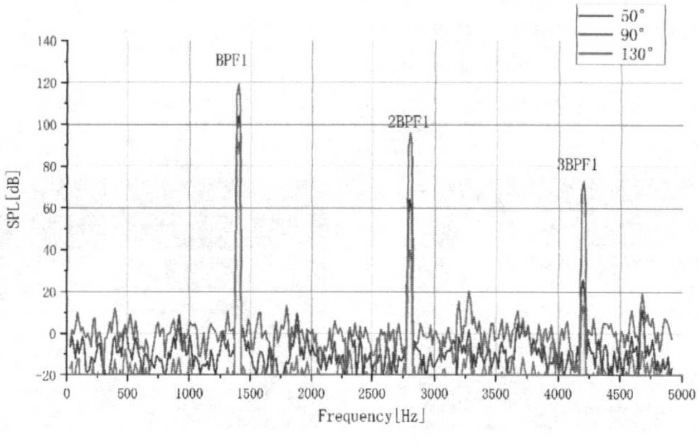

Fig. 6 Numerical predictions of the noise spectra at three different locations (single propeller)

Figure 7-10 shows the directivity distribution of BPF1, BPF2, BPF1+BPF2 and 2*BPF1+BPF2 noise of the open rotor and the present results are also compared with the literature. It can be seen that the overall calculated sound pressure level has the same trend as the literature [8], but the calculated value is slightly lower than that in the literature, which may be partly due to subtle differences in geometric models and grid differences. Figures 7 and 8 are the rotor alone noise generated by the front and rear blades respectively. It can be seen that the rotor alone noise has the maximum sound pressure level around the position of 90° to 100°. Fig. 9 and Fig. 10 show the directivity of the first interaction noise and the second interaction noise. It can be seen that the sound pressure level of the interaction noise is significant across a wide location range. At take-off condition, the interaction noise of the open rotor is very important which is also addressed by a lot of research.

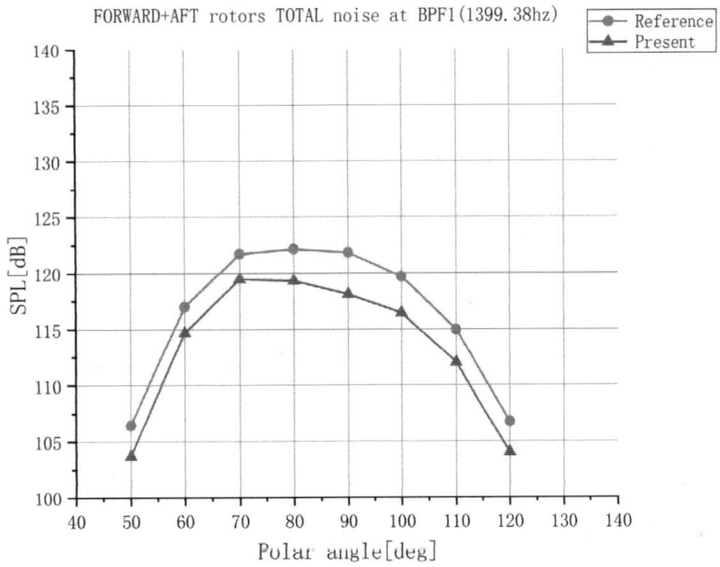

Fig. 7 Noise directivity at BPF1 frequency

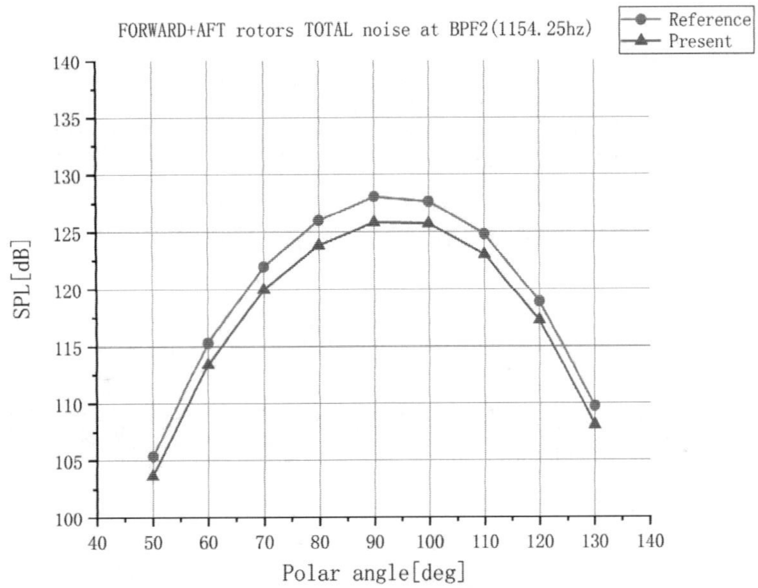

Fig. 8 Noise directivity at BPF2 frequency

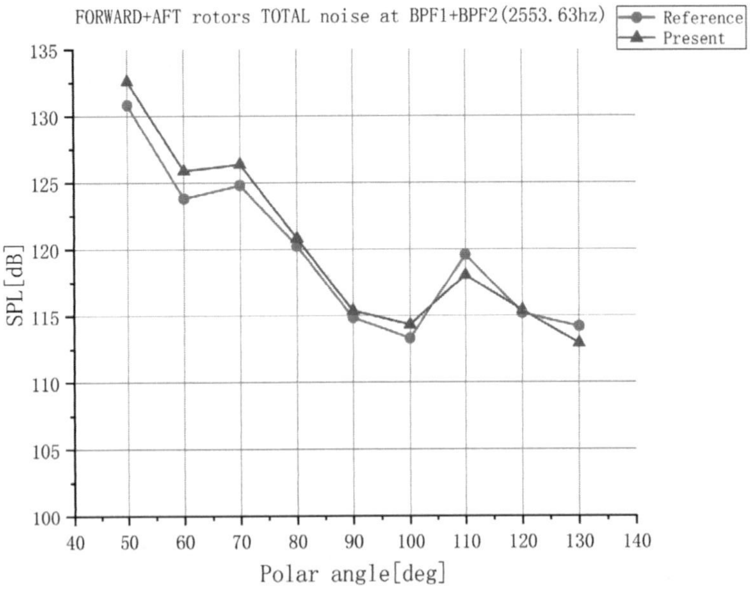

Fig. 9 Noise directivity at BPF1+ BPF2 frequency

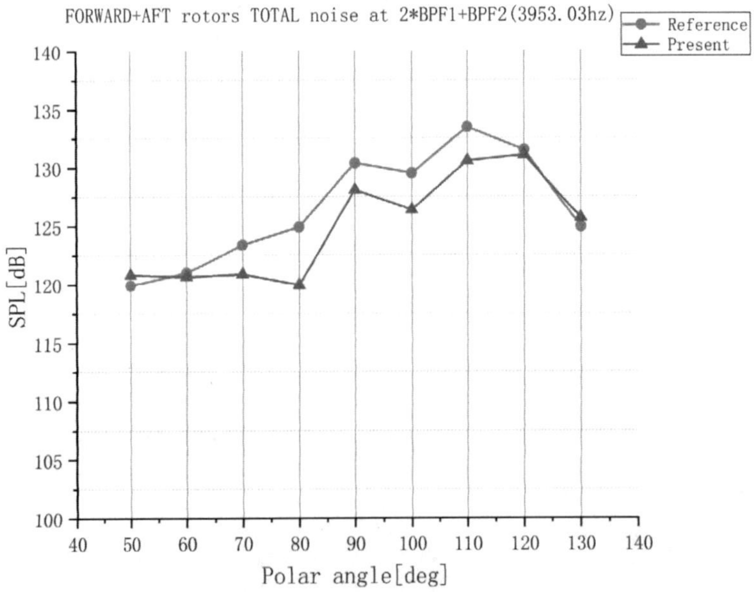

Fig. 10 Noise directivity at 2BPF1+ BPF2 frequency

The open rotor includes front and rear rows of high-speed counter-rotating propellers. The tip vortex, wake vortex, hub vortex and other vortex systems generated by the front row propeller will interfere with the downstream rear row propeller, leading to a variety of different complex noise sources. In order to deeply study the contribution of different noise mechanisms to the open rotor noise, the noise radiated from different regions of the rear rotor was calculated.

The noise radiated from 0%-12.5%, 12.5%-75% and 75%-100% of the rear blade was individually calculated. It is assumed that in the 0%-12.5% blade height, the noise is mainly generated by the interaction of the hub vortex and the rear blade. In the 12.5%-75% blade height, the noise is mainly caused by the interaction of the front propeller wake vortex and the rear blade. The noise radiated from 75%-100% of the blade height is mainly caused by the interaction of the front propeller blade tip vortex with the rear blade. It can be seen from Fig. 11 that the main noise source of the interaction noise at BPF1+BPF2 is blade tip vortex interaction noise and wake interaction noise, while the hub vortex interaction noise contributes less to the interaction noise. The investigation result may help to develop relevant noise reduction techniques for the open rotor. If we want to reduce the open rotor noise, blade tip vortex interaction noise and wake interaction noise need to be paid more attention.

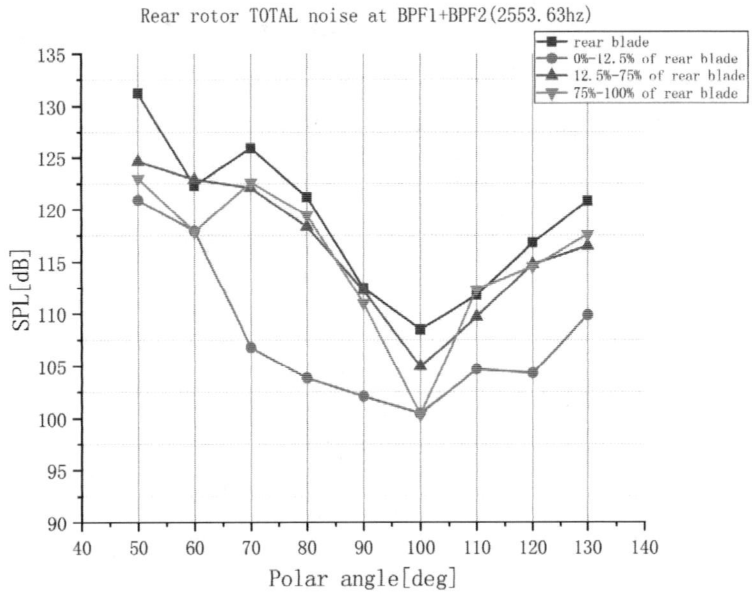

Fig. 11 Noise radiated from different zones of the rear rotor at BPF1+BPF2 frequency

4. Conclusion

The noise characteristics of an open rotor was studied by numerical calculation and compared with the noise characteristics of the single propeller. The contribution of different noise generation mechanisms to interaction noise was studied, which may provide some new ideas for the further noise reduction.

Comparing the noise of the single-rotor propeller and the open rotor, it can be seen that the noise spectrum of the open rotor is obviously more complex and higher than that of the single propeller. The open rotor will generate complex interaction noise in addition to the noise at blade passing frequency.

There is a significant difference between the directivity of the blade pass frequency noise and the directivity of the interaction noise. The sound pressure level of the blade pass frequency noise is the largest near 90°. For the BPF1+BPF2 interaction noise, the upstream noise is significantly larger than the downstream noise.

The noise produced by different zones of the rear blade was evaluated to study the contributions of different noise generation mechanisms. For the interaction noise, the main noise sources are the blade tip vortex interaction noise and wake interaction noise, while the contribution of vortex near the hub to the interaction noise is less obvious. When developing new noise reduction techniques in the future, designers need to pay more attention to the tip vortex interaction noise and wake interaction noise.

Acknowledgments

Funding: This research was sponsored by the National Natural Science Foundation of China, under the grant numbers of 12102451, 11802114 and 11602290.

References

[1] Kingan, M. J, Parry, A. B., Acoustic theory of the many-bladed contra-rotating propeller: the effects of sweep on noise enhancement and reduction[J]. Journal of Sound and Vibration, 2020, 468: 89-115.

[2] Parry,A.B., Kingan,M.J, Acoustic theory of the many-bladed contra-rotating propeller: physics of the wake interaction noise critical sources[J]. Journal of Fluid Mechanics, 2019,880:1-12.

[3] Broukaert J.F.,Mirville F,Phuah K, et al. Clean SkyReserch and Demonstration Programmes for Next-gener-ation Aircraft Engines[J]. The Aeronautical Journal, 2018,122(1254): 1163-1175

[4] HU Xiaoyu, Latest developments in engines for the next generation of narrow-body airliners[J].Journal of Avicaeroengine 1999, pp. 281-304.

[5] SHI WenBo, LI Jie, Numerical simulation of aerodynamic interference in flow field of Contra Rotoring Propeller[J].Journal of Aerospace Power, 2019, 34(04). 829-837.

[6] XIA Zhenfeng, YANG Yong. Characteristic analysis of unsteady aerodynamic interactions of control rotating open rotor[J]. Journal of Aerospace Power, 2014, 29(04): 835-843.Information on http://www.weld.labs.gov.cn

[7] YAN Wenhui, TANG Sijia, WANG Fengming, et al. Unsteady aerodynamic interactions of contra rotating propeller[J]. Journal of Aerospace Power, 2021, 36(7): 1398-1405.

[8] Ferrante P., Vilmin S., Hirsch C.. Integrated “CFD - Acoustic” Computational Approach to the Simulation of a Contra Rotating Open Rotor at Angle of Attack, May 27-29, 2013, Berlin, German, 19th AIAA/CEAS Aeroacoustics Conference, AIAA 2013-2242.

Experimental investigation on the heat transfer enhancement of steam condensation on tube with hydrophilic-hydrophobic hybrid surface

Huanmin Li, Zhuo You and Hao Zhang*

Wuhan Second Ship Design and Research Institute, Wuhan, China

*Corresponding author e-mail: zhyf1408@126.com

Abstract. The heat transfer enhancement of steam condensation on tube with hydrophilic-hydrophobic hybrid surface have been investigated comprehensively through experiment. It has been found that the steam condensation heat transfer performance on the tube with hydrophilic-hydrophobic hybrid surface has been enhanced significantly compared to that with hydrophilic surface. The results show that the steam pressure, cooling water velocity (flow rate) and non-condensable gas have influences on the steam condensation heat transfer performance on both tubes. The steam condensation heat transfer enhancement on the tube with hydrophilic-hydrophobic surfaces is due to the droplet drainage from hydrophobic stripes to hydrophilic stripes accelerating the discharge and formation of condensate from the tube. This work can be a guide for the design of enhanced condensing heat exchanger.

Keywords: heat transfer enhancement, steam condensation, hydrophilic-hydrophobic hybrid surface.

1. Introduction

Steam condensation at low pressure has been extensively applied in chemical industry, machinery, refrigeration, power generation, desalination and other fields due to its efficient phase change heat transfer. Depending on the wettability of the cold substrate surface, the phase change process during the steam condensation may result in either film-wise condensation (the formation of a continuous film) or dropwise condensation (the formation of droplet ensemble). Dropwise condensation heat transfer has attracted considerable attention because of its higher heat transfer coefficient compared to the film-wise condensation [1, 2].

In the process of dropwise condensation, the size and motion mode of the droplet are the key factors enhancing the condensation heat transfer. Through experimental investigation and theoretical analysis, the dropwise condensation heat transfer performance decreases with increasing the droplet radius on the condensation surface through experimental investigation and theoretical analysis, which indicates that regulating the droplet radius or its size distribution can strengthen and optimize the dropwise condensation heat transfer [3-6]. As their characteristic surface free energy, the gradient surface and the combined surface are used to control the droplet movement and size during dropwise condensation, then enhancing the heat transfer performance.

Daniel [7, 8] has found that the droplet coalescence drives the directional migration of droplets from the weak wettability region to the strong wettability region during dropwise condensation. This rapidly spontaneous droplet migration changes the droplet size distribution, thus improves the heat transfer coefficient by 3-10 times of film-wise condensation. In addition to the spontaneous movement of droplets driven by the wettability gradient surface to enhance heat transfer, with the development of surface structuring and functionalization technologies, the functionalized surface such as the hybrid surface has gradually become an important means for droplet removal, droplet radius regulation and heat transfer enhancement in the process of dropwise condensation. When the stripe width changes, the heat transfer performance of hydrophobic coating and hydrophilic stripes hybrid surface changes [9]. The finer the stripes are, the better the heat transfer performance with the hybrid surface is.

In this work, the experimental investigation has been done on the heat transfer enhancement of the tube with hydrophilic-hydrophobic hybrid surface compared to that with hydrophilic cooper surface. The aim of this study is to provide guidance for the design of enhanced condensing heat exchanger.

2. Experimental

2.1. The Experimental setup

Steam condensation on the heat exchange tubes with hydrophilic-hydrophobic hybrid surface and hydrophilic cooper surface is conducted on the experimental setup as shown in Figure 1. The experimental setup basically consists of steam generator, coolant circulating system, condensing chamber (test section), auxiliary condenser, image capture system and data acquisition system. During the condensation experiment, the team is generated by the boiler first then mixed with air, and then enters condensing chamber. The steam condensates in the condensing chamber and releases the latent heat of vaporization, which is cooled by the cooling water flowing through the coolant circulation system. The excessive steam is completely condensed by an auxiliary condenser. During steam condensation, the high-speed camera has been used to visualize and record the droplet dynamic properties such as droplet drainage, droplet radius and its size distribution on tubes with different surfaces.

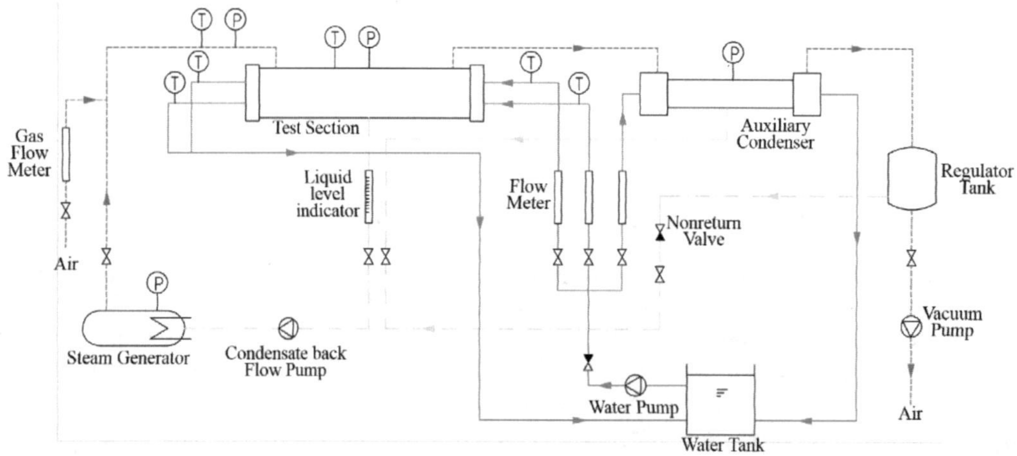

Figure 1. Experimental setup of steam condensation.

There are two parallel arranged heat exchange tubes installed horizontally in the condensing chamber. As shown in Figure 2, the heat exchange tubes are 16 mm in diameter,1.5mm in thickness and 1200 mm long with cylindrical condensing surface. The hydrophobic and hydrophilic stripes on the tube with hybrid surface are patterned vertically to axial direction of the tube. The widths of hydrophilic and hydrophobic stripes are 5 mm and 3 mm respectively. The white and black stripes on the tube with hybrid surface are hydrophilic and hydrophobic, respectively. The other tube is completely hydrophilic. Both of the tubes are made of purity copper.

Figure 2. The heat exchange tubes: (a) tubes in air; (b) hydrophilic tube and (c) hydrophilic-hydrophobic tube with steam condensation in the condensing chamber (white: hydrophilic tripes; black: hydrophobic stripes).

The temperatures of the steam, in condensing chamber and of the coolant circulation system are monitored by thermocouples. The accuracy of the thermocouple is ±0.05 K. The steam pressure is measured and monitored by pressure gauge with the accuracy of ±0.5 kPa. The electrical signals (voltage or current) of all the thermocouples and pressure gauges are collected by the Labview data acquisition system.

2.2. Data reduction
Calculation of heat flux q by sensible heat:

$$q = \frac{Q_c}{A_o} = \frac{m_c c_p (T_o - T_i)}{A_o} \tag{1}$$

Where, Q_c, A_o, m_c, c_p, T_o and T_i are total heat flow of cooling water, outside surface area of heat exchange tube, mass flow rate of cooling water, specific heat capacity of cooling water, outlet temperature of cooling water and inlet temperature of cooling water, respectively.

The heat flux can also be calculated according to Newton's cooling law:

$$q = K\Delta t_m \tag{2}$$

$$K = \frac{q}{\Delta t_m} = \frac{Q_c}{A_o \Delta t_m} \tag{3}$$

Where, K and Δt_m are total heat transfer coefficient and mean logarithmic temperature difference between steam and cooling water respectively. The mean logarithmic temperature difference between steam and cooling water can be calculated:

$$\Delta t_m = \frac{T_o - T_i}{ln \frac{T_v - T_i}{T_v - T_o}} \tag{4}$$

Where, T_v is the temperature of steam.

Calculation of the internal Heat transfer coefficient h_i in horizontal tube:

$$h_i = \frac{\lambda_c}{D_d} \cdot \frac{(f/8)(Re - 1000)Pr}{1 + 12.7(f/8)^{1/2}(Pr^{2/3} - 1)} \tag{5}$$

Where λ_c, D_d, Re and Pr are the thermal conductivity of cooling water, the diameter of heat exchange tubes, Reynolds number of cooling water and Prandtl number of cooling water, respectively. The resistance coefficient can be calculated:

$$f = (0.79ln(Re) - 1.64)^{-2} \quad (6)$$

The condensation heat transfer coefficient h_o on the outside tube can be calculated by extrapolation method:

$$h_o = \frac{1}{\frac{1}{K} - \frac{A_o}{A_i h_i} - \frac{A_o b}{A_m \lambda_w}} \quad (7)$$

Where, A_i, A_m, b and λ_w are inside surface area of heat exchange tube, mean area of heat exchange tube, the thickness of heat exchange tube and the thermal conductivity of heat exchange tube wall, respectively.

3. Results and discussions

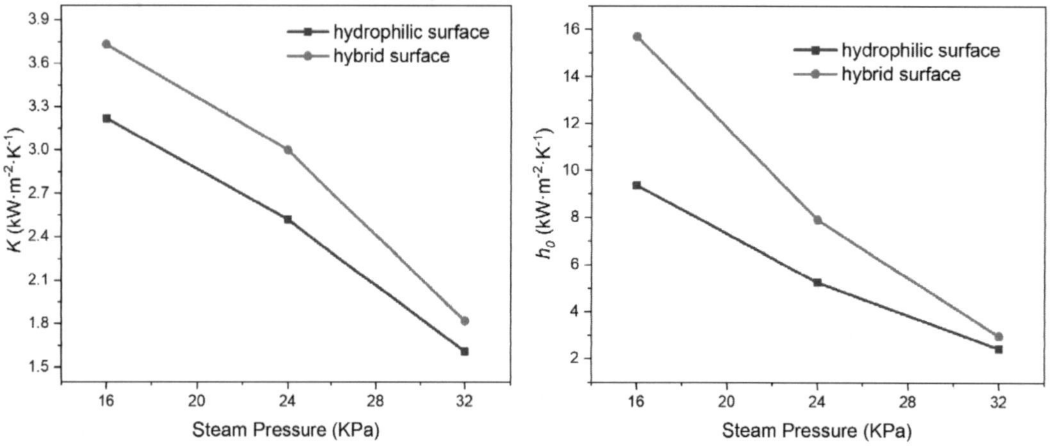

Figure 3. Influence of steam pressure on steam heat transfer performance (total heat transfer coefficient K and condensation heat transfer coefficient h_0 of tubes with hydrophilic surface and hybrid surface (hydrophilic-hydrophobic).

Under a constant cooling water velocity 2.5 $m \cdot s^{-1}$(flow rate) without non-condensable gas, the influence of steam pressure on the steam heat transfer performance of tubes with hydrophilic surface and hydrophilic-hydrophobic hybrid surface is depicted in Figure 3. The total heat transfer coefficient K and condensation heat transfer coefficient h_0 of the tube with hybrid surface are larger than that of tube with hydrophilic surface at the same condition. Compared to the tube with hydrophilic surface, the K and h_0 of the tube with hydrophilic-hydrophobic hybrid surface increase by 13.5%-18.9% and 22.01%-67.37% respectively. The lower steam pressure, the lower density and the higher flow rate of the steam leading to accelerated droplet detachment from heat exchange tubes. Thus, with increasing the steam pressure, the steam heat transfer performance of both tubes decreases.

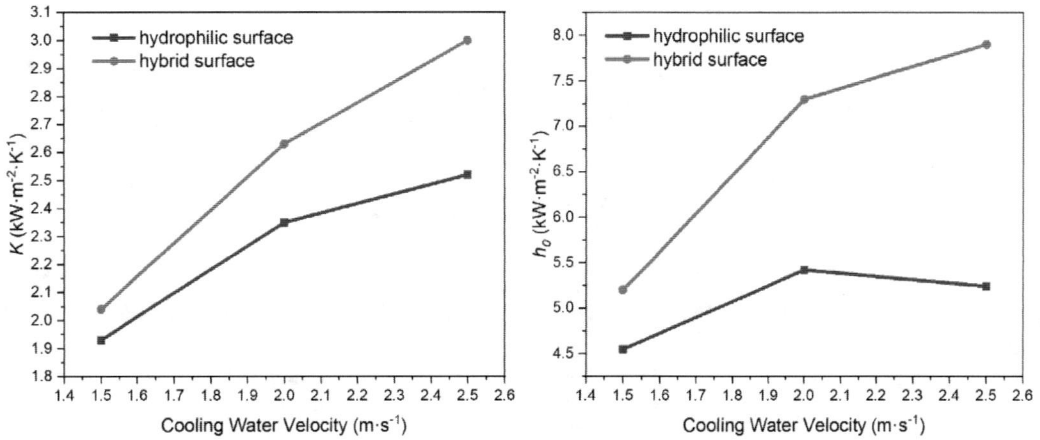

Figure 4. Influence of the cooling water velocity (flow rate) on steam heat transfer performance (total heat transfer coefficient K and condensation heat transfer coefficient h_0 of tubes with hydrophilic surface and hybrid surface (hydrophilic-hydrophobic).

Under a constant steam pressure 24 KPa without non-condensable gas, Figure 4 shows the steam heat transfer performance of both tubes with the change of cooling water velocity (flow rate). Compared to the tube with hydrophilic surface, the tube with hydrophilic-hydrophobic hybrid surface has larger K and h_0 increased by 11.6%-18.9% and 22.73%-50.01% respectively. Increasing the cooling water velocity means increasing the temperature difference between steam and cooling water, which leads to accelerated condensation droplet nucleation.

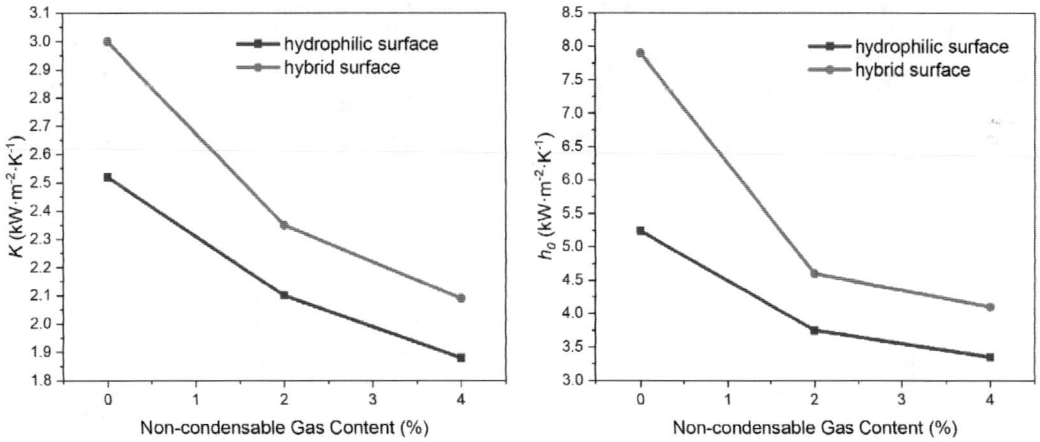

Figure 5. Influence of the non-condensable gas content on steam heat transfer performance (total heat transfer coefficient K and condensation heat transfer coefficient h_0 of tubes with hydrophilic surface and hybrid surface (hydrophilic-hydrophobic).

As shown in Figure 5, with constant steam pressure 24 KPa and cooling water velocity 2.5 $m \cdot s^{-1}$ (flow rate), the non-condensable gas content has influence on the steam heat transfer performance of both tubes. Compared to the tube with hydrophilic surface, the tube with hydrophilic-hydrophobic hybrid surface has larger K and h_0 increased by 11.5%-18.9% and 21.07%-50.01% respectively. The non-condensable gas hinders the steam molecular diffusion, which decreases condensation mass transfer and heat transfer efficiency. Therefore, with increasing the non-condensable gas content, the steam heat transfer performance of both tubes decreases.

Figure 6. The droplet dynamic property on the tubes with hydrophilic (left) and hydrophilic-hydrophobic (right) surfaces during steam condensation.

Under all the same conditions, the heat exchange tube with hydrophilic-hydrophobic hybrid surface has enhanced steam heat transfer performance compared to that with hydrophilic surface. To investigate the condensation heat transfer enhancement mechanism of the tube with hydrophilic-hydrophobic hybrid surface, the droplet dynamic properties such as droplet drainage, droplet radius and its droplet size distribution on both tubes are captured by high-speed camera. The droplet dynamic property on the tubes with hydrophilic (left) and hydrophilic-hydrophobic (right) surfaces during steam condensation are shown in Figure 6. The steam film-wise condensates on the tube with hydrophilic surface, which includes droplets nucleation, droplets growing up, droplets coalescence, flowing into the liquid film and falling off the tube. While, the steam condensates on the tube with hydrophilic-hydrophobic hybrid surface including film-wise condensation and dropwise condensation. The steam condensates on the hydrophobic stripes to form droplets. Then the droplets grow up, coalescence and are drained to flow into the liquid film on the hydrophilic stripes. The droplet drainage from hydrophobic stripes to hydrophilic stripes results in accelerating the discharge of condensate from the tube and the formation of new droplets, which significantly enhances steam heat transfer performance.

4. Conclusion

The heat transfer enhancement of the tubes with hydrophilic-hydrophobic hybrid surface and hydrophilic surface has been experimentally investigated thoroughly. The influence of steam pressure, cooling water velocity and the non-condensable gas content on steam heat transfer performance and the droplet dynamic property on the tubes with hydrophilic (left) and hydrophilic-hydrophobic (right) surfaces has been clarified. The following conclusions can be summarized.

(1) Compared to the tube with hydrophilic surface, the heat transfer performance on the tube with hydrophilic-hydrophobic hybrid surface has been enhanced significantly, with the total heat transfer coefficient and condensation heat transfer coefficient increased by 11.5%-18.9% and 21.07%-67.37% respectively.

(2) The steam pressure, cooling water velocity (flow rate) and non-condensable gas have influences on the steam condensation heat transfer performance on both tubes with hydrophilic surface and hydrophilic-hydrophobic hybrid surface.

(3) The droplet dynamic property on the tubes with hydrophilic and hydrophilic-hydrophobic surfaces during steam condensation indicates that the droplet drainage from hydrophobic stripes to hydrophilic stripes leads to the accelerated discharge and formation of condensate from the tube and then the steam condensation heat transfer enhancement.

Acknowledgments

We sincerely thank the financial support from the Hubei Provincial Natural Science Foundation (No. ZRMS2020002231) and the Young Top-notch Talent Cultivation Program of Hubei Province.

References

[1] Ji, X, Zhou, D, Chao Dai, & Xu, J, Dropwise condensation heat transfer on superhydrophilic-hydrophobic network hybrid surface, International Journal of Heat and Mass Transfer, 2019, 132, 52-57.

[2] N. Miljkovic, R. Engirght, Y. Nam, K. Lopez, N. Dou, & J. Sack., Jumping-droplet-enhanced condensation on scalable superhydrophobic nanostructured surfaces, Nano Letters, 2013, 13, 179-187.

[3] Peng, B., Ma, X., Lan, Z., Xu, W., & Wen, R., Experimental investigation on steam condensation heat transfer enhancement with vertically patterned hydrophobic–hydrophilic hybrid surfaces. International Journal of Heat and Mass Transfer, 2015, 83, 27-38.

[4] Baojin, Q., Li, Z., Hong, X., & Yan, S., Experimental study on condensation heat transfer of steam on vertical titanium plates with different surface energies. Experimental Thermal and Fluid Science, 2011, 35, 211-218.

[5] Kananeh, A. B., Rausch, M. H., Fröba, A. P., & Leipertz, A., Experimental study of dropwise condensation on plasma-ion implanted stainless steel tubes. International Journal of Heat and Mass Transfer, 2006, 49, 5018-5026.

[6] Hu, H. W., Tang, G. H., & Niu, D., Experimental investigation of condensation heat transfer on hybrid wettability finned tube with large amount of noncondensable gas. International Journal of Heat and Mass Transfer, 2015, 85, 513-523.

[7] Longley, J. E., Dooley, E., Givler, D. M., Napier III, W. J., Chaudhury, M. K., & Daniel, S., Drop motion induced by repeated stretching and relaxation on a gradient surface with hysteresis. Langmuir, 2012, 13912-13918.

[8] Macner A M, Daniel S, & Steen P H. Condensation on surface energy gradient shifts drop size distribution toward small drops. Langmuir, 2014, 30: 1788-1798.

[9] El Fil, B., Kini, G., & Garimella, S., A review of dropwise condensation: Theory, modeling, experiments, and applications. International Journal of Heat and Mass Transfer, 2020, 160, 120172.

AUTHOR INDEX